18th Process Plant Safety Symposium (PPSS)

Topical Conference at the 2016 AIChE Spring Meeting and 12th Global Congress on Process Safety

Houston, Texas, USA
10-14 April 2016

ISBN: 978-1-5108-2475-1

Printed from e-media with permission by:

Curran Associates, Inc.
57 Morehouse Lane
Red Hook, NY 12571

Some format issues inherent in the e-media version may also appear in this print version.

Printed by Curran Associates, Inc. (2016)

For permission requests, please contact AIChE
at the address below.

AIChE
120 Wall Street, FL 23
New York, NY 10005-4020

Phone: (800) 242-4363
Fax: (203) 775-5177

www.aiche.org

Additional copies of this publication are available from:

Curran Associates, Inc.
57 Morehouse Lane
Red Hook, NY 12571 USA
Phone: 845-758-0400
Fax: 845-758-2633
Email: curran@proceedings.com
Web: www.proceedings.com

TABLE OF CONTENTS

(17a) Beyond HAZOP and LOPA – Four Different Company Approaches 1
Karen Study, J. Wayne Chastain, Chris Devlin, Thomas Mueller, Paul Delanoy

(36a) High/Continuous Demand Hazardous Scenarios in LOPA 35
William Mostia, Hui Jin, Angela Summers

(36b) Advanced Analysis to Supplement HAZOP/LOPA for Effective Process Design 46
William R Banick, Cindy Wei

(36c) Identify SIF and Specify Necessary Sil, and Other IPLs, As Part of PHA/HAZOP, or Why It Is Not Necessary to "Boldly Go Beyond HAZOP and LOPA" 66
Arthur M. Dowell, William Bridges

(44a) One Company's Observations on the Implementation of LOPA 85
Anne Bertelsmann

(44b) Don't be Afraid of Conditional Modifiers 102
J. Wayne Chastain

(44c) Layer of Protection Analysis As a Multi-Functional Team Problem Solving Tool 103
Edward Cialkowski

(64b) Using Data Analytics to Improve Process Safety 120
Timothy J. Myers, Harri K. Kytomaa

(64d) Safety Culture and Leadership- Independent of Language 121
Kumar Israni, Yaneira Saud, Sarah Acton

(64e) Indulging in Invincibility (III)- Twelve Steps to Sobriety 122
Brian Kolodji

(64f) Case Study of Relief Analysis of Compressor Stations 123
Kumud Chaurasia, Aniket Patankar

(64i) An Approach to Reducing Risk during Online Isolation, Removal and Maintenance of Relief Devices Using Engineering Assessment 137
Aubry Shackelford, Jeff Heil, Ian Sergo

(64k) Common "Design Flaws" Identified during Pressure Relief System Analysis 138
Rafael Rocha

(64l) Considering Overfilling to be a Non-Applicable Overpressure Scenario 139
Natalie Doe, Dustin J. Smith

(64o) How Reliable Are Your Major Accident Hazard Barriers? 150
Mike Neill

(64q) Donde, Cuando Y Que Enseñar Y La Nueva Certification De Seguridad De Procesos Del CCPS 168
Nestor H. Sposito, Laura Acevedo Turci

(65h) PSM Legal Update 169
Mark Dreux

(65bb) Application of Computational Fluid Dynamics (CFD) for Liquefied Natural Gas (LNG) Pool Spreading and Vaporization on Water 170
Nirupama Gopalaswami, Konstantinos E. Kakosimos, Bin Zhang, Yi Liu, Ray Mentzer, Sam M. Mannan

(65aq) Modeling an Unconfined Liquid Spill on a Sloped Surface 171
Robert Deshotels

(65as) Apparent Discrepancies in Methods for Predicting the Explosion Energies of Bleves 198
Ronald Pape, Amy Richards, Michael Hanks

(65bk) Automated Optimization of Gas Detector Placement 199
Edward Marszal

(65d) Quantitative Risk Assessment in the Process Industries: Beyond Numerical Estimates of Individual and Societal Risk 200
Kehinde Shaba

(65ar) Quantifying the Importance of Accurate Physical Property Data in Dispersion Modeling 201
Devin Averett, Jef Rowley

(65l) Sustainable Risk Reduction with Dynamic Execution of Procedures for Any Operating Condition 213
Blair Morgan

(65x) Process Hazard Analysis of River-Crossing Natural Gas Pipeline 214
Yi Zhang, Xiaodong Ling, Wen Fang Zhao

(65u) Safe Depressurization of Large Thickness Vessels Under Pool Fire 215
Subhash S. Kale, George Maliakkal

(65c) Quantifying the Impacts of Human Factors on Functional Safety 221
Julia V. Bukowski, Loren Stewart
(65w) Using HAZOP/LOPA to Create an Effective Mechanical Integrity Program 228
Steven T. Maher
(65am) Reactive Chemical Hazards of Azide Compounds: a Case Study and Lessons Learned 229
Min Sheng, John Hull, Todd Tambling, Grant Von Wald
(65bd) The Application of Frequency Data in Quantitative Risk Assessment 230
Vinicius Simoes, Sarah Chambliss, Richard Gustafson
(65bi) Uncertainty in Quantitative Risk Assessment and Decision Making 231
Richard Gustafson, Vinicius Simoes, Sarah Chambliss, Stannia Sofyan
(65ao) Analysis of Process Systems during Accidental Fire 232
Ali Sari, Tohid Darvishzadeh
(65i) Assessing Safety Culture Via the Site Safety Index 233
Julia V. Bukowski, Denise Chastain-Knight
(65ai) Fire Hazard Associated with Polymers and Application of Polymer Nanocomposites for Flame Retardancy 242
Lubna Ahmed, Bin Zhang, Sam M. Mannan, Zhengdong Cheng, Logan Hatanaka, Qingsheng Wang
(65bh) Applying QRA Study to Select and Design Effective Release Prevention Safeguards 243
Stannia Sofyan, Vinicius Simoes, Cindy Wei
(65ap) Managing the Uncertainty in Blast Load Calculations with Enhanced Technical Safety Analysis 244
Ali Sari, Umid Azimov
(65at) Occupant Vulnerability for Process Plant Building Design Due to Secondary or Domino Events 245
Ali Sari, Umid Azimov
(65ah) Assessment of the Mechanical Integrity of a Pipeline with Metal Loss, Using Finite Element Method 246
Andrés F. Sánchez Akli, Alejandro Castellanos, Felipe Muñoz Giraldo
(65bg) An Improved Risk-Based Approach for Accidental Loadings 247
Ali Sari, Umid Azimov, Hieu Nguyen
(65bf) Occupant Vulnerability for Temporary Blast Resistant Modular Buildings 248
Ali Sari, Umid Azimov, Hieu Nguyen
(65aj) Occupied Building Blast Retrofit and Upgrade for Personnel Safety 249
Ali Sari, Umid Azimov
(65m) Using Hierarchy of Controls to Reduce the Likelihood of High Severity Scenarios 250
Alek Hamparian, Mardy Kazarians
(65e) Maintaining Consistency in the Determination of Consequence Severity Rankings for Process Hazard Analyses 264
Kirk Busby
(65z) Application of Open Source Software Tools in the Process Safety Education 276
Tianxing Cai, Qiang Xu
(65au) Beyond Quantitative Risk Analysis Results - Part I: Explosions and Blast Phenomena Characterization 277
Jordi Dunjó, Neil Prophet
(65av) Beyond Quantitative Risk Analysis Results - Part II: Fires and Domino Effect Characterization 278
Jordi Dunjó, Neil Prophet
(65o) A Leadership Vision for Accelerating Learning from Past Lessons to Create a Future Free of Incidents 279
David Jones, Zach Johnson
(65k) Analytics to Reveal Meaurable Factors to Process Safety Performance 280
Joseph Stough, Mei-Li Lin
(65ad) Coupling Dynamic Pressure Relief System Design with Experimental Vent Flow Regime of Dicumyl Peroxide / Toluene System 281
Surendra Singh, John Rynkar
(65be) We Did a Quantitative Risk Assessment Study. Now What Do We Do with It? 282
Irfan Shaikh
(65p) Effective Implementation of PSM to Reduce Operational Risks 283
Srinivasan Ramabhadran, Polshom Chan-Urai
(65az) Probit Models for the Determination of Damage Probability of Storage Tanks Due to Flooding Impact 284
Nicolás Villalba, Alejandro Castellanos, Fabio Ocampo, Felipe Muñoz Giraldo

(65ak) Analysis of Fire Extinguishing Methods in Engine Control Rooms ... 285
Julian Mendez, Nicolás Villalba, Fabio Ocampo, Oscar Mariño, Felipe Muñoz Giraldo, Alejandro Castellanos
(65af) The Study of HAZOP Revalidation Approach ... 286
Jing Xu, Chuntao Ge, Yi Zhang
(65a) Realizing the Importance of PSM: A Young Engineer's Perspective ... 296
Waheed Wakil, Travis Bruner
(65v) Scaling up Safely: Making Smarter Decisions about Rate Dependency ... 303
Zubin Kumana, Venkata Badam, Achilles Arnaez
(65g) Learning Lessons from a Big Chemical Disaster in Korea ... 315
Kwanghee Lee, Il Moon
(65aw) Study on Thermal Radiation of Enclosed Ground Flare ... 316
Peng Wang, Anfeng Yu, Wenyi Dang, Jiedong Zhang
(65bl) Simple Objects As Indicator of the Severity of Vapour Cloud Explosions ... 317
Vincent Tam, Bassam Burgan, Anqi Chen, Daniel Allason, Michael Johnson
(65q) Best Practices for Writing Operator Procedures and Trouble-Shooting Guides ... 318
William Bridges, Lauren Madden
(65ab) How Process Safety Information Affects the Quality of a Process Hazard Analysis ... 341
Michael Mayo, Dick Baum
(65al) Experimental Measurements of Flammability at Elevated Conditions Using Thermal Criterion ... 348
Ning Gan, Dragomir B. Bukur, Sam M. Mannan
(65f) Use of Process Incident Data for Accident Sequence Precursor Analysis ... 349
Erin Collins, Bijan Najafi
(65ax) Quantitative Risk Analysis of the Expansion of the Chlorine Room of the Industrial Water Plant of the Petrochemical Complex Jose Antonio Anzoategui of Petroquimica Venezuela Pequiven S.a ... 360
Jesus Corrales, Gerson Parpacen, Juan Duarte, Jesus Gonzalez
(65t) Assessing Cognitive Human Errors for the Process Industries ... 361
Erin Collins, Bijan Najafi
(65ac) Critical Considerations in the Vacuum Protection of Butane Spheres ... 372
Robert Siml
(65r) Assessing Risk in the Supply Chain Using the Bowtie Method ... 373
John F. Kill, Yaneira Saud, Mireille Busque, Radhika Bompelly
(65an) Reference Framework for the Application of Quantitative Risk Analysis for Hydrocarbon Pipelines, Coupled with Uncertainty Treatment Methods: Techniques for Analysis and Modelling Methods ... 374
Fabio Ocampo, Alejandro Castellanos, Nicolás Villalba, Felipe Muñoz Giraldo
(65j) Dynamics Operational Risk Management in Organizational Design, the Challenge for Sustainability ... 375
Salvador Ávila Filho, Ricardo Araujo Kalid, Carlos Roberto Oliveira Sousa, Jose Filipe Michel Gagliano Ferreira
(65ag) Availability Assessment of Water and Energy, a Proposition of Management to Chemical Industry ... 398
Salvador Ávila Filho, Jose Rafael Nascimento Lopes, Jean Marcel Prazeres Silva
(65ay) A Novel Open Source CFD Code for Explosion Modelling ... 422
Tatiele Ferreira, Sávio Vianna
(65ae) Compliance with Process Safety Management Elements ... 423
James Brigman
(65y) A Fuzzy Logic and Probabilistic Hybrid Approach to Quantify the Uncertainty in Layer of Protection Analysis ... 424
Yizhi Hong, Hans Pasman, Sonny Sachdeva, Adam S. Markowski, Sam M. Mannan
(65bj) Quantitative Risk Analysis of Nitrogen Fertilizer Plant, of Moron Petrochemical Complex ... 425
Jesus Gonzalez, Juan Duarte, Jesus Corrales
(65s) Management of Plant IPL's ... 426
Rick Stanley
(65b) Jack Rabbit II: Experimental Large-Scale Chlorine Release Trials ... 447
Shannon B. Fox
(65n) Novel Methodology for Inherent Safety Assessment in the Process Design Stage ... 448
Preeti Gangadharan, Ravinder Singh, Helen Lou
(79a) Creating a Culture of Operational Discipline within a Management of Change Process ... 449
Stacey Griffith
(79b) Managing Alarms to Support Operational Discipline ... 460
Denise Chastain-Knight, Todd Stauffer

(79c) **Achieving Operational Excellence through a Sustainable Approach within Process Safety** 482
Kumar Israni, Steve Hawkins
(83a) **Getting The Most From Your PHA – Lessons Learned From A Seasoned Facilitator** 492
Katherine E. Pearson
(83b) **LOPA Boot Camp – Basic Training with Practical Examples** 493
J. Wayne Chastain
(83c) **Hazards Inherent to Control Systems: Case Studies and Lessons Learned** 494
Brenton L. Cox, Stephen W. Garner, Andrew R. Carpenter, Mark Fecke
(102a) **Human Reliability and Process Safety Metrics: Considerations from an Abnormal Situation Management Perspective** 508
Peter Bullemer, Dal Vernon Reising
(102b) **Security and Risk Mitigation: Practical and Cost-Effective Solutions By the U.S. Nuclear Industry** 527
Gary Hayner
(102c) **Compressor System Performance at Abnormal/Non-Design Conditions** 535
Dylan Grosscup, Carl D. Ramlakhan, Christian Jarvis, Charles Lea
(121a) **Risk Criteria Selection and the Impacts on the LOPA Process: To Sum or Not to Sum, That is the Question** 570
Aaron Huberman
(121b) **A Call to Action: Understanding the Key Driver for Continual Process Safety Performance Improvement** 583
David Jones, Kay Modi
(121c) **Risk Based Process Safety Program in R & D Facility** 610
Monica Prasad, Praveen Kumar Bommaganti
(131a) **Restructuring of Process Safety at Air Products** 629
Brian E Farrell
(131b) **Becoming "Wiser" in Management of Change** 641
Sean J. Dee, Russell A. Ogle, Trey Morrison, Mark Fecke
(131c) **Successfully Managing a Changing Workforce - Application to Process Safety Management** 652
Charles A. Soczek
(64g) **Conducting a Human Reliability Assessment to Support PHA and LOPA** 658
Dave Grattan
(153b) **Quantitative Risk Analysis for Decision Support in Process Plant Operations** 676
Stein Haugen, Jan Erik Vinnem, Olav Brautaset, Nathaniel John Edwin
(153c) **Depressurization of CO_2 Rich Mixtures: Challenges for the Safe Process Design of CCS Facilities and CO_2 EOR Systems** 693
James Marriott, Apostolos Giovanoglou, Sathish Natarajan, Yemi Zaccheus
(161a) **Do You Feel Lucky? or Do You Want to Identify and Manage Safety - Critical Equipment?** 713
Mike Broadribb, Eric Freburger
(161b) **Using Past Incidents As Leading Indicators for Pipeline Integrity Management** 727
Hao Chen, Cynthia Spitzenberger, Lindsay Deal, Tone Rice
(161c) **Best Practices for Performing Data Mining to Support Asset Integrity Management** 748
Brent Stanley, Randal L. Montgomery
(168a) **Revitalizing Auditing Practices - One Company's Journey** 768
Robert Romano, Hope A. Luebeck
(168b) **Beyond Compliance Auditing: Drill 'til You Find the Pain Points and Release the Pressure!** 777
Laura Ankrom, Kathy Shell
(168c) **How Effective Are Your Safety Controls, Alarms, and Interlocks? The Importance of Functional Safety Auditing** 791
Monica Hochleitner, Eloise Roche, Angela E. Summers
(175b) **Leveraging PHA to Develop Requirements for Cyber-Security** 806
Edward Marszal
(175c) **Pipeline Risk Assessment Using Artificial Intelligence: A Case from the Colombian Oil Network** 817
Alexander Guzman Urbina
(179a) **Effective Operational Discipline Programs** 818
James Klein
(179b) **Best-in-Class MOC Implementation – Case Studies** 819
Rainer Hoff
(179c) **Facility Integrity: Learning from the Equipment Lifecycle** 820
Robert Wasileski

(189f) CSB Investigation of the West (Texas) Fertilizer Explosion and Fire 838
Johnnie A. Banks
(189a) Process Safety Incidents: Big Picture Revelations 840
Paul Baybutt
(189b) Accidents during Turnarounds, Cleanings, and Other Infrequent Operations 871
Michael C. Stern, Noah Budiansky, Vijay Somandepalli, Ali Reza, Timothy J. Myers
(189c) Biomass Industry Explosion Case Study 879
Amy Theis, Timothy Cullina, Kenneth N. Kurko
(189d) Atmospheric Tank Explosions: Mechanisms and an Unexpected Case Study 889
Michael S. Schmidt
(189e) Explosions in SMPO Hydrogenation Unit 910
Nick Gonzales
Author Index

12TH GLOBAL CONGRESS ON PROCESS SAFETY

Beyond HAZOP and LOPA - Four Different Company Approaches

J. Wayne Chastain, P.E.
Eastman Chemical Company
P.O. Box 511
B-18
Kingsport, TN 37662
(423) 224-0553
chastain@eastman.com

Paul Delanoy
The Dow Chemical Company
Estuary Road
Kings Lynn, PE30 2JD
UK
+44-155-36-94291
pdelanoy@dow.com

Chris Devlin, CPSP
Celanese
The Bishop Plant
Hwy 77 South
Bishop, TX 78343
(361)584-6394
CPDevlin@Celanese.com

Dr. Thomas Mueller
BASF SE
GUS/AB - M940, 67056 Ludwigshafen, Germany
+49-621-60-47677
thomas.h.mueller@basf.com

Karen Study, P.E.
The Dow Chemical Company
6519 State Hwy 225
Deer Park, TX 77536
(281)228-8679
kstudy@dow.com

Prepared for Presentation at
American Institute of Chemical Engineers
2016 Spring Meeting
12th Global Congress on Process Safety
Houston, Texas
April 10-13, 2016

UNPUBLISHED

Abstract

Abstract

For operations where application of standards, regulations and/or Recognized and Generally Accepted Good Engineering Practices may not be sufficient to address a particular company's risk, several options exist. For qualitative assessment of process hazards, Hazard and Operability Studies (HAZOP) and What-If reviews are two of the most common petrochemical industry methods used. Up to 80% of a company's process hazard analysis may consist of HAZOP and What-If reviews. [1] After the process hazard analysis, Layer of Protection Analysis (LOPA) is commonly used throughout industry to evaluate the required safety integrity level for instrumented protection layers in a semi-quantitative manner. [2] HAZOP, What-If and LOPA are all straightforward methods and are relatively easy to perform. However, much like a hammer, they are not always the best or most appropriate tool for a given job. At times, more advanced methodologies such as Fault Tree Analysis, Quantitative Risk Assessment, Event Tree, Failure Mode and Effects Analysis and Human Reliability Analysis are necessary to properly assess risk. However, these more advanced tools come with a price. They are often more expensive, time consuming and require a higher level of expertise. The decision to use these higher level methodologies is not taken lightly and different companies use different criteria for determining when to take this next step. This paper will present approaches by four companies, BASF, Celanese, The Dow Chemical Company, and Eastman Chemical Company. Each company will outline criteria used to determine when to go beyond HAZOP, What-If and LOPA and will present examples where more advanced techniques were used. The intent of this paper is to provide readers with real world examples that demonstrate the appropriate application of the 'right' tool and to illustrate what criteria can be used to make informed decisions regarding selection of a process hazard analysis methodology.

Keywords: HAZOP; LOPA; Quantitative Risk Assessment; Hazard; SIS; What-If; Fault Tree; Event Tree; Failure Mode and Effects Analysis; Human Reliability Analysis; HRA; FTA; QRA; FMEA

1. Introduction

The decision to use higher level methodologies such as Quantitative Risk Analysis (QRA) and Fault Tree Analysis (FTA) is not taken lightly due to the complexity, skill level, time and cost required. When lower level methodologies such as a qualitative assessment using HAZOP, What-If Reviews or Checklists and semi-quantitative techniques such as LOPA or a risk matrix suffice, they are generally accepted as a preferred method throughout many companies in industry. This paper will detail the risk assessment protocols used at 4 different companies in an attempt to demonstrate key considerations when choosing a risk assessment methodology.

2. Company Approaches

2.1. BASF Approach

2.1.1 History

In 2015 BASF celebrated its 150th anniversary. The history of BASF not only comprises leading innovations but also major incidents with a large number of fatalities and significant devastation, most notably the Ammonium Nitrate explosion in 1921 and the Dimethylether railcar explosion in 1948, both at BASF's largest site in Ludwigshafen.

While in the 1890s about 1% of the company's employees were killed per year and in the 1950s and 60s there were still yearly fatalities on larger sites, in the time between 1970 and present, the number of major incidents as well as the number of fatalities were significantly reduced. Although a high priority for process safety was already an integral part of BASF's "DNA" since the 1921 explosion, it took until the 1980s for a process safety organization to be formally established that functioned explicitly independent of operations and engineering.

Today, BASF's Process Safety Management is embedded in the RC® Management System. Process safety having priority over production targets, as expressed in the slogan "We never compromise on safety".

2.1.2 Current Process

2.1.2.1 Background

BASF constructs and operates production plants throughout the world, and aims to ensure that these plants meet high standards with regard to safety (both process safety and occupational safety), health and environmental protection. The overarching goal is to maintain these high standards throughout the life cycle of a production facility.

This is accomplished through the development and consistent performance of a comprehensive process safety management system within each production facility.

The first element of the process safety management system was formally introduced in 1982 and was targeted to the implementation of capital expenditure projects. The heart of this directive is a standardized BASF group-wide framework for a multi-step system of safety, health and environmental protection (SHE) reviews during the planning and construction of new process plants/facilities.

As a second element of the process safety management system, BASF launched a structured program for process safety reviews of all existing plants in 2010. Evolved from a less formal program in the mid-1990s for designated sites, the new approach focuses on standardizing the program and applying to all operations.

2.1.2.2 Safety, Health and Environmental Protection (SHE) Reviews

To ensure that all plants BASF builds and operates fully comply with the Corporate SHE requirements, a standardized group-wide framework for a multi-step system of SHE reviews during the implementation of capital expenditure projects has been set up. The SHE reviews are carried out by an interdisciplinary team of experts and serve to:

- ascertain the process-related hazard potentials,
- define the specific SHE Protection Concept and
- review the planning documents to ensure they conform to the specific SHE Protection Concept and meet the general safety and regulatory requirements, all as a function of progress of the project development.

Then, prior to start-up, the completed plant is checked to verify compliance with specifications and conformity with the planning documents. This system is firmly integrated with the project procedure in BASF's engineering division and corresponds to what is commonly referred to as the Pre-Startup Safety Review in many locations.

Additionally, for some selected projects as determined by the review team (e.g., in case of technologies new to BASF), a final review may be conducted sometime after plant startup involving a team from safety and operation to discuss any necessary adaptation of the SHE Protection Concept and seek lessons learned for the future. See Table 1 for a summary of the SHE review steps.

Table 1: The 6 Steps of SHE Reviews

Step 0	Process Development	Identification of basic EHS issues. Preliminary decisions on process technology, site selection and logistics concept.
Step 1	Conceptual Engineering	Review of the basic strategies devised for processes, site, civil/structural design and logistics; confirming/supplementing the solution strategies adopted.
Step 2	Extended Conceptual Engineering	Systematic Review of Piping & Instrument Diagrams (P&IDs). Approval of the SHE protection concept and any other essential requirements to be observed in detailed engineering.
Step 3	Detailed Engineering	Formalized Systematic review of P&IDs for Safety and Operability (HAZOP study; only high hazard potential).

Step 4	PSSR	Review of implementation of SHE concepts before start-up.
Step 5 (planned)	Post-Startup	Planned: Review of changes from start-up (only new technology with high hazard potential).

SHE Review Step 0

- Timing: After Project Requirements Definition and before Initial Approval of the project.
- Objective: Identification of basic SHE issues, particularly related to regulatory or authority approval. Preliminary decisions on process technology, site selection and logistics concept.
- Scope: Required for "large" projects especially if a decision shall be made between various alternative process technologies and sites or locations.

SHE Review Step 1

- Timing: Between Conceptual Planning and Extended Conceptual Planning.
- Objective: Reviewing SHE issues affecting the basic strategies devised for processes, site, civil/structural design and logistics, identification of the process hazards, development of a preliminary SHE Protection Concept including implementation check based on process flow diagrams.
- Scope: Required for all projects.

SHE Review Step 2

- Timing: After Extended Conceptual Planning and before Final Approval (appropriation) of the project.
- Objective: Completion and approval of the SHE Protection Concept including implementation check based on piping and instrumentation diagrams (P&IDs) to ensure functional safety of the process during all modes of operation (i.e., normal operation, startup, shutdown and non-routine operations).
- Scope: Required for all projects.

SHE Review Step 3

- Timing: As soon as final ("approved for construction") P&IDs are available and before mechanical completion of the project.
- Objective: Systematic review of the piping and instrument diagrams (P&IDs) for safety and operability, executed in the form of a HAZOP study.
- Scope: To be defined by the SHE Review team. A SHE Review Step 3 is usually required for plant sections which exhibit a particularly high hazard potential because of the substances and/or the process. Typically, this is true when handling large volumes/quantities of very toxic substances, or large volumes of flammable liquefied gases or where processes have an inherently high potential of runaway reactions occurring.

SHE Review Step 4 (= PSSR)

- Timing: Upon mechanical completion and prior to start-up of the plant.

- Objective: Verify prior to start-up that the SHE Protection Concept has been implemented in constructing the plant.
- Scope: Required for all projects.

SHE Review Step 5

- Timing: Between 6 months to one year after start-up of the new process/plant.
- Objective: Exchange about experiences and learnings with safety measures/equipment in the time between start-up and normal operation.
- Scope: Only for process/technology new to BASF or upon request of the business unit.

As can be seen in the description above, the SHE Protection Concept is a fundamental requirement used to help ensure the safe operation of every BASF chemical plant.

In order to develop and maintain organizational memory, it is crucial to document the hazards that have been identified, the risks associated with those hazards and the measures taken to eliminate or control those hazards for each process to ensure an acceptable level of risk in the plant. This documentation comprises the Process Safety portion of the SHE Protection Concept for a plant and is commonly referred to as the Plant Safety Concept. The Plant Safety Concept describes protection strategies, whose implementation in the plant is then checked in detail in the Diagram Reviews. The documentation of the Plant Safety Concept is standardized globally for all types of plants.

2.1.2.3 Risk Assessment with BASF Risk Matrix

A quantitative risk assessment for chemical plants may be theoretically possible, but it is not practical. BASF therefore uses a semi-quantitative risk assessment utilizing the safety review system and an internally developed risk matrix, known as the BASF Risk Matrix. The BASF Risk Matrix is a tool developed to assist in performing risk assessments. The BASF Risk Matrix is used for all risk assessments involving health effects, since it reflects BASF's philosophy on maximum acceptable risk and the determination if and what additional measures are necessary to eliminate or mitigate the risk

The BASF Risk Matrix is a 4x5 matrix with four severity classes (S1-S4) and five probability classes (P0-P4). The severity classes represent the magnitude of the consequences of an incident with decreasing effects. The probability classes represent the order of magnitude of how often the incident happens with decreasing likelihood. There are twenty potential risk fields within the matrix. These fields are broken down into six risk classes, designated by the letters A-F. These designations correspond with prescribed minimum safeguards, or design/control measures required in order to bring the risk to a level acceptable within BASF.

The risk assessment is executed by an interdisciplinary team within the P&ID-review of a Step Review or a Clean Sheet Review (see 2.1.3.1). For each scenario the severity of the effects of the incident and the probability of the occurrence of this event will be evaluated.

Table 2: BASF's Risk Matrix

	Risk Matrix			
	Severity			
Probability	S_1	S_2	S_3	S_4
P_0	**A**	**B**	**D**	**E**
P_1	**A/B***	**B**	**E**	**E**
P_2	**B**	**C**	**E**	**F**
P_3	**C**	**D**	**F**	**F**
P_4	**E**	**F**	**F**	**F**

*** Determined on a case-by-case basis decision whether A or B is needed.**

Probability:

P_0 Happened a couple of times (once per year or more often)
P_1 Happened once (Approx. once in 10 years)
P_2 Almost happened, near miss (Approx. once in 100 years)
P_3 Never happened, but is thinkable (Approx. once in 1,000 years)
P_4 Reasonably not to be expected (less than once per 10,000 years)

Severity: (Health Effects)

S_1 On site: Potential for one or more fatalities
S_2 On site: Potential for one or more serious injuries (irreversible)
S_3 On site: Potential for one or more lost time injuries
S_4 On site: Potential for minor injuries or irritation

Risk Class	Minimum Risk Reduction Measures
A	Process or design change preferred
B	Process or design change, one protective device equivalent to SIL 3 required (e.g. PSV, SIS)
C	Process or design change, one protective device equivalent to SIL 2 required (e.g. PSV, SIS)
D	One monitoring device of high quality with documented testing or administrative procedure of high quality
E	One monitoring device or administrative procedure
F	No technical measures needed

When these properties are assessed, one does not take credit for any of the existing or proposed safeguards (e.g., SIS interlock or PSV), using only the control system and the design features as they have been planned. This is done in order to obtain a "raw" risk value. From this "raw" value, the necessary level of protection can be determined.

Once the required level of protection is determined, the existing or planned safeguards can be evaluated to determine if they provide an adequate level of risk reduction

The letter designating the resulting risk classes refers to the Minimum Level of Protection necessary to make the risk acceptable. Further risk reduction may be desirable and/or necessary based on the individual situation. BASF has implemented a hierarchy of controls for risk reduction, which gives the order of preference for the types of control measures.

If the resulting risk is unacceptable (risk class A, B, or C), then implementing the minimum risk reduction measures will move the risk into the tolerable area. BASF, therefore, strongly recommends taking one or two additional measures ("layers of protection") to further reduce the risk associated with the deviation under consideration. If a letter D or E designates the resulting risk, then the risk is tolerable, but taking additional measures to further reduce the risk associated with this deviation should be considered As Low as Reasonably Practical (ALARP).

Through benchmarking studies against a number of varying risk evaluation tools, the BASF Risk Matrix has been demonstrated to not be overly conservative. Although BASF's Risk Acceptance Criteria is on the same level as industry-wide conventions, application of the BASF Risk Matrix places BASF's residual risk level on the conservatively low side.

2.1.2.4 Other Risk Assessment Tools

Generally, the BASF Risk Matrix is the only risk assessment tool used within BASF. The risk assessment is based on the Single-Failure principle, which states that only single, independent faults are considered as primary faults initiating the chain of events. However, this only holds true if the following assumptions apply:

- the primary faults are truly independent and
- the primary faults are rare events and happen by chance (highly infrequent) and,
- in addition, they do not persist for longer periods of time (short duration).

In cases of very high risk, BASF may decide to deviate from the single failure rule and to consider double jeopardy or a multiple failure scenario. These are case by case decisions at the discretion of the safety review team and/or operations.

In certain cases or technologies that may have a high number of "low frequency-high severity" scenarios, a Fault-Tree Analysis has been utilized as a means to challenge the adequacy of the planned safeguards by also taking into account multiple failure scenarios (e.g., dinitrotoluene process).

QRA is a tool that is generally not applied by BASF. However, it is used in countries where it is compliance driven due to regulatory obligations, e.g., Belgium and The Netherlands, as well as for some sites in North America supporting facility siting studies. Of course, quantitative calculations are needed and used for most facility siting studies, but these calculations are not considered to represent a full QRA.

To cope with these "low frequency-high severity" events, BASF's approach is to support the primary risk reduction measures by appropriate mitigating measures, e.g., secondary containment for highly toxic materials (containment chamber for processes handling highly toxic substances, e.g. Phosgene, Hydrogen Cyanide, Acrolein, etc.).

2.1.3 Revalidation Process

2.1.3.1 Plant Safety Concept and Clean Sheet Review

The program for process safety reviews of existing plants was introduced to eliminate safety relevant gaps which were not identified in previous reviews, were introduced by inaccurate or incomplete Management of Change Reviews (MOCs), or became apparent due to an incident.

Included in the reviews are "non-routine" activities, such as start-up and shut-down, plant sections not previously reviewed, e.g. off-gas systems, loading-unloading stations and tank farms, as well as acquired plants with varying safety standards. The program is based on multiple pillars:

- Each plant is obligated to have a documented Plant Safety Concept. As already described previously, the Plant Safety Concept documents the identified hazards of a process/plant and the measures for controlling those hazards. The Plant Safety Concept describes protection strategies, whose implementation in the plant is then checked in detail in a Clean Sheet Review.
- Periodic reviews (at a minimum of every 5 years for plants handling hazardous chemicals) are required to ensure that the Plant Safety Concept is up to date. This revalidation of the Plant Safety Concept includes consideration of revisions to the documents as a result of MOCs, small capital projects, changes to regulatory requirements, lessons learned from incidents (both within the unit/technology as well as applicable industry cases), or other changes in the process itself that might affect the accuracy of the existing documentation. During the revalidation of the Plant Safety Concept, any significant changes to the Plant Safety Concept are verified through a corresponding implementation check.
- A periodic Hazard Analysis (minimum of once every 10 years for plants handling hazardous chemicals) utilizing a Clean Sheet Review approach is required to reconfirm the implementation of the Plant Safety Concept in the plant. The implementation check itself is part of a hazard assessment and verifies that the Plant Safety Concept has been correctly implemented in the current plant. A number of different methodologies are used to fulfill the documentation of an implementation check, including P&ID review, HAZOP study, What-If analysis, etc., as approved for use. The selection of a methodology is based on the hazard potential and nature of the process or plant section being evaluated. Starting a safety review with a Clean Sheet is taking a fresh look at the hazards of the process. Revalidation of existing documentation does not always discover still hidden gaps. Clean Sheet is a term used to describe an approach whereby a new document is created starting with only the previously identified deviations (initiating events) and respective causes as the basis for the assessment, and any existing implementation check is only used as reference.

To manage the planning of the Clean Sheet Reviews/Revalidation of the Plant Safety Concepts as well as the respective personnel resources for carrying out the reviews a tool was developed, i.e. Global Plant List. For an even work load over the 10 year period (a periodic Hazard Analysis is required every 10 years; see above) the Global Plant List also contains the planned year of the review for each plant section. Over the last 4 years the program was ramped up and is now approaching the desired level of reviewing 10 % of the plants per year.

Table 3: Plant Safety Concept, Revalidation & Clean Sheet Review

Timeline		5th Year of Operation	10th Year of Operation	15th, 25th, ... Year of Operation	20th, 30th, ... Year of Operation
Plant Safety Concept Periodic Revalidation	Format	Plant Safety Concept	Plant Safety Concept	Plant Safety Concept	Plant Safety Concept
	Activity	Revalidation	Revalidation	Revalidation	Revalidation
Check of Implementation "Clean Sheet Review"	Check of	N.A.	Piping & Instrumentation Diagrams (PID) of plant sections with medium and high Hazard Potential	N.A.	Piping & Instrumentation Diagrams (PID) of plant sections with medium and high Hazard Potential
	Format		Diagram Review for medium Hazard Potential		Diagram Review for medium Hazard Potential
			Hazop for high Hazard Potential		Hazop for high Hazard Potential

2.1.4. Implementation Learnings

BASF has established an effective process safety management system to cope with today's challenges, including a large number of acquisitions having different safety cultures, mix of old (some plants are dating back almost 100 years) and new plants, as well as technologies.

The multi-step system of SHE reviews during the planning and construction ensures that all new process plants/facilities within BASF meet the highest process safety and SHE standards.

With the recently introduced program for process safety reviews of all existing plants, BASF now has a comprehensive system in place to maintain these high standards throughout the life cycle of a production facility.

With these two elements among others, BASF has a system in place that enables BASF's overall process safety performance to improve continuously while constantly and effectively being on guard to prevent major incidents in their plants.

The effectiveness of BASF's management system approach is validated to a certain extent by noting that there has been a continuous reduction of the average severity of process safety incidents over the last 10+ years. The reduced severity is reflected in lower release volumes, fewer injured workers and smaller financial damages from the incidents.

However, there is room for improvement that can be derived from analyzing incidents. Less than 20% of the process safety incidents today are attributed to design causes which are tackled in the SHE review process. The remaining 80% of the incidents have operational or mechanical integrity causes and these are not yet addressed as effectively as the design causes. Knowing these gaps allows for opportunity to strengthen the evaluation process to further elaborate operational and mechanical integrity issues that lead to process safety incidents.

Methods beyond HAZOP and LOPA might bring incremental value, but caution is needed in ensuring the boundary conditions and assumptions are valid. Otherwise, it is an exercise to create a lot of paper that makes people feel good but does not improve the level of protection.

2.2. Celanese Approach

2.2.1 History

In the early 1990s, Celanese began initial compliance activities with the then new OSHA Process Safety Regulation [3] requirement to perform process hazards analyses. These analyses were implemented as Process Hazard Analyses (PHAs), which are now called Hazard Identification Risk Assessments (HIRAs) at Celanese. The initial schedule for regulatory HIRAs included the more hazardous process units, such as oxidation reactions as first priority. These very first HIRAs necessitated a strong hazards identification method and as a result HAZOPs were performed almost exclusively. Recommendations and findings from these hazards identification studies were made on a consequence only basis using team consensus opinion.

The organization quickly realized that making these critical decisions on a subjective basis was not sustainable. Assessment of consequence via simulation tools and calculation of frequency of occurrence of that consequence via a defined method was needed. A migration towards risk based decision making resulted.

A multi-site team was formed and a consultant engaged (A. D. Little). The consultant proposed a simple LOPA method in 1994. Celanese piloted this method in which various layers were simply "counted" on several reaction systems. Due to the complex system interactions, common cause issues, enabling conditions, conditional modifiers and multiple scenarios in which the same consequence could be reached, the LOPA method did not provide the desired needed rigor.

An alternative approach was developed using simplified fault and event trees. The process was dubbed Risk Assessment/Risk Analysis, or RARM for short. It is still in use at present.

2.2.2 Current Process

The HIRA process in use at Celanese today includes up to 11 different elements depending upon the process being analyzed. These elements include: 1) process safety information audit, 2) preliminary consequence assessment, 3) hazard identification methodology selection, 4) preliminary consequence estimates, 5) hazard scenario grouping, 6) fault tree construction, 7) event tree construction, 8) risk matrix evaluation, 9) risk response evaluation, 10) managing findings and 11) facility siting. Each element will be described below.

2.2.2.1 Process Safety Information (PSI) Audit

All categories of PSI are audited prior to conducting the HIRA. Any deficiencies are corrected. Correcting PSI can impact the HIRA schedule, so the audit activity needs to be accomplished early. Celanese has guidance in this area to start the "HIRA" activities as much as one year prior to the publication date to allow for any significant PSI issues.

2.2.2.2 Preliminary Consequence Assessment

Using simulation tools, such as DNV's Phast™, Baker- Strehlow-Tang Vapor Cloud Explosion Model, ALOHA, and others, a range of possible release scenarios from the process are examined. A representative number of small, medium and large release scenarios of hazardous materials are evaluated. This is best done by section of the process unit. Sections would include areas like a tank farm, a reaction area, a purification area and so on. The HIRA team then uses the results as a calibration of consequence when performing the hazards identification studies. If individual models were run for each identified scenario, efficiency of the HIRA would be impacted. However, in some cases, additional models are needed.

2.2.2.3 Hazard Identification Methodology Selection

Recognized and Generally Accepted Good Engineering Practices (RAGAGEP) may cover some components of a process unit. For example, the burner management system of a power boiler is completely defined in NFPA 85 [4]. Given that a portion of the process unit is specifically covered by a RAGAGEP, then a simple audit of that equipment against the applicable RAGAGEP would suffice to identify hazards as well as concerns that require mitigation. Great care must be taken to understand the scope of equipment covered by the RAGAGEP. Celanese maintains developed a list of RAGAGEPs that HIRA teams need to prescriptively review and apply on an as-needed basis.

The current process safety regulatory framework excludes combustible solids. From a Celanese perspective, these hazards warrant our attention for worker safety. We have attempted to use the HAZOP tool in this domain, with unsatisfactory results. The HAZOPs took a long time and missed hazardous dust scenarios. A NFPA 654 [5] review of a process unit in combustible dust, such as a room full of extruders, is an ideal manner in which to identify and manage those dust hazards.

For the sections of the process not specifically covered by RAGAGEP, appropriate Hazards Identification (HI) Methodologies are employed. Celanese allows four HI methods at the discretion of the trained HIRA study leader: (1) HAZOP, (2) Failure Mode and Effect Analysis

(FMEA), (3) Checklist and (4) What-If. For complex process units involving highly hazardous chemicals, HAZOP is the best choice to identify hazards. FMEA can be used for equipment or instrumentation heavy systems; however, FMEA is weak in identifying human failure scenarios and is never used by itself. Checklists are used for specialized studies, such as Human Factors analyses and RAGAGEP studies. What-If is often used to analyze simple utility systems.

2.2.2.4 Preliminary Consequence Estimates

During the course of the HI studies, prior to moving on to the next hazard scenario a simple question is asked: "What is the "Ultimate Credible Consequence" if this scenario plays out and all protective functions fail?" If the consequence is operability or reliability only, then that is noted and a mechanism to transfer that concern to the correct functional area is employed. This typically would be to list the concern and hand off to the appropriate function to address, but not include in the HIRA report itself. Environmental or injury consequences are conservatively graded in a proprietary consequence scale using the preliminary consequence analysis information to aid in team decision making. This assessment need not be precise because when the consequence is considered to be high, it will be assessed further to better define that consequence.

2.2.2.5 Hazard Scenario Grouping

The risk criteria in use at Celanese assumes there are frequencies calculated for events, not scenarios. An event would be a reactor runway or uncontrolled exothermic reaction in which the reactor could burst. There may be 10 or more scenarios that can lead to that event. In order to manage the groups, rules are needed to prevent too much or too little grouping of scenarios. Grouping too little (as is often the case in LOPA) can lead to low resultant risk and no action. Grouping too much leads to alarmingly high risk and inappropriate action. Grouping at an "appropriate" level ensures event risk is controlled. Only scenarios leading to consequences of concern are grouped. Other minor consequence scenarios are eliminated screened from further analysis.

There are two types of failures that can be imagined which lead to hazardous releases:

- Passive Failures - Failures that can occur to equipment operating under normal pressure, temperature, flow, composition, etc. These failures can occur in any process at any time and are not unique. Examples include a flaw in a weld or a crack in a casting. Failures of this type are typically managed by the asset integrity function through defined equipment care plans. In other words, HIRA teams are not asked to march into the plant and measure pipe wall thicknesses or check weld seams with fluorescent dyes. Although not the primary function of HIRA, there are occasionally situations in which passive failures are assessed. For these situations the RARM process has a strict rule set to prevent over grouping. Those rules include, but are not limited to, requirements for the scenarios to be in the same process unit, involve the same "named equipment," operate under similar conditions with respect to temperature, pressure, and chemicals, be of the same magnitude regarding dispersion distances to levels of concern, and so on.

- Active Failures or Activity Based Failures - Failures specific to the process technology. These are related to controls, interlocks, human actions, and procedures that are specific to the process under study. It is not feasible to take the control system from a Methanol

Unit and duplicate those controls in an Alkylation Unit. Nor is it feasible to take the operating procedures for an Alkylation Unit and run a Methanol unit using those procedures. These controls and procedures are unique features of the process technology that must be examined in a HIRA as their failures can lead to release scenarios. Again, there are defined criteria for grouping these; however, these criteria tend to be more intuitive. An example, as previously mentioned, is an uncontrolled exothermic reaction. There may be many scenarios leading to this, all of which would be grouped together in a single fault tree, typically connected logically using an "or" gate. Scenarios may include: loss of cooling, too much Reactant A, not enough Reactant B, too much catalyst, agitation failures, and so on.

2.2.2.6 Fault Tree Construction

Fault trees in the RARM process are simplified in that the tree is built to a defined limit. In some fault tree implementation, the failure of a wire or corrosion at a screw terminal is discretely shown in the fault tree with other scenarios leading up to a transmitter failure. In the Celanese implementation of fault trees, all failures of that transmitter are considered in a single numerical value, therefore, the fault tree construction typically ends at a transmitter failure, or a controller failure, or an automatic valve failure, and so on.

The top event of the fault tree is defined as a release of material or energy from the process. That top event is defined by failures beneath, such as controls failures or human errors and failures of protective functions. The logic is present in a diagram which aids the user in visually understanding the interconnectivity of the components. Cut set analyses are preformed to test for logical errors in the construction.

Fault trees, once deemed logical by the team, are next quantified. The data used is derived from Celanese specific and industry experiences. It is maintained in a comprehensive format for internal use only. All facilities in Celanese use the same data set to calculate frequencies to ensure consistency in application of company resources to risk control. Facilities may increase failure frequencies and probabilities for local factors, such as environment, severity, etc., however they cannot reduce failure frequencies or probabilities without approval from the corporate Process Safety group.

2.2.2.7 Event Tree Construction

Event trees are used to analyze the range of possible consequences of a release from the process. For flammable releases, ignition probabilities are assessed. For toxic releases, wind directions are important to understand. In all cases the presence of people to receive the hazards is assessed.

Ignition probabilities are described in the RARM database and include factors such as rate of release, duration of release, area impacted by release and reactivity of the material. Nomographs have been developed for internal usage to determine probabilities of immediate and delayed ignition. Once ignited, the probability of fire versus explosion is assessed considering size of release and geometry of the volume of the plant impacted by the release. Dispersion modeling to ½ Lower Flammable Limit (LFL) is used to define the potential impact zone. Baker Strehlow

Tang vapor cloud explosion modeling helps define blast impact zones. Phast™ or ALOHA can be used to develop thermal radiation impact zones for a fire event.

Toxic impact zones are determined using dispersion models typically taken out to Emergency Response and Planning Guidelines -3 (ERPG-3) [6] isopleths for onsite injury concerns and to ERPG-2 for offsite health concerns. There is not a large population of EPRGs published; however, Westinghouse under contract to the Department of Energy developed temporary ERPGs, (called Temporary Emergency Exposure Limits or TEELs) that are intended to be used as ERPGs until such time as an official ERPG is published [7]. The expectations are that the published ERPG would be the same value as the TEEL, but this is not always true.

Worker impact probabilities are derived from two probabilities:

- For a release scenario which impacts workers outdoors, but does not impact workers in a safe haven building, the first probability calculated is the probability a worker assigned to that unit is outdoors. These probabilities are assessed via interviews with the workers and are calculated as the fraction of time outdoors.
- Given that the worker is outdoors, the probability that they are also in the impact zone is calculated. The impact zone of a flammable release is that area defined by the plume to ½ LFL, a thermal radiation level of concern, or a blast radius. The impact zone for a toxic release is defined to ERPG-3 or ERPG-2, for onsite or offsite effects respectively. To define a probability, some simple math is employed. The operating unit occupies a measureable area of real estate. The plume has a simulated footprint area. Dividing the plume footprint area by the unit area results in a dimensionless number which can be used to represent the probability that a worker outdoors is within the impact zone of the release. Internal simple criteria are used to define the consequence of various human exposures as derived from published thermal radiation data, Probit analysis [8] and other consensus values.

Calculations are made across the branches of the event tree. Multiple frequencies are possible for the same consequence. All frequencies calculated for the same consequence are summed together to represent the total frequency of that consequence from the release event. There may be multiple consequences of concerns, leading to multiple pairs of consequence and frequency for a given event.

2.2.2.8 Risk Matrix Evaluation

Celanese uses a five-by-six risk matrix. There are three regions defined in the risk matrix. High consequence and high frequency events result in "intolerable" risk which must be reduced. Very low frequency events fall into the "negligible" risk region that requires no further risk reduction. The region between "intolerable" and "negligible" is the As Low As Reasonably Practicable (ALARP) area of the matrix in which risk is reduced based on internal company requirements. The Celanese risk matrix incorporates many benchmarks including references from the Dutch, English, Hong Kong, and sharing session with other companies.

Once the various frequencies of consequences are calculated in the event tree, those pairs of data are used to enter the risk matrix. For that event, the highest risk calculated by any frequency and consequence pair defines the risk for that event.

2.2.2.9 Risk Response Evaluation

For "intolerable" risk, there is a stated company criterion for assessment and review of the calculations in a defined and managed time schedule. Approvals for further operation go to the highest level of the company. In the 20 year history of RARM, this level of risk has been infrequently calculated and in those few cases, the company has elected to not operate, even on a temporary basis until the risk is reduced. For "negligible" risk, no action is warranted. For risk in the ALARP region, the risk reduction criteria are scaled to the location of the controlling frequency / consequence pair in the matrix. The closer the calculated risk is to the "intolerable" region, timing and approval requirements become strict and allowable risk reduction activities are broader.

2.2.2.10 Managing Findings

HIRA reports and findings are reviewed by the HIRA team (of course), technology experts, site management teams and the law department. HIRA findings are required to be closed in a timely fashion as defined by company policy. Once a plan is agreed and documented, any change to that plan must be approved by corporate technical resources, corporate EHS, the law department, and ultimately, the Manufacturing Vice President.

2.2.2.11 Facility Siting

Part of the HIRA is a facility siting analysis in which the relative locations of the hazards to the populations are assessed. During a unit specific HIRA, Celanese applies one or more Siting Checklists to take a close up look at unit siting issues.

A facility wide study is done centrally for all Celanese operating facilities on a five-year schedule. In this study explosions, fire and toxic events are simulated for each processing unit. A facility often has many processing units. Then the impact of those events is determined on occupied buildings. Societal risk and individual risk are calculated and managed for existing facilities. For new facilities Celanese applies a stricter, harsher, consequence only, criteria.

2.2.3 Revalidation Process

The Celanese HIRA revalidation procedure follows the CCPS Guideline book on that topic [9] which requires engineering judgment as part of the procedure. If the last HIRA is acceptable and meets all requirements AND the MOC process is functioning specifically with high quality MOC safety reviews AND there have not been many significant or major Process Safety Incidents, then a Revalidation HIRA may be performed. Large issues in any of these three areas would force a redo of the entire HIRA. Often corporate resources are engaged to help the sites make the determination to redo or revalidate. In some cases where the process unit has changed significantly, such as added a third reactor, a redo may be suggested even though the project team would have performed a HIRA on the project. The following elements are included for the revalidation:

- PSI Review - All categories of PSI are audited and any deficiencies are corrected.
- Prior HIRA Review - A checklist is used to review the prior HIRA to see if a revalidation is possible. This is done by the HIRA team leader in a similar fashion as an audit. If the prior HIRA meets all company and regulatory requirements, a charter for a Revalidation Team is proposed to management. The HIRA team then reviews all Hazard

Identification Log sheets and all Risk Assessments. Any small problems noted are corrected and updated by the team. Large issues would force a complete redo of the HIRA.

- MOC Review - The HIRA team reviews MOCs performed since the last HIRA was performed specifically making quality judgments on the MOC safety review activities. Any small problems noted are corrected and updated by the team. Large issues would force a complete redo of the HIRA.
- Incident Investigation Review - The HIRA team reviews process safety incident investigations performed since the last HIRA and makes quality judgments on the root cause analysis. Any small problems noted are corrected and updated by the team. Large issues or repeats of significant or major incidents would force a complete redo of the HIRA.
- RAGAGEP Review - The HIRA leader works with corporate resources using the company RAGAGEP list to determine which RAGAGEPs are required to be reviewed in detail during the HIRA. If no RAGAGEP review was done in the prior HIRA, then in the course of the revalidation HIRA a RAGAGEP review is performed where applicable. If the prior HIRA included a RAGAGEP review, then that study is also reviewed during the revalidation. If the prior review was to an earlier version of the RAGAGEP, the revalidation HIRA would perform a new review using the most current version.

After the revalidation, a report is reviewed, then issued and any resultant recommendations are managed to closure.

2.2.4. Implementation Learnings

Since the initial OSHA PSM compliance activities in the early 1990s, many lessons have been learned as Celanese's process safety analysis and evaluation techniques evolved to their current form. Overall, within Celanese, the primary learning is that qualitative, consequence-based decision making dilutes the focus to improve process safety and overloads plant project resources. Rigorous consequence assessments using simulation tools such as Phast™, can lead to better focus on scenarios with the greatest potential impacts. During the move to risk based decision making at Celanese, as prior HIRA recommendations that were open at that time were evaluated. It was often the case that the qualitative consequence originally estimated was misrepresented and the recommendation was not valid. Given a rigorous consequence assessment, frequency assessments and the ultimate risk calculations performed on those prior HIRA recommendations, it was often the case that existing engineering and administrative controls adequately controlled the risk and the recommendation made on a qualitative basis was not needed. There were many examples where prior HIRA recommendations, if implemented, would not reduce risk. There were even a few examples where prior HIRA recommendations, if implemented, would increase risk. Those recommendations were revised with more robust solutions.

It was also observed within Celanese that simple techniques such as LOPA struggle with risk assessments of complex systems. For example, a pilot of a fault tree based risk process was performed on a complex reaction system that had four different chemical reactions that could credibly lead to an uncontrolled exotherm and toxic vapor release. With each of the four possible reactions there were multiple scenarios as causes. Usage of fault trees allowed proper assessment of that complex scenario logic and definition of mitigation to control risk. LOPA was attempted, but could not perform this analysis.

2.3. The Dow Chemical Company Approach

2.3.1 History

The late 1980s and early 1990s were periods of rapid risk assessment development. The Dow Chemical Company (Dow) already had a number of valuable process safety tools at this time such as its Loss Prevention Principles, Process Hazard Analysis (including a specific focus on chemistry), Fire and Explosion Index and Chemical Exposure Index. But it was apparent that there were new developments that offered opportunities to improve risk management. A number of different risk assessment methodologies were gaining popularity and these were evaluated for possible use within Dow. The conclusion of this evaluation was formally documented in 1994 as the Process Risk Management Standard (PRMS). The PRMS set out the four step risk review process detailed below but also, crucially, linked this process to the company risk criteria. The link to the company risk criteria ensures that all stages of the risk assessment process have the same risk management objectives. The PRMS has been updated several times since 1994 but the concepts and process outline have remained the same.

2.3.2 Current Process

Dow uses a four step risk review process to manage the risk of acute process hazards. Acute process hazards are those hazards that could result in instantaneous or short duration effects from fires, explosions or chemical releases. This approach provides criteria for screening, analyzing and prioritizing risk reduction opportunities. In all situations, if local laws or regulations are more conservative, the local laws or regulations will take precedence.

Once identified, risk must be evaluated to determine if the risk is considered tolerable. Dow uses two basic approaches to evaluating risk tolerability. The more traditional approach is qualitative in nature and typically involves implementation of design and operating standards derived from the company's collective experience, industry standards and regulations. This qualitative approach at Dow includes an assessment against Dow's Company Loss Prevention Principles (LPPs) and Business Most Effective Technology requirements (METs). Design, installation, modification and operating practices that relate to process safety and loss prevention are documented in these LPPs and METs. All projects and operations are assessed both initially and periodically against these requirements. The second approach is quantitative in nature and involves comparing the risk with numerical criteria. If the risk exceeds predefined and agreed upon companywide risk criteria, methods of reducing the risk must be implemented, the activity creating the risk must be discontinued, or the activity creating the risk must be reviewed with corporate management to agree upon appropriate actions. When the risk has been reduced to a tolerable level, the risk assessment is complete. The risk assessment is revalidated periodically or when significant changes are made to the process.

The Dow 4-Step Risk Review Process

An overview of the Dow Risk Review Process is shown in Figures 1 and 2. The philosophy incorporated into this risk review process is to escalate the levels of risk reviews for the chemical processes which are identified to have higher risk potentials based upon the more general risk reviews. Each level represented in the figures and acronym definitions will be discussed in subsequent text.

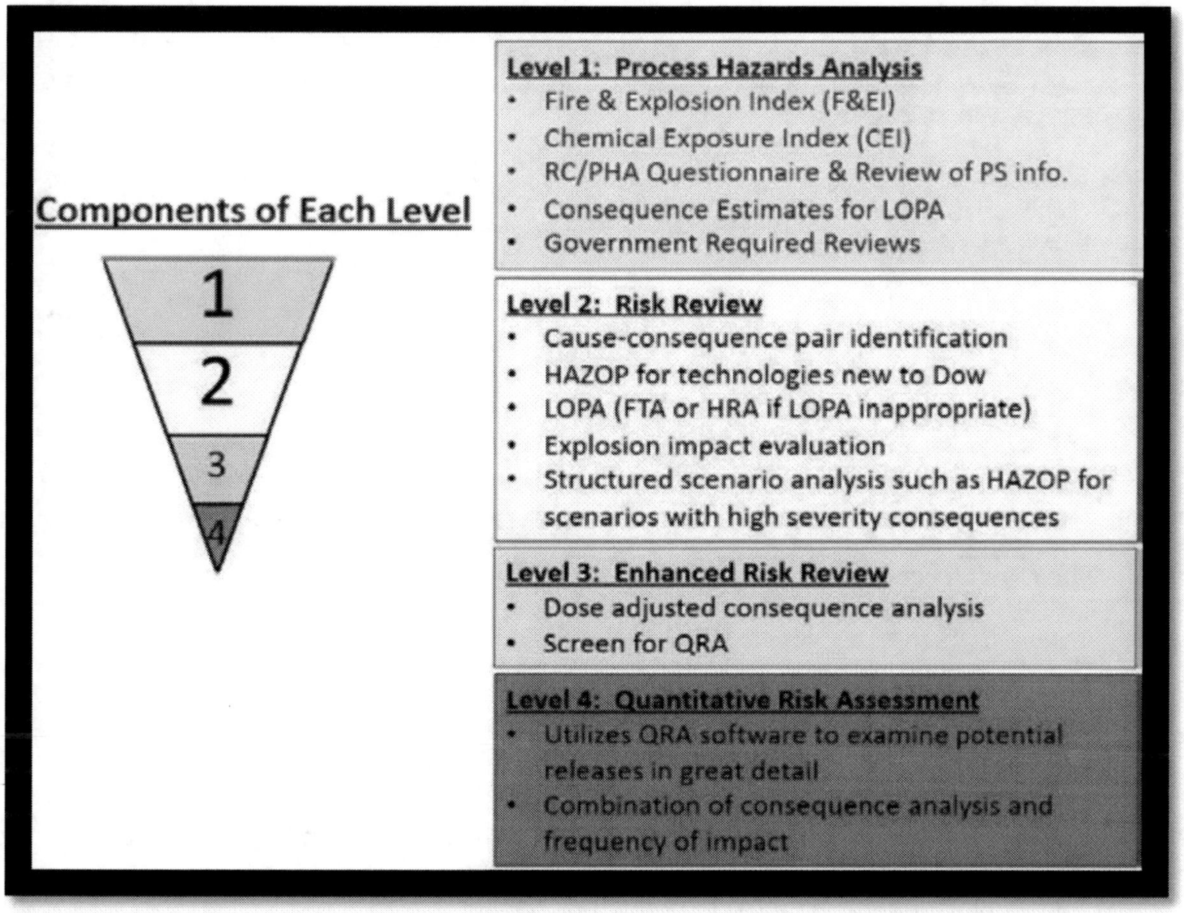

Figure 1: The Dow Process Risk Management Process – Components of Each Level

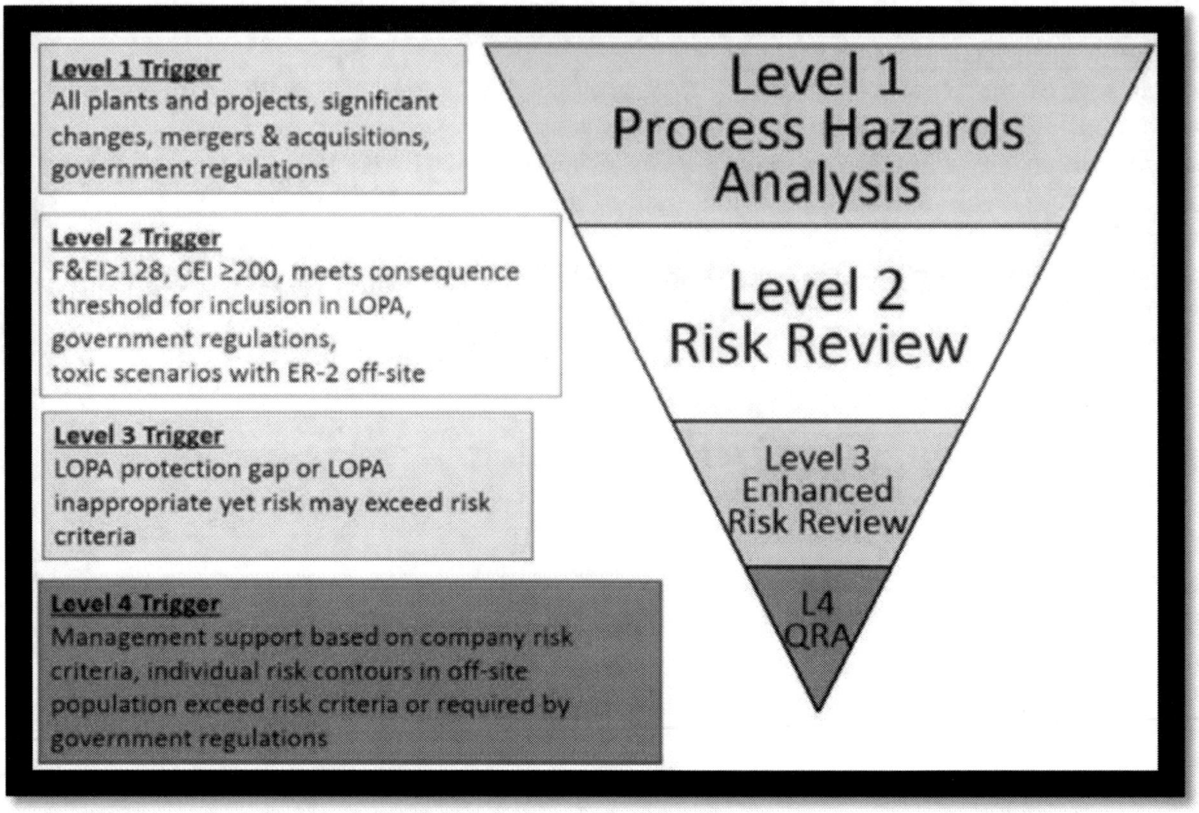

Figure 2: The Dow Process Risk Management Process – Triggers for Review Level

Level 1 Process Hazard Analysis – Applied to all processes

The first step of the Dow Process Risk Management process is to conduct a Process Hazard Analysis or Level I Risk Review. This is done using many of Dow's traditional risk review tools. For a chemical manufacturing facility, the primary tools will be:

- The Dow Reactive Chemical/Process Hazard Analysis (RC/PHA) Questionnaire and Protocol [10]
- Fire and Explosion Index (F&EI) [11]
- Chemical Exposure Index (CEI) [12]
- Consequence estimates for each LOPA scenario identified in the RC/PHA. [13]

The RC/PHA helps to identify scenarios and potential applications of company and external process safety standards and guidelines. The RC/PHA involves interaction with a multifunctional team including several diverse technical experts to study the chemistry, process safety information and processing activities, looking for any potential risk reduction opportunities. This is similar to a checklist approach for process hazard analysis. The CEI, F&EI and LOPA consequence evaluation provide a first estimate of the hazard potential. The CEI and F&EI provide a relative ranking for toxic and flammable impacts respectively. Higher numbers for the CEI and F&EI represent higher levels of risk potential. The LOPA consequence estimates are based on the maximum number of people potentially impacted by scenarios identified in the RC/PHA

Criteria for escalating to a Level 2 Risk Review

At this point in the process, a decision is made on the need for further risk review based on the following criteria:

- F&EI ≥128
- CEI ≥ 200
- Meets or exceeds the company consequence severity threshold for inclusion in LOPA
- New technology to Dow which is judged as presenting significant risk
- Management judgment regarding activities which warrant evaluation because of potential injury to people or damage to property or the environment.
- Government requirements specify Level 2 analysis tools be used
- CEI scenarios where the ER-2* concentration extends beyond Dow property

***ER-2** is similar to the ERPG-2 concentration as developed by the American Industrial Hygiene Association's Emergency Response Planning Committee. [6] Where an ERPG-2 value has been developed this is usually used as the ER-2 value. For those materials where no ERPG-2 exists, other values such as the Acute Exposure Guideline Levels from the US Environmental Protection Agency may be adopted.

Facility unit operations meeting any of these criteria must proceed to Level 2 of the risk review process. In reality nearly all chemical plants proceed to Level 2. The facilities that are only assessed at Level 1 are usually pilot plants that only handle hazardous materials in small quantities or plants that do not use materials with a significantly hazardous nature.

Level 2 Risk Review

A Level 2 Risk Review consists of more quantitative hazard identification and analysis than was carried out in Level 1. The Primary Tools utilized for Level 2 Risk Reviews include:

- Cause-Consequence Pair Identification - used to better develop/define scenarios for LOPA. For new technologies to Dow, HAZOP is the recommended tool to identify scenarios before the LOPA analysis is performed.
- LOPA - a semi-quantitative frequency analysis tool, evaluates risk tolerability of a scenario using consequence estimates from Level 1 and standard frequencies and probabilities.
- Explosion Impact Evaluation - determines construction requirements for occupied buildings near potential explosion sites.

For higher severity consequences, cause-consequence pair identification must be very thorough. To that end, a structured scenario analysis must be completed to assure that there are no other previously unidentified initiators of the scenarios or other high risk scenarios. The structured scenario analysis may use any of the following hazard identification and scenario development tools:

- HAZOP
- What-if Checklists
- Failure Mode and Effect Analysis
- Proprietary Scenario Screening Tool [14]

Fault Tree Analysis, Event Tree Analysis, Human Factors Analysis and dispersion modeling are other tools that may be used by the risk analyst as Level 2 tools for special circumstances or to address unlikely problems.

Criteria for escalating to a Level 3 Risk Review

If the LOPA analysis performed in the Level 2 Risk Review identifies one or more scenarios for which risk cannot be adequately mitigated to the company's tolerance of a particular outcome, or management judgment indicates additional risk analysis should be pursued, consideration is given to completion of a Level 3 Enhanced Risk Review. If LOPA is deemed to not be the appropriate tool to evaluate some risks (e.g., mechanical integrity failures) and the risks associated with that scenario are anticipated to exceed the company's risk tolerance - this would be another case when proceeding to a Level 3 Risk Review would be considered. Typical cases for risk assessment above Level 2 would be situations where the local population is close to the facility or plants that have a significant inventory of highly toxic materials.

Level 3 Enhanced Risk Review

The Level 3 Enhanced Risk Review will further refine the potential consequences of an event in the area surrounding that facility. There are two stages for a Level 3 review, 3A and 3B. These examine individual scenarios, typically from the previous LOPA study and apply Dose Adjusted Consequence Analysis (DACA) to achieve more accurate assessments of the risk. A 3A review is a screening tool for toxic impacts and a 3B review is used to more accurately assess the potential impact for events that do not screen out in the 3A process. The DACA studies use probit [15] data specific for the toxic chemical(s) being evaluated to estimate the probability of fatality based upon the chemical concentration and duration of exposure. Probit analysis is commonly used in toxicology to express numerically the relative toxicity of chemicals to living organisms. Dow only uses this type of analysis for reviews which proceed to Level 3 or Level 4 in the risk review process. The primary tools used for Level 3A and 3B Risk Reviews are the Phast™ and Safeti™ Software.

In Level 3A reviews, the first step is to examine the distance to potential public or on-site receptors compared to the toxic impact distance predicted by toxic dispersion modeling using dose analysis. Some scenarios may screen out at this point because the calculated release does not have the potential to impact the receptors. Level 3B reviews can be conducted to refine LOPA consequence estimates for toxic scenarios which do not screen out during level 3A reviews. In 3B reviews, the LOPA consequence is estimated by applying QRA techniques to individual scenarios and comparing this directly with the company's risk criteria. However, cumulative risk is not calculated.

A further option exists to perform a Level 3B review in which Individual Risk Contours (IRC) are developed for the facility. If the IRC that represents a generally acceptable risk can be shown to be entirely inside the site boundary then further escalation to a Level 4 risk study is not generally pursued. This decision is made by company and business management with engagement from process safety leadership.

Criteria for escalating to a Level 4 Risk Review

A Level 4 Risk Review, which is a complete QRA, is considered when any of the following situations occur:

- Level 3 Risk Review indicates that the risk posed to area(s) outside the site fence line that are normally populated by members of the public may be above predefined and agreed upon company risk criteria
- A QRA is required by local or national government regulations
- A QRA is required by some other external requirement (e.g., a business opportunity which involves another company and that company's risk review processes require the completion of a QRA)
- A QRA is supported by company management, business management and process safety leadership

Dow's LOPA process has inherent conservatism which normally ensures that controlling the risk for each individual scenario will result in an acceptable overall risk. However, at the extremes of the range of analysis with plants that have many potentially high hazard scenarios in their LOPA, it is possible that the inherent conservatism will no longer be sufficient to ensure that the overall risk is acceptable. Where this is deemed possible by company management, business management and process safety leadership, the assessment of the risk is escalated to level 4 where the overall cumulative risk is assessed using QRA.

Level 4 Quantitative Risk Assessment

A QRA utilizes highly sophisticated, probabilistic, mathematical modeling techniques to produce overall numerical estimates of risk. The primary tool utilized for Level 4 Risk Reviews is Safeti™ Software. Similar to Level 3A and 3B Risk Reviews, probit equations for the toxic chemicals under review are used to estimate potential consequences.

A QRA study begins by estimating the probability of occurrence and consequence. For each leak location, dispersion analysis is performed for many different wind directions per the site's specific weather data. Other details such as release height, direction and velocity are specified as well. For each scenario where the plume extends into populated areas, a mathematical estimate of fatalities is generated. The QRA software sums the probabilities of the initiating event times probability of wind direction times probability of population impacted in each area to arrive at an overall numerical probability estimate of potential impact on people. The estimated cumulative frequency of all remaining incidents which could cause a fatality are calculated. Once all applicable scenarios are modeled, a cumulative risk profile for the facility is compared to the company's established risk criteria to determine if any changes are needed.

Residual Risk

When it is determined that none of the Level 2, 3 or 4 Risk Review Criteria are exceeded, the risk analysis portion of the risk review is complete. Management procedures are then reviewed to ensure that appropriate and effective process safety management systems are in place and the residual risk is being monitored and controlled as planned.

2.3.3 Revalidation Process

All risk assessments are revalidated. Where an assessment is performed purely for regulatory reasons, then it is either revalidated at the frequency required by the regulator or more frequently. Assessments that serve both an internal and regulatory purpose are revalidated at either the internal or regulatory interval, whichever is the shortest. Internal revalidation intervals range from two to five years. Facilities classified as having a significantly higher than average potential hazard are assigned the shorter revalidation intervals.

Internal QRAs tend to represent the facilities with the higher potential hazards and are audited every few years. The audit examines:

- The assumptions in the original audit to ensure that they are still valid
- The operating conditions to ensure that they have not significantly changed
- The mitigation and protection layers in place at the time of the QRA to ensure that these are still providing the risk reduction assumed in the QRA
- General plant or site changes that might impact the QRA results
- Changes in population on or off the site
- The continued validity of the weather data used in the QRA

The QRA audit data collected is reviewed by the responsible process safety specialists and the QRA Subject Matter Expert. If warranted, a recommendation will be made to update the QRA study.

2.3.4. Implementation Learnings

All the risk management processes in use at Dow evaluate risks against the companywide risk criteria. Unlike risk matrices that generally have 1 or more of the potential severe consequences in the same matrix, the Dow risk criteria differentiates across a broad range of consequences. This application of the same risk criteria for all risk management processes allows Dow to consistently manage risk across the company regardless of the specific methodology employed.

An important finding from the implementation of risk management processes is that no single solution addresses all risk assessments. Detailed risk assessments such as QRA, which consume a lot of resources, are not a practical solution for all assessment requirements. However, less detailed methodologies such as LOPA rely on more assumptions and, if the assumptions are handled correctly, generally greater conservatism results. Spending additional time and money on the risk assessment has been shown generally to reduce the overall cost of a project because a more accurate assessment of the risk identifies more economically efficient risk reduction measures. Going forward, LOPA will continue to be the dominant risk assessment tool within Dow, but an increased use of more detailed tools such as dose-adjusted consequence assessment and QRA is very likely to occur. The more detailed tools will be used for specific situations such as for higher hazard facilities and locations with out-of-the-ordinary risks such as the site being near large populations.

2.4. *Eastman Approach*

2.4.1 History

Eastman Chemical Company's first foray into formalized process hazard identification was the development of the Hazard and Risk Assessment (HARA) process in the 1980s. This methodology specified the means of conducting HAZOP analyses on capital projects based on work by Ellis Knowlton at Chemetics [16] to both identify hazards and specify recommendations for improvement to the proposed process. In this same time period Eastman was conducting limited quantitative consequence analysis including developing internal programs for dispersion modeling of releases of toxic materials to assist in the design of new facilities.

With the promulgation of the OSHA Process Safety Management standard (29 CFR 1910.119) in 1992, Eastman evaluated their existing processes and developed a new methodology for conducting Process Hazards Analyses. The Eastman methodology focused on the development of unit operation and system component specific checklists and was documented in the Process Hazard Analysis and Risk Assessment (PHARA) procedure. These checklists included the primary scenarios and applicable deviation – guide word combinations for each type of system that were normally developed as part of a HAZOP analysis of each specific system. In addition, the checklists included questions which might be missed in a HAZOP analysis for specific unit operations.

To support Eastman's PHARA procedure, a computer program and associated work process was developed that allowed implementation of the specified methodology. This PHARA computer program became an enterprise solution for conducting and managing the data generated during PHAs. This software was rewritten in 2000 as an Oracle database application to be a comprehensive, flexible, and enterprise-wide solution. The new version of the software stored all PHA information for the entire corporation in a single database including all archived studies. Storing all of the data in a single relational database allowed recommendations to be managed across the entire organization or in any subset of the corporation.

Eastman identified specific scenarios which required a more in depth study to understand and determine if the controls were acceptable based on recommendations from incident investigations and scenarios with high loss potential and a perceived lack of traditional controls being in place. In some of these cases, Eastman enlisted the aid of a contractor to conduct detailed quantitative risk assessment. These analyses typically consisted of detailed fault tree analysis to determine the risk of the top event of concern.

In 2000, Eastman began the use of LOPA [2]. LOPA was used in a limited capacity initially as a means of achieving a relatively easy determination of the relative risk associated with specific scenarios. LOPA was used as a tool for comparing existing operations to more recently installed equipment to determine if additional risk reduction was to be recommended for the existing operation. To support the application of LOPA, Eastman developed risk elevation criteria to determine the appropriate level of management for the evaluation of acceptance of risk for the organization.

2.4.2 Current Process

Eastman Kingsport Site's overall risk assessment methodology is shown in Figure 3. Not all of the blocks within the overall risk assessment process are addressed in this paper but the flowchart does provide a view of the scope of the analysis for research, development, engineering, and operations at the Kingsport Site.

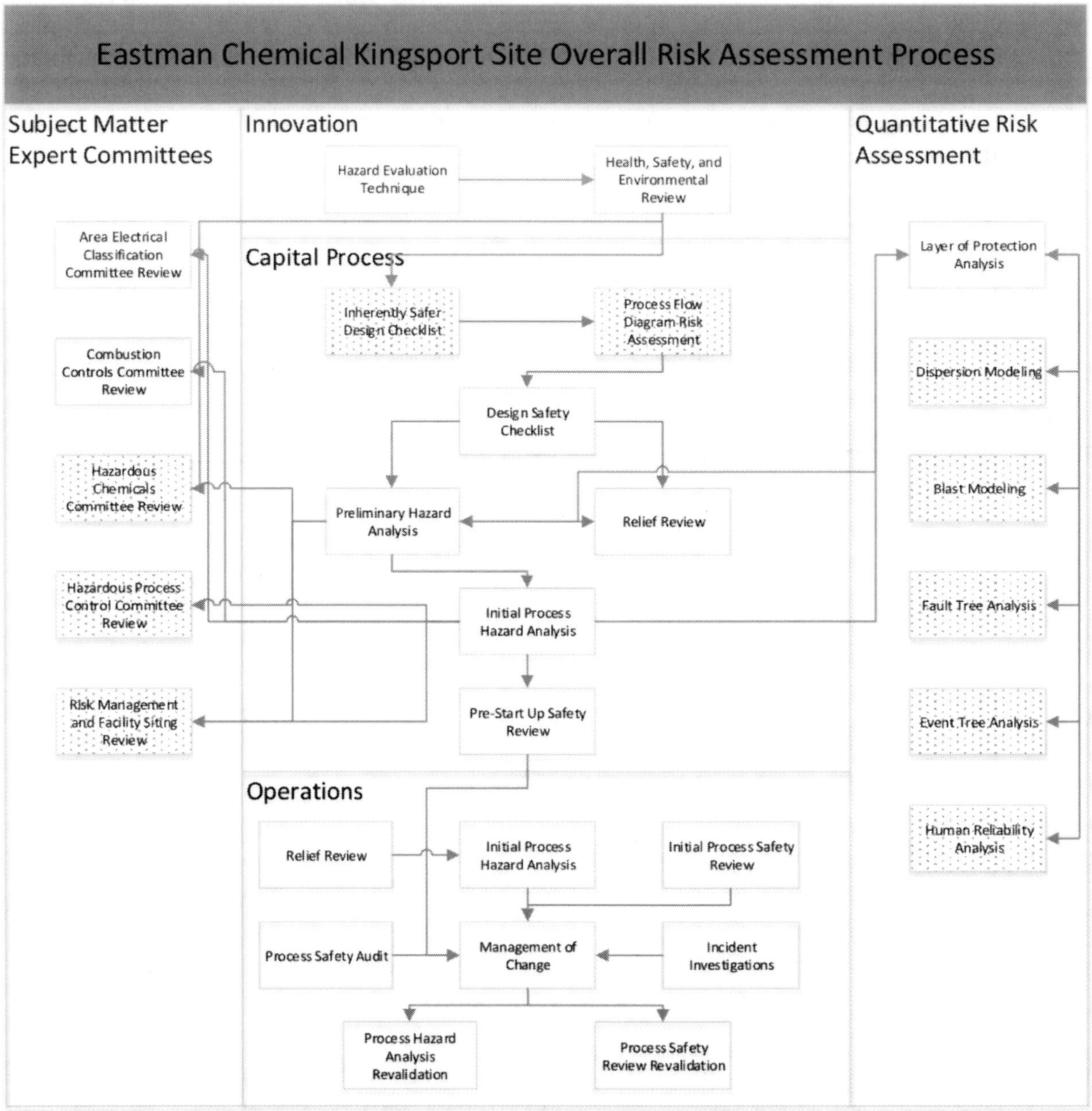

Figure 3: Eastman Chemical Kingsport Site Overall Risk Assessment Process

Eastman's policy requires hazard identification and risk assessment of all of the company's processes. The level of this analysis as well as the frequency of revalidation is dictated by the risk posed by the process as well as its coverage by the OSHA PSM regulation. All of the higher risk processes receive a full PHA and a revalidation every 5 years. Lower risk processes may have a less robust analysis of the risks.

As stated above, Eastman's primary tool for conducting a PHA is through a unit operation specific checklist. The checklist approach, as implemented by Eastman, requires that each question is addressed with a determination that the question is: a) not a concern for the equipment under consideration or b) requires documentation of the worst case credible consequences associated with the specific failure, a categorization of the consequence, a listing of the controls which prevent or mitigate the event, a qualitative determination of the adequacy of those controls, and a recommendation to correct any discovered deficiency. In cases where the PHA team is unsure of the adequacy of the controls, the recommendation is often to study the specific consequence, its associated controls and the adequacy of those controls using a more detailed methodology. If the team is unsure, consequence modeling may be recommended. Uncertainty about the controls adequacy will often result in a recommendation for more detailed frequency analysis typically in the form of LOPA.

2.4.2.1 Layer of Protection Analysis

LOPA is the primary quantitative risk analysis tool used within Eastman. Several different approaches exist throughout the organization due to different paths of historical development. The technique as practiced at the Eastman Kingsport Site (EKS) located in Kingsport, Tennessee is used as the basis for the discussion in this paper. LOPA can be initiated from a PHA recommendation, but is also often initiated due to a relief analysis determining that a relief contingency may be mitigated by the application of other controls, typically through application of safety instrumented systems. LOPA is the primary method to determine the required reliability and therefore, the safety integrity level of safety instrumented systems and can also be initiated when a SIS design is being evaluated.

The first step in a LOPA analysis at EKS is the development or the confirmation of the consequence used in the analysis since this determines the risk elevation criteria used as a target value in the LOPA. Often a qualitative evaluation of the consequence, such as that typically developed by a PHA team, is adequate for the selection of the risk elevation criteria. Situations which may require more detailed consequence modeling for the determination of the consequence category and therefore the risk elevation criteria include:

- Large releases of flammables that may result in a deflagration
- High pressure vessel failures which may result in significant damage
- Significant releases of toxic materials which could impact populations outside of the immediate area, both on-site and off-site

The common themes associated with these scenarios, often subjected to more detailed consequence analysis, are the potential for impact to personnel on the plant site at a distance from the failure or the potential for an event to impact the community. This impact can be due to fire, overpressure, or toxic exposure. If the impact of a particular consequence needs to be confirmed, the tools described in the consequence assessment methods section below are applied as appropriate.

Once the consequence and associated risk elevation criteria are confirmed, the company uses a standard LOPA methodology for the evaluation of the frequency of the event. EKS uses a LOPA methodology based on ultimate consequences and therefore the method does include conditional modifiers as well as enabling conditions as appropriate. One variation of EKS's method is the use of layers which can only grant half order of magnitude of risk reduction. Half orders of magnitude are only applied where the level of granularity supports the more precise values.

2.4.2.2 Detailed Risk Analysis

More detailed risk analysis techniques are applied when LOPA cannot address the scenario under consideration due to the complexity of the scenario or the weaknesses of the LOPA methodology. The broad areas which may require more detailed risk analysis techniques include:

- Fault Tree Analysis
- Human Reliability Analysis (HRA)

Fault tree analysis is used when multiple simultaneous announced failures are required for the occurrence of the scenario. If all but one of the failures that must occur for the event to progress are unannounced, then LOPA can be used. One of the basic limitations of LOPA is that only one initiating failure with a failure frequency can be used in a given analysis. All other factors that are used in LOPA are probabilities, typically probabilities of failure. When multiple failures are required for the event, and all but one are unannounced, the unannounced failures can be treated as enabling conditions with associated probabilities of failure in the LOPA analysis. The single announced failure is used as the initiating event.

In the case of multiple announced failures fault tree is an appropriate methodology. Experience with this type of event has indicated that normally the incidence rate of multiple independent simultaneous failures occurring is generally low enough that only events with very low target frequencies (i.e. catastrophic events) require fault tree analysis.

Fault tree analysis is also applicable when initiating events and independent protection layers are not truly independent. Fault tree analysis can evaluate the impact of common mode failure for the initiating event and the protection layers and allow the frequency of the scenario to be analyzed. Direct applicability of the rules of LOPA to this type of situation will often grossly overpredict the risk since the Independent Protection Layers (IPLs) and initiating event may not be independent, but the level of dependence may be limited. This type of analysis may be done on a group of related protection layers to determine a composite probability of failure on demand and this composite value may be used in a LOPA as a single IPL. This approach is analogous to the calculation of a probability of failure on demand for a safety instrumented system which accounts for common mode failure between components which is then used as an IPL based on its overall safety integrity level.

Because of the limitation of LOPA requiring independence of the initiating event and all independent protection layers, systems which are highly dependent on human failure will frequently result in the prediction of a failure rate that does not match the experience of the operating area. This is another situation in which initiating events and protection layers with some degree of common mode failure are eliminated from consideration by the rules of the LOPA methodology but can have value in prediction of the actual frequency of occurrence. In some of these cases Eastman has used the Technique for Human Error-Rate Prediction (THERP) [17] to evaluate scenarios with multiple human failures either in the initiating event or the protection layers for the event. This HRA technique is highly resource intensive and time consuming and due to the level of complexity and infrequency of use requires a significant amount of reinvestment of time in familiarization each time an analysis is conducted. Often, THERP and other HRA techniques are only evaluated as methods of last resort since a redesign of the system in question to eliminate the dependence on human error through the application of inherently safer design or active systems such as safety instrumented systems is the preferred path.

2.4.2.3 Consequence Assessment

Eastman conducts detailed consequence assessment to establish the consequence category for use in a PHA or to establish a risk elevation target for use in LOPA. These categories are established based on the number of personnel that may be impacted by the event on the plant site or the potential for the scenario to impact the community.

2.4.2.4 Blast modeling

The two primary scenarios evaluated for blast modeling are pressure vessel burst scenarios which can result in significant overpressure and shrapnel effects and vapor cloud explosions caused by the release of large quantities of flammable gases or liquids.

Pressure vessel bursts are typically modeled using the SuperChems® software with the failure point for the vessel falling into one of three main groups:

- Failure of the vessel under normal operating conditions. This event can occur due to an error in manufacturing of the vessel, fatigue, or corrosion.
- Failure of the vessel due to weakening of the vessel wall due to high temperatures. This scenario can occur when a runaway reaction causes high temperatures but not excessive pressures. A typical reaction that can exhibit this hazard is trickle bed hydrogenation.
- Failure of the vessel due to high pressure. High pressure can be caused by any of the relief contingencies that are normally evaluated in a relief design. If the relief system is not adequate for the scenario under evaluation or fails to perform when challenged, the vessel could fail at the high pressure. This pressure vessel burst is typically conducted at the ultimate yield point of the vessel (3.5 or 4 times the maximum allowable working pressure (MAWP) depending on the edition of the ASME code the vessel was designed by) or the maximum pressure that can be reached by the contingency if it is lower.

SuperChems® models the blast energy and properly accounts for the thermodynamic effects of the flashing liquid in the vessel when the pressure is dropped due to vessel failure. The program provides both overpressure and impulse as a function of distance.

Vapor cloud explosions can occur when a significant quantity of combustible vapor or gas is released and contained in a structure. Eastman uses tools from the BakerRisk Explosion Research Cooperative (ERC) to determine the predicted overpressure and impulse from such an explosion. Vcloud and Vapor Cloud Explosion Contours (VCEC) are software packages which use the amount of material in the explosive cloud, the properties of the material and the amount of confinement and congestion of the area of the VCE to evaluate the explosion. These programs implement the Baker-Strehlow-Tang methodology for estimating the blast wave from the vapor cloud explosion.

2.4.2.5 Blast Effects

Once the overpressure and impulse for an explosion are estimated, regardless of the source of the blast wave, the effects on buildings and standing structures are evaluated using another software package from the ERC. The Building Evaluation and Screening Tool (BEAST) program uses empirical effect models developed by Baker Risk to predict the level of damage to structures from the specified blast wave. The level of damage is used in the software to predict the occupant vulnerability.

2.4.2.6 Toxic Effects

Eastman primarily uses the SuperChems® software and internally developed software packages to establish the source term for dispersion modeling of released materials. Once the source term is established SuperChems® or the SAFER® program are used to model the dispersion. The dispersion models are used to determine the downfield concentrations of the chemical of concern. Eastman has used Emergency Response Planning Guidelines (ERPG) [18] concentrations, Acute Exposure Guideline Levels (AEGL) [19], Specified Levels of Toxicity (SLOT) and Significant Likelihood of Death (SLOD) Dangerous Toxic Loads (DTL) developed by the U.K. Health and Safety Executive [20], as well as probit functions for the toxic impact of chemicals on populations. Exposure to a toxic dose of material is the normal metric employed instead of a specified concentration level over time since the concentration is often changing. The typical approach takes into account not only the concentrations of material which could result in harm, but also concentrations which might impair the ability of individuals to escape from the cloud which can be significantly lower than the concentrations which might result in harm.

To better evaluate the impact to personnel in buildings to toxic exposure from exterior releases, Eastman has developed approaches to determine concentrations of toxics due to ingress of materials. These methods allow dosage to impacted personnel to be evaluated based on air handling equipment continuing to run during a release event or being properly shut down to minimize the ingress.

2.4.3 Revalidation ProcessEarly revalidations at Eastman were conducted by evaluating each change to the process that had occurred in the previous 5 year period (or since the last study if more recent) and updating the associated PHA documentation for the changes. Recommendations from the previous study were also addressed in the revalidation, updating the information based on any additional analysis or added controls. If the MOC process had not adequately addressed a potential hazard being introduced by the change, a recommendation might be made.

Over time additional process information was becoming available on existing processes and risk perception and acceptance was also changing. A determination was made that a more in depth revalidation process was needed to address these factors. Three new protocols (alpha, beta and gamma) were developed for conducting PHA revalidations. The alpha (α) protocol simply implemented the previous practice of a review of the MOCs implemented since the last PHA revalidation and any recommendations from the previous study. The beta (β) protocol included the alpha protocol but also required the team to reexamine all of the high consequence events identified in the previous study. The gamma (γ) protocol involved the PHA team revisiting every question in the PHA and reevaluating the consequences of the deviation, the controls, and their adequacy. The implementation of the beta and gamma protocols for revalidation led to later revalidations identifying new scenarios of concern and recommending additional studies and new controls to mitigate existing hazards. These more in depth protocols also served to improve the quality of the documentation of scenarios over time.

2.4.4. Implementation LearningsThe methods used by Eastman attempt to manage the risks associated with operations while minimizing the engineering costs by using more efficient tools and methods where possible. Eastman's culture and approach to risk management led to an outstanding process safety record and recognition by AIChE in 2015 as one of the companies setting the standard for excellence for process safety culture.

As is true for many companies, the quality of PHAs within Eastman has improved over time. The earliest PHAs tended to under predict risk while the LOPAs tended to be overly conservative. As more in-depth consequence analyses have been conducted, Eastman is seeing improvements in the accuracy of risk assessments. For these assessments, there is an optimum level of detail needed. In general, Eastman has determined that the greater the risk, the more detail that is needed to properly assess and manage the risk.

3. Summary and Conclusions

All four companies included in this manuscript, BASF, Celanese, Dow and Eastman have comprehensive hazard identification and risk analysis methodologies along with a corporate commitment to process safety. However, the methods used to assess and manage risk associated with operations are not identical. The companies are listed below in order of least to greatest use of higher level studies:

1. BASF relies on the Plant Safety Concept together with an Implementation Check which utilizes HAZOP and a semi-quantitative risk matrix. FTA is typically reserved for low-frequency/high-severity events and QRAs are performed where required by local regulations.
2. Dow and Eastman report a similar usage of higher level risk analysis tools such as FTA and HRA. The level of the analysis is dictated primarily by the risk posed from the process, suitability of LOPA for the analysis and local regulations.
    - Eastman: FTA is used when multiple simultaneous announced failures are required for the occurrence of the scenario or when initiating events and independent protection layers are not truly independent. HRA is used for systems which are highly dependent on human failure.
    - Dow: FTA and HRA are used in a similar manner to Eastman where special circumstances exist such as evaluating scenarios with potential high consequences and/or systems which are not properly assessed using LOPA methodology.

Additionally, Dow uses QRA for some of the higher risk processes in the company and where required by local regulations.

3. Celanese appears to have the broadest use of higher level techniques for processes which are not fully covered by RAGAGEP and/or prescriptive regulations. The Celanese HIRA methodology incorporates not only HAZOP, FMEA, Checklists and What-if, but also Fault Tree and Event Tree methodologies.

All four companies clearly recognize the importance of properly assessing risk. A few key learnings noted by each are detailed below:

- BASF
  - Methods beyond HAZOP and LOPA might bring incremental value, but caution is needed in ensuring the boundary conditions and assumptions are valid. Otherwise, it is an exercise to create a lot of paper that makes people feel good but does not improve the level of protection.
  - Another important issue and a current area of focus for BASF is consistency. Whether qualitative or quantitative techniques are utilized, the challenge has been consistently interpreting results which then drive the design and operation.
- Eastman – The earliest qualitative PHAs tended to under predict risk while the LOPAs tended to be overly conservative.
- Dow - Spending additional time and money on the risk assessment can reduce the overall cost of a project by more accurately assessing the risk and identifying more economically effective risk reduction measures.
- Celanese - Qualitative consequence based decision making dilutes the focus to improve process safety and overloads the plant project resources. Rigorous consequence assessments using simulation tools such as Phast™, can lead to better focus on scenarios with the greatest potential impacts. Simple techniques such as LOPA struggle with risk assessments of complex systems.

A key observation is that no single methodology addresses all risk assessments. Qualitative assessments were seen to lead to over-predicting risk at Celanese and under-predicting risk at Eastman. At Dow, Celanese and Eastman, it was observed that LOPAs generally led to more conservativism. However, detailed risk assessments consume a lot of resources and are not practical for all assessment requirements. So, the real task is to find the right level of detail for each study conducted. In general the most detail is needed for studies involving higher potential consequences. Additionally, the 'right tool' needs to be used for each study. The analyst must understand the relevant limitations of a given methodology when deciding how best to evaluate any given scenario or system.

4. References

[1] Nolan, D.P. (1994). Application of HAZOP and What-If Safety Reviews to the Petroleum, Petrochemical and Chemical Industries, p. 1. William Andrew Publishing/Noyes. Available at: http://app.knovel.com/hotlink/pdf/id:kt003X3L98/application-hazop-what/purpose Accessed November 3, 2015.

[2] Dowell, Arthur M. (1997). Layer of Protection Analysis: A New PHA Tool After HAZOP, Before Fault Tree Analysis, "International Conference and Workshop on Risk Analysis in Process Safety", 1997, CCPS/AIChE

[3] OSHA 1992. Process Safety Management of Highly Hazardous Chemicals, 29 CFR 1910.119. Occupational Safety and Health Administration. Federal Register, Vol. 57, No. 36, Washington, DC.

[4] NFPA. NFPA 85. Boiler and Combustion Systems Hazards Code, Massachusetts, USA (2015)

[5] NFPA. NFPA 654. Standard for the Prevention of Fire and Dust Explosions from the Manufacturing, Processing, and Handling of Combustible Particulate Solids, Massachusetts, USA (2013)

[6] ERPG. Emergency Response Planning Guidelines (ERPGs), Emergency Response Planning committee of the American Industrial Hygiene Association, http://response.restoration.noaa.gov/oil-and-chemical-spills/chemical-spills/resources/emergency-response-planning-guidelines-erpgs.html

[7] TEEL. Temporary Emergency Exposure Limits, U.S. Department of Energy Subcommittee on Consequence Assessment and Protective Actions (SCAPA), http://response.restoration.noaa.gov/oil-and-chemical-spills/chemical-spills/resources/temporary-emergency-exposure-limits-teels.html

[8] Methods for the determination of possible damage to people and objects resulting from releases of hazardous materials, The Netherlands Organisation of Applied Scientific Research (TNO), The Hague, Netherlands (1992)

[9] CCPS. Revalidating Process Hazards Analyses. Center for Chemical Process Safety, American Institute of Chemical Engineers, New York, NY, 2001.

[10] Integration of a Reactive Chemicals Program into the Process Hazard Analysis Process at The Dow Chemical Company, Phillips, Gary; Frurip, David J.; Powers, Bruce; and Overton, Tim as presented at the Second Annual Symposium, Mary Kay O'Connor Process Safety Center "Beyond Regulatory Compliance: Making Safety Second Nature", October 30-31, 20010, College Station, Texas. Available at http://pscfiles.tamu.edu/symposia/2001/Phillips%20Frurip%20and%20Powers%20DOW.pdf Accessed November 3, 2015.

[11] Dow's Fire and Explosion Index: Hazard Classification Guide 7th Edition ... ISBN-13: 978-0816904389. ISBN-10: 0816904383

[12] Dow's Chemical Exposure Index Guide, 1st Edition, American Institute of Chemical Engineers (AIChE) ISBN: 978-0-8169-0647-5, August 1998

[13] Center for Chemical Process Safety (2005) Use of Layer of Protection Analysis (LOPA) within the Dow Chemical Company, in Emergency Planning Preparedness, Prevention & Response, John Wiley & Sons, Inc., Hoboken, NJ, USA. doi: 10.1002/9780470924839.ch29

[14] First, Ken (2010) Journal of Loss Prevention in the Process Industries 23(6):705-718 · October 2010.

[15] Finney, D. J., Ed. (1952). Probit Analysis. Cambridge, England, Cambridge University Press.

[16] Knowlton, R.E. (1981). An Introduction to Hazard and Operability Studies. Vancouver, Canada. Chemetics International.

[17] Swain, A.D. & Guttmann, H.E. (1983). Handbook of Human Reliability Analysis with Emphasis on Nuclear Power Plant Applications. NUREG/CR-1278. USNRC.

[18] American Industrial Hygiene Association (AIHA). Emergency Response Planning Guidelines. Available at https://www.aiha.org/get-involved/AIHAGuidelineFoundation/EmergencyResponsePlanningGuidelines/Pages/default.aspx. Accessed March 1, 2016.

[19] United States Environmental Protection Agency (EPA). Acute Exposure Guideline Levels for Airborne Chemicals. Available at https://www.epa.gov/aegl. Accessed March 1, 2016.

[20] United Kingdom Health and Safety Executive. Toxicity Levels of Chemicals. Available at http://www.hse.gov.uk/chemicals/haztox.htm. Accessed on March 1, 2016.

High/Continuous Demand Hazardous Scenarios in LOPA

Hui Jin, PhD
Risk Analyst, SIS-TECH Solutions, LP
12621 Featherwood Drive, Suite 120
Houston, TX 77034
hjin@sis-tech.com

William (Bill) L. Mostia, Jr., PE, ISA Fellow
SIS-TECH Fellow, SIS-TECH Solutions, LP
bmostia@sis-tech.com

Angela Summers, PhD, PE
President, SIS-TECH Solutions, LP
asummers@sis-tech.com

Keywords: High demand; Continuous mode; LOPA; PFH; PFD; Safety instrumented system

Abstract

Layer of protection analysis (LOPA) has become one of the most important risk analysis techniques in the process industry. It is commonly used to determine the safety integrity requirements for protection layers, especially the safety integrity level (SIL) for safety instrumented functions (SIF). Once a SIL has been assigned to a SIF, the SIF is designed, installed, operated, maintained, tested, and managed according to IEC 61511. The standard requires that the SIL of the SIF be verified quantitatively against defined ranges based on its mode of operation. A key question is how the mode of operation impacts LOPA calculation.

One basic assumption in LOPA is that the safety integrity of the protection layers (including SIF) is given by the well-known average probability of failure on demand (PFD), which is the safety integrity measure for low demand mode per IEC 61511. However, what if the hazard scenario involved has a high(nominally defined as more than once a year) or continuous demand function? IEC 61511 explicitly defines the safety integrity measure for high/continuous demand SIF as the frequency of dangerous failures per hour (PFH), instead of PFD. In some scenarios, there is a mixture of safeguards operating in different modes, e.g. both low demand and high/continuous modes. Does LOPA still work? Is your SIL determination correct? Are your verification calculations correct?

A method is presented to allow both high demand and continuous mode scenarios to be assessed in LOPA without changing the general LOPA framework. The LOPA calculation of high demand and continuous mode functions is illustrated, showing how improper treatment of the operating mode results in excessive target SIL. A case study encountered in an actual project is used as an example to showcase the proposed calculation method.

1 Introduction

The defense-in-depth philosophy using multiple independent protection layers (IPL) is by far the most common risk analysis methodology utilized in the process industries. Layer of protection analysis (LOPA) is commonly used to analyze hazard scenarios and to assess the risk associated with the process. In addition to being a risk analysis method, LOPA is used as an allocation tool to determine what IPL are available, what estimated risk reduction is provided by the IPL, what the residual risk gap is, and what risk reduction is needed to close the risk gap.

The risk gap is often closed using a safety instrumented function (SIF) implemented by a safety instrumented system (SIS). The risk reduction allocated to the SIF determines the required safety integrity level (SIL) of the SIF. The SIS is then designed, installed, operated, maintained, tested and managed according to IEC 61511[1]. This standard requires in clause 11.9.1 that the design and management strategy be verified by calculation to demonstrate a performance better than the target criteria determined in the LOPA.

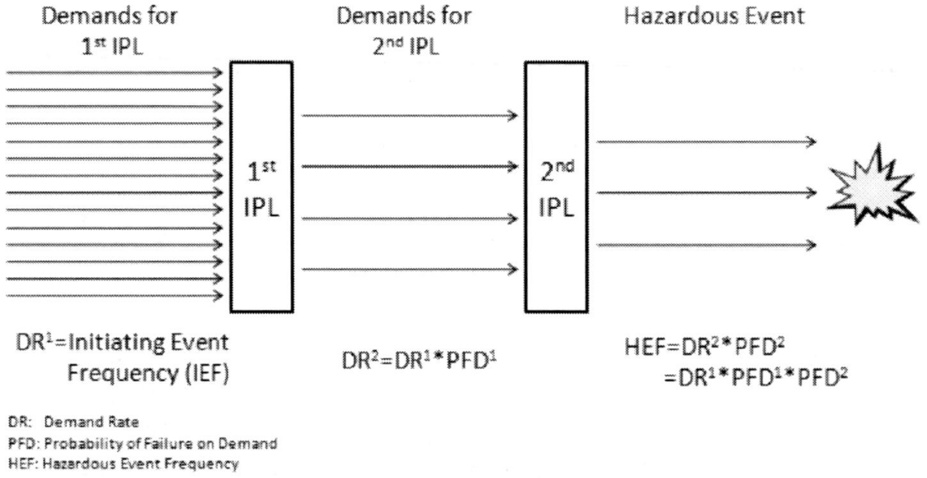

Figure 1. Defense-in- depth philosophy

The defense-in-depth philosophy may be illustrated by the incident sequence shown in Figure 1. A hazardous event starts with an initiating event that places a demand on the IPL, such as a process disturbance, human error, equipment failure, etc. IPLs are specifically designed and implemented to stop the initiating event from propagating into a hazardous event. When an initiating event occurs, a demand is placed on the first IPL. If this IPL functions properly, the hazard scenario stops. But if the first IPL fails to stop the initiating event, a demand is placed on the second IPL. If the second IPL fails, a demand is placed on the next IPL. If all of the IPLs fail to act properly, a hazardous event occurs. Therefore the hazardous event frequency (HEF) is a function of the initiating event frequency and the likelihood that the IPLs fail to work correctly.

The LOPA method [2, 3] is built upon the defense-in-depth philosophy and a fundamental assumption is that the safety integrity of an IPL is measured by the average probability of failure on demand (PFD). The LOPA method can be mathematically represented byEquation1.

$$HEF = DR^1 \times PFD^1 \times PFD^2 \times \ldots \qquad [1]$$

where:

HEF= Hazardous Event Frequency (per time)
DR^1= Demand Rate (Frequency, per time)
PFD^1= PFD of the first IPL (Probability, unitless)
PFD^2= PFD of the second IPL (Probability, unitless)

One condition that is often underappreciated is that for this calculation to be valid, the demand rate on the next IPL or HEF cannot be greater than the frequency of failure of the IPL under consideration. For example a demand rate of 10 times per year would greatly exceed the failure frequency achievable by most continuous interlocks of 1/10 years. If Equation 1 is used and the interlock PFD is 10^{-1}, the HEF = 10 times per year$\times 10^{-1}$ = 1 incident/year, which is greater than the failure frequency of the interlock. The hazardous event cannot happen at a frequency greater than the interlock failure frequency, which is 1/10 years. Mathematically, the limiting condition for this transition can be represented by Equation 2.

$$HEF\ or\ DR^{n+1} = DR^n \times PFD^n \leq PFH^n \qquad [2]$$

where:

DR^n = The demand rate of nth IPL
PFD^n= Average probability of failure on demand of the nth IPL
PFH^n= Frequency of dangerous failures per hour of the nth IPL.

As the demand rate increases, it is not uncommon that the limiting condition in Equation2 is violated. When the conditions in Equation 2 are not met, the PFD is no longer an appropriate safety integrity measure for the IPL, and the typical LOPA calculation represented by Equation 1 leads to an incorrect risk analysis and SIL requirement.

IEC 61511 differentiates its requirements according to the mode of operation: low demand, high demand and continuous. The standard further prescribes that PFD is only valid in low demand applications, and PFH is used for high demand and continuous applications. Then, is the typical LOPA calculation still valid when PFH is the appropriate safety integrity measure? And how should the LOPA calculation be modified if the current one is not adequate?

An example is used to show how the typical LOPA calculation in high demand and continuous applications leads to incorrect SIL requirement. To address this problem, a revised calculation is presented. The remaining paper is structured as follows. Section 2 discusses the classification of SIF modes of operation, and section 3 reviews the typical LOPA calculation. In section 4, an example of a SIF in high demand mode from a flare gas recovery and compression (FGRC) unit is used to show how excessive target SIL can be result from the typical LOPA calculation. A method to address high and continuous applications is proposed and illustrated using the same example. A brief discussion of PFH calculation of high and continuous mode SIFs is provided in section 5. Concluding remarks are given in section 6. Two important LOPA topics uncertainty and dependent failures are not covered in this paper. The readers are referred to Freeman and Summers [4] and Jin and Summers [5] for detailed discussions.

2 SIS Operation modes

IEC 61511[1]defines three modes of operation for a SIF:

- Low demand mode: *mode of operation where the SIF is only performed on demand, in order to transfer the process into a specified safe state, and where the frequency of demands is no greater than one per year.*

- High demand mode: *mode of operation where the SIF, is only performed on demand, in order to transfer the process into a specified safe state, and where the frequency of demands is greater than one per year.*

- Continuous mode: *mode of operation where the SIF retains the process in a safe state as part of normal operation.*

The mode of operation affects two key requirements in IEC 61511. The first is clause 11.4 on the minimum hardware fault tolerance (HFT) requirement for SIL 2. For a low demand SIF, the minimum HFT requirement is 0 to claim SIL 2, whereas for a high demand/continuous SIF, the minimum HFT requirement is 1 to claim SIL 2. The second is the safety integrity measure in clause 9.2. For a low demand SIF, the safety integrity requirement is measured by PFD, whereas for a high demand/continuous SIF, the safety integrity requirement is measured by PFH. See Table 1 for details of each SIL.

Table 1. Safety integrity requirements

Safety integrity level (SIL)	Low demand mode	High demand/continuous mode
	Average probability of failure on demand (PFD)	Frequency of dangerous failures per hour (PFH)
4	>=10E-05 to <10E-04	>=10E-09 to <10E-08
3	>=10E-04 to <10E-03	>=10E-08 to <10E-07
2	>=10E-03 to <10E-02	>=10E-07 to <10E-06
1	>=10E-02 to <10E-01	>=10E-06 to <10E-05

The HFT requirement has no impact on the LOPA process and hence is out of the scope of this paper. This paper focuses on the rationale for the differences in the safety integrity measure for low demand and high demand/continuous mode SIFs. Why are there different safety integrity measures for different operating modes?

Consider a SIF operating continuously as the last line of defense. A hazardous event occurs whenever the SIF fails dangerously. The HEF is equal to the dangerous failure frequency of the SIF. In IEC 61511, the SIL of a continuous mode SIF is determined by frequency of dangerous failures per hour (PFH) as shown in Table 1.

For a SIF operating in demand mode as the last line of defense, a hazardous event occurs when there is an initiating event (IE) demanding the SIF to respond and the SIF fails to do so. The HEF is equal to the IE frequency multiplied by the likelihood that the SIF is unavailable when the IE occurs. In IEC 61511, the SIL of a demand mode SIF is determined by PFD. This logic would seem to apply to any demand based operation. However at some rate of occurrence, the demands are so frequent that the SIF can no longer be judged by it probability of failure. The HEF is instead being limited by the SIF failure frequency. Where does this transition occur?

According to IEC 61511, the transition occurs at a demand rate of 1 per year. The rationale for this guidance is given in ISA TR 84.00.04 [6].ISA TR 84.00.04 uses a single channel (1-out-of-1) system as an example. For such an IPL with a test interval of TI, the PFD and PFH can be calculated as follows:

$$\mathrm{PFD} = \frac{\lambda_{\mathrm{DU}} \times \mathrm{TI}}{2} \qquad [3]$$

$$\mathrm{PFH} = \lambda_{\mathrm{DU}} \qquad [4]$$

If instead of average PFD, the maximum PFD at the time right before the testing is used for the purpose of being conservative, the PFD becomes

$$\mathrm{PFD} = \lambda_{\mathrm{DU}} \times \mathrm{TI} \qquad [5]$$

Inserting the above PFD and PFH into Equation 2, the limiting condition on the LOPA calculation becomes:

$$\mathrm{DR} \times \lambda_{\mathrm{DU}} \times \mathrm{TI} \leq \lambda_{\mathrm{DU}} \Longrightarrow \mathrm{DR} \times \mathrm{TI} \leq 1 \Longrightarrow \mathrm{DR} \leq \frac{1}{\mathrm{TI}} \qquad [6]$$

Assuming 1 year test interval, the limiting condition becomes a demand rate of less than or equal to 1 per year. When the demand rate is higher than 1 per year, PFH should be used as the safety integrity measure. This relationship is incorporated into the PFH and PFD ranges given in Table 1. The relationship of PFD and SIL was originally accepted in the 1980s and the PFD ranges were converted into PFH in the 1990s by assuming 1 demand per year with 1 year being comprised of 10,000 hours to keep the ranges as orders of magnitude. It should be pointed out that classification of a SIF as low demand or high demand is dependent on both the demand rate and the test interval. For example, if the test interval is 5 years, the limiting condition would be1demand every 5 years instead of 1demand per year.

3 Low Demand SIS System in a LOPA

When using LOPA, Equation 1 can be used to calculate the HEF for comparison with the tolerable event frequency (TEF) to determine whether the risk level is acceptable. When using LOPA to determine the safety integrity necessary to achieve the TEF, Equation 1 can be rearranged to as follow.

$$Residual\ Risk\ Gap\ SIS = \frac{TEF}{DR \times PFD^1 \times PFD^2 \ldots} \qquad [7]$$

As an example, consider a tower overfill scenario where the overfill may be initiated by a BPCS failure. The overfill can be prevented by 3 IPLs: 1) operator response to an alarm,2) a SIF and 3) a relief valve (RV). Apply typical LOPA calculation as shown in Equation 1, the scenario may be illustrated by Equation8:

$$TEF = BPCS\ Failure\ Frequency \times Alarm\ PFD \times SIF\ PFD\ \times\ RV\ PFD \qquad [8]$$

Assuming a TEF of 10^{-4} per year, a PFD of 10^{-1} for the alarm IPL, a PFD of 10^{-1} for the RV IPL, and 10^{-1}per year failure frequency for the BPCS, it is straightforward to yield a 10^{-1} risk gap and a SIL 1 requirement for the SIF.

$$Residual\ Risk\ Gap\ SIS = \frac{10^{-4}/\mathrm{year}}{10^{-1}/\mathrm{year} \times 10^{-1} \times 10^{-1}} = \frac{10^{-4}}{10^{-3}} \qquad [9]$$

$$= 10^{-1} ==> SIL\ 1\ SIF$$

It should be noted that the TEF and IE frequency are the only frequencies, which conforms to a basic reliability math rule that you can multiple probabilities and frequency but you cannot multiple frequencies together. It should also be noted that the expected operation sequence is:

Operation Sequence ==> Alarm ==> SIS==> RV [10]

This sequence determines the demand rate for each of the IPLs. In this particular example, with a BPCS failure frequency assumed to be 10^{-1} per year, the SIF implemented in the SIS are in the low-demand range. Hence, the LOPA for this scenario determines that the SIF has a SIL 1 requirement.

4 High Demand/Continuous Mode SIS in a LOPA

As discussed in Section 2,the transition between low demand and high demand/continuous mode is dictated by the limiting condition calculation. If the demand rate on the SIF exceeds the limit, the HEF is calculated using the frequency of IPL failure rather than the typical LOPA calculation, see Equation 11. *If the correct math is not used, the allocated SIL is likely to be higher than required.*

$$HEF = PFH^1 \times PFD^2 \times PFD^3 \times \ldots \qquad [11]$$

where:

HEF= Hazardous Event Frequency (per time)
PFH^1= Failure frequency of the first IPL (, per time)
PFD^2= PFD of the second IPL (Probability, unitless)
PFD^3= PFD of the third IPL (Probability, unitless)

There is folklore in the industry that continuous mode SIFs do not exist within the process industry. This is simply not correct. While unusual, they exist in a wide range of processes. For example, consider the dynamic positioning of a drillship or the control of the ethylene/oxygen ratio in an ethylene oxide reactor. Both of these functions must work continuously or else the hazardous event propagates.

High demand SIFs are commonly encountered. High demand mode may be caused by how the SIF prevents the hazardous event. Consider a valve interlock where the operator is prevented from moving a particular valve until a set of valves is properly aligned. Each time the operator does the lineup, there is the potential that the interlock may need to prevent the valve movement. If the operator does this activity as part of a normal batch routine, the valve interlock is likely in high demand mode. High demand mode may also be due to the number of potential sources of the hazardous event. For example, in a subsea production system where several subsea production trees feed the same platform, an abnormal situation in any of the subsea production trees may require a process shutdown on the platform. Whereas demand from each subsea production tree may be in low demand range, taken together, the demand rate from all sources can be high. Finally, high demand mode can come from unstable process control where process deviations occur more frequently than desired. These latter sources of demand should be addressed through improvements in the control system to reduce the demand rate rather than relying on IPLs to achieve the TEF.

4.1 Example system description

Multiple initiating events can occur in an FGRC unit that demands the IPLs to react. One of those initiating events is unintended gas compressor shut down in a string of flare gas recovery compressors that fed a flare. The potential consequence of gas compressor shutdown is rupture of a vessel and lines leading to fire and explosion. To avoid this consequence, two IPLs are used as part of the protection strategy.

- A SIF comprising 2-out-of-3(2oo3) voted pressure transmitters, a safety PLC, and a quick open valve controlled by a de-energize to trip solenoid.
- A pressure relief system comprising two redundant buckling pin valves.

4.2 Misapplication of LOPA to high demand SIF

LOPA was performed by the operating company to determine the SIL requirement for the SIF. This high pressure scenario is illustrated by Figure 2.If the SIF operates as required when demanded, the hazard scenario stops, otherwise a demand on the buckling pin valves occurs. If the buckling pin valves operate as required, the hazard scenario stops, otherwise a hazardous event will occur. Hence:

- The demand rate for the SIF (DR1) is the initiating event frequency
- The demand rate for the buckling pin valves (DR2) is the initiating event frequency time the PFD of the SIF
- The HEF is the product of the initiating event frequency, PFD of the SIF and PFD of the buckling pin valves

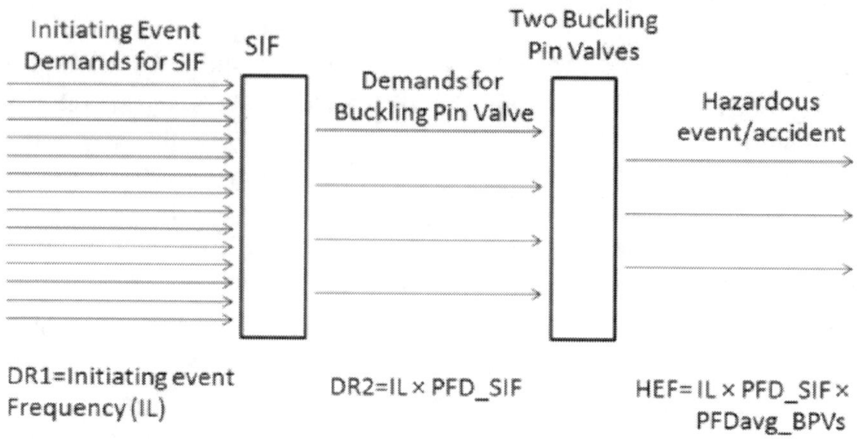

Figure 2. FRGU high pressure protection strategy low demand

In the LOPA, it is determined that the initiating event leading to high pressure occurs 50 times per year. With the two LOPA credits (PFD=10^{-2}) from the two buckling pin valves, it is determined that a SIL 3 (RRF =5000) SIF is required to achieve the desired TEF target of 10^{-4} per year. The SIL 3 requirement was determined based on the assumption that PFD was the appropriate measure for the proposed SIF. The typical LOPA calculation is given by Equation12.

$$Residual\ Risk\ Gap\ SIS = \frac{TEF}{IL \times PFD^{BPV}} \quad [12]$$

$$= \frac{10^{-4}/\text{year}}{50\ /\text{year} \times 1 \times 10^{-2}}$$

$$= 2 \times 10^{-4} ==> SIL\ 3\ SIS$$

$$==> Residual\ RRF = 5000$$

5 High Demand/Low Demand LOPA Methodology

For the FRGU example, the demand rate is 50 times per year, therefore the SIS is operating in high demand rather than low demand. For a SIF operating in high demand, we need to use PFH instead of PFD for the SIF, see Figure 3. Hence:

- The demand rate for the SIF (DR1) is the initiating event frequency
- The demand rate for the buckling pin valves (DR2) is approximately the PFH of the SIF
- The hazardous event frequency (HEF) is the product of PFH of the SIF and PFD of the buckling pin valves

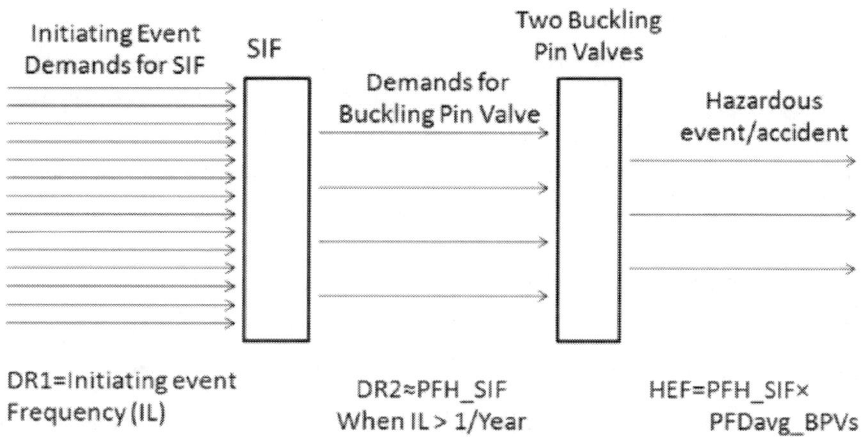

Figure 2. FRGU high pressure protection strategy high demand

We can modify the LOPA calculation to reflect this in Equation13, and using the

$$Residual\ Risk\ Gap\ SIS\ = \frac{TEF}{PFD^{BPV}} \quad [13]$$

$$= \frac{10^{-4}\ per\ year}{1 \times 10^{-2}} = 10^{-2}\ per\ year$$

$$= 1.14 \times 10^{-6}\ per\ hour ==> \text{SIL 1 SIF}$$

From the result, we can see that instead of a SIL 3 requirement as determined by the typical LOPA calculation, the actual requirement should be a SIL 1.

6 PFH Calculation

Lots of publications are available for PFD calculation, but not so many for PFH calculation [14]. To verify that a SIF operating in high demand/continuous mode meets the SIL requirement. The PFH of the SIF is calculated and compared with the values in the Table 1. Similar to PFD calculation, the PFH_{SIF} for a simple SIF is calculated as the sum of the PFH of the input subsystem (IS), logic solver (LS) and final element (FE):

$$PFH_{SIF} = PFH_{IS} + PFH_{LS} + PFH_{FE} \quad [14]$$

For the FRGU example, the input subsystem is 2-out-of-3(2oo3) voted pressure transmitters, the logic solver is a safety PLC, and the final element is a quick open valve controlled by a de-energize to trip solenoid.

The PFH of the logic solver is simply equal to the dangerous failure rate of the logic solver.

The PFH of the final element is equal to the sum of the dangerous failure rate of the solenoid and the quick open valve because the final element fails if either the solenoid or the quick open valve fails.

For the PFH calculation of the 2oo3 voted pressure transmitters, the calculation is a bit more complicated. The input subsystem fails when two or more transmitters fail. The input subsystem failure can be the result of independent failures or a common cause failure. The failures do not need to be simultaneous. The system failure could occur when the second transmitter fails before the first one is repaired. So, the PFH contribution from independent failures can be calculated as the product of the failure rate of one transmitter and the probability that one of the remaining two transmitters fails before the first transmitter is repaired. Mathematically, this is expressed as in Equation15, where 3 represents that the first failure can happen to any of the 3 transmitters, the first λ_{PT} is the dangerous failure rate of first failed transmitter, 2 represents that the second failure can happen to any of the 2 remaining transmitters, and $\lambda_{PT} * (MTTR_{PT} + \frac{TI_{PT}}{2})$ is the probability that one transmitter fails before the first failure is repaired.

$$PFH_{IS} = 3 * \lambda_{PT} * 2 * [\lambda_{PT} * (MTTR_{PT} + \frac{TI_{PT}}{2})] \qquad [15]$$

The contribution of common cause to the input subsystem failure can be calculated with the widely accepted beta factor method. The common cause PFH is the product of the beta factor and the transmitter dangerous failure rate.

Hence the PFH_{SIF} can be calculated by Equation16.

$$PFH_{SIF} = 3 * \lambda_{PT} * 2 * \lambda_{PT} * \left(MTTR_{PT} + \frac{TI_{PT}}{2}\right) + \beta\lambda_{PT} + \lambda_{LS} + \lambda_{SOV} + \lambda_{QOV} \qquad [16]$$

Using reliability parameters from SIL Solver® [8] in Table 2, the PFH of the SIF is calculated as 2.9×10^{-6} per hour. Compare with the PFH range for SIL in Table 1, the PFH_{SIF} meets the SIL 1 requirement for high demand mode.

Table 2. Example calculation parameters

Parameter	λ_{PT}(/yr)	λ_{LS} (/yr)	λ_{SOV} (/yr)	λ_{QOV} (/yr)	$MTTR_{PT}$ (hr)	TI_P(/yr)	β
Value	6.67×10^{-3}	2.10×10^{-4}	1.67×10^{-2}	8.33×10^{-3}	72	1	2%

It should be noted that from Equation16, it seems that the test intervals of the solenoid and the quick open valve do not affect the SIF PFH. This conclusion is premature, because in Equation16, the failure rates are based on the assumption that an appropriate mechanical integrity program is in place. Without a mechanical integrity program requiring inspection, testing and maintenance with appropriate interval, the failure rates used in Equation16 cannot be sustained. Therefore it is important that the valves and solenoids are tested regularly.

7 Concluding remarks

Typical LOPA calculations have the inherent assumption that the IPLs are in low demand operation. Assessing high demand or continuous IPLs using low demand assumptions results in excessive risk reduction requirements. An approach is proposed to correctly determine the SIL requirement when the SIF is operating in high demand or continuous mode. High demand and continuous mode SIFs are not uncommon in the process industry, so it is important to use appropriate calculation techniques to avoid incorrect application of LOPA and potentially excessive SIL requirements.

8 References (examples given to show format).

[1] IEC.IEC 61511 Functional safety: Safety Instrumented Systems for the Process Industry Sector - Part 1-3, Final Draft International Standard. Geneva: IEC. 2015.

[2] A. E. Summers. *Introduction to layers of protection analysis.* Journal of Hazardous Materials. Volume 104, Issue 1-3 (2003), Pages 163–168.

[3] CCPS. Layer of Protection Analysis: Simplified Process Risk Assessment. Center for Chemical Process Safety, American Institute of Chemical Engineers, New York, NY, 2001.

[4] R. Freeman and A. E. Summers. *Evaluation of uncertainty in safety integrity level calculations.* Process Safety Progress. Available online since November 2015.

[5] H. Jin and A. E. Summers. *Dependent, Independent, and Pseudo-independent Protection Layers in Risk Analysis.* Process Safety Progress. Available online since November 2015.

[6] ISA. ISA TR 84.00.04. Guidelines for the Implementation of ANSI/ISA-84.00.01-2004 (IEC 61511 Mod). The International Society of Automation, North Carolina, 2015.

[7] H. Jin, M.A. Lundteigen and M. Rausand. *New PFH-formulas for k-out-of-n:F-systems.* Reliability Engineering and System Safety. Volume 111 (2013), Pages 112–118.

[8] SIS-TECH. SIL Solver 7.0, SIS-TECH Solution, Houston, TX.

Advanced Analysis to Supplement HAZOP/LOPA for Effective Process Design

William R. Banick
Genesis
11750 Katy Freeway, Houston, TX 77079
bill.banick@genesisoilandgas.com

Cindy Wei
Genesis
11750 Katy Freeway, Houston, TX 77079
cindy.wei@genesisoilandgas.com

Prepared for Presentation at
American Institute of Chemical Engineers
2016 Spring Meeting
12th Global Congress on Process Safety
Houston, Texas
April 11-23, 2016

Advanced Analysis to Supplement HAZOP/LOPA for Effective Process Design

William R. Banick
Genesis
11750 Katy Freeway, Houston, TX 77079
bill.banick@genesisoilandgas.com

Cindy Wei
Genesis

Keywords: HAZOP, LOPA, human factors, fault tree, event tree, process design

Abstract

The HAZOP method is commonly used in identifying causes and consequences of hazard scenarios and the required safeguards in confirming the engineering design for process safety. The LOPA method is then often used to provide a semi-quantitative evaluation of the adequacy of the identified safeguards to reduce risk to an acceptable level. Where gaps exist, this method provides initial input into determination of required risk reduction factors or safety integrity levels for safety instrumented systems. Integration of HAZOP and LOPA techniques into a single study has been accomplished and this provides the practical advantage of utilizing the same team of subject matter experts to complete the evaluation of hazard scenarios, to identify the need for further risk reduction, and to recommend the specific safeguards required and confirm that the resulting residual risk will be acceptable.

While practical and efficient, the one-step comprehensive integrated HAZOP/LOPA method can have pitfalls. For example, the LOPA method isn't always appropriate for scenarios that may have high initiating event frequencies, such as repetitive manual or semi-automatic batch operations. In this case, use of standard initiating event frequencies for control failures or operator error could result in understating the risk. A human factors task analysis may be the more appropriate tool. Likewise, HAZOP/LOPA teams are usually instructed to provide concrete recommendations when additional safeguards are required, rather than just calling for further study. In some cases this could lead to costly overdesign. In particular, use of LOPA alone to confirm risk acceptability can lead to complex solutions to meet the relatively conservative order-of-magnitude threshold requirements for independent protection layers (IPLs). Quantitative methods are sometimes required beyond HAZOP/LOPA to obtain a more precise evaluation of the identified hazards. These quantitative methods are useful for design evaluations and recommendation resolutions when major capital expenditures need additional justification (e.g. when SIL-2 or 3 systems are required).

This paper presents a risk assessment approach incorporating human factors task analysis and fault tree and/or event tree analysis following the traditional HAZOP/LOPA analysis. This approach allows for a more complete evaluation of the engineering design and safe operation options. To illustrate this approach, an example is shown on the selection of appropriate engineering design options for cyclic regeneration of dense phase ethylene treaters due to their unique safety concerns. Using a combination of HAZOP scenario identification, LOPA, human factors task analysis and fault tree analysis, a cost effective design solution is developed to manage the unique risks associated with dense phase ethylene treater and regeneration systems. These techniques not only avoid the cost and complexity of implementing a SIL-3 instrumented safety system that are called for by HAZOP/LOPA, they also identify that the risks associated with routine opening of process equipment for blinding and un-blinding could be avoided as well.

1 Introduction

The HAZOP (Hazard and Operability) method was developed by Imperial Chemical Industries, Ltd. (ICI) in Britain during the 1960's with the primary objective of assuring that major incidents could be avoided during the lifetime of the facility through the completion of a thorough, impartial and adequate safety review. (Nolan, 2014). It is commonly used in identifying causes and consequences of hazard scenarios and the required safeguards in confirming the engineering design for process safety. The LOPA (Layers of Protection Analysis) method is then often used to examine the scenarios generated by the HAZOP and to provide a semi-quantitative evaluation of the adequacy of the identified safeguards to reduce risk to an acceptable level. Where gaps exist, this method provides initial input into determination of required risk reduction factors or safety integrity levels for safety instrumented systems. Integration of HAZOP and LOPA techniques into a single study has been accomplished and this provides the practical advantage of utilizing the same team of subject matter experts to complete the evaluation of hazard scenarios, to identify the need for further risk reduction, and to recommend the specific safeguards required and confirm that the resulting residual risk will be acceptable.

While practical and efficient, the one-step comprehensive integrated HAZOP/LOPA method can have pitfalls. HAZOP/LOPA teams are usually instructed to provide concrete recommendations when additional safeguards are required, rather than just calling for further study. In some cases this could lead to costly overdesign. In particular, use of LOPA alone to confirm risk acceptability can lead to complex solutions to meet the relatively conservative order-of-magnitude threshold requirements for risk reduction and the restrictive requirements for independent protection layers (IPLs).

While LOPA is a good risk screening tool, it may not always be the most effective design tool. Quantitative methods are sometimes required beyond HAZOP/LOPA to obtain a more precise evaluation of the identified hazards. Methods such as Human Factors Task Analysis and Fault Tree Analysis can provide valuable insight for detailed evaluation of operator error rates, operator intervention credits, instrumentation and mechanical component failure rates and overall protection layer and safeguard effectiveness. These quantitative methods are useful for design evaluations and recommendation resolutions when major capital expenditures need additional justification.

This paper presents an example where use of HAZOP/LOPA results alone could have led to costly overdesign.

2 Example – Dense Phase Ethylene Treater Regeneration

2.1 Background

In this case we examine the design of ethylene treaters or purification beds which are widely used in the petrochemical industry to remove trace impurities such as water, oxygenates, nitrogen compounds and acid gases (CO_2, H_2S, COS) from ethylene feedstock. Many ethylene treaters are operated under dense phase ethylene conditions at high pressures and ambient temperatures associated with cavern storage and pipeline delivery. Besides the highly flammable and reactive nature of ethylene, handling of dense phase ethylene presents additional hazards, including potential for self-polymerization and decomposition if temperatures are elevated, or shock chilling and brittle fracture if uncontrolled depressurization occurs. (Britton, 1994).

These treaters, which require periodic regeneration, are often connected to low pressure regeneration systems, where hazardous conditions can occur during operating transitions. Operating transitions are usually accomplished through manual operator steps.

Figure 1 provides a basic illustration of a dense phase ethylene treater with process and regeneration system connections. Two parallel treaters are typically provided with one in service and the other in regeneration/standby. Ethylene supply at 800-1200 psig from the process flows downwards in the treater during the purification/drying cycle, which typically lasts about one month. At the completion of the purification/drying cycle, treaters are switched, the spent treater is drained and depressurized, and then heated regeneration nitrogen flows upwards through the treater bed at low pressure in a purge, heating and cooling sequence. Once the regenerated treater is cooled, it is placed in standby or back in service by slowly pressurizing and filling. The valve manifolds illustrated in the figure are critical to providing positive isolation between the process and the regeneration systems. A double valve arrangement is commonly used in this service to provide this positive isolation.

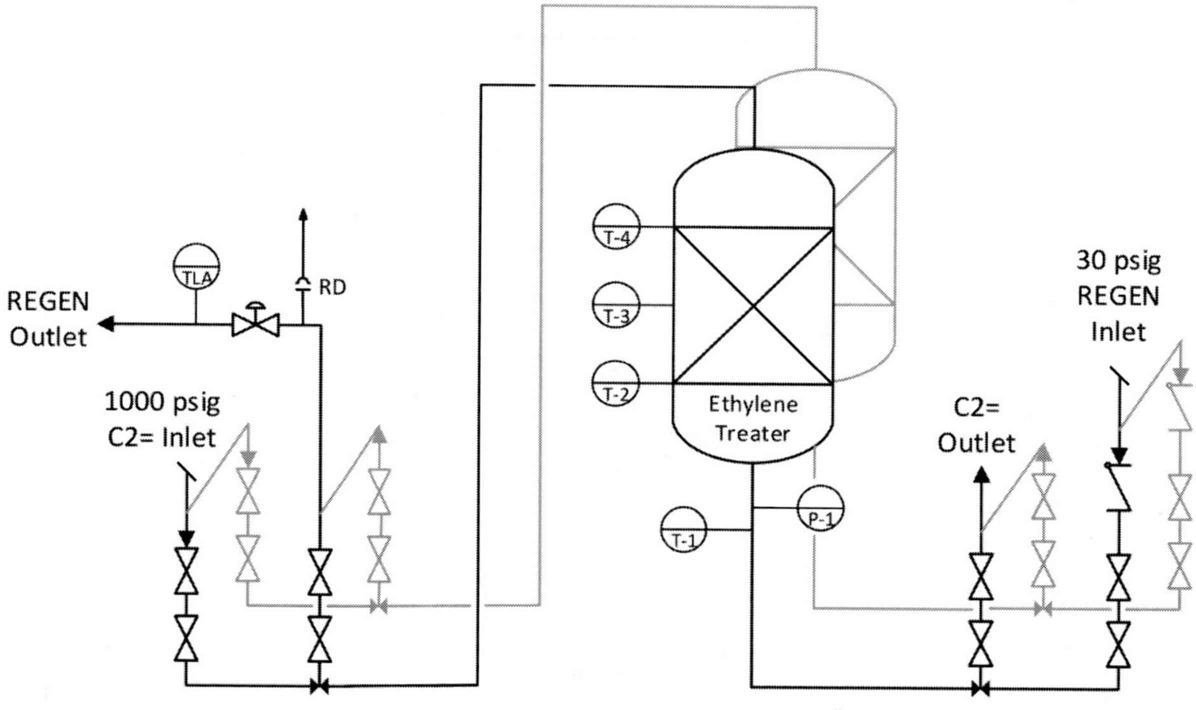

Figure 1: Schematic of Dense Phase Ethylene Treater

2.2 *HAZOP/LOPA Analysis*

Proper execution of operating transition steps is critical for these treaters and the potential for hazardous scenarios is carefully evaluated during the HAZOP/LOPA review. For example, the following scenarios were identified in the HAZOP/LOPA review:

1. Mis-alignment of a regeneration valve on the on-line treater, resulting in gross flow of ethylene into the low pressure regeneration system with loss of containment due to decomposition or brittle fracture
   - Decomposition can occur if the regeneration system outlet valve is closed in which case the high pressure dense phase ethylene acts as a piston and compresses the low pressure nitrogen resulting in high temperatures
   - Brittle fracture can occur if the regeneration system outlet valve is open, in which case sudden depressurization occurs with auto-refrigeration and over-chilling of the carbon steel piping and equipment
2. Opening a regeneration valve before the treater is drained and depressurized, resulting in sudden emptying of the treater contents into the regeneration system with loss of containment due to decomposition or brittle fracture
   - The treater contains enough ethylene that the same consequences that occurred in Scenario 1 can occur in this case.
3. Lining up ethylene into a hot treater after regeneration, if the cooldown step is missed, with loss of containment due to overheating from polymerization and decomposition

Each of these scenarios involves human error with initiating event frequencies, according to LOPA rules, of greater than once-in-10 years (> 0.1/yr). In each case the consequence is a major loss of containment which could lead to a fire or explosion and loss of life. Using typical industry risk acceptance criteria for a loss of life consequence, the frequency of occurrence would have to be reduced to the order of magnitude of between 10^{-4} and 10^{-5}/yr (or less than once-in-10,000 years). On this basis, the LOPA analysis identified the need for 4 IPL credits, as illustrated on the risk matrix in Figure 2. Table 1 provides a summary of the identified safeguards, LOPA gap and recommended additional safeguards.

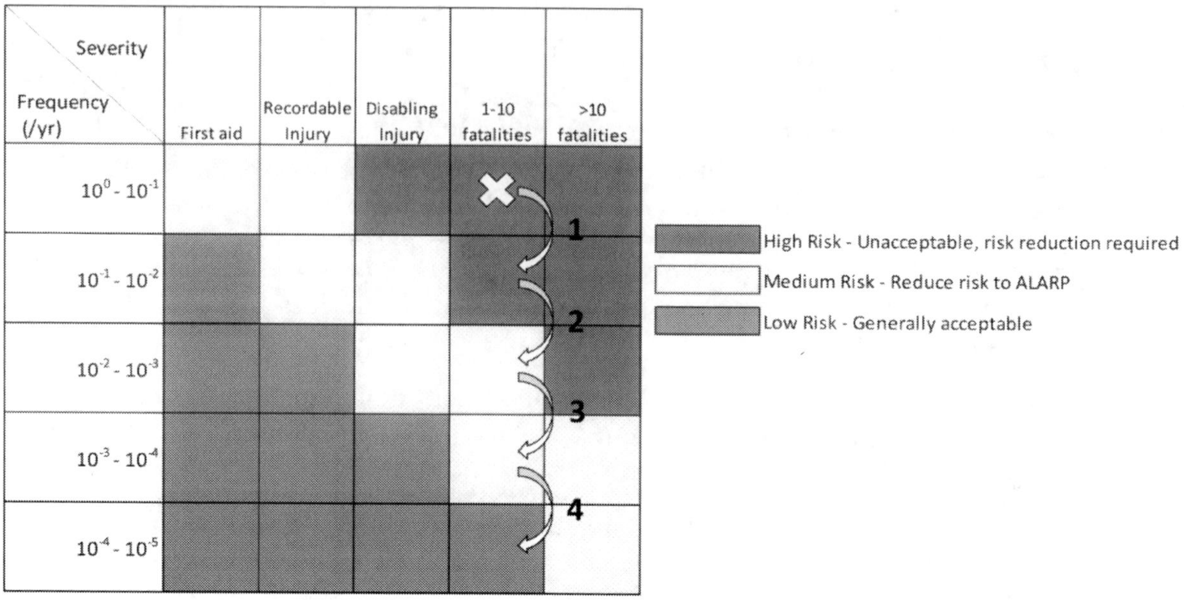

Figure 2: Typical Risk Matrix showing unmitigated risk and required layers of protection to reach acceptable risk level

Table 1: Ethylene Treater HAZOP Scenarios and LOPA Analysis

Scenario	Req'd IPL Credits	Existing Safeguards – IPL Credits	GAP	Additional Safeguards Required by LOPA – IPL Credits
1	4	Check valves, low temperature alarm or rupture disk on regeneration system – 1	3	Blinding Procedure – 1 Trapped Key Interlock – 2
2	4	DCS Sequence – 1 Check valves, low temperature alarm or rupture disk on regeneration system – 1	2	SIL-2 isolation system with pressure permissive – 2
3	4	DCS Sequence - 1	3	SIL-3 isolation system with temperature permissive – 3

As indicated in Table 1, for valve mis-alignment scenario 1, LOPA indicates the need for both a blinding procedure for 1 IPL (Independent Protection Layer) and a mechanical interlock system on the treater valves, such as a trapped key interlock for 2 IPL credits as per CCPS LOPA guidance (CCPS, 2015), to prevent inadvertent mis-alignments. These supplement the check valves, low temperature alarm and rupture disk on the regeneration system.

In missed step scenario 2, a SIL (Safety Integrity Level) –2 protective system involving automatic valves on the regeneration lines, with permissives for opening based on treater pressure, is suggested. This is required to supplement the 1 IPL that can be taken for the treater regeneration sequence logic, which is implemented in the DCS and the protection on the regeneration system. SIL-2 integrity is required since the large treater vessel contains enough mass to cause decomposition or brittle fracture if it is not drained and depressurized before either one of the regeneration valves is opened. In these scenarios, the blinding procedure and mechanical interlock system used in scenario 1 do not provide protection since the process valves are already closed and blinding/un-blinding of appropriate lines in preparation for a regeneration would be completed prior to the draining operation.

In the LOPA analysis for missed step scenario 3, a SIL-3 protective system with automated valves is suggested to prevent the operator from inadvertently placing the treater back in service prior to completing the cool-down or pre-load steps. This system would supplement the single IPL taken for the DCS sequence logic. The protective system would either prevent opening or would re-close the process inlet and outlet valves in the event of high bed temperature in order to stop the flow of the ethylene reactant and thereby limit the exothermic heating of the bed to below the point at which decomposition could be triggered. SIL-3 integrity is required since the temperature rise in this scenario would be too fast to allow alarms and operator response to be used as an effective IPL.

Together these results suggest that an automatic SIL-3 protective system may be the best solution since both the regeneration and process valves require automated action to cover all six scenarios. The treater design incorporating this automatic system is illustrated in Figure 3.

While this "fully automatic" SIL-3 system meets or exceeds the requirements of LOPA for all of the identified scenarios, it adds significant cost and complexity. And, as with any high integrity system, it would be a challenge to maintain this level of performance over the life cycle of the treater system. Full automation of the regeneration sequence may also provide some operational advantages, however, the added investment would not typically be justified by operational credits for a monthly operation. For these reasons, coupled with the fact that LOPA often provides a conservative analysis, it is prudent to conduct additional analysis. For this case, a human factors task analysis of the treater regeneration procedure was conducted and incorporated into a fault tree analysis for the safeguards identified in the HAZOP.

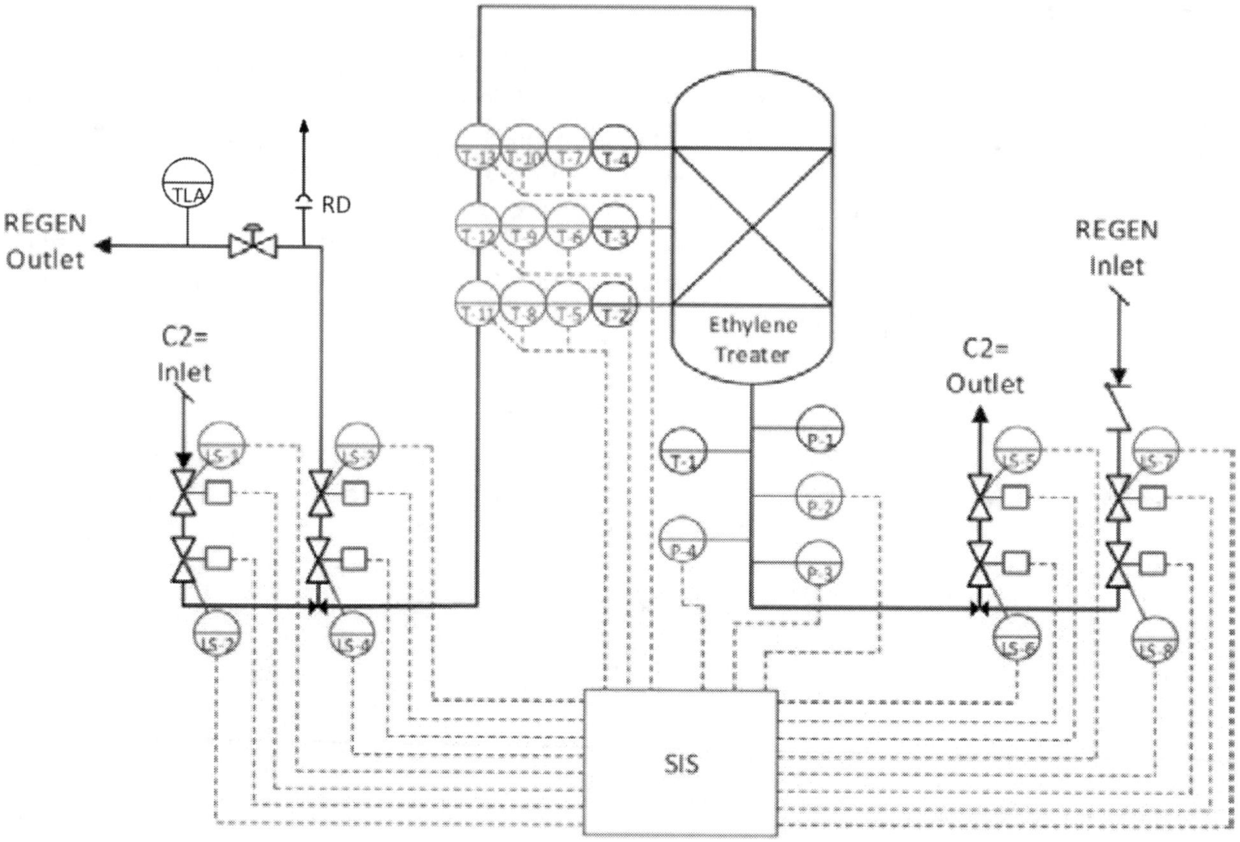

Figure 3: Schematic of LOPA Driven Design

2.3 Advanced Analysis

2.3.1 Human Factors Task Analysis

Where human error is the initiator for significant hazard scenarios, use of realistic error rates is critical. Nominal error rates may overstate, or worse, could underestimate the potential risk level. Likewise, where credit is to be taken for operator intervention in hazard scenarios, properly assessing these credits is critical and requires the use of a formal human factors task analysis approach. Human factors task analysis is a fundamental methodology in the assessment and reduction of human errors. It can be used to identify and eliminate the conditions that may potentially lead to human errors, and therefore, it is a useful tool in the design stage of a new system, or for the modification of an existing system.

A human factors task analysis workshop was conducted with experienced process and operations personnel and subject matter experts in the areas of human factors and process safety. A detailed treater regeneration sequence procedure was developed to provide a basis for the task analysis. Tables 14.15 and 14.16 in Lee's Loss Prevention Handbook (Mannan, 2012) were used as the basis for the human error rates. These tables cover "Potential Errors in Selection of Valve", "Omission of Items in Use of Written Procedures" and "Potential Errors in Checking", which were categories judged by the team to be highly representative of the operator actions being taken in a treater regeneration sequence, and specifically those actions leading to the identified hazard scenarios. Error multiplication factors of 3 to 5 were used to account for the upper limit of data uncertainty. The resulting overall operator error rates were judged to be realistic by the team.

Based on the workshop results, credit was taken for immediate operator recognition and correction of the regeneration valve mis-alignment on an on-line treater in scenario 1 in 3 out of 4 instances. This was based on the fact that opening the valve with excessive pressure drop would be more difficult, and the vibration and noise associated with sudden flow would be noticeable. No credit was taken for this correction in the valve mis-alignment for the isolated and undrained treater in scenario 2 since vibration and noise effects might be reduced given the fixed inventory in the isolated treater, which would depressurize quickly.

Additionally, credit was taken for the active verification of treater pressure and/or temperature by a second operator in missed step scenarios 2 and 3. Two cases were considered and discussed, one where the outside operator initiates action and the other where the console operator initiates action by radioing the outside operator. In each case, both operators are required to use written procedures, which call for verification of treater conditions, before proceeding to the next step. On this basis the team judged that either the console operator or the outside operator would provide a reliable second check of treater conditions prior to initiating valve line-ups.

The resulting operator error rates, or initiating event frequencies, are summarized in Table 2.

Table 2: Operator Error Rates

Scenario	**Error Description**	**Base HEP(1)**	**Multiplication Factors**					**Total per Regen**
			Grouped Valves	**No. of Valves**	**Opportunities per Regen**	**Shift Change**	**Un-certainty**	
1	Valve mis-alignment Error in Selection of valve	0.001	3	2(2)	2(3)	-	3	0.036
	Abnormal situation not recognized	0.25(4)	-	-	-	-	-	0.25
2, 3	Skipped Step Omission of Item Procedure without checklist > 10 items	0.01	-	-	1(5)	-	3	0.03
	2nd operator verification Checking that involves active participation	0.01	-	-	1(5)	3	4	0.12

Notes:

(1) HEP – Human Error Probability from the HRA Handbook, Source: After Swain and Guttmann (1983)
(2) Regeneration inlet or outlet valve
(3) Regeneration valves are manipulated twice per regeneration
(4) Based on judgement of task analysis team
(5) One depressurization (scenario 2) or one ethylene reintroduction (scenario 3) step per regeneration

2.3.2 Fault Tree Analysis Discussion

To quantify the human errors and evaluate the adequacy of safeguards, Fault Tree Analysis (FTA) was conducted after the completion of a task analysis workshop. Fault Tree Analysis is a technique for detailed safety and reliability analysis. It is a top down, deductive failure analysis in which an undesired event is analyzed using Boolean logic to combine a series of causes.

Fault tree analysis was conducted for the three identified HAZOP/LOPA scenarios using the results of the human factors task analysis along with the following data and assumptions:

- The design ethylene treater regeneration frequency is once per month.
- For the trapped key interlock mechanism, which would be attached to the manual block valves, a failure rate of 750 cycles with 10% failure probability was used based on the mechanical failure of the spring inside the interlock, as provided by the interlock manufacturer. Using the regeneration frequency of the treaters (one regeneration involving 2 cycles/month/valve), the mean time to failure (MTTF) was estimated to be 300 years (CEN, 2006).
- Supplemental safeguards, such as DCS sequence logic, alarms, check valves and pressure relief devices are included in the fault trees with nominal IPL failure rates assigned, consistent with those used in the LOPA analysis.
- In order to maintain a reasonable level of conservatism, the top event was assumed to be the process upset condition (high or low temperature) that could lead to piping or vessel failure. No credit was taken for probability of failure, or for post-release conditional modifiers, such as likelihood of ignition, presence of the operator or likelihood of injury.

Fault tree analysis for mis-alignment of the regeneration valve on the on-line treater in scenario 1 indicated that the trapped key interlock and the supplemental safeguard reduced the loss of containment frequency to an acceptable level of 3.7E-5/yr. As shown in Figure 4, this was achieved even without the use of the blinding safeguard. As such, for these scenarios, blinding would not be a required step in the regeneration procedure and this has the advantage of eliminating the need to open up equipment to install or remove blinds. Figure 4 also shows the combined operator error rate, or initiating event frequency, and the combined probability of failure (PFD) of the safeguards.

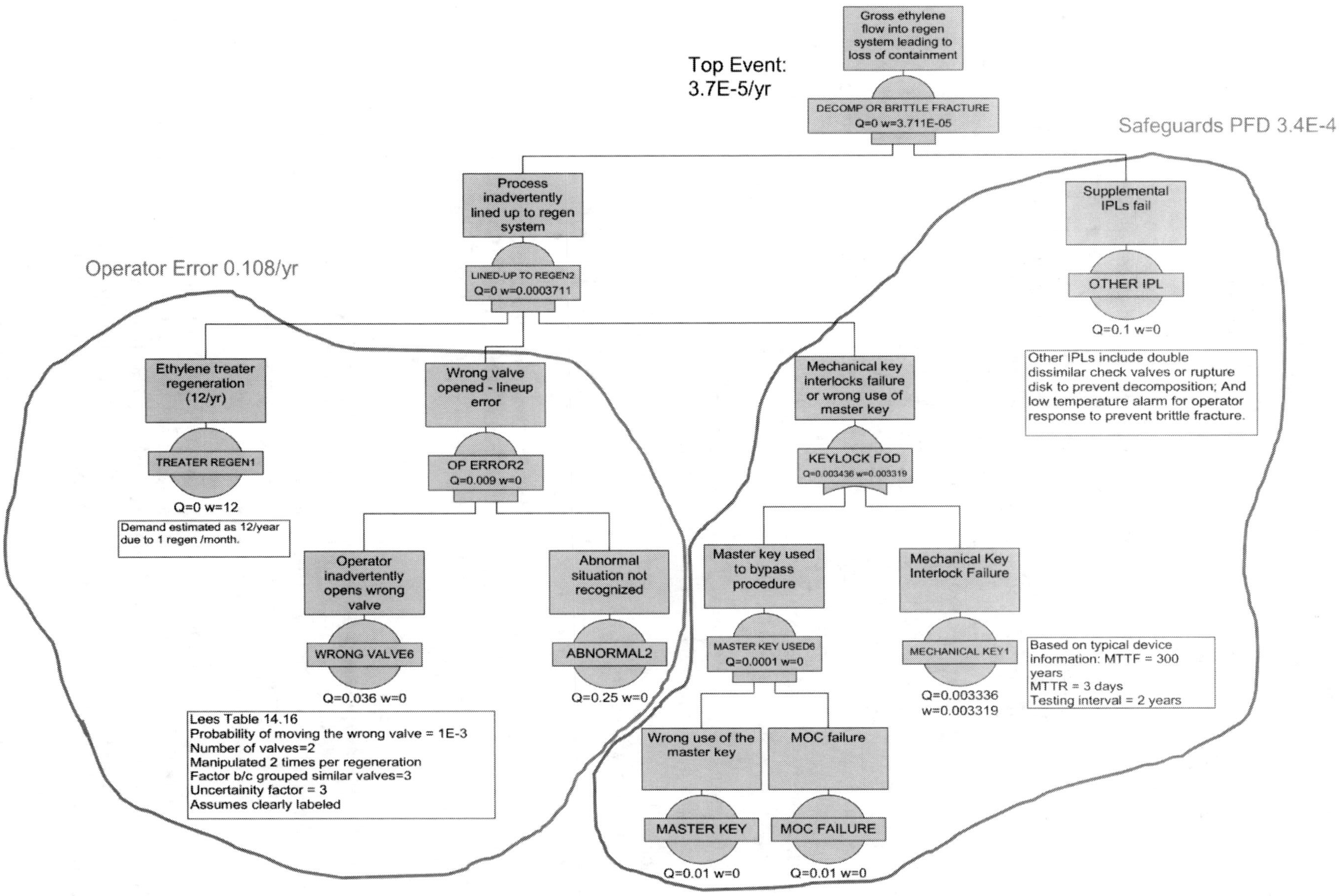

Figure 4: Fault Tree Diagram – Scenario 1

Given the risk reduction achieved by the trapped key interlock in scenario 1, which effectively provided "2+ IPL credits", use of this safeguard in a hybrid approach to address the other scenarios was suggested. The concept was to design the trapped key interlock system with appropriate process variable inputs to create a permissive using a key release solenoid. This would have the advantage of maintaining a simpler system utilizing manual block valves for the process and regeneration system isolation in lieu of the SIL-3 automated isolation valves suggested by LOPA.

A fault tree analysis was developed to verify the potential effectiveness of the hybrid concept for the missed step scenarios 2 and 3, using a pressure and temperature permissive input to the trapped key system. For example, in scenario 2, where regeneration valves are opened before the treater is depressurized, the key utilized to open the first regeneration valve can be trapped in a key exchange box, until the closed process valve keys are inserted and the treater pressure input meets a pre-determined low setpoint. The key required to open the first regeneration valve cannot be obtained, until both conditions are satisfied.

Similarly for scenario 3, where feed is re-introduced into the treater prior to cooldown, the key utilized to open the first process valve can be trapped in a key exchange box, until both closed regeneration valve keys are inserted and the treater temperature meets a pre-determined low setpoint, confirming completion of the cool-down step. If this hybrid arrangement achieves the required risk reduction factor, it would greatly simplify the overall design and avoid the need for a complex and costly SIL-3 instrumented system.

For the fault tree analysis of scenarios 2 and 3, the following additional data and assumptions for utilizing the trapped key interlock with a pressure and/or temperature permissive were utilized:

- Each ethylene treater will require multiple independent pressure and temperature transmitters, with one set providing input to a DCS treater regeneration sequence, another providing alarm inputs and a third providing input to an SIS as part of the key release solenoid pressure and temperature permissives.
- Generic equipment data from SERH (Safety Equipment Reliability Handbook) (Exida, 2015) was used for failure rates of the pressure and temperature sensors a logic solver and the key release solenoid to be utilized as part of the pressure and temperature permissives.

Fault tree analysis for scenario 2 confirmed that the trapped key interlock with the integrated low pressure permissive reduces the loss of containment frequency to 8.1E-5/yr. Likewise for scenario 3 the trapped key interlock with the low temperature permissive reduced the loss of containment frequency to 9.3E-5/yr. Both met the target top event frequency of less than 1E-4/yr, and therefore the hybrid approach provides the necessary risk reduction. The fault tree for scenarios 2 and 3 is illustrated in Figure 5.

Figure 5 also shows the combined operator error rate, or initiating event frequency, and the combined probability of failure (PFD) of the safeguards. Note that the fault tree shown includes the pressure permissive used specifically in scenario 2. The fault tree for scenario 3 is identical except that the pressure permissive is replaced with a temperature permissive, which has a slightly higher PFD. Safeguard PFDs and Top Event frequencies are shown for both cases.

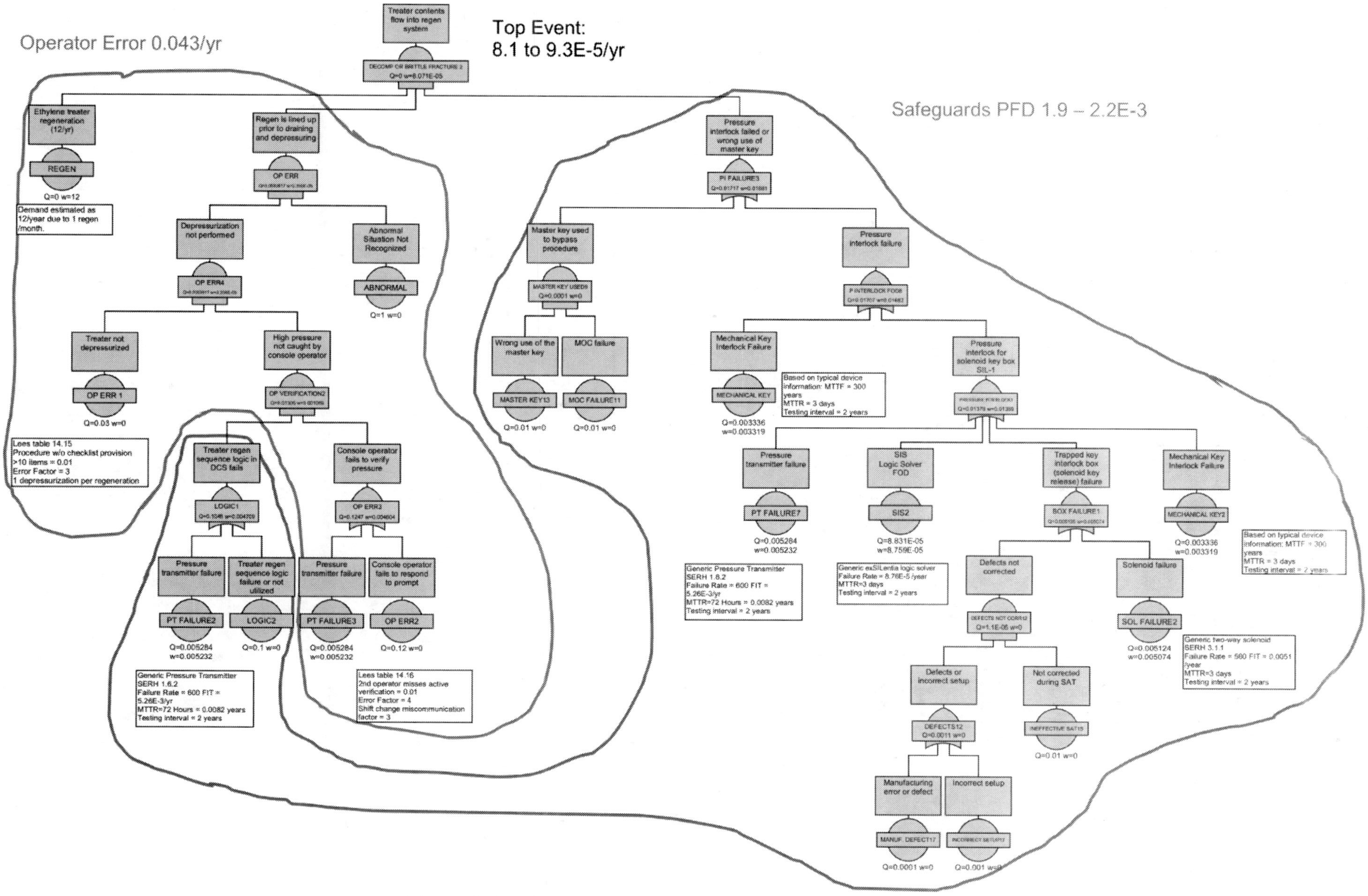

Figure 5: Fault Tree Diagram – Scenarios 2 and 3

A comparison of the LOPA results and the advanced human factors and fault tree analysis results is summarized in Table 3. The advanced analysis demonstrated that the risk tolerability criteria could be met by taking credit for operator intervention, using actual mean-time-to-failure (MTTF) data rather than order of magnitude numbers typically used by LOPA practitioners, and making use of the full benefit of the hybrid design incorporating process variable inputs into the trapped key interlock system.

The treater design incorporating the hybrid solution based on the additional analyses is illustrated in Figure 6. This simpler design solution reduces initial cost, reduces complexity and life cycle maintenance costs and provides fit-for-purpose risk reduction.

Table 3: Ethylene Treater HAZOP Scenarios and LOPA Analysis

HAZOP Scenario		**LOPA**			**Human Factors/Fault Tree Analysis**		
		IEF (/yr)	**IPL Credits**	**Freq. (/yr)**	**IEF (/yr)**	**Safeguards PFD**	**Freq. (/yr)**
1	Mis-directed Flow	0.1-1	4	<1E-4	0.108	3.4E-4	3.7E-5
2	Missed Step	0.1-1	4	<1E-4	0.043	1.9E-3	8.1E-5
3	Missed Step	0.1-1	4	<1E-4	0.043	2.2E-3	9.3E-5

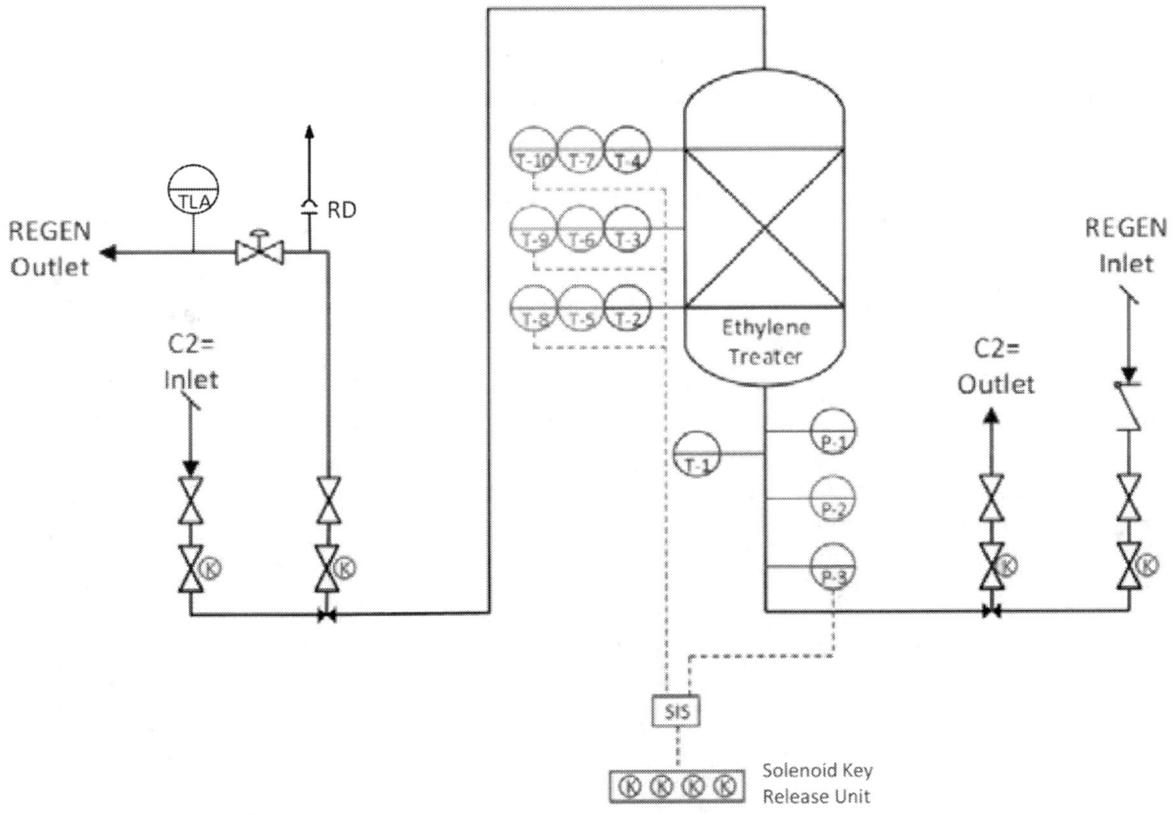

Figure 6: Schematic of the Hybrid Approach Based on Human Factors and Fault Tree Analysis

3 Cost Benefit Analysis

In the example, we "sharpened our pencil" on the LOPA analysis and using human factors and fault tree analysis demonstrated that a simpler design could provide the risk reduction required to yield an acceptable risk. In particular, for Scenario 3 the frequency was reduced just enough, at 9.3E-5/yr, to drop into the green generally acceptable box on the matrix. However, for high severity consequence scenarios, or where risk remains close to a medium risk yellow box, many operators now require that we also demonstrate that the risk has been reduced "as low as reasonably practicable" or ALARP.

According to the UK Health and Safety Executive (UK HSE 2016), making sure a risk has been reduced to ALARP is about weighing the risk against the sacrifice needed to further reduce it. The decision is weighted in favor of health and safety because the presumption is that the duty-holder (owner/operator) should implement the risk reduction measure. To avoid having to make this sacrifice, the duty-holder (owner/operator) must be able to show that it would be grossly disproportionate to the benefits of risk reduction that would be achieved. Thus, the process is not one of balancing the costs and benefits of measures but, rather, of adopting measures except where they are ruled out because they involve grossly disproportionate sacrifices.

Cost Benefit Analysis (CBA) can be useful as part of the ALARP demonstration process to compare the sacrifice (cost) and the risk reduction (benefits). The general rule is that risk reduction measures must be adopted unless the cost is grossly disproportionate to the risk reduction achieved. However, it is not straightforward to apply this rule because there is no standard criterion on judging "grossly disproportionate" based on how much costs outweigh benefits. Many companies must rely upon a company specific risk matrix for CBA and ALARP demonstration. Unfortunately, the typical risk matrix itself does not provide enough granularity in risk tolerability to provide assistance with this determination. And in our ethylene treater example, it could be argued that since we sharpened our pencil on the risk assessment that we should do the same for our risk tolerability criteria.

To help solve this problem the concept of risk units (RU) is introduced. Risk is the product of consequence frequency (or likelihood) and consequence severity and there are points on the matrix that are different pairings of frequency and severity but which represent the same statistical risk. As illustrated in Figure 7, since the colors of the risk matrix boxes represent distinguishable risk levels, two points of equal risk can be assigned at the yellow to green box breakpoints on the left and right sides of the matrix. Connecting these points defines a line of equal risk on the matrix, which has a slope roughly matching the color grouping. Furthermore, the risk represented on the matrix is usually annualized and on many matrices frequency is assigned a number, typically ranging from a low of once in 100,000 years, or 1E-5/yr., to a high of once per year, or 1/yr. By picking a point on the matrix and assigning it an arbitrary non-dimensional number of 10 risk units, a number can also be assigned to the severity by back calculating, where the product of the frequency and severity equals 10. Since the risk will be present for the life of the project or the facility, the annualized frequency is multiplied by 25 years (typical plant or project life span) to obtain a dimensionless risk unit number.

The resulting Risk Unit Matrix is illustrated in Figure 8. Note that the severity numbers assigned to each risk matrix box are in orders of magnitude consistent with the frequency numbers and consistent with the logarithmic nature of the risk matrix. Figure 8 also includes a matrix without the boxes, but with color shading in parallel with the lines of equal risk, which may provide a more realistic portrayal of risk acceptability breakpoints.

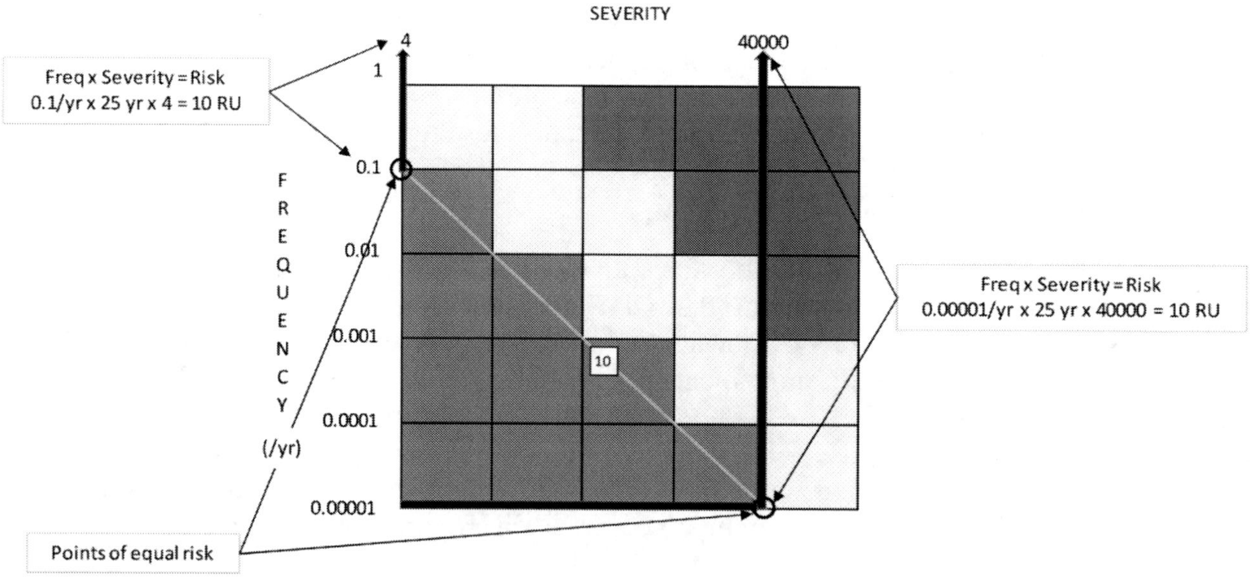

Figure 7: Derivation of Risk Units on a Typical Risk Matrix

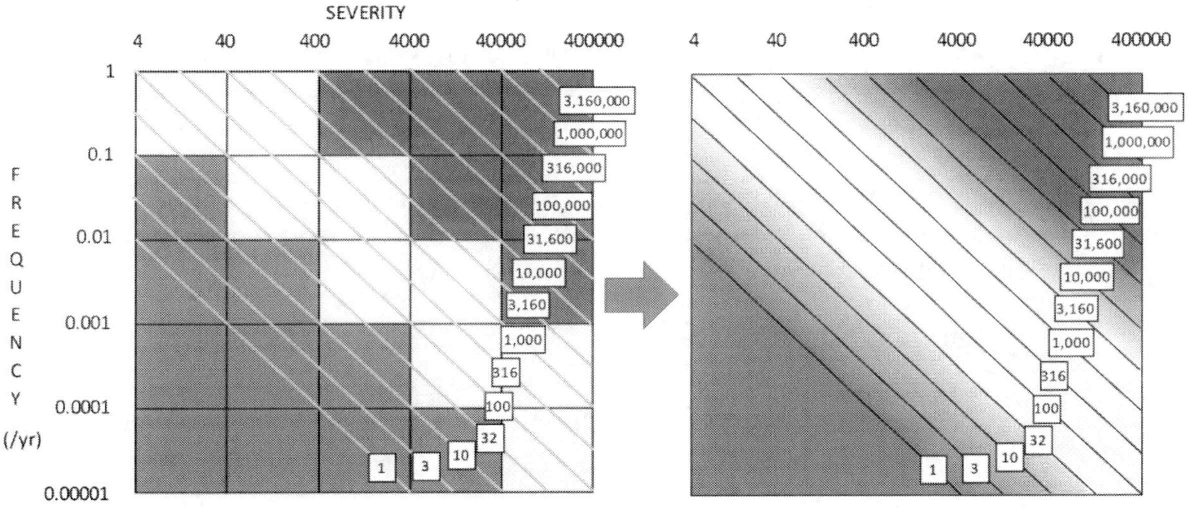

Figure 8: Risk Unit Matrix

Using this concept every point on the matrix can be assigned a risk unit number, essentially becoming an "infinite" matrix. It should be noted that the risk unit numbers are not intended to represent a measure of absolute risk and will only be used to measure relative risk.

The granularity provided by the Risk Unit Matrix provides a means to use cost-benefit analysis to compare different risk reduction options. This also provides a means to evaluate if the cost for incremental risk reduction is becoming disproportionate to the risk reduction achieved. In our example, the consequence of a major loss of containment event from an ethylene treater system could lead to a flash fire or vapor cloud explosion with the potential for 1-3 fatalities. Based on the Risk Unit Matrix this would be a severity number of 8620, obtained by multiplying the severity number at the left side of the 1-10 fatality column, 4000, by the cubic root of 10, representing the left hand third of the column on a log scale. For Scenario 3, the initiating event frequency is 0.043/yr (see Table 3) and the existing DCS sequence safeguard has a probably of failure of 0.1048 (see Figure 5, "Treater sequence logic in DCS fails"), for a combined frequency of 0.0045/yr. This results in a starting risk of 0.0045*25*8620 = 970 RU, assuming a 25 year life of the project or facility.

Table 4 provides a summary of the cost-benefit analysis comparison of the hybrid solution vs. a full SIL-3 protective system with automated valves. For each option the mitigated event frequency is multiplied by the same project or facility life span, and the same severity number since the end consequence remains the same, to determine the mitigated or residual risk units. Note that the cost utilized for the analysis covers the facilities required for the set of parallel treaters, since the risk assessment was based on the regeneration and error frequency associated with both treaters. Based on this analysis, the hybrid approach provides a risk reduction of 950 RU at a cost of $358 per risk unit. The SIL-3 system with automated valves provides a risk reduction of 969 RU at a cost of $1,589 per risk unit. The hybrid option clearly provides the most cost effective risk reduction. Note that this analysis only considers the installation cost of the system and if lifecycle costs for maintenance and testing were included the relative cost of the SIL-3 system would be even higher.

However, we still need to decide if the risk has been reduced to ALARP. For this example we are assuming that the only other risk reduction option that has been identified is the SIL-3 system and Table 5 provides the incremental cost-benefit analysis for implementing that system. As shown, the incremental cost for further risk reduction is $63,160 per risk unit. While there is no established benchmark breakpoint for $/RU, that high cost can be judged to be disproportionate to the risk reduction achieved and therefore the hybrid option meets the ALARP test. Just as importantly, it is likely that the project or facility involved will have better opportunities to utilize those resources more effectively to reduce risk elsewhere.

Table 4: Cost-Benefit Analysis for Hybrid and SIL-3 Designs

Design Option	Components	Cost ($)	Total Cost ($)	Mitigated Freq (/yr)	Mit. RU	RU Reduction	$/RU
Hybrid	Trapped key interlock on 16 valves @ $10K per valve	160,000	340,000	9.3E-5	20	950	358
	16 transmitters @ $10K per inst.	160,000					
	SIS system	20,000					
SIL-3	16 automatic valves @ $75K per valve	1,200,000	1,540,000	4.5E-6	1	969	1,589
	24 transmitters @ $10K per inst.	240,000					
	SIS system	100,000					

Table 5: Incremental Cost for SIL-3 Option

Design Option	Delta Cost ($)	Delta RU Reduction	$/RU
SIL-3 Incremental	1,200,000	19	63,160

4 Conclusions

Using a combination of HAZOP scenario identification, LOPA, human factors task analysis and fault tree analysis, a cost effective design solution was developed to manage the unique risks associated with dense phase ethylene treater and regeneration systems. The hybrid solution maintained a simpler system utilizing manual block valves for process and regeneration system isolation. A trapped key interlock system on the block valves, with treater vessel pressure and temperature permissives, provided fit-for-purpose risk reduction. These techniques not only avoided the cost and complexity of implementing a SIL-3 instrumented safety system, they also identified that the risks associated with routine opening of process equipment for blinding and un-blinding could be avoided as well. Additionally, the resources saved with implementation of the simpler design solution could be used more effectively to fund other risk reduction opportunities. The authors recommend that additional analysis be undertaken whenever LOPA calls for implementation of a high integrity (SIL-3) protective system and should also be considered where SIL-2 protection is suggested.

In addition to conducting a more detailed risk assessment to evaluate fit-for-purpose design solutions, a methodology to provide a more detailed evaluation of risk acceptability using the concept of risk units is also introduced. This provides the capability to make risk acceptance decisions consistent with level of detail provided by the risk assessment approach, and also provides a means to demonstrate ALARP.

5 References

API 579 (2007). *Fitness-for-Service.* (2nd Edition). (Section 3.3.3)

Britton, L. G. (1994). Loss Case Histories in High Pressurized Ethylene Systems, *Process Safety Progress, 13 (3),* 128-138.

CCPS (2015). *Guidelines for Initiating Events and Independent Protection Layers in Layer of Protection Analysis.*

CEN ISO 13849-1 (2006). *Safety of machinery – Safety-related parts of control systems.* (Part 1: General principles for design).

Exida (2015). *Safety Equipment Reliability Handbook.* (4th Edition).

Mannan, S. (2012). Lee's Loss Prevention in the Process Industries. (4th Edition). *Chapter 14, Human Factors and Human Error* (Table 14.15).

Nolan, Dennis P. (2014). *Safety and Security Review for the Process Industries, Application of HAZOP, PHA, What-If and SVA Reviews* (4th Edition). Gulf Professional Publishing.

UK HSE (2016). http://www.hse.gov.uk/risk/theory/alarpglance.htm

Identify SIF and Specify Necessary SIL, and other IPLs, as part of PHA/HAZOP – or - Why it is not necessary to "Boldly Go Beyond HAZOP and LOPA"

William G. Bridges, President
PROCESS IMPROVEMENT INSTITUTE, INC. (PII)
1321 Waterside Lane, Knoxville, TN 37922
Phone: (865) 675-3458
Fax: (865) 622-6800
e-mail: wbridges@piii.com

A. M. (Art) Dowell, III, PE
PROCESS IMPROVEMENT INSTITUTE, INC. (PII)
2437 Bay Area Blvd PMB 260
Houston TX 77058-1519
phone: 713-865-6135
e-mail: adowell@piii.com

Prepared for Presentation at
12th Global Congress on Process Safety
Houston, TX
April 10-13, 2016

UNPUBLISHED

12TH GLOBAL CONGRESS ON PROCESS SAFETY

Identify SIF and Specify Necessary SIL, and other IPLs, as part of PHA/HAZOP – or - Why it is not necessary to "Boldly Go Beyond HAZOP and LOPA"

William G. Bridges, President
PROCESS IMPROVEMENT INSTITUTE, INC. (PII)

A. M. (Art) Dowell, III, PE
PROCESS IMPROVEMENT INSTITUTE, INC. (PII)

Keywords: PHA, HAZOP, Hazard and Operability analysis, risk assessment, risk analysis, LOPA, IPL, independent protection layer, layer of protection analysis

Abstract

This paper shows how to apply the qualitative definition of IPLs within the setting of a process hazard analysis (PHA) to get most of the gain from *LOPA* without doing a LOPA (without using numerical values). The paper includes an implementation path to develop PHA leader competencies to guide the qualitative approach. We also show the way we use a PHA team to identify when a SIF is needed and to select the proper target SIL. This portion of the SIL evaluation and the identification and labeling of the IPLs during the PHA/HAZOP does not take any longer than a normal PHA/HAZOP, once the right habits are established. Note that this approach eliminates the need for a separate SIL Evaluation Study to identify the SIFs and select the target SIL. Finally, the paper ties together these two specific topics, along with the topic of making risk judgments, to show there is less than 5% need to go beyond HAZOP, and less than 0.01% need to go beyond LOPA.

Background

Identifying safety instrumented functions (SIFs) and other independent protection layers (IPLs) is important for any organization. Features that are designated as IPLs (including SIFs) are now being treated as "Safety Critical" or "Business Critical" or "Mission Critical" features.[1] These need to be maintained such that the predicted probability of failure on demand of the feature is actually achieved. These can be identified in a simplified risk assessment, such a Layer of Protection Analysis (LOPA).[2] But, these also can be identified with relative ease in a purely qualitative setting of a process hazard analysis (PHA) using hazard and operability analysis (HAZOP) or other PHA methods.

Similarly, a PHA team can usually identify in a qualitative sense if an instrumented safeguard loop meets the restrictions of a SIF. PHA teams can also be taught to assign qualitatively the necessary Safety Integrity Level (SIL) for the SIF. Per ANSI/ISA 84.00.01 Part 3, Section 3.8:

> *A qualitative method may be used as a first pass to determine the required SIL of all SIFs. Those which are assigned a SIL 3 or 4 by this method should then be considered in greater detail using a quantitative method to gain a more rigorous understanding of their required safety integrity*[3]

Many believe that only a quantitative analysis, such as LOPA, can identify the need for an SIF and assignment of the SIL necessary to reach tolerable risk. But, purely qualitative teams are permitted to make the same judgements and have been doing so for more than 50 years.

Once the IPLs (including SIFs) are identified, then the purely qualitative teams can make the final judgment on if the risk is a tolerable level or not. This last step is iterative and leads to the judgments of where additional or improved safeguards and IPLs are needed, including SIFs. The conclusion by the authors and most of our clients who follow this approach, is that once a PHA team is properly calibrated, they will be capable of judging tolerable risk and identify IPLs, including SIFs, without resorting to quantitative methods. Further, hundreds of PHAs have been performed using this approach, which together illustrate that less than 5% extra meeting time is required to identify and label the IPLs/SIFs.

This qualitative approach saves the time that it would take to do multiple LOPAs and yet still provides the comprehensive list of necessary SIFs (and their SILs) and other IPLs. The company can then use this information to assign inspection, test, and preventive maintenance (ITPM) activities and other management systems to ensure the necessary PFD against the accident scenarios are available.

Why is Qualitative Decision Making Normally Better than Quantitative?

As illustrated by the title of this conference session, "Boldly Go Beyond HAZOP and LOPA", many people believe that more quantification is better. But, a comparison of results of qualitative versus quantitative does not support this position. In the past, every quantitative analysis textbook, from *LOPA* (2001) to *Guidelines for Chemical Process Risk Analysis* (CPQRA; 1992), states that it is unrealistic to get an accurate estimate of risk for quantitative methods. Instead, QRA methods are mainly useful to compare risk of alternatives. The same is true of LOPA. Yet, more recently (CPQRA; 2000) most organizations use the risk estimated from LOPA or QRA as approximations of the actual risk of a scenario and base absolute decisions on such estimates. This is somewhat puzzling given that broad assumptions are built into such estimates about control of human factors, management systems, accuracy of the average failure rates used, and uncertainty (range of data that is used to provide the average) in the failure rates.

The authors are proficient in all of the risk assessment methods, from purely qualitative to full quantitative, and have supervised (and helped perform) more than 7000 PHA/HAZOPs, 3000 LOPA, and about 100 QRAs (and human reliability analyses [HRAs]). Assuming each analysis is performed by highly competent staff, the conclusions are that qualitative analysis is usually (more than 95% of the time) better than quantitative analysis at judging the risk of specific accident scenarios. By "better" we mean the PHA team can use more relevant data (their experiences at the site) which is nearly always more pertinent and more accurate than data from lookup tables provided in CCPS LOPA textbooks and other sources. Some may comment here that "highly competent staff" are difficult to come by on the PHA teams. We certainly understand that challenge. But that does not change the fact that without an excellent PHA, your organization will miss Many process safety scenarios and these scenarios are much more likely to not have sufficient safeguards in place. Having a highly competent PHA team is much more important than any other discussion point in this paper.

Note that CCPS textbooks on QRA and LOPA recognize the value of PHAs. As the newest CPQRA textbook (2000) concedes ***"… where the risks are clearly excessive and the existing safeguards are inadequate, corrective actions can be adequately identified with qualitative methods."*** And the LOPA book (CCPS, 2001)[2] lists and answers myths about LOPA being better than PHA/HAZOP, which it is not.

Based on analyses of these methods and their results, we believe a spectrum of methods is best, with qualitative methods being the first choice, and then other methods being used if the qualitative analysis team (PHA/HAZOP team) is confused on the risk. The summary below (Table 1) compares the three major risk analysis approaches:

Table 1: Comparison of Risk Analysis Approaches vs. Uncertainty and Usefulness

Approach	Assessment Methods	Risk Judgment Method	Estimated Range of the Results	Comments
Qualitative only (no numbers; not even a risk matrix).	What-If HAZOP of parameters HAZOP of steps FMEA	Voting of experts, from the specific site, but including at least one expert from outside of the process under review. **Fully capable of judging risk more than 95% of the time.**	Within one IPL of minimal value (so within about an order of magnitude)	Heavily focused on site-specific experience, with some consideration of general industry data and incidents
Simplified quantitative	LOPA RiskGraph	Multiplication of statistical averages of general failure rate data. **Was necessary to judge risk on about 5% of the scenarios.**	Plus and minus an order of magnitude, since the range of data for some to most inputs data is plus and minus an factor of 10	Typically does not use site-specific data for the PFD and IEF. Broad assumptions made about management systems at the site

Full quantitative	Fault Tree Event Tree Consequence modeling Human reliability event tree Bow Tie	Multiplication of statistical averages of general failure rate data; in some cases modified by expert opinion based on site-specific evidence. **Was needed for less than 0.01% of the scenarios.**	Plus and minus an order of magnitude, since the range of data for some to most inputs data is plus and minus an factor of 10	Typically does not use site-specific data for the PFD and IEF. Broad assumptions made about management systems at the site. See paper from Bridges and Dowell, 2015, for illustration of how QRA underestimated risk by three orders of magnitude in nuclear power PRAs.

General Approach to Qualitative Determination of IPLs and SIFs

One of most valuable outcomes of starting LOPA in the mid-1990s was the crystallization of the qualitative definition of an IPL. If the PHA/HAZOP leader was competent in the definition of an IPL, we found that IPLs could be just as easily identified in a PHA/HAZOP as in a LOPA. This was a significant outcome, since IPLs are what we need to focus on to maintain tolerable risk for each scenario; in other words, we can focus our reliability/maintenance efforts and operational efforts on IPLs to conserve resources while maximizing control of risk. SIFs are just one type of IPL and we found the same was true for identifying SIFs and setting their SILs. Also, as mentioned earlier, ANSI/ISA 84.00.01-2004 allows SIL to be set qualitatively.[3] The following sections explain the steps to achieve the competency necessary in the PHA/HAZOP team leader, scribe, and/or members to allow these qualitative judgments. COMPETENCY, as always, is King; the path to competency includes learning the rules for using this approach, so the rules are explained first in this paper. After a review of rules, the steps for **intentional competency development** are explained in more detail.

Review of Rules for an IPL

Below are the set of rules for each IPL. These rules can each be applied "qualitatively" to each candidate IPL. (These are similar to those found in Guidelines to Initiating Events and Independent Protection Layers.[1])

1. **Each protection layer must be truly independent of the other protection layers and independent from the initiating event (IE). That is, there must be no failure that can deactivate two or more IPLs and the IE cannot deactivate an IPL.**

   An IPL (or an IE) includes the ENTIRE sub-system, including any root valves, impulse lines, and bypasses. The other IPLs (nor IE) cannot share any of these (except for the mother board when using approach B for BPCS loops), and describe the textbook on LOPA.[2]

   A device, system, or action is **not** independent of the initiating event and cannot be credited as an IPL for either approach if either of the following is true:

   - ➢ Operator error is the initiating event and the candidate IPL assumes that the same operator must act to mitigate the situation. Human error is equivalent to the failure of a system and once a human has committed an error it is not reasonable to expect the same operator to act correctly later in the sequence of events. This approach is justified because the error may be due to fatigue, illness, incapacity (drugs or alcohol), distraction, work overload, inexperience, faulty operating instructions, lack of knowledge, etc., that are still present later when the action is required.

- Loss of a utility (electricity, air, cooling water, nitrogen, etc.) is the initiating event and a candidate IPL is a system that depends on that utility.

Other examples where the IPL is not independent include:

- Multiple flow meters, analyzers, etc., with a calibration error or other error during testing, due to human error, faulty calibration instruments, etc.
- Multiple units or SIF systems with a single source of power or a common circuit breaker unless it can be determined that fail safe action will always be initiated in the event of power loss—this is true for any other utility required for an IPL to reach a safe state;

2. **A control loop in the BPCS whose normal action would compensate for the initiating event can be considered as an IPL.**

   For example, an initiating cause for high reactor pressure could be failure of a local upstream pressure regulator; the normal action of the reactor pressure controller would be to close the inlet PV, thus providing protection against the impact event.

   NOTE: Under specific conditions, a BPCS can be used twice in the same LOPA scenario, as described later in detail in the LOPA book[2] and IPL book[1]. But, under no circumstance can a BPCS be used more than twice in the same scenario. So:

   a) a BPCS loop failure can be an initiating event

   b) a BPCS loop (the entire loop) can be an IPL

   c) a BPCS loop can give an alarm and then a human respond as part of a human IPL

   ...but in LOPA, the maximum use of a BPCS is one case of a and b, or two cases of b, or one case of a and c, or one case of b and c.

3. **The frequency reduction for an IPL is at least one order of magnitude, i.e., 10^{-1} PFD (that is, the availability is 90%).**

   - ***Example:*** If the operator has sufficient time to react, then the risk reduction for Operator Response to an Alarm is one order of magnitude, i.e., 10^{-1}

4. **The IPL is capable to prevent or mitigate the consequences of a potentially hazardous event. To make this judgment, the evaluator (such as the PHA team) must ask:**

   - Is the IPL valid for the mode of operation for the scenario (startup, shutdown, normal, batch, etc.)
   - Is the IPL applicable to the scenario under consideration? For instance: Is the PSV even designed for this scenario?
   - What are the maintenance/reliability practices and plant/company history? How much likelihood reduction credit will you take for a relief valve?
   - How good are the procedures and related training practices? Were the operators trained in specifics of how to respond to this alarm/indication?
   - Consideration of standards and certifications (PSV code stamp; IEC 61508 classification,

etc.) can help ensure safeguards qualify as IPLs.

5. **The IPL must be *Maintained* and *Validated* periodically; it must be proven that the IPL can be counted on to do what it was intended to do.**

   The IPL must be periodically maintained and it must be proven or validated. The site must have data that supports the reliability factor. The frequency and test method must comply with best industry practices for such IPLs. Also, the site must maintain a database for each IPL that statistically supports the PFD stated. For a component or instrumentation IPL, this requires maintaining a statistical failure rate database that justifies the PFD listed for each IPL. For a human IPL, the site must maintain data from "drills" of the action of the worker that statistically demonstrates that the worker(s) can indeed implement the required action (of the IPL) with the time specified in the IPL. For a PFD of 10^{-1}, the statistical data must support that 90% of the recorded data demonstrates the necessary speed and reliability. For a PFD of 10^{-2} the statistical data must support that 99% of the recorded data demonstrates the necessary speed and reliability.

6. **The IPL maintenance and validation must be *Audited*.** Auditing is required to ensure the validation, procedures, training, and resulting data are adequate. This is an administrative check. This auditing cycle is set frequent enough (typically 1 year for the first audit and then 5 year frequency after that) to ensure that validation is being carried out as planned and is sufficient to justify the IPL and its PFD.

7. **Specific errors (such as leaving an IPL bypassed) and systemic failures (such as the instrument tap plugging) that would impact the performance of the IPL (also applies to IEs) must be considered in the PFD (or IEF).**

   Example: When considering a PSV as an IPL; if there is a block valve (B/V) upstream or downstream of the PSV then the probability of leaving this closed (and perhaps even car-sealed-closed) must be included. Currently, site data indicates this probability is between 0.01 and 0.04, which means a PSV valve of 0.01 is Not valid.

   Example: When considering a SIL 2 or 3, systemic errors and failures can obviate the redundancy and increase the PFD (increasing the risk) since some of these errors and failures can themselves be 0.01 or higher. The SIL Verification for SIL 2 and 3 must account for these systemic errors and failures or else the assigned SIL is not valid; they may be no better than a SIL 1 in actual performance. (See later comments of SIS for LOPA).

8. **Related to 7, the boundary for the IPL (or IE) must include all relevant components upstream and downstream that could affect the performance of the IPL (or IE).** For example, for a PSV, the IPL includes the PSV as well as any inlet and outlet piping and any isolation valves upstream or downstream. Similarly, as shown on the Boundary B of Figure 1, an SIF boundary includes the root valves, impulse lines, and any bypasses associated with the SIF.

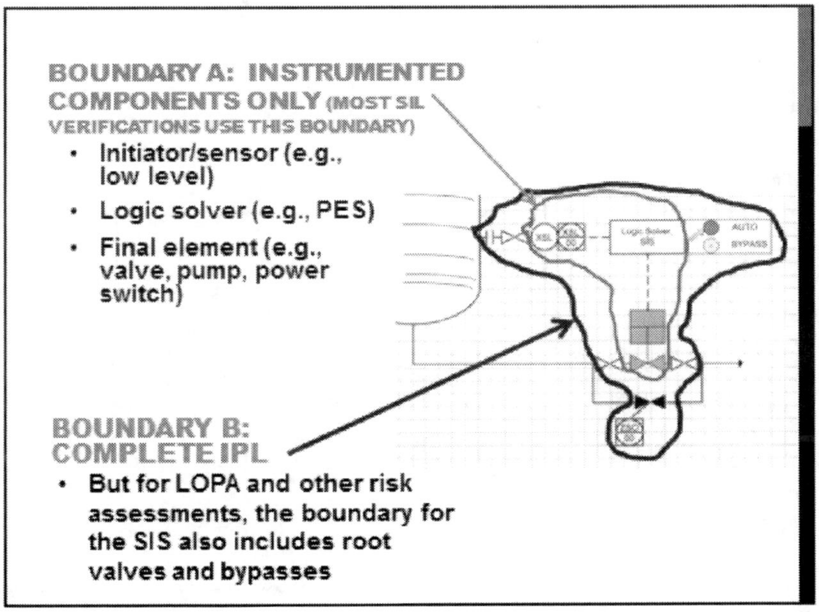

Figure 1. The True Boundary for an IPL includes all associated connections, isolations, and bypasses. Note how the traditional SIL Verification calculation boundary is much different that the true IPL boundary

Review of Specific Rules and Example Architectures of SIFs of Various SIL

The design of SIFs to meet a required SIL can be complex in certain applications, but in general a <u>**SIL 1**</u> SIF is a single loop that uses devices rated for use in a SIF of the appropriate SIL, including a safety PLC that is independent of the BPCS (control system), and that is rated for use in an SIF of the appropriate SIL. Figure 2 shows a typical example of a SIL 1 SIF configuration. These are tyically one sensor or switch, going to one fast acting final element (like a shutdown valve). These typically have a calculated PFD in the range of 0.03, depending on the proof test interval.

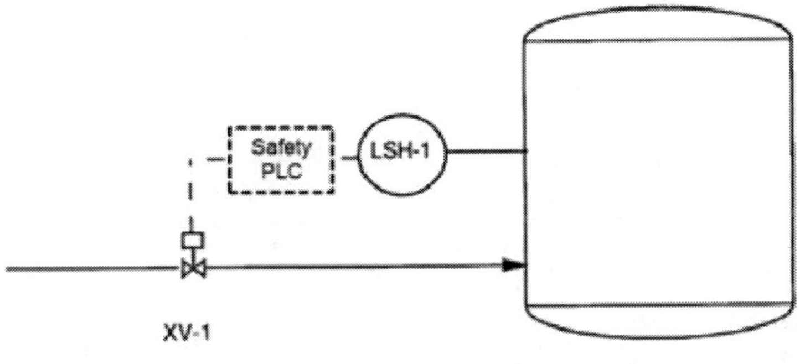

Figure 2. Typical configuration of a SIL 1 SIF

It should be noted that the achieved safety integrity level for an SIF is a function of the architectural constraints requirements for the particular SIL in addition to the PFD requirements for the SIL. The architectural constraints depend on the safe failure fraction of the devices, the voting architecture, and the diagnostics for the device.

The PFD of the SIF depends on the voting architecture, the failure rate of each device, the diagnostics to detect failures, and the proof test interval for the SIF.

In general a **SIL 2** SIF is a safety loop similar to a SIL 1, but which uses one or more redundant devices, that are voted one-out-of-two (1oo2) or better, for the safety functions. Depending on the architectural constraints requirement, the failure rate and diagnostics of the devices, the voting architecture, and the proof test interval, redundant devices may be required for the sensors and for the logic solver and/or for the final element. Figure 3 shows two typical configurations for SIL 2 SIFs – when redundancy is either be in the sensor/ switch/ transmitter devices (so, in the example shown here, two level swiches) or in the final element (such as two shutdown valves, where at least one needs to close for safety purposes). The calculated PFD for such systems, if the specific human erorrs of leaving the SIF in bypass or of miscalbrating of two devices is ignored, will be about 0.009, depending on the proof test interval.

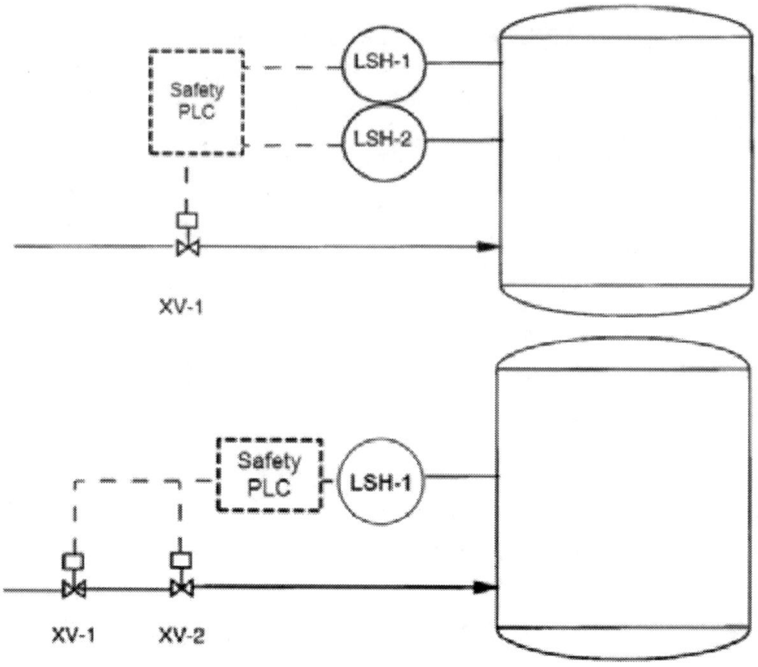

Figure 3. Typical configurations of SIL 2 SIFs

The required PFD for a **SIL 3** SIF is likely not possible to achieve in practice, as the human errors for maintaining such systems is greater than the target PFD of 0.001 to 0.0001. But, if such a system is designed to compensate for all such errors, then SIL 3 usually have both redundancy in the sensors and in the final elements. Figure 4 shows a typical example of a SIL 3 SIF configuration.

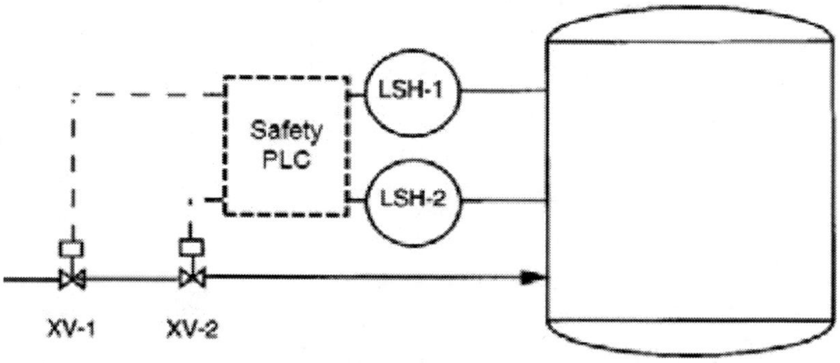

Figure 4. Typical configuration of a SIL 3 SIF

Note that there are architectural constraints and common mode issues for SIL 2 and SIL 3 that are beyond the scope of this paper. And of course there are more complex configurations for SIFs, but typically, they follow the patterns above.

Once the PHA leader and scribe understands the typical configurations of SIFs for the various SILs then they must learn the basic rules of SIS (which are listed in detail in IEC 61508 and 61511[ANSI/ISA 84.0 0.01-2004])[4], which are simplified below:

- Devices rated for use in a SIF of a particular SIL must be used (or "proven in use" devices must be used)

- All of the rules of an IPL must be followed for any SIF (e.g., the SIF cannot share any components with the BPCS)
- Only one SIF is allowed for a scenario per SIS (That is, all the safety instrumented functionality for a given scenario should be evaluated as one SIF, if contained within the same SIS.)
- The ITPM stated in the Safety Requirements Specification (SRS) must be followed and documented.

The PHA leader or scribe can now begin to judge qualitatively if a loop shown on a P&ID is intended to be a SIF or not, and if so, if it meets the basic rules of a SIF.

If the PHA team feels that a SIF is the best way to reach tolerable risk (given that the risk of the scenario is judged to be intolerable), and a SIF is not present or is not of a high enough rating, then the PHA team can recommend a SIL 1, 2, or 3 SIF (though at PII we do not typically go along with recommendations for a SIL 3 SIF, for the reasons mentioned earlier).

Implementation Path – Intentionally Achieving COMPETENCY in the Qualitative Definition of an IPL and SIF

Now that the rules and descriptions of SIFs (of various SILs) and other IPLs are defined, the next requirement for adequately using these in a qualitative PHA is for the **PHA team leader to understand fully these rules and definitions and to be competent in their application.** Many PHA leaders are not competent in even how to conduct a PHA; in fact, about 90% of the PHAs reports that we have reviewed around the world are woefully deficient, especially with respect to finding scenarios during startup, shutdown, and online maintenance. Per the authors experience, the path to the necessary competency is typically:

- Already be an experienced PHA leader, trained in all PHA methods, and capable of applying these methods to all modes of operation and capable to make sound qualitative judgments, along with the PHA team members, on when the number and type of IPLs is sufficent to control the risk. Achieving full competency as a PHA leader may require some remedial training on how to lead PHAs of startup, shutdown, and online modes of operation; or remedial training on how to uncover and discuss all plausible damage mechanisms. This assumes the PHA leader has the correct technical background, including many years in operations. *A poster from the classroom training from 2003 is shown in Figure 5.*

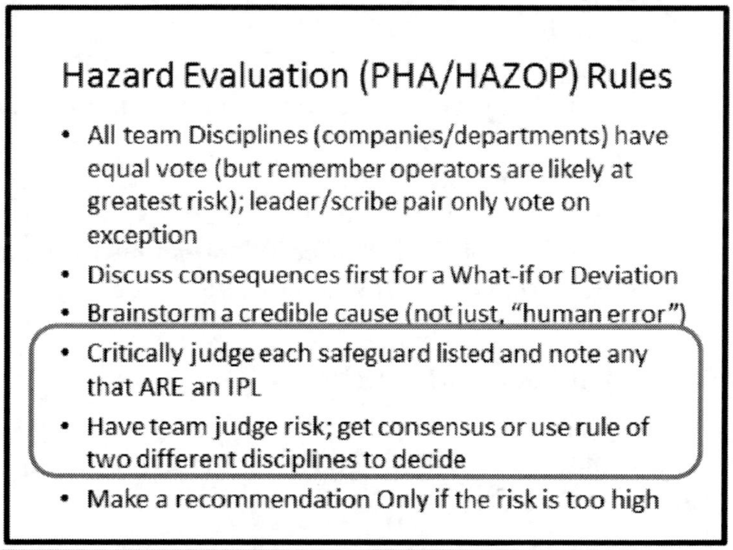

Figure 5. Wall Poster of PHA Rules; PII, 2003-2016 ©

- Attend a LOPA course to learn the basics of IPLs (including SIFs) as described in the previous sections. Or, attend a more progressive training course for PHA/HAZOP leadership that covers the definitions of IPLs and provides exercise time on how these determinations are made. The key on qualitative risk judgment is to know when there are enough IPLs for the accident scenario under review. *Another poster from the PHA Leadership classroom training from PII since 2003 is shown on the right* ➔

Independent Protection Layers (IPL)
must meet all of the following requirements

1. INDEPENDENT – of causes and other IPLs - there must be no shared components and no shared human
2. ENOUGH – big enough, fast enough, strong enough, survivable for the scenario under review
3. PROVEN – audited, tested, and/or validated routinely enough to provide the desire risk reduction factor
4. DOCUMENTED – design basis and proof results are recorded and available and demonstrate the desired reliability/availability

Figure 6. Wall Poster of IPL Rules; PII, 2003-2016 ©

- Get coaching (by someone already competent) during actual PHAs to learn how to help a team make judgments if safeguards meet the definition of an IPL (or SIF) or not and also on if there are enough IPls for the accident scenario (risk judgment). The PHA team leader must have either seen (or done) enough LOPA of similar nature to make such judgments or the PHA team must be able to judge when to go to LOPA or not. We know from thousands of PHAs over the past decades that a PHA team can make excellent risk judgments > 95% of the time, which also means that the IPLs and SIFs can be clearly identified > 95% of the time.
- Achieve competency, in the opinion of the competent coach on the skills above.
- In addition to the PHA team leader competency, the PHA team has enough understanding of either qualitatively risk judgment or LOPA risk judgment– just in time training by the PHA leader (we tend to accomplish this training across the first 5-10 accident scenarios we discuss):
  - IPL requirements – difference between just a safeguard and a safeguard that is an IPL
  - Initiating cause frequency
  - Consequence severity
  - Required mitigated consequence frequency to meet risk criteria for that consequence severity; or the Required IPLs qualitatively necessary to reach tolerable risk.
  - Risk reduction provided by existing IPLs

Implementation Path – Using the Qualitative Definition of an IPL and SIF

The competent PHA/HAZOP leader can now guide the PHA/HAZOP team through the following thought processes:

- IF (1) the safeguard meets the definition of an IPL and (2) if the team believes (qualitatively) this safeguard is critical to control risk to tolerable level (qualitatively),then add the designator " – IPL" to the right of the safeguard text (or else, turn on the "IPL Type" column in HazardReview LEADER [by ABS Consulting] and select the IPL type from that pulldown menu). If the safeguard is not going to be labeled an IPL, then it can be run to failure; unless the safeguard supports an IPL, such as when a sight glass support an LAH used in an Human Response IPL, in which case the sight glass will have some ITPM (such a periodic cleaning of the sight glass).

- IF an instrument is already in the ESD system or SIS and qualitatively meets the archetecture of an SIL 1, or SIL 2, or SIL 3, and also meets the definitions/rules for an IPL, then add the "- SIL-1" (or SIL-2, or SIL-3) to the right of the text (or else, turn on the "IPL Type" column in HazardReview LEADER and select SIL-1, 2, or 3 from that pulldown menu). *See the screen shot of a LEADER window of a PHA/HAZOP table in Figure 7.*

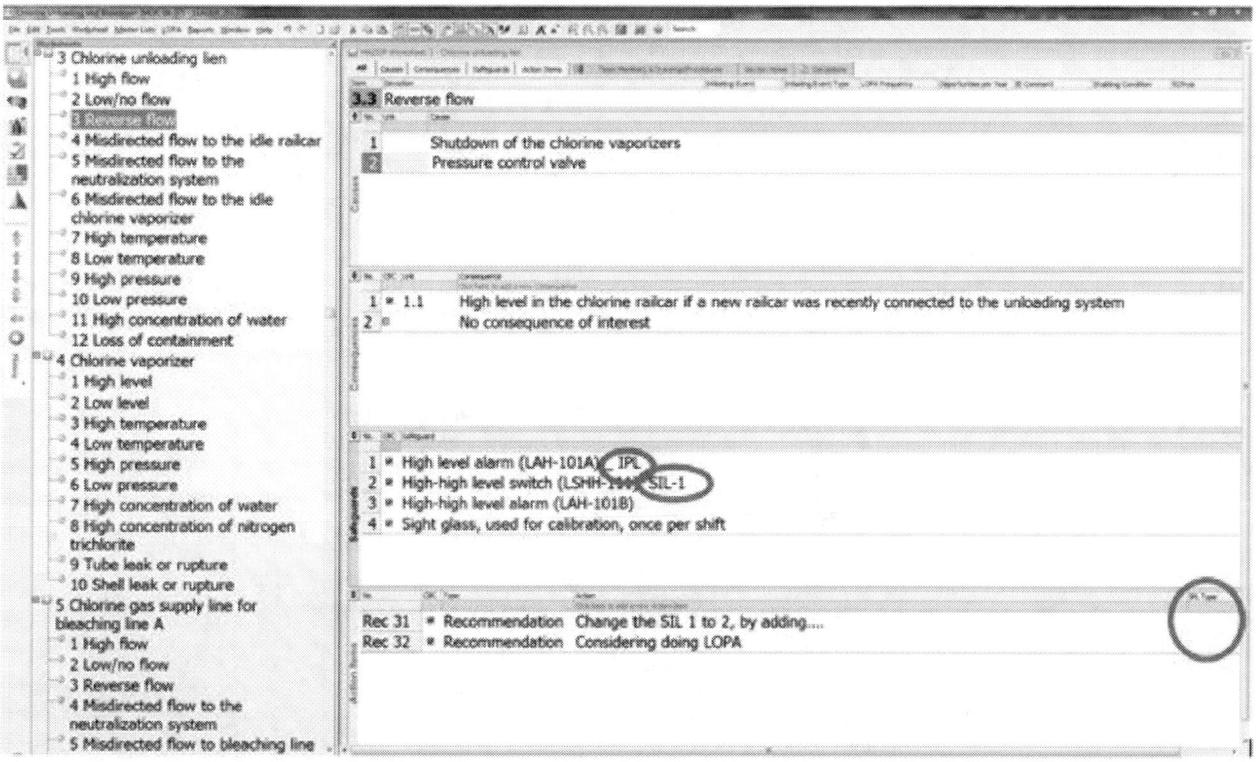

Figure 7. Screen Shot of LEADER software showing designation of IPL and SIL in the HAZOP Record

The non-human IPLs identified can be maintained as critical features in reliability/maintenance systems. For SIFs, the facility should have the SILs verified by calculations and have the SRSs developed. The human IPLs can be tested/drilled once a year. All of these activities are to ensure the IPLs/SIFs deliver the PFD anticipated, while still ensuring reliable operation/control by not cause too many spurious trips.

CASE STUDIES

Case Study 1 – Sinopec-SABIC Tianjin Petrochemical Company (SSTPC)[5]

PHAs of 9 full petrochemical plants were led and documented in 2013 - 2015 (see complete paper from 2015 GCPS).[5] The results were documented in English into *HazardReview LEADER*™ software; a Word report was generated for each unit's PHA. This report contains the typical entries, with considerable detail developed for each scenario. Excerpts from the report are shown in Table 1.

Table 2: Excerpts from Petrochemical Process PHA at SS-TPC[5]

No.: 2 XXXX storage spheres xxx-T-XX A/B/C/D/E/F/G/H/I/J/K/L (1 of 12)

#	Dev.	Causes	Consequences	Safeguards	Recommendations
2.1	High level	Too much flow to one sphere from XX Plant (through their pump; about 40 bar MDH)	High pressure (see 2.5)	High level SIF with level sensors voted 2oo2, to close inlet valve - **SIL 1** Overflow thru pressure equalization line to other spheres (through normally open [NO] valve) - **IPL**	
		Misdirected flow - Liquid from xxx Plant(s) to spheres (see 1.4)	Overpressure of sphere not credible from high level, for normal operating pressure of the column (which is 1.75 MPa), unless all spheres are liquid filled and then thermal expansion of the liquid could overpressure the spheres	High level SIF with level sensors voted 2oo2, to close inlet valve - **SIL 1** Overflow thru pressure equalization line to other spheres (through normally open [NO] valve) - **IPL** Spheres rated for 1.95MPa (19.5 Bar, approx) and the highest pressure possible from the column feeding the spheres is 1.75 MPa Level indication and high level alarm in DCS, used by operators to manually select which tank to fill - **IPL**	
			Overflow into the equalization line will interfere with withdrawal from the column, but this is an operational upset only		
			Excessive pressure on inlet of high pressure liquid pumps, leading to excess load on pumps and trip of pumps on high pumps, causing trips of xxx, xxx, etc. - significant operability issue		
2.2	Low level	Failing to switch from the sphere with low level in time (based on level indication)	Low/no flow - Liquid from spheres through high pressure product pumps to the vaporizer (see 4.2)	Level indication and low level alarm, inspected each year, per government regulation **(not IPL; part of the cause)** 9 other spheres with possibly enough level to switch to Feeding from two spheres at all times, so unlikely for BOTH spheres to have low level at the same time - **IPL** Two level indication from SIS level transmitter, with low level alarm, with more than 60 min available to switch tanks (SIF driven alarm and	Rec 4. Make sure the Human IPL of response to low level in all spheres and tanks is described in a trouble-shooting guide (like an SOP) and practiced once per year per unit operator. This will make this response a valid IPL.
			Low/no flow - Unqualified liquid from spheres back		

No.: 2 XXXX storage spheres xxx-T-XX A/B/C/D/E/F/G/H/I/J/K/L (1 of 12)					
#	**Dev.**	**Causes**	**Consequences**	**Safeguards**	**Recommendations**
			to Plant (see 6.2)	response) - **possible IPL, if action of the operator is quick enough**	
2.3	High temp.	Large area of damaged insulation Loss of cooling. when the tank is isolated from column	High pressure - vapor from spheres through condenser and return to liquid pump out line (only used when plant is shutdown) (see 3.7)		
2.4	Low temperature	Deviation during startup (see 2.9)	Loss of containment - due to sudden flashing to -100 °C; if there is also a sudden vibration (such as by the flashing from liquid to gas) (see 2.8)	Temperature indication	Rec 5. Consider adding an IPL, such as an interlock, to prevent opening of the isolation valves for liquid into a sphere (following maintenance or an outage), until the sphere has been pressurized with vapor, to prevent brittle fracture of the sphere and to prevent thermal shock of the sphere.
2.5	High pressure	Liquid filled and left blocked in	Loss of containment (see 2.8)	Pressure indication and high pressure alarm in DCS (1 on each of the 12 spheres); with operator response (with practice/drills) - **IPL** (Note: The pressure control valve to the flare is normally blocked in; and is only used when the standby cooling system is used)	
		High pressure in the gas equalization line from column	Loss of containment (see 2.8) More losses to flare from sphere - economic consequence	(Note: The pressure control valve to the flare is normally blocked in; and is only used when the standby cooling system is used)	

To help the teams make consistently good judgments on risk, each safeguard was judged to determine if it met the definition of an IPL or not. The team then qualitatively judged the risk (voted; sometimes with pressure from the team leader) and decided if the risk was controlled well enough; if not, then recommendations were made to reduce the risk to tolerable levels. As a cross-check of these judgments, SS-TPC required the PHA team also to score (using a calibrated risk matrix or LOPA) any scenario that had catastrophic consequences. The results turned out the same as purely qualitative risk judgment. The team also determined if a safety instrumented function (SIF) was needed (if a SIF was the best way to achieve tolerable risk) or if a SIF was intended in the design (based on the configuration of the dedicated safety instrumentation). If there was a SIF, the PHA team assigned the SIL based on either the risk reduction needed from the SIF or again by evaluation of the existing instrument configuration.

It was noted in some cases that the SIF that was installed for protection against scenarios during continuous mode of operation did not protect against even more catastrophic and much more likely consequences during startup or online maintenance. For such situations, additional IPLs, including SIFs specific to startup or online maintenance, were recommended by the PHA team.

The PHA covered all modes of operation for the units; besides a HAZOP or What-if of continuous modes of operation, the PHA team also used the Two Guideword or What-if approach to complete a PHA of startup, shutdown, and online modes of operation. The PHA of the non-routine modes of operation took about 25% of the total meeting time and was done at the end of the unit PHAs. This portion of the analysis followed the requirements in Chapter 9 of Guidelines for Hazard Evaluation Procedures[6], as improved in Bridges[7].

Many Other Case Studies

PII has led hundreds of PHA/HAZOP of entire process units over the past 13 years using the approach described above. In these cases, the PHA/HAZOP team, by itself, was able to make the determination of risk and the determination of IPLs and SIFs using only qualitative judgment about 97% of the time (95 to 100% of the time). In about 3% of the cases, a LOPA was recommended by the PHA/HAZOP team to help clarify the risk of a scenario. A full QRA (FTA, ETA, Consequence modeling, and/or HRA) was necessary and recommended for less than 0.01% of the scenarios. Clients for which qualitative judgment and qualitative identification of IPL and SIF/SIL were performed include:

- Advanced Turbine (ATTSSA; Saudi Arabia)
- CABOT (EU)
- ECOLAB (USA)
- Elementis (USA)
- Gulf Petrochemical (GPIC; Bahrain): 4 plants
- Irving Oil (Canada): 2 units
- MIDREX (USA)
- MOL Pakistan (Gas plant in Pakistan)
- PlusPetrol (Peru): 5 gas/oil well clusters, 5 gas processing trains
- Peru LNG
- QAFCO (Qatar): multiple ammonia, urea, and other plants; 7 process plants in all
- Roquette (USA)
- SABIC (Saudi Arabia): several affiliates; about 15 plant-size PHAs)
- SS-TPC (China, see Case Study 1): 9 petrochemical plants
- SUNCOR
- WISON (China)
- Many others – anonymous

In addition, we have trained hundreds of PHA/HAZOP leaders how to do the same, though we have only provided the follow-on coaching for about 20.

Summary

Based on the data above, the choice of methods for risk judgment is shown in Figure 8.

Figure 8: Flowchart for Risk Judgment Methods

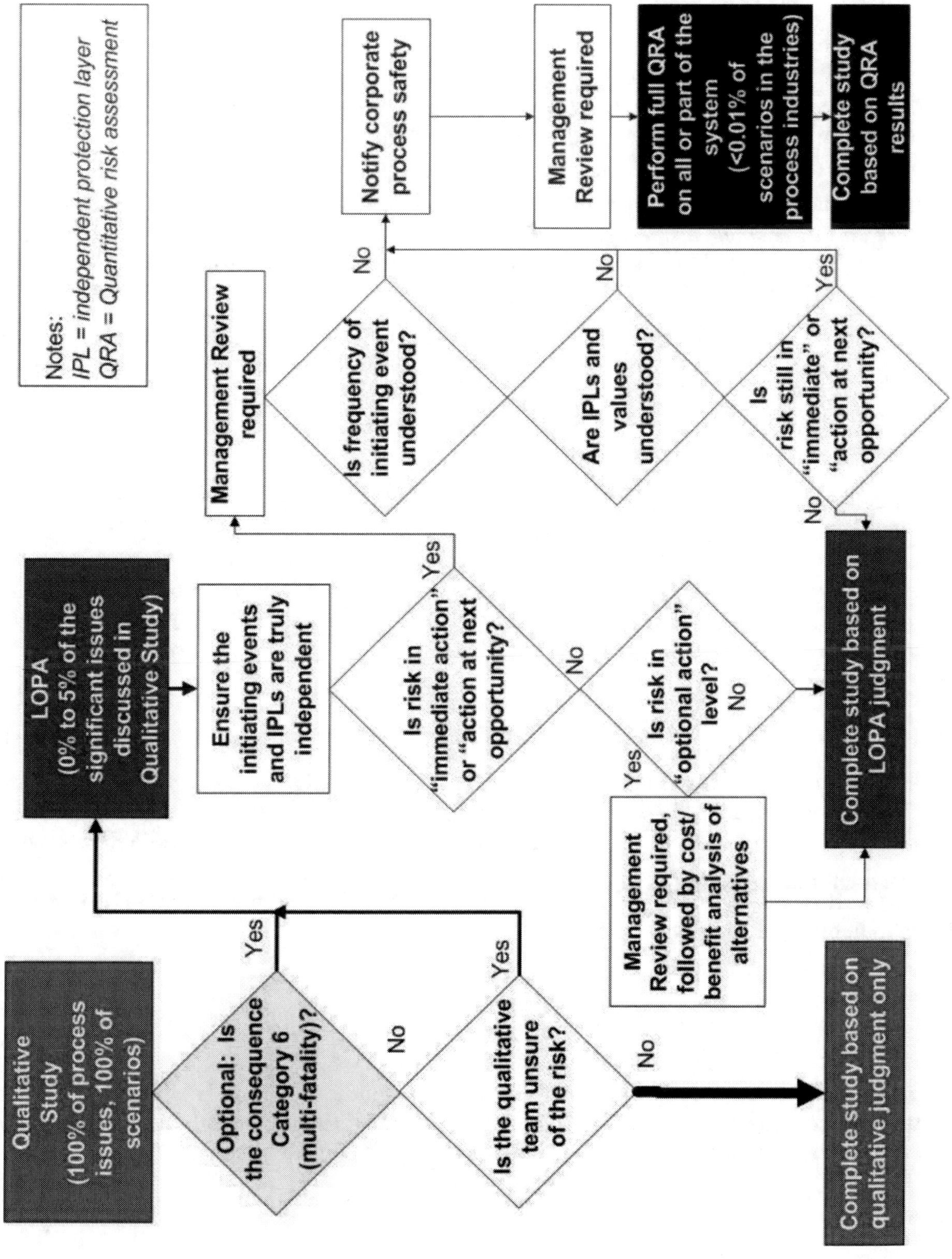

So, PHAs using qualitative methods has been shown to work effectively for making >95% of the risk judgments and for also judging which safeguards qualify as IPLs, including which qualify as SIFs and what SIL is warranted. And, others can learn to do the same with reasonable investment of time.

Further, adding these assignments to the task list of the PHA team leader/scribe increases the time required for the PHA by only about 0 to 5% (all of the increase, if any, is in extra documentation load).

Finally, our experience indicates that when companies want to over-use quantitative methods (starting with LOPA as the next step) for risk judgments and identification of IPLs and SIFs/SILs, they describe their needs in one of the following ways:

- "The PHA team leaders we use, coupled with the less-than-expert team members we provide them, are not competent to make good risk judgments." *Our suggestion is to work on this problem as a HIGH PRIORITY, since no amount of additional LOPA can fix this problem, and no other activity (other than near miss reporting) can find the accident scenarios you are missing.*
- "We want to get a Better estimate of the risk; HAZOP and other qualitative methods simply can't do that." *This is a myth; LOPA and QRA do NOT provide better risk estimates. In fact, the PHA methods likely have the lower error band on the risk estimate, because at most, we have found the qualitative teams off by 1 IPL (PFD = 1E-1) on their judgment of tolerability of risk.*
- "The PHA team leaders, and the teams we give them, do not make consistent risk judgments." *Our suggestion is to work on this problem by coaching of the PHA team leaders. Adding requirements for more quantification within the PHA meetings will lead less productive brainstorming, and therefore more accident scenarios being missed. And adding more LOPA outside of the meetings will result in a significant waste of time. The alternative of bolstering the skills of the PHA team leaders is much more productive.*

Conclusions

So, there is no need to **"Go Boldly Beyond HAZOP and LOPA."** The opposite is true. Organizations should instead **"Bring Hazard Identification and Risk Assessments Back to Earth".** There is plenty of practical evidence presented in this paper and available elsewhere than shows qualitative voting by a group of experts (the PHA team) does a better job >95% of the time than artificially defining the same scenario to fit within the bounds of LOPA or QRA and using "other people's data" as inputs.

It is critical that the PHA leader:

- Be trained and coached in LOPA and/ or in qualitative IPL definitions and in judging if there are sufficient IPLs.
- Be trained, coached, and confirmed to be able to guide the PHA/HAZOP team in the identification of true IPLs and the sufficiency of the IPLs (including SIFs) for the scenario; this can only be done by someone who is already fully competent already.

Acronyms Used

AIChE– American Institute of Chemical Engineers
BPCS – Basic Process Control System (such as a distributed control system [DCS])
CCPS – Center for Chemical Process Safety (of AIChE)
EHS – Environment, Health, and Safety (includes process safety)
ESD – Emergency Shut Down
ETA – Event Tree Analysis
FTA – Fault Tree Analysis
HAZOP – Hazard and Operability Analysis
HRA – Human Reliability Analysis
IE – Initiating Event
IEF – Initiating Event Frequency
IPL - Independent Protection Layer
LOPA – Layer of Protection Analysis
LSH – Level Switch High
MOC – Management of Change
PFD – Probability of Failure on Demand
P&ID – Piping & Instrumentation Diagram
PHA – Process Hazard Analysis
PLC – Programmable Logic Controller
PSI – Process Safety Information
PSM – Process Safety Management
PSV – Pressure relief valve or pressure relief valve
QRA – Quantitative Risk Analysis
SIF - Safety Instrumented Function
SIL - Safety Integrity Level
SIS - Safety Instrumented System
SS-TPC – Sinopec SABIC Tianjin Petrochemical Company

References

1. *Guidelines for Initiating Events and Independent Protection Layers*, 2015, CCPS/AIChE.
2. *Layer of Protection Analysis: Simplified Process Risk Assessment,* 2001, CCPS/AIChE.
3. *Functional Safety: Safety Instrumented Systems for the Process Industry Sector - Part 3: Guidance for the Determination of the Required Safety Integrity Levels*; ANSI/ISA 84.00.01 Part 3, 2004.
4. IEC 61511, *Functional Safety: Safety Instrumented Systems for the Process Industry Sector - Part 1: Framework, Definitions, System, Hardware and Software Requirements*, International Electrotechnical Commission.
5. "Implementation of Process Hazard Analysis at SINOPEC-SABIC Tianjin Petrochemical Company Ltd, China", Homoud Al-Maymouni, Yunzhong Gao (both from SS-TPC), and W. Bridges, *11th Global Congress on Process Safety,* Austin, TX, AIChE, April 2015.
6. *Guidelines for Hazard Evaluation Procedures*, 3rd Edition, 2008, CCPS/AIChE.

7. Bridges, W. and Clark, T., "How to Efficiently Perform the Hazard Evaluation (PHA) Required for Non-Routine Modes of Operation (Startup, Shutdown, Online Maintenance)," *7th Global Congress on Process Safety*, Chicago, AIChE, March 2011.

12TH GLOBAL CONGRESS ON PROCESS SAFETY

One Company's Observations on the Implementation of LOPA

Anne Bertelsmann, Ph.D., PE, FSE
Marathon Petroleum Company LP
502 10th Street South
Texas City, TX 77590
ABertelsmann@MarathonPetroleum.com

Prepared for Presentation at
American Institute of Chemical Engineers
2016 Spring Meeting
12th Global Congress on Process Safety
Houston, Texas
April 11-13, 2016

One Company's Observations on the Implementation of LOPA

Anne Bertelsmann, Ph.D., PE, FSE
Marathon Petroleum Company LP
502 10th Street South
Texas City, TX 77590
ABertelsmann@MarathonPetroleum.com

Keywords: LOPA, HAZOP, PHA, Risk Assessment, Risk Reduction Factors, Independent Protection Layers, Conditional Modifiers

Abstract

Layer of Protection Analysis (LOPA) has gained popularity as a semi-quantitative risk assessment tool, but often unexpected difficulties are encountered when rolling out LOPA. This paper discusses the practical aspects of LOPA implementation in Marathon Petroleum Company LP's Refining Organization (MPC Refining). Specifically, it shows how scenarios are selected from a Hazard and Operability Study (HAZOP) and how they are organized for conducting LOPA efficiently, and it discusses whether to conduct LOPA concurrently or after the HAZOP. The presentation also compares the [cause → consequence] or [consequence → cause] methodology and addresses the implications of combining multiple causes in a single scenario.

Another key aspect of LOPA implementation is the set of numerical values used for frequencies, modifiers and risk-reduction factors. The presentation shows how numerical values should be tied to the risk matrix and how internal consistency is achieved through the use of standardized values and restrictions on the use of frequency modifiers. It discusses potential solutions for cases where the LOPA outcome does not meet the risk target. Finally, the presentation shows how the use of application standards can supplement LOPA by defining minimum protections and providing standard LOPA scenarios.

1 Introduction

Layer of Protection Analysis (LOPA) is a semi-quantitative method for performing risk assessments. The primary purpose is to evaluate if sufficient protections against a hazard are in place. It was first developed in the 1990's and by now, several books (1) (2) (3) and many papers have been written on the topic – some general, others highlighting specific aspects. The overall methodology is based on probability calculations and is relatively straightforward – and it does not require advanced math (multiplication and division is generally sufficient). LOPA is mentioned as one of the techniques for Safety Integrity Level (SIL) determination (4 p. 51) (5 p. Annex F) and its benefits are discussed in the literature and at industry forums and conferences.

So, why is LOPA not universally implemented by now? The answer is probably that practical issues encountered during LOPA implementation can delay or even derail the roll-out in an organization. And while the methodology sounds deceptively simple, a good LOPA is so much more than a HAZOP combined with multiplication and division by ten. Key aspects for a successful LOPA are the identification and organization of the appropriate scenarios, proper selection of the numerical values and finally a clear path on how to handle LOPA outcomes.

This paper describes MPC Refining's experience with implementing LOPA and highlights many of the issues encountered along with the selected path forward. It does not imply that the solutions proposed here are the only ones or that they will work for everyone. But by highlighting the issues, the hope is that organizations can become aware of them before introducing LOPA, thus allowing them to anticipate and address potential pitfalls prior to implementation.

2 Connection between HAZOP and LOPA

LOPA is a semi-quantitative risk assessment tool, but it does not systematically identify potential hazards. The identification is done by other methods, such as HAZOP, What-If analysis or FMEA (Failure Mode and Effect Analysis). Thus, LOPA is based on a hazard identification study, most commonly a HAZOP. This raises the question of scope (i.e. which HAZOP scenarios should be selected for LOPA) and timing (when should the LOPA be performed relative to the HAZOP).

2.1 Selecting Scenarios from the HAZOP

Developing LOPA scenarios with a team is more time consuming than HAZOP scenarios since the failure progression, enabling events and conditions, safeguards and whether they qualify as Independent Protection Layers (IPL) are discussed in greater detail and probabilities have to be assigned for each. This takes additional team time and adds to the duration of a Process Hazard Analysis (PHA). It is thus important to select only those scenarios from the HAZOP for LOPA that will benefit the most from the methodical risk assessment.

PHA teams routinely perform a qualitative risk assessment as part of the HAZOP methodology – they describe the unmitigated consequence and assign a severity and then estimate how often such an event would occur with the safeguards in place. The combination of severity and frequency provides the risk ranking, using a company's risk matrix. Since frequency estimates are based on the experience of the team members, this method works well for low severity / high frequency events – i.e. the types of events that the team members have either seen themselves or that have been experienced by someone they know.

However, when it comes to the high severity / low frequency events – i.e. the catastrophic failures that very few have ever seen and hopefully most will never see – the HAZOP team does not have an experience base to make the frequency estimation. For these events, the LOPA methodology allows determination of a frequency based on standardized factors, not team experience, resulting in a more consistent (though not necessarily precise (1 p. 24)) frequency value. Each company will have to define what it considers "high severity events." MPC Refining performs LOPA for hazards that could result in one or more fatalities or severe injuries, have long-term off-site environmental impact or affect the community through shelter in place, evacuations, and/or road closures. This allows focusing the LOPA efforts on those scenarios with the highest potential impact. LOPA scenario selection is based on consequence severity, not on risk since risk requires a frequency and it is the purpose of LOPA to estimate that frequency.

Additionally, not every type of scenario lends itself well to LOPA. LOPA generally works best for events caused by equipment or control system failures, where the failure can be detected by instrumentation and automated protections can be put in place. However, LOPA is very conservative in the credits allowed for human intervention, giving no credit to general plant systems and activities, such as training, inspection, maintenance, procedures, safe work practices, etc. Some companies don't allow any credit for human intervention, but MPC Refining has found that too restrictive. HAZOP scenarios that rely predominantly on those types of safeguards (corrosion, maintenance activities, etc.) generally do not lend themselves to LOPA because the safeguards do not qualify as IPLs. Also, LOPA focuses on preventative protections, and scenarios where the consequence has already occurred with the failure (some types of fixed equipment failures, external impact, etc.) are not suited for LOPA.

Finally, purely financial consequences may be difficult to analyze with LOPA. The most common conditional modifiers used for health and safety consequences (e.g. occupancy) do not apply to financial consequences, resulting in higher frequencies. This could be addressed by setting different frequency targets for financial consequences, but this tends to lead to confusion when comparing results from different categories. Cost-benefit analysis may be a more appropriate tool than LOPA for scenarios with financial consequences.

Thus, a company's LOPA procedure should clearly define which HAZOP scenarios are applicable for LOPA. "Non-LOPA'ble" scenarios should be analyzed using the HAZOP methodology to confirm the adequacy of the existing safeguards. It is important to remind team members and management that just because a scenario is not taken to LOPA, it does not mean that it is insignificant. It only means that the LOPA methodology is not suitable – and that a different risk assessment method should be used.

2.2 Timing of the LOPA

LOPA can be done either simultaneously with the HAZOP or sequentially after the HAZOP, preferably immediately following the completion of the HAZOP. Either way is possible, but MPC Refining has found it beneficial do to them sequentially, as it was originally intended.

When LOPA is done concurrently with the HAZOP, the unit is analyzed node-by-node with the normal HAZOP methodology and when a scenario is identified that qualifies for LOPA, the team does the LOPA analysis right then before moving on to the next consequence. While this method has the advantage that the team is already familiar with the scenario, we have found that it has several significant drawbacks. HAZOP, with its focus on risk identification, uses inductive thinking, while LOPA uses deductive reasoning for the risk assessment. LOPA also analyzes scenarios in more depth and gives more detailed consideration to the qualifications of protection layers than HAZOP does. To analyze a scenario with LOPA, the team gets into the "LOPA-mindset," which takes some transition time. Then, when the scenario is completed and the next scenario uses the HAZOP methodology, the team is often still in the "LOPA-mindset" and tends to overanalyze the HAZOP scenario, taking additional time. And since the "LOPA mindset" is more focused on risk assessment than risk identification, HAZOP teams may miss identification of hazards when switching back and forth between the two methodologies.

When doing HAZOP and LOPA sequentially, the HAZOP is completed first, then the LOPA scenarios are selected, prepared for the LOPA and finally the LOPA is performed. There are advantages to this method. One is that it allows using a different facilitator for each study (not all good HAZOP facilitators are necessarily good LOPA facilitators and vice versa). It also makes it possible to vary the team composition slightly. LOPA, with its focus on instrumentation, can benefit from having a control engineer and/or instrument technician as part of the team, while the mechanical engineer or inspector may not be as important (their input was already captured during the HAZOP in the hazard identification).

When doing the studies sequentially, the team is trained in the LOPA methodology at the beginning of the study and can stay in "LOPA-mindset" throughout the study. A fresh look at scenarios developed in the HAZOP – even with the same team – often results in new insights. Doing the LOPA after the HAZOP also allows getting an overview of all LOPA scenarios first and grouping them for the LOPA study, since LOPA does not have to follow the flow of the process like the HAZOP does. In particular, the ability to re-arrange LOPA scenarios in a logical manner (see section 2.5) was found to improve efficiency and consistency significantly, and this is only possible when doing the studies sequentially.

Both HAZOP and LOPA generate recommendations, though LOPA recommendations tend to be more focused on providing a full layer (order of magnitude or factor of ten) of risk reduction, while HAZOP teams are more likely to include operability improvements or measures to reduce error rates – even if they don't provide a full layer of risk reduction. Conducting the LOPA simultaneously with the HAZOP may discourage such recommendations. When conducting the two studies sequentially, some reconciliation between the two sets of recommendations is required.

2.3 Cause → Consequence vs. Consequence → Cause

HAZOP uses [cause → consequence] methodology, i.e. the team identifies causes for the deviation under study in the current node and then develops the consequences, which could be outside the node under study. For a HAZOP, [cause → consequence] methodology generally works best, because we are used to "thinking forward." We can easily follow the failure progression forward and thus are more likely to identify hazards and consequences that have not yet been observed.

The purpose of LOPA is risk assessment, not risk identification, and thus it focuses more on the consequences. For that reason, MPC Refining prefers to group all the causes that lead to the same consequence together – even though each cause is analyzed separately. Protections are often similar for multiple causes of the same consequence, so the ability to use copy and paste from one scenario to the next improves LOPA efficiency. For LOPA, [consequence → cause] methodology flows better because it first looks at the consequence to select the risk tolerance criteria and then looks at initiating events (causes), probability factors and IPLs (safeguards). This means that the cause/consequence pairs identified in the HAZOP and recorded in the node of the cause will need to be rearranged such that they are grouped in the node of the consequence. This rearrangement helps to group all causes for one consequence almost automatically, but it is an additional step that has to be taken by the facilitator in preparation for the LOPA study.

The following graphic illustrates the re-grouping approach, using just a limited number of scenarios for illustration purposes.

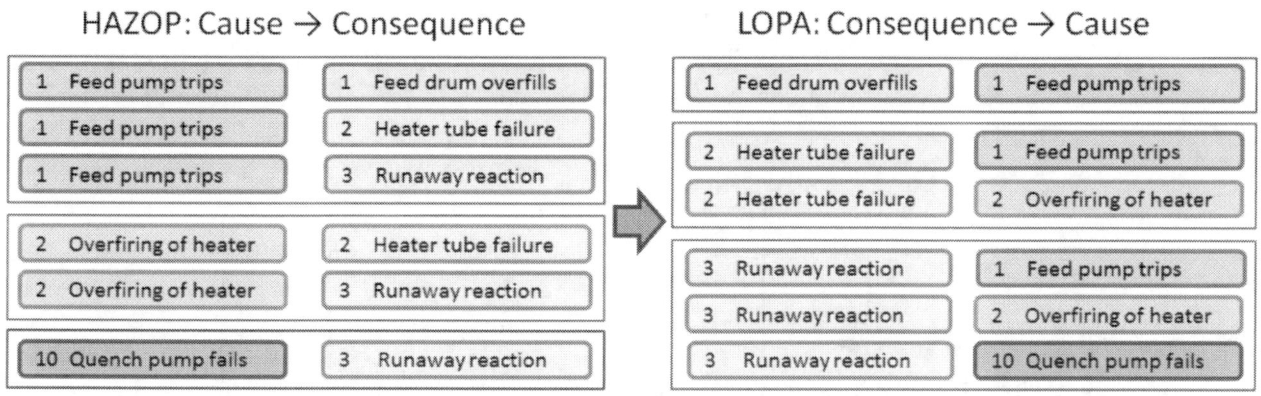

Figure 2.3-1: Comparison of [Cause → Consequence] and [Consequence → Cause Methodology]

With this approach, all causes leading to the same consequences are grouped together. Sometimes, this new view can help to identify causes that were missed in the HAZOP.

2.4 Individual Scenario vs. Cumulative Likelihoods

Once all the causes for one consequence have been assembled and each cause/consequence pair has been analyzed and the frequency for the consequence determined, it is possible to add the individual frequencies to determine the overall consequence frequency. This is often referred to as "cumulative likelihood" of a consequence.

The concept of adding the individual cause probabilities to calculate the overall event likelihood of a consequence is mathematically correct as long as the causes are independent of each other. However, when applying this math to LOPA frequencies, several issues were encountered. MPC Refining decided to base the LOPA analysis on individual cause/consequence pairs as recommended (1 p. 12) and does not determine cumulative frequencies for the basis of risk evaluation. Below are details on the issues that were encountered:

Accuracy of the LOPA numbers: LOPA makes simplifying assumptions with regards to the frequencies, and the values used in LOPA calculations are generally order of magnitude numbers. Thus, they are rounded and conservative (1 p. 7). By adding causes, the result for the consequence probability is no longer strictly "order of magnitude." For general engineering calculations, it is recognized that the digits *after* the decimal are often not significant. For order of magnitude calculation like LOPA, the digit *before* the decimal may not be significant. The approximate nature of LOPA does not support the accuracy implied in adding individual cause probabilities. This type of math is better suited for a detailed QRA (Quantitative Risk Analysis) rather than the simplified LOPA risk assessment.

Order of Magnitude and Risk Matrix: Generally, both the LOPA and the risk matrix are done in orders of magnitude. Thus, while risk rankings are normally shown in the center of a cell, a continuous frequency scale would actually place the risk on the edge of the cell (see figure 2.4-1 on the left). Adding any additional cause – no matter how unlikely – will move the dot into the next cell and result in higher risk rankings. This can exaggerate the impact of adding causes. One possible solution is to define the breakpoints for the risk matrix not as even orders of magnitude such as 10^{-5}, 10^{-4}, 10^{-3}, etc., but as half-steps, i.e. $10^{-4.5}$, $10^{-3.5}$, $10^{-2.5}$, etc. This shifts the breakpoints on the risk matrix by half a cell, so that a single order of magnitude cause/consequence pair will truly be in the center of the cell.

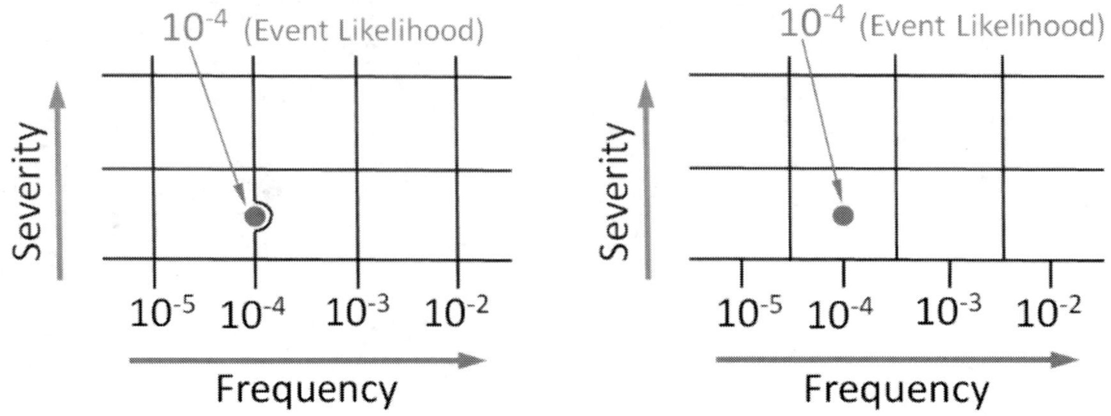

Figure 2.4-1: Risk matrix with boundaries at full and half orders of magnitude

Adding up to a total of three equally likely events will still result in the same risk ranking, while adding more than three events shifts the aggregated likelihood to the next higher risk. Intuitively, this makes sense, since events with many potential causes can be expected to be more frequent and thus might need more protections. However, this brings up the question of what constitutes "one cause."

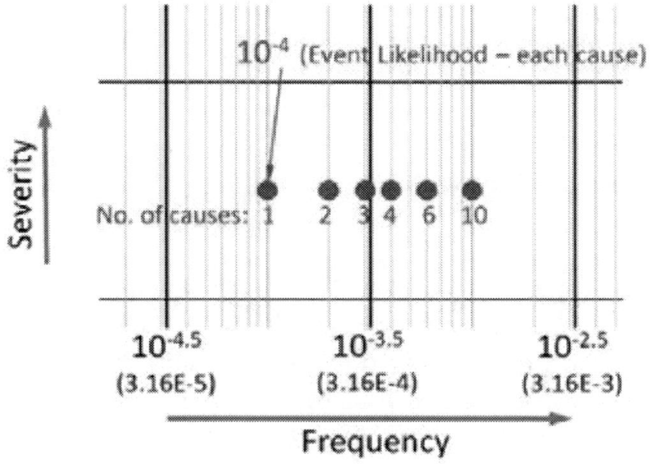

Figure 2.4-2: Cumulative likelihood for multiple causes in a risk matrix

<u>Cause definition:</u> Causes can either be individual failures or a combination of multiple failures. For example, the frequencies for control loop failure given in the LOPA literature (1 p. 70) generally include various failure modes, such as sensor failure, transmitter failure, valve failure, operator error, etc., and thus should be treated as one cause. However, this concept can become somewhat blurred in more complex cases, like cascaded control loops. Is each loop in the cascade a potential cause for failure with the

given frequency, or should the entire cascade be treated as one failure? Are five block valves in series in a line really five times more likely to be accidentally closed than a single block valve? When doing individual cause/consequence pairs, this does not matter, but when calculating cumulative likelihood, the selection of causes affects the outcome. It is thus one more source of variability for which clear guidance needs to be given in order to achieve consistent results if cumulative likelihoods are used.

2.5 Organizing scenarios for an efficient LOPA

A HAZOP follows the flow of the process, going from feed to final product to systematically cover all hazards. LOPA is not bound by the same requirements; it analyzes the hazardous scenarios identified by the HAZOP. For that reason, there is no need to perform the LOPA analysis in the same order as scenarios were identified in the HAZOP. Section 2.3 already discussed the benefits of combining causes for similar or identical consequences for the purposes of LOPA.

Taking this concept even further, MPC Refining found it beneficial to group similar equipment together for team discussions. For example, all fired heaters on a unit have similar hazards, whether it is a feed heater at the front of a unit or the reboiler on the final column. Arranging the order such that all fired heater scenarios are discussed during the same LOPA session can significantly improve the quality and efficiency of the LOPA. It allows the team members to familiarize themselves once with the equipment-specific hazards and the typical safeguards, as well as any applicable company standards, then apply that knowledge multiple times. The first scenario usually requires more team time to allow for the familiarization; subsequent similar scenarios discussed immediately afterward tend to be completed much more quickly. Grouping similar equipment or hazards together also improves consistency and allows involving the subject matter experts (either internal or external) on that equipment or process selectively. In addition to the fired heaters example, other potential groupings could be rotating equipment (pumps and compressors, either separately or together), loss of cooling events, reaction/mixing-related scenarios, reverse flow and/or blow through cases, etc.

This concept requires more preparation for the facilitator, since he/she has to first identify all LOPA scenarios from the HAZOP documentation, then group by consequence and group by equipment or hazard type. The facilitator has to plan the order in which to cover them and invite specialized team members as needed. During the LOPA, the scribe has to jump between nodes and keep track to ensure that all applicable scenarios are covered. However, the efficiency gains for the team time and the improvement in consistency are worth the extra preparation effort.

3 Frequencies, Modifiers and Risk Reduction Factors

Once everything is prepared for the LOPA session, the team will meet to discuss each scenario and assign numerical values to the initiating event frequency, the conditional modifiers and the probability of failure on demand of protection layers. The actual math of calculating the mitigated event likelihood and comparing it to risk targets is relatively straightforward and can be handled easily by most LOPA documentation software.

Standard values for the frequencies and probabilities are published in the literature (1) (2) (3) and the key to a good LOPA is knowing what values to select and when to apply them.

Traditionally, the values used represent frequencies and probabilities expressed in scientific notation (e.g. 10^{-4}). Alternatively, a "Risk Reduction Factor (RRF)" can be used, which is an inverse of the probability of failure; numerical values then are whole numbers (e.g. 10,000). LOPA can be conducted using either method, though whichever method is selected should be used exclusively. Mixing probabilities of failure on demand with risk reduction factors can lead to confusion. MPC Refining found that using Risk Reduction Factors works better with LOPA teams, since the whole numbers make it easier for team members not well-practiced in scientific notation to compare numbers.

3.1 Connection between Risk Matrix, Risk Tolerance and the Frequency Values

Most risk matrices follow the same principles in their design – they have an axis for frequency and an axis for severity. The cells of the matrix are used to define the need for risk reduction. Sometimes, a boundary is drawn into the risk matrix which defines the acceptable risk or the acceptable risk is defined through numerical values. However, while the general principle of a risk matrix is universal, the individual implementation with regard to severities, frequencies and the acceptance of risk varies between companies. The risk tolerance criteria defined either in the risk matrix or as a numerical value based on severity plays an important role in the outcome of the LOPA, since it determines how many protection layers are needed for each scenario. Risk acceptance criteria may vary between companies by several orders of magnitude, even for similar severities.

Yet, despite the variability in the risk criteria between companies, LOPA outcomes for the number of protection layers are often remarkably comparable. For example, a fuel gas trip for a fired heater on high or low fuel gas pressure is typically designed as a SIL2 – two layers of protection. If different companies with different numerical risk tolerance criteria arrive at the same conclusion, it indicates that they use different credits during the LOPA. Some companies may not allow certain credits at all, such as enabling events, conditional modifiers or human IPLs, and values for the initiating event frequencies and/or conditional modifiers (where allowed) may vary between companies. In the literature, recommended LOPA frequencies for initiating events are often defined as a range (1 p. 71). Rules for the applicability and types of conditional modifiers may vary. A company with a very low numerical value for the risk tolerance criteria will require more generous credits for frequencies and likelihoods to arrive at the same conclusion for the required number of protection layers.

This is illustrated in the following graph, which shows how two different companies might achieve their respective risk target through a combination of initiating event frequencies (I.E.), enabling conditions (E.C.), conditional modifiers (C.M.) and protections (IPLs).

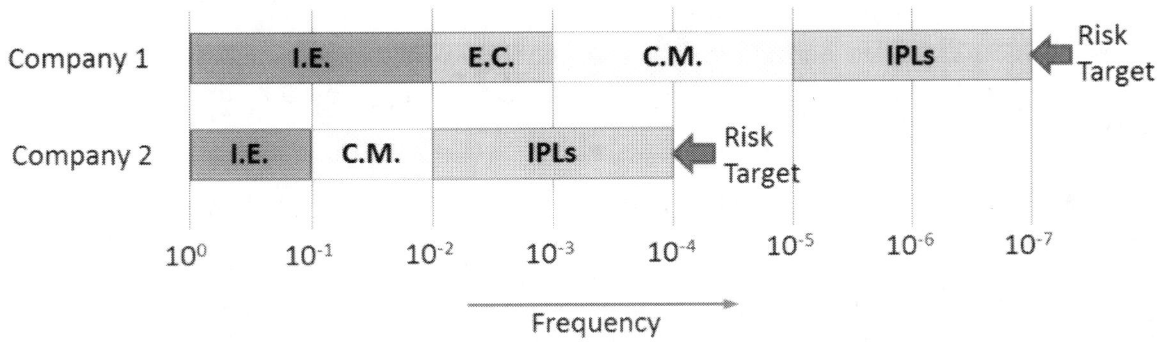

Figure 3.1-1: Comparison of risk targets and use of LOPA factors

Additionally, company internal guidance on frequencies and the applicability of modifiers can help refine certain areas of the risk matrix. For example, if the risk matrix does not distinguish between single-and multi-fatality scenarios, it is possible to define that a modifier such as occupancy shall only apply to single-fatality scenarios. This will result in additional protections required for multi-fatality scenarios. On the other hand, for companies that do make that distinction in their risk matrix, such a limitation may be too restrictive.

Thus, neither a company's risk criteria nor its LOPA procedures or guidance for use of data and numerical values should be evaluated in isolation – they have to be matched to work with one another.

3.2 Benefit of Clear Guidance and Company Specific Values

Recommended numerical values for the various LOPA factors (initiating events, IPLs, enabling events and conditional modifiers) are available in the literature – entire books have been written for that purpose (1) (2) (3). Most LOPA practitioners will be familiar with these numbers. So, why does MPC Refining define its own values and provide rules for their applicability? The short answer is consistency and alignment with the risk matrix. Recommended values from the literature are frequently given as a range, not a single number, so specifying a value and the conditions under which it applies helps to ensure consistent evaluations across the organization. Generally, the company-internal numbers are in alignment with the literature values, though there may be instances where we have chosen to deviate. Deviations from literature values are occasionally needed to reflect company experience or to better align with the risk matrix – the interconnectivity between the two is discussed in the previous section. The reason for the deviation from literature values is documented. The company-specific guidance does not just focus on the numerical values, but also on the conditions under which they apply.

At times, working out this detailed guidance ahead of time may seem overwhelming and it can be difficult to anticipate most of the potential cases. The temptation exists to limit the LOPA practice to broad guidance and general principles, rather than dictating the details. The downside of this approach is that the details will have to be worked out eventually – at the latest by the time the individual LOPA study is finalized. And if there is no clear guidance on which numbers to use, the local LOPA facilitator (often a contractor) will choose them. The numbers selected during the LOPA have a direct impact on the company's risk profile – they are used to determine where additional protections are needed and where they are not. Use of inappropriate values can result either in too many protections (unnecessary complexity and costs) or too few protections (increased risk).

LOPA drives risk decisions, mainly in the high severity scenarios. It seems self-evident that these types of decisions should be made centrally for a company, by the Subject Matter Experts (SMEs), and not left to the individual facilitators. If the SMEs with good access to internal and external resources cannot develop a detailed framework for LOPA, what are the chances that the local facilitators will make consistent risk decisions that align with a company's risk culture? Good facilitators are still required to conduct the LOPA and properly apply the right factors – but within the detailed guidance of the framework. At times, such guidance may seem overly prescriptive, but it helps in achieving consistency.

3.3 Limitation on the use of Enabling Conditions and Conditional Modifiers

Enabling Conditions and Conditional Modifiers (here collectively referred to as "modifiers") have the most variability of any of the LOPA factors. While initiating event frequencies and probabilities for the failure of protection layers are well understood and have clear criteria, the same is often not true for the modifiers. Mathematically, modifiers have the same impact on risk reduction and LOPA outcome as IPLs, though they have little of the rigor (independence, auditability, etc.) associated with IPLs. Yet, without modifiers, LOPA outcomes can be overly conservative and do not match observed event likelihoods (3 p. 14).

This makes modifiers the one aspect of LOPA that is most likely to be misused. Use of inappropriate modifiers can result in underestimating the number of protections needed and thus expose a company to a higher risk of a major incident. For that reason, MPC Refining has developed clear guidance on the use of modifiers that specifies their values and when they can be used. The guidance is conservative and the limitations may at times seem overly restrictive. However, the benefit is consistent application and a lower potential for misuse, even if a facilitator has limited knowledge or understanding of the finer points of their applicability.

The following examples show some of the guidance that MPC Refining has adopted for the use of frequency modifiers:

- Time at Risk – credit for time at risk cannot exceed a risk reduction factor of 10 (probability of 0.1) and does not apply to start-up. Without this restriction, teams may try to reduce the risk by specifying shorter durations rather than providing protections. Time at risk gives the impression that the shorter a hazardous activity is, the lower the risk. This is akin to saying that it is safe to run across a busy freeway as long as you are a fast enough runner. Restricting the maximum credit still allows some reduction where appropriate but prevents "tweaking" of this factor to achieve a certain risk target.

- Probability of Ignition – Ignition of flammables is a complex topic, since it needs to consider immediate and delayed ignition (where consequences may be different), area classifications, ignition sources, etc. While generalized values based on research and experiments are available (6), these values are often misapplied due to lack of understanding. For that reason, our guidance specifies a value of 1 (certain ignition) for any release within the battery limits of a process unit. This is conservative (by experience, not every release ignites), but it is straightforward to apply with little room for interpretation.
- Probability of Personnel Presence (Occupancy) – This modifier expresses the probability that a person is in the impact zone of the consequence, and is thus tied to presence patterns. However, to avoid the tendency of shortening presence times to achieve risk reduction, fixed values for the modifier are specified and teams can choose to either apply them or not (if the impact zone includes a frequently occupied area, such as the access way to the control room, a road or similar). Values used are order of magnitude – risk reduction factor of 10 (probability of 0.1) for a normal process unit and risk reduction factor of 100 (probability of 0.01) for restricted or limited access areas including elevated structures, flare yards and tank farms. Scenarios with a large impact area, such as a vessel rupture, are typically rated at the highest severity, and for these scenarios the probability of personnel presence is limited to a risk reduction factor of 4 (probability of 0.25), which is less than one full order of magnitude. This is done to reflect the larger footprint and the escalation potential. It results in more protections required for scenarios with large impact areas.

Additionally, MPC Refining's LOPA policy defines probabilities for certain types of equipment failures, limited to very specific situations. LOPA facilitators cannot "invent" additional modifiers, though there is a provision that allows for the development of new modifiers by the process technology SME and the LOPA SME. Finally, the combined risk reduction from all modifiers is capped at RRF 100 (probability of 0.01) to prevent over-use.

4 Supplementing LOPA with Application Standards

4.1 Potential Impact of LOPA Implementation

So, what happens when LOPA results in an outcome that does not meet risk criteria? The general answer is that more protection layers are needed. Often, that is the right choice because the LOPA highlighted a case where, either through a lack of independence or other reasons, the existing protections are not sufficient. However, there may be cases where the higher-risk LOPA outcome is more an artifact of the limitations on the frequency values and modifiers or of overstating the severity rather than an indication of an actual higher risk.

When LOPA is performed across a wide variety of processes and hazards, there will be scenarios where the risk outcome does not initially meet a company's risk criteria – yet engineering judgment, industry practice and incident history may indicate that the commonly existing safeguards are adequate. To determine whether the LOPA risk outcome may be overstated, a more detailed analysis (such as QRA or similar) can be performed to refine the frequencies and parameters. It is also possible to compare the event likelihood predicted by LOPA with incident rates across the industry. Such a comparison can provide a valuable reality check for the LOPA.

For example, HAZOP teams often identify the consequence of a complete seal failure on a pump operating above the flash point of the process as a potential fatality. The frequency of complete primary pump seal failure is given as 1/10 years (2 p. 92). Even if an occupancy of 10% and a tandem seal (2 p. 232) or fire suppression system (2 p. 244/246) as IPL are applied, the probability of a fatality predicted by LOPA is 1/1000 years per pump. Given the number of pumps in industry, LOPA would predict multiple fatalities due to pump seal failures per year – which does not match the observed frequency. The reality check just bounced.

Here are several options that MPC Refining considers on a case-by-case basis on how to deal with such a LOPA outcome. Each solution has potential drawbacks:

- Add additional protection layers – even though the risk determined by the LOPA is likely overstated and the additional protections would exceed what is common in industry, adding additional protections allows meeting the risk target. However, these additional IPLs can lead to higher maintenance costs and complicate operation of the equipment as well as add new failure modes.
- Credit additional modifiers – There may be some enabling events or conditional modifiers that apply to this scenario and would result in a LOPA outcome more in line with observed frequencies. However, data to justify these modifiers may be difficult to obtain, and opening the door to creativity with modifiers can lead to inappropriate use in other cases.
- Perform a QRA – It can be expected that QRA will give a more accurate result for the risk (and generally lower, due to the conservative nature of LOPA (1 p. 7)), but it is time consuming and information to substantiate the factors necessary for the QRA may be difficult to come by.
- Exclude the scenarios from LOPA – It is possible that LOPA is not a suitable tool for analyzing certain scenarios, in which case the HAZOP team needs to qualitatively evaluate the adequacy of safeguards. However, this solution requires a solid justification based on the type of scenario, not the risk outcome.
- Accept the higher risk outcome – With good justification, a LOPA outcome that falls short of the corporate risk criteria can be regarded as acceptable as long as a minimum number of safeguards are in place. The cases in which this is permissible need to be clearly defined and require approval by the executive management.

If a method for dealing with conservative LOPA outcomes is not established prior to the introduction of LOPA, it can be expected that LOPA teams will identify multiple scenarios in which, even though all the "typical" safeguards are in place, the risk does not meet the company criteria and they will generate recommendation to install additional protection layers. The organization is then faced with either implementing all the recommendations or declining them with proper justification – either way, work that requires additional resources.

4.2 Use of Application Standards

LOPA highlights the fact that safeguards can fail. For example, prior to LOPA, the general assumption was that if a vessel had an adequately sized relief valve, it was protected against overpressure. This is still true, *as long as the relief device works*. LOPA essentially adds the question, "Are you willing to take the risk that the relief device may fail?" and answers it based on probabilities and a company's risk matrix. In some cases, the answer may be, "Yes, that risk is acceptable," and then the relief device is sufficient. In other cases, the answer may be, "No, that risk is not acceptable" either due to the severity of the consequences or due to process conditions that can affect the reliability of the relief device. Then, a second, redundant method of protecting against overpressure needs to be available.

How can MPC Refining ensure that the question about the adequacy of existing protections is answered consistently? One part of the answer is to have clear definition of LOPA rules and company specific values for frequencies and modifiers. MPC Refining further improves consistency through the use of Application Standards, which define minimum protections for specific hazards. These standards are written for processes or units (for example, Fluidized Catalytic Cracking Unit or Sulfur Recovery Unit), for equipment (for example, heaters or compressors) and for typical hazards (for example, blow-through or reverse flow). They identify the typical consequences and their causes and list the required protections along with key design criteria. They take a tiered approach – identifying different severities based on process conditions and tailoring the number and type of protection layers to the severity.

The required protections defined in these standards are based on the experience of the company SMEs, licensor practices, guidance from industry organizations and risk considerations. However, it is not just the risk assessment alone that drives the need for protections – it is a combination of risk assessment, industry recommended practices and engineering experience. It allows weighing benefits and drawbacks of additional protection layers (each trip adds a potential new failure mode and complexity).

Since the application standards are written such that they align with the process hazard assessment process (identifying consequences and causes), PHA teams can readily use the information and find their specific situation in the standard. The standards detail the typical consequence severity, initiating events and their frequency, applicable modifiers and the proper risk reduction credit for the required IPLs. The team has the option to modify severity and frequency values if necessary, but will need to provide justification for any changes

MPC Refining also includes "standard LOPAs" in the Application Standards to provide detailed guidance to the LOPA team. Developing the standard LOPAs helps to identify upfront the cases where the LOPA outcome with the recommended IPLs may not meet the risk target so that these cases can be resolved once (through a combination of more detailed analysis, additional requirements and/or high level management approval), rather than relying on the local facilitators to resolve them in the course of each study.

LOPA teams check whether all the required protections from the standard are available in their process or unit. If the requirements are met, LOPA outcomes can be accepted without additional approvals and without creating a recommendation to add more protections. If not all required protections are in place, LOPA teams will create a recommendation to come into compliance with the standard, even if the LOPA achieves the targeted risk level. And LOPA teams always have the option of creating a recommendation, even if the risk target is met and all minimum protections are in place.

5 Conclusion

This paper discusses several of the detailed implementation issues that a company may encounter when introducing LOPA and highlights the path that MPC Refining is taking. In summary, these solutions are:

LOPA is targeted to the high severity consequences, and certain scenarios where LOPA is not a suitable methodology are excluded. The LOPA study is performed after the conclusion of the HAZOP to allow sufficient preparation, and does not follow the flow of the process but is grouped in a logical manner to improve consistency and efficiency. Cause/consequence pairs are analyzed individually with no summing of causes.

MPC Refining-specific values for frequencies, likelihoods and modifiers are matched to MPC's risk matrix and defined in the LOPA policy document. Detailed, prescriptive guidance helps to achieve consistent risk assessment. Clear rules for the use of enabling events and conditional modifiers are established, their use is restricted and values are capped. While these restrictions may be too conservative at times, they help to ensure consistency and prevent underestimating the risk.

Application standards supplement LOPA and define the required minimum protections for typical process hazards. These minimum protections requirements are determined based on a combination of risk assessment, industry guidance and engineering experience. When an installation is in compliance with these standards, it is ensured that the necessary protections are in place and that the LOPA outcome is deemed acceptable.

The solutions described in this paper may not work for every company and it should be remembered that there is often not one right way of doing LOPA. Hopefully, this paper helped to point out potential issues that may be encountered, and even if it does not provide suitable answers for all situations, it might at least help to raise the right questions. Working out the details prior to implementation will result in a smoother roll-out with fewer surprises.

6 References

1. **Center for Chemical Process Safety.** *Layer of Protection Analysis.* New York : American Institute of Chemical Engineers, 2001. 0-8169-0811-7.

2. **Center for Chemical Process Safety**. *Initiating Events and Independent Protection Layers in Layer of Protection Analysis.* Hoboken, New Jersey : John Wiley and Sons, 2015. p. Data table 4.14 (page 92).

3. **Center for Chemical Process Safety**. *Guidelines for Enabling Conditions and Conditional Modifiers in Layer of Protection Analysis.* New York : American Institute of Chemical Engineers, 2014. 978-1-118-77793-0.

4. **Center for Chemical Process Safety**. *Safe and Reliable Instrumented Protective Systems.* Hoboken, New Jersey : John Wiley & Sons, 2007. 978-0-471-97940-1.

5. *Functional Safety: Safety Instrumented Systems for the Process Industry Sector - Part 3.* **International Society of Automation.** 2004. ANSI/ISA-84.00.01-2004 Part 3.

6. *Development of Algorithms for Predicting Ignition Probabilities and Explosion Frequencies.* **Moosemiller.** New York : American Institute of Chemical Engineers, April 2009. 5th Global Congress on Process Safety .

441491 Don't be Afraid of Conditional Modifiers

Abstract:

Application of conditional modifiers as a factor in Layer of Protection Analysis is viewed by some experts in the industry as a potentially dangerous practice that can lead to underestimation of risk. While conditional modifiers can certainly be misused in a variety of ways, appropriate use of conditional modifiers in the LOPA methodology are well within a reasonable bound of practice. Conditional modifiers are factors which have a long history in the more detailed Quantitative Risk Analysis methods that predate and sired LOPA. There are general warnings that practitioners should be aware of regarding the use of conditional modifiers as well as specific restrictions on the use of each type of conditional modifier. However, as long as these factors are used consistently and appropriately, conditional modifiers can lead to a much clearer and more accurate picture of the risk associated with an activity. In general, users should not be afraid of using conditional modifiers in their LOPA but they should have a healthy respect for them, since misuse is a dangerous practice.

This paper will include examples from actual LOPAs which highlight appropriate use of conditional modifiers and some of the pitfalls of inappropriate use of these factors.

12TH GLOBAL CONGRESS ON PROCESS SAFETY

Layer of Protection Analysis as a Multi-Functional Team Problem Solving Tool

Edward Cialkowski
Air Products and Chemicals, Inc.
Allentown, PA
cialkoej@airproducts.com

Prepared for Presentation at
American Institute of Chemical Engineers
2016 Spring Meeting
12th Global Congress on Process Safety
Houston, Texas
April 11-13, 2016

Layer of Protection Analysis as a Multi-Functional Team Problem Solving Tool

Edward Cialkowski
Air Products and Chemicals, Inc.
Allentown, PA
cialkoej@airproducts.com

Keywords: LOPA review, problem solving, team

Abstract

Layer of Protection Analysis (LOPA) has been widely adopted as a method of organizing and quantifying hazard rates. Frequently, an initial LOPA indicates that a risk target has not been met. Unlike Fault Tree Analysis, the inherent simplicity of the LOPA format makes it conducive for use directly as a problem solving tool in a team environment. Teams at Air Products with representatives from operations, process controls, process design and process safety, have effectively used the LOPA format in problem-solving sessions to identify and improve components of the analysis involving human error, controls architecture, and inherently safer process designs. This multi-functional team approach has resulted in broader ownership of the safety analysis by the full team of stakeholders, and this has ultimately led to more cost effective solutions. The benefits of multi-functional team review are illustrated with a case study example.

1 Introduction

Air Products has greatly expanded the use of Layer of Protection Analysis (LOPA) for process safety risk assessment and management over the past decade. In transitioning from using primarily Fault Tree Analysis (FTA) to LOPA for frequency quantification, Air Products has observed both benefits and drawbacks of the LOPA approach.

Several of the drawbacks are widely understood and workarounds have been developed. For example, dependency between layers of protection or an initiating event and a layer of protection is not as obvious to spot with LOPA as it is with FTA. This is often addressed with appropriate training and quality assurance reviews. In addition, high demand scenarios, or initiating events with high frequency, may not be modeled in LOPA as correctly (or conservatively) as they would be in FTA with robust gate calculations. Checking high demand cases against an FTA or establishing certain rules for high demand cases in LOPA are strategies that can be used to manage this.

On the positive side, a LOPA takes less effort to construct and much less training to understand than an FTA does for the same scope of analysis. One of the benefits of this is that more risk assessments tend to be documented simply because the barrier to entry with LOPA is so much lower than FTA.

One of the most positive benefits Air Products has experienced in transitioning from primarily FTA to primarily LOPA for risk assessments is the ability to review the frequency quantification results with a multi-functional team. One does not need to be a process safety practitioner to understand a LOPA. Compared to a fault tree structure that often spans several pages, the LOPA format is simple, often fits on one page (or screen), and is highly intuitive once the basic rules of its structure are explained.

This paper presents a case study from an actual project in which LOPA was used to quantify and manage risk. As has been the case at Air Products for some time now, multi-functional teams review LOPAs developed by safety or systems engineers. This diverse team review approach may have started out as a means of assuring accuracy and quality, but there is increasing evidence that it has other distinct and substantial benefits as well.

By practicing multi-functional reviews of LOPAs, especially when initial results indicate that corporate risk criteria are not being met, Air Products has observed some encouraging benefits.

Multi-functional reviews of LOPAs have demonstrated the ability to do the following:

1. Broaden the base of process safety ownership beyond the process safety function

2. Achieve better solutions for the lowest asset life cycle cost

The purpose of presenting the detail of the case study below is to illustrate the above benefits of reviewing LOPAs with multi-functional teams. The case study is broken into two sections, with each section followed by a description of the detailed benefits observed fitting into one of the categories above.

The general structure of a LOPA is well documented elsewhere and is not the subject of this paper. However, there is one clarifying distinction that may help overcome some potential confusion with the case study that follows. Some LOPA practitioner guides suggest that each LOPA should have only one initiating event. At Air Products, the corporate risk criteria are defined based on a hazardous consequence, not an initiating event. Therefore, our practice is to include multiple initiating events in the same LOPA if the initiating events all have the same hazardous consequence. Each initiating event corresponds to a single row in the LOPA table, and each layer of protection corresponds to a single column in the LOPA table. In each row, the initiating event frequency is multiplied by the average Probability of Failure on Demand (PFDavg) for each Independent Protection Layer (IPL) that applies.

2 Case Study – Raw Material Storage Tank Containing Hazardous Material

A project to build a new chemical process facility includes a storage tank containing a raw material that is subsequently fed to a process unit. The liquid raw material has a high enough vapor pressure to generate a hazardous vapor cloud upon loss of primary containment. Any deviation that results in loss of containment has the potential to cause a severe consequence for which risk criteria have been set.

Figure A. Case Study Storage System Diagram

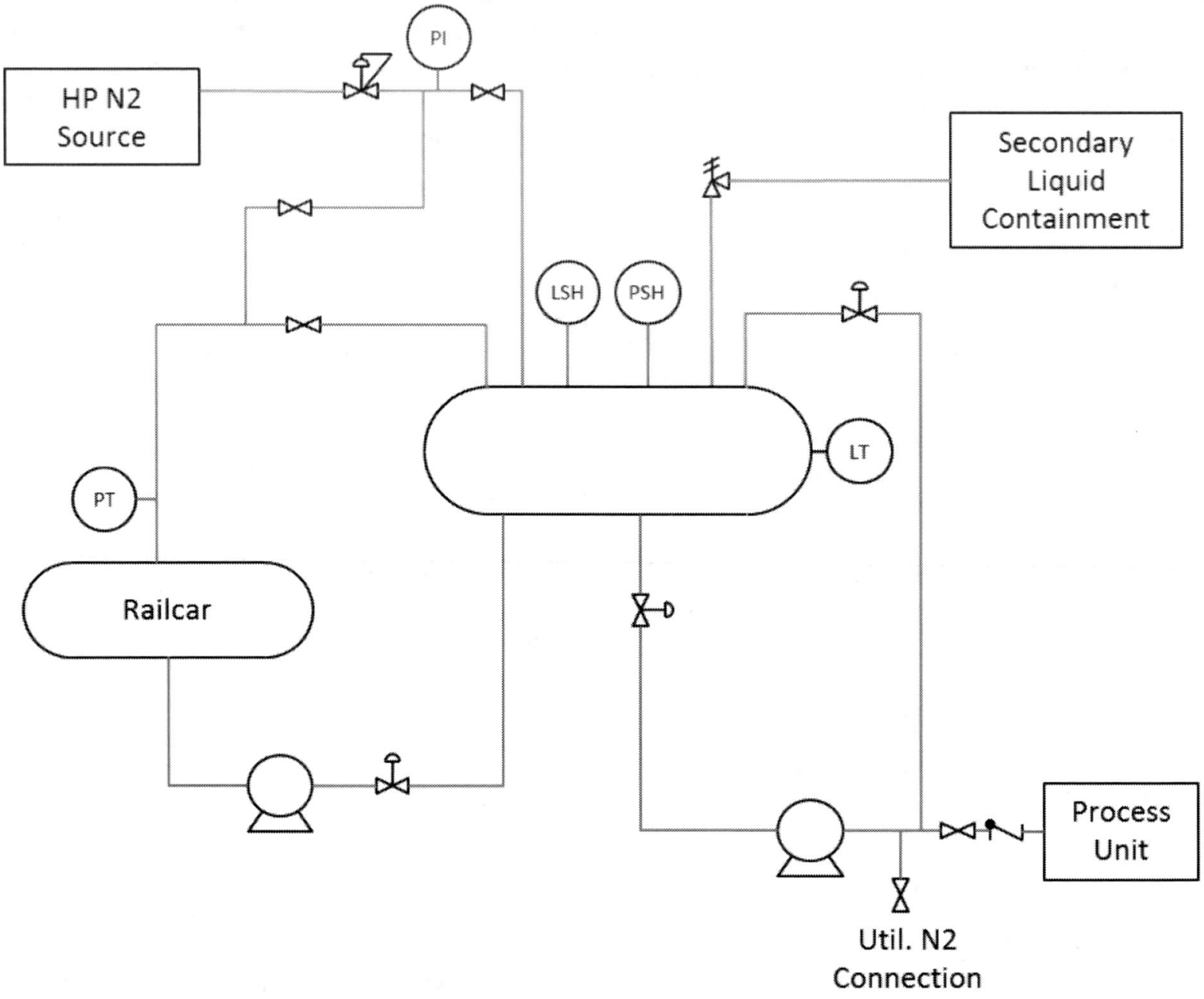

During the project hazard review, using HAZOP methodology, the team documents causes, consequences and safeguards for tank overpressure. The overpressure items from the hazard review are summarized in Table 1.

Table 1. Summary of Tank Overpressure Items Identified During Design Hazard Review

Cause	Consequence	Safeguards
Failure to open vapor return valve during offloading from railcar to tank.	Transfer pump deadhead pressure exceeds tank design pressure. Potential overpressure of tank.	PSH turns off railcar transfer pump. PSH closes railcar transfer valve. PSV adequately sized for this cause.
Level transmitter fails low.	Operator offloads material from railcar without sufficient capacity in the tank. Potential overfill and overpressure.	LSH turns off railcar transfer pump. LSH closes railcar transfer valve. PSV adequately sized for this cause.
N2 pad regulator adjusted too high.	Potential to overpressure tank since N2 pressure upstream of regulator exceeds tank design pressure.	PSV adequately sized for this cause.
N2 pad regulator fails open.	Potential to overpressure tank since N2 pressure upstream of regulator exceeds tank design pressure.	PSV adequately sized for this cause.
Operator error during line clearing with Utility N2	Potential to overpressure tank since Utility N2 pressure exceeds tank design pressure.	PSV adequately sized for this cause.
Process feed pump failure and reverse flow from process unit.	Potential overfill and overpressure of tank.	LSH closes pump suction and recycle isolation valves. PSV adequately sized for this cause.
Offload incompatible material from railcar due to spotting the wrong railcar for offloading.	Rapid reaction with residual material in tank resulting in heat and vapor generation with potential for tank overpressure.	PSH turns off railcar transfer pump. PSH closes railcar transfer valve. PSV adequately sized for this cause.

2.1 Initial LOPA

Note that the consequences for each of the overpressure events listed above are comparable: tank rupture with loss of containment and generation of a hazardous vapor cloud. Since the risk criteria is consequence based, the hazard rates for each row are summed in the LOPA structure to generate an overall consequence frequency and then compared to the risk criteria.

The LOPA, developed after the hazard review, would be structured by listing the causes of deviation as initiating events and the safeguards as layers of protection. Marks in the Table 2 below indicate which layers are effective in protection against each of the initiating events. Note that actual values for initiating event frequencies, PFDavg for layers of protection, and hazard rates are not presented because they are not essential to the topic of this paper.

Table 2. Initial LOPA Based on Hazard Review

	Layers of Protection				
Initiating Event	**Tank High Pressure Switch**	**Tank High Level Switch**	**Tank Relief Valve**		
Vapor Return Valve Left Closed	X		X		
Level Transmitter Fails Low		X	X		
N2 pad regulator adjusted too high			X		
N2 pad regulator fails open			X		
Operator error during line clearing with utility N2			X		
Process Feed Pump failure and reverse flow		X	X		
Offload incompatible material into tank	X		X		

When a risk analysis is first developed, such as in the LOPA depicted above, it is sometimes the case that the calculated hazard rate exceeds the risk criteria for the consequence.

What happens next is important.

Sometimes the safety analyst develops a risk assessment, determines there is a gap between estimated risk and the risk target (risk gap), and develops a list of design changes the team must accept. This "over-the-fence" or non-interactive method of design change has been used for years. It has been used for a number of reasons, a couple of which are mentioned below:

- Safety assessments may have been based on a set of esoteric rules that related the number of safeguards to the severity of a consequence in a non-quantitative way. The safety engineer was the only one who understood (rather, who needed to understand) these rules and the resulting effort tended to be conducted exclusively by the safety engineer. The tediousness and perception of arbitrariness of this method created a barrier to team participation.
- Safety assessments have been done using quantitative tools like fault tree analysis to generate hazard rates. Although robust in developing frequency estimates, these tools are not at all suited to communicating results or requirements to multi-functional audiences. The format of the analysis created a barrier to team participation.

What can often happen with non-interactive design change by the safety analyst is that the basis for the safety assessment remains mysterious to the team. If the proposed design changes increase project cost or add a burden to operations, those stakeholders have to accept these things without understanding why they are necessary, except for the reason that "safety says so." This can undermine team effectiveness and create a sense among the stakeholders of being disengaged from the safety objectives.

The intuitive structure of a LOPA, on the other hand, opens up the possibility of reviewing a safety assessment directly with the multi-functional stakeholders on the project team. Participants in the review could include project engineers, operations representatives, process engineers, controls engineers, asset managers, and others. The purpose of the review is to make the entire team responsible for the solution. Everyone on the team is responsible to help close the risk gap between the current LOPA result and the risk target.

The narrative of the group problem solving sessions below illustrate how this may work.

2.2 Group Problem Solving, Part 1

There are 3 initiating events for which the relief device is the only layer of protection. They all involve overpressure due to a high-pressure nitrogen source. And all 3 cases contribute significantly to exceeding the hazard rate target:

- Adjusting the tank headspace regulator set pressure too high for the tank
- Spurious fail open case for the tank headspace regulator
- Connecting utility N2 to the pump circuit for line clearing prior to maintenance

2.2.1 Utility N2 for Line Cleaning

If the line clearing case is done every couple years for maintenance, then that frequency coupled with the probability of human error results in a slightly lower hazard rate than the other N2 cases. As the team discusses the options for this case, the process engineer does some quick math and points out that, due to the large size of the tank, and the hydraulic limits of the N2 connection, it would take at least 2 minutes for the utility N2 line clearing flow to cause the tank pressure to go from the alarm point to a hazardous overpressure condition. The controls engineer responds that 2 minutes is enough time for the high pressure switch to respond and trip the isolation valves on the pump suction and return lines closed. Therefore, the high pressure switch, which was not identified as a safeguard for this case by the hazard review team, and is already in the scope of the design, can be claimed as a layer of protection.

Having dealt with the line clearing utility N2 case, the team moves on to consider the two N2 regulator cases for the tank headspace pad.

2.2.2 N2 Regulator Adjusted too High

Since the N2 regulator is used to both pressurize the railcar high enough to prime the transfer pump and then later adjusted lower to provide a pad for the tank, the operator needs to adjust the N2 regulator every time the tank is filled, which could be dozens of times per year. Let's say that this is the highest hazard rate in the table. It is now easy for the team to see that this particular operation is the greatest risk to tank overpressure.

Operations is now engaged. Can something be done with operating procedures or the operating environment to reduce the likelihood of human error? What mistake-proofing strategies could be employed? Are there visual aids that would help? How about creating procedural steps that allow for recovery such as checking and verification?

Process is now engaged. Can the process for priming the transfer pump be changed so the regulator does not need to be adjusted every time the tank is filled? It would add some cost, but a second regulator could be added to the scope that is used exclusively for priming the railcar offloading transfer pump.

With a little coaching from the safety analyst, the team recognizes that the separate regulator approach is an example of an inherently safe solution – it eliminates this particular cause of overpressure entirely. It is much more effective than adding procedural steps to reduce human error probability. It eliminates the possibility of human error by eliminating the procedure entirely. The team adopts the separate regulator design change and removes this line item from the LOPA.

2.2.3 N2 Regulator Fails Open

The initial LOPA indicates that the combination of regulator failure rate and probability of tank relief protection failing results in a frequency that exceeds the hazard rate criteria. As the team is discussing its options, the operations representative recognizes a new issue. The random failure rate of the regulator is shown in the initiating event frequency of the LOPA table. Although this is a relatively low value, it is high enough that one may expect it to occur sometime during the life of the plant (20+ years). The LOPA clearly shows that the tank relief is the only protection. Therefore, one would expect a demand on the tank relief device at the regulator failure frequency.

Due to the nature of the contents of the tank headspace, the consequence of venting through the relief could be significant and is therefore undesired. If an additional layer of protection is needed, it would be highly desirable to have that layer of protection function to prevent a demand on the tank relief valve.

The team identified two options:

- Adding an actuated isolation valve to the N2 supply line that would close on high tank pressure
- Adding a relief valve to the N2 supply line just downstream of the regulator and upstream of the tank

The controls engineer provides some feedback on the total installed cost of the isolation valve option. The project engineer had some data on small N2-service relief valves that shows the installed cost of the relief was a factor of 5 lower than the actuated isolation valve. The team adds the N2 relief valve with a set pressure 10% lower than the tank relief valve and moves on.

2.2.4 Summary of Scope Changes from Group Problem Solving, Part 1

The two changes made by the team during Problem Solving, Part 1are shown on the updated flowsheet in Figure B, and included in the LOPA in Table 3.

Figure B. Scope Changes from Problem Solving, Part 1

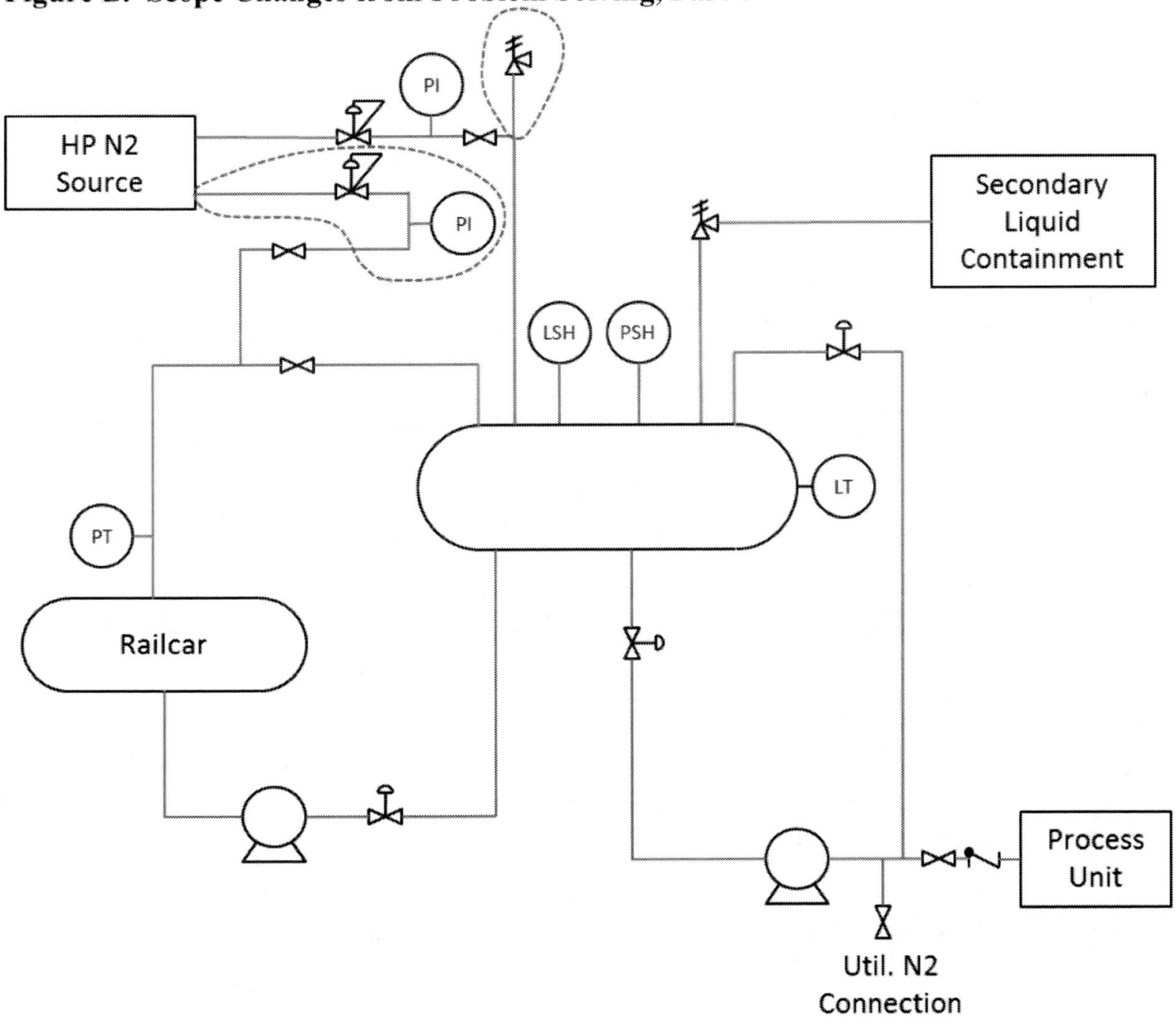

Table 3. LOPA Following Group Problem Solving, Part 1

	Layers of Protection				
Initiating Event	**Tank High Pressure Switch**	**Tank High Level Switch**	**Tank Relief Valve**	***N2 Regulator Relief Valve***	
Vapor Return Valve Left Closed	X		X		
Level Transmitter Fails Low		X	X		
~~N2 pad regulator adjusted too high~~			~~X~~		
N2 pad regulator fails open			X	*X*	
Operator error during line clearing with utility N2	*X*		X		
Process Feed Pump failure and reverse flow		X	X		
Offload incompatible material into tank	X		X		

2.2.5 Observed Benefits from Group Problem Solving, Part 1

The case study LOPA review is not yet complete, but this is a good time to summarize some of the benefits of the multi-functional review process that have been illustrated thus far.

Broaden the base of process safety ownership:

- Problem Visibility: LOPA presents a risk assessment in a format that is easy for a multi-functional team of stakeholders to understand. As a result, the team is able to actively contribute as reviewers and problem-solvers.
- Active Engagement: Once the team is asked to participate in the problem-solving process, the members' perception of their role transitions from being input providers to being active developers of the LOPA structure.

- Common Goal: The quantitative nature of the hazard rate criteria and the mathematical structure of the LOPA make it easy for everyone to see where risk gaps exist and where solutions are required.

Achieve better solutions for the lowest asset life cycle cost:

- Prioritize Effort: LOPA allows team members to identify the initiating events or protection weaknesses that are contributing most substantially to the total hazard rate. This improves execution efficiency by focusing effort on the issues with most leverage.
- Generate Viable Options: Having a diversity of expertise on the review team results in a wide variety of potential solutions to problems. The greater the number of viable proposals, the greater the likelihood the team will find the lowest cost solution.
- High Leverage Solutions: The LOPA structure clearly portrays which layers provide effective protection for each of the initiating events. As illustrated by the case study, teams become motivated to find ways to determine if a layer of protection for one initiating event can be effective in mitigating others.
- Option Sort in a Safety Context: LOPA development solutions may add an ongoing burden to operations. They may increase the installed cost of a facility. Some may do both to some degree. The multi-functional team review process creates a forum where those issues and costs can be discussed and minimized within a context where safety value is part of the analysis.

2.3 Group Problem Solving, Part 2

Returning to the case study, there remain two more initiating events that contribute to exceeding the hazard rate criteria.

- Vapor return valve left closed during railcar offloading
- Tank level transmitter fails low and tank overfilled

2.3.1 Vapor Return Valve Left Closed

Leaving the vapor return valve in a closed condition is a problem because this is a manual operation and the operator has an opportunity to skip the step of opening the valve every time the tank is filled. Despite the tank high pressure interlock protection, the hazard rate is still too high.

A problem-solving process similar to the regulator adjustment failure case occurs. Operations begins thinking about ways to reduce human error. They also ask if the valve could be automated and a batch sequence control program written to control it. It is a fair question, but when the controls engineer provides an indication of the equipment and engineering cost to set up such a system, the entire team agrees to keep looking for alternatives.

Then, looking through the railcar unloading flowsheet, the operations representative points out that, during the operability review, a pressure transmitter was put on the railcar vapor connection. This was done to provide a low pressure interlock in the event that the railcar headspace was blocked while the transfer pump was on. The interlock would alert the operator to the low pressure condition and stop the pump before it entered a cavitation regime. Could this same device be used as an indication that the vapor return valve from the tank was closed? The process engineer does some math comparing headspace volume in the tank vs. the railcar and determines that it would be possible to define a set point for the railcar low pressure interlock that protects both the pump and the tank.

At this point, the safety engineer emphasizes the importance of the LOPA rule that all layers of protection must be independent from each other. If the team is going to take credit for two instrumented layers of protection, those two layers must not share any common field elements. In this case, the team verifies that it is possible to identify two independent safety instrumented functions because there are two sensing elements (high tank pressure and low railcar pressure) and two final control elements (pump shutdown and isolation valve closure).

2.3.2 Level Transmitter Fails Low

After solving the closed vapor return valve issue, the only remaining line item that contributes significantly to exceeding the hazard rate target is the tank level indication failing low. There is a high level interlock on the tank already, and based on the prior experience of adding another pressure interlock with the vapor return valve closed case, the team is naturally drawn to the idea of adding a second high level switch to the tank.

There is a problem, however, with this approach. The safety engineer points out that both the initiating event and the first layer of protection involve the failure of a level device. If the team attempts to add another layer of protection based on a level device, the contribution of common mode failure will tend to dominate the frequency of failure.

This may be a difficult moment for the team, and especially for the safety engineer. Common mode failure is one of the more difficult concepts for stakeholders outside the safety profession to understand and therefore accept. Common mode failure is a real yet enigmatic contributor to the reliability of protective systems. Unlike the failure rate for an individual component, one cannot open a database and find a probability that two or more level transmitters will both simultaneously be in a failed state for reasons other than random failure. It is the safety engineer's responsibility to educate the team on this subject and steer them away from excessive redundancy and toward diversity in layers of protection.

In the case study, the team accepts the common mode failure limitation and moves on toward finding ways to determine when the tank is at risk of being overfilled without an additional level interlock.

Over in the process unit, the raw material from the tank passes through a flow control station into the destination vessel. The controls engineer points out that it would be relatively easy to totalize the flow through this meter, starting from zero each time the tank is filled. Since this is the only destination of this material, the totalized flow represents the quantity of material that has been removed from the tank. The control system could be configured to provide an alarm if the railcar offloading transfer pump is turned on when the current totalized flow value is less than the contents of a typical railcar. The entire team likes this approach because it diversifies the layers of protection while using equipment that is already in the scope of the project.

2.3.3 Summary of Scope Changes from Group Problem-Solving, Part 2

At this stage, the sum of all the individual row hazard rates in the LOPA is less than or equal to the hazard rate criteria for the loss of containment consequence. The work of the team is done. Figure C shows the updated flowsheet and the LOPA changes are reflected in Table 4.

Figure C. Scope Changes from Problem Solving, Part 2

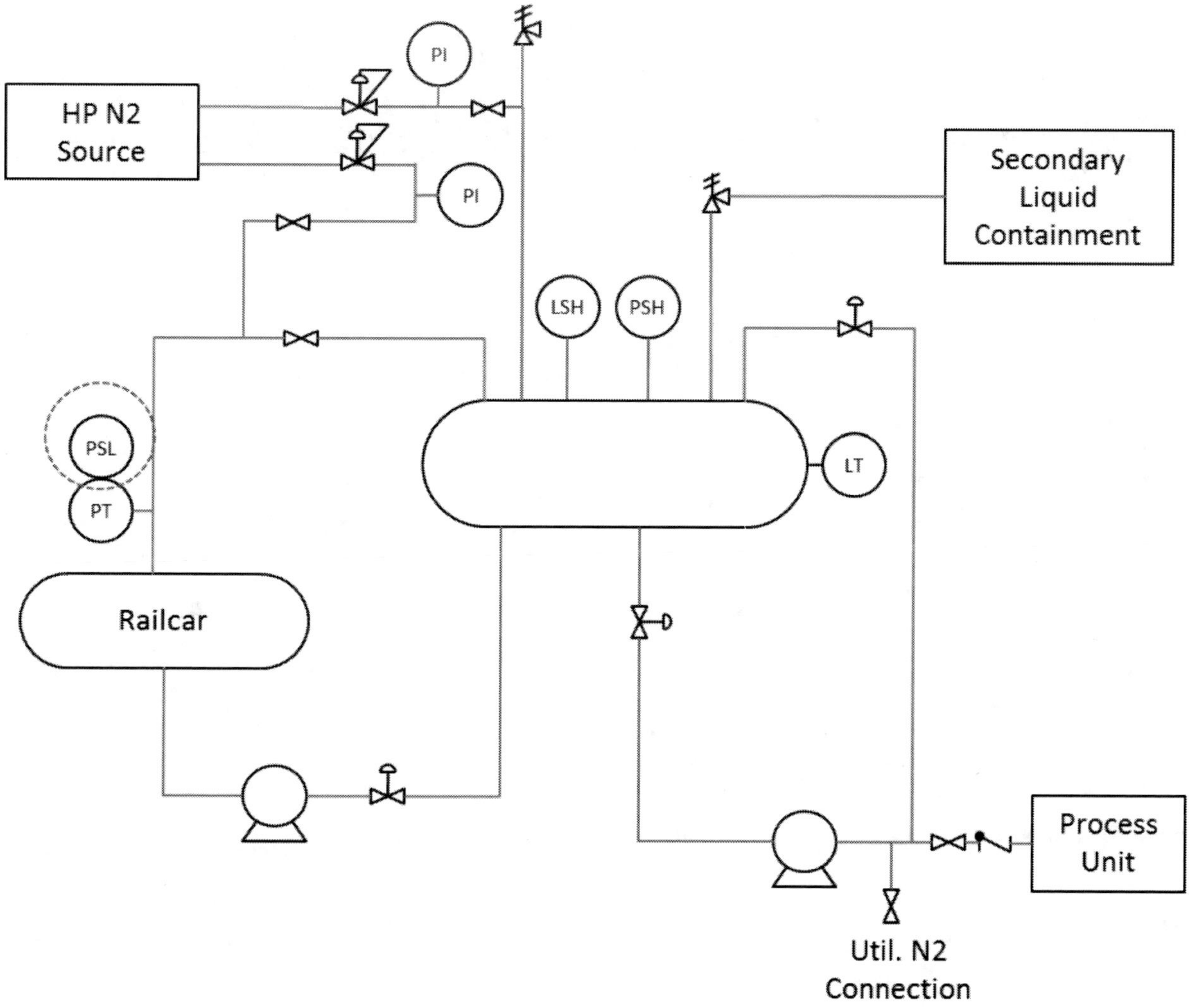

Table 4. LOPA Following Group Problem Solving, Part 2

	Layers of Protection					
Initiating Event	**Tank High Pressure Switch**	**Tank High Level Switch**	**Tank Relief Valve**	**N2 Regulator Relief Valve**	***Railcar Low Pressure Switch***	***Flow Tot. Alarm***
Vapor Return Valve Left Closed	X		X		*X*	
Level Transmitter Fails Low		X	X			*X*
~~N2 pad regulator adjusted too high~~			~~X~~			
N2 pad regulator fails open			X	X		
Operator error during line clearing with utility N2	X		X			
Process Feed Pump failure and reverse flow		X	X			
Offload incompatible material into tank	X		X			

2.3.4 Observed Benefits from Group Problem Solving, Part 2

This most recent work illustrated again some of the benefits described in section 2.2.5 above from the Part 1 problem solving session. But the Part 2 events just completed also illustrate some new benefits of multi-functional team LOPA review that are summarized below.

Broaden the base of process safety ownership:

- Education in Independence: The strict LOPA rule requiring protection layers be independent of each other teaches the team members what independence means with the practical examples immediately before them. Because independence is easy to understand in this context, team members are empowered to apply this knowledge in future LOPA reviews and even future project design hazard reviews.
- Education in Diversity: By taking a multi-functional team through a LOPA where common mode failure presents a limit on what types of protection layers can be applied, the team members begin to accept the concepts of dependent failure. They are far less likely to put 4 check valves in series when drawing flowsheets in the future.

Achieve better solutions for the lowest asset life cycle cost

- Creative Options for Diversity: A LOPA review team where members represent different disciplines or functions makes it easier to identify options to overcome common mode failure with equipment or method diversity. Often, equipment, instruments or procedures that are already required for operability reasons can be utilized to provide additional protections to overcome common mode failure with diversity. The key skill is understanding how the potential for diagnosing or correcting one problem can be positively inferred from the observation of something else.

3 Conclusion

The case study presented is illustrative of some of the positive experiences Air Products has had as it has utilized multi-functional teams to review LOPAs, especially in cases where risk criteria were not initially being met. In the circumstance where there is a risk gap in a LOPA between a risk target and the calculated hazard rate, Air Products generally has achieved better results using multi-functional teams for problem solving rather than having a safety analyst working in a non-interactive way. The better results have been observed to come in two forms: (1) broadened process safety ownership, and (2) better solutions to minimize asset life cycle cost

Interactive team review and problem solving of LOPAs has broadened the base of process safety ownership well beyond the process safety function by increasing:

- Problem Visibility – the ease with which team members are able to understand the problem
- Active Engagement – the level of impact each team member perceives they can have on the outcome
- Commonality of Goals – translating a corporate risk target to a specific objective within a process unit that everyone can understand and work together to solve
- Knowledge of Risk Quantification Basics – recognizing the importance of maintaining independence between layers of protection and the pitfalls as well as potential solutions for common mode failure

Multi-functional team problem solving of LOPAs has resulted in better solutions to achieve the lowest asset life cycle cost by:

- Prioritizing Effort – review team can easily assess the issues driving the results and focus their efforts on those items
- Generating Viable Options – having a diversity of expertise on the review team results in a wide variety of potential solutions to problems – this is especially valuable with common mode failure limitations are encountered
- Generating High-Leverage Solutions – teams quickly learn that the best layers of protection are those able to be applied to multiple initiating events
- Option Sorting in a Safety Context – solutions often create capital versus operating cost trade-offs that can be evaluated in the context of achieving safety goals

4 References

[1] CCPS. *Layer of Protection Analysis: Simplified Process Risk Assessment.* Joint Publication of the Center for Chemical Process Safety, American Institute of Chemical Engineers, New York, NY, 2001.

[2] CCPS. *Guidelines for Initiating Events and Independent Protection Layers in Layer of Protection Analysis.* Joint Publication of the Center for Chemical Process Safety, American Institute of Chemical Engineers, New York, NY and John Wiley & Sons, Inc. Hoboken, NJ, 2015.

[3] CCPS. *Guidelines for Enabling Conditions and Conditional Modifiers in Layers of Protection Analysis.* Joint Publication of the Center for Chemical Process Safety, American Institute of Chemical Engineers, New York, NY and John Wiley & Sons, Inc. Hoboken, NJ, 2013.

[4] IEC 61508 – 2010 Functional safety of electrical/electronic/programmable electronic safety-related systems Part 1-7.

[5] IEC 61511 Functional Safety: Safety Instrumented Systems for the Process Industry Sector.

[6] P. Gruhn. "Layer of Protection Analysis and Common Mistakes," 2013 Safety Control & Instrumented Systems Conference.

[7] NASA. *Fault Tree Handbook with Aerospace Applications. Version 1.1.* NASA Office of Safety and Mission Assurance, Washington, DC., 2002.

441507 Using Data Analytics to Improve Process Safety

Abstract:

Using Data Analytics to Improve Process Safety

Increasingly, data from multiple sources are compiled into central storage systems for entire facilities and in some cases entire corporations. The low cost of digital memory and the increasing process speed of modern microprocessors and computers continue to make it easier to analyze, manipulate and merge very large databases. Big Data and Data Analytics have been touted as allowing staff at all levels to make real time, data-driven business decisions. These concepts can be used to improve process safety at facilities by proactively analyzing sources of data to identify opportunities for improvement in operating practices, identify and track near misses, and in real time detect incipient instrument and equipment failures that could lead to catastrophic incidents. The aviation industry is one example that is voluntarily implementing similar analytics. Over the last decade the aviation industry has implemented flight operations quality assurance (FOQA) or flight data monitoring (FDM) programs which collect, store, and analyze recorded flight data with the goal of increasing overall safety.

In this presentation we show examples of analytics Exponent performed on data from facilities. These analyses were used to analyze alarm frequencies during specific types of operations, identify near misses, and proactively detect failures of instrumentation and equipment in order to prevent or mitigate catastrophic incidents.

442485 Safety Culture and Leadership- Independent of Language

Abstract:

The existence of safety in the workplace is mandatory for a sustainable business. Strong safety cultures are characterized by human factors, workforce engagement and fully committed safety leadership. With that being said, increased diversity in work environments and international recruitment can oftentimes complicate the process of implementing a common safety culture.

More specifically, those who come from different types of work environments than their current environment may have trouble interpreting the procedures and guidelines that are enforced at their new place of employment. Differences in languages and dialects may also lead to misinterpretations of intended safety guidelines. Lastly, there may be some instances where the workplace follows a non-flexible and hierarchical structure that is not very open to change.

In order to best develop safety practices in the workforce, the senior leaders need to first develop an environment based on the understanding of the different rationales and expectations of a multi-cultural workforce. Once that has been established, communicating with the workforce on safety expectations of the organization and optimizing interpretation of this message is critical.

This paper will continue the discussion from a previous publication; how safety leadership can consider linguistic differences when implementing a safety plan for a diverse workforce. The discussions will emphasize the learnings of the Bradley Curve, showing how important it is to follow safe practices to reach the proximity of an almost 100% incident free area, and will provide real examples from international work assignments and diverse work teams, highlighting this need and offering potential approaches

439897 Indulging in Invincibility (III)- Twelve Steps to Sobriety

Abstract:

<>8/3/2015 <>Indulgence in Invincibility (III)- Twelve Steps to Sobriety <>by Brian P. Kolodji, PE

Could Bhopal, Macondo, or Texas City have been prevented? Directly related examples are presented where PS cultural indulgences contribute to a false sense of invincibility. Light is shed on where the perennial dilemmas of misplaced loyalty, unachievable goals, and belief in corporate higher power contribute to and enable denial of process safety culture failure. Successes and a program are presented where these failed practices are brought out of the shadows and recovery begins.

12TH GLOBAL CONGRESS ON PROCESS SAFETY

Case Study of Relief Analysis of Compressor Stations

Kumud Chaurasia
Process Engineer
Lloyd's Register Energy Americas, Inc.
1330 Enclave Parkway, Suite 200, Houston, TX 77077, USA
Kumud.chaurasia@lr.org

Aniket Patankar
Senior Process Engineer
Lloyd's Register Energy Americas, Inc.
1330 Enclave Parkway, Suite 200, Houston, TX 77077, USA
Aniket.patankar@lr.org

Prepared for Presentation at
American Institute of Chemical Engineers
2016 Spring Meeting
12th Global Congress on Process Safety
Houston, Texas
April 11-23, 2016

Case Study of Relief Analysis of Compressor Stations

Kumud Chaurasia, Process Engineer

Aniket Patankar, Senior Process Engineer

Lloyd's Register Energy Americas, Inc.

1330 Enclave Parkway, Suite 200, Houston, TX 77077, USA

kumud.chaurasia@lr.org, aniket.patankar@lr.org

Keywords: Relief analysis, Compressor Station, case study, concern identification

Abstract

A compressor station is a facility which helps the transportation process of natural gas from one location to another. Compressor stations include several key component parts, the primary being the actual compressor unit, either centrifugal or reciprocating. They also typically include scrubbers, strainers or filter separators which remove liquids, dirt, particles, and other impurities from the natural gas. This study will give an overview of the relief analysis conducted for multiple compressor stations. The relief analysis was system based as opposed to relief device based so as to identify any unprotected systems that may need overpressure protection. It will discuss the overpressure scenarios identified for these systems. It will provide some general statistics on the total number of applicable overpressure scenarios. It will give a detailed breakdown of the various concerns found including, but not limited to, undersized devices, unprotected systems, installation concerns, data needs, and documentation discrepancies. The study will discuss possible mitigation solutions for these concerns.

1. Introduction

Compressor stations are typically located along a natural gas pipeline to facilitate the compression of the gas to a specified pressure, which allows the gas to flow along the pipeline to a target location. Since these pipelines typically extend for several hundreds of miles, the gas tends to lose pressure frequently. This necessitates the requirement to compress it frequently during the transportation process.

The size of the station and the number of compressors (pumps) varies, based on the diameter of the pipe, and the volume of gas to be moved. Compressor stations include several key components, the primary being the actual compressor unit, either centrifugal or reciprocating, scrubbers, and strainers or filter separators which remove liquids, dirt, particles, and other impurities from the natural gas.

It is critical to ensure that the natural gas transported to its intended target is free from any contaminants. This is accomplished through four mains steps. The first step is contaminant removal, which occurs upstream of a compressor station near the well head. Here, the raw natural gas, at very high temperature and pressure, is passed through separators in order to remove oil, water, sand etc. It is then passed through a slug catcher, typically located within the compressor station, to remove any leftover condensate. The third contaminant removal step occurs downstream of the compressor unit (after the gas has been pressurized) via a contactor unit that predominantly removes liquid water and carbon dioxide using triethyleneglycol (TEG) as the absorbent. The TEG that separates out at the bottom of the scrubber unit, carrying the water and carbon dioxide with it, is typically regenerated in a glycol regenerator unit before it is recirculated back into the scrubber unit. The final step prior to delivery of the gas to its intended target is to remove any leftover TEG in the gas using a scrubber unit. There are also several tanks within the compressor station for storage of the condensates and TEG. Additionally, the compressor stations contain several components that are part of the glycol regenerator unit including filters, flash tank separators, pumps, heat exchangers, etc. A general process flow diagram of the compression station is shown in Figure 1.

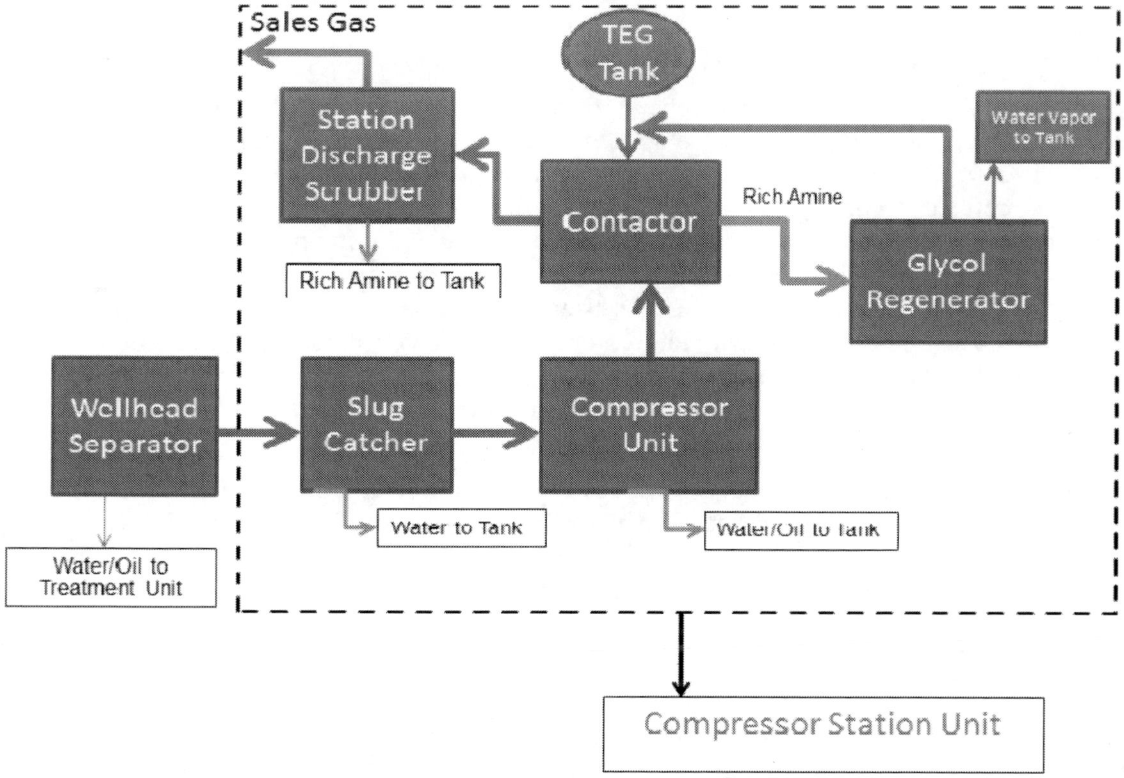

Figure 1: Process flow diagram of Compressor Station

2. Relief Analysis

For this case study, twenty separate compressor stations were analyzed. Specifically, the steps required for performing the analyses, different applicable scenarios, and all the concerns identified with mitigations options are presented. The analysis was conducted using API 520 Part 1 and Part 2 standards, while considering the client's design philosophy as well.

2.1. Steps involved in Relief Analysis

The relief analysis, which is a system based analysis performed for each equipment in the unit, is a methodical four step approach including data and design documentation collection, contingency analysis, pressure relief calculations, concern identification and mitigation. These steps are discussed in detail below.

Data and Design Documentation Collection: Since the client for whom the relief analysis is being performed may not always provide all the information necessary to conduct the analysis, it is imperative that all necessary data and documentation required is collected at the client site prior to commencing the project. Examples of such data collected include isometrics for PSV inlet and outlet piping, PSV and control valve information, equipment and relief device name plate information, U-1A forms, equipment drawings etc. Usually, the piping and instrumentation diagrams (P&IDs) and heat and material (H&M) balance are provided by the client in order to complement the other data collected. All the data collected is then entered into a client project database.

Contingency Analysis: Prior to commencing the contingency analysis, the compressor station is divided into various systems, with each system consisting of several equipment. The interface of systems is defined based on where block valves are situated within the P&ID. Once the systems have been defined, a contingency analysis is performed where all scenarios that could possibly result in an overpressure condition are identified. For the 20 compressor stations that were analyzed in this case study, the following eighteen contingencies were identified: Power Failure, Cooling or Reflux Failure, Steam Failure, Instrument Air Failure, External Fire, Blocked Outlet, Isolated with Heat Input, Equipment Failure, Inadvertent Valve Operations, Failure of Automatic Controls, Gas Breakthrough, Internal Boundary Failures, Reverse Flow, Run Away Reaction, Accidental Mixing, Change in Conditions, Vacuum, Other (applicable scenarios other than listed above).

Calculations: Once the contingency analysis is finished, the required flow rate of the applicable overpressure scenario and the capacity of the relief device (based on the relief device ASME orifice area) are calculated. The inlet and outlet pressure drops, Mach number and sound power level of the piping is also calculated based on the dimensions of the relief device.

Concerns Identification and Mitigation: After the calculations are completed for a system, all the concerns are identified and documented in the database. This is as per the guidelines provided in API 520 Part 1, 2 and the design philosophy provided by client. The concerns are identified based on the calculation results and discrepancies and assumptions made during the analysis. Once all the concerns are identified, recommendations for implementation of mitigation strategies are provided to the client.

3. Results of Case Study

3.1. Relief Device Identification

For the twenty compression stations that were analyzed as part of this case study, the relief devices included all the relief valves, pressure and vacuum vents (PVRVs), goosenecks, and rupture disks. Usually, the relief valves and rupture disks were situated on the pressure vessels while the PVRVs and goosenecks were situated on the low pressure tank. A total of 1099 relief devices were identified for the 20 compressor stations. Depending on the type of relief device, the method of calculation may vary. For example, in the case of PVRVs and goosenecks, it is not necessary to calculate the pressure losses.

The breakdown of the percentage and number of the relief devices, by type, that were analyzed in this study are shown in Table 1 and Figure 2 respectively.

Table 1: Percentage of different relief devices by type

Relief device Type	Percentage
Relief valve	78%
Gooseneck/PVRV's	21%
Rupture Disk	1%

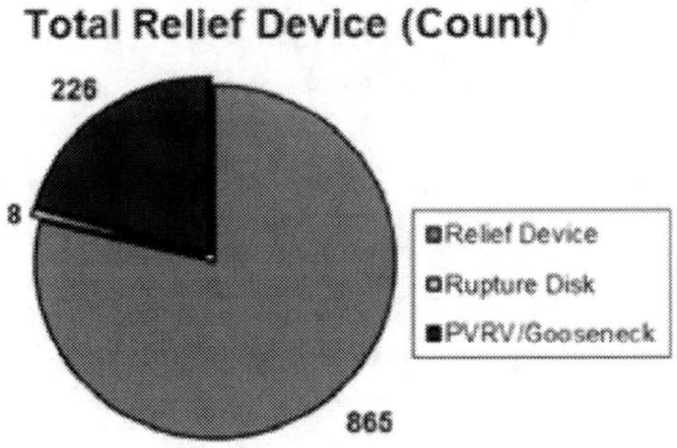

Figure 2: Actual number of different relief devices, by type, for the twenty compressor stations

3.2. Relief Valve Type

There are three types of relief valves that are typically used in the oil and gas industry: conventional, bellows, and pilot valve. Based on the type of valve, the outlet pressure drop limits vary; hence it is important to identify the relief valve type. As shown in Table 2, out of a total of 865 relief valves identified, the majority of the valves in the compression stations analyzed were conventional relief valves,. The breakdown of the number and percentage of the relief valve that were analyzed in this study are shown in Table 2 and Figure 3 respectively.

Table 2: Percentage of different relief valves, by type

Relief Valve Type	Percentage
Conventional	68%
Pilot	23%
Bellows	9%

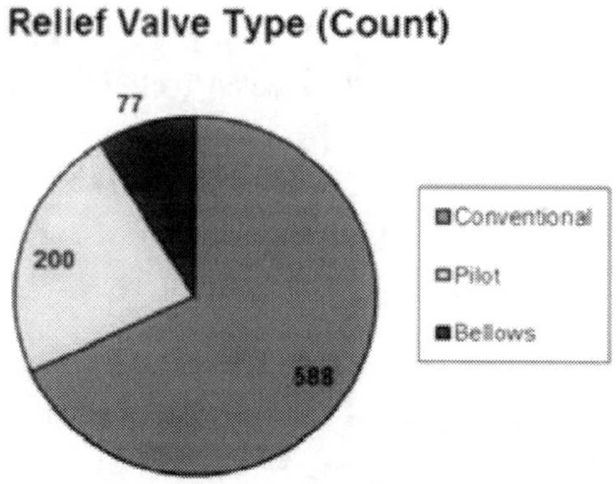

Figure 3: Actual number of relief valves, by type for the twenty compressor stations

3.3. Equipment Type

Based on the equipment type, the overpressure scenario may be applicable or not and the data required to do the analysis will be different. Hence, it is very important to identify the type of the equipment. A total of 2014 equipment were identified in all the compressors stations, of which majority were pressure vessels.

The breakdown of the percentage and number of the equipment type that were analyzed in this study are shown in Table 3 and Figure 4 and respectively. Note that the "Other" type of equipment includes compressors, glycol regenerators etc.

Table 3: Percentage of different equipment, by type

Equipment Type	Percentage
Pressure Vessel	48.5%
Tower	8.1%
Pumps	8.9%
Exchangers	16.6%
Fired Heater	0.1%
Low Pressure Tanks	6.1%
Piping Component	6.4%
Others	5.3%

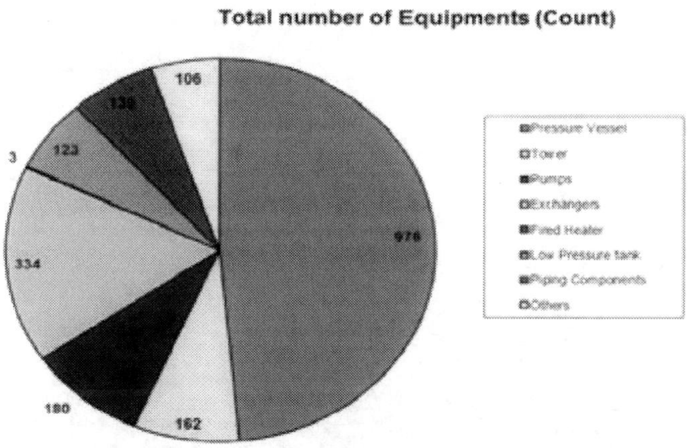

Figure 4: Actual number of equipment, by type, for the twenty compressor stations

3.4. Applicable Overpressure contingencies:

All the applicable scenarios, for each system, were identified after reviewing the P&IDs and other available data. This section shows all the applicable scenarios identified. The breakdown of the percentage and number of the applicable contingencies that were analyzed in this study are shown in Table 4 and Figure 5 respectively. Of a total of 3026 applicable contingencies, the majority of the contingencies in the compression stations analyzed were external or pool fire and blocked outlet.

Table 4: Percentage applicable contingencies by type

Contingency Type	Percentage
External fire	22.8%
Cooling or reflux failure	0.1%
Blocked Outlet	22.6%
Failure of automatic controls	13.3%
Gas breakthrough	8.5%
Inadvertent valve operation	5.6%
Instrument Air failure	2.9%
Internal boundary failure	1.6%
isolated with Heat Input	6.4%
Power failure	0.9%
Reverse flow	3.2%
Vacuum	7.1%
Other	5.0%

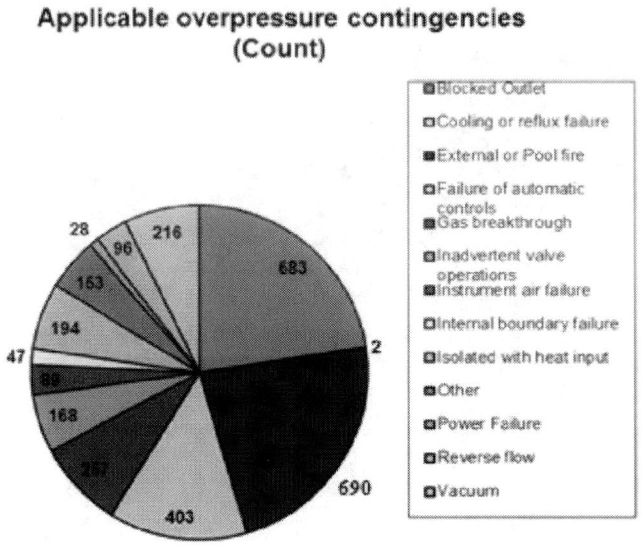

Figure 5: Actual number of applicable overpressure contingencies, by type, for the twenty compressor stations

3.5. Concerns Identification:

This was one of the most important steps in the relief analysis. Based the calculation results, assumptions, discrepancies etc., all the concerns were identified and documented based on the guidelines. The concerns were divided into five major categories: Undersized Concerns, Installation Concerns, Unprotected Systems, Data Requirement, and Documentation Updates. It is important that assumptions made during this stage be documented, verified and approved by the client. The undersized concerns, installation concerns and unprotected system concerns were assigned a higher priority since they could have a direct impact on the operation and failure of the compressor station. Hence, it was recommended that these concerns be mitigated immediately.

The breakdown of the percentage and number of the concerns that were analyzed in this study are shown in Table 5 and Figure 6 respectively. Of a total of 3278 applicable concerns that were identified in this case study, the majority of the concerns in the compression stations were related to Data Requirements.

Table 5: Percentage of concerns identified, by type

Concern Type	Percentage
Undersized	10.4%
Installation	24.6%
Unprotected Systems	9.5%

Data Requirement	50%
Documentation Updates	5.5%

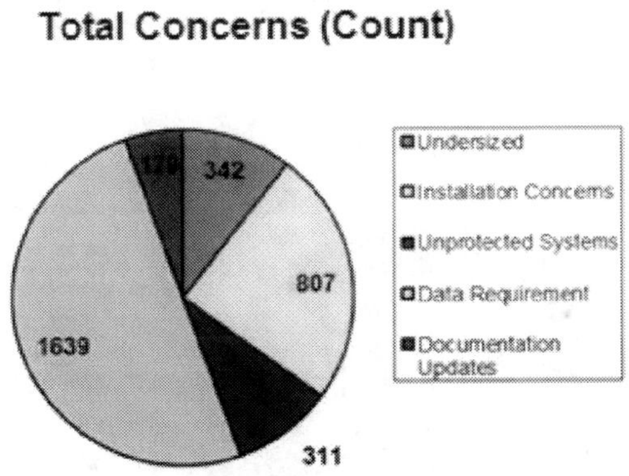

Figure 6: Actual number of different concerns, by type, for the twenty compressor stations

Undersized Concerns:

When the required relief flow rate is more than the capacity of the relief device, the relief device is defined to be undersized. This section shows the overpressure scenarios which were undersized. This includes all the contingencies for which the relief devices were undersized. The most common reasons for relief devices to be undersized were identified as being because of incorrect upstream pressure, incorrect composition of the relief stream, or that the contingency was not identified at all in the previous study.

The breakdown of the number and percentage of the undersized concerns that were analyzed in this study are shown in Table 6 and Figure 7 respectively. Of a total of 342 undersized concerns that were identified in this case study, the majority of the contingencies in the compression stations were Vacuum, Gas Breakthrough, External or Pool Fire and Failure of Automatic Controls.

Table 6: Percentage of type of contingency within undersized concerns

Contingency Type within Undersized Concern	Percentage
Blocked outlet	11.1
External fire	15.2
Failure of Automatic Controls	15.2
Gas breakthrough	15.5
Inadvertent valve operation	1.1
Internal Boundary failure	8.2
Reverse flow	8.5
Vacuum	24

Other	1.2

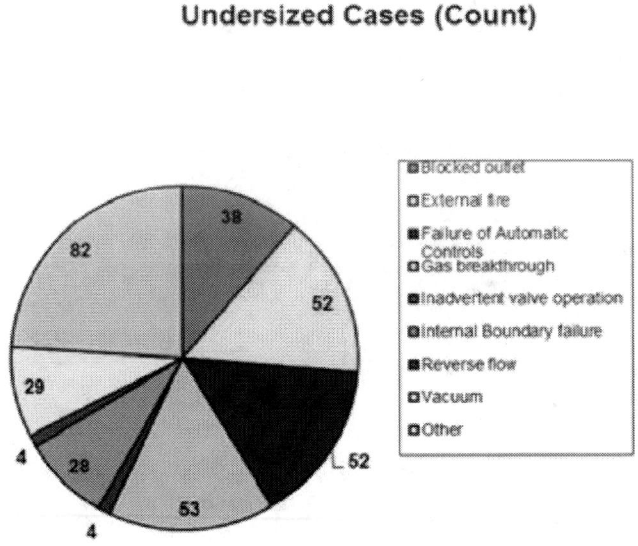

Figure 7: Actual number of different contingency types within the undersized concerns for the twenty compressor stations

<u>Installation Concerns:</u>

The installation concerns (total of 807 were identified) are divided into eleven sub categories as shown below. This is based on the guidelines used to perform the analysis.

Inlet Pressure Drop: The inlet pressure drop is calculated based on the relief device capacity. This includes all the scenarios for which the inlet pressure drop is greater than 3% of set pressure. As per API 520, if the inlet pressure drop exceeds 3%, the potential for capacity reduction and/or mechanical damage may exist for the relief valve. Note that inlet pressure drop concern was not identified for the relief devices which were undersized for one or more overpressure scenarios.

Outlet Pressure Drop: The outlet pressure drop is calculated based on the capacity of the relief device. This includes all the scenarios for which the outlet pressure drop is greater than 10% of set pressure for single conventional valves, or greater than 16% of set pressure for multiple conventional valves except for fire case, greater than 21% of set pressure for conventional valves for external fire, and greater than 50% of set pressure for bellows valves. As per API 520, if the outlet pressure drop exceeds the limit, the potential for capacity reduction and/or mechanical damage may exist for the relief valve. Note that outlet pressure drop concern was not identified for the relief devices which were undersized for one or more overpressure scenarios.

High Exit Velocity: The exit velocity and sound power level of the relief valve piping can affect its stability. All overpressure contingency with Mach number greater than 0.8 and sound power level greater than 155 was documented. Note that the high mach number concern was not identified for the relief devices which were undersized for one or more overpressure scenarios.

High set pressure: The relief device with set pressure greater than maximum allowable working pressure (MAWP) of the equipment was identified since the relief valve can no longer protect the system if this condition exists.

Isolation valve: This includes the block valves that are not chain locked open as required (the most common ones were the block valve on relief device inlet and outlet piping).

Pocketed Piping: In the event of liquid release and if the relief device is pocketed, the liquid can accumulate in the piping and hence the stability of the piping can get affected; hence it is very important to correct this concern.

Restricted line: In the event the inlet segment of the relief device piping is greater than the inlet nozzle as per API 521, these segments needs to be identified and documented.

The breakdown of the number and percentage of the installation concerns that were analyzed in this study are shown in Table 7 and Figure 8 respectively. Of a total of 807 installation concerns that were identified in this case study, the majority was the high outlet pressure drop concern.

Table 7: Percentage of installation concerns by type

Installation Concern	Percentage
High inlet Pressure drop	17.2%
High Outlet pressure drop	24.8%
High Exit Velocity	5.9%
Pocketed Piping	5.3%
Restricted Line	1.7%
Isolation Valves	11.8%
High Set Pressure	0.9%
No Weep Hole	20.8%
Liquid to Atmosphere	1.5%
Valve Stem Rotation	4.0%
Other	6.1%

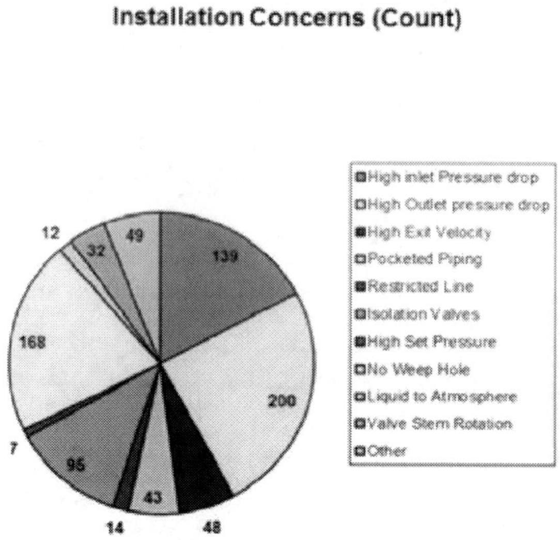

Figure 8: Actual number of installation by type for the twenty compressor stations

Unprotected Systems

These concerns were identified only for those systems that had applicable scenarios without any relief device protecting them. Contingency analysis for all the systems which didn't have any relief device protecting them was performed. As the case study was conducted on system based analysis, it was easier to identify these systems. The systems which had an internal relief device were considered to be unprotected. There were few cases in which there was a block valve present between the system and the relief device; hence these systems were considered unprotected. This list also includes the overpressure contingencies which resulted in over temperature of vessel in the event of external fire. Of a total of 3278 concerns identified, 311 were categorized as unprotected concerns.

Data Requirement

All the assumptions which were made during the analysis were documented under this category. During the analysis, a data need log was created if a proper supporting document was not provided and any follow on assumptions made due to the lack of documentation was approved by the client. Some general assumptions which were made were liquid levels, MAWP of the system (usually was assumed to be same as relief device set pressure), and relief valve information. Of a total of 3278 concerns identified, 1639 was data requirement concerns.

Documentation Updates

During the analysis, some discrepancies were found between different documents such as the U1A-form and the P&ID and these discrepancies were documented, with the correct information needing to be updated on the relevant documents. Documentation updates are usually a low priority concern as they may not impact the operation and safety of plant significantly. Of a total of 3278 concerns identified, 179 were documentation updated.

3.6. Mitigation of Concerns

Once the concerns were identified and documented, different recommendations were provided in order to mitigate the identified concerns. Usually, the clients choose to implement mitigation strategies predominantly for the higher priority concerns such as the unprotected concerns, installation and undersized concerns. An example list (not comprehensive) of the recommended mitigation strategies that were provided to the client for undersized, installation concerns and unprotected systems is detailed below.

Mitigation Strategies for Undersized Concerns

The provided recommendation to resolve undersized concerns depends on the overpressure contingency for which the relief device was undersized.

External Fire: In the event the relief device is undersized for external fire, it was checked if the vessel is insulated and whether credit of insulation could be taken. Usually, in order to overcome this concern, it was recommended that a bigger relief device be installed.

Blocked Outlet: In the event the relief device is undersized for blocked outlet, it was recommended that a bigger relief device be installed.

Isolated with Heat input: In the event the relief device is undersized for "isolated with heat input", it was recommended that a bigger relief device be installed. An alternate recommendation was to see if the undersized concern could be mitigated using administrative controls.

Tank Draining (Vacuum): This was a major concern identified in low pressure tanks. This overpressure scenario was missed in previous studies and hence, a lot of undersized concerns were found. It was recommended that a restriction orifice be installed on the drain line. Another option recommended was to install a bigger PVRV.

Failure of automatic control: For this undersized scenario, it was recommended that a restriction orifice be installed on the line, while still ensuring that it can allow the normal flow through it; however it can regulate the maximum flow such that the relief device is adequately sized. Another option recommended was that a restriction orifice be installed so that the maximum C_v of the control valve can be reduced, which ensures that the relief valve is adequately sized. It was rarely suggested that a bigger relief device be installed for this case.

Inadvertent valve operation: These cases usually included the inadvertent opening of bypass valves around the control valve. For these cases, usually the pipe capacity calculation was reperformed based on the equivalent length of piping from source to destination to check if the concern still existed. If the concern still exists, it was recommended that the bypass valve be chain lock closed or a restriction orifice be installed.

Gas breakthrough: Majority of the time, the failure of automatic controls and inadvertent valve operations resulted in gas breakthrough. For these cases, usually the pipe capacity calculation was reperformed based on the equivalent length of piping from source to destination to ensure if the concern still existed. If the concern still exists, it was recommended that the bypass valve be chain lock closed or a restriction orifice be installed.

Reverse flow: This is one of the most common overpressure contingencies that can be missed while doing the analysis. That is one the reason why the relief device was undersized for this case. In order to overcome this concern, it was recommended that multiple check valves be installed in the direction of flow. An alternate option recommended was the installation of a bigger relief device.

Mitigation Strategies for Installation Concerns

Inlet & Outlet pressure drop: For pressure drop concerns, it was recommended that the concerned piping be modified. The equivalent length of the piping was provided such that it was within the prescribed limit. Alternately, if the relief device was oversized, it was recommended that a smaller relief valve be installed. In this scenario, it is important to still ensure that the relief device is still adequately sized. For inlet pressure drop, it was recommended that a pilot valve with a remote sensing line be installed. For outlet pressure drop, it was recommended that bellows or a pilot valve be installed.

Mitigation Strategies for Unprotected System

If the system is unprotected, the most common recommendation was to install a new relief device. An alternate option recommended was to chain lock open any open block valve to take credit for other relief device. The third option recommended was to mitigate the scenario with administrative controls. This option was common for heat exchangers which have only thermal expansion as the applicable overpressure contingency.

4. Summary

This case study provides an overview of the relief analysis conducted for multiple compressor stations. As stated previously, the relief analysis was system based as opposed to relief device based so as to identify any unprotected systems that may need overpressure protection. It is very important to use the correct operating conditions and fluid composition to perform the relief analysis. The study provided an overview of the various types of overpressure scenarios identified for these systems, with a statistical breakdown of the scenarios by type. It was found that external fire and blocked outlet were the most commonly identified overpressure scenarios. In addition, this case study provided a breakdown of the types of relief devices, a further breakdown of types of relief valves, and the different types of equipment in the compressor stations. A breakdown of the type of concerns identified during the compressor station revalidation was provided along with a general idea of the mitigations that can be implemented to resolve the concerns. When performing relief analysis for compressor stations, it is very important to look out for any contingencies previously not considered, possibly due to changes in facility design or changes in the guidelines.

440451 An Approach to Reducing Risk during Online Isolation, Removal and Maintenance of Relief Devices Using Engineering Assessment

Abstract:

Overpressure protection must be provided for pressurized equipment by means of pressure relief devices, which sometimes require isolation and removal for inspection or repair. It would be advantageous to have a means of isolation without shutting down the process, especially in light of the additional risks associated with shutdown and startup; however, doing so may put the pressurized system at risk of being unprotected in the event of overpressure. Current recognized practices of providing dedicated personnel for monitoring and responding to an overpressure event, or installation of an independent spare, are non-ideal in many instances. An engineering assessment can identify specific ways for providing temporary overpressure protection, including elimination of root causes of overpressure events, provision of alternative pathways to overpressure protection, modification of process operations, and/or establishment of a means of equivalent or better risk reduction for the duration of the isolation. Guidelines and examples are provided to illustrate the analysis, along with recommendations for documentation.

441392 Common "Design Flaws" Identified during Pressure Relief System Analysis

Abstract:

Common "design flaws" identified during a pressure relief system analysis.

It is the goal of a pressure relief system analysis to identify possible gaps in a pressure relief system. In many oil and gas refining and petrochemical plants, it's common to encounter certain "design flaws" which can lead to overpressure scenarios where the system is not adequately protected.

Within these design flaws there are a few which frequently occur, regardless of the process. These can be just simple mistakes or related to consideration not being given to the whole system as part of the design. This paper will focus on five common design flaws:" unusually large bypass valves around a (smaller) control valve"; "liquid head between relief device and equipment being protected"; "how does the suction of the feed pump behaves during the scenario", "tube rupture can affect an entire system and not just the exchanger"; and "common API 650 tank concerns". For each "design flaw" a few common mitigations will also be discussed.

The goal of this presentation is not to discuss the issues in detail, but rather just bring to the industry's attention a few concerns being frequently identified.

Considering Overfilling to be a Non-Applicable Overpressure Scenario

Natalie Doe
Smith & Burgess
190 S. LaSalle St., Ste 550, Chicago, IL 60657
Natalie.doe@smithburgess.com

Dustin Smith, P.E.
Smith & Burgess
7600 W. Tidwell, Ste 600, Houston, TX 77040
Dustin.smith@smithburgess.com

Prepared for Presentation at
American Institute of Chemical Engineers
2016 Spring Meeting
12th Global Congress on Process Safety
Houston, Texas
April 11-23, 2016

Considering Overfilling to be a Non-Applicable Overpressure Scenario

Natalie Doe
Smith & Burgess
190 S. LaSalle St., Ste 550, Chicago, IL 60657
Natalie.doe@smithburgess.com

Dustin Smith, P.E.
Smith & Burgess
7600 W. Tidwell, Ste 600, Houston, TX 77040
Dustin.smith@smithburgess.com

Keywords: PSV, Risk mitigation, liquid overfilling, Safety Instrumented Systems

Abstract

The consequences associated with liquid overfilling are serious for plant safety, as illustrated by many recent incidents (2009, Bayamón, Puerto Rico; 2008, Chesapeake, VA; 2005 Hertfordshire, England; 2005 Texas City, Texas). In many cases, providing overpressure protection solely with pressure relief devices may not protect the facility. Preventing the uncontrolled release of liquid may be the only viable safe option. This paper will present an abbreviated analysis of several incidents that could have been avoided with proper preventive measures. Engineering methods will be presented to eliminate the need to release liquids from a closed system. For instance, the system can be designed such that upstream pressure sources are limited below the allowable downstream pressure. A designer can provide a safety instrumented system with an appropriate Safety Integrity Level (SIL) rating to reduce or eliminate the credibility of overfilling. Alternatively, consideration may be taken for alarms and operator response, with consideration for RAGAGEP, to reduce the credibility of overfilling to an acceptable risk. Systems can be redesigned to be inherently safer by minimizing contained liquids. This paper then gives a few examples of the use of system design, shutdown systems, alarms/operator response combination and inventory controls to eliminate or reduce the potential for liquid release scenarios.

1 Introduction

The consequences of an overfilling overpressure scenario can result in increased risk to the facility when mitigated with pressure relief devices. This could occur due to many different root causes. In order for liquid overfilling to require overpressure protection from a relief device, the following must all be true:

1. The source pressure must be capable of exceeding the system MAWP. Since the relief device is supposed to be set at or below the system MAWP, if the relief device is set lower than the MAWP, a liquid release may occur at a lower pressure.
2. The flow of liquid through the system can be stopped by closing a valve on the system outlet or the loss of a discharge pump (etc). In general, system designers ensure that liquid flows can be stopped, so this is usually a given.
3. There are no acceptable administrative or instrumented controls in place to prevent the scenario.

When reviewing the consequences of an overpressure scenario, the designer must check to ensure that the overfilling event may also not occur during a larger plant upset (power failure, instrument air failure, or other global utility disruption). This would add a simultaneous load to the disposal system, which should be included in the disposal system design.

Overpressure protection

The system designer should consider the root cause of the overfilling scenario to determine how to protect the system or eliminate the scenario. Relief devices are often used to prevent excessive vessel overpressure due to liquid overfilling. For many systems, overfilling is not expected to be the controlling case for relief device sizing; however, a review of incident history shows that many different systems have been overpressured by overfilling with liquid, resulting in relief. Therefore, when relief devices are used to provide protection for overfilling, it is important that proper consideration is given to the effluent handling system.

1. For a system with several pieces of protected equipment, consideration must be taken for the static head throughout the column to ensure equipment at grade is adequately protected. This is particularly an issue for low pressure tanks.
2. When systems overfill, there is the potential for a large amount of liquid to be released to the effluent handling systems. Unless liquid holding drums are sized for the potential liquid rate, there could potentially be an unplanned liquid release with significant consequences.

This paper will outline recent incidents and their consequences due to overfilling, and will overview methods to protect the vessels from overpressure and the system from the consequences of large liquid releases.

2 Overview of recent incidents

The following incidents from the past decade illustrate the consequences of overpressure due to overfilling and the handling of liquid releases.

2009, Bayamón, Puerto Rico;

Per the CSB report, "Tank 409 overflowed with gasoline, resulting in a vapor cloud that encompassed 107 acres of the CAPECO tank farm." [1] The report also states that "Tank 409 overflowed for an estimated 26 minutes before the vapor cloud ignited. The CSB determined nearly 200,000 gallons of gasoline… rushed out of six vents in the tank." These six vents responded to the upset conditions to relieve the excess pressure in the tank; thus, the overpressure protection provided was sufficient to prevent a catastrophic vessel failure. The action of the "engineered" pressure relief devices for this scenario worked and then became the initiating event for this incident. The final result was that the company went bankrupt and 17 of 48 tanks were burned.

2008, Chesapeake, VA;

Per the CSB report, "At about 2:20 pm, as the tank reached a level of 26.72 feet, a vertical split started midway up the shell and rapidly extended to the floor and roof of the tank on the side opposite the workers." [2] In this case overpressure protection to prevent catastrophic failure of the tank would not have worked because the tank was not "full". This incident could not have been protected for by pressure relief devices - only instrumentation, procedures or the like.

2005 Hertfordshire, England;

Per the aforementioned CSB report, "Similar to the CAPECO incident, the vapor cloud explosion and multiple tank fires occurred after a tank was overfilled with gasoline." [1] Thus another incident of overpressure protection for the overfilling scenario working; as such, it operated as designed. The "engineered" overpressure mitigating system was the initiating event for an incident that caused more than a billion dollars' worth of damage to the surrounding commercial and residential property and injured 43 people.

2005 Texas City, Texas

Per the CSB report, "The incident occurred during the startup of an isomerization (ISOM) unit when a raffinate splitter tower was overfilled; pressure relief devices opened, resulting in a flammable liquid geyser from a blowdown stack that was not equipped with a flare. The release of flammables led to an explosion and fire." [3] The pressure relief devices opened, and prevented excessive overpressure of the ISOM tower. As such, the engineered pressure relief devices in this case also worked, preventing a catastrophic overpressure-related loss of containment, but the design for the disposal of the fluid created an additional hazard with significant consequences.

Many of these incidents involved tanks, but that is inconsequential as in each and every case either a relief device was insufficient to prevent overpressure, or the engineered design of the pressure relief device and disposal was the root cause for these scenarios. Not only do the pressure relief devices have to be designed to prevent loss of containment from overpressure, the disposal system has to be designed to ensure the effluent has a safe location to be stored or is adequately destroyed. This can be particularly difficult for liquid releases.

3 Means of overpressure protection

Relief Devices and Disposal

While relief devices are an acceptable method to protect from overpressure due to liquid overfilling, complete risk mitigation requires proper consideration of the ability of the disposal system to handle the liquid effluent. As mentioned previously, additional pressure due to static head should be considered when protecting multiple pieces of equipment. For column systems with relief devices located on the top of the column, this can have significant impact on associated equipment located at grade. It is common practice to perform overfilling relief device calculations at 10% overpressure rather than 16% overpressure when multiple devices are installed, in an attempt to account for the static head. For tall column systems, or for systems with a heavy relief fluid, this may not be sufficient to ensure the allowable accumulation on the equipment located at grade is not exceeded. The relief device set pressure may need to be lowered below the system limiting MAWP to ensure adequate protection for all equipment.

Additionally, the disposal system must be capable of handling the liquid flow from the relief devices. The disposal piping should be evaluated to determine if it can handle the weight of the liquid. A knockout drum must be incorporated as part of the disposal system to ensure liquid is not sent to the flare or vent stack. For large liquid releases, it may not be practical to size one knockout drum to obtain the required particle separation. Multiple knockout drums in parallel can be used to ensure adequate particle separation for these cases. Failure to size the knockout drum(s) correctly can lead to liquid being sent to a system which cannot handle it, which can have serious consequences. The CSB report which reviewed the Texas City incident [3] unduly criticized atmospheric blowdown drums; many of their preferred disposal systems, i.e. flare systems, have had similar incidents [4] with the authors aware of countless near misses from flare systems.

Overpressure protection by system design

The need to design a disposal system for the liquid effluent from relief devices that protect a system of equipment from overpressure by overfilling can be eliminated by designing the system such that relief devices are not needed. The acceptable way to do this is to ensure that the overfilling scenario is not credible. For vessels designed to ASME VIII, UG-140 (formerly known as code case 2211) states "A pressure vessel does not require a pressure relief device if the pressure is self-limiting… and this pressure is less than or equal to the MAWP of the vessel at the coincident temperature…" [5]. This can be achieved in many ways.

Equipment Design

A designer can utilize equipment design to ensure that the disposal system does not need to accommodate effluent from relief devices installed for liquid overpressure protection.

Example 1 – Pump Selection

Instead of feeding a system with a positive displacement pump, a centrifugal pump may be used that is not capable of exceeding the downstream system MAWPs. This is illustrated in Figure 1 below, as P-1 cannot exceed the MAWP of V-1. The defining quality of centrifugal pumps is that their flow will decrease with increased head. Therefore, as the pressure in the downstream system increases, the capacity of a centrifugal pump will decrease. Thus, pump selection can be an effective method of limiting the overfilling rate.

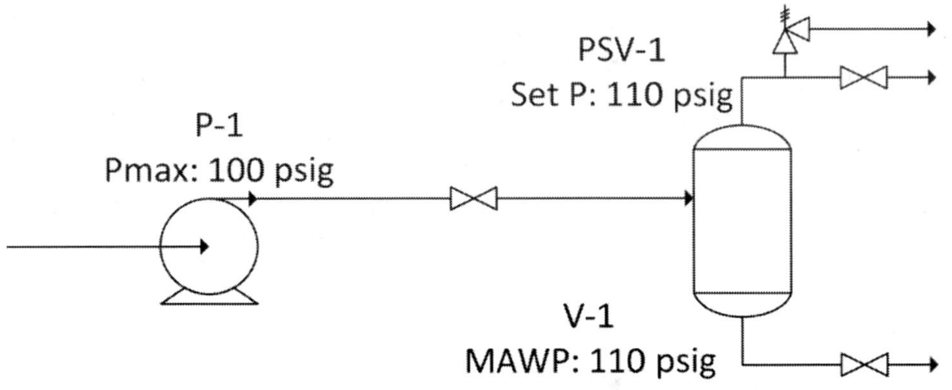

Figure 1: Upstream sources incapable of overpressuring the system

In contrast, positive displacement pumps are constant volume machines that are capable of maintaining high volumetric flow even at increased pressures. Therefore, alternative methods can be applied to protect from overpressure due to positive displacement pumps.

<u>Example 2 – PRV Discharge to Process</u>

If a relief device is installed on the pump instead of the vessel, the discharge piping can be routed to the pump suction. This allows for the liquid relief to be sent to process, rather than to a disposal system. However, the designer is cautioned that this will stop the forward flow of liquid so equipment upstream must be checked for this upset condition. An example is shown in Figure 2 below, as PSV-2A reduces the maximum feed pressure and eliminates the overfilling scenario on V-2.

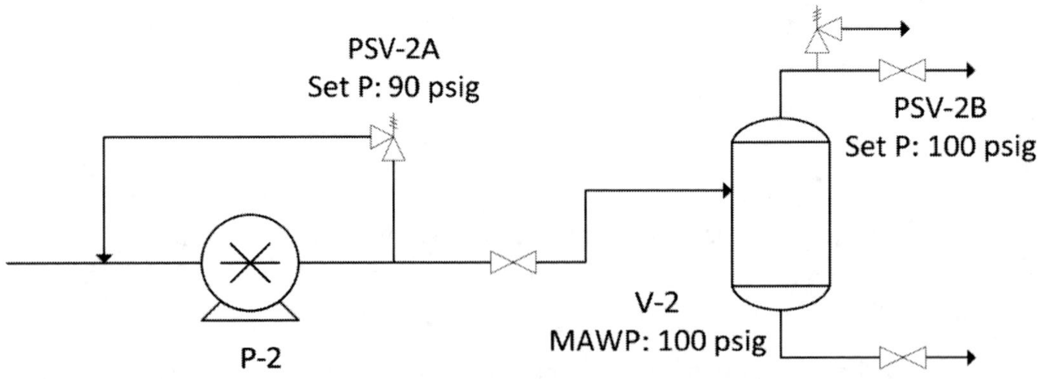

Figure 2: Protecting upstream pressure sources

In other cases, protecting for overfilling by pump selection or upstream protection may not be possible. Inventory controls may be utilized in these situations to ensure overfilling is not credible.

Example 3 - Inventory Controls

If possible, a designer can minimize the liquid inventory in a system. This is also a tactic in inherently safer process design. If the total liquid inventory of a system is insufficient to fill the vessel(s) vapor space then overfilling cannot occur. In this case overpressure does not occur irrespective of the maximum upstream pressure. An example of this is illustrated in Figure 3. The positive displacement pump (P-3) cannot overpressure the vessel (V-3) because the liquid inventory in the upstream tank (T-3) is less than the vessels vapor space.

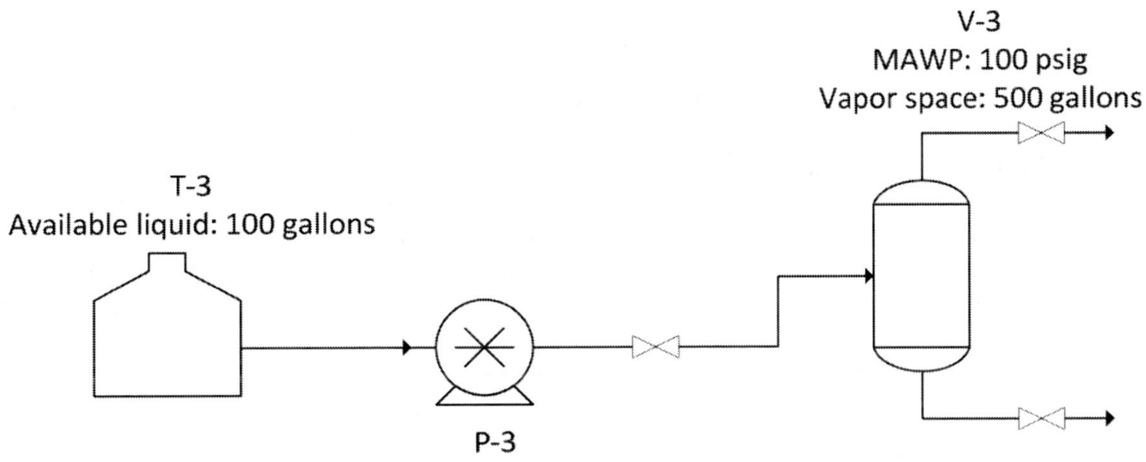

Figure 3: Inventory Controls

Instrumented Solutions

For installations where it is not practical to remove the credibility of the overfilling scenario by equipment design, safety instrumented systems (SIS) can be implemented. There are two ways to protect a facility from the effects of liquid effluent: remove the relief device entirely, or mitigate the consequences of the liquid effluent escaping from the disposal system.

Example 4 – Replace Relief Devices with instrumentation

The system can be designed such that instrumentation is used instead of overpressure protection. A designer can find guidance in API STD 521 which states in Annex E that "High Integrity Protection Systems (HIPS) are designed to avoid overpressure incidents by removing the source of overpressure or by reducing the probability of an overpressure contingency to such a low level that it is no longer considered to be a credible case."[6] ASME VIII UG-140 allows for the use of these systems, provided that detailed description of the safety system is documented, and includes "the identification of all truly independent redundancies and a reliability evaluation... of the overall safety system" [5] The ASME requirements for these systems are further explored elsewhere [7]. Figure 4 below shows an example of a system protected by instrumentation. Three separate level taps are used to determine the liquid level in V-4. Two independent emergency shutdown devices (ESD), each with an independent controller, are used to shut off the feed on high liquid level. The redundancies expressed in this example would be integral to determining the safety integrity level (SIL) of this system.

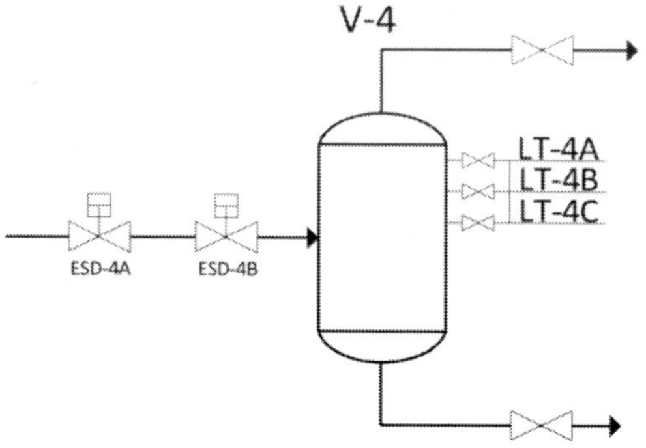

Figure 4: Protection through instrumentation

Example 5 – Add a layer of Protection

Alternatively, the design of a SIS does not require that overfilling is mitigated to a point at which it cannot occur; rather, that the probability is low enough that the total system risk is at an acceptable level. Depending on the existing system, the most cost effective solution may be to mitigate the effluent system release with a SIS but rely on a the relief device to protect the equipment. Due to the uniqueness of each system a risk assessment (such as a LOPA) will be required to determine the reliability of the SIS. The designer may choose to put in a less reliable safety instrumented system in combination with an existing relief device, or the safety instrumented system may replace the relief device. As mentioned above, these redundancies are used to determine the safety integrity level (SIL) of the system. For most cases, a SIL-3 System is needed to remove the credibility of the overfilling scenario. However, a SIL-1 or SIL-2 system may be acceptable when reviewed in a risk analysis [such as Layers of Protection Analysis (LOPA)] to mitigate an incident caused by a release of liquid to the disposal system. Either way, a detailed risk assessment should be performed to determine the SIL rating required to mitigate the hazard [8][1]. An example of a SIS that could provide overfill protection would be level indicators independent of the process control instrumentation from a separate vessel tap feeding a dedicated PLC that triggers a shutdown valve to shut off the feed to the system on a high level alarm.

Operator Response to mitigate overfilling

For many facilities, a combination of instrumentation with an operator response is a reasonable mitigation of risk from overpressure. Research has shown the following factors tend to determine the reliability of operators to take the intended action [9][10]:

1. Trigger – there is a clear indication to the operator that something is wrong triggering the need to take action. It is unreasonable to assume that the operators would have mitigated the overfill scenario at Bayamón PR as they had no indication of level or procedures to know when the tank overfilled [1].
2. Time – Per the work performed by the NRC [9], an individual in a stressful situation with limited time to perform a moderately complex task has a 100 times likelihood of failing compared to the exact same task with sufficient extra time allotted for execution. Others, [10] have put the reliability of an operator completing a clearly understood task on which he/she is trained, with adequate time under stressful conditions, is equivalent to a SIL-III system.
3. Consequences – if the mitigation action causes (or is perceived to cause) a chain reaction that is worse than the failure being mitigated there may be reluctance to perform the action. Additionally, if the operator assumes (or is told) that they will be punished for stopping production, then the likelihood of following the procedure decreases greatly.
4. Training – without the knowledge and training to act, the probability that an action will be taken diminishes greatly, especially if the action is not obvious to the individual operating the facility.

As safety instrumentation systems can be complex and expensive, many facilities choose instead to take credit for operator response to high level alarms. There are several important factors to consider when taking these credits. API 521 4.4.7.3 [6] outlines several factors to consider when relying on level alarms:

a) Whether the level instruments used for safeguards against overfilling are on separate process taps from the process control system
b) Whether level instruments used for safeguards against overfilling are susceptible to the same common mode device failures as those used for the basic process control system.
c) Whether the programmable level transmitter is set to show a low or zero level when the level exceeds the instrument range
d) Whether an open system differential pressure cell is used that can show a false low level due to accumulation of liquid in the upper impulse line if it is overfilled or collects condensed liquid
e) Whether density changes due to temperature effects or composition changes impact readings from differential pressure transmitters
f) Whether the instruments are proven in use for the specific process applications
g) Whether the range of at least one of the level measurement(s) can indicate a valid level reading of the full range between the high critical alarm point and any shutdown or interlock point
h) Whether operating characteristics of the level measurement during off-design, start-up, and shutdown operations are considered in display of level, physical properties, setting alarms, trip points, operator training, and operating procedures
i) Maintenance and testing frequency required for the instrumentation

Even if all of these factors are considered, and the reliability of the operator is evaluated, it is important to analyze the risk associated with human error, and the operator not responding in the allotted time. As with instrumentation systems, relying on level alarms and operator response is not a guarantee overfilling will not occur; rather, it lowers the risk to an acceptable level.

Overfilling of low pressure tanks present a unique challenge as compared to pressure vessels. API has created a specific document to provide further guidance to protect for the overfilling of low pressure storage tanks [11].

Example 6 – Alarms with subsequent operator response

As with Example 4 above, Figure 5 shows a vessel (V-6) with three independent level taps. These level taps are connected to two independent high level alarms (HLA). The designer performed a LOPA, and determined the operator can reliably address the alarm in 30 minutes. This is less than the fill time of V-6 from the time when the alarm sounds (1 hour), so credit may be taken for the operator actuating HV-6A and HV-6B to shut off the feed and prevent overfilling.

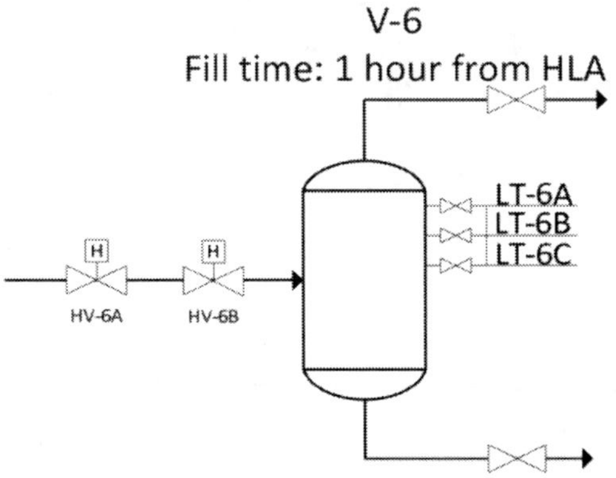

Figure 5: Operator Response

4 Conclusion

Liquid overfill can have serious safety and financial consequences, as seen in the recent incidents outlined in this paper. It is important for the designer to consider both the means to prevent overfilling, as well as methods to respond to an overfilling event. Appropriately designed relief devices and disposal systems are an acceptable method to react to an overfilling event. However, if this is to be the sole means of protection, careful consideration must be given to the large liquid effluent rate sent to the disposal system. Alternatively, the event may be prevented through the proper application of equipment design, instrumented solutions, alarms and operator response, or some combination thereof.

There are a large number of incidents involving overfilling that occurred despite the fact that the installed overfill protection reacted as designed. The protecting relief devices opened and were adequate to prevent overpressure of the overfilled vessel, but the disposal system was not designed for that case. The incidents mentioned in this paper resulted in billions of dollars' worth of damage, and even the bankruptcy of one company. When protecting for overfilling, it is imperative to consider the ramifications of allowing large liquid releases to occur. Overfilling protection design that encompasses consideration of all of the options presented above can prevent an incident with catastrophic consequences.

5 References

[1] U.S. Chemical Safety and Hazard Investigation Board. "Caribbean Petroleum Refining Tank Explosion and Fire" Available at http://www.csb.gov/caribbean-petroleum-refining-tank-explosion-and-fire/. Accessed on 2/5/2016.

[2] U.S. Chemical Safety and Hazard Investigation Board. "Allied Terminals Fertilizer Tank Collapse" Available at http://www.csb.gov/allied-terminals-fertilizer-tank-collapse/. Accessed on 2/5/2016.

[3] U.S. Chemical Safety and Hazard Investigation Board. "BP America Refinery Explosion" Available at http://www.csb.gov/bp-america-refinery-explosion/. Accessed on 2/5/2016.

[4] HSE (Health and Safety Executive) (Eds.), 1997. The explosion and fires at the Texaco Refinery, Milford Haven, 24 July 1994. A report of the investigation by the Health and Safety Executive into the explosion and fires on the Pembroke Cracking Company Plant at the Texaco Refinery, Milford Haven on 24 July 1994, HSE Books, London.

[5] American Society of Mechanical Engineers, 2013 ASME Boiler & Pressure Vessel Code: Section VIII, Division I. s.l.

[6] API Standard 521, "Pressure-Relieving and Depressuring Systems," American Petroleum Institute, 6th ed., 2014.

[7] Smith, Dustin J. "Safety instrumented systems in lieu of pressure relief valves." *Process Safety Progress* 33.4 (2014): 345-349.

[8] A.E. Summers. "Overfill Protective Systems – Complex Problem, Simple Solution," 12th Annual Symposium, Mary Kay O'Connor Process Safety Center, October 2009.

[9] Blackman, Harold S., David I. Gertman, and Ronald L. Boring. "Human error quantification using performance shaping factors in the SPAR-H method." *Proceedings of the Human Factors and Ergonomics Society Annual Meeting*. Vol. 52. No. 21. SAGE Publications, 2008.

[10] Mannan, Sam, ed. Lees' Loss Prevention in the Process Industries: Hazard Identification, Assessment and Control. Vol. 1. Butterworth-Heinemann, 2004.

[11] API Standard 2350, "Overfill Protection for Storage Tanks in Petroleum Facilities," American Petroleum Institute, 4th ed., 2012.

How Reliable Are Your Process Safety Barriers?

Mike Neill
Petrotechnics USA Inc.
10497 Town & Country Way, Suite 700, Houston, TX 77024
mike.neill@petrotechnics.com

Prepared for Presentation at
American Institute of Chemical Engineers
2016 Spring Meeting
12th Global Congress on Process Safety
Houston, Texas
April 11-23, 2016

How Reliable Are Your Process Safety Barriers?

Mike Neill
Petrotechnics USA Inc.
10497 Town & Country Way, Suite 700, Houston, TX 77024
mike.neill@petrotechnics.com

Keywords: Process Safety Management; Process Safety Barriers; Operational Risk Management

Abstract

Using Process Safety Management techniques, we design and implement multiple layers of protection for our hazardous plants. For these barriers to remain effective over the life of the asset is a challenge, particularly with aging assets. We know barriers deteriorate over time without proper, preventative maintenance and inspection. Often a lack of focus on critical barriers can result in activities being deferred or cancelled without recognizing the potential consequence.

There is generally a lack of understanding in our organizations of process safety techniques and approaches. While the criticality of these barriers maybe well-understood in the Process Safety department, this is unlikely true through other organizations, including Operations and Maintenance.

With so much equipment being tagged "safety critical" on our facilities; it is often difficult to ensure safety barriers receive the focus they need to ensure their effectiveness at protecting against high consequences events when needed. Operators need to manage the health of their barriers in near real-time and mitigate the impact arising from any impairments.

By understanding dynamic aspects of cumulative risk, all members of an organization can easily identify barrier impairments. This level of visibility helps make better operational activity management decisions on account of operational activities' potential impact on Major Accident Hazard barriers.

This paper will examine the traditional approach to barrier design and assumptions on reliability, the approach to maintenance, and the reality of operational dynamics. It will introduce the concept of cumulative risk impacts on safety barriers and examine operational dynamics which impact barrier performance.

1 Introduction

Confucius, the Chinese Philosopher from around 500BC is quoted as saying that "a man who lacks reliability is utterly useless." Well the same can be said for our Process Safety barriers, whose sole purpose is to prevent major accidents or mitigate the consequences of such events. These barriers enable us to operate our hazardous facilities, exposing people to potential danger, since we satisfy ourselves that the calculated risk of incident is acceptable, assuming reliable performance. Unfortunately, when we investigate incidents, we see again and again examples where multiple barriers fail to perform. Occasionally we see evidence where the design was at fault and the assumptions made were incorrect, and we see plenty of evidence where Management Systems designed to prevent barrier failure, fail themselves.

All too often the barriers do not work as intended for a variety of reasons ranging from corroded pipe, seized valves, inoperable switches, blocked flowlines to a simple case of responsible people failing to follow operating procedures.

We focus a lot of attention on making sure we design our barriers properly in the first place but do we do enough during the life of the facility to make sure the barriers continually perform per design? Are our organizations set up to optimally manage these barriers and are our frontline workers engaged sufficiently and armed with the right knowledge to play their role in identifying barrier impairments and to make appropriate operating decisions? These are questions this paper will address and will conclude with examining new solutions which are becoming available to help the industry with this challenge.

2 Evidence of Barrier Failure

The failure of barriers designed to protect or mitigate the consequences of Major Accident Hazards (MAH) often results in an incident such as a significant gas release or an over pressure situation. The failure of multiple barriers can escalate into a major accident such as an ignition barrier failure coincident with a gas release which results in an explosion or fire. The term "barrier", refers not just to equipment hardware such as piping, vessels, instruments, actuated valves, etc., but also includes procedural systems, where plant staff follow agreed practices to operate the plant and respond to emergencies.

The latest Marsh and McLennan report (ref1) provides a current update to property damage losses in the hydrocarbon industry. This report, in its 23rd edition, outlines the 100 largest losses across our industry over the past 40 years. A direct quote from that report

> *"Accidents that result in major losses, such as those reported in this document, generally occur because of the failure of a number of barriers within the process-safety management system — all occurring at the same time; typically, none of these losses are the result of the failure of a single barrier or protection measure."*

Furthermore, in the same report, other quotes reflecting on the incidents,.

"N*one of the losses listed in this document should be considered 'black swan' events.*"

("Black Swan" events are those regarded as inconceivable and impossible to consider as a credible threat.)

The implication of this statement is that the scenarios and failure modes were anticipated so either the design of the barriers was inadequate to meet conditions or the design of the barrier was adequate, but its performance did not match expectations. It could also be a combination of both.

In another report, (ref2) Ron Jarvis from Swiss Re looked at a sample of 45 incidents in the Oil, Gas, and Petrochemical fields over the prior 15 years. Despite this being a random sample and not necessarily a statistical analysis, it is worthy of consideration, not least because its aim was to prioritize those risk control elements whose failure resulted in a majority of incidents.

The study concluded 52% of primary incident cause was failure due to corrosion, 27% failure of operating practices and procedures, and 14% was attributed to unsafe maintenance, equipment isolation and permit to work management. Considering secondary and tertiary causes, poor design and/or inadequate hazard identification is another cause which becomes more prominent. The author goes on to identify inadequate piping and equipment inspection programs resulted in 68% of corrosion failures, the remainder being due to incorrect material selection or defective materials when first installed. Regarding the failure of operating practices and procedures, Jarvis highlights poor management of deliberate by-passing of Safety Critical Equipment (SCE) and the management of abnormal situation on the plant. 57% of all incidents occurred during abnormal conditions on the plant, including start-up, shutdown, or maintenance.

The evidence from these two studies is conclusive that Major Accidents occur as the result of multiple barrier failures. You may ask why it is important to recognize this seemingly obvious fact! This Author believes it is having attended many Process Safety Conferences over the years where the emphasis is on barrier design and static risk scenarios with little focus on what happens to barriers over the life of the asset. This, and working in an industry where there seems only to be a tacit recognition of a dynamic component to MAH risk in which a 5 year review of safety management systems seems adequate. The fact is, barriers fail and over time most, if not all, will fail without periodic inspection, testing, and preventative interventions.

The challenge we face is the data on Major Hazard Barriers, which can indicate an impaired status or diminished integrity, is not easily shared with the people who need to know and who routinely make decisions on daily operations based on assumptions that everything is in good health.

3 Barrier Design Versus Barrier Performance

3.1 Design Principles

When designing Oil, Gas, and Petrochemical plants, many of the principles used are determined by regulatory requirements in addition to generally accepted engineering design standards. Wall thicknesses and material specifications for pipes and vessels will be based on the expected fluid properties and operating conditions as will the specifications for valves and rotating equipment selection. HAZOP and HAZID studies look at potential hazard scenarios and failure mechanisms and identify methods for minimizing risk and/or mitigating the outcome. Risk-based studies look at ways to reduce MAH risk to a level which meets the criteria of acceptance. When the risk levels are excessive, additional barriers are designed.

Unfortunately, barriers themselves are not 100% reliable. This can be the result of a material defect in piping or vessel walls, or corrosion of those same walls, a stuck relief valve, a deluge system plugged with scale, or a standard operating procedure which is not followed. A single safety barrier cannot eliminate risk but it can reduce the likelihood of an event happening or mitigate the magnitude of the consequences if it does occur. As such, the traditional design approach is to apply multiple barriers, some preventative and some mitigative. Together they reduce the overall consequential risk.

Modern techniques of Inherently Safe Design, pioneered by Trevor Kletz, seek to build plants where hazards are removed rather than controlled and where volumes of hazardous fluids are minimized to reduce the magnitude of consequences. This is paving a way towards safer plants. At present, the fundamental needs of Oil, Gas, and Petrochemical plants continue to demand processing flammable, explosive, and toxic fluids often at high pressures and high temperatures and in large volumes. As such, the approach of multiple barriers to reduce and mitigate risk is still essential for plants - old and new.

When a hazardous scenario has been identified, the risk of the consequences occurring is calculated using statistical and Quantitative Risk Analysis (QRA) techniques. Risk acceptance criteria is often prescribed or advised by the regulator or set by internal corporate policy for an Operator. For example, the threshold may be a broadly acceptable risk of death of 1×10^{-6}. If a scenario results in the calculated risk of this outcome being higher than the threshold, additional barriers need to be designed to reduce the risk level to one which is acceptable.

Each barrier is credited with a level of risk reduction. This is calculated using performance characteristics of the barrier which, for hardware, is provided by the manufacturer based on empirical data such as "mean time between failure," etc. The assurance from the manufacturers of this performance is founded on a manufacturing quality system designed to provide reproducible results.

For vessels and piping the wall thickness and material specification are the principle drivers of strength and the ability to contain the process fluids. Once again, manufacturers' quality assurances assume manufacturing standards are followed. Mill certificates demonstrate material specification compliance. Hydrostatic testing of vessels and fabricated flowlines / pipelines, at pressures below the material yield points, are part of commissioning new plant and verifying a high level of integrity prior to start-up.

This process of risk reduction through designing additional barriers where necessary is carried out across the facility until the overall level of risk is reduced to an acceptable level.

Unlike hardware barriers, Procedural Barriers typically are less reliable since they require human interventions to be consistent and repetitive and with a discipline to follow standard procedures. Since most Operators are not robotic by nature, reliability is less than for engineered systems. On balance, this is a good thing since, unlike robots, Operators are free-thinking, unconfined by a narrow scope, and are able to identify a wide spectrum of developing situations and respond appropriately to regain control.

Reliability for procedural barriers is improved if the procedures themselves are well-written and workable so have been repeatedly tested and improved over time. The use of checklists provides a simple go/no-go approach, removing ambiguity in the step-by-step procedure. Training, and of course, experience are the other strong determinants of the reliability of procedural barriers.

3.2 Barrier Categories

The diagram below shows an example of fundamental barrier groupings. These are categories of barriers based on their function. While shown as the familiar "Swiss Cheese" model, the layout is similar to a "bow-tie" diagram where the left hand barrier groups are "preventative" and the right hand are "mitigative." It is useful to understand barriers in the context of multiple layers of protection and to appreciate that the term "multiple layers" does not necessarily indicate multiple redundancy.

As an example, if there is a gas leak on a plant than the failure in itself is a serious incident and has the potential for fire and explosion. Another group of barriers would be ignition control, preventing the escaping gas from igniting. Gas detection is another barrier and fire suppression systems a further barrier categorized under Protection. So each of the multiple layers maybe performing different roles to either prevent an event happening or to mitigate the resulting consequences or preventing escalation.

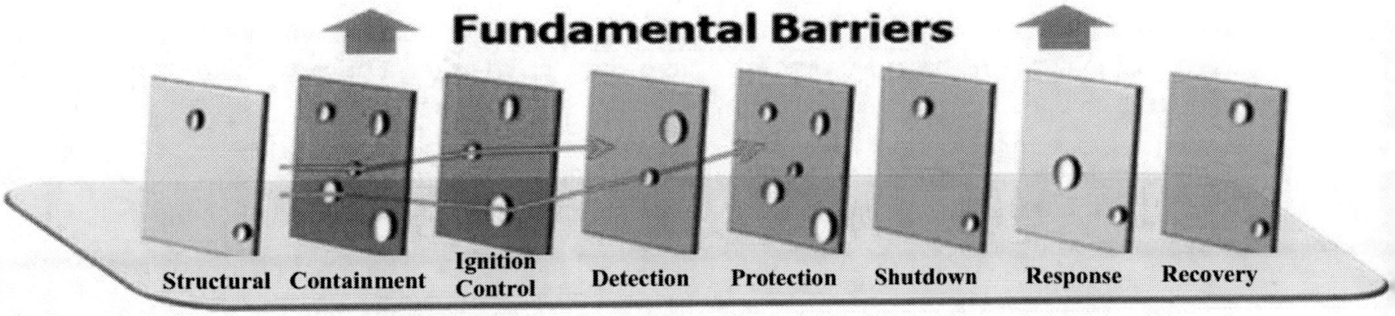

Figure 1. Fundamental Process Safety Barriers

Another example of categorizing is shown in the table below was proposed by Blacklaw, et al. (ref 3), detailing some of the specific barriers for offshore installations:

Table 1. Cumulative Risk Assessment Barrier Model

Barrier System	Example Components
Structural Integrity - (offshore structure)	Jacket, Topsides, Cranes
Process Containment	Vessels, Pipelines, Rotating Equipment, Pressure Systems, Relief Valves
Ignition Control	Ignition Prevention Systems, Flare Systems, Ventilation, Positive Pressure Systems, Gas Seals, etc.
Detection Systems	Fire & Gas Detection and Alarm Systems
Protection Systems	Fire Water Pumps, Deluge & Sprinkler Systems, Safe Refuge, Foam Systems, Blast Walls, etc.
Shutdown Systems	Pressure Trips and Shutdown Valves, ESD Systems, etc.
Emergency Response	Escape Routes, Emergency Lighting, Fire Water Mains & Portable Equipment, etc.
Life-Saving	Life Rafts, Personal Survival Equipment, etc.

3.3 Barrier Failure and Maintaining Barrier Performance

The reliability of engineered barriers will deteriorate over time, eventually to a level where they will just not work. This is simply a factor of material changes due to environmental actions such as corrosion or the rigors of service, erosion, wear, etc. To prevent or slow this deterioration, regular inspection, testing, and preventative maintenance is necessary.

As was explained in section 3.1, the reliability of engineered barrier performance determines the level of risk reduction for which credit is taken resulting in a residual risk which meets the acceptance criteria.

The data used for barrier reliability is based on manufacturer's information but assumes an ongoing schedule of inspection, maintenance, calibration and function testing to maintain that performance. What this means in practice is to take risk reduction credit for a shutdown valve, all the components of the system, from the pressure switch sending the signal to the controller, the actuator, valve stem, valve body and valve seat all have to be periodically inspected, tested, and undergo preventative maintenance. If these activities fail to occur within a specified timeframe, the_only consequence is the risk of major accident increases. Without these sustaining activities, dependability of the system declines, reducing or eliminating the ability to claim a risk reduction credit- thus the consequential risk has to increase.

This simple logic is important because there are often temptations to defer inspection and maintenance of such systems due to other operational priorities. Such deferrals are justified by arguments that it is preventative work and that there have not been reports of any failures…..so there is no reason to assume they will not still work!

The single most important set of barriers on a plant are those classed as "primary containment." Primary Containment includes pipework and vessels, plus a myriad of valves and flanges which connect them all together. It is not only the main flowlines but also includes by-pass lines, impulse lines for pressure switches, temperature gauges, sight glasses, etc. plus small bore pipework for instrument connections, bleed lines, chemical injection, etc.

The performance of these barriers is essential to keep the process fluids contained and under control. They are simpler than instrumented systems, not relying on moveable parts but instead on wall thickness and material specification plus the devices used to couple such as flanges, gaskets, bolt tensions, etc. Corrosion, both internal and external, are the most common failure mechanisms. Erosion, scale deposits, chemical damage and accidental damage will play their part. Inspection programs monitoring for losses in wall thickness are typically in place across the facility. These are supplemented with preventative programs such as protection coatings, corrosion inhibitors, anodic and cathodic protection, etc. Periodic hydraulic testing and recertification of vessels and pipework is also a standard requirement from most regulatory authorities and / or insurance companies.

3.4 Management Systems

Whether specified by the regulator or by internal policy, Management Systems are typically in place to manage the processes described in the sections above. As part of an organization's governance, the suitability of these Management Systems will be periodically reviewed, and their effectiveness tested by audit to ensure they are properly applied. Performance indicators provide a measure of the effectiveness of such systems with API RP 754 (ref. 5) being a standard most commonly used in the O&G & Petrochemical industry for process safety KPI's. In this standard, Tier 1 and 2 measures are known as "lagging indicators" based on historical events such as loss of containment. Tiers 3 and 4 are referred to as "leading indicators" since they are typically based on auditing of management systems in place to prevent or mitigate process safety events. In the UK, the Health & Safety Executive issued guidance, HSG 254, on developing process safety indicators (ref 6) a few years before API published their standard. Here they promoted the identification of MAH scenarios, barrier identification and identifying the "Risk Control System" or RCS which is managing the performance of that barrier. Leading and lagging indicators, similar to API 754 are also recommended in this guide.

While it is essential to the overall safe management of a facility that barrier / risk management systems are in place, both API 754 and HSG254, generate indicators which inform on the system alone and not on specific pieces of equipment or areas of the plant. Furthermore, what are called "leading indicators" are based on data collected months if not years in advance so may not represent what is current. What is clear is that they do not help day to day decisions on risk levels in specific areas of the plant or working on specific equipment.

4 The Reality of Operations and Dynamic Risk

4.1 Deferred Maintenance of Safety Critical Equipment

Maintenance Management Systems are programmed with time-based schedules for inspection and preventative maintenance of all equipment on a facility. Where this equipment forms one of our major accident hazard barriers, it is flagged as "safety critical." A rolling schedule is managed based on different time horizons. As equipment breaks down, repair work is added in based on priority. At two weeks or seven days, the schedule is typically "frozen" and handed over to Operations.

Executing the plan as designed is often a difficult challenge for a variety of reasons and maintenance backlogs, or maintenance activities beyond their planned completion dates, become common as activities slip on the schedule. This becomes more of a concern when the backlog contains safety critical equipment. In such instances, Operators need to justify why the maintenance is not being carried out and a "deferment" review process involving technical and engineering authorities should be conducted.

As we have seen earlier, when inspection and preventative maintenance of safety critical barriers is delayed, the inevitable outcome is that MAH risk levels increase.

Deferred safety critical maintenance is an unfortunate reality in the O&G & Petrochemical industry, especially with old facilities and infrastructure. The magnitude of this issue is difficult to judge since data on the topic is not easy to access as few Operators want to share data which shows them in a bad light; however, there is one exception.

The UK sector of the North Sea is a region where Operators have, for several years, voluntarily submitted data on the maintenance backlog for Safety Critical Equipment (SCE). Oil and Gas UK, the cross-industry body recently published this data in their annual Health & Safety Report in 2015 (ref 4). The diagram below shows how the growth in SCE backlog over the last 4-5 years. The report mentions the regulator has raised concerns the industry has increasingly deferred maintenance backlogs rather than addressing them.

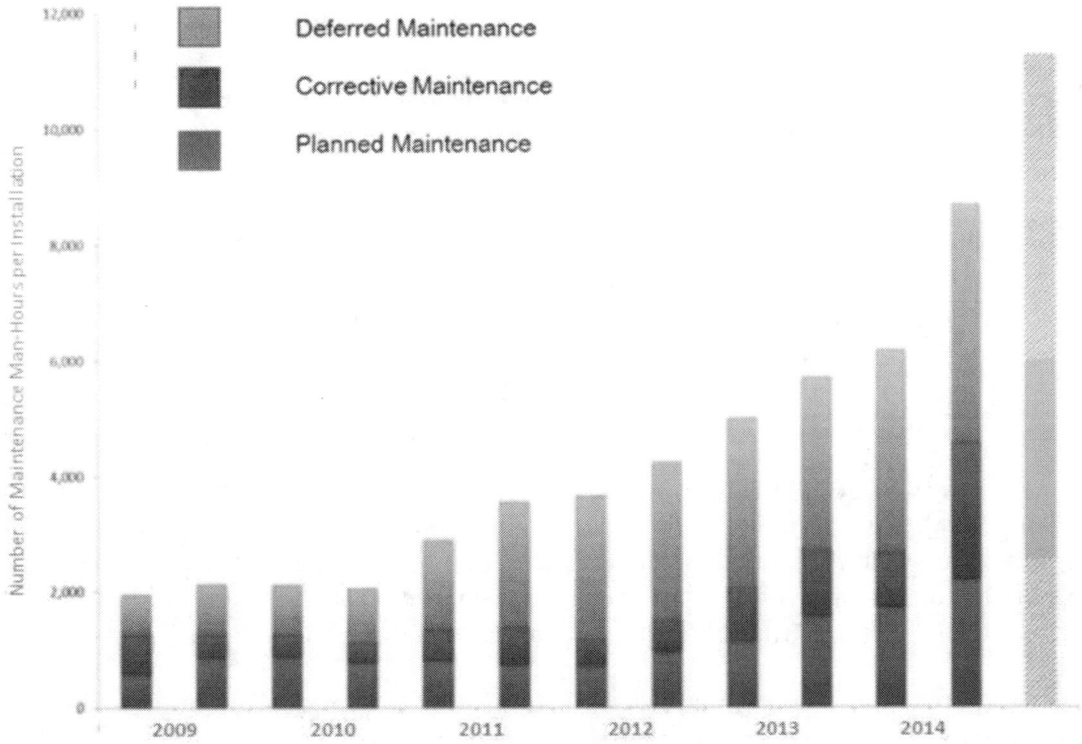

Figure 2. Oil & Gas Health & Safety Report 2015 (ref 4) – Growing Backlog of Safety Critical Maintenance

What this picture says is there are a number of barrier systems on the facility which, with the deferment of inspection and preventative maintenance, are knowingly "impaired. In other words, they are not providing the level of risk reduction that was intended by design. Add to this, the barrier systems which have been inspected but have failed to perform adequately and are therefore also in need of corrective action.

4.2 Deliberate defeat of barrier systems

Another category of compromised barriers must also be considered. Every day, many barriers on a facility are deliberately defeated. This is a necessary on a plant where ongoing maintenance and repair activities have to take place with the plant in a live state or condition. Detection systems, such as gas and fire detectors, have to be deliberately defeated to carry out work in a vicinity which could accidently trip them. Alarms and process trips have to be bypassed for work to be carried out. Ignition sources have to be brought onto the live plant in order for work to be carried out.

The terminology used to describe these situations include: "temporary defeats (of safety system), "instrument override", "interlock by-pass" but they amount to the same thing. A formal, managed process should be followed during this action, often accompanied with a risk assessment. This will often focus on the following critical parameters: **time** – the longer a defeat is in place the greater the risk; **reinstatement** – ensuring the temporary defeat does not get forgotten and becomes permanent; **communication** – ensuring all necessary parties are aware of the status.

Even our most important barrier, primary containment, has to be compromised fairly regularly in order to replace and repair equipment such as valves, pumps, filters, etc. Work involving "breaking containment" is, again, normally managed with a formal process involving energy isolation, draining and purging and gas testing.

All of these situations where safety barriers are deliberately defeated involve an additional level of risk. While the defeats maybe managed with formal process and additional barriers or controls maybe put in place, there will be a residual risk which needs to be considered along with the other sources of risk on the facility.

4.3 Operational Decisions and Corporate Silos

On a daily basis, plant Supervisors make the decision, typically informally and often unconsciously, on whether the plant is safe to operate. They also decide whether it is safe to carry out simultaneously activities - such as maintenance and repair work, project construction work and normal Operational activities. These decisions are often based on incomplete information, specifically with regard to the status of barriers on the plant.

Unfortunately it is quite common within our organizations, that data and information is stored in areas often difficult to access by anyone other than the people who created the data. While this may not be intentional, the result of such "discipline silos" is that updated reports from these discipline groups tend to be too generalized to support specific decision-making.

Given the wide distribution of data, it is not surprising that daily decisions are often made on the basis of exception, i.e. unless told otherwise and in the absence of any visible concerns the assumption is that the barriers are in good health. Visual evidence of barrier impairment cannot be relied upon other than very obvious external corrosion or physical damage to equipment and "local memory" of ongoing status rely on individuals being on shift or very good handover notes and Operations logs. A good supervisor will make great efforts to find the most current information before making decisions but as always when trying to prove a negative….how do you know when to stop looking?

The table below shows some random examples of the types of discrete risk management processes and disconnected data systems which are common in our organizations. This is a short list to demonstrate the principle. The point is that our organizations and discipline groups tend to be very insular and the data they collect is difficult to access from outside the group. This is partly because the work carried out to identify deviations or verify conformance is fragmented and generally covers the whole facility. The output reports tend to be summarized around performance indicators (examples of which are included in the table). These KPI's are useful to judge the effectiveness of the Risk Management Process (RMP) averaged over sizeable time steps (per month , per quarter or per year). They are not necessarily helpful for supporting real time decisions in specific areas of the plant. For that, decision makers need to know the status of specific barriers in particular areas of the plant to identify whether planned activity or plant changes are safe to instigate.

Table 2. Examples of Corporate "Silos" of data and information

Lead Discipline	Risk Management Process (RMP)	Data "silo" system	Performance Indicator
Maintenance	Identified barrier impairment	MMS (Maintenance Management System)	% impaired vs not impaired
Engineering	Corrosion Inspection	Inspection Systems	% systems inspected % failed
Engineering	Safe Operating Limit excursions	DCS / Data Historian	Number of SOL excursions
Instrumentation	Relief Valve testing	Testing Systems	% tests carried out
Instrumentation	Instrument Malfunction	DCS / Data Historian	% malfunctioning instruments
Engineering	Management Of Change	MOC system	% MOC's outstanding
HSE	Incident Reporting	Accident / Incident Database	Tier 1 – 4 reporting statistics
HR	Competency Management	Competency Management / Training database	% workers trained
Process Safety	HAZOP / HAZID	Risk Analysis Database	% Hazops complete
Management	Management Audits	Audit / Action Tracking System	Audit Actions complete

4.4 Cumulative Risk & Dynamic Risk

Two factors of risk management which are gaining some attention in our industry are cumulative risk and dynamic risk. Dynamic risk is perhaps the most simple concept since it is just a recognition that the risk situation is not static in any of our facilities. It changes with a variety of time dependent mechanisms. We have covered extensively the way in which barrier systems fail over time which results in a change in the consequential risk of the considered scenarios. The degree of risk elevation will depend on the length of time the barrier(s) is impaired. Other time dependent components to risk include environmental conditions particularly those at the limit of design such as hurricanes, tornadoes, floods, earthquakes, etc. Ongoing activities on the plant will potentially have an impact on risk either through introduced hazards or introducing potential casualties if an event occurred. Community expansion over time outside the perimeter fence is also a dynamic which changes the risk profile of a facility.

Built into some of our standards and regulatory practice is a 5 year review of hazard and risk scenarios across the facility. This is a suitable time frame for reviewing major equipment items and to ensure that appropriate barriers and risk control systems are in place as well as an opportunity to implement new attitudes towards risk and risk reviews. It does not, however, address the reality of day to day changes in our barriers, temporary situations and the changing hazards of a typical work day and as a result, the changes in overall risk. Instead we rely on informal processes and diligent front line supervisors to recognize when risk levels are reaching unacceptable levels.

Cumulative risk is another component of this dynamic picture which is gaining some recognition in areas, notably, offshore in the North Sea. Oil and Gas UK have recently issued in draft form guidance on the management of Cumulative risk (ref 7). In this document they define Cumulative Risk as follows:

The Cumulative risk on an installation is considered to be the total risk at a particular point in time due to a combination of:

- The risk described in the Safety Case
- Changes to the above caused by one-off operations or other operational changes
- Additional risk from deviations (from a normal state), the interactions between them and the influence of other factors on these risks

The North Sea is a mature basin area with many of the operating facilities having been first installed in the 1970's. Maintaining these facilities is a challenge and, as we saw in section 4.1, a backlog of safety critical maintenance now exists on many facilities. The regulator has a concern on how operators manage the cumulative effect of multiple "deviations", i.e. impaired barriers and how decisions are made.

There is nothing special about the North Sea and it probably represents what Deepwater Gulf of Mexico will look like in 10 to 15 years. Similarly, onshore refineries, chemical plants, storage and distribution facilities suffering the effects of ageing equipment and infrastructure will need to understand and manage the cumulative and dynamic impact of multiple impaired barriers and simultaneous operations. To achieve this successfully, we need to have a better set of tools to help decision makers, in real time, assess overall risk from a number of disparate sources and enable them to more safely and efficiently manage activities and the operation of their plant.

5 A New way to Manage Risk, Barrier Health, and Activity

5.1 New Category of Enterprise Operational Excellence Software

New software platforms are emerging under the category of Enterprise Operational Excellence Management enabling companies to make better decisions in all areas of activity and risk management. These systems are designed to manage activity with the benefit of understanding risk from a number of sources. Barrier impairment, in such systems, is easily tracked and very visible when activities are planned and executed.

These systems help manage some of the following challenges:

- They provide a unified approach to assessing operational risk and governing the status of Process Safety barriers
- They provide opportunities to produce meaningful real-time and predictive risk indicators, to manage the sequence and execution of activities.
- Real time and predictive cumulative risk modeling in time and space dimensions
- Safe execution of frontline activities across the whole planning-to-execution business process.
- Data exchange with other enterprise platforms such as Maintenance Management Systems (CMMS), Inspection databases, Process Data Historians, etc. integrating across many of the data silos in our organizations
- Visualize risk and activity in space and time
- Prioritize work scheduled based on optimizing future reduction of risk
- Easy identification of barrier impairments and management of deviations in risk control systems
- Ability to manage, in real time, all risk and activity in the one place using easy simple graphical representation of the facility.
- Management of all activities including permitted work, energy isolations (LOTO), task risk assessments / Job Safety Analysis, confined space entries, routine activities, shift handover, Operator Duties, etc.
- Efficient and safe execution of work schedule
- Real time and predictive KPI dashboards shwing risk, productivity, plan attainment, time on tools, audit compliance, etc.

The screen capture below shows an example of how graphical views of the plant can indicate where MAH barriers are impaired. Icons show the different categories of impaired barriers. Other icons show ongoing or planned activities. The view can be real-time or predictive depending on the shift date selected. The future position is based on planned activities and estimates for the duration of impacted barriers. The system predicts the overall level of risk for that time period and displays with a simple traffic signal red, orange, green pattern to indicate high, medium, or low level risks. It can also highlight the specific area on the plant where the highest risks are predicted.

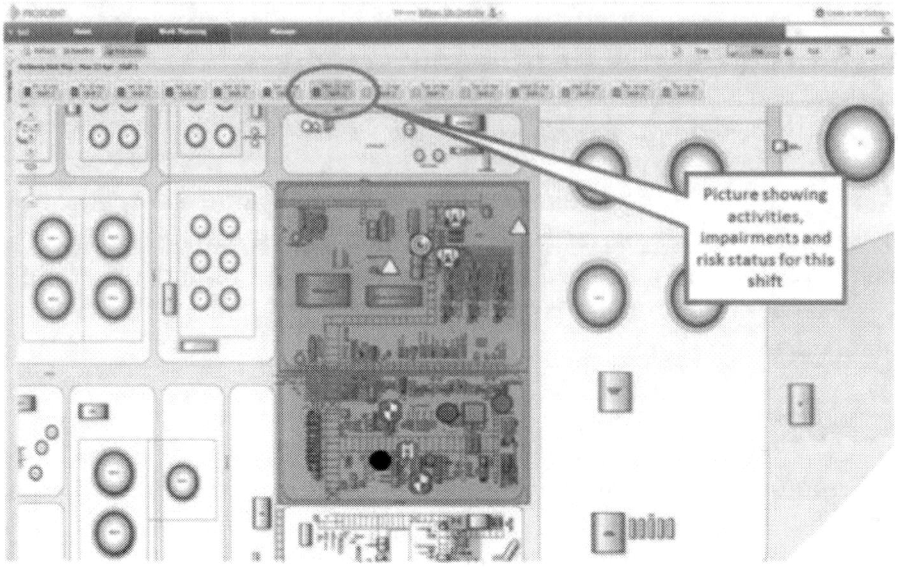

Figure 3. Graphical View of the Plant to Display Operational Risk Impairments and Deviations

Specific risk view screens for each area of the plant allow drill-down into the data to understand which activities or conditions are driving higher risks. This view is also available for current real-time or for future shift periods, with the traffic signals indicating cumulative risk for the facility during that period. This allows an operator to see the current or future state of fundamental safety barriers in each area of the plant.

Figure 4. Risk Views Against Specific Plant Areas for Current or Future Shifts

The impact on risk is driven by specific work activities, plant statuses, or barrier conditions. Each of these is captured as a "line item" and displayed in a Gantt chart style. Much of this data is "sucked in" from the disparate silo stores and is displayed in a single place showing all future work and conditions. This screen automatically shows conflicts depending on configuration rules established. For example, if there was a containment issue in an area and an operator was trying to schedule hot work in the same area, the system would prompt a conflict and shows the activities with a red boundary. Even if it was not a conflict rule, the high risk status for the shift would alert the situation.

Simple drag and drop features allow Frontline Leadership and Planning to simply move conflicts or compounding risk events so that they do not coincide and in doing so, reduce the levels of predicted peak risks.

Digital displays help Operators identify prolonged barrier impairments which are elevating risk levels in specific areas of the plant to unacceptable levels. Simple simulate/fix functionality empowers operators to see the risk reduction impact of fixing an impaired barrier earlier than scheduled.

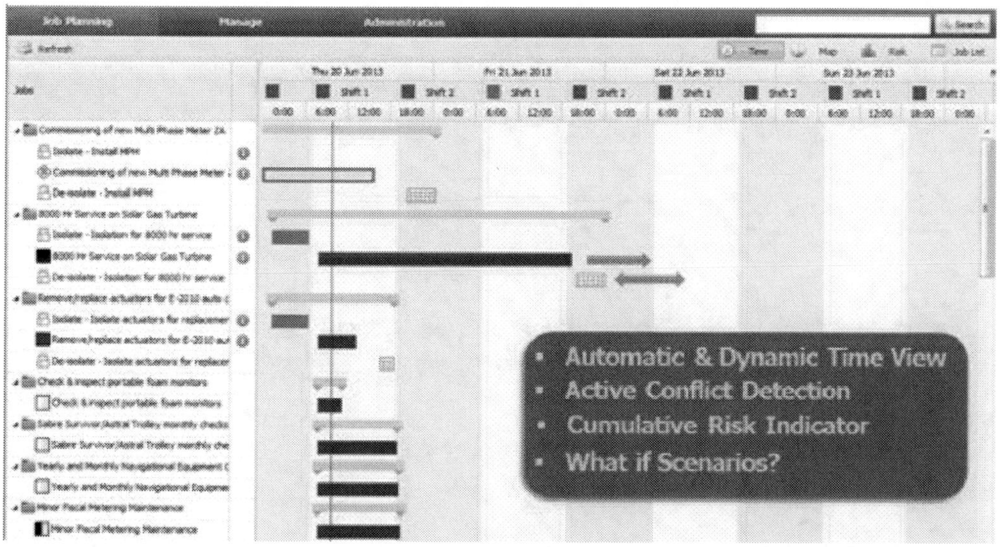

Figure 5. Time View showing Durations of all Planned Activities

A digital risk assessment engine enables Operators to evaluate the risk involved in their planned activities. At the same time, everyone across the plant can see the status of impaired barriers and assess the cumulative risk impact of their activities and those of other departments on all barriers, in all areas, at all times. This means medium-term planning and scheduling (90+ days) can be prioritized using MAH risk criteria. Rather than just blanket labeling of work priorities with the term "safety critical," the true impact on MAH risk can be assessed and activities prioritized accordingly.

With this data being collected centrally and used to assess the overall impact on risks, the quality of analysis and reporting surpasses that which has been achievable to date. The screen below show some of the types of reporting possible such as tracking overdue deviations (i.e. deviations or impairments which are not resolved within an expected timeframe), risk control systems impacts and trends and even trending of MAH risks over time.

These systems are designed for enterprise deployment so reporting can cover a whole organization of assets and identify where barrier impairments and resulting MAH risks are highest.

Figure 6. Examples of Dashboard Type Reports

6 Conclusions

We have seen how major accidents occur due to the failure of several of our protective / mitigative barriers. Simply put, the barriers did not perform as designed. Their reliability is dependent on continued inspection and preventative maintenance which, as we have seen, can often be deferred resulting in an increase of MAH risk. This is because the risk reduction credit for the barrier, taken during design, is no longer applicable. Another problem we face is that the data and information regarding these impaired barriers is often not easily accessible, being stored within corporate silos. Frontline Supervisors are making day to day decisions on the basis of assumed barrier status and cannot easily access information which could contradict some of these assumptions. They also need to take into account the dynamic nature of risk which can change from day to day, hour to hour and the cumulative impact of multiple impaired barriers.

Our frontline supervisors and decision makers have the responsibility to manage a safe plant so they need this information in order to make fundamental decisions such as whether the plant is safe to operate and whether the planned program of work is safe to execute. Currently they do not have easy access the relevant data in order to make these decisions.

Fortunately new technology platforms are emerging where the data on impaired barriers, temporary defeats, equipment isolations, work activities, etc. can be sourced through a single place. In this same place, activities can be planned and executed. Planned maintenance schedules can be imported and work carried out safely. Cumulative risk indicators show the aggregate impact of multiple activities and impaired barriers on risk in specific areas.

These new platforms represent an opportunity for a significant change in the way we manage risk on our facilities and should lead to safer and more efficient operations.

7 References.

[1] "The 100 Largest Losses 1974-2013 – Large Property Damage in the Hydrocarbon Industry", 23rd Edition, Marsh Energy Practice 2014.

[2] "An analysis of common causes of losses in the Oil, Gas & Petrochemical Industries" Presented at IChemE Safety and Loss Prevention Special Interest Group, Ron Garvis Swiss Re Risk Engineering Services, July 7th 2015

[3] The Cumulative Risk Assessment Barrier Model, Alan Blacklaw, Al Ward, Kevin Cassity, BG Group, SPE, 146255, 2011

[4] Health & Safety Report 2015 – Oil & Gas UK

[5] API Recommended Practice 754 - Process Safety Performance Indicators for the Refining and Petrochemical Industries, API, April 2010

[6] HSG 254: Developing Process Safety Indicators, UK Health & Safety Directorate, 2006

[7] Cumulative Risk Guidance (Issue Draft for Comment), Oil & Gas UK, February 2016

445034 Donde, Cuando y Que Enseñar y la nueva certification de seguridad de procesos del CCPS

Abstract:

El primer objetivo de este paper es el de tratar de desmistificar que la Seguridad de Procesos debe enseñarse solamente o fundamentalmente en la carrera de Ingenieria Quimica. Basandonos en el concepto de los 20 elementos propuestos por el libro del Center for Chemical Process Safety "Seguridad de Procesos Basada en el Riesgo " (RBPS por sus siglas en ingles) se puede demostrar que ciertos elementos, como integridad mecanica y planificacion de emergencias, cruza transversalmente a otras disciplinas del conocimento academico como Ingenieria Mecanica, Ingenieria Electrica, etc. y ademas, otras no pertinentes al ambito de la ingeniera, como es el caso de divulgacion a la comunidad. Se presentaran ejemplos de ciertas universidades de los Estados Unidos, Canada, y Colombia. Como segundo objetivo, y conscientes del hecho de que no existe en el mundo actual, un modelo unico y probadamente eficaz para implementar la ensenanza de la Seguridad de Procesos, los autores buscan presentar diferentes modelos y llamar a la accion a las universidades de paises hispanoparlantes para desacelerar la brecha entre lo que provee la Universidad y lo que necesita la industria.

Ademas, este paper buscara explicar los requerimientos del nuevo proceso de certificación de seguridad de procesos implementado por el CCPS, que es una nueva herramienta para demostrar competencia en seguridad de procesos, que permitirá a los profesionales agregar a sus credenciales "Certified Process Safety Profesional" or CPSP.

440342 PSM Legal Update

Abstract:

Over the last year, there have been many changes to the PSM Standard. There have been two court cases explaining key terms in the PSM Standard such as "change," "process," and "process equipment" and discussing key issues such as the time to implement PHA and audit findings. In addition, OSHA has issued four memoranda expanding the boundaries of PSM by defining RAGAGEP, creating a new mixture rule for Appendix A substances, completely revising the retail exemption and defining "proximate" for the purpose of establishing the boundaries of a process. OSHA issued a Request for Information to gather information for 14 potential changes to the PSM Standard to modernize it. After reviewing all of the responses to the RFI, OSHA has just initiated the review process under the Small Business Regulatory Enforcement Fairness Act (SBREFA) in which a draft rule with the proposed changes to the PSM Standard will be published and small business will have an opportunity to comment on it. The SBRFA process will provide a framework for additional changes being developed for the PSM Standard. This presentation will provide an overview of the key changes in the PSM Standard that OSHA has implemented and a summary of the changes being considered by OSHA.

440951 Application of Computational Fluid Dynamics (CFD) for Liquefied Natural Gas (LNG) Pool Spreading and Vaporization on Water

Abstract:

A marked increase in LNG utilization is forecast with an increase in LNG marine operations. A ship-ship collision of an LNG marine tanker or a rupture of the LNG loading and unloading lines can result in accidental spillage of LNG on water. Upon release, a spreading liquid can form a pool with rapid vaporization, leading to the formation of a flammable vapor cloud. Safety analyses for the protection of the public and property involve the determination of consequences of such accidental releases. The evaluation of consequences resulting from an accidental spill of LNG on water involves the determination of the rate (vaporization rate/source term) at which flammable hydrocarbon vapor is produced and the dynamics of the spreading pool. A specific offshore scenario of rupture of an offloading line is considered resulting in an LNG spill on water, between the space enclosed by the offloading deck and the LNG ship. This scenario was studied experimentally by releasing LNG in a narrow concrete trench filled with water. The experimental study involves the determination of the vaporization rate for a continuous release of LNG in a trench. Along with the vaporization rate, the spreading rate, pool area and pool height were measured. The computational work involves development of a methodology to model the pool spreading and vaporization behavior in Computational Fluid Dynamics (CFD) for the defined scenario. The overall aim of this research is to improve the understanding of pool spreading and vaporization of LNG releases on water.

Modeling an Unconfined Liquid Spill on a Sloped Surface

Robert Deshotels
Retired
21492 Pedroso, Mission Viejo, California, USA
R.L.Deshotels@cox.net

Prepared for Presentation at
American Institute of Chemical Engineers
2016 Spring Meeting
12th Global Congress on Process Safety
Houston, Texas
April 11-23, 2016

Modeling an Unconfined Liquid Spill on a Sloped Surface

Robert Deshotels, PE
Retired
21492 Pedroso, Mission Viejo, California, USA
R.L.Deshotels@cox.net

Keywords: liquid spreading, hydraulic radius, pool evaporation, friction factor, liquid spill modeling, simulation of liquid spreading

Abstract

When determining the consequences of a vapor cloud or fire from a liquid spill, the area of the liquid pool is a key variable. If the pool is contained by walls or berms, dispersion or flame parameters can be evaluated directly from the pool area using correlations that are well known to consequence analysts. For a spill onto an unconfined paved area or sloping deck, there is no simple formula for the pool area. For an unconfined pool, area must be calculated from flow modeling. The progressive movement and size of a liquid spill onto a sloped surface can be accurately modeled with well-known fluid equations using an empirical friction factor. The fluid flow formulas for piping can be applied to a sloping flat spill surface by including a correction for the equivalent hydraulic radius, the ratio of flow area to wetted area of the frictional surface. For uncomplicated flow surfaces, spreadsheet calculation is sufficient. Example calculations point out the need for improved knowledge of friction factors in terms of quantitative surface characteristics, and an improved formula for heat transfer during film boiling of cryogenic liquids.

1 Overview

The surface area of a spill of a toxic or flammable liquid spill is crucial to the analysis of hazardous consequences. The evaporation rate from a pool of volatile liquid, and hence the dispersion rate, depends on liquid properties, meteorological conditions, and the surface area of the pool. In the case of a pool fire, the heat flux radiating to nearby structures or people depends on the height and width of the flame, which depend on the size and shape of the pool. Usually, a liquid spill is confined by berms or curbs, so that the worst-case spill area is known from confinement dimensions. For an unconfined spill, there isn't any obvious maximum pool area, even if the source of the spill has a limited volume, because the pool will tend to continue to spread as time progresses. Therefore, unconfined spills require careful analysis to determine whether the spill can affect sensitive areas or lead to accident escalation. As well as environmentally sensitive areas, sensitive areas include escape paths, structures used for the shelter of employees or the public, and areas where fire or explosion could lead to escalation of the consequences.

Equations for the rate of spread of a liquid pool have been published in standard references on hazard evaluation [Ref. 1, 2], but friction loss was not considered in these reference guides. Friction loss is a major factor in the speed of liquid flow, especially for a thin pool depth. For a fixed-volume release, frictional loss can be the most important factor in whether a pool spreads into sensitive regions.

Much of the recent research on liquid spreading across a surface involves the effects of capillary action and surface tension on the motion of a thin film [Ref. 3 - 5]. There is much to be learned about the interaction between spilled liquid and the surface below in terms of capillary action and the hydrophilic/hydrophobic nature of the surface. However, the complexity surrounding surface tension applies only for very thin films of liquid. It will be shown that very thin films are not usually a significant part of industrial-sized releases.

For liquid spills that present a design challenge to industrial-size facilities, equations that are commonly used for design of piping and open channels, such as the Darcy equation, can be adapted to the purpose of modeling the time-dependent flow of an unconfined spill [Ref. 6 - 8].

For volatile liquids, the open channel flow equations give relatively accurate estimates for dispersion consequences. However, the concern may be that the spill will flow into a sensitive area. For non-volatile pools, the following question will be answered: "At what point is the spill thickness too low to produce a significant hazard?"

With a methodology for evaluating the minimum significant pool thickness in place, the basic steps for modeling an unconfined spill are:

1. Determine an appropriate release scenario.
2. Evaluate flow parameters for a generalized frictional flow formula.
3. Estimate the endpoint minimum significant pool thickness.
4. For sequential time increments, calculate pool dimensions, taking evaporation into account.

At each stage of modeling, uncertainties in the models and parameters should be considered to ensure that conservative choices are made. Uncertainty is especially significant when postulating a release scenario, estimating temperature for the liquid and surroundings, and appropriate modeling for heat transfer and evaporation. The largest pool area and evaporation rate produce the worst toxic or flammable release to the atmosphere. When the concern is distance of pool travel, minimum evaporation will create a worse case. There may be several potential worst case scenarios that must be quantified to determine which one is actually the worst.

Two example cases of unconfined liquid spills are provided in this article:
Example A. A leaking tanker truck located on a large paved area;
Example B. A piping leak onto the sloped deck of a floating oil production platform.

2 Release scenario

Generally, the source of the liquid spill will be a vessel or piping system with many potential leak points and spill trajectories. One possible way to address the multitude of possible release scenarios is to choose a worst credible case. If the worst case results in tolerable risk and acceptable mitigation expense, there is no reason to evaluate any less conservative cases. Sometimes, a release scenario may be so conservative as to be speculative. For example, if a release is conjectured to be the total contents of a tank due to accidental impact, there may not be any good basis for determining the speed at which the tank breaks, the size of the break, or the splashing that would occur upon impingement against nearby objects. In this case, simple assumptions should be made, and there should be a discussion of the sensitivity to uncertainties.

Any individual spill case may require a different approach to conservatism, but in general, the most conservative release scenario involves the greatest volume and flowrate. Concerning scenario parameters, pool spread distance is worse when the slope is high, and evaporation is low. Higher liquid temperature decreases viscosity and increases evaporation. When the spill poses multiple hazards or affects multiple possible targets, several potential worst-cases may have to be run to find the actual worst. When risk acceptability or mitigation cost are sensitive to the choice of release scenario, quantitative risk assessment tools such as event tree analysis should be used to pick the appropriate set of release cases. Data from offshore platform leaks [Ref. 9] can be used to estimate the probability of specific leak sizes from common process piping and equipment items.

Standard fluid dynamics methods [Ref. 7, 10] can be used to calculate the flow for a release of a given opening size and specified liquid conditions. The volumetric flow rate through an orifice, q, can be expressed in terms of the following parameters (in order of mention) flow coefficient, flow area of the orifice, gravitational acceleration, and differential pressure across the orifice, expressed in height of liquid:

$$q = C\,A\,(2g\,h)^{1/2}$$

The value of the flow coefficient, C, is a function of the Reynolds number (discussed in the section "Formulas for Flow of a Liquid Pool") and the ratio of the orifice diameter to the diameter of the pipe or vessel that is leaking. Plots of C are readily available [Ref. 7].

For liquids that flash upon release to atmospheric pressure, classical thermodynamics theory [Ref. 10] will accurately predict the vapor-liquid split and liquid properties after flashing. The mass fraction of the leaking fluid that vaporizes after release, f_v, can be expressed in terms of the exponential of the following parameters (in order of mention) liquid specific heat capacity, initial temperature of the liquid, atmospheric boiling point, and latent heat of vaporization. The heat capacity and latent heat should be the mean values over the temperature range T_i to T_b.

$$f_v = 1 - \exp\{-C_p(T_i - T_b)/\Delta H_v\}$$

Depending on release conditions and liquid physical properties, the liquid release may reach the pool surface as a steady stream or a spray of droplets. If the release is in the form of an undispersed stream, the trajectory can be evaluated using classical Newtonian physics, including the effect of wind pressure on the stream. In some cases, the release will immediately begin breaking into droplets, and in all cases a liquid stream will disperse into droplets if it falls enough distance.

Empirical formulas have been developed to predict the distance that a liquid release will travel before breaking-up into droplets. Droplet formation is characterized by the dimensionless Weber number, which is the ratio of shear force on the liquid surface, $\frac{1}{2}\rho_g v_d^2$, to surface tension pressure on the drop, σ/d. The Weber number is defined in terms of the following parameters (in order of mention) gas density, velocity of the drop relative to the air, drop diameter, and surface tension :

$$W_e = (\tfrac{1}{2}\rho_g v_d^2 d) / \sigma$$

For subcooled liquids, the length between jet release and breakup into droplets, L_b, can be expressed in terms of the following parameters (in order of mention) jet diameter at the narrowest point, Weber number, and Reynolds number [Ref. 11]. This formula applies to leaks large enough to cause a serious hazard, but not to a 2 mm diameter hole or less. The units of measure must be consistent with the dimensionless Weber number.

$$L_b = 540\, D\, W_e^{1/2}\, R_e^{-5/8}$$

For superheated liquids [Ref. 11],

$$L_b = 2.75 \times 10^{10}\, D\, W_e^{1/2}\, R_e^{-2}$$

Based on experimental results for liquids without superheat, the typical droplet size, d, can be expressed in terms of the following parameters (in order of mention) surface tension, gas density, and jet velocity relative to the wind. The drop size depends on the liquid, and could be double the typical value [Ref. 1].

$$d = 24\, \sigma / (\rho_g v_d^2)$$

For superheated liquids, the droplet size depends on the difference between the liquid temperature and the atmospheric boiling point [Ref. 1],

$$d = 3 \times 10^4 / (10 + 4.0\, \Delta T_{sh})$$

The terminal velocity of a droplet in air is a complicated function of the liquid density and drop diameter, which can be read from a plot in a standard reference [Ref. 12].

Comparison of the terminal velocity to the wind speed will show whether droplets will rain out onto the pool or drift away like a cloud. The direction of release and Newtonian physics determine the trajectory that the droplets will follow from the liquid breakup point to the point where the pool begins growing. Different wind conditions may have to be considered. The probability of the combination of accident scenario, wind and temperature should be consistent with a tolerable risk for the accident.

3 Liquid pool growth and shrinkage

The pool volume will increase as long as the liquid release rate into the pool exceeds the evaporation rate. The release rate changes with time, depending on source inventory, pressure and effectiveness of process shutdown controls. The evaporation rate depends on pool area and changing temperatures.

It is possible to determine a worst-case maximum pool area using the initial leak rate and assuming a minimum evaporation rate. In this case, the maximum pool area equals the maximum pool volume created over the interval of concern (e.g., maximum escape and rescue time) divided by the minimum significant pool thickness. If the hazards evaluated from this bounding case give a tolerable risk, there may be no need to actually model the growth and configuration of the pool. However, even if the worst case results in tolerable risk, dynamic modeling of the pool may provide insight that would be valuable for planning and training.

Dynamic modeling consists of stepwise calculation of pool characteristics at small increments of time. At each increment, the following variables are computed:

1. Based on process conditions of pressure, temperature and composition, determine the flow properties and vapor pressure of the liquid just upstream of the release point. Using classical thermodynamics, determine the fraction of vapor flash after release into the atmosphere.

2. For the release scenario chosen, calculate the liquid flow rate into the pool using the orifice flow equation that is appropriate to the scenario. Calculate the liquid trajectory and dispersion from the jet breakup formula that applies to the orifice size, Weber number, and flashing superheat. It can be assumed that the liquid volume will be in a conical shape, with the target area being circular on a flat substrate, or elliptical on a sloped substrate. While a liquid release is in the momentum-driven region of flow (about 6 to 100 orifice diameters from the release point), a liquid jet will diverge at an angle of 10° from the centerline [Ref. 6]. From the flowrate, trajectory and divergence angle of the release, the target area of the liquid added to the pool is determined, as well as the volume increase in the target area during the time interval.

3. From a predicted Reynolds number, or the Reynolds number from the previous time increment, the applicable flow equation for pool spread is chosen. The volume flow and velocity at the edge of the pool can then be calculated, from which pool edge growth can be calculated. The equations for flow of the pool are given in the section "Formulas for Flow of a Liquid Pool".

4. A conservative calculation of pool evaporation is done, based on the new pool area calculated in the previous step. Depending on the purpose of the modeling, the evaporation calculation should be either a low estimate or a high estimate. An exact calculation is generally not possible because of uncertainty about rates of diffusion and heat transfer. An approach to providing a conservative estimate of evaporation is provided in the section "Evaporation". Evaporation may result in a cooling effect on the pool. The evaporation calculation is expressed as a uniform decrease in pool thickness, which will result in an adjustment in the pool surface area.

5. Knowing the inventory volume and process conditions, new process conditions are calculated from the liquid loss through leakage flowrate calculated in step 2 above. As the process pressure decreases, any evaporative cooling inside the process containment should be considered in the revised process conditions and composition.

Considering the accuracy of the postulated release scenario and air dispersion models, a simple model for pool thickness should suffice. It may be reasonable to model the pool volume as a cone with an elliptical base (because of the slope of the substrate), with the peak of the cone located at the target point of the liquid release trajectory.

4 Minimum significant pool thickness

The driving force for pool flow is the difference in height between different parts of the pool. For a pool moving down a sloped surface, the uphill part will become thinner as the pool spreads downhill. Furthermore, it is common for the leak rate from the process to stop or diminish after a while, causing the rate of evaporation rate from the pool to exceed the fill rate. At some point, part or all of the pool may become too thin to pose a significant hazard. To see whether this is possible in a specific case, it helps to categorize the different hazards posed by a pool, as follows.

Pool liquid that has not evaporated can pose a hazard of toxic exposure by direct contact, or a fire hazard.

For liquids that are not flammable, but are combustible upon ignition by flame, the pool may be too thin to maintain combustion. This happens when heat transfer to the substrate will quench any fire that starts by cooling the liquid below the flash point. This thickness could be calculated, but instead of solving unsteady-state heat transfer calculations using uncertain heat transfer coefficients and uncertain burning rates, the minimum fire hazard thickness for a non-volatile liquid should be determined by a field test. For one rough example, in a backyard test in a one-foot square confined pool of kerosene on concrete sidewalk at 20°C in breezy air, it was found that the minimum liquid thickness that would maintain a flame was between 2 and 3mm.

Where a liquid is highly toxic to people or the environment (e.g., aqueous ammonia) OSHA Standard 1910.119 has established a threshold quantity. A pool below this quantity no longer poses a catastrophic level of hazard, but mitigation is still imperative. Consequently, there is no insignificant pool quantity or thickness in the case of toxic liquids.

Vapor that is evaporating from a pool can create a toxic or flammable cloud leading to toxic exposure, flash fire, or explosion. The severity of exposure of a person to a toxic cloud is measured as the integrated dose of concentration over time. Therefore, for toxics, the quantity of material in the vapor cloud is more important than the peak rate of evaporation. As a result, the thickness of a pool does not affect the severity of a toxic release much, especially for a thin film that will evaporate soon. For a flash fire or explosion, severity is measured as heat flux or blast pressure, which depend on the size and concentration of the flammable cloud. Thus, for a spill of volatile liquid, the pool thickness just prior to complete evaporation does not have much impact on accident consequence.

When there is a concern that a combustible spilled liquid will flow into a sensitive area, the pool can be considered non-hazardous, provided there is no continuous source of ignition, if the entire pool depth is less than the minimum fire hazard thickness, determined from field testing of flame propagation.

5 Formulas for flow of a liquid pool

All practical formulas for the flow of liquids are derived from Bernoulli's equation, an energy balance that includes all of the involved forms of energy: potential energy (elevation difference), pressure, velocity, and frictional loss. The Bernoulli equation for the energy change for a fluid flowing from point 1 to point 2 can be expressed in terms of the following parameters (in order of mention) elevation, pressure, density, velocity, gravitational acceleration, and frictional loss expressed as height of fluid [Ref. 6, 7]. Consistent units of measure must be used. With U.S. customary units of measure, the gravitational acceleration factor of 32.2 ft/s^2 is needed to convert between pounds force and pounds mass.

$$Z_1 + P_1/\rho_1 + v_1^2/2g = Z_2 + P_2/\rho_2 + v_2^2/2g + h_L$$

Because the entire pool is at atmospheric pressure, there is no pressure energy term, P/ρ, to consider for a pool. The change in the elevation term Z is the only driving energy for pool flow, unless the release into the pool has a significant velocity due to the release trajectory at the target point. Further explanation follows for the velocity head term and friction head term.

Flow of a liquid pool on a sloped surface has two components. The first component of pool flow is the downhill motion of the entire pool. The driving force for downhill motion is the gravitational force multiplied by the slope. To determine downhill motion, the downward velocity due to the slope must be added to any velocity head imparted by the trajectory of the release flow into the pool, and the friction head must be subtracted. The downhill momentum increases as time continues. The second component of pool flow is the spreading of the pool outward from the center of mass, which can be modeled from the difference in height between the liquid and the surrounding surface.

Unlike flow through a pipe or in an open channel, liquid spreading on a sloped surface does not have a constant flow depth or cross-sectional flow area. Consequently, velocity varies through the pool. Therefore the value of the velocity used in the velocity head term, $v^2/2g$, and friction head term, h_L, must be considered carefully. For a pool modeled as a cone with a round or elliptical base, spreading begins at the high point as gravity-induced downward flow, then turns outward to be parallel with the base at the outer edge. To determine pool growth, the velocity head due to the elevation difference must be decreased by the friction head to yield the effective elevation change from the high point to the pool edge. From the effective elevation change, a new pool radius can be calculated for a chosen time differential. The new height can be calculated from the mass balance of the pool.

When the substrate is sloped or there is a velocity head imparted by the release stream, the growth in the pool can be evaluated by considering the cone to have four quadrants: uphill, downhill, and two equal sideways quadrants. Each quadrant is broken into differential elements at intervals of radial length along the distance from center of the ellipse outward to a point on the edge in order to evaluate a velocity that applies to each differential element of the quadrant. This allows a calculation of the friction head contribution of each element, from which the radius increase for each quadrant is calculated, and the new pool height determined as discussed in the previous paragraph.

There are different liquid flow formulas (e.g., Darcy, Fanning, Manning) that can be used to estimate the empirical term for frictional loss, h_L.

Regardless of the formula used for friction loss, there are four regimes of flow that can apply to an unconfined liquid spill:

1. Frictionless flow
2. Turbulent flow
3. Laminar flow
4. Thin film flow

5.1 *frictionless flow*

When the flow path has a low elevation gradient and a high pool depth, frictional loss may be insignificant. Also, it is possible for frictional loss to be very low in the case where the pool is heated by film boiling at the substrate interface [Ref. 4]. Otherwise, in actual spill cases that pose a serious risk to a facility, frictional loss is significant. When significant frictional loss is ignored, the simplified formulas for liquid flow rate [Ref. 1, 2] will overestimate the speed of liquid travel. Because friction loss is not difficult to evaluate, there seems to be little incentive to ignore friction.

5.2 *turbulent flow*

Flow turbulence is characterized by the Reynolds number. The Reynolds number is dimensionless, and can be expressed in any consistent units of measure, in terms of the following parameters (in order of mention) pipe diameter, velocity, density, and absolute viscosity.

For pipe, $R_e = D\, v\, \rho\, /\mu$

Commonly used formulas for liquid flow through a pipe or channel [Ref. 6 - 8] express frictional loss in terms of a friction factor, f, which is determined from experiments. However, formulas for pipe and channels cannot be used directly; the formulas must be modified to account for the different geometry of flow.

In an unconfined liquid spill, the flow often transitions from turbulent flow to laminar flow as the pool spreads thinner. Although the Reynolds number can be calculated exactly, the transition between the turbulent and laminar flow regimes is not consistent from one case to another. The transition is usually gradual, and can depend on flow disturbances and flow history. Turbulent flow is ordinarily exhibited when the Reynolds number is above 4000, with laminar flow exhibited below 2100. Laminar flow has less friction loss than turbulent, resulting in a higher velocity for a given slope. However, laminar flow results in a lower heat transfer rate. In the transition range between flow regimes, the flow should be modeled conservatively as either fully turbulent or laminar. The conservative choice will depend on the situation.

Unconfined liquid flow on a horizontal or sloped surface is analogous to gravity-induced flow in a pipe, except that wall friction only occurs along the bottom of the spill. The concept of hydraulic radius, R_H, [Ref. 6, 7] must be applied.

$$R_H = \text{cross sectional flow area /wetted perimeter}$$

For flow through a round pipe full of fluid, the flow diameter equals $4R_H$. For unconfined pool flow, the hydraulic radius is the pool height, z, at the point where velocity is measured, and the equivalent flow diameter equals $4z$. The Reynolds number for a round pipe full of liquid is expressed in terms of the inside pipe diameter. For full flow through a narrow slot, the Reynolds number is expressed in terms of the height of the slot. For open flow across a flat surface, we typically assume that there is no friction at the top surface of the liquid or the edges, although this is not strictly true due to relative motion between the air and liquid surface. With negligible air friction, the zero-shear level is at the top of the liquid pool, rather than at the centerline as in a round or rectangular pipe. Therefore, the equivalent hydraulic radius used to evaluate the Reynolds number for a round liquid pool, with a flow area (circumference multiplied by height) of $2\pi\ r\ z$ and a perimeter of $2\pi\ r$, should be corrected to twice the usual value, or $2z$. In the Reynolds number and friction head equations, the D used for full-flow round pipe should be replaced by $8z$ for a pool.

$$\text{For a pool, } R_e = 8z\ v\ \rho\ /\mu$$

Wind friction can affect pool motion, but the effect is expected to be minor. Weather stations report wind speed at a reference height of 10 m above grade. If the spill is at grade, the wind is reduced to about 8% of the reference value in the daytime, or 5.5% at night (the difference is due to air stability) [Ref. 4]. It should be noted that wind wave height cannot exceed the pool thickness; therefore, the surface of the liquid will be smooth, even under the influence of wind. Wind drag can be evaluated from low-speed aerodynamics. The drag pressure caused by air friction is the drag coefficient multiplied by the wind velocity head. For wind blowing parallel to a flat surface the drag coefficient, C_D, is 0.04 [Ref. 13]. To convert into liquid head, the wind head is multiplied by the ratio of gas density to liquid density. If the wind is in the opposite direction of flow, the wind head is added to the friction head. If the wind is in the same direction as flow, the wind head is subtracted from the friction head. The wind head, h_{drag}, can be written in terms of the following parameters (in order of mention) drag coefficient, air density, air velocity, and liquid density.

$$h_{drag} = C_D\ \rho_G\ v_G^2\ /2\rho_L$$

Table 1

Formulas for Turbulent Flow Friction Loss

Darcy (Moody)	$h_L = f\,L\,v^2 /2D$
Fanning	$h_L = 2f\,L\,v^2 /D$
Manning	$h_L = f\,L\,v^2 /2R_H$

Note that the Moody friction factor is four times larger than the Fanning friction factor. This can be confusing because plots of friction factor do not always mention which friction factor is given. For the Darcy and Fanning equations, the friction factor is a function of the Reynolds number and relative roughness of the pipe wall, ϵ/D. For a pool having a height of z, D should be replaced by $4R_H$, or 8z. Roughness is determined by experimentation, rather than surface measurement. Based on the limited data available and the difficulty in defining roughness, the friction factor is the major cause of uncertainty in flow rate calculations. The Manning equation is an empirical equation developed from data for water flow in deep open channels. For the Manning equation, a roughness parameter n is used to correlate the friction factor to the hydraulic radius [Ref. 6, 8].

$$\text{Manning } f = 29n^2 /R_H^{1/3}$$

Table 2

Roughness Factors for Turbulent Flow

Surface	**ϵ ft**	**n ft$^{1/6}$**
Clean commercial steel	0.00015	
Galvanized iron	0.0004	0.012
Cast iron	0.00085	0.011
Concrete pipe	0.001 - 0.01	0.015
Riveted steel	0.003 - 0.03	0.014
Earth, straight and uniform		0.019
Brick open channel		0.017
Concrete open channel		0.022
Gravel open channel		0.033

5.3 laminar flow

Unlike turbulent flow, fully laminar flow along an unobstructed path exhibits no eddies in the axial or radial direction. Friction loss is due to viscous effects alone, and the friction factor is a function of Reynolds number alone, not surface roughness.

Darcy laminar $f = 64/R_e$

Fanning laminar $f = 16/R_e$

The value of f for laminar flow is substituted into the Darcy or Fanning equation for friction head. However, the laminar friction factor was developed for round pipe containing full flow, and will be somewhat in error for open flow on a flat surface, due to the parabolic velocity profile that exists between the pipe wall and zero-shear point in laminar flow [Ref. 6]. Although this paper proposes a correction in the hydraulic radius for the location of the zero-shear point, there is no published data to verify that this proposed correction is accurate or conservative.

Like turbulent flow, as long as there is a height difference between the top of the liquid pool and the surface of the substrate, the Bernoulli equation and derived Darcy formula tell us that the pool will continue to spread without limit, although at an ever decreasing rate. For turbulent or laminar flow, the Darcy formula does not consider the effects of surface tension, capillary action, or hydrophilic/hydrophobic nature of the surface under the spill.

5.4 thin film flow

Recently, investigators have been able to explain why liquid pools do not actually continue to spread endlessly [Ref. 5]. When the liquid height is less than around 100 nm (0.01 cm), long-range intermolecular forces (like van der Waals forces) begin affecting the ability of the liquid to wet the surface of the substrate. There is a minimum thickness of liquid film, beyond which the film no longer wets the substrate; instead, the film breaks into individual droplets. In the case of a level substrate, flow stops. The minimum film thickness is affected by substrate surface properties such as roughness, cleanliness, porosity, and surface tension between the solid and liquid. There are many cases for which the proposed model has not been tested (including a sloped surface), but two conclusions can be drawn:

1. For thin pools, the laminar flow model gives a higher spreading rate than any model that considers the effect of surface tension.

The thin film flow model does not apply for liquid pools thicker than about 0.01 cm.

6 Evaporation

Evaporation will occur if the liquid has a significant vapor pressure at the pool temperature, and the atmosphere above the pool is not saturated with the liquid vapor. The temperature of the pool can change because of heat transfer with the substrate or air, which would affect the vapor pressure.

Evaporation rate from the pool is described in terms of classical thermodynamics [Ref. 10]. The driving force for evaporation is the difference between the vapor pressure of the liquid and the partial pressure of the liquid in the air. The mass evaporation rate, W, can be written in terms of the following parameters (in order of mention) molecular weight, mass transfer coefficient, flow area, vapor pressure of the pure liquid, partial pressure of the vapor in the bulk stagnant air above the liquid, gas law constant, and temperature of the liquid. Pressure and temperature are in absolute units.

$$W = M\,K\,A\,(P_{sat} - P_v)\,/(R\,T)$$

At atmospheric pressure and practical operating temperatures, ideal gas can be assumed. Except in the case of a boiling liquid, the rate of evaporation is restricted by the rate of diffusion from the surface of the liquid. Diffusion coefficients have been measured or estimated for common chemicals [Ref. 14]. The resistance to mass transfer from a homogenous liquid to air is mainly due to the resistance of the air. The resistance of the air depends on the transport properties of the air (expressed as the Schmidt number) and flow regime (expressed as the Reynolds number). There are many theoretical and empirical correlations for evaporation, depending on the geometry of the surroundings, flow regime, concentration of the evaporating component in each phase. A great deal of judgement is required in choosing an appropriate correlation. Key issues involve friction modeling for a boiling pool, wind effects, and heat transfer from the substrate [Ref. 5]. It seems that nearly all of the correlations were developed for equipment design and were tested in laboratories, not outdoors in industrial-sized facilities where nearby objects can obstruct or funnel the wind. Therefore, correlation accuracies claimed in literature may not apply to evaporation from a real pool [Ref. 5]. An empirical correlation, rather than theoretical-based, should be applied when the flow is parallel to the interface through which mass transfer occurs (such as pool evaporation) [Ref. 14]. For a calculation that overestimates evaporation, one can assume that there is no resistance to mass flow within the liquid or across the liquid-air interface, which results in an air concentration just above the interface that is equal to the partial pressure of the liquid relative to atmospheric. When the resistance to diffusion is essentially due to air convection alone, the diffusivity is in the range of 10^{-4} to 10^{-6} m^2/s. A simple but crude estimate of diffusivity, K, based on molecular weight, M, may be sufficient [Ref. 10], by assuming the ratio between the diffusivity and cube root of molecular weight is constant for the liquid of interest and some reference liquid with known diffusivity, K_0.

$$K = K_0\,(M_0/M)^{1/3}$$

An evaporation correlation specifically for a pool of volatile liquid has been developed, accounting for the wind speed and stability class of the air [Ref. 2]. This formula does not consider evaporative cooling of the pool. The evaporation rate, W, involves two parameters, λ and n, which depend on the Pasquill-Gifford air stability classification. The equation is written in SI units in terms of the following parameters (in order of mention) λ, wind velocity, vapor density, vapor pressure, ambient pressure, wind velocity squared, pool diameter, and n.

$$W = \lambda\,v_w\,\rho_v\,P_v\,/P\,\{2/(v_w^2\,D\}^n$$

Table 3

Pool Evaporation Parameters

Pasquill Stability Class	λ	n
A	1.0×10^{-3}	0.078
B	1.2×10^{-3}	0.091
C	1.5×10^{-3}	0.111
D	1.7×10^{-3}	0.130
E	1.8×10^{-3}	0.149
F	1.8×10^{-3}	0.180

To calculate a high estimate of evaporation for non-cryogenic spills, assume the spill occurs on a windy, hot summer day; for a low estimate assume a still, cold winter night.

A separate formula applies to a spill of a cryogenic liquid on land, explicit as a function of time [Ref. 2]. This equation is written in SI units in terms of the following parameters (in order of mention) a surface correction factor, thermal conductivity of the substrate, heat of vaporization, thermal diffusivity of the substrate, substrate temperature, liquid temperature, and time. The surface correction factor, *X*, is assigned a value of 1 for a smooth substrate, or 3 for a rough substrate such as gravel. Note that this equation crudely assumes that heat transfer is proportional to the temperature difference between the substrate and liquid; this assumption does not apply at all beyond the point of maximum nucleate boiling, when partial or full film boiling occurs.

$$W_{cryo} = 2X\, k\, /\Delta H_v\, (\pi\, \alpha)^{-1/2}\, (T_S - T_L)\, t^{1/2}$$

If the pool is boiling, the evaporation rate is usually limited by heat input. If the liquid flashes upon release from the process, the liquid and vapor reach equilibrium soon after release, usually before the liquid touches the ground. The liquid will not continue to boil unless heat is added from the air or substrate. The main uncertainty involves heat transfer from the substrate under the pool. Different modes of heat transfer are possible, nucleate boiling, film boiling or transition [Ref. 15, 16]. The transition from nucleate boiling to film boiling occurs at a critical temperature difference between the substrate and liquid, which depends on liquid properties. Heat transfer rates for boiling LNG and NPG have been documented from field experiments with liquefied gas inside level confinement [Ref. 16]. In the case of film boiling, which often occurs in cryogens (especially LNG), the friction factor at the substrate interface is markedly decreased. Equations for the friction factor during film boiling of LNG spilled over a water surface have been published [Ref. 4]. If the pool is supported by the ground, transfer through the substrate depends on several factors including moisture content. One bounding option that overestimates evaporation is to assume that the substrate temperature remains constant. A bounding assumption that underestimates evaporation is that for fully developed film boiling, the heat flux equals the minimum for film boiling for that liquid.

Several references show an "S" curve for typical boiling liquid heat flux as a function of temperature difference between the substrate and liquid. The typical curve shows a low flux at low temperature difference, rising like the front end of a sine wave to reach a maximum flux at the point of fully developed nucleate boiling. When the temperature difference increases beyond this critical point, the heat flux decreases like the back end of a sine wave, reaching a minimum flux point caused by a layer of vapor that separates the substrate and liquid. Beyond the minimum point, the "typical" flux curve curves up again at a slightly lower slope than the initial curve for nucleate boiling. This increase is attributed to the increase in radiative heat transfer to the liquid from the hotter substrate. The data that supports this "S" curve is mainly drawn from data from experiments with silicone oil, which boils at a high temperature. In the typical experiment, the substrate is heated beyond temperatures where radiant heat transfer is significant. This data may apply to film boiling of cryogenic liquids inside pressure vessels, but an atmospheric spill of LNG is different. If the increase in film boiling heat transfer is due to radiative transfer, it should not be expressed as a linear relationship with temperature. Radiative heat transfer is a function of absolute temperature to the fourth power. The equation for transfer between parallel plates, or a flat substrate to a liquid suspended by a flat vapor layer is [Ref. 15]:

$$Q/A = \epsilon \sigma T_1^4 - \alpha \sigma T_2^4$$

where Q/A is heat flux, ϵ is emissivity, σ is the Stefan-Boltzmann constant, T_1 is the absolute temperature of the substrate, α is the absorptivity of the liquid, and T_2 is the absolute temperature of the boiling liquid. Units of measure must be consistent with the Stefan-Boltzmann constant, which is 1.713 $\times 10^{-9}$ Btu/(ft^2 h °R^4) or 5.67 $\times 10^{-8}$ W/(m^2 °K^4). The emissivity and absorptivity of common materials is given in standard references [Ref. 15]. It would seem that the radiative effect on film boiling heat flux could be calculated, but there has been no experimental data to validate this conjecture.

7 Conclusions

Use of the Bernoulli equation for liquid flow provides a method for calculating the spread of an unconfined liquid pool on a sloping surface. The Bernoulli equation is an expression of the fluid flow energy balance, so there is no need for the user to characterize the flow as gravity-driven or momentum-driven in order to choose an appropriate model. Friction loss can be calculated the same as for piping or open channel flow after flow geometry is accounted for. Although flow can be calculated exactly if all parameters are known, the surface roughness parameter can be uncertain by an order of magnitude. The most important variable, release flow rate into the pool is generally determined by accident conditions that are difficult to predict. Furthermore, the evaporation rate from the pool can be affected by heat transfer rates and mass diffusion coefficients that are somewhat difficult to predict outside of a laboratory. Therefore, care should be taken to assure conservative assumptions, methods and data.

The calculation accuracy of liquid flow across a sloping surface is limited by roughness data and heat transfer knowledge. Within this current limitation, one-dimensional spreadsheet calculations can provide useful bounding results. However, when accurate data has been developed, computational fluid dynamics will provide more accurate results.

The non-skid checkered deck plating used on the decks of land-based processing facilities and floating production platforms has well known surface dimensions, but no documented friction factor for flow calculations. Rather than relying on field testing of qualitatively defined surfaces, isn't it time that someone develop a method for estimating friction factors based on dimensions and some measurable index of roughness?

Spills of cryogenic fuels such as LNG and LPG pose a high risk of catastrophic accident worldwide. The heat flux during film boiling as a function of differential temperature could be more accurately calculated as function of the difference of the fourth power of the absolute temperatures, like the Stefan-Boltzmann equation, rather than a first-order difference. Experiments with cryogenic liquids such as liquid nitrogen could test this possibility and be used to develop a better correlation for film boiling, especially for liquids with low boiling points. Furthermore, experiments with cryogenic liquids flowing on sloped surfaces would be useful for accident analysis.

8 References

1. J.L. Woodward, *Estimating the Flammable Mass of a Vapor Cloud*, "Discharge Characteristics", p 49-59, Center for Chemical Process Safety of the American Institute of Chemical Engineers, 1999.

2. A. Cox, F. Lees, and M. Ang, "Classification of Hazardous Locations," pp 46-47, Inter-Institutional Group Classification of Hazardous Locations of the Institution of Chemical Engineers, Rugby, UK, 1990.

3. M. Apel-Paz and A. Marmur, "Spreading of liquids on rough surfaces," J Hazard Mater, 146, pp 273-279, 1999.

4. S. Brambilla and D. Manca, "Accidents involving liquids: A step ahead in modeling pool spreading, evaporation and burning," J Hazard Mater, 161, pp 1265-1280, 2009.

5. A. A. Pahlavan, et al., "Thin Films in Partial Wetting: Internal Selection of Contact-Line Dynamics," Phys. Rev. Lett. 115, 034502, July, 2015.

6. R. H. Perry and D. W. Green, *Perry's Chemical Engineers' Handbook,* 7th ed., "Incompressible Flow in Pipes and Channels," pp 6-9 through 6-16, 6-20, 6-21 McGraw-Hill, New York, 1997.

7. Crane Co., *Flow of Fluids, Technical Paper 410*, Ver. 15, pp 1-1, 1-4 through 1-7, www.flowoffluids.com, 2009.

8. S.D. Lin, *Water and Wastewater Calculations Manual,* pp 242-252, McGraw-Hill, 2001.

9. UK Health and Safety Executive, "Hydrocarbon Releases Database," https://www.hse.gov.uk/hcr3/, 2005.

10. D.A. Crowl and J.F. Louvar, *Chemical Process Safety Fundamentals with Applications,* 3rd ed., pp 122-140, Prentice Hall, 2011.

11. J.H. Lienhard and J.B. Day, "The Breakup of Superheated Liquid Jets," J Basic Eng. Trans., ASME, pp 515-521, 1970

12. R. H. Perry and D. W. Green, *Perry's Chemical Engineers' Handbook,* 7th ed., "Particles Dynamics," pp. 6-50 through 6-54, McGraw-Hill, 1997.

13. B. Jones, *Elements of Practical Aerodynamics,* 2nd ed., p 17, Wiley & Sons, 1939.

14. R. H. Perry and D. W. Green, *Perry's Chemical Engineers' Handbook,* 7th ed., "Mass Transfer," pp 5-42 through 5-79, McGraw-Hill, 1997.

15. R. H. Perry and D. W. Green, *Perry's Chemical Engineers' Handbook,* 7th ed., "Heat Transfer," pp 5-8 through 5-29, McGraw-Hill, 1997.

16. J.L. Woodward and R.M. Pitblado, *LNG Risk Based Safety: Modeling and Consequence Analysis,* pp 142-152, Wiley, 2010.

17. Western Regional Climate Center, "Cooperative Climatological Data Summaries," http://www.wrcc.dri.edu/climatedata/climsum/ Feb. 2016.

18. ASHRAE, *Fundamentals Handbook,* ASHRAE, Ch. 22, 1977.

Example A

Leaking LNG tanker truck located on a large paved area

A-1 Accident scenario

The concern is that a tanker truck at a LNG terminal could spill LNG into a storm sewer, with the potential that the natural gas could ignite in a sewer line in a public area. A calculation must be performed to see whether the contents of the truck would evaporate before reaching the nearest storm sewer drain.

At a terminal used to fill city vehicles powered by compressed natural gas, the cryogenic storage tanks are filled by a 10,000 g tanker truck. The tanker is double-walled for insulation and impact protection. The tanker contains LNG at about 70 psig and -210°F. In this example, the LNG is assumed to consist of 100% methane. At an actual facility, the LNG composition would be known. At atmospheric pressure, methane boils at -259°F. The tanker is unloaded through a nominal 3" flexible hose.

The terminal equipment and storage tanks are located inside a protective building. The tanker unloading hose is connected to a double-walled pipe that goes through the building wall. To reach the unloading station, the truck must drive along a concrete driveway that is 59 ft wide. The driveway slopes at a 1% grade toward a storm drain that is located a little more than 400 ft down the driveway.

A-2 Release modeling

The intent is to model a bounding case, to see if the worst-case spill of LNG can reach the storm drain. Leaks from tanker trucks seldom occur, and when they do, it is usually due to accidental impact or failure of the unloading hose or coupling. Failures involve some degree of human error, and therefore are difficult to assign a probability. The failure chosen for modeling is a sudden loss of the total contents of the tanker, which may not be credible, and certainly is a bounding case.

The LNG spill quickly flashes upon release to the atmosphere. The flash can be assumed to be adiabatic. After the initial flash, the LNG continues boiling during the life of the pool. The longest spill travel occurs when there is minimum LNG evaporation or lowest surface friction. Evaporation is minimized by cool air, cool pavement, and poor heat transfer. The lowest evaporation would occur on a cold winter night. However, friction is minimized during film boiling, which is favored by a hot summer day. Calculations that follow show that the summer temperature case is worse for pool travel. At the terminal site, the lowest annual temperature is 39°F (4°C), and the highest annual temperature is 96°F (36°) [Ref. 17]. The maximum temperature typically occurs on a clear, breezy afternoon, consistent with a pavement surface temperature of about 180°F (82°) due to solar gain. Frictionless flow is assumed when the temperature difference between the substrate and liquid indicates the possibility of film boiling. The summer case is average for about 60 days a year, or 16% of the year. Therefore, the summer temperatures could be exceeded 16% of the time. It turns out that the ability of the LNG pool to travel 400 ft is not sensitive to surrounding temperature within the range that occurs at the site.

The modeling covers two phases of pool spreading: unconfined and partially confined. For simplicity, the first phase begins with the volume of the tanker in the shape of an upright cylinder, the same height as the center of the tanker, located at the tanker unloading station. After the cylinder expands to the width of the driveway, which takes about 3 seconds, the pool shape changes due to partial confinement imposed by the curbs on each side of the driveway.

A-3 Pool spread methodology

The boiling regime and heat transfer rate depend on the temperature of the pavement, which will vary with time due to contact with the LNG. A preliminary step of analysis is to correlate pavement surface temperature as a function of contact time with boiling LNG. The heat transfer from the pavement to the LNG depends on the temperature profile of the pavement and sub-pavement as a function of depth. The starting point profile was created from typical day and night temperatures as a function of time, using thermal properties of concrete and aggregate and a plot of air-concrete heat transfer coefficient as a function of wind speed [Ref. 18]. The summer daytime case requires data on solar radiation, which varies with location [Ref. 17]. The profile of temperature versus pavement depth depends on the recent history of air temperature and solar radiation. For the winter night case, there is radiative heat loss on a clear night from the pavement to the sky which is at a temperature of nearly absolute zero. However, the winter night radiation effect is insignificant. For the summer day case, there is no historical data of pavement surface temperatures, so a pavement profile was calculated by solving the one-dimensional differential equations for conductive heat transfer, starting with the temperature profile at sunrise at an average daily minimum summer temperature, allowing for night solar radiation. The air temperature and solar gain varied until six hours later, when air temperature and solar gain reached the average summer maximum. Several trial calculations showed that the daily temperature cycles did not have a significant effect on the profile below a depth of about 12". The summer temperature at 12" depth is assumed to be the average of the summer minimum and summer maximum air temperatures for the month. The equations for one-dimensional heat transfer in to a thin section of substrate at the surface of the pavement are:

$$Q_{radiation}/A = F_R\ \Delta t$$
$$Q_{convection}/A = h\ (T_{air} - T_1)\ \Delta t$$
$$Q_{conduction}/A = -k\ x\ (T_1 - T_2)\ \Delta t$$
$$\Delta T = (Q/A)/m\ C_P$$
$$m = \rho\ A\ x$$

where Q is the heat transfer rate
A is the transverse heat transfer area
F_R is the solar heat flux
Δt is the differential time step
T_1 is the temperature in the 1st section of pavement
k is the thermal conductivity of the pavement
x is the thickness of the section of pavement
ΔT is the temperature change in a section during time dt
m is the mass of the section
C_P is the specific heat capacity of the pavement

ρ is the density of the pavement

By taking a small enough time step, the heat transfer equations can be solved accurately enough by using the temperatures from the previous step. The exception occurs at the first section, where the temperature changes rapidly with time. This section was solved by obtaining the difference between the first and second sections by trial and error. Trial and error is easily done by manually adjusting the temperature difference in the spreadsheet for each time step. This procedure was used to calculate the temperature profiles for the summer case influenced by solar gain, and for the profile when pavement is exposed to LNG.

A summary of the summer temperature profile at different pavement depths is tabulated below. The bottom row shows the profile at the hottest part of a typical summer day:

Table A-1
Pavement Temperature Profile before an LNG Spill
Pavement sections are 1" thick. Section numbering begins with the section on top.
The profile for the hottest hour (hour 6) is used for spill calculations.

Time h	T_{air} °F	Solar Btu/h-ft^2	Section 1	Section 2	Section 3	Section 4	Section 5	Section 6	Section 7
0	60.4	0	59.1	61.1	63.1	65.1	67.1	69.1	71.1
1	66.4	50	69.9	68.2	68.1	68.5	69.3	70.3	71.4
2	72.5	100	85.4	80.2	77.6	75.9	74.9	74.4	74.4
3	78.5	150	105.1	96.4	91.1	87.0	83.9	81.7	80.1
4	84.5	200	128.6	116.3	108.0	101.2	95.8	91.4	88.0
5	90.6	250	155.3	139.3	127.9	118.3	110.2	103.4	97.7
6	96.6	250	178.6	161.2	147.7	135.9	125.5	116.4	108.4

When LNG is spilled onto pavement in the summer or winter, the pool stays in the film boiling regime until the surface of the pavement cools to within 20°C (36°F) of the LNG boiling point. In the film boiling regime, there is virtually no friction loss. As the tanker spill spreads to fill the area between the curbs of the driveway, the liquid is modeled as an upright cylinder in film boiling, non-frictional flow. For lack of better information, it is assumed that the vapor film under the liquid is of uniform thickness. Energy loss due to internal turbulence is ignored, which overestimates velocity and distance travelled. In the case of frictionless, gravity-induced open flow, the change in elevation equals the resulting velocity head: horizontal velocity squared divided by twice the acceleration of gravity, where the velocity is the average value across the cross-sectional area of flow.

$$z = v^2/2g$$
$$v = (2g\ z)^{1/2}$$

The velocity head equation applies to both components of pool motion – downhill motion of the entire pool, and radial spreading. In the case of downhill motion, z_d is the horizontal distance travelled, Δy, (including the downhill radius of the cylinder) multiplied by the slope of the surface.

$$v\,\Delta t = \Delta y = (2g\,z_d)^{1/2}\,\Delta t$$

In the case of radial spreading, z_r is the height of the pool at the center

$$\Delta r = (2g\,z_r)^{1/2}\,\Delta t$$

Neglecting viscous forces, the initial radial increase equals $(2g\,z)^{1/2}\,\Delta t$, where Δt is the time interval. Prior to considering evaporation loss, the cylinder volume is the same as before, but the cylinder is shorter by Δz. Therefore, the new pool height is the volume of the pool divided by the area calculated from the new radius.

In the beginning, the pool is deep, and the time required to reach the curbs is short, so it is not yet important to model the effect of the driveway slope. However, when the elevation difference between the uphill and downhill portions of the pool becomes significant relative to the pool thickness, the slope should be considered. The cylinder can be partitioned into four quadrants: uphill, downhill, and two sides. Without surface friction, the top of the cylinder is level, so each point on the circumference has a different height above the pavement. The sloped z values for the uphill and downhill quadrants are used to calculate new radii of the uphill and downhill quadrants. The net effect of the longer downhill radius is that the cylinder will spread farther in the downhill direction and become slightly elliptical.

The evaporation loss is the combination of evaporation from the surface of the liquid plus the vapor generated from boiling. The boiling heat transfer rate for LNG can be conservatively calculated by using the minimum heat flux value of 20 kW/m^2 when the temperature difference between the substrate and LNG exceeds 20°C [Ref. 16]. Evaporation from the surface is due to the combined effects of heat input from solar heat and convective heat transfer with the air. Evaporation is assumed to be limited by heat flux, not diffusion into the air. The heat flux terms for solar heat and air convection are the same as used for calculating the pavement temperature profile for a summer day.

As the pool spreads thinner, there is a time when Δz exceeds the pool thickness on the uphill edge of the pool. When this happens, the uphill edge is moved so that the thickness becomes zero at the uphill edge.

A-4 Results

Upon release to atmospheric pressure, 17.3% of the LNG flashes, based on an adiabatic flash, taking data from the methane pressure-enthalpy diagram.

The one-dimensional, time-dependent heat transfer calculations showed that on an example hot summer day, concrete that is in contact with boiling LNG can take almost 10 minutes before the liquid is in frictional contact with the pavement. This is because a film of natural gas vapor insulates the concrete, maintaining a high temperature difference between the boiling LNG and the pavement surface. Film boiling occurs for about 8.7 minutes in the summer example. For the next minute, boiling becomes more nucleate, finally reaching the point of maximum heat flux. The same spill on a cold winter night showed that the duration of film boiling was about half as long at this site, but still greatly exceeding the time required for the pool to evaporate. Other factors, such as difference in evaporation rate between summer and winter are not important. Therefore, because flow will be essentially frictionless during film boiling, the hot summer case is the worse case for distance travelled by the moving pool of LNG.

Example B

Piping Leak onto the Sloped Deck of a Floating Oil Production Platform

B-1 Accident scenario

There is a potential for a large fire on a lower deck of a floating oil production platform. To assess the potential for this fire to cause escalating damage to the platform and possibly interfere with escape, the size of the pool must be estimated. The area of the oil spill is the major factor affecting the height and heat flux of the potential pool fire. This particular spill case is one possible event of an event tree performed for a quantitative risk assessment. The length of the spill path is known, but the width of the pool as it flows down the deck must be estimated by flow calculation.

The spill source is a manifold of 12-inch pipe on a deck made of metal checker plate of the kind typically used in industrial applications. The piping is supported every few meters, so that the piping does not interfere with the flow of oil across the deck. The deck plates are galvanized metal with tread ridges to provide a non-skid surface. The deck has a slope of 2.5% from the keel line down to the edge of the deck, 31.5 m away. The lower deck is shielded from sunlight, and there is little difference in temperature between the air, deck, piping, and contained oil.

The potential leak consists of light crude oil that has been separated from produced water and gas. The crude oil contains enough dissolved gas to be flammable, but the oil loses an insignificant volume of light components when flashed to the atmosphere. The crude oil is contained at a maximum pressure of 8 barg. The transport properties of the oil at flowing temperature are:

Density	889 kg/m^3
Viscosity	11 cp, or 0.00112 kg s/m^2
Mol Wt	250 kg/mol
Surf tension	27 dyn/cm, or 0.027 N/m

Based on the number of potential leak points and the probability of leakage from each type of potential leakage, the probability of a serious leak is on the order of 10^{-3} per year for a leak equivalent to a 50-mm diameter hole. The most important contributor to serious leakage is gasket failure at a flange or valve. Nearly all flanges are arranged vertically, with the opening aimed downhill. Ordinarily, any leak onto the deck will be quickly detected by flammable gas detectors, resulting in a manual shutdown of the piping system. In the accident case for the fault tree branch of interest, the detector alarm system or operator response fails, which is estimated to happen less than 10% of the time, which lowers the probability to 10^{-4} per year. The probability of ignition is about 5% to 10%, which places the accident scenario in the barely credible range. The average wind speed is 4.3 m/s, directed crosswise to the slope of the deck. Extreme weather conditions at the time of the accident would make the scenario beyond credible, and consequently need not be considered.

B-2 Release modeling

The maximum credible leak, equivalent to a 50-mm hole is likely to occur as a narrow gap rather than a round opening, either as a tear in piping due to impact or a gap between flanges due to improper bolt tightening or gasket failure. Round holes often result from corrosion, which is not a problem with crude oil in this case. The gasket thickness is 2 mm, so a 2 mm gap is assumed. A full blow-out of a 2 mm gasket from a 12-inch line results in an opening with an area of 1920 mm^2, which is about equal to the area of a 50-mm hole.

The flow rate through the leak opening can be calculated from the flow coefficient [Ref. 7], which is a function of the Reynolds number for the pipe. The Reynolds number for the pipe is 17,400 and the ratio of orifice diameter to pipe diameter is less than 0.2, which gives a square-edge flow coefficient, C, of 0.60. The standard equation for flow through an office gives the leak flow rate. Based on the pressure drop, density, and equivalent orifice size, the leakage flow is 0.0500 m^3/s. The liquid velocity at the release point is the volumetric flow rate divided by the flow area, or 26.0 m/s.

For holes less than 2 mm wide, the drop diameter is about 1.9 times the narrow dimension (2 mm in this case) [Ref. 1, 11]. This correlation gives a drop diameter of 3.8 mm. A second method for calculating drop size is based on the Weber number being between 12 and 22 [Ref. 1]. For a Weber number of 12,

$$d = 24\,\sigma/(\rho_g\, v_d^2).$$

For the site air density of 1.204 kg/m^3, surface tension of 0.027 kg/s^2, and drop velocity of 26.0 m/s, the estimated drop diameter is between 0.80 mm and 1.46 mm. The free fall velocity for a drop between 0.80 mm and 3.8 mm is between 3 m/s and 10 m/s [Ref. 12]. The chosen values are 2.3 mm and 2 m/s.

The release will break up into droplets at a distance of L_b, which is a function of the Weber number and Reynolds number. For a drop diameter of 2.3 mm, and an air density of 1.204 kg/m^3, the Weber number is 34.7. The Reynolds number at the release point is 47,500. The distance from release to break-up into drops is

$$L_b = 540\, D\, W_e^{1/2}\, R_e^{-5/8}$$

In this case, L_b = 0.0874 m.

The trajectory of the release begins as a circular sheet 2 mm thick sprayed from a pair of vertical flanges. Before the sheet contacts the deck below, it breaks into droplets and the velocity slows to 2 m/s, relative to the air. For the portion of the spray directed straight down, that liquid lands directly under the leak point. For the portion of the spray directed up, the droplets will be in the air for about 0.36 s, calculated from the free fall drop velocity. The wind force on a droplet F_d, is a function of the drag coefficient C_d, projected drop area facing the wind A_d, air density ρ_a, and relative velocity v between the drop and air [Ref. 12]:

$$F_d = \tfrac{1}{2}\, C_d\, A_d\, \rho_a\, v^2$$

From Newtonian physics, acceleration of the droplet is force divided by mass:

$$a = F_d/(V_d \rho_d)$$

Distance travelled by a droplet is:

$$y = \frac{1}{2} a t^2$$

At the free-fall velocity, the droplet Reynolds number is 12,900. For droplet Reynolds numbers exceeding 1,000, C_d has a value of 0.445. The volume of a spherical droplet is:

$$V_d = \pi d^3/6$$

Solving the above, the wind will move liquid droplets downwind around 0.2 m, crosswise to the slope of the deck. At the same time, the trajectory of the leak will move droplets up to 2 m downhill, depending on the position on the vertical flange from which the droplet was released.

B-3 Pool spread methodology

The exact direction of leakage (and splashing against objects) cannot be predicted, but the model should be plausible. The release of crude oil is assumed to be continuous and at a constant flow of 0.0500 m^3/s. The starting point of the pool is directly below the leak point, located 31.5 m from the downhill edge of the deck. The initial pool area is 2 m in radius, based on droplet dispersal radius from the assumed leak due to gasket failure. After the release lands on the deck, the effect of wind is insignificant compared to gravity. Evaporation is not significant. The fluid properties of the oil remain constant. Friction loss is calculated from the Darcy equation for laminar flow. The pool flow is in the laminar regime at the edges of the pool, and is a conservative bound on areas that are in the turbulent or transition flow regime.

The pool is modeled as a conical volume with the peak located directly under the leak point. In the case of gravity-induced open flow, the change in elevation equals the resulting velocity head plus head loss from friction. The velocity head is the horizontal velocity squared divided by twice the acceleration of gravity. The friction loss is a linear function of the area of contact and the Reynolds number.

$$z = v^2/2g + h_L$$

$$v = (2g\,(z - h_L))^{1/2}$$

To include the effect of the sloping deck, the conical pool is partitioned into four quadrants: uphill, downhill, and two sides. The height of the peak is the same for all quadrants, but in each quadrant the vertical distance between the peak and the edge can be different. The radius for the uphill quadrant is adjusted as needed so that the uphill edge is never higher than the peak. As pool spreads faster in the downhill direction, the pool model changes from a right cone with a circular base to an angled cone with an elliptical base. The contact area for a quadrant of a cone is $\frac{1}{4} \pi r^2$. In this example, the quadrants are calculated as sections of cones of different radii, which do not match at the edges. The result is a pool area that looks like a wind rose. This mismatch results in an error in pool area calculated as an ellipse. It would have been possible to calculate the area more correctly or increase the number of sections from four if it had seemed that this error was important.

Because the friction factor is a linear function of Reynolds number in laminar flow, the average friction factor occurs at the radius associated with half of the area, r divided by the square root of

2, or approximately $\underline{r} = r/1.414$. The Darcy equation for friction loss, with the hydraulic diameter, $8z$, substituted for pipe diameter, is

$$h_L = f\,L\,v^2 /(16\underline{z}\,g)$$

where $\underline{z}$ and v are evaluated at $r/1.414$, and the quadrant radius, r, is substituted for L.
For laminar flow, the Darcy equation for friction factor is

$$f = 64/Re$$

where

$$Re = 8\,\underline{z}\,v\,\rho\,/\mu$$

giving the following for laminar flow friction head

$$h_L = \mu\,r\,v\,/(2\underline{z}^2\,\rho\,g)$$

For turbulent flow ($Re > 4{,}000$), the Churchill equation for the Fanning friction factor can be used [Ref. 6]

$$1/f^{0.5} = -4\,\log\{0.27\epsilon/D + (7/Re)^{0.9}$$

where ϵ is the surface roughness. The roughness for galvanized steel is 0.152 mm. The variable D is replaced with the equivalent value of $8z$. The Fanning f must be multiplied by 4 for use in the Darcy equation.

After the average friction loss is calculated at $\underline{r}$, the velocity at that point can be calculated. Then the outward spread at that point can be calculated by

$$.\ \Delta\underline{r} = v\,\Delta t$$

The outward spread at the radial edge of the pool, Δr, is calculated by assuming that the ratio between $\underline{r}$ and r does not change:

$$\Delta r = 1.414\,\Delta\underline{r}$$

The equation for Δr and even the proposal that pool spreading can be modeled with a conical shape has not been validated by computational fluid dynamics or actual experiments.

The deck material is assumed to be flat after installation, and has typical checker plate ridges. The ridges rise from the base at an angle of 45° to a height of 1.5 mm. The ridges occupy about 25% of the surface area. Each ridge has about 1.4 times the surface area of the flat base, which increases the overall area of the plate by 8%. Therefore, the ridges increase the hydraulic radius by a factor of 8% above that of flat plate. The roughness value, ϵ, is estimated to be the same as for galvanized steel pipe.

B-4 Results

Table B-1

Oil Spill Spread on a Sloping Ship Deck from a Constant Continuous Leak

The radii for the side, downhill, and uphill quadrants are labelled ***a***, ***b***, and ***c***, respectively.
The height of the conical peak, **z**, is relative to the deck directly below.
The velocities for the quadrants are calculated at the centroid.

Time s	V m^3	a m	b m	c m	z m	v(a) m/s	v(b) m/s	v(c) m/s
0	0.005	2.00	2.00	2.00	0.001	0.004	0.790	0.088
1	0.055	2.28	3.65	0.33	0.012	0.271	1.136	0.175
2	0.105	2.67	5.80	0.44	0.012	0.269	1.409	0.123
3	0.155	3.03	8.40	0.44	0.011	0.238	1.675	0.000
4	0.205	3.35	11.46	0.44	0.010	0.204	1.944	0.000
5	0.255	3.62	14.98	0.44	0.009	0.175	2.207	0.000
6	0.305	3.85	18.96	0.44	0.008	0.150	2.486	0.000
7	0.355	4.05	23.40	0.44	0.007	0.130	2.759	0.000
8	0.405	4.26	28.30	0.44	0.006	0.141	3.034	0.000
8.6	0.435	4.43	31.46	0.44	0.006	0.152	3.199	0.000

As an overall characterization of a spill on a sloped surface, the spill can be viewed as having a roughly triangular footprint. In this case, with a surface slope of 2.5%, the spill could be viewed as a triangle 31.5 m long and 8.9 m wide at the base. This corresponds to a divergence angle of about 8° from the centerline of flow.

Uncertainties in the results include the question of whether the conical model is representative of the action of an actual liquid spill at an industrial site. There is no published data that could be found containing a friction factor for checker plate decking. The checker plate ridges may have an effect on flow pattern, such as deflecting the flow. In this case, the flow pattern that has the largest area (greatest width before stopping at the edge of the deck) gives the worst consequences. There was no study done for this report to investigate the effect of various conditions affecting the width of the flowing pool. However, results are based on recognized flow equations and should be viewed as useful in the absence of experimental or computational fluid dynamics results.

440778 Apparent Discrepancies in Methods for Predicting the Explosion Energies of Bleves

Abstract:

The authors have used calculation methods for BLEVE blasts as described in the CCPS book *Guidelines for Vapor Cloud Explosion, Pressure Vessel Burst, BLEVE and Flash Fire Hazards*, and most recently similar methods in the PHAST process industry hazard analysis/consequence analysis software tool, developed by DNV-GL. Predictions made using PHAST appeared to be overly conservative, which instigated a comprehensive comparison of PHAST to other methods in the literature. There is not universal agreement concerning the dominant mechanisms driving the air blast pressure wave from a BLEVE, and there are major differences in methods used to predict the explosion energy driving the blast wave. The CCPS guidelines book and PHAST consider the explosion energy driving the blast wave to be from both the vapor and liquid phases present at the time of loss of containment. For each phase the contribution to explosion energy is the difference in internal energy between the thermodynamic states of the liquid and vapor at the time of loss of containment and the final states of the materials when isentropically expanded to atmospheric pressure. Recently, Birk compared various prediction methods to experimental data for propane BLEVEs and concluded that the flash vaporization of the liquid may be too slow to contribute significantly to the blast, therefore the blast energy is primarily from the expansion of the vapor in the vessel at the time of loss of containment. Other authors also base the explosion energy on the gas expansion and others on the liquid superheat. Ogle and coworkers have discussed the available energy of a BLEVE (exergy) and compared this to the BLEVE explosion energy predicted using several methods in the literature. These various approaches can lead to drastically different predictions of explosion energy and the resultant blast pressures and impulses. BLEVE theories and models are discussed in this paper and predictions are compared to limited experimental data. This paper is meant to provide comparisons between the many various approaches, including many more that are not listed above, and summarize their differences.

441089 Automated Optimization of Gas Detector Placement

Abstract:

Chemical process facilities are moving toward the use of performance-based design techniques for fire & gas detector placement. These approaches are displacing methods for layout of gas detection based on uniform "grids" that date back more than 25 years. First generation performance-based design techniques utilized 'gas volume detection' principals that are valid in terms of detecting and mitigating a large explosion event, but can result can result in an excessively large number of detectors and a costly design. This paper presents a superior alternative to the first generation technique of volume detection that uses sophisticated modeling of gas release scenarios with computational fluid dynamics (CFD). The techniques described in this paper align with ISA TR 84.00.07 guidelines for scenario-by-scenario basis. Furthermore, the paper breaks the paradigm that CFD can only be used sparingly on complicated gas dispersion problems. With the proper definition of the process, optimal selection of boundry criteria, and automated scripting of multiple scenarios and data collection from those scenarios, the thousands of CFD runs required to develop results can be efficiently and cost effectively performed. This paper will demonstrate practical application of CFD modeling and probabilistic risk analysis in order to improve on the performance of gas detection systems while at the same time, reducing unnecessary conservatism in the number of detectors. The paper will present methodology for determining scenario-based gas detector coverage and include a worked application example for an offshore gas processing platform.

441414 Quantitative Risk Assessment in the Process Industries: Beyond Numerical Estimates of Individual and Societal Risk

Abstract:

The Chemical Process and Oil and Gas industries (both on and offshore) have largely adopted Quantitative Risk Assessment (QRA) with open arms following its application in the nuclear industry. Currently, the methodology not only forms a core aspect of the risk management programs of numerous organizations who design, build, own, manage or operate high hazard facilities (refineries, LNG terminals etc.). It also occupies a central role in the legislative and regulatory framework in many jurisdictions across the world. It has become a tool of choice for both the regulators and the regulated parties.

As its application has become more widespread, it has been the subject of extensive critical inquiry, much of which has revolved around the primary product of the process, namely numerical estimates of risks to exposed people. Advocates highlight the utility and rationality such numbers offer in the often challenging decision making process associated with risk management. Critics make reference to the wide variability and uncertainty associated with such predictions as well as the implicit value judgments they embody. Less emphasis has been placed on the wealth of other insights (often latent) afforded by a well scoped out, executed and documented QRA.

The aim of the paper is to examine and explore the value of QRA over and beyond the well-known orthodox output of individual or societal risk estimates. It identifies and outlines what these added benefits are and elaborates on their significance - particularly in the context of achieving risk management objectives. It argues that such latent benefits should be considered to be equally or more important to the typical output and that more attention should be paid to these whilst conducting a QRA. A popular adage goes thus "sometimes, it's the journey that teaches you a lot about the destination"; so it is with QRA.

12TH GLOBAL CONGRESS ON PROCESS SAFETY

Quantifying the Importance of Accurate Physical Property Data in Dispersion Modeling

Devin Averett
Baker Engineering and Risk Consultants, Inc.
11011 Richmond Avenue, Suite 700
Houston, TX 77042
daverett@BakerRisk.com

Jef Rowley
Baker Engineering and Risk Consultants, Inc.
11011 Richmond Avenue, Suite 700
Houston, TX 77042
jrowley@BakerRisk.com

Prepared for Presentation at
American Institute of Chemical Engineers
2016 Spring Meeting
12th Global Congress on Process Safety
Houston, Texas
April 11-23, 2016

Quantifying the Importance of Accurate Physical Property Data in Dispersion Modeling

Devin Averett
Baker Engineering and Risk Consultants, Inc.
11011 Richmond Avenue, Suite 700
Houston, TX 77042
daverett@BakerRisk.com

Jef Rowley
Baker Engineering and Risk Consultants, Inc.

Keywords: Dispersion Modeling, Physical Properties, DIPPR

Abstract

Modeling discharge and dispersion of materials from processing vessels is an important part of a facility siting study. In this paper, we investigate the dependence of dispersion modeling on physical property data accuracy, including the impact of uncertainty in the lower flammability limit, heat of vaporization, liquid density, and vapor pressure data. Vapor pressure is further analyzed by comparing correlations obtained from the Design Institute for Physical Properties (DIPPR) to the Clausius-Clapeyron equation using the normal boiling point as the reference.

Overall, the predicted dispersion distances are relatively resilient to uncertainty in physical properties. However, errors in dispersion distances can be large under certain conditions. It is important to have accurate data for these cases. To ensure data accuracy, we recommend using a modern, critically reviewed database such as the latest DIPPR database, which contains thermophysical data for a large number of compounds and routinely employs rigorous analysis and consistency checks to assure the reliability of the data.

1 Introduction

Dispersion modeling can be a main driver for many major economic and safety decisions. Modeling of a toxic release may help dictate where and what type of personal protective equipment (PPE) is needed, where to place high-occupancy buildings and work areas, and whether risk to personnel or the surrounding community warrants investment of capital into mitigating the effects or decreasing the likelihood of such a release. Modeling of a flammable vapor cloud may provide similar information regarding safety from the risks posed by flash fires and explosions, but it can also aid in estimating the economic impact from explosion events by determining which buildings and equipment are likely to be damaged. This impact is not just in the cost of replacing equipment, but also in the loss of production during the repair downtime. This information may in turn determine insurance rates and proper amounts of coverage.

With such important decisions hinging on the results of a model, it is vital that the model is reliable and provides as accurately as possible a representation of what will happen during an event, especially when there is already so much inherent uncertainty due in large part to the natural variability in the weather. So selecting a good model that will make the best of a difficult situation is vital. When modeling a dispersion scenario, there are many models available. Typically the most accurate models are based on computational fluid dynamics (CFD), as most other models do not explicitly account for the presence of obstacles; however, use of CFD is computationally expensive and impractical if it is necessary to model numerous release scenarios subject to various weather conditions. Other models have been developed that sacrifice some accuracy in the presence of obstacles to gain computational speed.

Whatever the model, similar inputs are typically required, many of which are scenario-dependent (e.g., atmospheric conditions such as wind direction and speed, weather stability class, etc.). The inputs that remain constant across all different scenarios are the physical properties of the material being released. Unlike the weather, a material's chemical properties do not exhibit large and unpredictable fluctuations. So if there is any input to the model in which one can place a reasonable level of confidence, then surely it is the physical property data.

Of course there is disagreement in the literature pertaining to physical properties. Aside from the more obvious sources of error (human error, bad equipment calibration, etc.), different experimental techniques may yield differing results, and there is even variability in experiments employing the same or similar techniques. The trouble, then, is determining which literature value is likely to be the most correct, which may not be an easy task.

In this paper, we explore whether the effort of finding accurate material properties for consequence modeling is worth the effort and cost. If the physical properties are not exact, does it make a significant difference in dispersion model results? We take a quantitative approach to answer this question by utilizing the dispersion model currently used in SafeSite$_{3G}$®, BakerRisk's in-house software for consequence modeling. By varying a number of different properties, we can determine the impact they have on the final dispersion results. Conclusions can then be drawn on the importance of obtaining the best physical property data in dispersion modeling.

2 Methodology

Dispersion model results can't be summarized in a single number. There are many points of comparison that could have been used in our analysis. In this study, we use for our basis of comparison the horizontal distance at which the centerline of our release reaches a concentration equal to the lower flammability limit of the material.

The dispersion model used is detailed by Rowley [1]. In short, it solves the conservation of mass, momentum, and energy in one dimension, and then concentration profiles in the vertical and cross-wind directions are obtained based on similarity concepts. The model itself actually draws from several other ancillary models for handling phenomena such as droplet trajectory, pool formation, and droplet and pool evaporation. Physical properties for materials used in the model were obtained from the most recent version of the DIPPR® 801 database [2].

Several different physical properties were varied to monitor the effect any variance in the values would have on dispersion distances. These were chosen based on which properties intuitively would seem to have the greatest impact on dispersion results; namely lower flammability limit (LFL), heat of vaporization (HVP), liquid density (LDN), and vapor pressure (VP).

Several chemicals were chosen for the analysis, selected to give a wide range of normal boiling points (NBP) to ensure materials of different volatilities were adequately explored. The chemicals, in order of increasing normal boiling points, are methane, ethylene, propylene, ethylene oxide, methanol, *n*-octane, phenol, 1-decanol, glycerol, and *n*-eicosane. The intent was to select fairly common chemicals that differed in normal boiling point by approximately 50 K.

Dispersions were run at two different weather conditions, both at atmospheric pressure of 1 atm and an ambient temperature of 70 °F. The first weather condition used a wind speed of 1.5 m/s and a Pasquill stability class of F, and the second used a wind speed of 3 m/s with a D stability class. In addition, different dispersion runs were performed for different storage temperatures and pressures. Pressures of 10, 100 and 1000 psig were run at temperatures ranging from -275 to 675 °F in 50 °F intervals. Releases where the material exists as a solid at those storage conditions were ignored.

To establish a baseline set of results, dispersions were run for the different materials with physical properties set equal to the accepted values from DIPPR. Additional dispersion runs were then gathered where one physical property was altered slightly. Each property was altered by 5% in either direction. (Note that if altering the property resulted in a phase change at the storage conditions, these cases were omitted from the analysis, the presumption being that even if a physical property may not be accurately known, it is known in which phase the material is being stored.) We were then able to obtain an estimate as to how much the dispersion distances change in relation to a standard step change in a specific physical property. The vapor pressure analysis was further extended by looking at how dispersion results differed when using the Clausius-Clapeyron equation using the normal boiling point from DIPPR as the reference point. This was done to see how important it is to obtain accurate data across a broad range of temperatures rather than simply at a single point and extrapolating from there numerically with a basis on theory.

3 Results and Discussion

Taking one property at a time, results from the dispersion runs are summarized below, accompanied by some discussion on implications that can be drawn from the data. All results are plotted as:

$$\frac{\%\Delta(Distance\ to\ LFL)}{\%\Delta(Physical\ Property\ of\ Interest)}\ vs.\ \frac{Process\ Temperature\ (K)}{NBP(K)}$$

where $\%\Delta$ refers to a percent change in the pertinent value.

3.1 Lower Flammability Limit (LFL)

The flammability limit has the most straightforward effect on dispersion distances. This is because it has no effect on the dispersion model whatsoever. It merely changes the target concentration we are looking for when calculating the extent of the plume to LFL. Figure 1 shows the results for all runs.

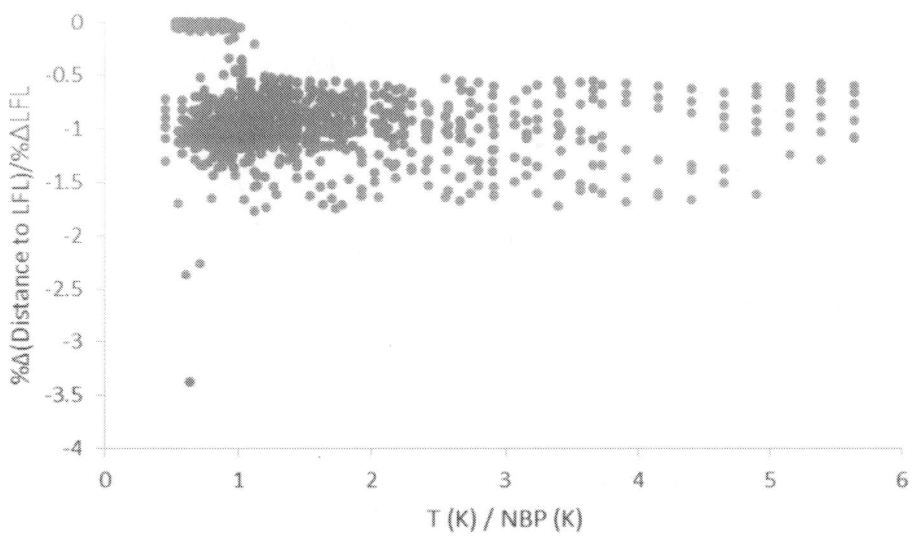

Figure 1. Ratio of change in dispersion distance and change in LFL vs. ratio of process temperature and NBP.

The uppermost cluster showing virtually no change in dispersion distance represents those scenarios that form highly non-volatile pools. Otherwise, all scenarios show a decrease in dispersion distance with an increase in LFL, which intuitively makes sense. The largest deviations occur for materials near saturation (T/NBP = 1). On average, dispersion distances changed only slightly, with an average change of -0.81 % in dispersion distance for a 1 % increase in LFL and a standard deviation of 0.45 %. While this may not appear to be very significant, the trouble with LFL is the inaccuracy in measuring LFL. Literature values may vary widely, and these large discrepancies will result in large differences in dispersion distance. For example, within the DIPPR database, looking at only methane, the accepted value of 5 % vol [3] differs by as much as 12 % from other experimental values (a value of 4.4 % vol [4]). A previous study has shown that large discrepancies can occur simply due to the method of measurement. Data obtained using the US Bureau of Mines tube method can give dispersion distances to LFL that differ by as much as 200 % from data obtained using the EN 1839 tube method [5]. Indeed, the same study discusses the reality that rather than a discrete value, the LFL exists as a range of concentrations that may yield varying degrees of flame propagation, and different methods of measurement may give measurements biased towards one end of this range or the other. Fortunately, in this study we found that small differences in LFL generally didn't have a large impact on dispersion distances, but certainly caution should be taken in cases where the LFL is low or wide ranges of LFL data are available in the literature.

3.2 Heat of Vaporization (HVP)

For heat of vaporization, we examine a much narrower range of conditions. This is because for HVP to play any significant role at all, an appreciable amount of both liquid and vapor phases must be present. Thus high-temperature vapor releases and low-temperature liquid releases would give virtually the same results for large changes in HVP because a negligible amount, if any, of vaporization is occurring. Figure 2 shows the results from the analysis.

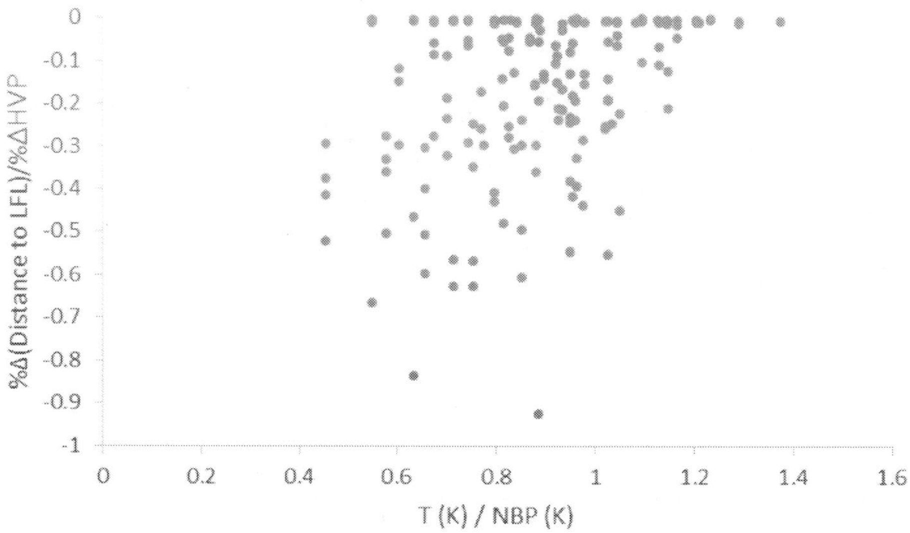

Figure 2. Ratio of change in dispersion distance and change in HVP vs. ratio of process temperature and NBP.

Similar to LFL, we see that small changes in HVP result in approximately the same order of magnitude of change in dispersion distances. For every one percent increase in HVP, the distance to ½ LFL will on average change by -0.17 %, with a standard deviation of 0.19 % and a minimum of -0.93 % occurring in the cases we analyzed. And this, remember, is only for cases where any appreciable phase change is taking place. If at least somewhat reliable HVP data can be obtained, then we can be confident that it will not affect dispersion distances to any significant degree, and especially not if the scenario does not include a phase change.

3.3 Liquid Density (LDN)

The effect of liquid density on dispersion models is much more complex than the effect of LFL or HVP. First of all, liquid density will affect the discharge rate. It would seem logical that a higher density would lead to a material that would discharge at a greater rate from a specific orifice, suggesting greater dispersion distances. However, this would affect other characteristics of the flow as well such as the choke pressure. The change in choke pressure can cause additional issues, particularly in cases where the liquid flashes upon release. Any change in pressure will then have a large effect on the vapor fraction of the discharge, and hence will cause a large swing in the overall density of the initial discharge.

The second issue to consider is of plume buoyancy. For plumes of high liquid fractions, an increase in liquid density will cause a decrease in plume buoyancy, resulting in the liquid raining out at a nearer distance. This would likely result in shorter dispersion distances, which could either supplement or compete with any effects from the change in discharge.

Although there are other issues affected by LDN, the last issue we will mention here is that momentum must be conserved. Given two identical droplet-filled plumes (both plumes start with the same velocity and have the same number of droplets with equal diameter) with the only difference being the density of the material in the droplets, and assuming an equal air entrainment rate between both plumes, conservation of momentum dictates that the denser plume will decelerate at a slower rate than the less dense plume. So with these simplistic plumes, the liquid in the denser plume will rain out at a farther distance, exactly the opposite conclusion that was reached when considering plume buoyancy. We see that with LDN, trying to reason out what effect changing LDN might have on a single scenario without running the model can become very complicated.

We turn to the model, then, for some answers. Figure 3 shows results for LDN dispersion runs. Again, some data was omitted because it was not relevant to the analysis, namely the cases where no liquid was present throughout the scenario. Note that unlike previous properties, both negative and positive deviations in dispersion distances are seen, and the differences become more pronounced the more the temperature is decreased. The point that gives the largest deviation (a surprising 13.1 % increase in dispersion distance per one percent increase in density) is phenol at P = 1000 psig and T = 275 °F for the F-classified weather. Under these storage conditions, phenol exists as a liquid. Looking at this specific case in more detail, we see that the droplets in the denser plume rain out more quickly (less than a second compared to about 11 seconds), but the velocity in the less dense plume is knocked out very quickly, causing rainout to occur at about 28 ft compared to the denser plume at about 60 ft.

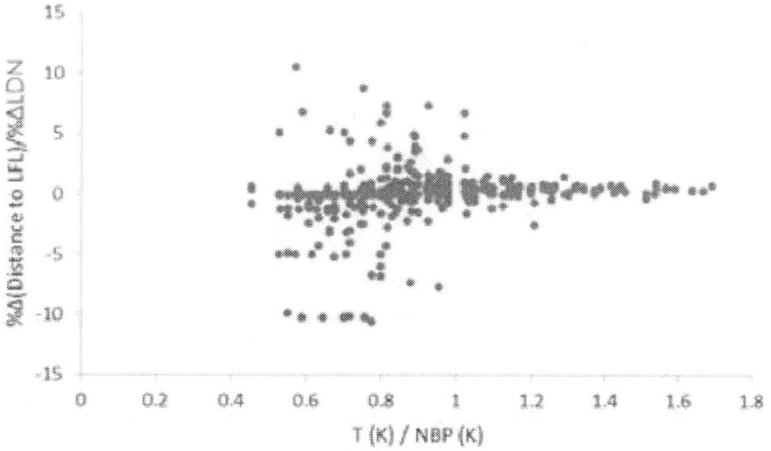

Figure 3. Ratio of change in dispersion distance and change in LDN vs. ratio of process temperature and NBP.

On average, a one percent increase in LDN results in only a 0.08 % increase in dispersion distances, but that has a standard deviation of 2.21 %. Thus, having inaccurate LDN data can cause large differences in dispersion results, and those differences can be difficult to predict. Fortunately LDN is a relatively simple property to measure, but if the scenario deals with a wide range of temperatures it is imperative to have accurate temperature-dependent data for LDN.

3.4 Vapor Pressure (VP)

3.4.1 Deviation in Vapor Pressure

Like HVP, vapor pressure can really only affect releases where two phases are possible. Hence, pure vapor releases are ignored, as well as releases low enough in temperature to form pools with no significant amount of evaporation. Also ignored are cases where the change in VP would result in a phase change at process conditions, the assumption being that the initial phase is known.

Like LDN, VP may have competing effects on dispersion modeling. An increase in VP would cause a flashing liquid upon discharge to have an increased vapor fraction and thus a decreased mass flow, likely suggesting shorter dispersion distances. However, it also means droplets and pools evaporate at higher rates, serving to lengthen dispersion distances.

Figure 4 shows the results from the VP runs. As would be expected, most of the variability occurs when the temperature of the fluid is near the normal boiling point where a two-phase release is much more probable. The most extreme result (an 11 % increase in dispersion distance per 1 % increase in VP) come from *n*-eicosane at P = 100 psig and T = 625 °F for the D-classified weather. In this case, the difference is largely buoyancy driven. For higher VP, the resulting excess vapor is sufficient to increase the buoyancy of the plume and significantly delay rainout, keeping the concentration and mass of material in the plume at higher levels for a longer time.

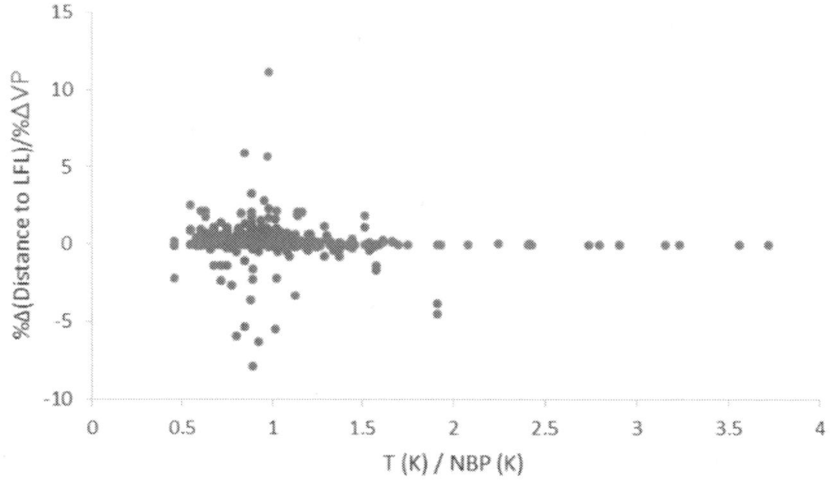

Figure 4. Ratio of change in dispersion distance and change in VP vs. ratio of process temperature and NBP.

The average result is only a 0.18 % increase in dispersion distance per 1 % increase in VP, but with a standard deviation of 1.19 %. While it does not tend to affect dispersion distances to the same degree as LDN, VP does have the potential to significantly alter dispersion results.

3.4.2 Clausius-Clapeyron Equation

It was also of interest to see if one could still get reliable results using the Clausius-Clapeyron equation, or at least the well-known approximation of it, shown here:

$$\ln\left(\frac{P_1}{P_2}\right) = \Delta H_{vap}\left(\frac{1}{T_2} - \frac{1}{T_1}\right) \qquad (1)$$

where P_1, T_1 and ΔH_{vap} are the vapor pressure, temperature, and heat of vaporization, respectively, at a reference state, and P_2 and T_2 are the vapor pressure and temperature, respectively, at some new state. Hence, at the reference state, the equation is exact, but errors will develop and worsen as the desired temperature moves farther from the reference temperature. Often, heat of vaporization data is available at the normal boiling point of a material. To establish a basis for this study, then, we select the normal boiling point of the material to be our reference point, using the HVP value obtained from the DIPPR database at this temperature. The DIPPR database base correlations for HVP and VP are based on rigorously evaluated experimental data, so we can be confident that the DIPPR correlation will always result in a more accurate answer than the Clausius-Clapeyron equation, regardless of the temperature. Hence, in the following discussion we assume any deviation from results derived from DIPPR data are unfavorable. Again, pure vapor releases, releases low enough in temperature to form pools with no significant amount of evaporation, and cases where the change in VP would result in a phase change at process conditions are ignored.

Earlier results relating to vapor pressure would suggest that using the Clausius-Clapeyron equation would likely have little effect on dispersion distances since significant deviations seen in Figure 4 are concentrated around the NBP, the area in which the Clausius-Clapeyron (as we have defined it) should be the most accurate. Following the same procedure as the other analyses (testing several materials at several temperatures and pressures), there surprisingly were some significant deviations in dispersion distances. Figure 6 and Figure 6 show the results. Note that because we are not dealing with a strict percentage deviation in vapor pressure, the y-axes in these figures are not presented as ratios, but rather Figure 5 shows straight percent differences in dispersion distance and Figure 6 shows absolute differences in dispersion distances. This makes it difficult to draw a comparison with the results from earlier in this study. Nevertheless, it is clear that particularly at temperatures just below the NBP, significant deviations can occur from the use of Clausius-Clapeyron. Over all runs, there was an average change of 10.0 %, with a standard deviation of 40.4 %.

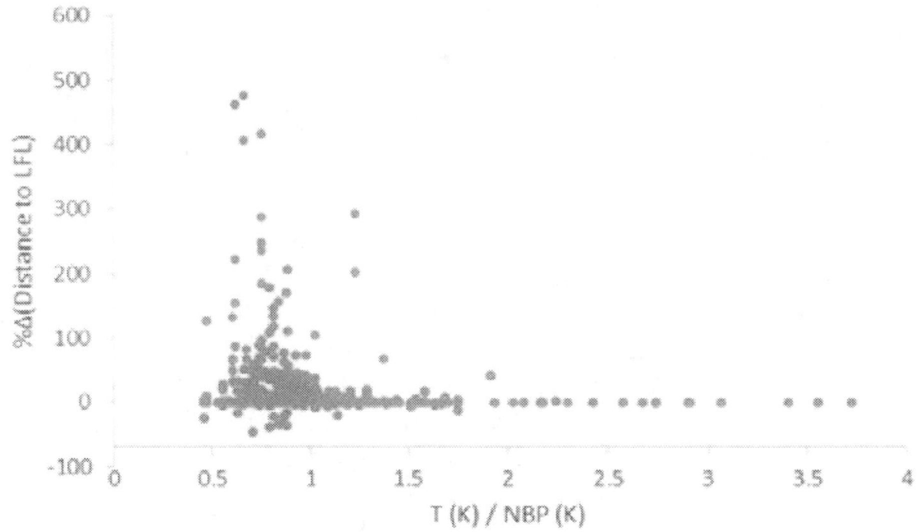

Figure 5. Percent change of dispersion distance vs. ratio of process temperature and NBP.

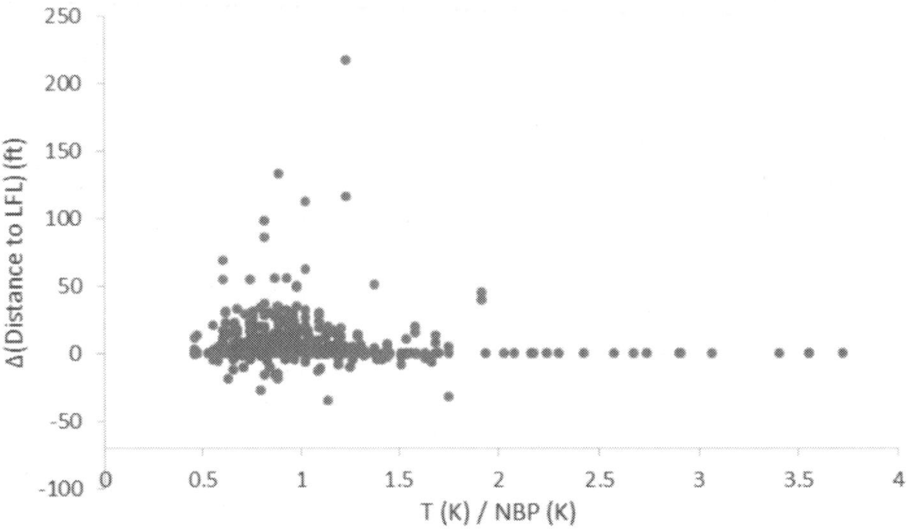

Figure 6. Absolute change in feet of dispersion distance vs. ratio of process temperature and NBP.

Still, the magnitude of these deviations are at times so extreme that it warrants further inspection to determine the reasons behind it. Figure 7 helps make results from Figure 5 and Figure 6 more relatable to results from Figure 4 by showing exactly how much VP as calculated from Clausius Clapeyron differs from VP calculated from the DIPPR correlation, once again plotted against T / NBP. Figure 4 shows deviations per percent change in VP, and Figure 7 shows that using Clausius Clapeyron will frequently result in a deviation much higher. The general trend seen in Figure 7 is that higher-boiling components reach greater deviations at lower T/NBP ratios. (Note that if plotted against T - NBP, the data would look much more uniform across all compounds.) It is the high-boiling compounds that make up the majority of the higher deviations in Figure 5. So not only do we find that Clausius Clapeyron can give very different results from correlations based on temperature-dependent experimental data, but we also see that the magnitude of those differences have a positive correlation with the NBP of the material. For heavier compounds, then, it becomes even more important to use accurate VP data in dispersion modeling.

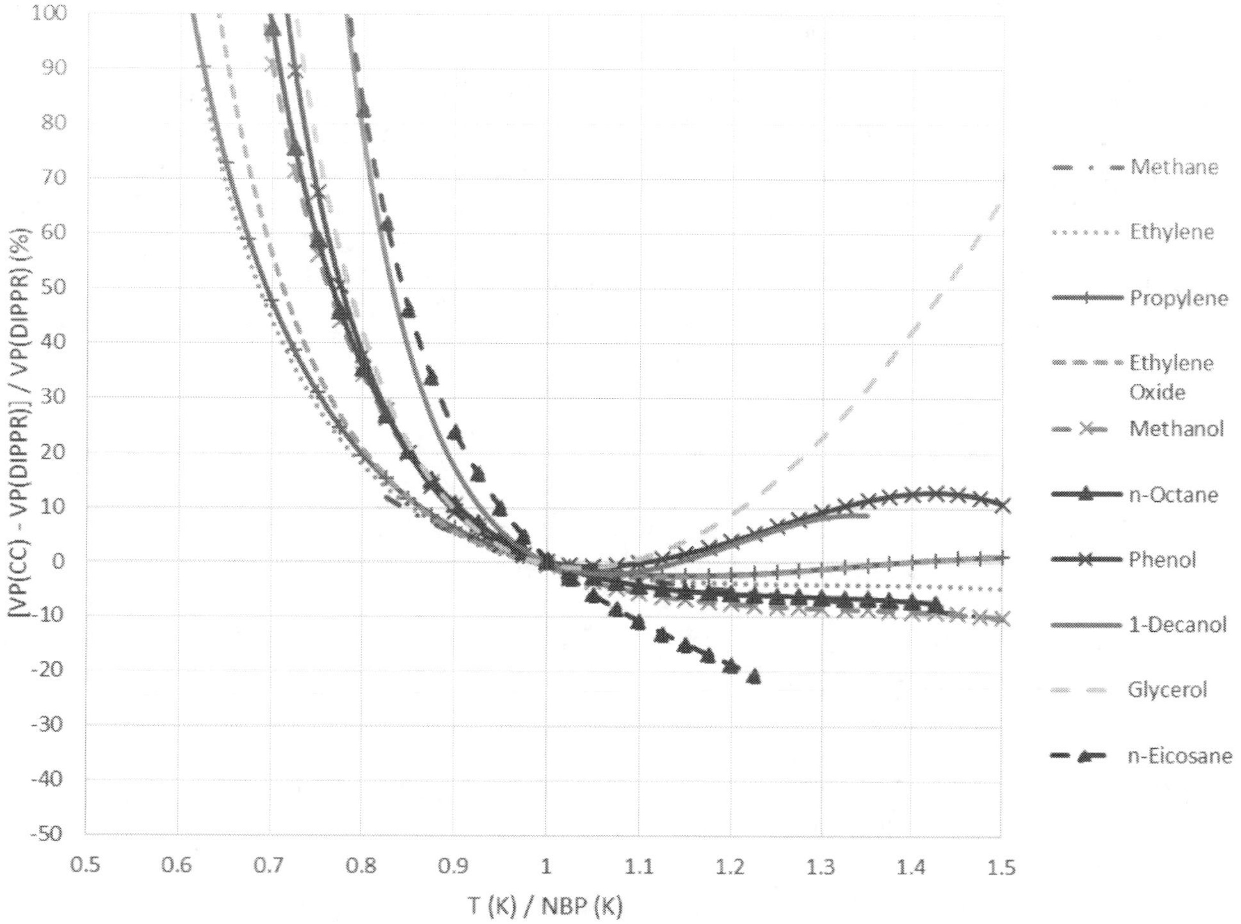

Figure 7. Percent difference between VP as obtained from the DIPPR correlation and the VP as given by Equation 1 plotted against T / NBP.

4 Conclusions

Overall, dispersion distances turned out to be quite resilient to deviations in physical properties. Minor changes in HVP or LFL resulted in only slight deviations in dispersion distances. In many cases, minor deviations in LDN and VP resulted in only minor deviations of dispersion distances. However, under the right conditions (far into the liquid regime for LDN or near the NBP for VP) large deviations in predicted dispersion distances can occur with only minor changes in physical property data. Consequently, great care should be taken as to where one obtains the physical property data used in dispersion modeling.

Furthermore, the Clausius-Clapeyron equation may not work well across the wide range of conditions that may be encountered during even a single dispersion. The processing temperature may be different from the temperature upon discharge, the temperature of the droplets in the plume, or the temperature of any pool formed, and we see in this study that it is vital to have reliable temperature-dependent vapor pressure data rather than rely on an extrapolation from a single point.

Obtaining accurate physical properties may be difficult and time-consuming, and we can often be tempted to settle for any values we can find, regardless of the source. Particularly in the case of LDN and VP, this study has shown that all physical property data should be met with some level of scrutiny when attempting to model dispersions.

We recommend using a critically reviewed database such as the current version of the DIPPR database [2]. The DIPPR database contains thermophysical data for a large number of compounds, and it routinely employs rigorous analysis and consistency checks to assure the reliability of the data. The database is continually updated and reviewed by experts in thermodynamics to ensure the recommended data reflect the latest advancements in thermophysical property measurement and evaluation. For properties where literature values are not available, state-of-the-art prediction methods are utilized to offer a best-guess value for situations where an estimate is needed.

5 References

[1] J. R. Rowley, "A new integral dispersion model based on one-dimensional turbulence theory," in *IChemE's 24th Hazard Symposium*, Edinburgh, UK, 2014.

[2] R. R. Rowley, W. V. Wilding, J. L. Oscarson and N. F. Giles, DIPPR® Data Compilation of Pure Chemical Properties, New York, NY: Design Institute for Physical Properties, AIChE, 2015.

[3] M. G. Zabetakis, "Flammability Characteristics of Combustible Gases and Vapors," in *Bulletin No. 627*, U.S. Bureau of Mines, 1965.

[4] K. Nabert, G. Schön and T. Redeker, Sicherheitstechnische Kenngrößen brennbarer Gase und Dämple, Deutscher Eichverlag, 2004.

[5] J. R. Rowley and J. E. Bruce-Black, Proper application of flammability limit data in consequence studies, Southport, UK: IChemE's 23rd Hazard Symposium, 2012.

[6] G. H. Tryon, Fire Protection Handbook, 12th ed., Boston: National Fire Protection Associations, 1962.

441423 Sustainable Risk Reduction with Dynamic Execution of Procedures for Any Operating Condition

Abstract:

Managing the volume of new information created daily is a challenge. Add to that the complexity of new equipment, employee turnover, and third party contractors working on-site, documenting clearly and concisely is critical to the operation. To truly mitigate risk we need look to operating companies focused on leading indicators of risk management where improved work conditions and profitability of the operation are the by product.

- Reduce human error
- Improve user adoption
- Improve operator efficiency
- Optimize SME time
- Quick resolution incident investigation
- Improve time "on the job"
- Extend equipment life

These objectives, met through the implementation of procedure best-practices to significantly reduce risk and improve the reliability of the operation.

During this presentation we will share the results of our procedure best-practices benchmark report and review the impact of asset level operating procedures for our clients.

As operating companies evolve through the knowledge management maturity model, they not only standardize best-practices but leverage that knowledge to optimize performance with a collaboration hub and dynamically generated procedures to fit any situation.

We will demonstrate how asset level operating procedures, designed for routine operator and facility maintenance, directly relates to improvements in asset life and performance, site-safe incident reduction based on human factors, and improve equipment start-up predictability when returning from planned maintenance.

This is so much more than simply documenting procedures; this is about sustainable adoption to improve operating discipline. Innovatia will introduce a complete procedure management solution to facilitate knowledge transfer in real-time to mitigate risk and improve reliability of the operation.

441155 Process Hazard Analysis of River-Crossing Natural Gas Pipeline

Abstract:

In order to control the process hazard of the river-crossing natural gas pipeline, HAZOP analysis is performed to qualitatively analyze its potential risk. Based on the results of HAZOP analysis, high risk scenarios are identified. So as to calculate the risk caused by pipeline's catastrophic rupture, FLACS software which based on CFD data simulation is used to quantitatively calculate the flame propagation and explosion overpressure of the explosion of natural gas. With the results of HAZOP analysis and quantitative calculation, safety measures are put forward to ensure the long-term safe operation of river-crossing natural gas pipeline.

Safe Depressurization of Large Thickness Vessels Under Pool Fire

Subhash S. Kale and George Maliakkal [(1)], Bechtel Corporation

(1) To whom all correspondence should be addressed

Abstract:

Vessels fail at pressures lower than the Maximum Allowable Working Pressure (MAWP) during fire due to degradation of strength at high temperatures. Analytical solution of the vessel wall temperature profiles as a function of fire exposure time available in API 521 Annex A for thicknesses of 3.2 mm (0.125 in.), 12.7 mm (0.5 in.) and 25.4 mm (1 in.) was extended for vessel thicknesses up to 152.4 mm (6 in.) to include high pressure/large diameter vessels encountered in gas processing. For fire depressuring, plots of ratio of the maximum vessel pressures to avoid failure to MAWP were developed for various wall thicknesses and compared to a plot of the ratio of vessel internal pressure to MAWP, both as a function of time. If the vessel internal pressures throughout fire depressuring were less than the reduced vessel failure pressures, then vessel rupture is avoided. The methodology is illustrated for the case of a high pressure vessel of wall thickness 57.15 mm (2.25 in). in natural gas service.

Introduction:

Large vessels, for e.g. vertical compressor suction KO drums, have a substantial flammable hydrocarbon liquid inventory, which if leaked can form a source of pool fire. The vessels usually have a large diameter 3 - 6 m which requires vessels with thicknesses higher than 25.4 mm (1 in.). API 521 [1] Section 4.4.13.2.4.1 gives information on wall temperatures for plates 3.2 mm (1/8 in.), 12.7 mm (1/2 in.) and 25.4 mm (1 in.). Also, Appendix A of API 521 gives information on maximum pressure as % of Maximum Allowable Working Pressure (MAWP) for plates of thickness up to 25.4 mm (1 in.). No data is available in API 521 for plates of thicknesses greater than 1 in. This work uses the analytical method given in Appendix A of API 521 to extend the calculations for vessels of thickness higher than 25.4 mm (1 in.) - for e.g. 50.8 mm (2 in.) to 152.4 mm (6 in.). thicknesses. The analytical equation given in Appendix A of API 521 is solved to develop wall temperature profiles for various wall thicknesses including those beyond 25.4 mm (1 in.). The temperature versus allowable stress data given in Table A-7 of the Appendix A can then be used to develop temporal profiles of maximum allowable pressure to avoid failure as a function of MAWP. The vessel internal pressure experienced during fire case depressurization is calculated using HYSYS v 7.3 depressurization utility. The internal pressures during depressurization, expressed as a ratio of the internal pressure to MAWP, is overlaid on the failure pressure temporal profiles to ascertain whether the vessel can avoid rupture during a pool fire scenario. The following describes the assumptions used in the calculations, the methodology to compute the temporal profiles of wall temperature and the maximum pressure to avoid failure.

Assumptions:

1. A carbon steel vertical vessel with hemispherical heads is assumed.
2. The surfaces are assumed to fully engulfed in flames hence both convection and radiation terms in equation A-1 of API 521 Appendix A apply.
3. The parameters given in Appendix A are slightly modified to better fit the temperature profiles given in Figure C-9 of Appendix C of API 521.
4. Since the vessels have thicknesses above 25.4 mm (1in.), the depressurization is from the design pressure, which is assumed to be same as MAWP, to half the design pressure in 15 minutes per API 521 guidelines. For these calculations, a design pressure of 80 barg is assumed.
5. The calculations are illustrated for a vessel of controlling thickness 57.2 mm (2.25 in.) for the hemispherical heads. The methodology can be extended to larger thicknesses from 25.4 mm (1 in.) to 152.4 mm (6 in.).

Wall Temperature Calculation by Analytical Method:

The analytical equation given in A-1 of Appendix A of API 521 is stated as:

$$q_{absorbed} = \sigma\ (\ \alpha_{surface}{}^{*}\varepsilon_{fire}{}^{*}T_{fire}{}^{4} - \varepsilon_{surface}{}^{*}\ T_{surface}{}^{4}\ ) + h\ (T_{gas} - T_{surface})\quad (1)$$

Also, by a heat balance,

$$q_{absorbed} = m\ c_p\ (\text{Delta } T_{surface} / \text{Delta time})\qquad (2)$$

The following gives a description and the values used for the various parameters in this equation per API 521 Table C-3:

$q_{absorbed}$, Heat absorbed by the wall

σ, Stefan Boltzmann Constant = 5.67 E-8 W/m^2.K^4 (1.713E-09 BTU/H-ft^2-°R$^{4)}$

$\alpha_{surface}$ is the equipment absorptivity = 0.8

ε_{fire} is the fire emissivity = 0.85

$\varepsilon_{surface}$ is equipment emissivity = 0.75

T_{fire} is the fire temperature = 904 °C (1652 °F)

$T_{surface}$ = equipment temperature initially assumed to be 59 °C (122 °F or 582 °R)

The surface temperature increases as the surface heats up due to radiation from the fire

h is the convective heat transfer coefficient 20 W/m^2-K (3.52 BTU/hr-ft^2-R)

Plate area is assumed to be 2.3 m^2 (24 ft^2) (per API Section 4.4.413, Figure 1)

Density of carbon steel is 7850 kg/m^3 (490.1 lb/ft^3)(from internet)

Plate thickness varies from 3.2 mm (1/8 in.), 12.7 mm (1/2 in.) and 25.4 mm (1 in.).

Heat capacity of carbon steel = 502 J/kg°-K (0.12 BTU/lb-°F)

Additionally, larger thicknesses 50.8 mm (2 in) and 76.2 mm (3 in) are also used.

From equation (1), $q_{absorbed}$ is calculated.

The plate volume is calculated for each thickness from the given area.

Using the given carbon steel density, the plate mass is calculated.

From equation (2) Delta $T_{surface}$ is calculated.

Starting with the initial surface temperature, the change in the surface temperature is calculated from equations (1) and (2) for different times in 0.5 second increments. The resulting wall temperatures are plotted and compared to values given in API 521 Appendix C (Figure C-9) as shown in Figure 1.

Figure 1 Calculated wall temperature profiles as a function of time

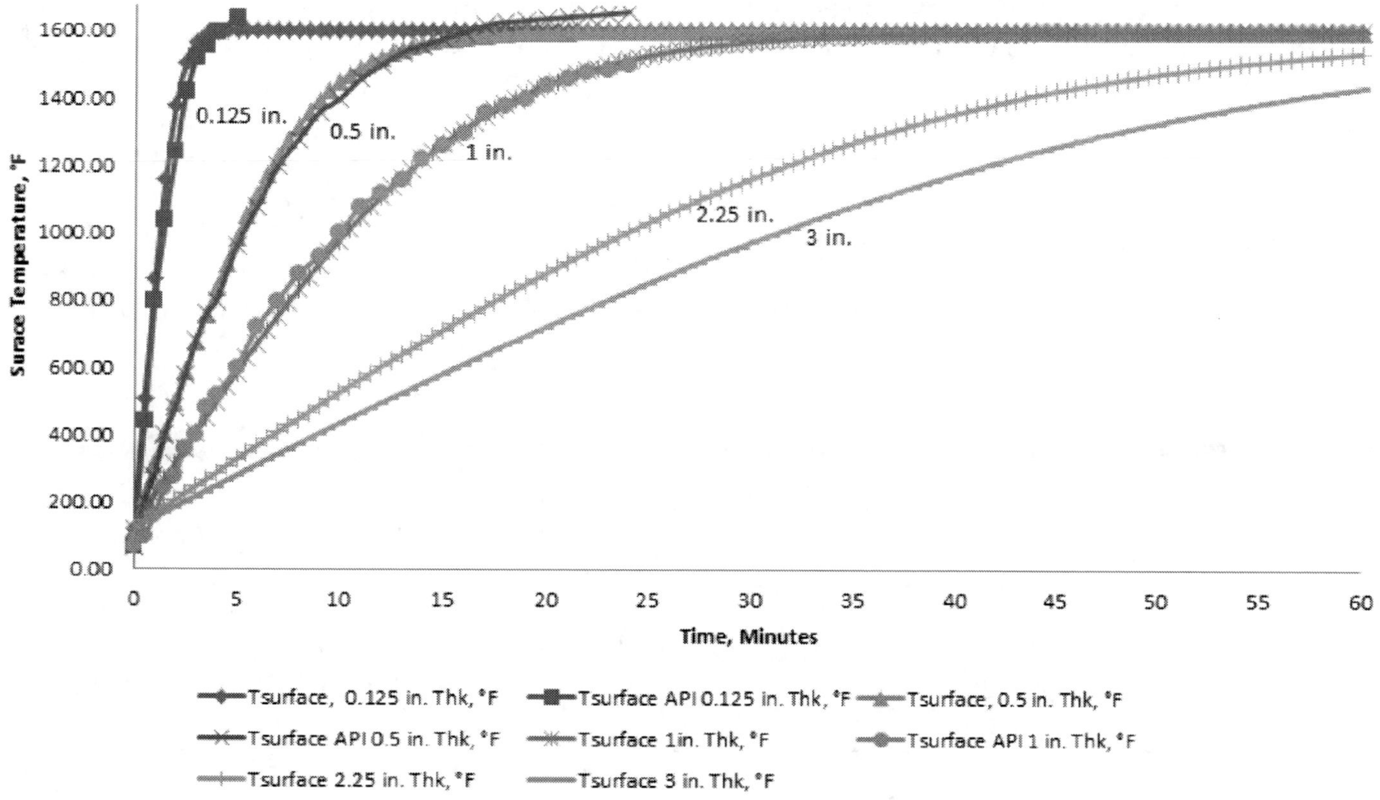

Calculation of the maximum pressure to avoid failure for various wall thicknesses:

As described in Appendix A Section A.3.5.4.6, the wall temperatures can be combined with tensile strength versus temperature data for carbon steel to determine the tensile strength, or the maximum pressure to avoid failure of a vessel, as a function of time in a pool fire scenario.

The tensile strength versus temperature data for carbon steel from Table A-7 in API 521 Appendix A is given in Table 1 for carbon steel.

Table 1 Tensile strength versus temperature for Carbon Steel (SA-515, SA-516)

Temperature (°C)	Temperature (°F)	Tensile Strength (MPa)	Tensile Strength (psi)
482	900	313.7	45,500
538	1000	251.7	36,500
593	1100	187.5	27,200
649	1200	137.9	20,000
704	1300	93.1	13,500
760	1400	62.2	9025

As the wall temperature heats up to 482°C (900°F), the tensile strength or the maximum pressure to avoid failure, decreases. These values are normalized with respect the tensile strength of 313.7 MPa (45,500 psi) and a safety factor of 0.75 is applied to generate Figure A-5 in Appendix A of API 521.

These values are extrapolated to larger thickness greater than 1 in using regression to determine the times to reach failure pressures expressed as a ratio of maximum pressure to avoid failure to MAWP for various thicknesses.

Table 2 Vessel rupture strength as % of MAWP versus tire exposure times

% of MAWP	100%	80%	60%	40%	34%	30%	26%	24%	
Thickness									
in	Time (mins)	Time (mins)	Time (mins)	Time (mins)	Time (mins)	Time (mins)	Time (mins)	Time (mins)	Comment
0.125	1.2	1.5	1.9	2.4	3.1	6.0	12.0	17.5	API 521
0.5	3.6	3.9	4.5	6.1	7.6	10.5	17.0	28.0	API 521
1	12.6	13.1	14.1	17.0	20.0	23.1	31.0	37.0	API 521
2	25.19	25.91	27.58	33.24	38.83	42.15	52.16	59.66	Extrapolated
2.25	28.52	29.28	31.14	37.48	43.74	47.11	57.68	65.18	Extrapolated
3	38.49	39.44	41.81	50.20	58.44	62.00	74.22	81.72	Extrapolated

Figure 2 shows a plot of the maximum allowable pressures to avoid failure for plates of different thicknesses from 25.4 mm (1 in). to 76.2 mm (3 in.)..

Figure 2 Failure pressures for various thicknesses as a function of time

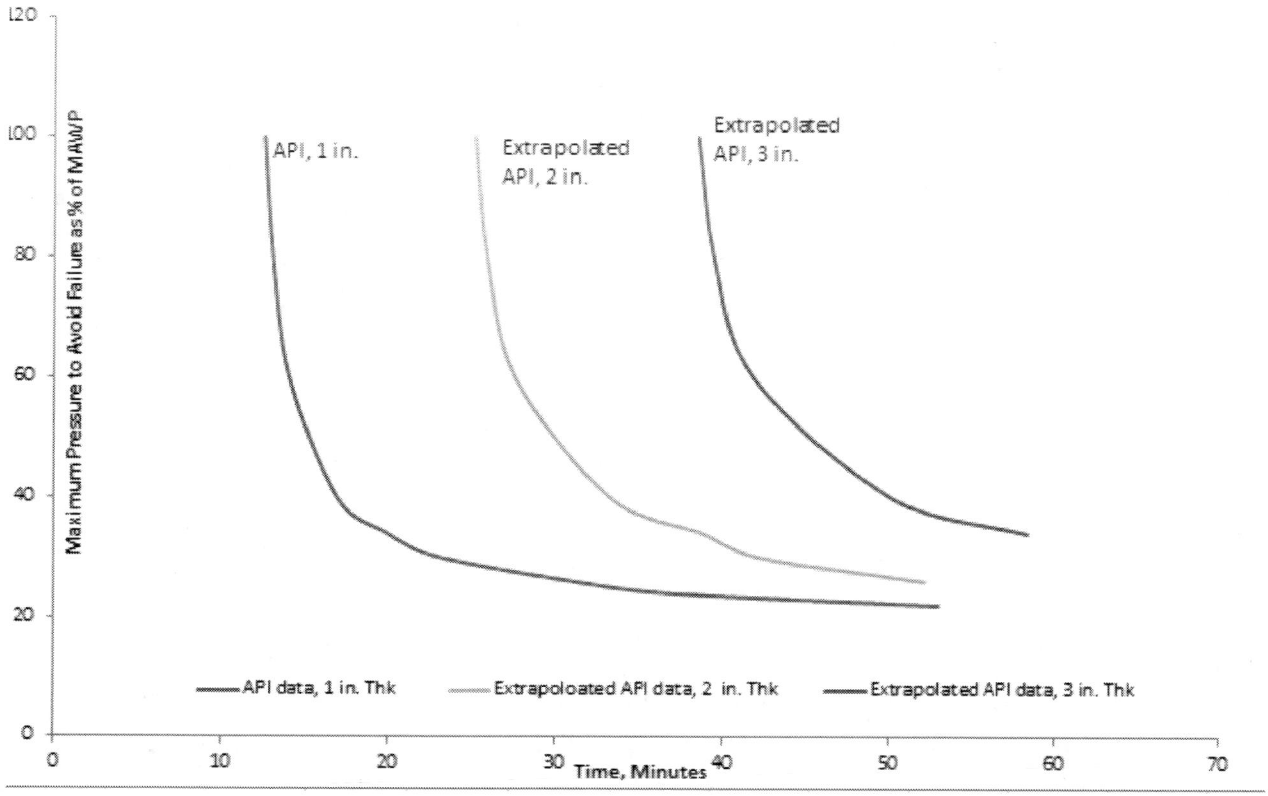

A hypothetical vessel of 4 m diameter and 7 m height with a vessel thickness of 101.6 mm (4 in.) having a hemispherical heads of thickness 57.2 mm (2.25 in.) is depressuerd from a design pressure of 80 barg to a pressure of 40 barg in 15 minutes using HYSYS v 7.3 depressuring utility. The resulting profile of the vessel internal pressure as a function of MAWP is overlaid on the temporal profiles of the maximum allowable pressures to avoid failure for various thicknesses. Data for about 60 minutes is plotted for comparison with the profiles of internal pressures as shown in Figure 3.

Figure 3 Plot of failure pressures for various thicknesses and the vessel internal pressures as a function of time

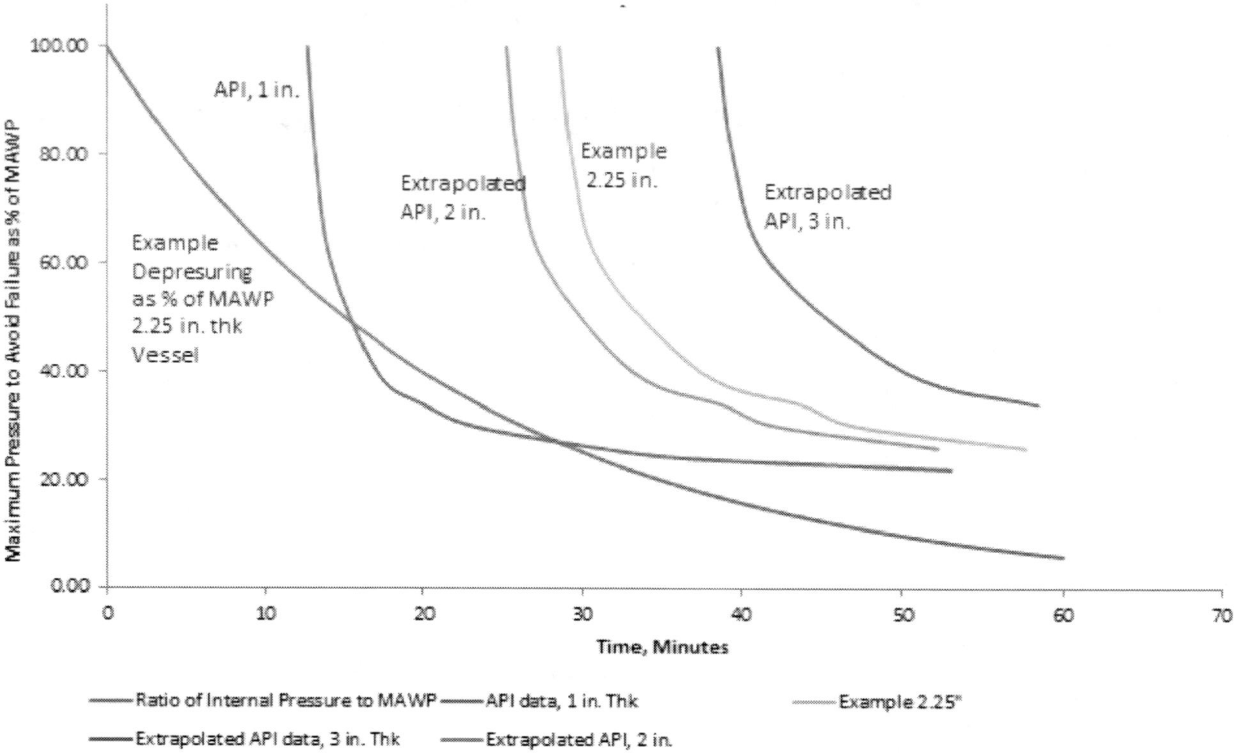

Discussion:

The kinks in the plot are due to round off errors and errors in the statistical regression. However, as seen from Figure 3, the vessel internal pressure for depressuring in a pool fire scenario is less than the maximum allowable pressure to avoid failure for thicknesses greater than 1". Thus, a methodology is developed for calculating wall temperature profiles during pool fire for vessels of various thicknesses and to check if the vessel internal pressures during depressurization exceed the failure pressures for vessel thicknesses greater than 1in.

References:

1. Pressure-relieving and Depressuring Systems, API Standard 521, Sixth Edition, January 2014.

12TH GLOBAL CONGRESS ON PROCESS SAFETY

Quantifying the Impacts of Human Factors on Functional Safety

Julia V. Bukowski, Ph.D.
Dept. of Electrical & Computer Engineering
Villanova University, Villanova, PA 19085
Julia.Bukowski@villanova.edu

Loren Stewart, CFSP
exida
64 N. Main St., Sellersville, Pa 18960
lstewart@exida.com

Prepared for Presentation at
American Institute of Chemical Engineers
2016 Spring Meeting
12th Global Congress on Process Safety
Houston, Texas
April 11-23, 2016

Quantifying the Impacts of Human Factors on Functional Safety

Julia V. Bukowski, Ph.D.
Dept. of Electrical & Computer Engineering
Villanova University, Villanova, PA 19085
Julia.Bukowski@villanova.edu

Loren Stewart, CFSP
exida

Keywords: systematic failure rate, product failure rate, site-generated failure rate, FMEDA, PFDavg

Abstract

It is not difficult to think of numerous ways in which human factors might impact functional safety. For example, imperfect repair, improper calibration, faulty installation, etc. can all contribute to decreases in functional safety performance. However, if actions are to be taken to mitigate these impacts then it is essential to have some quantitative measure of the impacts so that the actions' effectiveness can be assessed. This paper describes how the Site Safety Index™ (SSI) is used to adjust safety metrics, computed under the assumptions that human factors play no part in safety system performance, to reflect the effects of human factors on safety system performance on a site by site basis. Examples of the application of the method showing the safety performance changes attributable to human factors are provided.

1. Introduction

A safety instrumented function (SIF) has an inherent product failure rate based on its design and implementation and the assumption that the end user will take all necessary actions (i.e., meet all IEC 61508 [1] and/or IEC 61511 [2] requirements) to insure that the inherent product failure rate will be achieved. In practice, end-user activities and practices often fall short of this ideal and, consequently, the safety performance achieved by the SIF is often less than the ideal [3]. The SSI measures, on a scale of 0 to 4, the extent to which end-user actions support the achievement and retention of ideal SIF safety performance. The SSI is described in detail in [4]. This paper explains how end-user actions as measured by SSI affect various parameters that are used to calculate SIF safety performance and show how different levels of SSI impact various safety metrics.

2. Notation

FFD	field failure data
FMEDA	failure modes, effects and diagnostic analysis

IEC	International Electrotechnical Commission
MTTFS	mean time to fail safe
OREDA	Offshore and Onshore Reliability Data
PFDavg	average probability of dangerous failure on demand
RRF	risk reduction factor
SIF	safety instrumented function
SIL	safety integrity level
SRS	Savannah River Site
SSI	Site Safety Index^{TM}
1oo1	1-out-of-1 architecture

3. Using SSI to Measure Impacts of End-User Activities and Practices on Safety Performance of SIFs

3.1 Why a Site Safety Index^{TM}?

Studies [5] have shown that the failure rates estimated from field failure data (FFD) for identical modules in the same industrial application exhibit a difference in estimated failure rates for the same product from site to site. Typically, when compared to inherent predicted failure rates for the same products, the FFD estimates are approximately 1.2 to 3 times higher depending on product type. Site audits have attributed these differences to dissimilarities in end-user activities and practices which vary from site to site. This variation in end-user activities and practices can be captured by the SSI [4]. Further, these variations in activities and practices directly impact many variables used in models which assess the safety performance of SIFs.

3.2 The Key Variables Impacting PFDavg and Other Metrics

A key metric for measuring the safety performance of SIFs is average probability of failure on demand (PFDavg). Other closely related metrics are risk reduction factor (RRF), safety integrity level (SIL), and mean time to fail safe (MTTFS). Based on studies of FFD and proof test reports, nine key variables impacting PFDavg have been identified [6]. These nine variables are

1. Failure rates of each device including failure modes and any diagnostic coverage from automatic diagnostics (an attribute of devices chosen and of end-user practices);
2. Proof Test Effectiveness (an attribute of the proof test method and, to the extent that there is more than one possible proof test method, an attribute of end-user practices);
3. Proof Test Intervals (assignable by end-user practices);
4. Mean Time to Restore (an attribute of end-user practices);
5. Probability of Initial Failure (PIF) (an attribute of end-user practices);
6. Mission Time (an attribute of end-user practices);
7. Proof Test Duration (an attribute of end-user practices);
8. Site Safety Index^{TM} (an attribute of end-user practices); and
9. Redundancy of Devices including common cause failures (an attribute of SIF design).

3.3 How Key Variables Are Modified to Quantify the Impacts on PFDavg and Other Metrics

Clearly, redundancy of devices is not an attribute of end-user practices or activities. The SSI itself is not modified in safety performance calculations; rather it is used to modify many of the other variables which impact PFDavg and other metrics. Specifically, whenever SSI is less than 4, the following five key variables are modified:

- failure rates of each device
- proof test effectiveness
- proof test intervals
- mean time to restore
- PIF

Failure rates of each individual device in the SIF are increased by a specific multiplier which is determined by the SSI value and the device itself. It is known that final elements are more likely to be negatively impacted by less than ideal end-user practices than are sensors or logic solvers. By increasing device failure rates on an individual device basis, it is possible to more accurately account for the effects of site practices on safety performance. Proof test effectiveness, i.e., the likelihood that all failures are discovered and completely repaired, is also decreased with decreasing SSI.

Proof test intervals, mean time to restore, and PIF are increased by specific multipliers determined by SSI value. It is common to find that less than ideal end-user practices include delaying proof testing, reducing proof test coverage, requiring additional restoration time, and introducing initial failures through imperfect pre- or post-installation testing. Final elements that may develop failures simply due to lack of motion of the moving parts over an extended time (a few months) are especially vulnerable to initial failures in excess of 1% [7, 8, 9].

The two variables not modified are mission time and proof test duration. Mission time is not modified because its impact is generally small unless the mission time is sufficiently beyond the SIF useful life so as to cause the SIF to see failures due to ageing, i.e., failures no longer modeled by a constant failure rate, without renewal during the mission time. Proof test duration has no impact if proof testing is performed with the process off-line. For proof testing performed on-line, proof test duration generally has only a very small effect unless proof testing frequency is very frequent, i.e., greater than or equal to once per month – an impractical situation for most industrial processes.

3.4 Calibrating the Multipliers

As with the development of the Failure Modes, Effect and Diagnostic Analysis (FMEDA) failure rate predictive technique [10] and its associated validated databases [11, 12, 13], it is necessary to calibrate the database of multipliers for the various modifications and to provide on-going validation against actual FFD. The initial calibration phase is still in progress and uses the extensive FFD of the Offshore and Onshore Reliability Data Project (OREDA) [14, 15] as well as the detailed proof testing and root cause failure records of the Savannah River Site (SRS) [8].

4. Examples of Quantifying the Impacts of SSI on Various Metrics

The same simple 1oo1 SIF was analyzed under five different levels of SSI. Figure 1 shows the impacts on PFDavg and SIL of these variations in SSI. For the same 1oo1 system, Figure 2 shows the impacts on RRF and SIL of variations in SSI. Finally, Figure 3 shows the impact on MTTFS of variations in SSI.

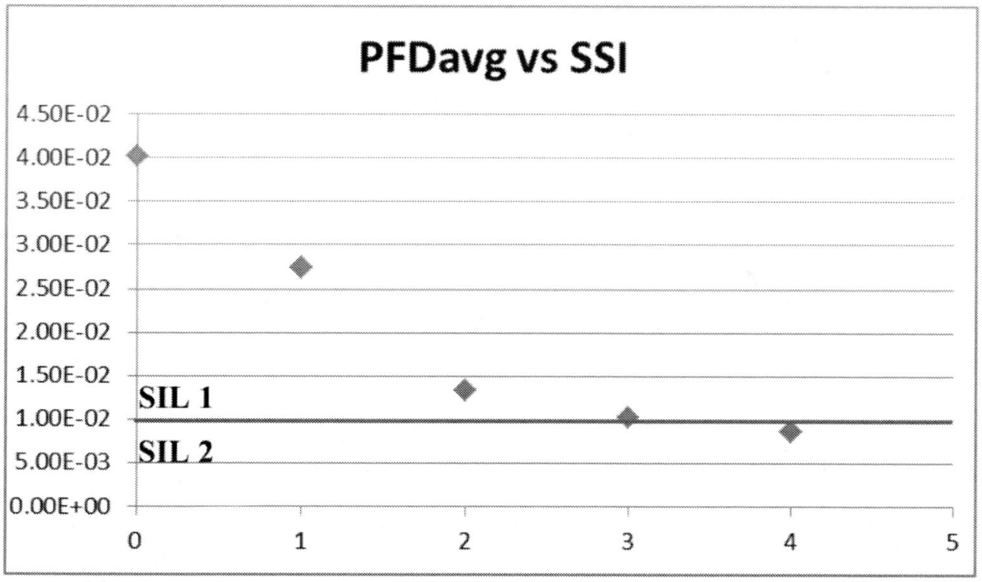

Figure 1. Impacts on PFDavg and SIL of Variations in SSI.

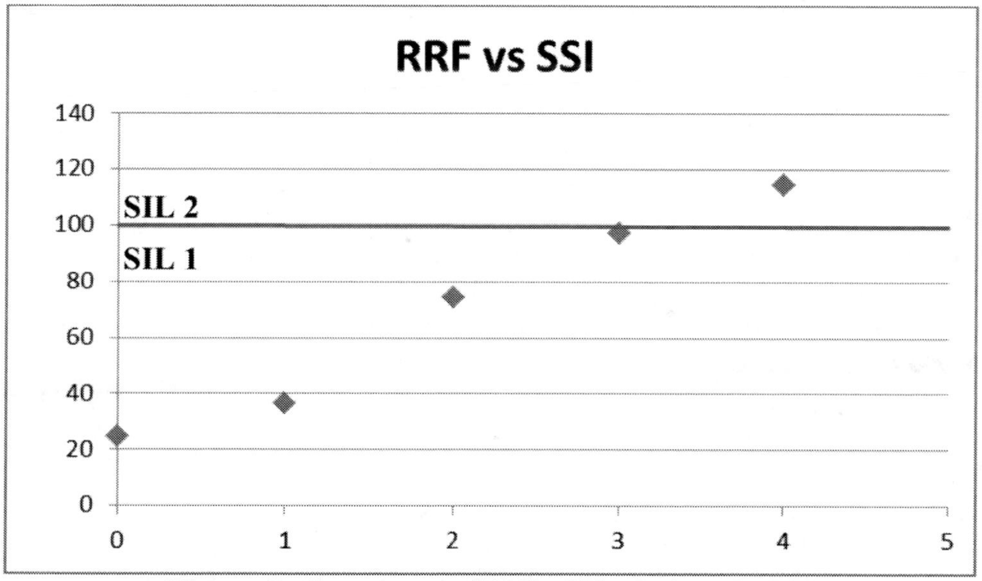

Figure 2. Impacts on RRF of Variations in SSI.

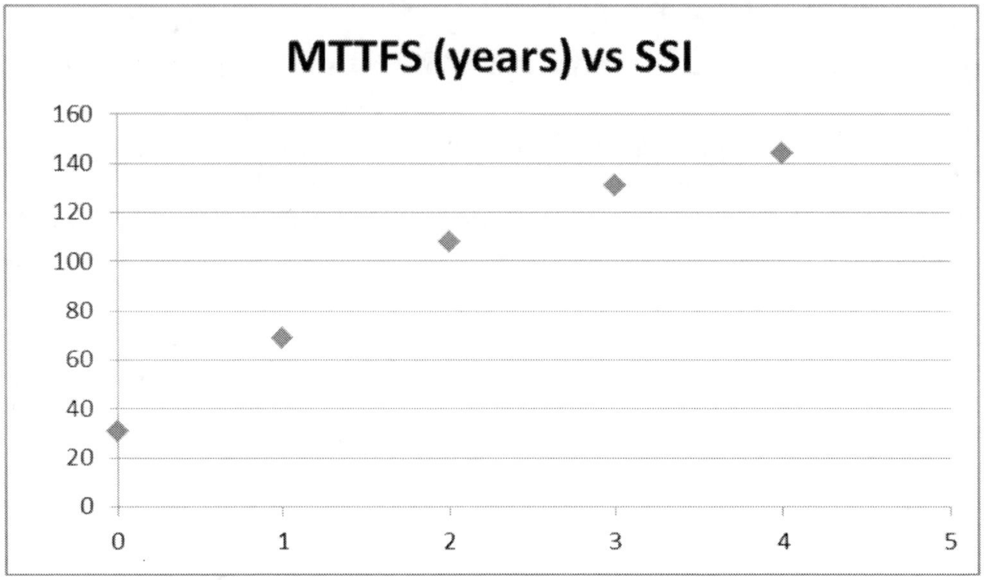

Figure 3. Impacts on MTTFS of Variations in SSI.

5. Summary

This paper clearly shows that the impacts of end-user activities and practices on safety performance of a SIF can be quantified through a simple model like the SSI. It also explains how the SSI modification multipliers are being calibrated and will be updated as increasing amounts of FFD become available. The methodologies presented here and in [4] allow end-users to measure quantitatively their own activities and practices on a site-by-site basis and to measure the impacts of those activities and practices on SIF safety performance.

6. References

[1] IEC 61508, *Functional Safety of electrical / electronic / programmable electronic safety-related systems*, Geneva, Switzerland, 2010.

[2] IEC 61511, *Application of Safety Instrumented Systems for the Process Industries*, Geneva, Switzerland, 2nd edition, 2004.

[3] J. V. Bukowski, W. M. Goble, and I van Beurden, "Product Failure Rates vs Total Failure Rates at Specific Sites: Implications for Safety," *Proceeding AIChE 11th Global Congress on Process Safety*, Austin, TX, April 2015.

[4] J. V. Bukowski and D. Chastain-Knight, "Assessing Safety Culture via the Site Safety Index™," *Proceeding AIChE 12th Global Congress on Process Safety*, Houston, TX, April 2016.

[5] W. M. Goble, "Maintenance capability: Does it matter in the long term?" *Hydrocarbon Processing*, June 1, 2013.

[6] "The Key Variables Needed for PFDavg Calculation," White Paper, exida 2015, http://www.exida.com/Resources/Whitepapers/the-key-variables-needed-for-pfdavg-calculation

[7] J. V. Bukowski, R. E. Gross and W. M. Goble, "Probability of Initial Failure for Spring Operated Relief Valves," *Proceedings of the ASME 2011 Pressure Vessels and Piping Division Conference*, Baltimore, MD, July 2011.

[8] J. V. Bukowski, R. E. Gross and W. M. Goble, "The Adhesion Failure Mode in Stainless Steel Trim Spring Operated Pressure Relief Valves," *ASME Journal of Pressure Vessel Technology*, Vol 135 No 6, December 2013.

[9] L. Stewart, J. V. Bukowski and W. M. Goble, "Improving Reliability & Safety Performance of Solenoid Valves by Stroke Testing," *Proceedings AIChE 9th Annual Global Conf on Process Safety – Process Plant Safety Symposium*, San Antonio, TX, April-May 2013.

[10] W. M. Goble and A. C. Brombacher, "Using a Failure Modes, Effects and Diagnostic Analysis (FMEDA) to Measure Diagnostic Coverage in Programmable Electronic Systems," *Reliability Engineering and System Safety*, Vol. 66, No. 2, Nov 1999, pp. 145-148.

[11] W. M. Goble and J. V. Bukowski, "Development of a Mechanical Component Failure Database," *Proceedings of the 2007 Annual Reliability and Maintainability Symposium*, Orlando, FL, January 2007, pp. 451-455.

[12] J. V. Bukowski and, W. M. Goble, "Validation of a Mechanical Component Constant Failure Rate Database," 2009 *Proceedings Annual Reliability and Maintainability Symposium*, Fort Worth, TX, January 2009, 338-343.

[13] J. V. Bukowski and L. Stewart, "Comparing Failure Rates for Safety Devices: FMEDA Prediction vs OREDA Estimation," exida White Paper available at www.exida.com

[14] SINTEF, *OREDA Offshore and Onshore Reliability Data Handbook, Vol 1. – Topside Equipment and Vol. 2 – Subsea Equipment*, 6th Ed, OREDA Participants, 2015.

[15] SINTEF, *OREDA Offshore Reliability Data Handbook, Vol 1. – Topside Equipment and Vol. 2 – Subsea Equipment*, 5th Ed, OREDA Participants, 2009.

441468 Using HAZOP/LOPA to Create an Effective Mechanical Integrity Program

Abstract:

Many people view the conduct of a HAZOP/LOPA to address regulatory requirements as a chore, and stop there. However, the implementation of a quality HAZOP/LOPA has the potential to provide a framework for addressing numerous safety and operational optimization issues at plants, including the formulation/refinement of the Mechanical Integrity Program. The purpose of this paper is to focus on the mechanical integrity program, illustrate how a quality HAZOP/LOPA can support the effective implementation of some of the new Damage Mechanism Review requirements for California Refineries (e.g., 5189.1(k)), and optimize key elements of an effective Mechanical Integrity Program, e.g.:

- inspection/testing methods
- testing intervals
- maintenance outage periods
- repair prioritization and allowable outage
- identification of low priority equipment

441486 Reactive Chemical Hazards of Azide Compounds: a Case Study and Lessons Learned

Abstract:

While a researcher was servicing a vacuum pump an unintended detonation occurred. This pump had been used in a sulfonyl azide synthesis ($R\text{-}SO_2\text{-}N_3$). The detonation occurred when he was applying effort to the hose barb from the outlet of the pump to connect it with a rubber hose. Differential Scanning Calorimetry (DSC) and headspace Gas Chromatography/Mass Spectrometry (GC/MS) were employed for the root cause investigation. The DSC showed a large exothermic decomposition (-1388J/g) for the residual sample from the pump fitting. Headspace GC/MS identified the presence of hydrazoic acid in the pump oil. The investigation concluded that either the azide precursor (sodium azide) or hydrolysis of the product (sulfonyl azide) was responsible for generating hydrazoic acid, which escaped from the dry ice trap and appeared in the downstream equipment, and probably reacted with metal parts (brass) to form a highly shock sensitive material (copper azide).

This paper reviews the reactive chemical hazards for azide compounds involved in this case study. Sodium azide is not particularly sensitive to mechanical impact, but its propensity to form hydrazoic acid in water is well-known, even at neutral pH. Likewise, the test data on sulfonyl azide product showed that it is not particularly sensitive to mechanical impact, but it can generate hydrazoic acid upon hydrolysis with water. Metal azides which are formed by the reaction between hydrazoic acid and metal surfaces of equipment parts are highly hazardous due to their extreme impact and friction sensitivity.

Based on lessons learned, this paper also describes further steps for synthesizing organic azides using aqueous sodium azide. These steps include, but are not limited to: (1) Utilizing a dedicated hood and equipments to limit potential HN_3 vapor streams from contacting any reactive metal surfaces. (2) Using a dual caustic trap system designed to scrub the outlet gases swept from the azide reaction step, instead of dry ice trap. (3) Using a 5% aqueous solution of ceric ammonium nitrate as a decontaminating agent for metal surfaces inside the fume hood at the completion of the campaign. (4) Eliminating all brass or copper fittings from equipment used in the synthesis, due to the potential formation of the unstable explosive copper azide that can form from exposure of brass or copper parts to HN_3.

441495 The Application of Frequency Data in Quantitative Risk Assessment

Abstract:

Quantitative Risk Assessment (QRA) is a vital resource for estimating risk in the process industry. There are four primary steps involved in conducting a QRA: hazard identification, frequency analysis, consequence analysis, and risk calculation (the combination of frequency and consequence to determine the risk). Frequency analysis is a key component of the QRA and can have a major impact on the risk profile.

The primary method of estimating event frequency for QRA utilizes historical leak frequency data, of which the Hydrocarbon Release Database (HCRD) is the most widely used leak frequency database. One main issue with applying historical frequency is not having transparency of the causes of the events. Using the historical data approach, a single value is generated for the frequency of an event occurring, with no association to the cause (e.g. corrosion, dropped object, etc.). It has also become widely accepted to apply these offshore databases to onshore processes. There are many concerns in applying these databases to onshore processes, as there are differences in the types of equipment, the hazards the equipment is exposed to, and the safety management systems in place.

When QRAs were first conducted during the 1980s, they were an exhaustive and detailed process. In today's fast track industry, often frequency analysis is conducted without undertaking an evaluation to determine if the frequency being applied is appropriate for the hazard at hand. Leak frequency analysis has become an isolated activity, rather than relying on frequencies established during HAZOP and reliability assessments such as fault tree analysis.

An alternative method to employing historical databases to estimate event frequency is conducting fault tree analysis. Fault tree analysis involves dissecting a hazard into its causes, using a "top down" approach, beginning with hazard identification and working down to a level in which applicable frequency data is available. Fault trees provide a logical relationship between the hazard and its causes.

If historical leak frequency data is employed, which is independent of the cause, the resulting risk is also independent of the cause. Therefore, mitigation measures and the decision making process tend to focus on the back-end consequence of the event, rather than the cause of the event.

Fault tree analysis, although it has had limited application in onshore and offshore risk assessment, provides a major advantage of providing a link between the cause of an event and the associated frequency. This transparency between the event frequency and cause enables us to identify major contributors to the risk and provide risk reduction measures based on what is actually causing the event to occur.

The objective of this paper is to provide an alternative to employing offshore historical frequency data to QRA and demonstrate that the alternative method can provide major advantages in mitigations by linking the estimated risk to the cause and focusing the decision making process on the front-end of the problem rather than solely on the back-end.

441528 Uncertainty in Quantitative Risk Assessment and Decision Making

Abstract:

Including uncertainty in Quantitative Risk Assessment (QRA) is recommended in most published Best Practice Recommendations for QRA, including those published by TNO, CCPS, and Norwegian authorities. Unfortunately typical commercial QRA techniques and models used in Process Risk Analysis do not include the capabilities of uncertainty within the model formulations. Effects on uncertainty may be based on qualitative estimates or ignored, or at most accounted for by sensitivity analysis by "swinging" variables to estimate individual variable sensitivity. By not presenting the uncertainty, the risk assessment results may be misrepresented as possessing an accuracy greater than is warranted.

Failure to account for uncertainty also tends to support centrist thinking and poor quality decision making. The natural tendency of users of risk analysis is to believe in their results, especially if much effort has been put into developing the results. If the result is a single number, that number may be assumed by the analyzer and users as "right". In addition, there is a tendency in people to assume that by picking the most likely values for the inputs into a model, the end result is also the most likely result. This may not be true if the model is complex and non-linear. The most likely results must be determined by including variable distributions and uncertainty modeling in the analysis.

Decision analysis is also deceptively simple when single point criteria or simple curves are used for decision making purposes. A challenge in risk assessment is utilizing results and presenting them in a straight forward way that capture the uncertainty and complexity while allowing clear, high quality decisions by end-users.

This paper presents sources of uncertainty, analysis methods, examples of potential errors introduced by ignoring uncertainty, and illustrates how one can include uncertainties in your quantitative risk assessment (QRA) and benefit from it in your risk assessment decisions.

441529 Analysis of Process Systems during Accidental Fire

Abstract:

In the past 40 years, there have been over 60 major fire accidents reported on offshore platforms. The thermal effects imposed on process equipment and systems (such as pressure vessels, flare scrubbers, etc.) has the likely potential to result in catastrophic consequences such as explosions given that these process systems typically operate at very high pressures. In the event of an accidental fire, blowdown valves are engaged to relieve the excess pressure of the process system to acceptable values in a specific time period. During that time, however, the increased temperature of the blowdown system due to the impinging fire may result in the degradation of its structural capacity. The combination of a degraded vessel shell and a high internal pressure typically leads to sudden ruptures and explosions which could result in the escalation of the initial fire or injuries to personnel. Therefore, it is critical to accurately capture the pressure and temperature of the process system in a fire blowdown event.

A Blowdown analysis is typically carried out using simple thermodynamic software. While these software provide cost-effective approaches to studying process systems, they are unable to accurately compute the local effects of heat transfer and fluid behavior. Moreover, stress concentrations on the pressure vessel due to nozzle effects are not captured correctly using classical approaches and advanced analysis is required. Therefore, to accurately capture the rupture time of pressure vessels, scrubbers, etc., during a fire scenario, advanced approaches such as Computational Fluid Dynamics (CFD) are required.

In this paper, we will demonstrate a new approach in analyzing onshore and offshore process systems in an extreme fire scenario. Heat transfer through the vessel shell as well as the internal fluid will be accounted for by computing the rates of conduction, convection, and radiation. The time history of temperature and pressure as well as the change of phase of the fluid inside the pressure vessel will be captured in the analysis. A degradation study of the structural steel will be performed to predict the time of rupture of the process system to internal pressure using advanced analysis. Moreover, the effect of jet fire momentum on the vessels and their supports will be computed according to industry guidelines.

The methodology is validated by comparison with experimental data. The rupture time and rupture location will be computed and design recommendations will be provided to increase the strength of the process system at rupture locations. The current approach could provide accurate results to compute the response of process systems such as high-pressure vessels, scrubbers, etc., during offshore or onshore fire scenarios.

Assessing Safety Culture via the Site Safety Index^{TM}

Julia V. Bukowski, Ph.D.
Dept. of Electrical & Computer Engineering
Villanova University, Villanova, PA 19085
julia.bukowski@villanova.edu

Denise Chastain-Knight, P.E., CFSE
exida
64 N. Main St., Sellersville, Pa 18960
dchastainknight@exida.com

Prepared for Presentation at
American Institute of Chemical Engineers
2016 Spring Meeting
12th Global Congress on Process Safety
Houston, Texas
April 11-23, 2016

Assessing Safety Culture via the Site Safety Index™

Julia V. Bukowski, Ph.D.
Dept. of Electrical & Computer Engineering
Villanova University, Villanova, PA 19085
Julia.Bukowski@villanova.edu

Denise Chastain-Knight, P.E., CFSE
exida

Keywords: Safety Lifecycle, safety culture, systematic failure rate, site-specific safety performance, PFDavg

Abstract

How can a company establish a baseline measurement of its safety culture against which to gauge improvement? The Site Safety Index™ (SSI) quantifies in part (on a scale of 0 – 4) the degree to which a company's end-user practices support the attainment and retention of an appropriate safety culture for operations and maintenance. Further, the SSI can be used to appropriately adjust parameters that directly impact measures of safety such as probability of failure on demand (PFDavg). Thus, the impacts of safety culture can be further quantified and the effects of changes to safety culture can be assessed.

1. Introduction

Consider the following hypothetical scenario. A company assesses the risk intrinsic to one of its plant processes (given its current level of mitigation) to be greater than the tolerable risk dictated by company policy or regulatory standards. Further mitigation to reduce the process risk to an overall risk at or below tolerable risk level will require a safety instrumented function (SIF) which will result in a risk reduction factor (RRF) of at least 110, i.e., a SIF of safety integrity level (SIL) 2. The company may choose to design and implement the required SIF itself or to deploy an existing SIF design and implementation.

In either case, the company will need to ascertain that the SIF meets its risk reduction requirements. There are several ways to do this. For example, it may rely on proven-in-use data to claim the company has met its risk reduction goal. This proven-in-use data may come from another company or the database of a consortium of companies. Alternately, the company may predict the risk reduction and SIL level by using manufacturers' published data or other available databases perhaps in conjunction with a commercial tool. If by either or both means of risk reduction/SIL verification the company determines the risk reduction goal is met by the SIF, is this alone sufficient for the company to state that, when implemented in its plant, its process risk will be at or below tolerable levels? **No!**

When the company implements the SIF in its process, the actual risk reduction achieved by the SIF will be determined not only by the SIF attributes taken into account by the verification methods but also by the company's safety culture, i.e., by the overall impact of end-user practices that take place within the company in general and especially within the plant/site itself. Further, these practices may differ from site to site even in the same company.

For example, if the risk reduction/SIL was estimated from proven-in-use data, unless "systematic" failures were eliminated, these data inherently contain the impacts of the end-user actions/omissions which took place at the plant or plants during the time period in which the proven-in-use data were collected. When the company implements the SIF, it may actually experience the same, higher or lower risk reduction than that estimated from proven-in-use data depending on whether the company's end-user current practices are the same, better or worse than those practices in the plants *when* and where the proven-in-use data were collected. Especially for consortium-based data, it is not generally known exactly from where the proven-in-use data were gathered much less what the specific end-user practices were at those plants. Thus, estimated risk reduction/SIL based on proven-in-use data is difficult to adjust to more accurately represent the actual risk reduction/SIL the company will experience.

If the risk reduction/SIL was predicted based on the manufacturers' published failure rates or other failure rate data, these rates are likely predicated on the assumptions that the end-user faithfully performs all required actions to sustain the safety integrity of the equipment. Such actions include, but are not limited to, insuring that

- the equipment implementing the SIF is appropriate to the task and environment,
- the equipment is properly installed and calibrated,
- the equipment is correctly functioning when installed,
- all in-service maintenance is correctly and completely performed on schedule,
- the equipment is maintained so that no ageing occurs prior to proof testing,
- proof testing is completely and correctly performed,
- the equipment is replaced when its useful life can no longer be extended by maintenance and refurbishment.

Note that the list extends beyond the usual concepts of maintenance alone. Any required action that is omitted, delayed, or performed incompletely or imperfectly will reduce the risk reduction (and perhaps SIL) achieved in the SIF's actual use in the plant. However, in the case of predicted risk reduction/SIL, which is based on manufacturers' data or Failure Modes, Effects and Diagnostics Analysis (FMEDA) [1], it is possible to adjust the initially calculated risk reduction/SIL to account for a company's end-user practices. This adjustment occurs through the use of a SSI.

This paper explains the way in which the SSI is measured and illustrates how the SSI can impact typical safety parameters such as PFDavg. A second paper [2] explains how the SSI is used to adjust the initially calculated risk reduction/SIL.

2. Notation

FMEDA	failure modes, effects and diagnostic analysis
IEC	International Electrotechnical Commission

PFDavg	average probability of dangerous failure on demand
RRF	risk reduction factor
SIF	safety instrumented function
SIL	safety integrity level
SSI	Site Safety Index^{TM}

3. Measuring a Site's Safety Culture

3.1 Why a "Site" Safety Index?

The reader may wonder why the terminology "Site" Safety Index is used as opposed, perhaps, to "Company" Safety Index. It is well documented that a safety product may exhibit a safety performance less than that specified by the manufacturer or by a FMEDA [3, 4] and that even within the same company, failure rate estimates derived from field failure data can vary significantly from site to site [5]. Thus, while company-wide policies and procedures will impact the safety achieved, the possibly different ways company-wide directives are implemented on a site-by-site basis requires that a safety index be measured at the site level.

3.2 How is the Site Safety Index^{TM} Evaluated?

There are currently two ways that SSI can be evaluated: via an external audit, which provides the most accurate measure of SSI, and via a self-administered questionnaire, which relies on less information than the audit and provides an approximate measure of SSI. These two evaluation methods are detailed next.

3.2.1 Via an External Audit

The international standard IEC 61511 [6] lists the requirements that must be met for the entire Safety Lifecycle process. In an on-site audit, it can be verified that the site meets these requirements in combination with additional detailed requisites for the operations and maintenance activities. Additionally, all safety and lifecycle procedures are audited to be sure: a) that the procedures are adequate to support the requirements and b) that they are being correctly followed at the site. The outcome of the audit is a report which includes the SSI score obtained as well as recommendations for changes and improvements. The interpretation of the measured SSI level itself is discussed below.

3.2.2 Via a Self-Administered Questionnaire

A site that is interested in acquiring an approximate measure of its current state of safety culture may choose to complete a questionnaire of about 40 items which represents a subset of the information gathered in a complete audit. These questions cover activities and practices spanning a portion of the lifecycle of the SIF and include installation and commissioning, safety validation, operations and maintenance, SIF modification and decommissioning. Table 1 lists the questions for the operations and maintenance phase of the lifecycle. The complete set of questions is available at [7]. The questions, which are answered Always, Often, Never, or Not

Applicable, are subject to weighted scoring. The total score received determines an approximate SSI level for the site.

Table 1 Questions Regarding Site Activities and Practices during the Operations and Maintenance Phase of the SIF Lifecycle

Do your operation and maintenance procedures include competency requirements for those performing tasks to maintain the "as designed" functional safety of the SIF?
Do your operation and maintenance procedures include the information which needs to be maintained on device failures, proof test results, false trips, demand rates and restore times on the SIF?
Do your operation and maintenance procedures include procedures for fault diagnostics and repair after a failure in the SIF?
Do your operation and maintenance procedures require instruments to be rebuilt or replaced at the end of their useful life?
Do your operation and maintenance procedures require the use of a computer tool like SILStat™ to track demands, false trips, failures and maintenance performance?
Do your operation and maintenance work tasks proceed only in accordance with the relevant procedures?
Are operators that work with the SIF in their area trained to ensure they understand how the SIF functions and what hazards are being protected against?
Do you perform periodic proof-tests using the written procedure?
Is your frequency of proof tests decided using the PFD_{avg} calculation?
Are scheduled proof test performed on schedule?
Do you repair any deficiencies found during the proof testing in a safe and timely manner?
Do you maintain records of proof tests including descriptions of tests and inspections performed, equipment tested, dates, and names of personnel performing work?
Do you maintain records of proof tests including "as found" results of the tests and inspections?

The basis of the validity of using a questionnaire for determining SSI is the accepted practice per IEC61508 of using a questionnaire to determine the β factor that describes the extent of common cause failure to be modelled when evaluating the safety performance of redundant SIF architectures [8].

3.3 Interpreting the SSI Level

The SSI is a number from 0 – 4 which is an indication of the level of site activities and practices that contribute to the safety performance of SIF's on the site. Table 2 details the interpretation of each SSI level. Note that the levels mirror the levels of SIL assignment and that SSI 4 implies that all requirements of IEC 61508 and IEC 61511 are met at the site and therefore there is no degradation in safety performance due to any end-user activities or practices, i.e., that the product inherent safety performance is achieved.

Table 2 Details of the Interpretation of Each SSI Level

SSI Level	Description
SSI 4	Perfect - Repairs are always correctly performed, Testing is always done correctly and on schedule, equipment is always replaced before end of useful life, etc.
SSI 3	Superior - Repairs are usually correctly performed, Testing is usually done correctly and on schedule, equipment is usually replaced before end of useful life, etc.
SSI 2	Typical - Repairs are mostly correctly performed, Testing is mostly done correctly and on schedule, most equipment is replaced before end of useful life, etc.
SSI 1	Good - Repairs are often correctly performed, Testing is often done and often on schedule, some equipment is replaced before end of useful life, etc.
SSI 0	None - Repairs are not performed, Testing is not done, equipment is not replaced until failure, etc.

4. An Example of the Impacts of SSI

The specific ways in which SSI modifies key variables (e.g., failure rate, proof test effectiveness, etc.) which impact PFDavg, SIL and RRF, as well as other safety metrics, is explained in detail in [2] where additional examples are presented. For the purposes of this paper, Figures 1 and 2 provide a simple example of how SSI can impact safety performance. In Figure 1, computations were performed under the assumption that SSI level is 4 and thus end-user activities and practices do not degrade the safety performance of the SIF. Note that in this case the SIF SIL level is computed as SIL 2 based on PFDavg, and the RRF is computed as 113 – slightly greater than the requirement of the opening hypothetical example. In Figure 2, the analysis is repeated but this time the SSI level is assumed to be 3. Now, based on PFDavg, the SIL performance is a SIL level 1 and the RRF is computed as 98 – insufficient for the requirement of the hypothetical example. This shows how the impacts of site activities and practices can be quantified and how effects of changes to these activities and practices can be measured.

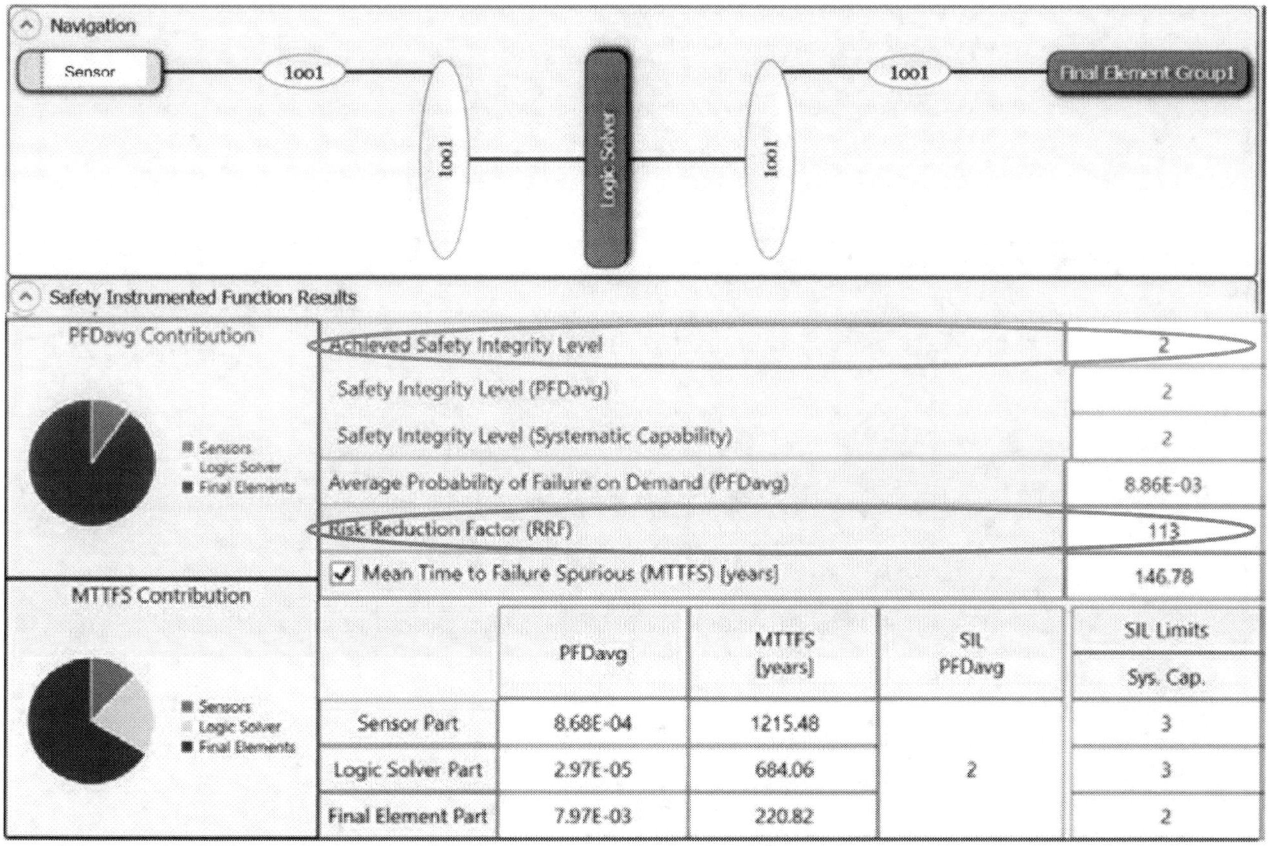

Figure 1. SIL verification/RRF computations performed under assumption SSI = 4.

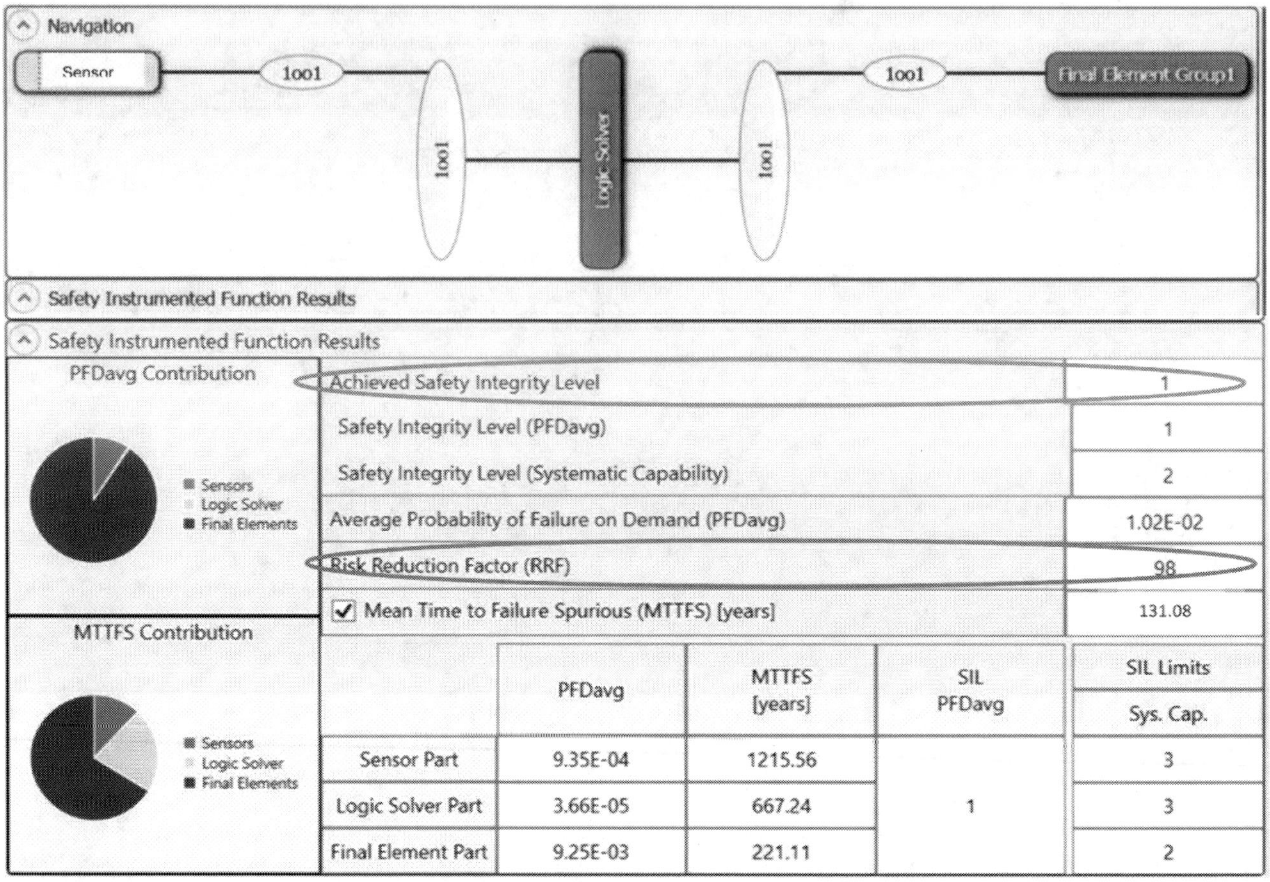

Figure 2. SIL verification/RRF computations performed under assumption SSI = 3.

5. Conclusions

The SSI provides a simple means to measure how well site-specific end-user activities and practices support the achievement and retention of SIF maximal safety performance. Further they can be used to adjust metrics of SIF safety performance from the ideal achievable to a more realistic measure of what is actually attained at a specific site given the site's activities and practices. This last concept is explored more fully in a separate paper [2].

6. References

[1] W. M. Goble and A. C. Brombacher, "Using a Failure Modes, Effects and Diagnostic Analysis (FMEDA) to Measure Diagnostic Coverage in Programmable Electronic Systems," *Reliability Engineering and System Safety*, Vol. 66, No. 2, Nov 1999, pp. 145-148.

[2] J. V. Bukowski and L. Stewart, "Quantifying the Impacts of Human Factors on Functional Safety Failure Rates," *Proceeding AIChE 12th Global Congress on Process Safety*, Houston, TX, April 2016.

[3] J. V. Bukowski, W. M. Goble, and I. van Beurden, "Product Failure Rates vs Total Failure Rates at Specific Sites: Implications for Safety," *Proceeding AIChE 11th Global Congress on Process Safety*, Austin, TX, April 2015.

[4] J. V. Bukowski, W. M. Goble, R. E. Gross and S. P. Harris, "The Effects of Maintenance Actions on the PFDavg of Spring Operated Pressure Relief Valves," *Proceedings of the ASME 2014 Pressure Vessels And Piping Division Conference*, Anaheim, CA, July 2014.

[5] W. M. Goble, "Maintenance capability: Does it matter in the long term?" Hydrocarbon Processing, June 1, 2013.

[6] IEC 61511, *Application of Safety Instrumented Systems for the Process Industries*, Geneva, Switzerland, 2nd edition, 2004.

[7] www.exida.com/SSI

[8] IEC 61508, *Functional Safety of electrical / electronic / programmable electronic safety-related systems*, Geneva, Switzerland, 2010.

441381 Fire Hazard Associated with Polymers and Application of Polymer Nanocomposites for Flame Retardancy

Abstract:

Due to advances in polymer science, their widespread application is found in daily products including vehicles, electronics, clothing, furniture *etc*. However, most of the items that first ignite in any fire scenario includes mostly polymeric materials according to NFPA reports which clearly elucidates the importance into looking for flame retardancy approach for polymers. Polymer nanocomposites is a relatively new approach undertaken for flame retardancy and has the potential to bring multi-functionality including improved thermal behavior, physical barrier effect, catalytic charring effect and better material performance. The objective of this paper is to initially provide the current state-of-the-art review on polymer nanocomposites in terms of their fire retardancy. The paper describes the hazards associated with burning of polymers, flame retardant mechanism, synthesis and characterization of polymer nanocomposites with focus on thermal analysis. Different kinetic model has been studied and applied for analysis of thermogravimetric analysis data. The kinetic parameters of the polymer degradation have been identified. Primarily polystyrene-silica nanocomposite has been prepared by in-situ method and were characterized for their morphology and thermal behavior using characterization methods, such as Scanning Electron Microscopy (SEM), X-Ray Diffraction (XRD), Thermogravimetric Analysis (TGA), and Differential Scanning Calorimetry (DSC). Initial results show nanocomposites with promising fire-retardancy properties.

441530 Applying QRA Study to Select and Design Effective Release Prevention Safeguards

Abstract:

In the past two decades, Quantitative Risk Analysis (QRA) has become a staple safety analysis in offshore oil and gas industry. A comprehensive QRA study could be a powerful tool if the results are understood and changes are implemented to improve safety in the engineering design. Conventional QRA study relies on generic leak frequency and parts count method. One major limitation of conventional QRA study is that it is focused primarily on the loss of containment of hazardous fluids and what happens when they are released. To reduce the risk to acceptable level, usually post-release mitigation measures are addressed. It is a challenge to tackle the effectiveness of release prevention safeguards. This paper presents an approach to address the dominant risk contributors identified in conventional QRA study. The aim is to increase confidence and facilitate decision making in the design selection of loss of containment prevention safeguards.

Sensitivity analysis is usually conducted as part of a QRA study to evaluate the effect of input parameters. However, in most cases, sensitivity analysis focuses on the mitigation measures, such as safety system locations or ratings, personnel distribution or modification to the QRA model itself. Means of release scenario prevention is rarely investigated to the point that it is beneficial to the operation or to the design project. This paper proposes that in order to maximize the use of QRA results, decision makers should investigate means of scenario prevention by performing reliability analysis (e.g., fault tree analysis, human reliability analysis) on the top contributing scenarios. This paper presents some examples of reliability analysis performed on top risk contributors. This paper also presents suggestions on how the results can potentially be applied for the benefit of the operation or design project. The outcome from the QRA study in combination with reliability analysis will increase decision makers' understanding of the top risk contributing scenarios. This approach can be utilized to identify opportunities for improvement in the prevention safeguards.

441536 Managing the Uncertainty in Blast Load Calculations with Enhanced Technical Safety Analysis

Abstract:

The safety of personnel working within onshore and offshore oil and gas processing facilities with respect to vapor cloud explosion (VCE) scenarios revolves around the risk associated with the event. Conducting a risk assessment is quite typical within the offshore and onshore loss prevention communities. The consequence of a VCE event is a very complex phenomenon however. It varies based on several blast load parameters. With improved understanding of blast loading parameters such as blast overpressure, impulse, time lag, net blast loading, etc. structural designers have gained increased confidence in their design and have reduced conservatism in their blast designs. However, there is a significant gap in the risk assessment of onshore and offshore facilities between the process safety and the structural design groups. Using structural insight in combination with rigorous process safety risk evaluation techniques we can reduce the gap between the risk and consequence of a blast design, providing the stakeholders with a more accurate understanding of the risk and consequence of a blast event.

This paper will attempt to address the impact of uncertainty in pressure exceedance curves and;

- how the confidence level impacts the design blast loads;
- how the slope of the exceedance curves impacts the blast design loads;
- two different blast events can result in similar design blast loads for a given risk;
- how a small change is frequency can lead to a significant change in design blast loads;
- how the negative phase of a blast scenario could lead to failure of some specific type of structures;
- how the response of long wall or piping structure is impacted by time-lag of the design blast load;
- how the design blast pressure and design blast impulse impacts a structure;
- how non-structural members, typically not considered in blast design, impact the blast design;
- how simultaneous application of blast load on different parts of structure, taking into account time-lag, affects the blast design.

Currently for a blast event, both blast pressure and blast impulse are defined as independent variables whereas in reality they are coupled. The correlation between the impulse and pressure will also be discussed in the paper. Defining appropriate pressure time-history is critical for structural design. Rise time, positive-phase and negative-phase duration are unknowns in a blast design. How an appropriate time-history is established will also be addressed.

Prior knowledge of implications of a blast event for an associated risk is helpful during structural blast design. Early involvement of structural engineers within the process safety procedure is a recommended solution to such questions. Enhanced loss prevention of offshore and onshore structures can only be established by always keeping the consequence for a blast event for a given risk in mind.

441537 Occupant Vulnerability for Process Plant Building Design Due to Secondary or Domino Events

Abstract:

Statistics indicate that it is in fact not uncommon to have an accidental explosion on an offshore platform, refinery or petrochemical installation. The focus for several years has been on occupied buildings and their resistance to blast loads from accidental releases of hydrocarbons. Common practice dictates that domino effects may not need to be considered given that a large catastrophic release is equivalent in consequence. However, the catastrophic events are filtered out in a risk assessment due to their unlikelihood and as a result, domino effects are not captured. For example, a small release of highly reactive fuel, can lead to a small explosion which can easily damage nearby piping and vessels which can in turn become catastrophic events. The design and assessment of piping systems and vessels against blast loads are of paramount importance and require advanced analysis capabilities. In order to have a safe, i.e. blast-resistant process piping system design, an advanced nonlinear analysis in time domain is required so that we may properly capture the dynamic response of the piping and vessels subjected to blast loading. This type of analysis, in general, should take into account (1) Blast drag load and blast overpressure; (2) Nonlinear material properties, i.e. thermal and strain-rate dependence; (3) Effects of non-structural masses and adjacent piping systems; (4) Pipe supports; (5) Failure of flanges and piping detail; (6) Effects of Operating Temperature on Material Properties; (7) Blast direction/Ignition location/Attenuation sensitivity; (8) Shielding effect, e.g. pipes behind large objects such as vessels; and (9) Effects of pipe insulation, i.e. increased pipe diameters. This paper first discusses the methodology/techniques to account for such effects in finite element analysis for a safe design. The paper then uses case studies to demonstrate the methodology we develop to show how advanced simulation techniques are applied as tools for piping blast response to minimize the risk of secondary failure due to initiating events. These case studies involve a piping system that is simulated from global to detail modeling. The goal of the paper is to discuss how to achieve a safer design for process piping systems by using advanced analysis.

441629 Assessment of the Mechanical Integrity of a Pipeline with Metal Loss, Using Finite Element Method

Abstract:

Following the guidelines proposed by the ASME B31G and API579 standards, for the assessment of systems that compromise its mechanical integrity, given the existence of defects associated with the presence of corrosion damage, the probability of failure of a pipeline, which transports hazardous materials associated with the Oil & Gas industry, and is exposed to pitting corrosión damage was calculated when it overcomes its yield stress. For this, a methodology that combines the corresponding structural analysis of the system and subsequent reliability analysis was proposed. Through the first, the calculation of the von Mises stress present in areas where the pitting defects was made; for this, the corrosion growth model to be used in the present work was defined by implementing the correlation proposed by the Southwest Research Institute (SwRI) to find out the corrosion rate growth in annual terms. Subsequently the pipeline section where the study would be carried out was defined, based on the results obtained in a previous study, which analyzed information related to corrosion damage present on the pipeline, collected through the In Line Inspection (ILI) technique. This allowed to restrict the test area to a total length of 80 cm and 1/8 section of pipe, lastly simulating the system and obtaining the loads associated with the von Mises stress affecting corrosion defined areas. For reliability analysis it was necessary to find the probability distribution function associated with the von Mises stress on the system studied; defining an experimental design which sought to evaluate different values of the random variables defined on the model proposed by SwRI, thus simulating the system a sufficient number of times in order to find the distribution of von Mises stress for this case study. With the distribution function defined, system reliability analysis was carried out through Monte Carlo simulations; thus, it was established that the limit state function would have a conventional structure comparing resistance with loads. Thus, the resistance was set as the yield stress of the material, while the load would be the von Mises stress, and thus the random variable from which random numbers governed by the probability distribution found previously were generated. As mentioned previously, it was defined that the present system fails when the von Mises stress exceeds the yield stress of the material; finding that for the time range evaluated (year one, five, ten and fifteen), the probability of failure is behaved as expected, increasing year by year if no maintenance is not performed to the pipeline section evaluated.

441539 An Improved Risk-Based Approach for Accidental Loadings

Abstract:

Accidental loading such as fire, blast, ship impact, and dropped objects constitute the great majority of potential and actual fatalities in both onshore and offshore operations. In order to help prevent human loss and for increased safe design or assessment of an onshore/offshore asset, the risks of these accidental loadings should be quantified, in terms of both probability/frequency and consequence aspects. In this paper we propose a unified risk-based approach for the assessment of all accidental loadings mentioned above.

This approach will address both human and asset risk assessment against accidental loadings. Technical safety and process safety inputs such as hazard curves, frequency exceedance curves, and return periods are where the process all begins. Based on these inputs, an event frequency analysis is carried out to determine how often one should expect the event to happen, i.e., the frequency aspect of the assessment. Depending on the type of the accidental loadings and the subject of the risk assessment (human or asset), the output can include a fragility curve, damage frequency matrix, and individual risk matrix. Additionally, a consequence analysis is carried out to determine the severity levels of the event. In a risk-based consequence analysis, the severity levels are fully determined based on the risk associated with the event. Based on both event frequency and consequence, the risk ranking is identified using a risk matrix. If the risk is not acceptable according to the risk matrix, mitigation/repair/strengthening options are then developed.

As an example, a risk-based structural fire analysis is compared with the conventional approach which is widely used in the industry. In a conventional Passive Fire Protection (PFP) analysis using Ductility Limit Analysis (DLA) according to API RP 2FB, fire loads are deterministically applied to a structure whose response is then analyzed. The initial PFP scheme is developed based on the analysis and then optimized based on input from all disciplines (e.g. technical safety, operations, structures, etc.). This approach is sometimes misinterpreted as a "risk-based" approach; however, it does not take into account the frequency aspect of the risk assessment. In a risk-based PFP analysis using DLA, fire scenarios are developed in a particular target zone based on exceedance curves (i.e., input from technical safety discipline). The ductility level analysis is performed to determine the structural consequence. If personnel safety is of interest, the consequence of the structure is then linked to individual risk to determine fatalities. The amount of PFP to be applied on the structure is fully based on the risk that is produced by the fire scenarios in the target zones or areas.

Followed by the approach, a new perspective on safe design of onshore/offshore structures for accidental loadings is outlined to estimate the associated risk to potential targets such as human personnel as well as asset components. The proposed assessment methodology will contribute towards identifying the mitigation measures and safety-critical procedures and equipment, and toward a safer design.

441540 Occupant Vulnerability for Temporary Blast Resistant Modular Buildings

Abstract:

Human fatality and injury in the process facility due to overpressure effects can be categorized as (a) direct injury to body due to pressure change, (b) secondary injury resulting from fragment or debris impact, and (c) tertiary injury as a result of body being thrown by explosion drag on stationary objects. Secondary and tertiary effects are typically the dominant causes of fatalities in the process facilities. Direct injuries may result from change in internal pressure as a result of wall/roof response or penetration of blast overpressure into the building. Existing studies allow for an estimation of occupant vulnerability to blast on typical default building construction types. There exists no such building occupant vulnerability calculation method for modular buildings. This paper seeks to develop an improved understanding of the impact of blast loads on modular building and a more comprehensive understanding of the vulnerability to personnel who are within this type of building if they are subjected to a blast load. The study will be carried out in analytical basis which will include basic and advanced numerical methods along with engineering judgment and accident investigation for estimation of vulnerability. A numerical model can include deformable crash test dummies as well as many types of non-structural components in order to estimate occupant vulnerability and maximum individual risk. The outcome of this study will include new response criterion limits related occupant vulnerability in modular buildings as well as design and retrofit guidelines for critical non-structural components.

441541 Occupied Building Blast Retrofit and Upgrade for Personnel Safety

Abstract:

Accidental blast events at petrochemical facilities such as BP Texas City have shown that the occupied buildings near the process areas can be vulnerable to blast and collapse even under low blast overpressures. Additionally, existing permanent and portable buildings may exhibit insufficient capacity to resist blast loading during facility siting studies per API RP 752 and 753. Although it is not possible to completely eliminate risk from blast loading, it is possible to minimize the individual risk, which is a probability an occupant will experience fatality due to an explosion event, by applying retrofit and upgrade strategies to increase the blast-resisting capacity of an occupied building and meet the safety performance criteria.

There are several ways to decrease individual risk due to explosion with respect to occupants in a building:

1. Moving the building from a high blast region near the process area to a lower blast region, and design a new building;
2. Retrofitting existing occupied building;
3. Replacing existing occupied building with a blast resistant modular building; and
4. Replace the existing occupied building with a new blast resistant building.

The first and forth options can be inefficient from construction cost and time perspective. Additionally, the first option may require purchasing an additional land and obtaining permit. The second and third options are significantly cost and time efficient, and causes less interruptions for the operations. However, these options require analysis and engineering judgment from a highly skilled and experienced engineer. In this paper, case study will be presented where a variety of retrofit and upgrade options will be discussed from perspectives such as design, constructability, cost and individual risk.

Implementing Engineered Controls to Reduce the Likelihood of High Severity Scenarios

Alek Hamparian
Kazarians & Associates, Inc.
100 West Broadway, Suite 970, Glendale, California
Alek.Hamparian@Kazarians.com

Mardy Kazarians
Kazarians & Associates, Inc.
100 West Broadway, Suite 970, Glendale, California
MKazarians@Kazarians.com

Prepared for Presentation at
American Institute of Chemical Engineers
2016 Spring Meeting
12th Global Congress on Process Safety
Houston, Texas
April 11-13, 2016

Implementing Engineered Controls to Reduce the Likelihood of High Severity Scenarios

Alek Hamparian
Kazarians & Associates, Inc.
100 West Broadway, Suite 970, Glendale, California
Alek.Hamparian@Kazarians.com

Mardy Kazarians
Kazarians & Associates, Inc.

Keywords: Hazard Identification and Risk Analysis, Process Safety

Abstract

The Center for Chemical Process Safety (CCPS) *Guidelines on Inherently Safer Chemical Process A Life Cycle Approach* (2009) defines a hierarchy of controls that can be used to decrease the risk of a system. These guidelines include a hierarchy of layers of protection that can be used in Process Hazard Analyses (PHAs) to determine adequacy of the existing safeguards. This approach can be particularly useful for companies that have potentially high severity scenarios for which only a procedural layer of protection is available.

Releases of hazardous materials have occurred that could have been prevented by implementation of engineered safeguards. From a review of these incidents, it had become evident that the owners and operators of the affected facilities relied on procedural safeguards to prevent occurrence of those events. Examples of such cases have involved gasoline, chlorine, and anhydrous ammonia, which are three commonly used hazardous materials at facilities in the United States.

Process Hazard Analysis (PHA) provides a thorough and efficient method for systemically reviewing complex systems for safety concerns. A commonly applied PHA method is the Guide-Word Style Hazard and Operability (HAZOP) approach. HAZOPs are a team effort and risk determination varies based on the experiences and safety culture of team members. Some teams may determine that procedural safeguards are adequate to prevent against a scenario that could lead to a severe event (e.g., liquid release of a volatile toxic material such as chlorine). However, accidental releases have been experienced demonstrating inadequacy of procedural safeguards to minimize the risk of a severe event.

The above noted CCPS guidelines consider procedures as the least effective protection layer. An active engineered safeguard is considered a more reliable protection layer. The purpose of this paper is to point out the potential for a pitfall in risk acceptance where a high severity consequence scenario is concluded to be of acceptable risk based solely on operator action. It is recommended for owner and operators of such facilities to require mitigation feature other than procedural controls for high severity scenario regardless of the risk ranking.

1 Introduction

Releases of hazardous materials have occurred in a wide range of systems with safeguards in place. A number of these events have occurred in systems that relied on procedures and operator intervention as the primary safeguard. The Center for Chemical Process Safety (CCPS) [1] and the American Industrial Hygiene Association (AIHA) [2] have each published guidelines on the topic of hierarchy of controls which can be used as a qualitative measure to establish the effectiveness of risk reduction measures. Both CCPS and AIHA have ranked engineered controls as a higher level safeguard than human intervention. The October 23, 2009 explosion at the Caribbean Petroleum Corporation's facility in Puerto Rico shows the importance of implementing engineered controls in addition to human intervention. The Chemical Safety Board (CSB) [3] investigated that event and recommended for similar facilities to add engineered controls such as "automatic overfill protection systems" to prevent future accidents.

Facilities that contain certain hazardous materials above a certain quantity must comply with the Process Safety Management (PSM) [4] and Risk Management Plan (RMP) [5] process safety regulations that require a Process Hazard Analysis (PHA). The purpose of a PHA is to identify potential hazards posed by a system, the safeguards that are in place to minimize the likelihood or severity of the event and risk rank the scenarios that can lead to adverse consequences. In some instances, the probability qualitatively assigned to an adverse scenario with only procedural safeguards may be underestimated. This underestimation of probability can lead to a lack of urgency to implement engineered safeguards to prevent an adverse consequence.

In addition to PHAs, some companies voluntarily perform Layer of Protection Analysis (LOPA), which is a more detailed analysis than PHA. LOPA is used to semi-quantitatively assess whether sufficient safeguards exist to meet a target risk level. In some instances, teams may use LOPA in such a way as to show that adverse consequences are unlikely by crediting conditional modifiers that may not be applicable to the scenario. This can lead to a determination that additional safeguards are not necessary as risk tolerance has been met.

It may be prudent to establish rules to implement engineered safeguards in potentially severe scenarios that rely only on procedural safeguards. The purpose of this paper is to describe the establishment of a criterion in which potentially high severity scenarios are identified that rely primarily on procedural safeguards and how to add an engineered safeguards to reduce the likelihood of the scenarios occurrence.

2 Industry Incidents

Many incidents have occurred in the chemical process industry where a procedural safeguard was the main layer of protection.

The CSB has performed many incident investigations and provides recommendations to different organizations with the main purpose of learning from past events to enhance safety. A recent investigation focused on the CAPECO incident of October 23, 2009 when a gasoline storage tank was overfilled leading to the formation of a vapor cloud and explosion [3]. In Appendix B of that investigation report [3], CSB lists twenty-two major accidents that have occurred in the process industry since 1962 that involved hydrocarbon storage tank overfill. A majority of these accidents involved processes that did not have an automatic overfill protection system and only relied on human intervention to prevent an overfill.

Similar events have occurred at facilities that handle anhydrous ammonia, a highly hazardous material. Ammonia is a toxic material with a low boiling point that vaporizes upon release into the atmosphere. Inhalation of ammonia vapors can result in irritation of the mucous membranes of the nose, throat, and lungs. If a plume of ammonia vapors with a concentration above 5,000 ppm is inhaled, sudden death may occur [6]. Many refrigeration systems use anhydrous ammonia as the refrigerant in the United States. In ammonia refrigeration systems, compressors provide the motive force to circulate the ammonia. Compressors require oil for lubrication, which by design leaks into ammonia side of the system. Entrained oil is removed from the system from the oil pots where oil separates from ammonia and settles in the pot. In order to ensure proper ammonia refrigeration system cooling efficiency, these oil pots must be drained periodically to remove the oil on the ammonia side of the system. This is achieved by opening a drain valve into an open pale. System pressure moves the oil. The operators are required to be vigilant and close the valve as soon as oil composition starts to change indicating that the pot is drained of oil and ammonia may follow.

Releases of ammonia have occurred from ammonia refrigeration systems due to oil drain operations. An engineered safeguard that could prevent this type of release is a self-closing emergency stop valve. It is designed such that an operator must hold the valve open by holding on to it and when operator's hand is removed the spring loaded self-closing emergency stop valve immediately closes. If an emergency stop valve is installed on an oil drain valve, it can limit the amount of ammonia released to atmosphere in the event that the drain operation is performed too long and ammonia starts to release once the oil has been drained from the pot.

The accident at CAI, Inc. and Arnel Inc., [7] is another example of an accident involving a system without an engineered safeguard. The CSB performed an incident investigation on this accident and issued recommendations [7]. According to the CSB, an initiating cause of the explosion was determined to be that a steam valve was left open to a tank containing flammable liquid. This steam valve was periodically used to heat the contents of a tank that stored ink and then manually shutoff by an operator once the desired temperature of the material was reached. On November 22, 2006 the steam valve was forgotten open and the ink in the storage tank overheated and vaporized. The vapor cloud found an ignition source and a large explosion occurred. The CSB recommended that CAI, Inc. install safety controls that would prevent overheating of flammable or combustible liquids.

3 Hierarchy of Controls and its Application

3.1 Hierarchy of Controls

CCPS has published "Inherently Safer Chemical Processes, A Life Cycle Approach", which describes the importance of making chemical processes inherently safer [1]. The CCPS states the following, "A chemical manufacturing process is inherently safer if it reduces or eliminates the hazards associated with materials and operations used in the process and this reduction or elimination is permanent and inseparable" [1]. The CCPS has described methods of making chemical processes safer and an important concept of the guidelines is levels of inherent safety. The CCPS [1] describes the levels of inherent safety as follows from most effective to least effective:

First Order Inherent Safety Measures –
Avoid Hazards
Second Order Inherent Safety Measures –
Reduce Severity;
Reduce Likelihood
Layers of Protection –
Apply Passive Safeguards;
Apply Active (Engineered) Safeguards;
Apply Procedural Safeguards

It is important to note that the least robust layer of protection listed by the CCPS is the application of procedural safeguards. This means that the application of active safeguards is a more robust layer of protection than procedural safeguards. The highest or first order of inherent safety listed is avoiding hazards. This is defined as elimination of a hazard. The two second orders of inherent safety listed are reducing severity and reducing likelihood. The severity of a hazard can be reduced by a number of methods including reducing the quantity of the material in a system, reducing the pressure or temperature within a process, substituting the material for a less hazardous material, etc. The likelihood of a hazard's occurrence can be reduced by a number of methods including using more corrosion resistant material, redesigning a process from four vessels to requiring two vessels, etc. The next order of inherent safety after the first and second orders are layers of protection. Layers of protection are ranked in order as passive safeguards, active safeguards, and procedural safeguards with procedural safeguards being the least robust and passive safeguards being the most robust. An example of a passive safeguard is designing the vessel within a system with a maximum allowable working pressure that is higher than the highest pressure that can potentially be reached within the system. An example of an active safeguard is an automatic overfill protection system that shuts off flow to a tank when high level is reached in the tank. An example of a procedural safeguard is relying on operators to shut off the flow to a tank when a certain level is reached.

The AIHA also addresses hierarchy of control from the standpoint of occupational safety. The "Z10-2012: Occupational Health and Safety Management Systems" Standard, defines hierarchy of controls which can be summarized as follows:

A. Elimination;
B. Substitution of less hazardous materials, processes, operations, or equipment;
C. Engineering controls;
D. Warnings;
E. Administrative controls; and
F. Personal protective equipment

There are many similarities between the CCPS's inherent levels of safety and the AIHA hierarchy of controls. Similar to the CCPS, the AIHA hierarchy of controls considers engineering controls as more robust than human intervention, warnings and administrative controls. The AIHA hierarchy of controls is written for occupational safety whereas CCPS's hierarchy of controls is written specifically for process safety applications. In both hierarchy of controls, the first item listed is eliminating a hazard. The second order of inherent safety listed by the CCPS is "reduce severity, or reduce likelihood", which is very similar to the second hierarchy of control listed by AIHA. Substituting a less hazardous material translates to reduced severity. Substituting less hazardous operations can be considered as a measure that reduces likelihood. The third hierarchy of control listed by AIHA is "engineering controls", which can be considered the same as the first two layers of protection listed by the CCPS, "Apply passive safeguards, Apply active safeguards". The third layer of protection listed by the CCPS, "Apply procedural safeguards" can be considered a combination of the fourth and fifth hierarchy of controls listed by AIHA, "Warnings; Administrative controls". The one major difference is that the CCPS does not call out alarms specifically as the AIHA does with the *warnings* hierarchy of control. The CCPS's inherent safety measures refer to alarms in the procedural controls as all alarms require personnel intervention.

An example of an engineered control referred to by AIHA in regards to occupational safety is a shutoff system for an automated laser welding machine that shuts off immediately upon detection of limbs within the welding area. This is an example of an engineered control designed to prevent personnel exposure to laser burns. An automatic overfill protection system that is considered an engineered safeguard by the CCPS for process safety applications is also considered an engineering controls according to the AIHA hierarchy of controls.

Regulatory bodies involved with PSM and RMP have taken notice of the importance of hierarchy of controls. Current drafts of California's Process Safety Management [8] and Risk Management Plan [9] regulations for refineries will specifically require hierarchy of hazard controls to be formally addressed. The draft regulations state that an analysis must be performed on all existing processes at a refinery that pose a process safety hazard. The analysis team must document the possible inherent safety measures for each hazard that can be implemented to reduce risk. The analysis team must review the feasibility of each level of inherent safety and must document why an inherent safety measure cannot be implemented for the process if that decision is made.

3.2 *Application of Hierarchy of Controls*

The most preferable time period to apply the hierarchy of controls to a system is during its design phase when changes can be made without severe economic impact. In addition to the design phase of a system, the hierarchy of controls can be applied throughout a system's life cycle when changes are considered. However, the upcoming regulations in California will be requiring application of the concept to built and operated systems.

To be able to apply hierarchy of controls to a system, it is only rational to establish what hazardous scenarios are possible and what safeguards are in place as protection. Process Hazard Analysis (PHA) provides a systematic method to identify and rank potential hazard scenarios and to identify the layers of protection in place to minimize the risk. By conducting the PHA, one can claim that all potential hazard scenarios that a team of experts could identify have been addressed systematically.

The hierarchy of controls can be used during a Process Hazard Analysis (PHA) to determine the adequacy of available safeguards to prevent each release scenario identified. This can be completed by identifying the available safeguards for scenarios that are postulated to have severe consequences and questioning whether or not inherently safer measures can be implemented. In addition, the hierarchy of controls can be applied when a recommendation is made during a PHA. The hierarchy of controls can act as guidance to choosing a higher level control measure to reduce the risk of a scenario. The hierarchy of controls can be reviewed while making any major changes to a system to assess if there are inherently safer options than the change being proposed. The hierarchy of controls can also be reviewed during an incident investigation to determine if a higher order of control should be implemented to prevent the recurrence of an incident.

The thought process of applying hierarchy of controls for a gasoline producer may be the following:

1. Currently there is a need for gasoline as vehicle fuel since a majority of automobiles sold in the United States require gasoline as the energy source. Therefore, this is a viable business venture that society at large needs. Eliminating gasoline from the line of products produced and sold by the company is not a prudent business decision because its safety risk is relatively well understood and has historically been accepted by society. Therefore, eliminating the hazard (i.e., moving out of gasoline business) as the very first item of the hierarchy of control is not a prudent option.
2. The next hierarchy of controls addresses reducing severity and likelihood. In order to reduce the severity of a gasoline storage tank related mishap (e.g., overfill and vapor cloud explosion), the hazard radius of a vapor cloud would need to be reduced. This can be done by reducing the pool surface of released material, as an example. Pool area is a function of the berm walls around the storage tanks and height of the walls. As another example, the owner may decrease the flow rate into a tank allowing the operators more time to discover and shutoff the inflow. These changes may not be feasible or the risk reduction gained from them may be smaller than other options (e.g., automatic shut-off on high level).

3. In order to reduce the likelihood of a gasoline storage tank overfill, the number of times the storage tank filling operation is performed can be reduced. This will require larger storage tanks which may increase the quantity of gasoline release in case of tank failure. Thus, the risk reduction gained from decreasing frequency of fills may be offset by the increase in risk due to larger tank size.
4. Engineered controls are the next available hierarchy of control. The first type of layer of protection based on CCPS's hierarchy of controls is a passive safeguard. One commonly used passive safeguard for large hydrocarbon storage tanks is secondary containment (a berm). The purpose of a berm is to contain a spill. However, in the case of gasoline, the light ends within the mixture would vaporize resulting in a vapor cloud formation with the potential for an explosion. As discussed above, reducing the pool size may reduce the evaporation rate but better risk reduction can be obtained by other means.
5. The next hierarchy of control is an active safeguard such as an automatic overfill protection system that will shut off the pump filling the storage tank when high level is detected. For a built and operated system, this option may provide the best risk reduction.
6. Procedural safeguards are the next hierarchy of control after active safeguards. The procedural safeguard for a gasoline storage tank filling operation would be operator monitoring of the level on the storage tank, installation of a high level alarm, and shutting off the feed pump manually at a certain level.

4 Process Hazard Analysis

4.1 Use of Process Hazard Analysis

A Process Hazard Analysis (PHA) is a hazard analysis tool that must be performed for systems covered under the PSM and RMP regulations [4 and 5]. PHA is a systemic tool for identifying the hazards of a system. PHA also provides a means to establish the potential severity of each hazard scenario, the available safeguards to prevent the hazard and likelihood of the scenarios occurrence. By applying risk ranking protocols, the risk gap of each scenario is identified in the PHA process.

The most common PHA method is HAZOP, which is typically performed in a brainstorming session with the participation of team members with experience in different fields necessary to properly analyze a system.

An important part of a PHA is establishing the risk of a hazard scenario. Risk is defined as "the combination of the expected frequency (events/year) and severity (effects/event) of a single incident or a group of incidents" [10]. Before a PHA study commences, the management responsible for the system must decide what risks they are willing to accept and what risks they will consider unacceptable. The acceptable and unacceptable risks are then translated into a risk tolerance matrix. The risk tolerance matrix contains severity and likelihood pairs that are acceptable and severity and likelihood pairs that are considered unacceptable and require risk reduction. This risk matrix is then used by PHA teams in all the PHA studies owned and operated by the management. The risk matrix provides a basis to identify risk gaps and whether or not system upgrades should be considered to improve safety.

4.2 *Augmenting PHAs with Hierarchy of Controls*

PHA results can vary from team to team. People working in the chemical process industry have different career experiences, which can result in different risk estimations during a PHA. It is possible and the authors' experience in leading PHA studies corroborates this observation that the likelihood of a severe consequence may be underestimated by PHA Teams. For example, operators who work with a system on a daily basis and have not experienced any incidents may be under the impression that the probability of a specific high severity scenario occurring is very low. Operators may feel that a system has been performing well for many years and if an issue did arise, the operators are knowledgeable enough to stop the progression to the bad event. Also, some PHA team members may believe that certain initiating causes of hazardous consequences are very unlikely based on their experience. Many of these issues can lead to underestimation of the risk of a severe consequence. Therefore, it is possible that the risk of a high severity scenario is estimated within the acceptable range allowing reliance on procedures based safeguards with no other safeguard. Due to the fact that risk ranking issues such as the one described above exist, it may be prudent for companies to use hierarchy of controls as the bases to require that all high severity scenarios to include at least one independent layer of engineered safeguard protection.

5 Layer of Protection Analysis

Layer of Protection Analysis (LOPA) is a tool that can be used to semi-quantitatively estimate the likelihood of a high severity scenario. Typically, LOPA is applied to the high severity scenarios identified during a PHA in order to enhance the accuracy of scenario likelihood. As part of the LOPA process, typically the LOPA Team estimates probabilities on an order of magnitude basis. The values that are estimated are the initiating cause likelihood, failure probabilities of the safeguards, occupancy factor, ignition probability, time at risk, and other conditional modifiers [11]. The initiating event likelihood is an estimate of how likely it is that the initiating cause occurs that can initiate the postulated chain of events. For example, according to the CCPS [12] the probability of a basic process control system (BPCS) failure is 0.1 per year. Also, if the safeguard has special features (e.g., it is designed to be of high reliability) the team may assign 0.01 probability of failure.

Occupancy factor refers to the amount of time personnel are within the hazard zone. For example, the CCPS [13] states that the probability of one person being present in a hazard zone is 0.1 for an area where operators spend 17 hours on average per week. Ignition probability refers to the probability that a released flammable material finds an ignition source resulting in a fire or explosion. Ignition probabilities will vary based on many factors such as the properties of the released material in terms of the released temperature, hot equipment near the release point, etc. Time at risk factor applies to non-normally operating modes of a system. Other conditional modifiers may also be used depending on the scenario. For example, some companies may take a conditional modifier stating that a catastrophic compressor seal failure due to liquid being sent to the suction side may occur once every ten times.

LOPA teams may credit additional conditional modifiers and enabling conditions resulting in a low risk level within the acceptable range. One example is crediting a time at risk factor for scenarios that can only occur during startup. Another likelihood reduction can be achieved through the application of occupancy factor for scenarios where generally personnel are not present within the hazard zone. Application of these factors raises an important question about the possibility of reliance on procedural safeguard as the only safeguard against a high severity consequence. Similar to the PHA, management may mandate that for all high severity scenarios, the LOPA team should consider an engineered safeguard regardless of the likelihood and risk ranking.

Startup is an interesting case as a safety risk contributor. The CCPS states that time at risk is not advised to be used during startup, "Use of a time-at-risk enabling condition in such a situation may lead to a conclusion that the average risk meets a facility's risk criteria; however, the peak risk imposed on persons exposed to the hazard may be inappropriately high" [13].

6 Limitations and Benefits

6.1 Increased Complacency Potentially Leading to Adverse Consequences

A potential adverse outcome of adding engineered safeguards to a system is an increase in operator complacency, which may lead to an adverse consequence. Operators involved in tasks such as storage tank filling may become complacent if an engineered safeguard is available as a backup in case they do not shut off the filling operation in a timely manner. Many engineered safeguards are meant to serve as a final barrier to prevent the occurrence of a release. The knowledge that a final barrier such as an engineered safeguard is available may lead to personnel relying on that safeguard and not being as attentive to system operation as before the addition of the engineered safeguard. Increased complacency may unintentionally lead to an adverse consequence as typical engineered safeguards have a reliability of approximately 90%. This means that approximately 10% of the time that they are required to prevent an adverse consequence, they may not work. If operators become complacent and rely on engineered safeguards to prevent incidents, an incident may occur if the engineered safeguard does not function as required. For this reason, it must be stressed to personnel prior to installation of engineered safeguards that they are not meant to be a replacement of a safeguard (i.e., engineered safeguard replacing a procedural safeguard). Adding engineered safeguards to a system must be understood to be a measure to reduce risk significantly, which it cannot accomplish if it only replaces a procedural safeguard.

6.2 Engineered Safeguards Are Not Perfect

Engineered safeguards should not be considered as means of eliminating the potential for an adverse consequence. Engineered systems are not guaranteed to work every time they are needed. Incidents have occurred in the chemical industry in which an engineered safeguard did not work. One such incident occurred on August 14, 2002 in which a chlorine hose ruptured during a chlorine transfer and resulted in a release of over 48,000 pounds of chlorine [14]. The chlorine transfer system did have automatic shutoff valves that were supposed to activate if chlorine was detected at nearby chlorine detectors. The chlorine detectors did sense chlorine and sent a signal to the automatic shutoff valves to close. However, all automatic shutoff valves did not close as they had experienced some internal corrosion. The automatic shutoff system was not tested properly as part of the Mechanical Integrity Program. Management at the site was not aware of the corrosion issue within the system and did not take proactive measures to ensure the automatic shutoff system activates properly in the event that it was needed.

Another example of an engineered safeguard failing to work is the explosion at the Buncefield Oil Storage Depot on December 10, 2005 [15]. A gasoline storage tank was overfilled and the automatic overfill protection system did not activate as it was not placed into service properly after its previous test. Maintenance errors such as this can happen and it is important for personnel working with engineered safeguards to understand its functionality, and test and maintenance procedures in order for it to activate when needed.

The above two examples of an engineered safeguard failing to activate should not lead to a conclusion that adding engineered safeguards to a system does not reduce risk. Engineered safeguards do indeed reduce risk if the Maintenance Program of the system to which it is added is updated to include the proper procedures and testing frequency recommended by the manufacturer of each piece of equipment associated with the engineered safeguard. It should also be noted that personnel must be aware of the functional details of the engineered safeguard system to ensure that it is operable so that it can prevent an accident when needed as it is designed to.

6.3 Resources

Adding engineered safeguards to a system requires resources from many departments. Engineered safeguards added to an existing system may need to be designed according to the Instrumentation, Systems, and Automation Society (ISA) Standard 84.00.01-2004 [16]. The ISA 84 Standard has many detailed requirements for installing Safety Instrumented Systems (SIS). In order for these requirements to be properly met, input is needed from many different groups within a company.

6.4 Financial Aspect

Adding engineered safeguards may prevent financial losses incurred due to major accidents. Some may argue against adding engineered safeguards to a system that has been found to be of low risk but high consequence severity due to the immediate financial strain it may have on a company. It is true that adding an engineered safeguard requires some financial investment and the system must be included in the periodic test and maintenance program. However, a major accident can cost millions of dollars in monetary losses and other non-tangible impacts (e.g., personnel injuries and reputation impact). For example, in 1999 an explosion occurred due to a gasoline storage tank overfill at the Thai Oil Company that cost $22.3 million and resulted in seven fatalities [3]. The storage tank only had a high level alarm, which was not heard by onsite personnel. The automatic overfill prevention system would have cost a small fraction of that monetary loss, not to mention the loss of life that was experienced.

An opinion that is increasingly being accepted by industry management and safety professionals is that safety is an investment, not a cost. According to OSHA's Business Case for Safety Webpage [17], "Employers that implement effective safety and health management systems may expect to significantly reduce injuries and illnesses and reduce the costs associated with these injuries and illnesses, including workers' compensation payments, medical expenses, and lost productivity." We can extrapolate this statement to process safety. "Z10-2012: Occupational Health and Safety Management Systems" Standard [2] is generally accepted as being a very good guidance document to implementing an effective safety and health management system. A very important section within that standard is the one on hierarchy of controls. Therefore, implementing a system in which engineered safeguards are added for high severity scenarios where only procedural safeguards exist would be a step towards improving process safety. This can result in a decrease in incidents at a chemical processing facility, which can have a direct positive impact on facility profitability.

Adding engineered safeguards to a system does not necessarily mean that future accidents will be prevented. However, it will reduce the likelihood of a major accident occurring at a facility. The investment required to add and maintain an engineered safeguard can be significantly less than the costs associated with an accident.

7 Conclusion

In the chemical process industry, many potentially high severity tasks exist for which procedural safeguards are the main layer of protection. The hierarchy of controls guidance developed by the process industry ranks engineered safeguards as a more robust layer of protection than procedural safeguards. Multiple chemical process industry accidents have occurred in which the procedural safeguard layer of protection failed and an engineered safeguard did not exist. These accidents could have been prevented if the associated system had an engineered safeguard as a layer of protection in addition to the procedural safeguard.

The PHA and LOPA methodologies are tools that can be used to identify high severity scenarios and the existing safeguards. It is possible, to apply these methods properly and arrive at a low risk conclusion for high severity scenarios that solely rely on a procedural safeguard. It is recommended in this article that these methods be augmented by owners and operators responsible for highly hazardous operations with a mandate that all high severity scenarios to include at least one engineered safeguard. This should ensure that regardless of the estimated likelihood of the scenarios, the operators are not the only safeguard preventing the high severity release.

References

1. *Inherently Safer Chemical Process: A Life Cycle Approach.* Center for Chemical Process Safety, American Institute of Chemical Engineers, New York, NY, 2nd Edition, December 2008.

2. *Z10-2012: Occupational Health and Safety Management Systems*. American Industrial Hygiene Association, Falls Church, Virginia, June 2012.

3. *Final Investigation Report – Caribbean Petroleum Tank Terminal Explosion and Multiple Tank Fires, Bayamon, Puerto Rico, October 23, 2009*. U.S. Chemical Safety and Hazard Investigation Board, Report 2010.02.I.PR, October 2015.

4. *Process Safety Management of Highly Hazardous Materials*. Occupational Safety and Health Administration, Federal Code of Regulations, 29 CFR, §119.119.

5. *Risk Management Programs for Chemical Accidental Release Prevention.* U.S. Environmental Protection Administration, 40 CFR Part 68.

6. *Safety Data Sheet (SDS) for Anhydrous Ammonia*. Hill Brothers Chemical Company, HBCC SDS No. CA10000, Revision Issued: 5/14/2015.

7. *Investigation Report – Confined Vapor Cloud Explosion – CAI, Inc. and Arnel Company, Inc. – Danvers, Massachusetts*. U.S. Chemical Safety and Hazard Investigation Board, Report 2007-03-I-MA, May 2008.

8. *Process Safety Management for Petroleum Refineries*. California Occupational Safety and Health Administration, California Code of Regulations, Title 8, §5189.1, Draft Version 5.0, September 14, 2015.

9. *California Accidental Release Prevention (CalARP) Program.* California Governor's Office of Emergency Services. Title 19 of CCR Division 2, Chapter 4.5, Draft, September 24, 2015.

10. *Guidelines for Hazard Evaluation Procedures*. Center for Chemical Process Safety, American Institute of Chemical Engineers, New York, NY, 3rd Edition, April 2008.

11. *Layer of Protection Analysis – Simplified Process Risk Assessment* Center for Chemical Process Safety, American Institute of Chemical Engineers, New York, NY, October 2011.

12. *Guidelines for Initiating Events and Independent Protection Layers in Layer of Protection Analysis*. Center for Chemical Process Safety, American Institute of Chemical Engineers, New York, NY, February 2015.

13. *Guidelines for Enabling Conditions and Conditional Modifiers in Layer of Protection Analysis*. Center for Chemical Process Safety, American Institute of Chemical Engineers, New York, NY, November 2013.

14. *Investigation Report – Chlorine Release – DPC Enterprises, L.P.* U.S. Chemical Safety and Hazard Investigation Board, Report No. 2002-04-I-MO, May 2003.

15. *Buncefield: Why Did It Happen?*. United Kingdom Health & Safety Executive, February 2011.

16. *ANSI/ISA-84.00.01-2004: Functional Safety: Safety Instrumented Systems for the Process Industry Sector*. Instrumentation, Systems, and Automation Society (ISA), September 2004.

17. https://www.osha.gov/dcsp/products/topics/businesscase/. Occupational Safety and Health Administration, Business Case for Safety, Accessed on February 8, 2016.

12TH GLOBAL CONGRESS ON PROCESS SAFETY

Maintaining Consistency in the Determination of Consequence Severity Rankings for Process Hazard Analyses

Kirk Busby
Kazarians & Associates, Inc.
100 West Broadway
Glendale, CA 91210
Kirk.Busby@kazarians.com

Prepared for Presentation at
American Institute of Chemical Engineers
2016 Spring Meeting
12th Global Congress on Process Safety
Houston, Texas
April 11-13, 2016

Maintaining Consistency in the Determination of Consequence Severity Rankings for Process Hazard Analyses

Kirk Busby
Kazarians & Associates, Inc.
100 West Broadway
Glendale, CA 91210
Kirk.Busby@kazarians.com

Keywords: Hazard Identification and Risk Analysis, Process Safety, Consequence Modeling

Abstract

A Process Hazard Analysis (PHA) is an important tool for identifying and managing the hazards associated with a chemical process. One of the most critical aspects of the analysis is the risk ranking of scenarios, which is used to determine which events pose a high risk and may require further mitigation. The risk ranking of a scenario is built on two factors – severity and likelihood. The severity of a scenario is a measure of the potential worst-case impact on facility operation, facility personnel, the public, or the environment, while the likelihood is an estimate of the probability with which an event may occur, given the presence of the safeguards acting to prevent or limit the severity of the event.

Severity of a scenario is generally expressed in qualitative terms. While a typical approach to assigning consequence severity rankings involves categorizing events by impact (e.g., recordable injury, permanent disability, or fatality), this approach relies on the subjective judgment of the analysis team in determining the exact impact of a specific event. Experience has shown that, because of the subjective nature of this step, different PHA study teams have had varying opinions on the potential severity level of a given scenario. It is the aim of this study to provide guidelines to improve consistency of severity rankings both within a single analysis, and across multiple analyses with separate PHA study teams. It is possible to increase the consistency in consequence severity rankings through application of standardized guidelines. A method has been developed to aid PHA facilitators based on dispersion modelling of various conditions typically encountered in a PHA. The underlying data is based on groups of chemicals with similar properties and other conditions (e.g., break size, break location, etc.).

1. Introduction and Background

Process Hazard Analysis (PHA) is an important tool for identifying and managing the hazards associated with a chemical process. The goal of a PHA is to use a systematic methodology to identify each potential hazard source within the process. A typical PHA methodology is the guideword-style Hazard and Operability (HAZOP) study. To ensure completeness, the HAZOP methodology breaks down a complex system into individual nodes, making it easier to thoroughly analyze each section of the process. Further, within each node, hazards are postulated according to a standard combination of guidewords and process deviations. For example, when the guideword 'No' is paired with the deviation "Flow," all scenarios within the node which lead to a loss of flow are considered.

Once a scenario is identified, the risk posed to the facility must then be determined. This is normally achieved through the risk ranking process. To determine the risk ranking, each scenario is assigned a severity and a likelihood level. This pair is then used in a risk ranking matrix, which defines the risk targets developed by company management prior to the start of the PHA. Assigning a risk ranking is one of the most important facets of the PHA process, as it is used to determine which risks are acceptable given the safeguards currently in place, and which require further actions to ensure that the risk is maintained at a tolerable level.

Currently, there is a large potential for variance in risk ranking determination, specifically with regard to consequence severity. While severity determination is meant to be a qualitative ranking, there are a number of benefits to improving the consistency of severity rankings among similar release scenarios. Kazarians & Associates, Inc. has developed a method that has proven effective in reducing variance among consequence severity rankings through the use of pre-determined dispersion modeling results.

2. Current Methods for Consequence Severity Determination

It is common practice to estimate the worst case consequence severity for the identified potential release scenarios. For this, it is assumed that all existing active safeguards are unavailable. Typically, the severity ranking is a qualitative determination made by the PHA study team based on their understanding of the potential impact a given scenario may have on the safety of facility personnel and the public, as well as the environment or business operations. The text provided in defining each severity level is generally used to determine the consequence severity. Table 1 provides an example of severity level definitions available to the PHA study team.

Definitions such as those in Table 1 can provide a good starting point for consequence severity selection. However, use of these limited definitions can result in inconsistent severity rankings, both within an individual PHA, as well as among PHAs. This variation stems from the fact that the PHA study team generally carries the chain of events to the release condition (i.e., flange leak, pipe rupture, etc.) whereas the definitions of consequence severity level is often in terms of level of harm (i.e., first aid injury, fatality, etc.). This then requires the study team to envision how the release will ultimately result in harm, which means that they have to take into consideration factors such as source conditions, dispersion, and ignition, which are often complex phenomena by themselves. Therefore, this leads different PHA study teams to hold varying opinions of what the impact of a certain event may be. Typically, these opinions are based on the experience of the personnel present during the PHA.

Table 1. Typical Consequence Severity Guidance

Consequence Severity	Impact on Personnel	Impact on the Environment	Impact on business
1	Minor first aid injury	Slight spill onto soil, release to flare	< $10,000
2	Recordable injury	Small spill onto water, moderate spill onto soil	< $100,000
3	Permanent disability	Moderate spill onto water, large spill onto soil	< $500,000
4	Fatality	Moderate spill into an environmentally sensitive area	< $1,000,000
5	Multiple fatalities	Large spill into an environmentally sensitive area	< $5,000,000

The following example is taken from PHAs conducted for chlorination systems at two facilities operated by the same organization. The example demonstrates the difference operational experience can make in the severity ranking of a given scenario. During PHA sessions for the two systems, the PHA team encountered the same scenario in which a storage tank is exposed to an external fire, resulting in a pressure increase within the tank, ultimately resulting in tank rupture. In this scenario, the two teams have arrived at an identical endpoint (i.e., release scenario). While the release conditions (i.e., temperature and pressure) will be the same for both cases, the first team concludes that a single fatality may result, the second believes that multiple fatalities may result. The two teams may record the scenarios according to Figure 1.

In each case, the PHA study team has arrived at the same chain of events leading up to an integrity failure of the storage tank, and release of chlorine into the atmosphere. However, the final consequence statement and therefore risk rankings vary based on the estimated severity of the release.

Inconsistency in severity rankings can cause a number of problems following a PHA. Most importantly, they can lead to an over- or under-estimation of the risk a scenario poses. A consequence with the potential to cause severe injury that has only been identified as a first aid issue could result in insufficient safeguarding to protect against the event. This could ultimately lead to major safety incidents during the operating lifetime of the process. Safeguards are typically added when the risk ranking of a scenario does not meet the pre-determined corporate risk target. A lower severity ranking may indicate that the appropriate safeguards are in place, when in reality an additional protection may be necessary.

Team 1:

Causes	Consequences	Safeguards	S	L	RR	HAZOP Recommendations
Greater than 140°F in the Chlorination Room caused by a fire	▪ Pressure buildup in the Chlorine Storage Tanks. ▪ Potential for loss of containment of a Chlorine Storage Tank if the relief valve fails to lift. Release of chlorine into the Chlorination Room. Potential for personnel exposure.	▪ Relief Valves will lift to prevent overpressure. ▪ Detectors alarm locally and on SCADA. ▪ Scrubber activated by detectors at 3 ppm. ▪ Plant housekeeping practices prohibits storage of any combustible materials in the Chlorination Room.	4	D	M	

Team 2:

Causes	Consequences	Safeguards	S	L	RR	HAZOP Recommendations
Fire inside the Chlorine Building.	Potential for increased pressure in the storage tanks. Potential to rupture the storage tank. Potential for release of chlorine inside the building. Potential for personnel and public exposure.	1. PSVs on the storage tanks set at 225 psig. 2. Chlorine detectors alarm at 1 ppm. Scrubber system will start at 3ppm on any one detector. 3. Good housekeeping practices.	5	E	M	No recommendations necessary. Safeguards are considered adequate.

Figure 1. Identical PHA Scenarios with Differing Consequene Statements and Risk Ranking

A second and more subtle consequence of inconsistent severity ranking is the potential discrepancy between processes or facilities. If two PHA study teams assign different severity rankings to similar events on two similar systems, there is a potential for one system to require an extra safeguard to be installed which is not in place on the second system. This discrepancy between units could potentially lead to confusion later in the life of the plant. If a group of operators is responsible for both systems, there could be confusion as to which setup is correct, as well as the incorrect assumption that both systems are configured in the same manner. Additionally if an incident occurs in the system with fewer safeguards, there is a potential for increased regulatory scrutiny and fines due to the fact that the second system is not as well protected as the first.

3. Improving Consistency through Dispersion Modeling

To reduce variation among consequence severity rankings, it is possible to provide PHA teams with improved guidance on the potential impacts of a release. Dispersion modelling is a tool which is commonly used to determine the hazardous area resulting from a specific chemical release scenario. There are a number of sophisticated dispersion models and software packages currently on the market which can provide a picture of the hazard distance presented by various chemical release events. Many of the commercially available packages allow the user to input specific data such as the temperature, pressure, quantity, and release orifice size, and calculate the released quantity, pool size, vaporization rate, and resulting hazard distance. Depending on the model chosen, the user may be able to customize the scenarios to a greater degree, including such factors as terrain type, degree of confinement, explosive characteristics, etc.

Currently, some PHA teams already use dispersion modeling to assist in performing a more in-depth analysis for select scenarios with complex release characteristics. These studies are useful, but require a large amount of input data to be determined by the PHA team in order for them to generate accurate results. Therefore, it is much too time consuming to use such an analysis for every release scenario postulated during a PHA.

To combine the time and cost efficiency of PHA with the increased accuracy afforded by the use of dispersion modeling, the dispersion calculations can be performed generically prior to beginning the PHA discussions. To utilize this method, a database is developed for each hazardous chemical which may be released, and dispersion calculations are performed for the range of temperatures, pressures, phases (liquid or gas), and break sizes which may be encountered during the PHA. Every possible combination of parameters does not need to be considered, as the PHA team may quickly interpolate between results when needed. With a pre-developed matrix in front of them during the PHA sessions, participants can quickly relate a set of known release conditions to an approximate hazard zone, which will allow them to choose a more refined consequence severity. Some large chemical processing and manufacturing corporations have already begun to adopt such methods to improve the consistency of their PHAs.

The method developed by Kazarians utilizes PHAST modeling software by Det Norske Veritas [Reference 1] to create databases for both flammable and toxic chemicals. Specifically, the common alkanes (i.e., methane through hexane), hydrogen, hydrogen sulfide, anhydrous ammonia, and chlorine have been analyzed, in both liquid and gaseous form. In each case, a generic outdoor release event is postulated for varying temperatures, pressures, spill quantities, and release sizes. From this basic input data, PHAST is capable of generating information on the maximum dispersion distance, the flammable range, the range of explosive overpressure, and toxic effects. For each release event, the maximum distance of each hazard type is compared, and the greatest distance is selected. For example, for a specific set of release conditions, the maximum distance at which an injury may occur due to thermal radiation is 200 feet, while the maximum injury distance due to explosive overpressure is 250 feet. For these release conditions, 250 feet will be selected and recorded as the worst-case injury hazard distance. The final result is a set of eight tables for each chemical, with four liquid phase tables and four vapor phase tables. Tables 2 and 3 below show the four hazard distance tables for methane in the liquid phase.

Table 2A. Methane Liquid Injury Hazard Distance (ft)

	Source Temperature (F)			
Release (lbs)	**-250**	**-200**	**-150**	**-117**
10	110	100	140	160
100	350	230	320	350
1,000	610	510	710	780
10,000	1,520	1,130	1,570	750
100,000	4,060	2,480	3,350	3,350
500,000	7,690	4,260	5,360	5,360

Table 2B. Methane Liquid Injury Hazard Distance (ft)

	Source Pressure (psi)				
Break Diameter (in)	**50**	**150**	**300**	**400**	**450**
0.25	20	30	30	30	30
0.5	110	110	110	110	110
0.75	180	210	210	210	210
1.0	210	280	310	310	310
1.5	280	410	480	510	510
2.0	380	540	640	670	700
3.0	510	770	970	1,070	1,100
4.0	610	1,000	1,330	1,430	1,460
6.0	840	1,430	2,020	2,180	2,250

Table 3A. Methane Liquid Fatality Hazard Distance (ft)

	Temperature (F)			
Release (lbs)	**-250**	**-200**	**-150**	**-117**
10	40	40	60	70
100	240	90	130	150
1,000	530	210	310	340
10,000	1,390	470	690	1,690
100,000	2,990	1,070	1,500	1,500
500,000	5,290	1,860	2,400	2,400

Table 3B. Methane Liquid Fatality Hazard Distance (ft)

	Pressure (psi)				
Break Diameter (in)	**50**	**150**	**300**	**400**	**450**
0.25	20	20	20	20	20
0.5	40	50	50	50	50
0.75	110	140	140	140	140
1.0	140	210	240	240	240
1.5	210	340	400	440	440
2.0	300	470	570	600	630
3.0	440	700	890	990	1,030
4.0	530	930	1,260	1,350	1,390
6.0	760	1,350	1,940	2,110	2,170

Tables 2 and 3 present two sets of hazard distances. One represents the distance inside which an injury is expected to occur, and one inside which it is possible for fatalities to occur. Additionally, the sets have been subdivided into Tables A and B, which provide dispersion information for Release Quantity vs Temperature, or Break Diameter vs Pressure. The PHA team may elect to use either of the tables based on what information is available for a given scenario. For example, Break Diameter vs Pressure tables may be more useful for a pipe break scenario, while Release Quantity vs Temperature may be easier to use in the event of a seal failure or liquid overfill scenario.

There are a few notable trends which can be identified from Tables 2 and 3. Specifically, it can be seen that in the tables relating Break Diameter and Pressure, the maximum hazard distances reach a constant value as pressure increases. This is due to the release reaching what is known as "sonic velocity." [Reference 2] When a material under pressure is released through an orifice, the velocity of the release is initially dependent on the pressure within the vessel. Higher release pressure will typically result in a higher release velocity. However, once the release velocity has reached sonic velocity, further increases in pressure will no longer cause the velocity to increase. Put simply, no more material can fit through the release orifice, no matter how much force is behind it.

4. Defining Exposure Thresholds

The final dispersion distances reported in Tables 2 and 3 can vary greatly depending on what thresholds values are chosen for injuries and fatalities. This analysis used three potential sources of impact to the population: overpressure due to explosion, thermal radiation due to fire, and exposure to toxic vapors. Each threat has a unique hazard zone that is determined by its effects on both the human body and surrounding structures.

There is a great deal of research performed on the various effects on the human body caused by exposure to overpressure, thermal radiation, or toxic chemicals [References 3 through 10]. From this research, threshold values can be estimated to determine the exposure limits for injuries and fatalities. Table 4 presents the threshold values used in this article to differentiate between effects.

Table 4. Exposure Threshold Definitions

Explosion Overpressure	
Effect	**Threshold Value**
Injury	0.5 psi
Fatality	2 psi
Thermal Radiation	
Effect	**Threshold Value**
Injury	5 (kW/m^2)
Fatality	10 (kW/m^2)
Toxic Exposure	
Effect	**Threshold Value**
Injury	ERPG-2
Fatality	ERPG-3

It is important to note that these values represent conservative estimates of the potential impact an event may have on a person. For example, the threshold values for thermal radiation exposure are time dependent. Exposure to 10 kW/m^2 will not result in an immediate fatality, but prolonged exposure of approximately 60 seconds could result in death. This value is chosen as it represents the possibility of a fatality since it requires rapid evacuation from the hazard zone.

For each scenario modeled in PHAST, the distances to each threshold value described in Table 4 are compared, and that which results in the largest hazard distance is selected as the worst-case hazard zone. Toxic effects are only calculated for those chemicals which pose a significant risk when inhaled or contacted.

5. Using Hazard Distance Tables During PHA Sessions

Once developed, hazard distance tables can be used during PHA discussions to quickly determine an appropriate severity ranking based on the potential impact to facility personnel and the public. To convert the hazard distance tables into a form which may be more easily used during the PHA, the PHA facilitator may review the company risk matrix, as well as the facility layout prior to starting the PHA sessions. During this time, the facilitator should convert the dispersion distances in Tables 2 and 3 into charts providing guidance on the potential worst-case severity of different release events.

For example, a hypothetical company has developed a risk matrix with the consequence severity rankings described in Table 1. The system under analysis is an LNG storage tank and associated piping, which are located in an isolated part of the facility, but approximately 700 feet from the closest residences. Using this information, the PHA facilitator may translate the worst-case hazard zones in Tables 2 and 3 into a set of tables with suggested consequence severity rankings which are tailored to the system under review. Table 5 provides an example of these consequence severity tables for the specific LNG system described above.

Table 5A. Methane Liquid Consequence Severity Rankings

	Source Temperature (F)			
Release (lbs)	**-250**	**-200**	**-150**	**-117**
10	1	1	2	2
100	3	2	2	3
1,000	4	3	3	4
10,000	4	4	4	5
100,000	5	5	5	5
500,000	5	5	5	5

Table 5B. Methane Liquid Consequence Severity Rankings

	Source Pressure (psi)				
Break Diameter (in)	**50**	**150**	**300**	**400**	**450**
0.25	1	1	1	1	1
0.5	1	1	1	1	1
0.75	2	2	2	2	2
1.0	2	2	3	3	3
1.5	2	3	4	4	4
2.0	3	4	4	4	4
3.0	4	4	4	5	5
4.0	4	5	5	5	5
6.0	5	5	5	5	5

When translating the calculated hazard zone to consequence severity, it is important to note that the dispersion modeling calculations are performed generically, and their results may need to be adjusted based on the specifics of the system under review. The relationship between hazard distance and similar consequence categories (i.e., first aid injury, reportable injury, and permanent disability) should be developed with the primary decision maker for the facility (i.e., company management). To accomplish this, the facilitator should record the decision making process used to relate hazard distances to consequence severity levels, and present these decisions to company management prior to beginning the PHA sessions, and adjust the severity tables based on comments received. .

Additionally, the model calculations make a number of assumptions which may or may not apply to the system being analyzed. Most importantly, the calculations assume that releases occur outdoors, and outside of any secondary containment areas. Indoor releases can have a significantly reduced hazard zone outside of the building. This may depend on the ventilation capabilities of the building, or doors normally kept open. Toxic material may build to higher concentrations within the building than those reported in the hazard distance tables. Similarly, the presence of secondary containment limits the surface area of a liquid pool, which may result in a smaller vapor cloud than an uncontained liquid spill. To adequately assess the severity of a chain of events, these deviations from the assumed model must be accounted for. In cases such as these, the hazard distance tables should be used as a starting point, but adjust their consequence severity ranking according to the physical systems present in the field.

During the PHA discussions, once a scenario has been identified and the release conditions are established, the consequence severity tables can be consulted to assist in selecting the most appropriate consequence severity ranking. To use the tables, a process engineer or other person familiar with the design of the system under consideration should determine the physical parameters of the release. It must first be determined whether the released material is in the liquid or the vapor state. Subsequently, the release conditions must be specified in terms of temperature, pressure, quantity spilled, or breach diameter, depending on what information is most readily available to the team. This determination is not intended to be a rigorous calculation, but rather a quick estimation of the conditions of the release.

Once the release parameters have been agreed upon, the team may then use the consequence severity tables to determine the worst-case severity ranking. For example, in the LNG system described above, it is postulated that a full break occurs in a one inch diameter pipe containing LNG at 170 psig. Given the information at hand, the team consults Table 5B, and identifies 150 psig as the closest pressure rating found on the tables. From the Table, the team selects a severity ranking of 2, indicating that a recordable injury is the anticipated worst-case consequence.

6. Conclusion

Kazarians & Associates, Inc. has developed a method which uses pre-developed consequence severity tables to refine the judgement of PHA teams with regard to the impact that a scenario may have on personnel. This method has been applied, resulting in a more accurate determination of consequence severity. When applied across multiple teams and studies, this increase in accuracy will lead to consistency in severity ranking among similar PHA scenarios. In turn, this will lead to consistent application of safeguards across multiple systems, resulting improved plant operability and avoiding the potential for confusion among operating staff.

7. References

[1] Phast Modeling Software, version 7.11.33.0. DNV GL AS.

[2] *Guidelines for Hazard Evaluation Procedures*, published by the Center for Chemical Process Safety of the American Institute of Chemical Engineers, 3rd Edition, April 2008.

[3] *OGP Risk Assessment Data Directory: Vulnerability of Humans*. The International Association of Oil and Gas Producers. Report No. 434-14.1. March 2010.

[4] Toups, Harry J., *Fires and Explosions - Fundamentals and Design Considerations* [PowerPoint Slides]. Retrieved February 26, 2015 from www.sache.org / links / Pike21Jul2004 / Fires%20and%20Explosions.ppt.

[5] Clancey V J, 1972. Diagnostic features of explosion damage, 6th Intl. Meeting on Forensic Sciences, Edinburgh, Scotland.

[6] ALOHA 5.4.3 Modelling Software, Office of Emergency Management – U.S. EPA and Emergency Response Division – National Oceanic and Atmospheric Administration.

[7] *Handbook of Chemical Hazard Analysis Procedures*. Federal Emergency Management Agency, U.S. Department of Transportation, U.S. Environmental Protection Agency.

[8] Hockey, S M, and P J Rew. *Review of Human Response to Thermal Radiation*. HSE Contract Research Report No. 97/1996. Health and Safety Executive. July 2, 1996.

[9] Raj, Phani K. *A Review of the Criteria for People Exposure to Radiant Heat Flux from Fires*. Technology & Management Systems, Inc. October 25, 2006.

[10] National Fire Protection Association. (2013). *NFPA 59A: Standard for the Production, Storage, and Handling of Liquefied Natural Gas (LNG).*

441564 Application of Open Source Software Tools in the Process Safety Education

Abstract:

Process safety professionals have used, developed, and contributed to open source software tools. They have fascinated more and more notice from the society. The free software will enhance the collaboration and motivate the scientific and technical development. The software users can use and even modify the software modules. A lot of applications can be found in the real industrial practice so the engineering learning process should follow this trend. Before starting the career, the students should get to know how to take the full advantage of these tools in their future jobs because the future engineers should know how to find the free software packages in order to solve process safety problems.

This presentation will tell the audience how to integrate the open source software into process safety education or learning practice. The examples of modeling projects which can be practiced by the students in the classroom will also be introduced. The modeling will give the students more opportunities to recall their knowledge of mass and energy balance, thermodynamics, heat transfer, reaction kinetics, transport phenomena, process safety, statistics and numerical methods. These projects will also help them to learn how to conduct the risk assessment and process safety management through different kinds of model development. These chemical engineering education activities will allow them to explore computer aided tools such as computing and process simulators throughout the curriculum.

441688 Beyond Quantitative Risk Analysis Results - Part I: Explosions and Blast Phenomena Characterization

Abstract:

The present paper illustrates the results of a Quantitative Risk Analysis (QRA) case study of a refinery. The main purpose of a QRA is to evaluate the risk levels (both individual and societal risk) of a process due to Loss Of Containment scenarios (LOCs). A QRA can be used to optimize the risk reduction measures to be implemented to comply with a given tolerable risk level. Moreover, the development of a detailed QRA is the best opportunity to perform a specific analysis addressing structural response due to overpressure impacts caused by explosions and also estimating the Time To Failure (TTF) of structures and equipment due to thermal radiation of fires.

Part I of this paper is focused on illustrating how a detailed analysis of QRA results can provide valuable information for characterizing the blast phenomena parameters for a given location. The analysis gathers three key sources of information: (1) peak overpressure, maximum and minimum positive phase durations and impulses as a function of the cumulative frequency of occurrence; (2) total number of LOC scenarios which contribute to the impact of explosions (including vapor cloud explosions due to both immediate and delayed sources of ignition); and (3) the identification of each LOC scenario and associated individual frequency of occurrence. This information is valuable for ensuring the blast loading function for a subsequent structural response analysis (e.g., equivalent elasto-plastic Single Degree Of Freedom analysis), and also provides specific information regarding which LOCs contribute most on overpressure impact.

441690 Beyond Quantitative Risk Analysis Results - Part II: Fires and Domino Effect Characterization

Abstract:

The present paper illustrates the results of a Quantitative Risk Analysis (QRA) case study of a refinery. The main purpose of a QRA is to evaluate the risk levels (both individual and societal risk) of a process due to Loss Of Containment scenarios (LOCs). A QRA can be used to optimize the risk reduction measures to be implemented to comply with a given tolerable risk level. Moreover, the development of a detailed QRA is the best opportunity to perform a specific analysis addressing structural response due to overpressure impacts caused by explosions and also estimating the Time To Failure (TTF) of structures and equipment due to thermal radiation of fires.

Part II of this paper is focused on illustrating how a detailed analysis of QRA results can provide valuable information for characterizing potential domino-effects due to thermal radiation of fires at a given location. The analysis gathers three key sources of information: (1) thermal radiation as a function of the cumulative frequency of occurrence; (2) total number of LOCs which contribute to the impact of fires; and (3) the identification and characterization of each LOC scenario and associated individual frequency of occurrence. This information not only is valuable for ensuring the heat radiation for a subsequent analysis intended to estimate the TTF and associated conditions remaining in the system (e.g., Wall Dynamics Analysis), but also provides specific information of which LOCs contribute most to the heat radiation impacts. This information ensures which LOC scenarios need to be addressed for optimizing the risk reduction decision-making process. Additionally, the accurate estimation of potential equipment failure due to thermal radiation provides added value for predicting when and how impacts such as BLEVEs (Boiling Liquid Expanding Vapor Explosions) are expected to occur.

441700 A Leadership Vision for Accelerating Learning from Past Lessons to Create a Future Free of Incidents

Abstract:

The practice of process safety management (PSM), involves the analysis and understanding of what can go wrong, what the effects could be, and what should be done to prevent or mitigate an incident. That is: (1) what are the causes (human errors and equipment malfunctions)? (2) What are the potential consequences and impacts (to people, property, process, reputation)? and (3) what are the risk controls needed to prevent, detect, control, mitigate, or respond to such events (safeguards, layers of protection)?

Lessons learned from industrial disasters, as well as successes by industry leaders in preventing such incidents, form the basis for a practical framework for a PSM system as a holistic management tool for managing hazards and their risks. Quality design/development and effective implementation of a robust risk management system driven by a proactive culture of leadership to reduce operational risks and improve performance, enables organizations to avoid future incidents.

However, a conventional approach of longer-term continual improvement of operational risk performance could easily be outpaced by the rapid changes in technology advances, organizational and human factors, and the dynamics of the market. Operating companies, service providers and regulators alike are challenged by such changes and their potential impacts. Lessons of disaster investigations from the process industry have been full of learning opportunities. In fact, many of the lessons and their root causes are common across most process safety incidents and near misses ... yet, 15-20 years has proven to be an unacceptable period for gradual learning.

This presentation will present some common lessons as a priority for establishing improvement strategies as organizations strive for operational excellence. Available proven risk management methodologies and best practices for operational risk reduction will be summarized. Given that the global industry has made remarkable progress toward developing proven solutions aimed at prevention, organizations must now commit to accelerating the adoption of these solutions before at-risk operations experience near future incidents.

The focus of the paper will be the challenge to accept a leadership position to resolve existing but perhaps unidentified risks, then to commit to action to proactively identify hazards and their associated risks, and implement risk reduction measures before losses occur. A framework of a robust risk-based management system, incorporating all of the necessary elements based on lessons learned and including supporting leadership and management processes to achieve risk reduction performance objectives, will be presented.

Attendees will benefit from practical takeaways of how each management system element can be developed with "leadership designed in" to control the process safety risks at three opportune times: before the incident (prevention), during the incident (emergency management) and after the incident (incident investigation, root cause analysis and lessons learned).

The presentation will pose a challenging question: "Knowing what we already know, why would we want to take years to gradually reduce our risks, when we could accelerate the learning from past lessons and prevent incidents using available best industry practices"?

This paper will draw upon the authors' extensive experience in designing, developing and implementing PSM systems.

441698 Analytics to Reveal Meaurable Factors to Process Safety Performance

Abstract:

Analytics on a Large Industry Data Set Reveal the Leadership and Organizational Factors that Drive Down Risk and Reduce Incidents

With access to a multi-year data set of over a million incidents from 1000's of operating facilities worldwide in the Energy and Chemicals industry, the research team mined the data set to explore the differences in risk management practices and behaviors which most effectively explain the variation in process safety outcome performance. By modeling and examining predictive relationships, the study identifies four attributes - worker engagement, process discipline, risk sensitivity, and learning-mindedness – as critical for reducing incidents.

Findings from this study support a progressive and integrated approach and suggest that companies that are proficient in driving process safety performance are companies that, to a greater degree, have successfully engaged their workforce, strengthened their operating discipline, sharpened their risk sensitivity, and become learning-minded. A learning-minded organization manages each incident and responds to each failure deliberately to identify and fully leverage improvement and learning opportunities.

While this study provides statistical evidence to support the strong correlation between leadership, organizational behavior attributes and safety performance, it also emphasizes the organizational factors that are necessary to enable and sustain effective risk mitigation efforts. By deriving these insights from a data set of routine field-level activities, operational measurements may be created to measure progress and drive focus on engagement, discipline, risks, and learning.

This paper will conclude by sharing examples of how companies have applied these findings to establish quarterly and monthly measurements to drive leadership and organizational focus in these critical areas and ultimately to improve outcome performance.

441960 Coupling Dynamic Pressure Relief System Design with Experimental Vent Flow Regime of Dicumyl Peroxide / Toluene System

Abstract:

Abstract

A safe process design requires knowledge of the chemical reactivity of the desired reactions as well as the undesired reactions due to process upset conditions. Techniques used for sizing pressure relief systems for reactive systems under thermal runaway conditions include direct empirical scaling of experimental data obtained from a vessel with very low thermal inertia, analytical methods, and computer simulations using reaction kinetic models developed from adiabatic calorimetry data.

Pressure relief size estimation is based on the required flow capacity through the relief device at a specific set pressure, in order to keep the pressure in the vessel below the maximum allowable accumulated pressure. The flow capacity of the relieving device is highly dependent on the quality of the vent flow regime, whether it is liquid, all-vapor/gas, or two-phase. This requires knowledge of the physical properties of the contents inside the vessel at the time of venting. For a reactive system, these properties can be difficult to determine.

In this paper, the pressure relief size of a reactive system containing 50% dicumyl peroxide in toluene will be estimated using the dynamic simulation method. The vent flow regime of the reactive system will be determined from measured data, according to the methods described in the DIERS (Design Institute of Emergency Relief Systems) project manual. An automatic pressure tracking adiabatic calorimeter (APTAC^{TM}), which is a low thermal inertia calorimeter, will be used to conduct vented tests. The pressure relief size estimates of the above dynamic simulation method will be presented and compared with the DIERS simplified method.

441701 We Did a Quantitative Risk Assessment Study. Now What Do We Do with It?

Abstract:

Quantitative Risk Assessment studies have been extensively used in risk based regulatory regimes and they are becoming more common as a replacement for facility siting requirements in Process Safety Management (PSM) covered regulatory regimes and as a corporate risk management tool. Besides assessing risks to society and personnel, it can also assist in practical risk reduction implementation in design projects and operating assets. This paper intends to discuss these practical risk reduction implementation to show value to projects and operating assets, and how implementation can be achieved to using specific examples. The following examples will be discussed in the paper and will outline a broader set of QRA implementation improvements:

- Operating modes input to isolatable inventories and failure criteria
- Failure mechanisms from Hazards and Operability studies, and corrosion or mechanical integrity surveys
- Survivability criteria for manned and unmanned buildings for upgrades or siting challenges
- Specialized consequence analysis such as uncommon release material, indoor releases, and elevated receptors.
- Spill containment application
- Thermal protection and passive fire protection application
- Personnel and structural cryogenic protection application
- Risk based fire and gas detection mapping
- Escape, evacuation and rescue impacts
- Operational risk reduction using performance standards

The objective of this paper would be to better understand how to utilize QRA studies by improving their content and by practical implementation (as necessary) of mitigating risk reduction measures in design projects and to existing assets.

442063 Effective Implementation of PSM to Reduce Operational Risks

Abstract:

Ultimate success of process safety management (PSM) is measured through effective operational risk reduction. Usually, there is a lot of emphasis on putting the various elements of PSM in place through adoption of established international standards and best practices. Many of these standards and best practices are very well documented and element leaders/champions are appointed. Unfortunately, many of these elements and element champions are in silos and thereby not contributing to the overall reduction in operational risk. The risk is not reduced until the documentation is converted into reality on the ground.

Also, there is a lot of emphasis on doing high quality technical studies and quantitative process hazards analysis but very little is done to follow up on implementing the recommendations and results of these studies. The operational risks are not reduced until the recommendations are actually implemented. So, despite significant amount of effort, the improvement in the effectiveness of PSM implementation is limited.

Learning from this experience, SCG Chemicals has adopted a practical, bi-focal approach that is driven with a specific purpose in mind — operational risk reduction. The short term focus was to address deviations from design intent and established procedures in the field. If these items are not addressed today, they can contribute to a potential catastrophic incident tomorrow. This enhanced operational discipline and prevented 'normalization of deviations'. The longer term approach focuses on redoing structured PHAs and using the risk matrix to establish the levels of acceptable/unacceptable risks. Management decisions were made to address priority of the unacceptable risks with a rigorous follow up on implementing recommendations. This process has contributed to better outcomes in increased effectiveness of PSM while optimizing the money, time, and effort required. This paper will share the experience of SCG, along with DuPont as a partner, and the journey in implementing PSM effectively to reduce operational risks.

442151 Probit Models for the Determination of Damage Probability of Storage Tanks Due to Flooding Impact

Abstract:

Global climate changes may significantly influence temperatures and precipitation patterns, increasing the frequency of severe meteorological events which may lead to flooding. Flooding may trigger severe accidents on population among urban as it may do on chemical process industries too, causing the damage of equipment and the loss of containment of hazardous materials which might transform into toxic dispersions, fires, explosions or severe environmental contamination. This type of events are known as Natech events (Natural-Technological). This type of events may lead to extremely severe consequences bringing up the need to develop approaches for the risk assessment of Natech scenarios. Recently important efforts have been dedicated to develop tools that include Natech events in the quantitative risk assessment framework. In order to allow a quantitative risk assessment of a Natech event, a key point is the definition of tools for the estimation of equipment damage probability on the basis of intensity parameters of flooding. Past accident data analysis evidenced that horizontal and vertical storage tanks were often damaged in Natech events triggered by floods. Because of this the present study presents specific probit models for the damage probability of horizontal and vertical storage tanks due to flooding intensity parameters (flooding water height and flooding water).

Probit models in this study were derived on the basis of a mechanical model used for the assessment of the vulnerability of storage tanks due to the impact of flooding. For the determination of damage probability Montecarlo simulations were applied over the mechanical model in order to obtain a more accurate value and a probabilistic distribution of this parameter in dependence of the flood impact vector. Once the damage probability distribution is determined the probit method was applied using a logarithm approach with the impact flood vector (water height and water velocity) as the dose variables and the probability of damage as the response variable. A logarithm regression using the minimum square method was utilized to determine the probit constants for the model. Since the mechanical model used as basis has a huge dependence on the geometry of the storage tank being studied, it was find the damage probability will be affected and its value will change. To make sure this affections will be taken into account in the probit modeling a relationship between the geometry of the tanks and the probit constants was established using a polynomic regression. Once again the minimum square method was used and these two variables were related using a second order polynomic approach. Probit constants were determined for two substances (gasoline and LPG) and two types of storages tanks (vertical and horizontal).

The results of this study are a useful tool for the development of the quantitative risk assessment of Natech events due to flood impacts in storage parks. Also contributes to the reduction of the uncertainty in the framework for the QRA of Natech events triggered by floods.

442178 Analysis of Fire Extinguishing Methods in Engine Control Rooms

Abstract:

The study of electrical fires is a major concern for fire safety in the industry and more particularly for oil industry. In petrochemical facilities, the fire hazard may cause severe consequences on safety functions inside the fire compartment and furthermore may lead to the loss of one or several compartments, as just as structural damage on the facilities. From the OECD Fire Database a majority of the fire ignition is due to electrical sources (close to 50%, with a significant contribution attributed to electrical or electronic cabinets, with an occurrence of more than 10% of all fire events). Because of the above mentioned, it is necessary to implement tools to study these phenomena without using an experimental method. For example, the simulation of the dynamics of combustion and fire spread, allows an estimation of the effects on the structure of the case study preemptively. This project addresses the comparison of different fire protection philosophies in engine control rooms, using Fire Dynamics Simulator as support tool. The project studies three different cases, one of them without a protection system, another a clean agent as extinguishing tool and a final one, which implements an inherently safe design. For the first case study, the obtained data for the room temperature and the heat release ratio indicate a potential risk for the personnel and the equipment of the facility. It is expected that the study of the remaining cases show the significant advantages of each fire extinguishing philosophy.

12TH GLOBAL CONGRESS ON PROCESS SAFETY

The Study of HAZOP Revalidation Approach

Jing Xu
SINOPEC Research Institute of Safety Engineering
No.218, Yan'an san Road,Qingdao,China
xuj.qday@sinopec.com

Chuntao Ge
SINOPEC Research Institute of Safety Engineering
No.218, Yan'an san Road,Qingdao,China

Yi Zhang
SINOPEC Research Institute of Safety Engineering
No.218, Yan'an san Road,Qingdao,China
Zhangyi.qday@sinopec.com

The Study of HAZOP Revalidation Approach

Jing Xu
SINOPEC Research Institute of Safety Engineering
No.218, Yan'an san Road,Qingdao,China
xuj.qday@sinopec.com

Chuntao Ge
SINOPEC Research Institute of Safety Engineering

Yi Zhang
SINOPEC Research Institute of Safety Engineering

Keywords: HAZOP, HAZOP revalidation, update HAZOP,PHA, MOC

Abstract

Abstract: HAZOP evaluation method is used to analyze the risk of the process by the worldwide chemical enterprises and design units. However, Standard Specification for many countries stipulate the enterprise needs to carry on the process hazard analysis once again (revalidating) in several years later. The goal of periodic revalidation is to identify the cumulative changes of risk over time and to improve understanding of risk. The ultimate goal is to avoid accidents. But there is no systematic approach to guide the HAZOP revalidation at present. If we redo HAZOP blindly (performing a new HAZOP similar to the initial one) will only be time-consuming. This paper introduces the concept of HAZOP revalidation which is based on the analysis of the initial HAZOP and discusses many of the factors that need to be considered for the revalidation such as the change management, accident analysis and HAZOP report quality. At the same time, the specific procedure and approach of the HAZOP revalidation are proposed. The panel of experts scoring method is used to determine the revalidating approach and then decides whether to redo or update the initial HAZOP. If the scoring results suggest redo the HAZOP, and then conduct the Hazard and Operability Analysis in accordance with the complete HAZOP procedures; if the scoring results suggest update the initial HAZOP and then modify and supplement previous HAZOP analysis results to meet enterprise current status. This approach is not only time-saving and more targeted which can guide enterprises to conduct better and faster HAZOP revalidation.

1 Introduction

Hazard and Operability Study (HAZOP) is a kind of process safety analysis methods which is invented by Britain's imperial chemical industries (ICI) in the 1960s[1]. It is carried out by a suitably experienced multi-disciplinary team (HAZOP team) during a set of meetings to identify potential risk and operational problems in the system.

However, the safety status of enterprise is not invariable after the HAZOP.Many changes may occur in the later production process inevitably. The safety status of enterprise will change if the technological process, equipment, safety equipment, operating procedures, personnel, organization structure, and environment changes in the entire life cycle. So getting through the HAZOP analysis is not once for all. HAZOP Revalidation is to conduct a comprehensive evaluating for process hazards situation of the system during the next few years after the first time HAZOP analysis. To eva`luate the validity of the analysis results and determine whether meet the requirements of laws and regulations,and improve the process safety management level of enterprises at the same time. The goal of regular revalidation is to identify the change of the risk as time goes on and to improve the understanding of the risk. The ultimate goal is to avoid the accident. However, HAZOP analysis of a typical chemical process takes a few weeks to a few months. When in the HAZOP revalidation, whether can take advantage of the existing HAZOP analysis report to make the HAZOP revalidation more simplified and efficient and guarantee the results conform to the enterprise current situation. More and more companies try to simplify the method of revalidation, rather than redo HAZOP blindly[2].

2 Status analysis of HAZOP revalidation

HAZOP technology has developed more than 50 years getting a lot of project application, and become one of the general evaluation method on process safety domain at present. Petrochemical companies and design companies worldwide use HAZOP method to analyze the risk of petrochemical process. Such as Dow chemical, Rhodia company, BP, Bayer, Royal DSM, SINOPEC, CNPC, CNOOC and so on[3].

OSHA(the Occupational Safety & Health Administration) put forward this revalidation concept at Process Safety Management in 1992. Claim that “After the first process hazard analysis, revalidation should be done at least every five years, to ensure consistency of process hazard analysis and the current process” [4]. In the risk management procedures of EPA(Environmental Protection Agency) and aEuropean SEVESO Ⅱ have corresponding requirements in the instruction. CCPS(Center of Chemical Process Safety) published a book Revalidating process hazard analyses Defined that the revalidation process attempts to protect this investment by identifying and building upon the still pertinent portions of the prior PHA .Corrections are made and new content is added to the PHA as required,and the result are documented to serve as the new PHA[5]. In the United States, the large-scale petrochemical enterprises have done 2 ~ 3 times HAZOP, identifying the potential risk in the system, found a lot of potential safety problems. The study of HAZOP technology started late in China. Now a lot of enterprises done HAZOP before will conduct HAZOP revalidation. Guidelines for process safety management of petrochemical corporations (AQ/T3034-2010) standard requires: Guidelines for process safety management of petrochemical corporations (AQ/T3034-2010) standard requires: Companies should conduct a process hazard analysis in the process during construction and should to finish the process hazard analysis to confirm again and updated every three year[6]. The difference between Revalidation and first HAZOP is the former is conduct on the basis of the practical operating experience. The key for the first time HAZOP is to perfect the design and the retrieval permissions[7]. HAZOP Revalidation is to compare the results of the first HAZOP whether still applicable to the current environment, more dependent on site operation actual situation and record filing, etc. There is no systematic method to guide how to HAZOP review again. However, blindly redo HAZOP is not only time-consuming, also need a lot of resources. More and more companies expect to revalidate on the basis of the first HAZOP analysis.It can be more simple and efficient, and at the same time guarantee the quality of process hazards analysis.

3 The Study of HAZOP Revalidation Approach

3.1 The key factor of HAZOP revalidation

HAZOP revalidation using redo or update way is decided by many factors, such as whether there is a big defect of the previous process hazard analysis, whether change management system is perfect, whether there is a major accident happened, whether the information is complete, new standards, etc.

The way for HAZOP revalidation depends on the following three aspects:

a. The design change number, the size of the problem, process complexity, change process hazard analysis quality and management of change since the last HAZOP analysis;
b. The quality of Previous HAZOP analysis, including process safety information comprehensiveness, the comprehensiveness of HAZOP analysis, HAZOP record content integrity, HAZOP results application effect, etc;
c. Previous accidents and device status: the type of process safety accidents, statistical accident frequency, Consequences severity and the reason for the accident , etc.

3.2 The process of HAZOP revalidation

a. Establish revalidation team, the composition of team members are as follows[8]:

- the group leader: trained, rich experience in organization HAZOP analysis; Responsible for communication between HAZOP team and project management personnel; Review plan
- Process equipment technician: familiar with the past project, past and present design/operational problems and change the content
- Site operator: familiar with the situation, the operation experience
- owner (user) : familiar with the operating environment, the consequences of deviation and the degree of danger of deviation
- Experts: providing related professional knowledge. Technology, automatic control experts shall be invited to assist analysis group for part analysis
- Recorder: review record, record the related changes, and update or supplement HAZOP records, to assist the analysis team leader for planning
- Maintenance personnel (if required)

b.Collect information, including the HAZOP report, change management (MOC) data, the previous accident report, HAZOP recommendations, process engineering and safety information, operation procedures, etc

c.Expert assessment team scoring method. Details see Table 1 - the marking table for determining HAZOP revalidation methods and Table 2 - HAZOP revalidation method
d.According to the examination way to implement. Redo HAZOP or update HAZOP.

The flow to determine the specific revalidation way is shown in figure 1

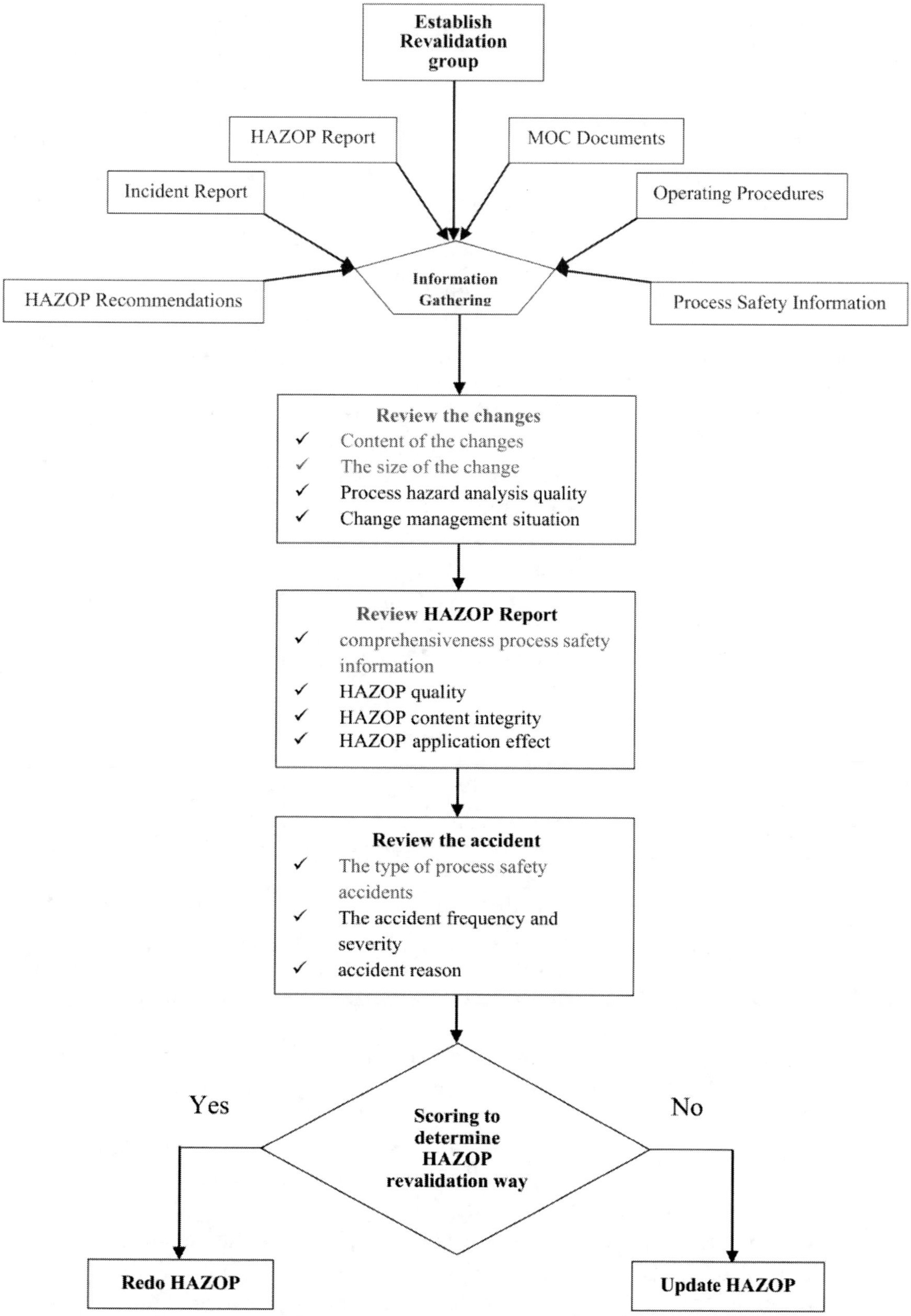

Fig.1 HAZOP revalidation approach decision diagram

Redo HAZOP is not introduced in the paper, the most effective way to update the HAZOP is to collect all data and accident information related to the change, to review the preliminary HAZOP Suggestions, to view the process pipeline flow chart meters (PID), understanding the changes and the defects of the system[9]. Such as process flow, equipment, safety equipment, operating procedures, operating conditions, location and environmental changes, and keep record by the recorder. According to the situation of change size, change management and change of process hazards analysis quality lead revalidation group to carry out different levels HAZOP for change unit by the team leader. Update the HAZOP results that conform to the enterprise the status quo. Update the HAZOP is time saving and maximum use of existing resources compared to redo HAZOP.

Table 1. HAZOP revalidation reference score table

Review items		**Index description**	**Reference scores**
Change Review	Change content (max)	Change of Process/tank type/operating procedure	15*
		Change of safety facilities	10
		Change of pipeline, pump, valve	5
		No change	0*
	Scale of change	Change of working hours more than one month/ 5 types of job post	10*
		Change of working hours about 2 weeks/3 types of job post	5
		Change of working hours about one week/2 types of job post	2
		No change	0*
	Quality of hazard analysis of change	No hazard analysis of change	10*
		Bad hazard analysis of change(lack of pertinence, record, etc)	5
		Good hazard analysis of change(More targeted, detailed record)	2
		Eexcellent hazard analysis of change(effective method, comprehensive analysis)	0*
	Management of change	Records and files are deficient and not combined with operating procedure or job training partly after change	5*
		Records and files are complete and combined with operating procedure or job training partly after change	2
		Complete and electronic records and files,which combined with operating procedure or job training after change	0*
HAZOP quality review	Comprehensiveness of process safety information	Deficient data or big change of site layout	5*
		Partly complete data or small change of site layout	2
		Complete data and less or no change of site layout	0*

	Integrity of HAZOP record	Deficient node description, design intent, drawing name, date, participant and the content of the standard form, bad reference	5*
		Partly complete node description, design intent, drawing name, date, participant and the content of the standard form, good reference	2
		Complete node description, design intent, drawing name, date, participant and the content of the standard form, wonderful reference	0*
	Rationality of HAZOP	Deficient deviation and guide word, unreasonable nodes, bad suggestions and measures	15*
		Partly complete deviations and guide words, reasonable nodes, common suggestions and measures	10
		Complete deviations and guide words, more reasonable nodes, good suggestions and measures	5
		Perfect deviations and guide words, most reasonable nodes, appropriate suggestions and measures	0*
	Application effect	Weak effect, less use of risk identification	5*
		Good effect, general use of risk identification	2
		Perfect effect, effectively use of risk identification	0*
Review of accidents		Vast or serious process safety accidents	30*
		Some specific process safety accidents	15
		Some accidents which has nothing to do with the process	5
		No accidents	0*

* reference point， according to the actual situation within the range of its rated arbitrary integer (interval)

Table 2. HAZOP revalidation approach determination table

Scores	Review way
⩾60	Redo HAZOP
<60	Update HAZOP

4 Application examples

A spherical tank area of PetroChina has four 3000 m3 spherical tank, two 5000 m3 spherical tank, storage for butene 1, butadiene. The company has completed the HAZOP analysis in 2010. Device in service conduct HAZOP analysis every four years according to the company policy. HAZOP revalidation use the proposed method, completed HAZOP analysis efficiently.

Table 3. Application of HAZOP revalidation approach determination table

Review items		practical situation	**Scores**
Change Review	Change content (max)	No change	0
	Scale of change	No change	0
	Quality of hazard analysis of change	Good hazard analysis of change(More targeted, detailed record)	2
	Management of change	No change	0
HAZOP quality review	Comprehensiveness of process safety information	Partly complete data or small change of site layout	2
	Integrity of HAZOP record	Perfect deviations and guide words, most reasonable nodes, appropriate suggestions and measures	0
	Rationality of HAZOP	Complete deviations and guide words, more reasonable nodes, good suggestions and measures	5
	Application effect	Good effect, general use of risk identification	2
Review of accidents		No accidents	0
Scores			11

5 Conclusion

This paper introduces the concept of HAZOP revalidation which is based on the analysis of the initial HAZOP and discusses many of the factors that need to be considered for the revalidation such as the change management, accident analysis and HAZOP report quality. At the same time, the specific procedure and approach of the HAZOP revalidation are proposed. The panel of experts scoring method is used to determine the revalidating approach and then decides whether to redo or update the initial HAZOP. If the scoring results suggest redo the HAZOP, and then conduct the Hazard and Operability Analysis in accordance with the complete HAZOP procedures; if the scoring results suggest update the initial HAZOP and then modify and supplement previous HAZOP analysis results to meet enterprise current status. The method has been applied for a spherical tank area of PetroChina. Conclusion is spherical tank area further implement the last countermeasures of HAZOP in view of the actual operation situation, do not need the modified or added. To satisfy the needs of the enterprise for better and faster process hazards analysis.

6 References

[1]Dunjó J, Fthenakis V, Vílchez J A, et al. Hazard and operability (HAZOP) analysis. A literature review[J]. Journal of Hazardous Materials, 2010,173(1):19-32.

[2]Smith K E, Whittle D K. Six steps to effectively update and revalidate PHAs[J]. Chemical engineering progress, 2001,97(1):70-88.

[3]WAN Gujun, LI Qi, ZHANG Guangwen, et al. Implementing HAZOP to Control Major Accidents (Ⅰ)[J]. Safety Health & Environment, 2009,9(011):2-4.

[4]ZHAO Jianmin. Discussion About HAZOP Revalidation [J]. Modern Occupational Safety, 2012(134):83-85.

[5]L. F W, K. W D. Revalidating process hazard analyses[M]. Center for Chemical Process Safety, 2001.

[6]State Administration of Work Safety. AQ/T 3034-2010 Guidelines for process safety management of Petrochemical corporations[S]. 2010.

[7]FU Jianmin, ZHAO Dongfeng, CHEN Guoming, et al. Quantitative study over the probability of HAZOP analyzing technique in petro-chemical plants[J].Journal of Safety and Environment,2008,8(6):130-134.

[8]Herbert I. Learning the lessons-retrospective HAZOPs[J]. Offshore Europe, 2011.

[9]Wagner T, Champion J. A work process for revalidating LOPAs and other risk analyses[J]. Process Safety Progress, 2012,31(2):122-129.

12TH GLOBAL CONGRESS ON PROCESS SAFETY

The Importance of Process Safety:

A Young Engineer's Perspective

Waheed Wakil
Smith & Burgess
Waheed.Wakil@smithburgess.com

Travis Bruner
Smith & Burgess
Travis.Bruner@smithburgess.com

Prepared for Presentation at
American Institute of Chemical Engineers
2016 Spring Meeting
12th Global Congress on Process Safety
Houston, Texas
April 11-23, 2016

The Importance of Process Safety:

A Young Engineer's Perspective

Waheed Wakil
Smith & Burgess
Waheed.Wakil@smithburgess.com

Travis Bruner
Smith & Burgess
Travis.Bruner@smithburgess.com

Keywords: Process Safety Management, Young Engineers, Overpressure, OSHA, LOPA

Abstract

Most young engineers experience a defining moment or moments in their career that shape their attitude towards importance of Process Safety Management (PSM). Experiencing a plant upset first-hand knowing that there were actual capital or human life impacts or both from an incident at the site, one soon realizes early in their engineering career why process safety is so important. It becomes evident that the most successful corporations are those who are willing to invest large capital costs into PSM documentation projects or projects that have come as a result of findings identified during a study, allowing them to prevent incidents from occurring. The findings generated during the course of a study may result in simple documentation update action items, or they may require a multi-million dollar investment. For young engineers, this exposure puts into perspective the value of their work, and emphasizes the importance of a thorough and adequate study.

1 Introduction

Many young engineers have a point early in his or her career when process safety hits home. Different companies develop their own unique strategies towards safety that they are comfortable with, which results in considerably different safety cultures within industry. Unfortunately, it is all too common for engineers to treat process safety as an afterthought, with priority given to efficiency and profitability. However, as we will see, treating safety with the utmost importance can prevent costly and life-threatening incidents.

2 First Hand Experiences

2.1 Organic Peroxides Plant Background

Two major upsets at a chemical plant within a year may very well serve as a defining "moment" in a young engineer's career that forces him to reflect on and realize the importance of dedication to process safety. A fellow Process Safety Consultant at Smith & Burgess who was a process engineer at a plant that produces organic peroxides was fortunate enough to live through two such accidents unscathed. His story is a lesson and reminder that even the most disastrous accidents can be prevented with due diligence, a safety-focused company culture, and proper attention to basic Process Safety Management.

Based on the nature of the chemicals produced, a Process Hazard Analysis for this particular plant would have revealed scenarios with severe consequences. However, the safety culture at this plant was not adequately focused in preventing and mitigating upsets. Not only were poor preventative maintenance practices used, but lack of written emergency procedures and poor training prevented adequate mitigating protection layers after the initiating events. A few relevant OSHA requirements are worth noting here as failure to meet these requirements was likely the reason for loss of containment:

- Per OSHA 29 CFR 1910.119(f)(1), "The employer shall develop and implement written operating procedures that provide clear instructions for safely conducting activities involved in each covered process consistent with the process safety information...".
- Furthermore, per OSHA 29 CFR 1910.119(j)(2), "The employer shall establish and implement written procedures to maintain the on-going integrity of process equipment."
- Lastly, per OSHA 29 CFR 1910.119(n), "The employer shall establish and implement an emergency action plan for the entire plant in accordance with the provisions of 29 CFR 1910.38. In addition, the emergency action plan shall include procedures for handling small releases."

If these basic elements of PSM would have been implemented at this plant, both accidents may have been prevented or mitigated.

Organic peroxides are highly unstable – their O-O bond breaks very easily, and that is why their decomposition properties are so useful in initiating polymerization reactions. This useful property can also be very dangerous. When an organic peroxide is in its pure form, the rate at which its exothermic decomposition generates heat is likely to be greater than the rate at which the heat can be dissipated to its surrounding environment if proper heat transfer equipment are disabled, resulting in a situation called self-accelerating decomposition in which the reaction continues to accelerate due to its own generated heat.

The organic peroxides produced in this particular plant are characterized by very low Self Accelerating Decomposition Temperatures (SADT), defined as the minimum temperature at which an organic peroxide would experience self-accelerating decomposition after one week of storage. As such, glycol cooling jackets on vessels and piping in this plant maintained operating temperatures in the ballpark of -30 °C (-22 °F) to prevent the possibility of self-accelerating decomposition. For this heat transfer to work properly, the organic peroxides must be continuously flowing in the process piping and equipment and cannot become stagnant. At the same time, the glycol that cools the organic peroxide must also continuously flow. Stoppage of flow for either component may be disastrous. As we will see in the following incident descriptions, loss of this cooling results in rapid increase of temperature and decomposition reaction rate, resulting in a deflagration that cannot be mitigated by a typical relief device such as a safety valve or a rupture disk. To prevent loss of containment and subsequent damage to surrounding structures, this scenario must instead be prevented or mitigated using written operating procedures, regular inspection and maintenance of equipment, and employee training.

2.2 Organic Peroxides Incident #1

What went wrong?

Organic peroxides are so thermally unstable that precautions must be taken when packaging and storing the material. Packaging must occur continuously to prevent the material from becoming stagnant in piping and the temperature must be monitored and controlled at or below the recommended storage temperature. If material remains stagnant without sufficient cooling, organic peroxides can reach their SADT temperature and undergo a violent reaction, which usually results in rupturing the packaging, dispersing peroxide, as well as discharging liquid and gaseous decomposition products considerable distances. The heat generated during this reaction has the potential to auto ignite flammable vapors.

An operator was transferring organic peroxides to 55 gallon drums, which are immediately transported by forklift to an icehouse for storage. An incident occurred which caused the operator to abandon the task in the packaging building to provide assistance elsewhere. While the operator was away, the product remained stagnant in the filter where glycol cooling was inefficient. This caused the material to exceed its SADT. The material began to decompose as temperature continued to increase. When the operator returned the material suddenly exploded propelling the head of the drum through the air. The remains of the drum were embedded in the ground and lights in the facility were shattered. The operator was behind a blast shield, but still suffered severely burnt hands and damage around the ear. Although no life threatening injuries were inflicted, the plant suffered considerable downtime and costly repairs to the filter, packaging equipment, and process piping.

How could it have been prevented?

Proper implementation of PSM standards likely could have prevented this incident from occurring - in particular, OSHA 29 CFR 1910.119(f)(1), mentioned previously. Documented operating procedures that clearly state not to leave the packaging unit unattended during operation to avoid stagnant organic peroxides, inefficient cooling, and the potential for explosion could have prevented (or, at the very least, limited) the disastrous effects of this incident.

2.3 Organic Peroxides Incident #2

What went wrong?

This plant utilizes large plug flow reactors that are jacket-cooled. On one particular day, the site was hit by a thunderstorm with wind speeds high enough to break a tree branch that would damage a power line, resulting in total loss of electricity to the plant. Electric power was vital to this plant as it kept peroxide flowing through the reactors and piping and it kept glycol flowing through the jacketing to keep the peroxide temperatures down to safe levels. The plant had a backup generator, but only a single operator was aware of the location of its switch. It just so happened that it was this operator's day off.

In the event of an emergency such as this, interlocks were in place to purge out residing organic peroxides in the equipment using control valves with actuators regulated by instrument air. Once glycol and peroxide flow was lost in the event of a power failure, the peroxides would rapidly exceed their SADT. The purging action would dump the peroxides into waste water drains that are gravity fed to open tanks, where it is much safer for the peroxides to reach their SADT. However, the instrument air that controlled the valves was supplied from air storage, and during the purging, the instrument air ran out. Nitrogen was used as backup for air during emergency situations, but due to poor inspection and maintenance for cost savings, there was damage to the backup nitrogen header that had gone unnoticed. This damage caused a leak in the header when it was exposed to the nitrogen pressure, rendering the header useless as there was not enough pressure to actuate the purge control valves properly. At this point, purging was halted. The catastrophic failure was now inevitable. With stagnant peroxides in piping and equipment, and no jacket cooling, fluid temperatures rapidly sored past the SADTs and eventually resulted in a deflagration. The reactor top heads failed first, and the building in which the reactors were located was designed to fail at the ceiling and the outer wall. The building and its contents were destroyed, and the plant suffered 6 months of downtime. Fortunately, operators were protected on the other side of a blast wall, and no one suffered major injuries.

How could it have been prevented?

Looking at the number of Independent Protection Layers (IPLs) that were in place at the time of failure, namely the operator's response to activate the backup generator and interlocks that functioned to purge out residing peroxides, there were not enough layers of protection in place to prevent a deflagration. Table 1 summarizes the highest possible credits that could be taken for each IPL based on CCPS's LOPA book [1].

Table 1: IPL Credit Summary

Initiating Event	Initiating Event Frequency	Number of IPL Credits Required for Category IV Consequence	Maximum Number of IPL Credits Available
Loss of Cooling	Frequency $\geq 1X10^{-2}$	**2.0**	Operator Response for Backup Generator 0.5
			SIL 1 SIF 1.0
			Total **1.5**

Like many such accidents, it is not uncommon for multiple independent failures to cascade into a catastrophic event, especially when a key to a successful emergency response from operators is proper training and a key to proper function of mechanical equipment is proper maintenance and inspection. To pinpoint PSM measures that would have likely prevented this accident, we can start by calling on OSHA 29 CFR 1910.119(n) referenced previously in this paper. The first broken link in the series of events leading up to the accident was the lack of training amongst operators with respect to activating the backup generator in the event of loss of power. This should have been a basic emergency response action that every operator at the plant should have been trained on and been aware of, but unfortunately, no one at the plant at the time of the failure was trained on how to activate the generator. The failures cascaded into the second phase of emergency response when the instrument air was exhausted at exactly the wrong time. To address the issue of the backup nitrogen header leak, we call upon OSHA 29 CFR 1910.119(j)(2), "The employer shall establish and implement written procedures to maintain the on-going integrity of process equipment." To cut costs, plant management had reduced regular inspection and maintenance of the backup nitrogen header, and it proved disastrous on this day when they needed it the most.

3 Conclusion

As illustrated in the incident descriptions, a poor safety culture is likely to result in poor preventative inspection and maintenance cycles and lack of clear emergency operating procedures and training. In most process plants, inadequacies in PSM elements can result in disastrous consequences. For a young engineer, it sometimes takes a real-life experience to change their perspective on PSM from one of passive compliance to one of active responsibility. Each element of PSM is important and critical, from the most obvious and simple, like written procedures on activating a backup generator, to the most expensive and extensive, like redesigning a flare system. It makes every detail in every process safety analysis that much more important and that much more rewarding when the benefits are apparent.

References

[1] CCPS. *Layer of Protection Analysis: Simplified Process Risk Assessment.* Center for Chemical Process Safety, American Institute of Chemical Engineers, New York, NY, 2001.

Scaling Up Safely: Making Smarter Decisions about Rate Dependency

Zubin Kumana, P.E.
Smith & Burgess LLC
7600 W. Tidwell Rd, Ste 600
Houston TX 77040
Zubin.Kumana@smithburgess.com

Venkata Badam, P.E.
Smith & Burgess LLC
Venkata.Badam@smithburgess.com

Achilles Arnaez
Smith & Burgess LLC
Achilles.Arnaez@smithburgess.com

Prepared for Presentation at
American Institute of Chemical Engineers
2016 Spring Meeting
12th Global Congress on Process Safety
Houston, Texas
April 11-23, 2016

Scaling Up Safely: Making Smarter Decisions about Rate Dependency

Zubin Kumana, P.E.
Smith & Burgess LLC
7600 W. Tidwell Rd, Ste 600
Houston TX 77040
Zubin.Kumana@smithburgess.com

Venkata Badam, P.E.
Smith & Burgess LLC
Venkata.Badam@smithburgess.com

Achilles Arnaez
Smith & Burgess LLC
Achilles.Arnaez@smithburgess.com

Keywords: Process Safety, Safe Design, and Venting and Emergency Relief

Abstract

Over the lifetime of a facility, it is almost a certainty that the throughput of a process will be increased, often repeatedly. Rate-dependency is frequently used as a way to identify pressure relief system sizing that would be affected by the scaling up of process throughput. This paper will define rate-dependency as it relates to pressure relief systems and describe the various ways in which rate-dependent classifications may be affected by the mechanical and thermodynamic constraints of the process and equipment. In addition, examples of when the "rate-dependent" designation may no longer apply due to common assumptions and methodological choices will be provided.

1 Introduction

In the refining and chemical processing industries, increasing throughput through a processing unit is often considered as a means of maximizing value with existing capital or minimal additional investment. Often, the throughput will be increased many times over the lifetime of the facility. The pressure relief system for a process is designed in part based on the throughput of the unit; therefore, the change in the throughput may affect the adequacy of the pressure relief system design.

Verifying the design of the process unit's pressure relief system for these increased rates is typically done by identifying overpressure scenarios as either rate-dependent or rate-independent. A rate-dependent overpressure scenario is one whose required relief rate varies with changes in process throughput. A common example is a blocked outlet scenario on a separator. With higher the process throughput, the quantity of mass that must be relieved to limit the pressure to the allowable accumulation is expected to increase.

Conversely, rate-independence refers to overpressure scenarios whose required relief rate is independent of process throughput. A common example is exterior fire, where the required relief rate depends on the heat of vaporization of the liquid mixture inside the vessel, the density of the vapor and liquid phases, and the wetted area of the vessel the fluid is residing in. Changing the feed rate to the vessel is not expected to result in a change to the required relief area for the exterior fire scenario. In some circles, these scenarios may also be termed "equipment-dependent". While the equipment and piping remains constant, the required relief rate remains constant. However, as rate scale-up increases, the equipment and piping may be changed to accommodate the additional capacity.

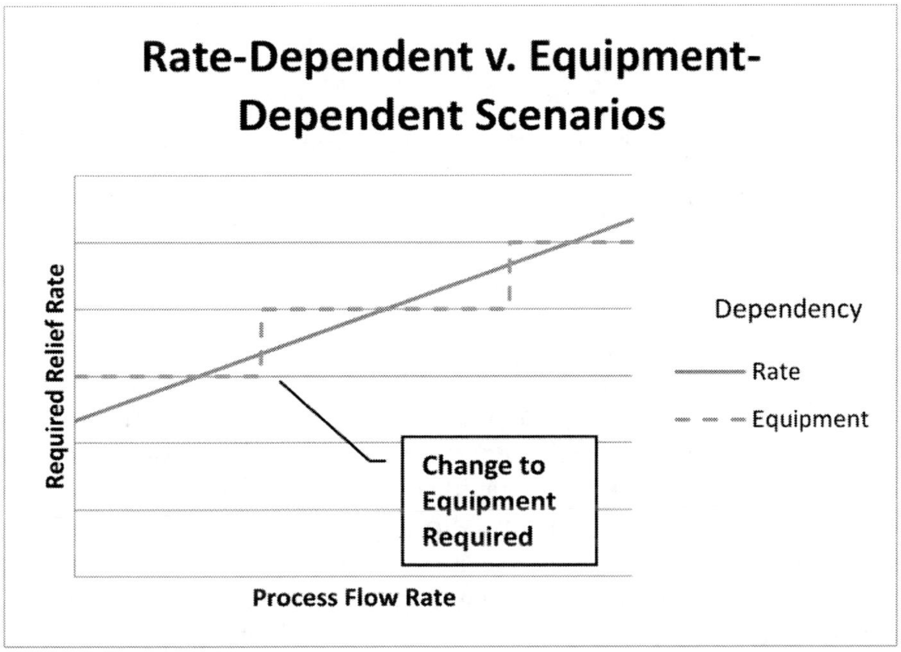

Figure 1: Rate-dependent scenarios increase in required relief rate (and therefore area) with process flow rate increase, while equipment-dependent scenarios are effectively constant until a change to equipment is required.

1.1 How Rate Dependency is used in Pressure Relief System Design

The common practice in pressure relief system design is to select a unit process basis that the user believes represents both the actual operation of the process as well as the worst case for the relief system design – specifically, the case that will result in the largest required areas and piping diameters. Since process units can have a range of operating modes and rates, selection of these cases tends to favor the following:

- Modes of operation with lighter feed, as this can result in more vaporization and therefore larger vapor and two-phase relief loads
- Higher process throughput rates; often the peak rate achieved, or some percentage above the peak rate

The conscientious user might separately specify a worst-case heavy liquid relief load basis in order to properly design the liquid handling capabilities of the disposal system.

As the pressure relief system design is completed, the engineer performing the design will typically identify overpressure scenarios as rate-dependent as the analysis proceeds. This designation is then used when the throughput rate is modified in the future. This allows engineers to minimize the amount of recalculation that needs to be performed, by only performing new calculations for the overpressure scenarios that are dependent on rates; the scenarios that had been identified as rate-independent are believed to be static and are left unchanged.

A generalized expression for pressure relief device sizing is given in API Std 520 Part I (9th ed.) Annex B, equation B.5, rearranged to solve for required area [1]:

$$A = \frac{W}{G\prod[K]} \qquad \text{(Eq. 1)}$$

Where

G	is the theoretical mass flux through the nozzle
W	is the mass flow through the PRV
A	is the effective discharge area
Π[K]	is the product of all applicable correction factors

The scale-up of the pressure relief system design is typically done using the assumption that the only change is to the flow rates through the unit; variables such as fluid composition, operating pressures and temperatures are assumed to remain constant. This effectively limits the relationship between the throughput rate and the relief system required areas to a linear one; as the fluid mass flux is considered constant, and the applicable correction factors are generally constant (exceptions are always possible based on Reynolds' number and actual backpressure resulting from flow), the values of A and W are directly proportional.

For the evaluation of the pressure relief system design, this has led to the following widespread practice:

1. The user takes the heat and material balance at the previous throughput rate
2. The user obtains a list of calculations that had been identified as rate-dependent
3. The user takes the percentage increase in flow and factors this value into the required relief rates
4. The user re-evaluates the pressure relief device requirements vs. capacity (along with other associated concerns where relevant, e.g. pressure losses if calculated at required relief rates)

This process rarely involves revalidating the process unit simulation to confirm the heat and material balance. Typically this practice has been followed for small changes to throughput rates. However, once the procedure has been implemented, the potential exists for repeated applications to increase the degree of deviation from a validated baseline even further. This poses a danger, as excessive extrapolation and repeated application of the scale-up process without carefully considered heat and material balance revalidation can quickly lead to larger and larger deviations, and the potential for undersizing the pressure relief system.

This paper focuses on rate scale-ups that do not involve any change to the as-built system. The rate scale-ups considered are therefore limited to the maximum throughput that the existing equipment can support. Debottlenecking the equipment in the process unit (rotating equipment, exchanger heat transfer area, and pipe sizing) may be required to accommodate further rate scale-up. As a capital project, this effort typically includes development of a new project baseline heat and material balance along with a review of the pressure relief system design.

2 Proper Identification of Rate-Dependency

2.1 How to Identify Rate-Dependency

Smarter decision-making with regard to rate dependency begins with understanding the ways that the traditional definition of rate-dependency may be invalidated by subsequent modeling choices in the determination of the required relief rate.

A general rule of thumb often passed around among novices in pressure relief system design is that overpressure scenario required relief rates that result from equations are rate-independent, and required relief rates that come from a heat and material balance (HMB) are rate-dependent. However, while it may be generally true for many facilities, there are aspects that are commonly overlooked in this approach, and some engineers may take subsequent steps in the required relief rate calculation that can change the classification. These additional steps are often taken as a means to "sharpen the pencil" when the initial determination of the required relief rate results in an undersized pressure relief device. These may include such actions as:

- Taking credit for an outlet that would not be blocked as a result of the scenario (including secondary causes like controller action and shut off systems)
- Reducing heat transfer into the system due to elevated bubble point temperatures
- Limiting flow rates based on the hydraulic limits of the piping, including valves and fittings
- Limiting flow rates to the maximum capacity of rotating equipment that can deliver the stated pressure at the point of relief

For example, if determining the required relief rate in the event of a blocked liquid outlet on an exchanger by taking the normal feed rate as the required relief rate is found to result in a required relief area greater than the currently installed relief area, the engineer may review the inputs and determine that when discharging against the relief pressure, the capacity of the feed pump is reduced below the normal feed rate, potentially making the existing installed relief area adequate. In this example, the overpressure scenario has changed from rate-dependent to equipment-dependent.

Table 1 shows common sources for determining the required relief rate for an overpressure scenario, and potential additional modeling choices that can change the classification.

Table 1. Examples of Dependency and Factors That May Change Classification†

Overpressure Scenario	Rate Commonly Derived From	Dependency for Common Derivation	Factors that may change Classification
Closed outlets	HMB/Simulation, Exchanger duty	Rate-Dependent	Limited by pump/compressor, valve, or pipe capacity
Reflux failure	HMB/Simulation, Exchanger duty	Rate-Dependent	
Cooling failure	HMB/Simulation, Exchanger duty	Rate-Dependent	
Overfilling	HMB	Rate-Dependent	Limited by pump or valve capacity
Inadvertent opening of manual valve	Line capacity	Equipment-Dependent	Credit taken for volume outflow
Failure open of inlet control valve	Valve capacity	Equipment-Dependent	Credit taken for normal minimum flow or outflow
Check valve failure	Valve or line capacity	Equipment-Dependent	
Abnormal heat input	Valve or exchanger capacity, HMB/Simulation	Rate-Dependent	
Exterior fire	Fluid properties, Vessel area	Equipment-Dependent	
Hydraulic expansion (pipe)	Fluid properties, Pipe area	Equipment-Dependent	
Hydraulic expansion (exchanger)	Fluid properties, Exchanger duty	Equipment-Dependent	Exchanger duty driven by hot side flow rate
Tube rupture	Exchanger properties, fluid properties	Equipment-Dependent	
Tube leak	Exchanger properties, fluid properties	Equipment-Dependent	
Plate weld failure	Exchanger properties, fluid properties	Equipment-Dependent	
Column power failure (Complex)	HMB/Simulation, Exchanger duty	Rate-Dependent	
Column power failure (Simple*)	Fluid properties, reboiler duty	Rate-Dependent	
Loss of Absorbent	HMB/Simulation	Rate-Dependent	

*Simple column failure includes overpressure scenarios where the required relief rate can be determined as a function of the reboiler duty and heat of vaporization of the residing fluid.
†Not a comprehensive list

3 Rate-Dependent Factors Influencing Pressure Relief System Design

Although the common practice is to size the pressure relief system as a linear function of process flow rate, the reality is more complex, as other factors that are themselves influenced by process flow rate can also impact the pressure relief system design, primarily through their impact on pressure relief device sizing. These factors can impact the pressure relief device sizing by causing additional deviations in relief rate calculations for rate-dependent overpressure scenarios (i.e., non-linear relations), as well as creating changes in required relief rates for overpressure scenarios previously categorized as equipment-dependent.

A quick review of three factors influencing pressure relief device sizing will aid in understanding the limits of scaling up the pressure relief system design basis: composition, operating conditions, and heat exchanger duties. Note that this list is intended for illustrative purposes of the effects and is not an all-inclusive list; there are other factors that can influence the design of the pressure relief system that may change during a process rate scale-up.

3.1 Composition

3.1.1 Composition and Pressure Relief System Design

Fluid composition impacts the pressure relief device sizing in two significant ways.

1. Firstly, it can directly impact the determination of the required relief rate; for example, fire calculations use the latent heat of vaporization for a mixture, and depending on the particular modeling choices selected (e.g. inclusion of sensible heat, vaporization interval selected) may include further dependencies on heat capacity, bubble point temperature, vapor to liquid volume ratio, and other variables. Orifice flow calculations (such as are found in tube rupture calculations) depend on the density or molecular weight, compressibility, and specific heat ratio for the fluid.

2. Secondly, the composition of the fluid mixture directly impacts the relationship between the pressure of the fluid and the specific volume of the fluid, which governs the determination of the critical mass flux across the relief device nozzle (G in equation 1).

3.1.2 How Scaling Up the Process Can Affect Composition

It has been commonly observed that the compositional data in a heat and material balance is rarely adjusted when a scale-up of the unit throughput is performed. This assumption can be invalidated by any of the following effects:

1. Changes in the operating pressure and temperature – Changes in operating temperature and pressure can have an impact on the composition by affecting the separation of the vapor and liquid phases within the process equipment. Increasing the operating pressure can raise the bubble point temperature of the fluid mixture, requiring additional heat to achieve the same separation point. Decreasing the operating pressure can reduce the bubble point temperature of the fluid mixture, possibly resulting in more vapor flashing off, with a higher molecular weight.

2. Limitations on feed sources – In order to maximize the throughput for the specific process unit, feeds from multiple sources may be required. If the process unit has excess capacity but insufficient feed is available from on-site units, feed may be supplemented from offsite or from tankage. These feeds may have variations in compositions, and as the new feed is added, the blend ratio can change. For example, a gas oil hydrotreating unit (GOHT) may have additional capacity, but due to changes in product demand the crude column system may be producing a higher proportion of lighter products. To take advantage of the slack

in the GOHT, additional gas oil may be imported, resulting in a change in composition of the feed.

3.2 Operating Conditions

3.2.1 Operating Conditions and Pressure Relief System Design

The operating conditions within equipment in the process unit can directly impact the pressure relief device sizing primarily through the temperature's effect on the fluid conditions (for overpressure scenarios that model the required relief rate using its operating temperature). Temperature also influences the maximum mass flux across the pressure relief device. Since the relief device is sized at the relief pressure, there is no similar direct effect of operating pressure.

More significantly, changes in the normal operating temperature and pressure can affect the required relief rate calculations for overpressure scenarios that require these values as inputs, such as modeling flow across an orifice, flow through piping or valves, or where separation processes are occurring. For example, inadvertent opening of a valve into a separator may involve flow of the fluid mixture across a valve and series of pipe fittings, followed by a separation where the vapor is expected to overpressure the downstream vessel, while the liquid will slowly fill the vessel. The change in overall vapor quality can both affect the required rate to be relieved while the vessel is filling as well as affect the available time for operator intervention – in some cases making the required relief rate better, and in other cases making it worse. Another example is the potential for changes in composition, temperature, or pressure to change the net positive suction head available (NPSHA) for pumps, requiring adjustments to liquid levels in order to maintain proper pump operation, thereby affecting equipment-dependent overpressure scenarios.

It is even conceivable that such a change may alter the applicability of an overpressure scenario, where the existing operating conditions are already near the threshold of applicability – from either direction. Changing the operating pressure may affect the suction pressure of a pump or compressor, thereby altering the deliverable pressure to a downstream system, whose determination of applicability for a control valve failure or inadvertent opening of a manual valve overpressure scenario may depend on the comparison of this pressure to the downstream system maximum allowable working pressure.

Changes to operating conditions can also affect the specification of the pressure relief device. Operating pressures that were previously at the limit of the acceptable range that may now exceed the operating limits of the pressure relief device may require alternative relief device specification (e.g., if the operating pressure exceeds 90% of the PRV set pressure, this may require replacing a conventional PRV with a pilot operated PRV). Significant changes to operating temperature may also affect the set pressure of the pressure relief device. For a spring loaded valve, a new cold differential set pressure may be required.

3.2.2 How Scaling Up the Process Can Affect Operating Conditions

The operating conditions can be impacted in the following ways during a scale-up:

1. The mechanical characteristics of rotary equipment – For example, for a given installed impeller diameter, as the flow rate is increased, the operating point of the rotating equipment must move out on the pump performance curve, resulting in a lower delivered pressure. (This can be countered by changing out the pump impeller, but that is outside the scope of this paper).
2. Piping hydraulic limitations – As flowrate is increased, the pressure drop is expected to increase. With changes in the pressure drop between equipment, the operating pressure profile may be affected. For flow in the turbulent region, for the same diameter of piping, the pressure drop normally follows a trend of

$$\frac{F_1\,v_1}{F_2\,v_2} = \frac{Q_1}{Q_2} = \sqrt{\frac{\Delta P_1}{\Delta P_2}} \quad \text{(Eq. 2)}$$

   Where

F	is the mass flow rate
v	is the specific volume of the fluid
Q	is the volumetric flow rate of the fluid
ΔP	is the pressure drop through the pipe

3. Heat transfer – As flowrates change, when running up against the limits of the heat transfer equipment, changes in temperature and mass quality may result. For example, applying the same duty to a fluid entering at a higher feed rate will soak up more sensible heat and result in a smaller vapor fraction exiting the equipment.

3.3 Heat Exchanger Duties

3.3.1 Heat Exchanger Duties and Pressure Relief System Design

Heat exchanger duties can potentially cause deviations in the relief device sizing basis if the process duties are used as inputs in determining the required relief rates. Note that some users choose to use the design duties of the exchangers in lieu of the process duties. This may mitigate the impact of scale-up (up to the point where new heat transfer equipment is required). Common examples of overpressure scenarios impacted by this factor are blocked outlets on heat exchangers, thermal expansion, and simplified column calculations (i.e. boil-up).

The other route through which heat exchanger duties can impact the pressure relief system sizing is by impacting the operating conditions of the process (see section 3.2.1).

3.3.2 How Scaling Up the Process Can Affect Heat Exchanger Duties

Scaling up the process flow rate through an exchanger can significantly impact the exchanger heat duty if it is limited by flow instead of limited by heat transfer area and overall heat transfer coefficient ("UA-limited"). For the same heat exchanger, the flow pattern correction factor F_T is expected to be constant. In these cases an increase in process fluid flow can result in increased duty, up to the point where the exchanger UA becomes limiting. Note that for forced convection, since the Nusselt number is a function of the Reynolds number, the value for the overall heat transfer coefficient U is not constant; rather, it is itself a weak function of flow rate (as well as composition and operating conditions). If the exchanger duty is limited by UA, then the exchanger heat transfer duty can be still changed by differences in log mean temperature difference (ΔT_{lm}). The temperatures changes may come about as described in section 3.2.2.

3.4 Extent of Deviations and Overall Impact on Pressure Relief System Design

The above discussions have focused on how rate scale-ups can impact the pressure relief system design in ways beyond the typically assumed linear relationship between mass flow rate and required relief area, as well as impacting scenarios that might otherwise not be classified as rate-dependent. While it can be expected that minor changes in flowrate would have a correspondingly minor impact on composition, temperature and pressure, and heat exchanger duties, and therefore a minor impact on the pressure relief system design, it is important to recognize that the extent of deviation from linear behavior can increase with the extent of the process throughput scale-up. Additionally, repeatedly performing scale-ups on a previously scaled-up process, without revalidating the heat and material balance, can lead to the same result in a less obvious manner.

4 Examples

4.1.1 Examples of Compositional Impact on Pressure Relief System Design

For equipment and systems where relief rate is determined by calculating a rate of vaporization, the relief rate typically increases as the proportion of lighter components within the feed increases. Even where the relief rate is not determined by calculating a rate of vaporization, a lighter feed may not change the required relief rate mass flowrate but might instead decrease the relief device capacity, on a mass basis. For example, crude units can be operated with significant variations in the crude slate. These changes in feed composition could cause different effects for relief device design within the crude unit, even with no change in overall unit throughput.

The sizing scenarios for crude column systems commonly include partial power failures that result in continuing crude feed and heat input from the fired heater, with loss of all overhead and pumparound cooling, among other effects (such as backflow and desalter water carryover). In these scenarios, the relief stream is based on the lighter portion of the crude feed, which typically consists of the offgas, liquefied petroleum gas (LPG), naphthas, kerosene, and/or diesel, etc., depending on the heating capability of the feed heater relative to the feed. Therefore, a change in crude slate that increases the proportion of naphtha cuts in the crude while decreasing the proportion of atmospheric gas oil (AGO) cuts can result in a direct increase on the relief rate of the crude column, whereas a change that only changes the heavier portions of the crude feed may not have a significant effect on the crude column. Note that composition changes that do not affect the crude column may affect another system, such as the vacuum column.

4.1.2 Examples of Changes in Operating Conditions on Pressure Relief System Design

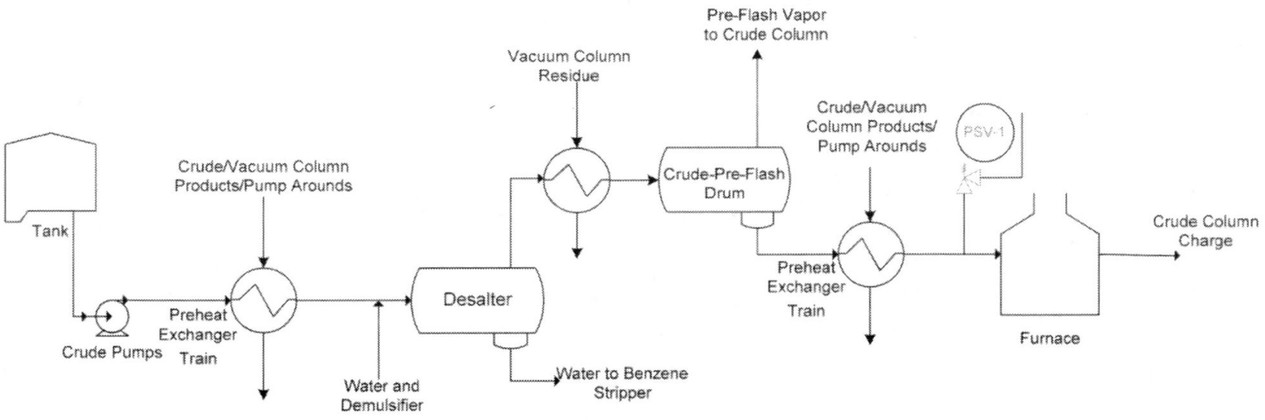

Figure 2: Crude Feed Pre-heating System Process Flow Diagram

The process flow diagram in Figure 2 represents the feed pre-heating section of a Crude Column system. Feed from tankage is pre-heated in a series of exchangers against hot fluid from Crude/Vacuum Column products and pump arounds; the crude feed is then treated for salt content in the Desalter before sending to the Crude Pre-Flash Drum. The Crude Pre-Flash Drum operates significantly lower than the Desalter, allowing the crude to flash. The flashed vapor is directly sent to Crude Column, while the liquid from the drum is sent through another train of exchangers, the furnace, and then to the Crude Column.

a. Effect of Operating Pressure Change on Composition

PSV-1 on the Flashed Crude pre-exchangers is sized for Blocked Outlet, and this case is usually selected as rate-dependent. However, a change in operating pressure of the Crude Pre-Flash Drum, with or without the throughput change, affects the composition of flashed liquid and vapor at the Crude Pre-Flash Drum, because the bubble point of the crude stream components is directly related to the operating pressure. In this case, as the fluid composition is changed, an assumption of a linear relationship between changes in throughput and required area may not be accurate.

b. Effect of Operating Temperature Change on Composition

Similarly, a change in the throughput may result in change in product rates, Crude Column product/pumparound rates and Vacuum Column product/pumparound rates, which could result in change in the heat input at the Crude Preheat exchangers. Similar to pressure, a change in Pre-Flash Drum inlet temperature, may affect the composition of Crude Pre-Flash Drum liquid and vapors, and the rate dependent assumption may no longer apply for Blocked Outlet case for PSV-1.

5 Conclusion

The economic driver for maximizing the production from process units is ever-present. However, it is vitally important to ensure that the techniques used to evaluate the pressure relief system design are appropriate and suitable to the task. Awareness of the assumptions inherent in rate-dependent classifications, and the potential to invalidate this classification with further "pencil-sharpening", is a starting point. Understanding the ways in which process throughput scale-ups can impact other factors in determining the design of the pressure relief system, such as affecting the fluid composition, operating conditions, and heat exchanger duties, and their impact on required relief rate determination or even scenario applicability, helps to illuminate potential problem areas when evaluating a process throughput scale-up, and informs the decision on whether a revalidation of the heat and material balance is appropriate. Altogether, these considerations should enable engineers to make smarter decisions about rate-dependency and ensure safer operation during process throughput scale-ups.

6 References

[1] American Petroleum Institute, *API Standard 520: Sizing, Selection, and Installation of Pressure-relieving Devices Part I – Sizing and Selection*, 9th edition, July 2014.

442298 Learning Lessons from a Big Chemical Disaster in Korea

Abstract:

Learning lessons from a big chemical disaster in Korea

Kwnaghee Lee and Il moon*

Department of Chemical and Biomolecular Engineering, Yonsei University, 50 Yonsei-ro, Seodaemun-gu, Seoul 120-749, Republic of Korea

On September 27, 2012 a massive chemical dispersion disaster occurred at the Gumi National Industrial complex in the southern city of Korea. About 8 tons of hydrogen fluoride gas leaked as the valve of tank lorry was opened due to a worker' mistake. The operators at the factory had never experienced such a big chemical disaster, and the fire station in the area did not have enough equipment at the time of chemical accident, so that an initial counter measurement could not be conducted smoothly. Due to this accident, 5 workers were killed and about 3,600 local residents had to receive a medical treatment at hospitals due to the accident. And the damage proceeded to the second and third stage, causing 17.7 billion KRW of damage to enterprises and spoiling the crops on 323.8ha of land. After the Gu-mi disaster, law and regulations, government organization, and industrial safety system were completely revised and culture have leaded positive effects towards managing and responding to chemical disasters. The governmental action has been led to make more practical plan and manual to control and respond to chemical accident. The governmental reorganization against chemical disaster such as Joint inter-agency Chemical Emergency Preparedness Center which is a specialized agency focused on responding to chemical disasters has also made to increase technicalities of chemical accidents response. In addition, the relevant laws and regulations has clarified the different acts and programs on control, management and assessment of chemicals along as well as its accident. Moreover, private and public enterprises have been also making efforts to systematically manage safety issues and build high-level safety culture. In this paper, we discuss the changes attitudes against chemical accidents from various viewpoints and the successful case in which the relevant agencies responded to major chemical accidents more efficiently after those changes. In order to improve the initial response system against chemical accidents continuously, it is required to strengthen the role of on-site coordinator, establish the Chemical Accident Investigation Committee on a national level. This study covers the big systematic changes in laws, regulations, business culture for the chemical safety management after a big chemical release accidents.

442301 Study on Thermal Radiation of Enclosed Ground Flare

Abstract:

Flare system is an important part of the safety systems in refineries or petrochemical facilities, and on oil and gas platforms. There are generally two types of flares: the elevated flare and the enclosed ground flare. Enclosed ground flares get more and more applications because of its high efficiency, flame invisibility and low nose. During an emergency flaring situation, a portion of the heat produced by the flame will radiate to its immediate surroundings.The amount of radiation heat emitted by flares is important in the design and location of flare systems. But there are no theoretical or empirical methods for predicting the radiation of enclosed ground flares in current literatures. Predicting the thermal radiation becomes one of the greatest challenges in designing a groud flare system. To investigate the radiation of the enclosed ground flare, the flaring process of a series of flare gases were simulated using CFD method.The temperature and radiation field was calculated.Results shows that the radiation flux at ground level were below 1.58kW/m2; at the height above 9m, the 1.58kW/m2 radiation isopleth can reach 60-70 meters radially from the enclosed ground flare axis,so the design of towers or other elevated structures containing operating platforms should consider radiat.

442313 Simple Objects As Indicator of the Severity of Vapour Cloud Explosions

Abstract:

Simple objects such as instrument boxes and oil drums are often found on petro-chemical plant sites. In the event of an accidental vapour cloud explosion (VCE), they serve as reliable indicators of the explosion overpressure at their positions provided there is a good understanding of their behaviour in a VCE.

Such objects were used to great effect in estimating the overpressure magnitude in the Buncefield incident in 2005. In large scale tests conducted since the Buncefield incident (both as part of a joint industry project and subsequently), such objects were located at varying distances from the cloud, as well as some within the cloud. Overpressure measurements were made at or adjacent to the location of these objects and many have been analysed using advanced finite element analysis. The objectives were to (i) gain a better insight into the response of these objects when subjected to different types and magnitudes of pressure loading, (ii) validate finite element models; (iii) assess the suitability of different finite element modelling techniques for predicting the response of these structures and (iv) develop pressure impulse (PI) diagrams which can be used to provide a rapid assessment of explosion magnitude through an examination of the damage level.

This paper presents an overview of the tests performed and the analysis results with comparisons with the actual behaviour of the objects.

Best Practices for Writing Operating Procedures and Trouble-Shooting Guides

Lauren Madden, Process Safety Manager
REC Silicon
Moses Lake, WA 98837
e-mail: Lauren.Madden@recsilicon.com

William G. Bridges, President
PROCESS IMPROVEMENT INSTITUTE, INC. (PII)
1321 Waterside Lane, Knoxville, TN 37922
Phone: (865) 675-3458
Fax: (865) 622-6800
e-mail: wbridges@piii.com

Prepared for Presentation at
12th Global Congress on Process Safety
Houston, TX
April 10-13, 2016

UNPUBLISHED

12TH GLOBAL CONGRESS ON PROCESS SAFETY

Best Practices for Writing Operating Procedures and Trouble-Shooting Guides

Lauren Madden, PSM Manager
REC Silicon

RECSiLICON

William G. Bridges, President
PROCESS IMPROVEMENT INSTITUTE, INC. (PII)

Keywords: Procedures, Human Factors, Risk control, process safety management, PSM

Abstract

There is no complete, best practice guideline or textbook for writing operating procedures and trouble-shooting guides. This paper presents the proven, best approach for developing accurate operating procedures and for ensure the format of the page and steps are optimized to reduce human error rates. This approach and the 35 rules established in this paper are based on the foundations set by Swain and others in the 1970 for control of human error rates, but uses experiences from more than 60 sites where this approach has been successfully followed. The approach and rules for developing procedures for responding to process deviations (such as those needed for Human IPL) are again the best approaches found and have been applied successfully since the early 1990s; these are many times called operational trouble-shooting guides. The paper also discusses how and when to integrate trouble-shooting guides into distributed control systems. Several case studies are provided that show the gains from following this approach. The guidelines in this paper build upon ones presented in 1999 at CCPS.

Background

Operating procedures have always been crucial to the safety, quality, and productivity of process systems. And with the advent of new safety and quality standards such as OSHA's process safety management (PSM) regulation, OSHA's personal protective equipment (PPE) regulation, EPA's risk management program (RMP) regulations, and the ISO 9000 quality standard, many companies are facing the daunting task of developing or upgrading their procedures to satisfy varied and sometimes complex or conflicting requirements.

Although there are overlapping characteristics that the various regulations and standards share, each approaches the procedure-writing process from a somewhat different perspective (e.g., quality or safety, protection of workers or protection of the public/environment), and the required level of detail for documenting the procedures differs greatly between these regulations and standards. For instance, the ISO 9000 quality standard advocates documenting procedures and work instructions that impact quality and suggests keeping the procedures as simple as possible, while the OSHA PSM regulation requires detailed procedures that address all operating modes (startup, shutdown, etc.) and that address operating limits, consequences of deviations, means to avoid hazards, safety and health considerations, and safety systems and their functions. How can companies develop and maintain procedures that ensure productivity and simultaneously satisfy the different regulations and standards? The key is to remember that the ultimate goal of the regulations and standards is consistent: reduce human errors that can impact quality, productivity, and/or safety.

Procedure-related errors are errors that occur because some characteristic of the procedure caused task performance to fail. ***This is currently the most critical human factor at most sites since 90% of accidents have at least one root cause related to mistakes within procedures.*** Reducing these procedure deficiencies can reduce human error rates by a factor of 2 to 10, normally. [1, 2]

The concepts and rules presented in this paper are applicable to all industrial settings. Most importantly, this approach **is effective in satisfying both safety and quality concerns.** With this approach, procedures can and have been developed that provide a sound basis for training and that provide the users and supervisors with an understanding of what can go wrong and how it can go wrong. This approach also provides management with an understanding of the inherent risk (to quality, productivity, safety, etc.) of an activity (i.e., the activity addressed by the procedure) and an appreciation of the safeguards that are being relied upon to mitigate this risk.

An extensive case study is presented by REC Silicon, which illustrates this approach, along with supporting the benefits of the accurate and clear procedures. But the same is true at more than 60 sites where the same approach has been followed.

Fundamentals of Writing Effective Procedures

When developing procedures, or any written material, always remember that the burden of written communication is on the writer, not on the reader. It should never be up to the reader to decipher the meaning of a message. Engineers who write procedure manuals may "tend to include information that is more relevant to them than it is to the operators actually running the unit" [Sutton, 1992].[3] When writing procedures, your guiding principle should always be: if the written procedures are difficult to read, difficult to locate (e.g., if titles are do not have a clear meaning), or inconvenient to use, they will seldom be used [Swain, 1983].[1]

Deficient Procedures are the most prevalent problem in process industries since procedures have not traditionally been developed from the perspective of optimizing human factors; instead, procedures have been traditionally developed to meet a compliance requirement to have written procedures. Examples of procedure deficiencies (inaccuracies) include:

- Incorrect/incomplete/nonexistent (most procedures we have audited have been only 70-75% accurate – the inaccuracies include missing critical steps, steps as written are not what needs to be done, or the steps are out of sequence)
- No/misplaced warnings (for a example, a warning should Never **contain** the action to take; it should instead **emphasize** the action to take)
- Poor format and presentation rules

Figure courtesy of Process Improvement Institute, Inc. (PII), copyright 2003-2016[4]

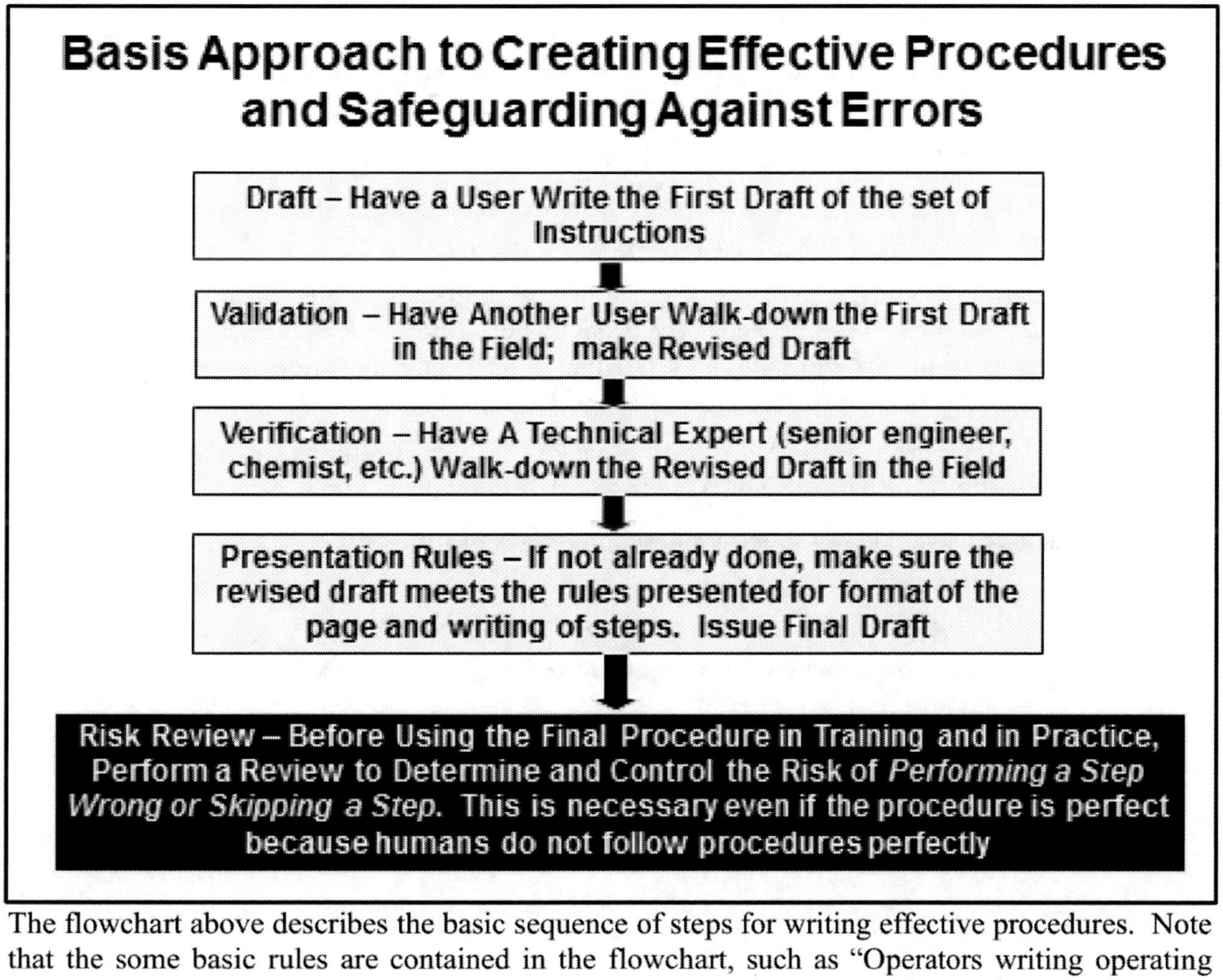

The flowchart above describes the basic sequence of steps for writing effective procedures. Note that the some basic rules are contained in the flowchart, such as "Operators writing operating procedures" and "Maintenance craft-persons write maintenance procedures" and "Lab technicians write lab procedures." Procedures are Not written by engineers or department superintendents. This is a critical rule because it ensures the procedures are written in the common language of the users and is written at the right level of detail.

Universal Purposes of a Procedure[4]

1. A written guide to accomplish initial training of employees
2. A quick reference manual for experienced (trained) employees

The Right Information to Write – Content is KEY[4]

The Content of the procedure controls the human error related to "procedure-based" error – if and content is wrong, the trainee is very likely to learn the step wrong (if the trainer is using the procedure as a basis or when the trainee refers the to the procedure step that is wrong). So:

If the content of the step is wrong it does not matter if the presentation is clear; but if the content is perfect, you can reduce error rates a further 50 to 70% by presenting the information properly.

Ask experienced mechanics or operators to describe a simple task that they perform. You will quickly see that even simple tasks can involve many steps.

You should consider the following questions when writing each procedure step:

- How is the step performed? Does the action involve interacting with a computer terminal, an automatic controller, or simple devices (gauges, valves, micrometers)?
- Can the actions be performed as written and in the sequence written? Can the equipment be operated as specified? Can the steps be physically performed? Does the user have the training or experience to understand and carry out the action using the information available, or is additional information needed?
- Does the user need to be alerted of any potential hazards (Cautions or Warnings) or need any supporting information (Notes) before performing the action?
- Does the user need to know specific operating ranges or limits to:
    - Perform this action
    - Recognize the successful completion of the action
    - Recognize an actual or potential problem
    - Make a decision

For example, does the user have to verify temperature or pressure visually? Is the user waiting for a computer command or verification? How should the process react? Will exceeding a limit result in an accidental chemical release, personal injury, or quality concern? What is the chance of not operating within safe operating limits, and what would be the consequences? If operation of the component is critical and a deviation will probably cause a failure, state the corrective actions. If failure is unlikely, keep going.

- Is needed information found on an instrument, panel, or monitor or is it in the procedure or another source such as a graph, table, drawing, or specification sheet? Should this information be included in the procedure or be referenced?
- What is the next logical step? How is the next step affected by what is performed in the current step?
- What are the results of improper task performance? If an action is critical, spell it out in detail. For example, when starting a special pump that may bind or cavitate if not vented and primed properly, the procedure may require more information than simply stating "start the pump."
- Is the action frequently performed? Is it an action that is easily overlooked? Is this a complex piece of critical equipment that has a low frequency of use? These questions can help determine the level of detail in your procedure.
- Is the action performed so infrequently or is it so complicated that the user is not sure how to do it? Is the action so complicated that nobody is ever certain it's done right the first time?
- Is the decision point clearly defined if a decision is required? Unclear decision points can cause arguments and delays in performing actions.

To address the concerns listed above, a few simple rules are needed for developing content:

- The procedures should be written at the level of someone who has just completed the basic training for that task. Do not write the procedure for someone just hired or for the 10+ veteran.
- Clearly identify ahead of time what activities need procedures to help ensure error rates or controlled low enough, and identify which activities are considered common "skills" of all of the staff.

  Example: Starting a pump is typically considered a "skill." This means a procedure step can simply be "Start the benzene recycle pump (P-119)." The procedure does not need to explain how to start the pump. But, starting a pump is a skill that nonetheless must be learned; it requires doing several sub-steps related to positioning of intake and discharge valves, how the discharge pump is throttled at startup, checking local pressure gage, etc. So, a training module is needed for starting pumps of various types, and the operator needs to learn this necessary skill. Once the skill is learned, it will be applied to so many various cases and so often, that a procedure is no longer needed as a "refresher" on how to do "start the pump."

- The first draft of the procedure must be walked-down in the field by another user. (Simply reading through the procedure does not catch the enough of the mistakes.)
- The revised draft should be reviewed in the field by a technical staff person, such as a process engineer (for operating procedures).

Next:

- The procedure needs to follow the page format and step writing rules described in the next section. This ensures the accurate steps are clearly presented.

The target is to reach an accuracy of 95% or better (so, no more than one wrong or missing step out of 20 steps). The following have been observed in the field by PII staff by direct data collection at more than 110 sites/plants around the world, based on walk-down review of several to dozens of procedures at each unit/plant:

- At 95% accuracy or better and when the same procedures follow 80% of the rules for procedure clarity (presented next), then most users will follow the written procedures and will try to keep the procedures up-to-date.
- At about 85% accuracy or less, about half of the users stop using the procedures
- At about 70% accuracy or less, less than 10% of the users will refer to the procedure or will try to keep it up-to-date. ***So, the written procedures are useless.***
- Unfortunately, the typical Operating procedure walked down by PII staff (accompanied by senior operators, a process engineer, and a shift supervisor) is about 75% accurate (so one step in four is missing or wrong). Usually these inaccurate procedures also follow less than half of the best practices for procedure format, presented next.

PHA/HAZOP of procedures

Just as critical as getting the content accurate and presenting the steps clearly, is performing a PHA/HAZOP of the written steps to determine if you have enough independent protection layers (IPLs) against errors that the "imperfect" operators will eventually make when using the "near perfect" written procedures. This analysis is described in detail in Chapter 9 of the *Guidelines for Hazard Evaluation Procedures*, 3rd edition, 2008[5]. **For an update on the methods for such analysis and for a discussion of the extra emphasis on PHA of procedures by US OSHA, refer to the more recent paper (Bridges, Marshall; 2015).**[6 and 7, 8]

Procedure Clarity

The importance of the accuracy of the procedure steps were discussed above. The clarity of the procedure steps (how they are written and how the page is formatted) is also important. Following best practices for step and page format reduce human errors by a multiplying factor of 3 to 5.[1, 9]

The best practice rules for writing and validating procedures have been published for many years *(see Bridges & Williams, 1997).*[11] These rules include best practices for formatting of the pages and steps; these have been gradually improved over the past decades and now are incorporated into PII's training materials.[4]

The Table on the next is a checklist (in auditing format) based on the current set of best practice rules for developing accurate and clear operating, maintenance, and other work-instructions (procedures).

PROCEDURE QUALITY CHECKLIST *(courtesy PII, copyright 2003-2016)*[4]

#	Issue	Response
	Procedure Content Checklist	
1	Is the procedure drafted by a future user of the written procedure? (Engineers should not author procedures to be used by operators or maintenance staff.)	
2	Is the procedure validated by a walk-down in the field by another future user of the procedures?	
3	Is the procedure reviewed and commented on by technical staff (engineers or vendors)?	
4	Is the procedure checked versus the Page and Step format rules below?	
5	Is a hazard review of step-by-step procedures performed to make sure there are sufficient safeguards (IPLs) against the errors that will occur eventually (when a step is skipped or performed wrong)?	
6	Is the content measured using "newly trained operators" to judge the % of steps that are missing, steps that are confusing or wrong, and steps that are out-of-sequence? (A score of 95% accuracy of content is good; 98% should be the targeted average.)	
	Page Format Checklist	
1	Is the title of the procedure the largest item on the page?	
2	Is the procedure title clear and consistent with other titles, and does it uniquely describes the topic?	
3	Are the document control features the smallest items on the page?	
4	Are temporary procedures clearly identified?	
5	Is white space used effectively? • Is there one blank line between each step? • Does indentation help the user keep their place? • Are the margins large enough to reduce page congestion?	
6	Is type size is 12 pt font or larger?	
7	Is mixed case used for words of steps, with ALL CAPS used only for special cases (such as IF, THEN, AUTO, and WARNING)?	
8	Is the step number very simple (single level of number used)? Only an integer?	
9	Have sections or information not necessary to performing the steps been moved to the back or to another part of the manual or training guide?	
10	Are section titles bold or larger than the text font? Do sections have clear endings?	
11	Is the decision on electronic presentation versus hard copy correct? Are electronic linkages to procedures clear and accurate and easy to use? If paper is chosen, is the color of the paper appropriate?	
12	Is the overall page format (such as Outline format or Two-Column [T-Bar] format) appropriate to the use of the procedure?	
13	Are play script features added for tasks that must be coordinated between two or more users? • Play script is normally used when there are two or more hand-offs of responsibility for steps.	
14	Are rules followed for formatting of Warnings, Cautions, and Notes? (See annotated rules, such as Warnings are for worker safety and Warnings must clearly standout from rest of page.)	
	Step Writing Checklist	
1	Is each step written as a command?	
2	Is the proper level of detail used throughout? This is judged based on: • *Who* will use the procedures • Same level of detail used in similar procedure steps	
3	On average, is there only one implied action per instruction? Best practice is to average 1.2.	
4	Does the procedure indicate when sequence is important? • If sequence matters, each step should be numbered (with an integer or letter) • If sequence does not matter, bullet lists should be used	
5	Are only common words used? Apply "education" level test (5 grade reading level is best)	
6	Do all acronyms, abbreviations, and jargon aid understanding? • Develop a list of such terms for use in procedures *and* communication. • Use terms that users use (within reason)	

7	Is each step specific enough? No room left to guess/interpret: • The meaning of a word or phrase (Check vs. Make sure) • The intent of a step or series of steps • A desired quantity or value • To what equipment the step applies	
8	Is the procedure free of steps that require in-your-head calculations? • Values expressed as ranges rather than targets with error bands • Conversion tables, worksheets, or graphs provided where needed	
9	Are graphics to the user's advantage? • No explanatory paragraphs or lengthy instructions that could be replaced by a picture • No impressive graphics that provide no real advantage	
10	Are references to the user's advantage? • No lengthy explanations or instructions that could be replaced by branching to a reference • No references to a procedure that references still another • No gaps or overlaps between this procedure and a referenced document • If branching, must branch to a procedure, not to a specific step in a procedure	
11	Are rules followed for writing warnings, cautions, and conditional steps? • No more than 2 per page • No actions within a warning or caution (actions must always be numbered steps) • Warnings and Cautions contain descriptions of potential consequences	

Each of these rules is explained in detail in the course *Writing Effective Operating and Maintenance Procedures*, by PII.[4] (Visit www.piii.com for more details.) Each rule is backed by experimental or observed data that indicates there is a clear advantage in error reduction by using the rule. Below are a couple of examples from others on procedure writing rules.

> ***Proper Page Format:*** *In one study, Haney took a procedure presented in narrative format and converted it to column format. He then asked experienced technicians who had been completing the procedure using the narrative format to complete it as presented in the column format. He found that when the technicians used the column format, their errors were reduced by a factor of 3 compared with the narrative format [Haney, 1969].*[10]

> ***Including Supplemental Checklists:*** *Checklists can be effective aids to performance and often should be used to supplement procedures presented in any format. Checklists can be particularly effective in terms of preventing errors of omission. Some of the literature supports using checklists only for simple procedures that do not need to be performed under time or other constraints [Sutton, 1992].*[3] *However, there is also support for using checklists for emergency procedures in which a core set of steps is used to respond to emergency situations, particularly when the actions follow a simple sequence of steps for shutting down a process [Connelly, 1992].*[12] *Overall, experimental data have shown that the probability of errors of omission and commission can be reduced by a factor of 3 to 10 when checklists are used as reminders of key steps in a work instruction [CMA's A Managers Guide to Reducing Human Error, 1990].*[9]

For more data on the history and development of these rules, see Bridges & Williams (1997)[11] and especially refer to originating data and discussion for most of the rules (Swain, 1983).[1] **The procedure on the next page follows the rules**; this is a T-Bar format. Several other page formats are also valid.

Unloading Monomer from Tanker to Storage

STEPS	DETAILS
1. Wear standard PPE, plus rubber gloves and full-face organic respirator.	Standard PPE includes hard hat, safety glasses with side shields, and steel-toe shoes.
2. Weigh in the tanker.	Record weight on GROSS line of form.
3. Check bill of lading.	... to verify correct type of material is in tanker.
4. Sign in tanker driver.	Driver must sign in as a visitor and be escorted at all times.
5. Take Certificate of Analysis to QC.	***CAUTION: To avoid contamination, DO NOT unload until you receive approval from QC.***
6. Spot tanker.	A) Direct driver to location. B) Verify brake is set. C) Chock the wheels on at least one side between the two rear axles. D) Ground the tanker (attach grounding strap to an unpainted metal surface).
7. Have the storage area operator make sure the storage tank can hold contents of tanker.	***WARNING:*** *Failure to perform this step can result in an overflow and spill of hazardous Monomer.* Perform **Tank Gauging Procedure (SOP-01-804)** or check the storage tank load cell readout. Storage tank operating limits are stenciled on the control panel.
8. Place 5-psig nitrogen pad on tanker.	A) Connect from local 5-psig nitrogen drop to aft bottom nitrogen connection on tanker. B) Open nitrogen supply valve. C) Then open valve at tanker.
9. Connect Monomer unloading hoses.	A) Verify hoses and gaskets are in good condition. (Hoses are normally stored on hose rack at unloading spot.) B) Connect from tanker to air pump. (Use aft bottom connection on tanker. Remove cap.) C) Connect from air pump to storage tank. (Connect to storage tank line labeled "from TW.")

Trouble-Shooting Guides (Procedures for Responses to Critical Deviation Alarms) [13]

Trouble-shooting guides are a special form of operating procedure; they are written for the actions that we want the operators to take to recover from a process deviation, *before an emergency situation occurs.* They are called guides since rarely can we predict the process conditions at the time the action is required. Trouble-shooting guides (and training and drills of the action) are required for any action that is considered a Human IPL. The Action Limit is what we show as the Min or Max in a Trouble-Shooting Guide. The action limit triggers the demand to use the trouble-shooting guide.

- The response is typically still possible, but it is time dependent. The time available is called the process safety time (PST). The operator must complete the diagnosis, make the necessary change(s), and make sure they are out of harm's way by the end of the Maximum Allowable Response Time (MART). [13]
- We usually set an alarm or a pre-alarm to trigger this action.
- This is usually before the shutdown triggers (ESD occur automatically) or release points (PSV set points) are reached
- The Min and Max shown in the Trouble-shooting Guide table is not the absolute safety limit for a system, but is instead some value that leaves us some time to take action to prevent from reaching the absolute limits.
- There is still time to prevent or avoid the final consequence that could occur if we reach the ultimate limits of the process. Usually, we want the MART to be ½ or less of the PST, and we want MART > 10 minutes for trouble shooting in the field/plant and MART > 5 min for trouble-shooting only from the control room. [13]

If the unit has a good PHA/HAZOP, then it is best to extract information from the HAZOP (or What-If) analysis tables to start the development of each guide. The guide is then finished with input from the process experts.[4]

Examples of Creating a Trouble-Shooting Guide from a HAZOP Table

HAZOP Table Entry	**Trouble-Shooting Guide Entry**
Cause: *Bypass valve is open or passing*	Make sure the bypass is tightly closed
Safeguard: *Isolation valves for the vessel*	Isolate the vessel, if necessary
Safeguard: *Relief valve*	Make sure the relief valve block valves/relief path are open

The key categories of information needed in a trouble-shooting guide are:[4]

- IMMEDIATE ACTION (by system or by operator)
- DECIDE IF ALARM is REAL
- FINDING and FIXING the CAUSE
- FIX or BYPASS PROBLEM

The example below is of a trouble shooting guide for one critical alarm/action.

Trouble-Shooting Guide

Alarm or Indicator:	**PAL 4446 – Low Pressure Alarm for Suction of Organic Feed Pump 40-PM-18.445**

Action Limit:	5 kPa		
Consequence:	Possible pump seal failure, releasing or spraying organic waste into the berm.		
Process Area:	FB&D Incinerator; Liquid Organic Liquid Feed	**Oper. Mode:**	Normal
Drawing #s:	D-400-PI-013		

IMMEDIATE ACTION (by system or by operator)

- DCS should shut down the organic feed pump (40-PM-18.445).
- From the DCS display, MAKE SURE the organic feed pump is shutdown.
- HAVE the field operator check for leaks near the organic feed pump.
- IF there is a large leak/release, THEN use the ESD switch to shutdown the unit and then follow/complete the shutdown and isolation procedure, OPS-ESD-117.
- IF there is a minor leak or no leak, THEN:
  - COMPLETE the rest of the trouble-shooting,
  - and DECIDE how to contain the leak for now,

DECIDE IF ALARM is REAL

- From the DCS, CHECK the pressure and feed tank level trends. IF the trends indicate the alarm if valid, THEN continue with finding the cause or fixing or bypassing the problem.

FINDING and FIXING the CAUSE

- CHECK valves upstream of the organic feed pump to see if any are closed too far, including checking ESD valves.
- CHECK, by feel with hand, if the heat tracing is on; IF Not, then TURN ON or open heat trace valves
- MAKE SURE nitrogen to the pump seal is at the normal operating pressure.
- CHECK if the line if plugged or frozen (skill)

Trouble-Shooting Guide

Alarm or Indicator:	**PAL 4446 – Low Pressure Alarm for Suction of Organic Feed Pump 40-PM-18.445**

- CHECK if the line if plugged or frozen (skill)
- CHECK if the level is actually low, use the Organic Feed Tank.
- IF the cause is low level in the feed tank, THEN resolve the problem if necessary based on cause that is found (skill)

FIX or BYPASS PROBLEM

- IF necessary, SHUT DOWN the Unit to allow fixing of the problem.
- FOLLOW the proper procedure to resolve problems (repair procedure, line clearing procedure, etc.)
- IF the decision is made to continuing operation without all equipment in normal condition, THEN:
  - FOLLOW Temporary Operating mode, if there is a temporary procedure already written for this possible problem/condition
  - FOLLOW MOC procedures to obtain approval for any non-standard temporary operating procedure or mode

END

Optimal presentation of trouble-shooting information.

Ideally (or ultimately) such trouble-shooting information should be imbedded in the basic process control system (DCS) so that the operators can access the information on demand, with the click of mouse or key.

Comments on Current Industry Consensus Guides for Writing Procedures

None of the consensus textbooks available for procedure writing cover all of the major points of this paper. The best "guidelines" for writing procedures currently exist in the course notebooks used by PII (and likely in the course notebook used by ABS Consulting, as they had similar origins).

Notably, the AIChE/CCPS textbook, *Guidelines for Writing Effective Operating and Maintenance Procedures*, 1996[14]:

- has NO description at all on the development and use and format of Trouble-Shooting Guides. This should have been a major chapter in the textbook.
- mentions only about half of the rules listed in this paper (and the related PII course notebook) for formatting of pages and steps
- emphasizes technical staff or professional procedure writers writing procedures, rather than the Users themselves writing the procedures (it violates the first rule of procedure writing)
- Fails to cover the competency needed by procedure writers.
- recommends and emphasizes having subsections in each work instruction/procedure (between the Title of the task and Step 1) rather than relegating such information either to other sections of the operating manual or rather than turning this information into steps.
- fails to mention the need for a PHA of procedures that focuses on hazard review of deviations from the steps; this is required by best practices and by US OSHA for PHAs under PSM.

This guideline textbook is not worth the investment.

As a better example, but one that nonetheless still falls short of what is needed, the ASM Consortium *Guidelines for Effective Procedural Practices,* 2010[15]:

- has good discussion on the need and thought processes related to handling abnormal situations (trouble-shooting or response to critical deviations), but there is little description on the format and detailed content of the trouble-shooting guides; in fact, the guideline fails to differentiate between routine procedures for startup, shutdown, and online maintenance, and those used for response to critical alarms (trouble-shooting)
- mentions only about half of the rules listed in this paper (and the related PII course notebook) for formatting of pages and steps
- mentions that experience users themselves should write their own procedures, but it does not emphasize this rule and leaves the authorship open to engineers or a committee. And it specifically mentions use of retired senior operators for writing operating procedures. While this is one good way to write the procedures (while also capturing the knowledge from this experienced group is prior staff), we have found that the current users do a better job overall, once they have achieved competency in how to write procedures.
- mentions the competency needed for procedure writers in general.
- mentions the need for a PHA of procedures, but does not reference the method for this in Chapter 9 of *Guidelines for Hazard Evaluation Procedures*, CCPS/AIChE, 3rd edition, 2008,[5] or other related papers on the topic. Also, the description of the PHA is misleading because it focuses on how PHAs should be used in writing procedure content (which is fine for trouble-shooting guides) but fails to emphasize that a step-by-step analysis of deviations from the steps are needed to ensure that the process has enough IPLs against mistakes that can occur when imperfect humans use the procedure.

Though this guideline has deficiencies, it is worth the investment and provides useful insight not found elsewhere. Just be sure to address the deficiencies noted above.

Implementation of this approach

PII staff has helped more than 60 sites implement the approach above. The typical approach for successful implementation is:

1. Perform a gap analysis to determine the average accuracy of procedure content, average clarity (as judged versus the procedure format rules, and identify any examples of excellent procedure writing.
2. Convince management of the problem and get their buy-in on the solution.
3. Train (1.5 to 2 days for operators; 1 day for lab or maintenance staff) and coach the users of procedures (normally for 3 to 8 days, to have them achieve full competency) so they can be effective authors. Normally, 20 to 25% of the users (hands-on operators, hands-on maintenance and inspection staff, lab staff) need to be trained as authors. Not all of those trained will make good writers; the organization needs to judge who does the best jobs. About 15 to 20% of the staff should end up being authors.
4. Get organized, like a project, for tracking and scheduling of the effort. Use master spreadsheets such as those used at REC Silicon (shown later).
5. Use a Traveler form (described later) to make sure each new procedure gets a proper amount of review by other users and by technical staff (such as engineers)
6. Reward procedure writers for their efforts. Especially ensure there are no disincentives for taking the extra effort to write procedures.
7. Track the progress on the effort; fix problems as they arise; enforce schedules (to show management cares).
8. Perhaps use to specialized software make it easy to output the content of procedures into different user-friendly formats such as training tools, training competency records, and field checklists. WordPerfect (formerly from Corel) made this relatively easy with the use of Primary and Secondary merge files; MS Word makes this more difficult. As a result, some specialized software have been developed to help keep formats consistent and to help reformat the content, on demand.
9. Issue procedures through a document control system that does not add extra workload.
10. Of course, ensure there is JSA performed for procedures that have "personal injury" potential and perhaps a PHA of the steps in the procedures whenever there is "process safety" potential.
11. Train all affected workers in the new procedures.
12. Identify the best writers to become the trainers of new authors, over the long run.

Overview of REC Silicon

REC Silicon is one of the world's largest manufacturer and suppliers of

- Silane Gas and specialty silicon gases.
- Granular Polysilicon from Fluid Bed Reactor technology for solar photovoltaic applications
- Ultra-pure float zone Polysilicon for the electronic industry

REC Silicon's Product and Service Offerings

Solar Grade
Polysilicon with various applications for the solar industry

Silane Gas
Product used in production of semiconductors, flat panel displays & thin film technologies

Electronic Grade
Polysilicon delivered to the semiconductor industry as well as other electronic market segments, such as optics & MEMS

Float Zone
High-purity polysilicon delivered to the power electronics industry

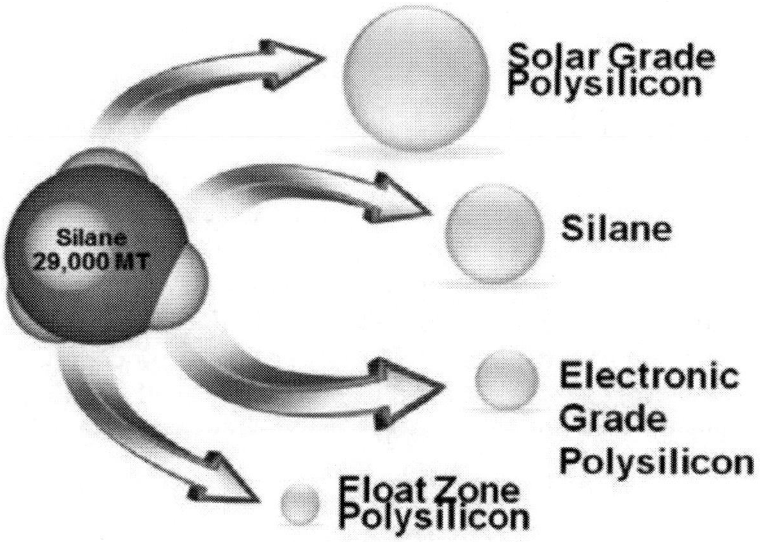

Moses Lake, Washington, USA 396 Employees	Silicon III, IV FBR	Silane gas: 22,000 MT Granular silicon: 16,300 MT
Butte, Montana, USA 255 Employees	Silicon II	Silane gas: 7,000 MT Siemens grade: 3,600 MT

Implementation at REC Silicon

REC production facilities in the USA have been actively implementing process safety for many years, with a renewed interest in 2013 in meeting industry best practices (which goes well beyond US OSHA PSM requirements). To that end, REC commissioned gap analyses versus best practices at both the Butte and Moses Lake production sites. One common finding was that the operating procedures and maintenance procedure for both sites needed improvement with respect to accuracy of the steps as well as clarity (page format and step format). The following were some of the issues with the procedures and related management systems, per the audit results:

- Procedure Inaccuracy (procedure content was less than 80% accurate)
- Technical Staff (such as process engineers) were viewed as owners, not shop floor
- Annual review process ineffective
- Procedure formats were poor; did not follow best practices for reducing human error
- Inconsistent use of Warning, Caution, and Note
- Procedure Naming – not consistent
- Tried to take on all modes of operation in one procedure title
- Procedure accessibility was difficult in the legacy database; multiple searches needed to find tool for job
- Procedure Safety Reviews – no formal methods were applied for procedure safety reviews; i.e., there were no PHA of procedures with process hazards or JSA of tasks with personal hazards

<u>**Procedure Author Training** – ***the first step in any procedure writing effort***</u>

In early 2014, REC Silicon adopted Process Improvement Institute's (PII's) procedure author training module (*Writing Effective Operating and Maintenance Procedures*[4]) with minor customizations. The training consists of 2 days of classroom lecture and about 6 exercises, followed by practical procedure development and field validation steps in the coaching phase. The classroom course covered all aspects of procedure writing with the focus on optimizing the aspects of the procedures that will lower the likelihood of human error when using the procedures. Topics included:

- human factors overview
- getting the content right
- page format rules and selection
- step format rules
- addressing operating limits (including creating trouble-shooting guides
- nuances of emergency shutdown and emergency operating procedures

- implementation guides and high level procedures for the task of getting all of the work instructions written
- examples/exercises on each topic

The coaching consisted of the instructors from PII, who are expert in organizing procedure writing efforts, helping the procedure writers organize and write operating and maintenance procedures, and trouble-shooting guides. Coaching was for 2+ days. About 20 staff from the Moses Lake and Butte sites attended the initial author training and coaching session.

REC Silicon has revised its management systems to require this competency for all personnel developing or changing procedures and to reach full operator certification. Operators must demonstrate an active role in the procedure MOC process to qualify for recertification every 3 years, otherwise they must repeat the classroom and field training. The unit trainers have taken over leading the procedure author training.

The benefits we have seen from these changes include transfer of ownership of procedures from the technical day staff to the operator since these are the people performing the work in the field. This is necessary to have accurate procedures. **The emphasis on intentional competency development for what is required to write effective operating procedures has greatly improved the quality of our procedures and our operator training and certification program.** The quality of the annual reviews and audits has improved tremendously with the rewritten procedure format and content changes.

Safe Operating Limits and Consequence of Deviation

REC Silicon has developed an engineering standard that governs the content and format for this type of special safety troubleshooting procedure. *CCPS Guidelines for Engineering Design for Process Safety*[16] is the basis for determining the safe operating limits. Production engineers are the owners of these procedures. Engineering competency is required to develop the safe operating limits table based on the available process safety information. A consequence of deviation is then developed by the Process Hazard Analysis team. Finally involvement of certified operators is required to develop steps to correct or avoid exceeding the safe limit. This combined effort produces the safety troubleshooting guide, which complies with the approach listed earlier in this paper.

REC silicon has risk ranked its process areas and is focusing on developing safety troubleshooting guides for high risk process first. To date, for the Moses Lake Facility, thirty six (36) guides have been developed. These have been developed in conjunction with PHA updates to the safe operating limits. We have not yet implemented the formal training and competency testing of these safety procedures. This is planned for the next phase of the initiative.

How REC Silicon approached the re-write effort:

To begin this initiative, each operating department listed all of the existing operating procedures into the procedure development tracking form. Each department also developed a list of "skills" that are used across multiple procedures and that operators or maintenance staff are required to attain before using the written procedures that were developed for more complex tasks (such as starting a pump, or hookup of a chemical transfer hose).

The workers (users) then screened each procedure for safety, equipment damage, or quality risk. This assisted in determining whether 2, 3 or 4 peer walk-downs were required on the validation traveler workflow. Since most of the REC Silicon procedures covered more than one mode of operation, the term "parent" was developed for the original procedure and each procedure was rewritten and broken out into a separate "child" procedures. The procedure re-write assignments were divided up between all four rotating shifts and the superintendent assigned certified authors (e.g., operators trained/coached as procedure writers) from their pool. Each department is managing their pool of qualified authors. For example, our fluidized bed reactor (FBR) unit has 11 qualified authors. Silane, which is a much larger department with 3 production units, has 56 qualified authors. Cross shift validation was utilized to ensure all operating approaches were considered. The assigned author was responsible for consolidating red line comments.

The traveler process introduced by PII was customized to consider information that is marked for removal from the parent procedure. **A Traveler form shown to the right accompanies each new or revised procedure through the writing and editing/review process.** When complete, the traveler form is then kept in the QA file for the procedure to validate long-term how the initial procedure was developed.

Procedure Development Traveler - Moses Lake			
New Document Number:		Area:	
Procedure Title:			
Procedure Type:			
Author:			
Initial Draft Completion Date:			
Author Walkdown Date:			
Author Revision 1 Completion Date:			
Peer 1:		Shift:	
Peer 1 Walkdown Date:			
Optional Peer 2:		Shift:	
Peer 2 Walkdown Date:			
Optional Peer 3:		Shift:	
Peer 3 Walkdown Date:			
Optional Peer 4:		Shift:	
Peer 4 Walkdown Date:			
Author Revision 2 Date:			
Engineer/Verifier:			
Verification Completion Date:			
Author Revision 3 Completion Date:			
QA Person:			
Author Revision 4 Completion Date:			

Images and layouts copyrighted by PII, 2008-2016[4]

A color coding process has been developed to indicate if the content can be permanently deleted, or must be moved into the unit training manual. The unit trainers have been responsible for tracking and reporting weekly progress on the initiative. **See an example of the overall tracking form below.** A total of 184 "parent document" procedures for Moses Lake were tracked through the development, review, and revision process (along with about 200 child documents).

Images and layouts copyrighted by PII, 2008-2016[4]

Procedure Development Tracking Form - Moses Lake

							Initial Development			Validation				
Item	New Doc #	Old Doc #	Area	Procedure Title	Proc Type	Author(s)	Initial Draft	Author Walkdown	Author Rev 1	Peer 1 Walkdown	Opt. Peer 2 Walkdown	Opt. Peer 3 Walkdown	Opt. Peer 4 Walkdown	Author Rev 2
1	FBP.3.xxx	FBP.3.160	FBR	Hydrogen Booster Compressor Cleanup	Cleanup	LT/FG	3/19/14	3/19/14	3/20/14	3/21/14	3/21/14			
2	FBP.3.xxx	FBP.3.160	FBR	Hydrogen Booster Compressor Operation and Op Restrictions	Operation	LT/FG	3/20/14	3/20/14	3/20/14	3/21/14	3/21/14			
3	FBP.3.xxx	FBP.3.160	FBR	Hydrogen Booster Compressor LOTO	LOTO	BCA/OL	3/19/14	3/19/14	3/21/14					
4	FBP.3.xxx	FBP.3.160	FBR	Hydrogen Booster Compressor Standby	Standby	BCA/OL	3/20/14	3/20/14	3/21/14					
5			4.0	STC/TCS from Receiving Tank to STC/TCS Storage Transfer Through Metering Line	Transfer	ME/RR	3/19/14	3/19/14	3/20/14	3/20/14	3/21/14	3/21/14	3/21/14	
6			4.0	STC/TCS from Receiving Tank to STC/TCS Storage Transfer Through Bulk Line	Transfer	ME/RR	3/19/14	3/19/14	3/20/14	3/20/14	3/21/14	3/21/14	3/21/14	
7			4.0	H2 Recycle Compressor Switching to Standby	Standby	ME/RR	3/20/14	3/20/14	3/20/14	3/20/14	3/21/14	3/21/14	3/21/14	
8			4.0	H2 Recycle Compressor Cleanup	Cleanup	ME/RR	3/20/14	3/20/14	3/20/14	3/20/14	3/21/14	3/21/14	3/21/14	

From PII's procedure author training,[4] the Quality and Content checklist was modified and used as quick reference guide to make sure that all the content development and formatting rules are being followed during the re-write process. We used Microsoft Word and made templates for the Tbar, Flat and Consequence of Deviations.

Procedure Management of Change

Through this project REC Silicon has adopted formal safety review methods for procedures. These methods standardize the approach to determine if a procedure is safe to use as written or if revisions are required. These details are outlined in a completely new REC Silicon policy - Procedure Management of Change PSM.30.004.01. The REC Silicon IT department developed a custom tool to execute these requirements in an electronic workflow. The benefits from these changes are formally transferring competency to lead and document job safety analysis to the shop floor supervisors. Prior to this effort, JSA was traditionally a review lead by our safety specialists.

Overall Summary of the procedure writing initiative:

For the Moses Lake facility, 398 procedures (of which 184 were unique and 214 were similar to the 184 unique ones) for normal and routine operations, various maintenance tasks, and trouble-shooting guides have been developed through this project (new and rewrites combined.) Similarly, for the Butte facility, 116 procedures (of which 24 were unique and 92 were similar to the 24 unique ones) for normal and routine operations, various maintenance tasks, and trouble-shooting guides have been developed through this project (new and rewrites combined.) In addition, about 56 existing procedures (that were in "okay" state) were updated and the format improved.

We do not have any data yet for reduction in error rates or increase in production, quality or safety; however we have seen great improvements with the new formats with the ability to conduct job observations in the field. Our job observations assess accuracy of the work being performed, accuracy of the written procedures, and an opportunity for on the job operator training.

This type of project can seem overwhelming for a manufacturing site. It is essential to form a project team with key decision makers from all the correct disciplines and have a project sponsor that can remove barriers. The way REC Silicon was able to accomplish this type of effort was dedicating time to meet each week, setting short term goals, and tracking KPIs to meet those goals. The procedure writing initiative at REC Silicon is still very much "a work in progress". Leaders of the project have to make time to support the staff to keep the momentum going on this initiative and also on the related culture change, until it becomes "the way the work always gets done."

Quotes from workers involved in initiative:

- Operators are receiving better work instructions
- Sequence of steps is well understood
- Much easier to audit operator compliance to procedure
- Standardizes how we do work – protects our equipment from damage, protects our people, protects the environment
- No longer need 20 year operator tribal knowledge and interpretation to complete procedure
- Procedures are not cluttered anymore, they just have the necessary information
- Procedure element now requires involvement of many shop floor workers
- Procedures are really safe to use
- JSA technique is alive on the shop floor
- Procedure can be trusted
- Technical staff no longer doing all procedure work. Freed up to optimize processes and realize other cost saving benefits.

Conclusions

Having clear and accurate procedures is a must. Poor procedures show up as one of the root causes in 90% of the accidents. The approach described in this paper has worked for more than 35 years for one of the authors, and continues to work today. Controlling risk is mostly about controlling human error rates, and having accurate and clear procedures are one key to this control. One other gain from the approach described in this paper that the users become the owners, and therefore they have a vested interested to keep the procedures up to date and clear.

No doubt, writing all of the procedures for a facility is a large task. But, it can be accomplish by site staff, with support from process safety or quality control leadership, after enough users become competent authors; attaining this competency takes less than 2 weeks of work, on average.

Acronyms Used

AIChE– American Institute of Chemical Engineers
ASM – Abnormal Situation Management consortium
CCPS – Center for Chemical Process Safety (a division of AIChE)
CMA – Chemical Manufacturer 's Association, now American Chemical Council (ACC)
FBR – Fluidized Bed Reactor
HAZOP – Hazard and Operability Analysis
IPL - Independent Protection Layer
JSA – Job Safety Analysis
LOPA – Layer of Protection Analysis
MART – Maximum Allowable Response Time
MOC – Management of Change
NUREG – Nuclear Regulation, US Nuclear Regulatory Commission
OSHA – Occupational Safety and Health Administration, US Department of Labor
PHA – Process Hazard Analysis
PII – Process Improvement Institute, Inc.
P&ID – Piping & Instrumentation Diagram
PSI – Process Safety Information
PSM – Process Safety Management
PST – Process Safety Time
SOP – Standard Operating Procedure

References

1. Swain, A. D., Guttmann, H. E., *The Human Reliability Handbook - with Emphasis on Nuclear Power Plant Applications*; NUREG/CR-1278, U.S. Nuclear Regulatory Commission, Division of Facility Operations, 1983.
2. Gertman, D., Blackman, H., Marble, J., Byers, J. and Smith, C., *The SPAR-H Human Reliability Analysis Method,* NUREG/CR-6883, U.S. Nuclear Regulatory Commission, Office of Nuclear Regulatory Research, 2005.
3. Sutton, Ian S., "Writing Operating Procedures," *Proceedings of the International Conference on Hazard Identification and Risk Analysis, Human Factors and Human Reliability in Process Safety*, AIChE, Center for Chemical Process Safety, New York, NY, 1992.
4. *Writing Effective Operating and Maintenance Procedures*, Course Notebook, Process Improvement Institute, Inc., 2003-2016.
5. *Guidelines for Hazard Evaluation Procedures*, 3rd Edition, 2008, CCPS/AIChE.
6. "Necessity of Performing Hazard Evaluations (PHAs) of Non-normal Modes of Operation (Startup, Shutdown, & Online Maintenance)", W. Bridges and Mike Marshall (US OSHA), *18th Annual International Symposium, Mary Kay-O-Connor Process Safety Center, College Station, TX, October 2015.*

7. "How to Efficiently Perform the Hazard Evaluation (PHA) Required for Non-Routine Modes of Operation (Startup, Shutdown, Online Maintenance)," W. Bridges and T. Clark, *7th Global Congress on Process Safety*, Chicago, AIChE, March 2011.

8. Bridges, W.G., et. al., "Addressing Human Error During Process Hazard Analyses," *Chemical Engineering Progress*, May 1994

9. *A Manager's Guide to Reducing Human Error*, 1990, Chemical Manufacturers Association (CMA)

10. Haney, R. W., "The Effects of Instructional Format on Functional Testing Performance," *Human Factors*, 11, 1969, pp. 181-187.

11. "Effective Procedure Writing – A Two-Part Series (Part 1: Reducing Risk of Worker Errors; Part 2: Addressing Process Safety Management Requirements for Operating Procedures)," *Chemical Engineering Progress*, 1997, W. Bridges with T. R. Williams (both of JBFA).

12. Connelly, C. S., "Emergency Operating Procedure Writing for Oil Refinery Process Units," *Proceedings of the International Conference on Hazard Identification and Risk Analysis, Human Factors and Human Reliability in Process Safety*, AIChE, Center for Chemical Process Safety, New York, NY, 1992.

13. *Guidelines for Initiating Events and Independent Protection Layers*, 2015, CCPS/AIChE.

14. *Guidelines for Writing Effective Operating and Maintenance Procedures*, 1996, CCPS/AIChE.

15. *ASM Consortium Guidelines – Effective Procedural Practices*, 2010, ASM Joint R&D Consortium.

16. *Guidelines for Engineering Design and Process Safety*, 2nd Edition, 2012, CCPS/AIChE.

12TH GLOBAL CONGRESS ON PROCESS SAFETY

How Process Safety Information Affects the Quality of a Process Hazard Analysis

Michael Mayo
Smith & Burgess LLC
7600 W. Tidwell Rd, Ste 600
Houston TX 77040
mike.mayo@smithburgess.com

Dick Baum
dick.baum@smithburgess.com

Prepared for Presentation at
American Institute of Chemical Engineers
2016 Spring Meeting
12th Global Congress on Process Safety
Houston, Texas
April 11-23, 2016

How Process Safety Information Affects the Quality of a Process Hazard Analysis

Michael Mayo
Smith & Burgess LLC
7600 W. Tidwell Rd, Ste 600
Houston TX 77040
mike.mayo@smithburgess.com

Dick Baum
dick.baum@smithburgess.com

Keywords: PHA, PSI, MI

Abstract

This paper is a tutorial that shows the importance of how Process Safety Information affects the quality of a Process Hazard Analysis. The Process Hazard Analysis (PHA) is the one requirement of the PSM standard that has the facility stop and review a process in its entirety to ensure safe operation. However, a high quality PHA is dependent on accurate process safety information (PSI) that is consistent with the current equipment state (Inspection History, Changes from original design, operating conditions etc.). A PSI sub-element, relief system design basis documentation, requires significant amounts of other PSI and if organized with forethought, will greatly improve the quality, effectiveness and, potentially, speed of a PHA. If the relief systems documentation were performed in a system that stored all of the other relevant PSI and was cross referenced with current inspection records (MI systems), then this subset of PSI would contain much of the critical information needed for a PHA. The system would have to be designed such that it was accessible to the PHA team and organized to match the PHA workflow. This paper will present a proposed organization of PSI and then show how with a trained PHA facilitator and organized and accessible PSI/MI data, specific items found in PSM audits of PHAs would have been avoided.

1 Introduction

From defining the Causes, developing the Consequence scenarios and ending with Safeguards, accurate PSI information is required to provide the most credible assessment of risk. The best practice of minimizing the assumptions used during a PHA study has two significant benefits;

- Management can be assured (feel good) that the PHA process utilized all relevant and up-to-date PSI information thereby reducing potential inaccurate assumptions and,
- It will reduce time spent for discussions between participants to come to common conclusions and speed up the PHA process while maintaining the integrity of the study.

As learned from many PSM audits, PSI information is typically spread throughout many locations and maintained by numerous groups;

PSV files - Engineering

Mechanical Equipment Data/Inspection reports - Engineering, Maintenance, Inspection Departments

Critical Instrument Logs - Engineering or Controls

In a perfect world, maintaining a common database to store all up-to-date PSI information would be ideal, but the delegation of maintaining and access would be cumbersome. There are many people involved with PSI information and each has their own responsibilities and actions for validating the data. One way to reduce the roadmap for locating PSI data is to maintain relief systems documentation in such a way that it provides most all relevant data in a simplified format and links to other sources of PSI information.

This poster will walk through the PHA process and how readily available PSI information can be located with robust relief systems documentation.

2 Causes

Typically, the detailed cause of a PHA would not reference back to any PSI data. The failure or misalignment of valves are always credible occurrences that are documented. Where PSI information becomes relevant is in the cause category. The example shown in figure 1 indicates the following;

- Inlet PCV-001 on methane supply to V-001 reactor fails open

Cross-referencing to the P&ID indicated a figure 2, it is noted there is a PSV-002 upstream of PCV-001, a vessel, V-001 with its own relief device, PSV-001. A quick check during the PHA study of the relief systems documentation for PSV-002 will show that the set pressure of 400 psig is indeed above the MAWP of V-001, 285psig.

This validates the cause category of high pressure. Had the set pressure of PSV-002 been documented at 250 psig, this cause probably belongs in another category, i.e. more flow.

Processes: 1. V-001 Reactor

Nodes: 1. Feed streams

Deviation	Causes	Consequence	Risk			Safeguard
			S	L	RR	
1. High Pressure	1. Inlet PCV-001 on methane supply to V-001 reactor fails open	1. Potential for high pressure in V-001 reactor, potential **flange leak**, loss of containment, explosion, jet fire, employee injury, damage to equipment and loss of production				1. PSV-001 on V-001 reactor
		2. Potential for high pressure in V-001 reactor, potential **rupture**, loss of containment, explosion, jet fire, employee injury, damage to equipment and loss of production				

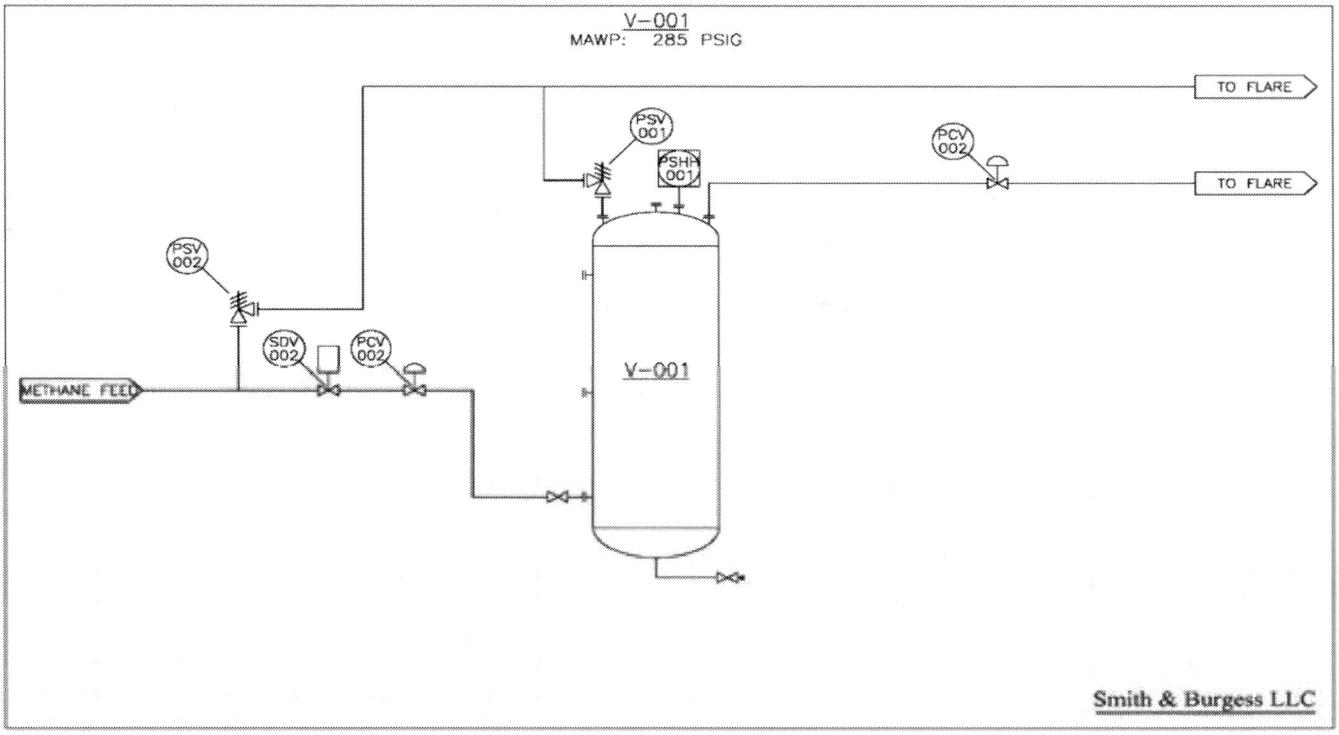

3 Consequences

The consequences developed during the PHA can benefit greatly from accurate and readily accessible PSI information. In the high pressure deviation we can use the PSI information in establishing the overpressure ratio to assist in pinpointing the consequence severity levels. Using the PSI information for PSV-002, a 400 psig supply into vessel V-001 with a MAWP of 285 psig, results in an overpressure ratio of 1.404. Based on your client's guidelines for overpressure and probabilities, this could result in either a flange leak or vessel rupture.

The consequence severity of these two scenarios is quite different but with the PSI data well documented and easily obtained, accurate severities can be obtained during the PHA.

Processes: 1. V-001 Reactor

Nodes: 1. Feed streams

Deviation	Causes	Consequence	Risk			Safeguard
			S	L	RR	
1. High Pressure	1. Inlet PCV-001 on methane supply to V-001 reactor fails open	1 Potential for high pressure in V-001 reactor, potential **flange leak**, loss of containment, explosion, jet fire, employee injury, damage to equipment and loss of production				1. PSV-001 on V-001 reactor
		2 Potential for high pressure in V-001 reactor, potential **rupture**, loss of containment, explosion, jet fire, employee injury, damage to equipment and loss of production				

PSV-002 set pressure can help determine consequence severity

PSI data for V-001 can help determine the consequence severity

4 Safeguards

The safeguards listed during a PHA have the most critical requirement for validation and accuracy during a PHA. They are the protection layers that mitigate or prevent major accidents from occurring and thus leave no room for assumptions or quantitative assessments. PSI information in relief system documentation should be current and accurate for the current operating parameters. A well-documented and arranged system will allow for easy access during the PHA. Information from other PSI data such as P&ID's does not always reflect the purpose of the safeguard. For example, many P&ID's show the governing case of a pressure relief device, which may or may not satisfy the scenario release and could lead to inaccurate assumptions being made during the PHA. (Fire case vs overpressure)

Processes: 1. V-001 Reactor

Nodes: 1. Feed streams

Deviation	Causes	Consequence	Risk			Safeguard
			S	L	RR	
1. High Pressure	1. Inlet PCV-001 on methane supply to V-001 reactor fails open	1 Potential for high pressure in V-001 reactor, potential **flange leak**, loss of containment, explosion, jet fire, employee injury, damage to equipment and loss of production				1 PSV-001 on V-001 reactor set at XXX PSV-001 set pressure can determine the likelihood
		2 Potential for high pressure in V-001 reactor, potential **rupture**, loss of containment, explosion, jet fire, employee injury, damage to equipment and loss of production				

5 Conclusion

Well documented PSI information is critical to the PHA process and provides the validation needed to ensure that the process is being operated and maintained in the best possible safe state. The examples shown indicate how comprehensive pressure relief files can provide much of the needed PSI data or arranged to provide the roadmap for data locations through links to other files. The additional benefit for having accessible PSI information is a smooth transition during the PHA and reduced costs and time for the PHA process.

442025 Experimental Measurements of Flammability at Elevated Conditions Using Thermal Criterion

Abstract:

Accurate measurements of flammability limits for combustible gases and vapors are necessary for the prevention of fire and explosion. The flammability limit, which is the maximum or minimum fuel concentration at which a gas mixture is flammable in a given atmosphere, has been studied for decades. However, due to the measurement conditions and apparatus set difference, the flammability values provided by different experimental methods are inconsistent, especially at elevated conditions. In this study, the flammability limits of methane-air were measured in an 8-L closed cylindrical vessel with upward flame propagation at maximum initial temperature of 300°C and maximum initial pressure of 2 bar. During the experimental measurement, a new thermal criterion was used to define if the gas is flammable and the experimentally determined flammability limits were compared with existing data reported in the literature. Finally, the temperature and pressure effect on flammability was investigated and compared with the modified Burgess-Wheeler Law.

Use of Process Incident Data for Accident Sequence Precursor Analysis

Erin P. Collins
JENSEN HUGHES
333 E. 46th Street, #11L
New York, NY 10017
ecollins@jensenhughes.com

Bijan Najafi
JENSEN HUGHES
2105 S Bascom Ave, Suite 350
Campbell, CA 95008-3278
bnajafi@jensenhughes.com

Prepared for Presentation at
American Institute of Chemical Engineers
2016 Spring Meeting
12th Global Congress on Process Safety
Houston, Texas
April 11-23, 2016

Use of Process Incident Data for Accident Sequence Precursor Analysis

Erin P. Collins
JENSEN HUGHES
333 E. 46th Street, #11L
New York, NY 10017
ecollins@jensenhughes.com

Bijan Najafi
JENSEN HUGHES

Keywords: accident sequence precursor, data, risk analysis

Abstract

Investigations of recent process incidents have identified multiple process safety management deficiencies that led to the accident. Furthermore, audits of the Safety and Environmental Management System (SEMS) program under the U.S. Department of the Interior's Bureau of Safety and Environmental Enforcement (BSEE) have shown that the "management of risk" elements such as Hazards Analysis and Management of Change (MOC) are not consistently implemented as tools to manage risk. In addition, the requirement to perform hazards analysis is frequently documented and often referenced as a corrective action, rather than as a "system driver". There are several pieces to risk assessment and risk management that allow facility operators to evaluate and measure what is actually occurring in their facilities, including the review of historical event data to identify what went wrong and for near misses, what almost happened. This paper describes techniques that have been used in the commercial nuclear power and other industries to collect and apply historical experience data to improve safety and operability. An approach using accident sequence precursor analysis of incident data and risk analysis techniques is recommended to identify potential accidents before they occur as part of a comprehensive treatment of historical data.

1 Introduction

There are several pieces to risk assessment and risk management that allow facility operators to evaluate and measure what is actually occurring in their facilities, including the review of historical event data to identify what went wrong and for near misses, what almost happened.

For example, the U.S. Chemical Safety Board (CSB) investigation of the February 18, 2015, explosion at the ExxonMobil Refinery in Torrance, California, identified multiple process safety management deficiencies that led to the accident and a serious near miss. In the case of the near miss, "...large pieces of debris from the explosion were thrown into other units of the refinery…One of these pieces of debris hit scaffolding in the refinery's alkylation unit, narrowly missing a tank containing tens of thousands of pounds of modified hydrofluoric acid, or HF. The CSB determined that had the debris struck the tank, a rupture could have been possible, resulting in a potentially catastrophic release of extremely toxic modified HF into the neighboring community." [Ref. 1]

Such near misses are not unique to the chemical process industry, but have been documented in the offshore oil industry as well. As described by EnergyWire [Ref. 2], "Partial failures happen regularly on oil rigs, with a worker catching the errors and correcting them before a major blowout. These are not always reported to regulators, but they are logged carefully and stored as raw data on rigs and onshore in Houston."

The Discussion Paper on Precursor Analysis by the research organization Resources for the Future [Ref. 3] identifies the specific challenge faced by the oil and gas industries of "how to reduce the risk of low-frequency oil spill events, and especially high-consequence events like the Deepwater Horizon accident, when historical experience contains few oil spills of material scale and none approaching the significance of the Deepwater Horizon." Based on a review of successful programs in the nuclear power, aerospace and other industries, the paper recommends the consideration of accident precursor analysis as a means for providing "a risk-informed oversight capability focused on preventing catastrophic oil spills and based on data from ongoing operating experience on the Outer Continental Shelf (OCS)."

The ASP approach is similar to Layer of Protection Analysis (LOPA) [Ref. 4] or Bowtie Analysis in that it postulates accident scenarios and the paths that could lead to a potential accident, but differs in that it uses insights from actual historical near-miss event data to identify the branch points which, if different, could have led to a severe accident.

This paper discusses the collection of historical event data, various programs that undertake such data collection, and proposes steps for use of this data toward risk assessment and management using ASP methods.

2 Collection, Evaluation and Application of Historical Events Data

While HAZOP and FMEA are useful tools for postulating incidents and their consequences, historical event data provides important input to the risk assessment process beyond the development of accident frequencies. Incident data collection and analysis can provide multiple qualitative insights, such as information for developing and refining risk models and a better understanding of system operation and failure (including the human-machine interface and efficacy of procedures and practices).

The CSB investigates incidents at chemical process facilities, produces investigation reports such as the one for the West Fertilizer Company Fire and Explosion in April 2013 [Ref. 5] and conducts public meetings to present and discuss their findings, as with the February 2015 ExxonMobil Refinery explosion [Ref. 6].

Following the Deepwater Horizon accident, the Deepwater Presidential commission report [Ref. 7] found at the time that "Although the raw incident data are online, MMS last produced an analysis of offshore incidents—critical data for promoting the safety of offshore operations—for calendar year 2000."

Probably for this reason, incident analysis was highlighted in the discussion of the Workplace Safety Rule in 2009 [Ref. 8] and one of the key areas where it was expected that the U.S. Department of the Interior's Bureau of Safety and Environmental Enforcement (BSEE) would be focusing their attention during the second Safety and Environmental Management System (SEMS) audit cycle in 2015 was on "Developing representative leading indicators for SEMS performance, while continuing to track industry accepted lagging indicators for evidence that SEMS programs are driving an improvement in safety performance." [Ref. 9] The development of these representative indicators by its nature must come from the review of historical evidence.

The narratives and timelines of previous accidents and near-misses can indicate what triggered the event, the automatic system response and failure, subsequent system failures and their causes, and worker involvement either to cause, exacerbate or mitigate the event. These insights in turn feed back into the data collection process by indicating what trends are of interest to monitor and, therefore, what categories of data should be collected.

The Workplace Safety Rule announcement by MMS in 2009 [Ref. 8] indicated that "As a result of MMS research conducted on accident panel investigations and reports, incident analysis, and [Incidents of Non-Compliance] INCs, it appears that equipment failure is rarely the primary cause of the incident or accident. This is due to technological advances which have provided industry with very efficient and reliable equipment for finding, producing, and transporting offshore oil and gas. However, in most cases, accidents and oil spills can be traced to human error and/or organizational failures." This is consistent with the Deepwater Presidential commission report [Ref. 7], which identified several human error-related issues, including the following:

- Numerous changes to the temporary abandonment procedures for securing the well were made in the two weeks leading up to the operations plan for the worker activities, but "[t]here is no evidence that these changes went through any sort of formal risk assessment or management of change process"

- "The drilling crew and other individuals on the rig also missed critical signs that a kick [unplanned influx of gas or fluids] was occurring. The crew could have prevented the blowout—or at least significantly reduced its impact—if they had reacted in a timely and appropriate manner. What is not now clear is precisely why the crew missed these signals."

- "The crew should have diverted the flow overboard when mud started spewing from the rig floor. While that ultimately may not have prevented an explosion, diverting overboard would have reduced the risk of ignition of the rising gas. Considering the circumstances, the crew also should have activated the blind shear ram to close in the well. Diverting the flow overboard and/or activating the blind shear ram may not have prevented the explosion, but likely could have given the crew more time and perhaps limited the impact of the explosion."

While the human involvement in incidents cannot be denied, the ATHEANA method of human reliability analysis (HRA) developed by the US Nuclear Regulatory Commission [Ref. 10] discusses the situation of "error-forcing contexts" or EFCs in which the plant conditions and so-called "performance shaping factors" (context-specific influences on reliable human performance) conspire to make human error nearly unavoidable. The ATHEANA document provides examples of these EFCs that have occurred in past US nuclear plant accidents, such as mismatches between the actual plant conditions and the operator job aids (e.g., training, experience) during the Three Mile Island event. EFC elements were also identified by ATHEANA for each of the four stages of information processing (i.e., detection, situation assessment, response planning, and response implementation) that describe the range of human activities required to respond to abnormal or emergency conditions. The review of past events and near-misses considering these EFC elements may indicate that organizational factors and gaps in procedures or training may in fact be the root cause.

An independent investigation of the Deepwater Horizon by the UC Berkeley Center for Catastrophic Risk Management stated that "[T]he Macondo system's good record of lost time accidents could not provide effective precursors for the major impending system accident. Efforts to prevent system accidents must address a broader set of organizational issues and hazards associated with all of the elements that comprise a given system. The good record of lost time accidents plausibly led to complacency that contributed to the disaster." [Ref. 11]

A program such as ASP looks at historical events from the perspective of commonality of potential outcomes and identifies the organizational issues that can cause near misses to become severe accidents.

3 Chemical Process and Nuclear Power Industry Experience

Examples of the projects that exemplify the type of incident reviews that factor industry experience into lessons learned are the Accident Sequence Precursor (ASP) program [Refs. 12, 13 and 14], the AIChE CCPS Process Safety Incident Database (PSID) [Ref. 15] and Process Equipment Reliability Database (PERD) [Ref.16], the EPRI Fire Events Database (FEDB) project [Refs. 17 and 18], the EPRI Guidelines for PRA Data Analysis [Ref. 19], the AIChE CCPS Guidelines for Process Equipment Reliability Data [Ref. 20], and the EPRI/NRC-RES Fire HRA Guidelines [Ref. 21]. Jensen-Hughes has been an active participant (lead in some cases) in the development of all of these projects except PERD and PSID.

3.1 Accident Sequence Precursor (ASP) Program

The ASP Program began in the early 1980s under the auspices of the U.S. Nuclear Regulatory Commission in response to the Three Mile Island accident. ASP is concerned with the identification and documentation of operational events that have involved portions of accident sequences and with the estimation of associated frequencies and probabilities. Identification of precursors requires the review of operational events for instances in which plant functions that provide protection against severe accidents (core damage, in the case of nuclear plants) have been challenged or compromised. The intention is to understand how these 'near misses' could have propagated into a severe accident had the circumstances been slightly different and thereby inform the risk modeling, data collection and operational insights.

[NOTE: The following explanation of the US NRC's ASP program is extracted from Reference 11.]

Reports of plant incidents are reviewed to identify those potentially involving failures in systems that provide protective functions against severe accident-related initiating events or the initiating events themselves. Those events selected undergo one- or two-engineer review(s) to determine if the reported event should be examined in greater detail. Events determined to be potentially significant as a result of this initial review are then subjected to a thorough, detailed analysis. This extensive analysis is intended to identify those events considered to be precursors to potential severe accidents, either because of an initiating event or because of failures that could have affected the course of postulated off-normal events or accidents. A variety of other data sources about the same event are also reviewed and compared to thoroughly understand the contributors to each event.

Quantification of accident sequence precursor significance involves the determination of a conditional probability of a subsequent severe accident given the failures observed during an operational event. This probability is estimated by mapping failures observed during the event onto the ASP accident sequence models (event trees and linked fault trees modified to reflect the event), which depict potential paths to the severe accident, and by calculating a conditional probability of the severe accident.

The information gained from the ASP process not only provides quantitative data, but provides insights into the types of events that should be included in risk models. In addition, events that provide insights into unusual failure modes with the potential to compromise continued safety functions but are not considered to be precursors are documented by ASP as "interesting" events.

Building upon this effort, the National Academy of Engineering Program Office undertook the Accident Precursors Project in 2003 to examine the complex issue of accident precursor analysis and management. This seven-month project was designed to document and promote industrial and academic approaches to detecting, analyzing, and benefiting from accident precursors, as well as to understand public-sector and private-sector roles in using precursor information. [Ref. 10]

3.2 AIChE CCPS Process Safety Incident Database (PSID)

The purpose of the CCPS PSID is to pool process safety incident experience among participating companies so they can learn from the experiences of others without suffering the consequences of failures, while minimizing corporate liability. PSID includes process safety incidents with a potentially important lesson to be learned from incidents that did or could have resulted in a fire, explosion, fatality, multiple injuries, significant release of hazardous materials, or any other unique process safety incident (including near-misses).

3.3 AIChE CCPS Process Equipment Reliability Database (PERD)

The equipment reliability database is intended to provide high quality, valid, and useful data to the Hydrocarbon Processing Industries (HPI) and Chemical Process Industries (CPI). This type of data provides input to analyses of availability, reliability and equipment design improvements, maintenance strategies and life cycle cost determination. The database includes Lessons Learned and Contributing Factors, which are data input requirements.

3.4 Fire Events Database (FEDB) Project

The FEDB is a collection of fire events that occurred in nuclear power plants since 1967. These events were reported to the U.S. Nuclear Regulatory Commission or the industry groups such as the Institute for Nuclear Power Operations (INPO) and the Nuclear Energy Institute (NEI). These events were reviewed to extract information on the equipment involved in the fire initiation, response of fire prevention and mitigation systems, response of the fire brigade and the consequences of the fire (degraded incipient or catastrophic). Guidance was also provided for the interpretation and categorization of fire event data. Since its publications the event history has been used, at time in combination with tests and a process called Phenomena Identification and Ranking Table or PIRT, to help define and characterize accident initiation, progression, mitigation and consequences, including:

- Types and characteristics, e.g., timing, of accident initiators,
- Sequence of events, including prevention/mitigation and human interface that helps define accident scenarios (or sequences) with risk implications,
- Severity of the event, e.g., heat release rate of an electrical panel fire or zone of influence from a high energy arc fault in a high voltage switchgear,
- Mitigating systems response, e.g., reliability and effectiveness of the fire brigade response to various types of fires,
- Cable damage and quantification of effects from fire, such as spurious indication of instrumentation and control systems leading to failure of accident prevention/mitigation systems or inadequate/erroneous operator response, and
- Qualitative insights on the types of fires that are most likely to occur, their duration, severity and how the plants address them.

3.5 EPRI Guidelines for Data Analysis for PRA

The EPRI Data Guidelines discusses the nuclear power industry approach to collecting, analyzing and providing summary statistics for several major categories of data types: initiating event data, component reliability data, unavailability (downtime) data and common cause failures. Initiating events are quantified using a combination of historical evidence of event occurrence, both from the specific plant and across the nuclear industry.

3.6 AIChE CCPS Guidelines for Process Equipment Reliability Data

Many of the aspects of data analysis discussed in the EPRI Data Guidelines were identified thirteen years earlier for the chemical process industry in the AIChE CCPS Process Equipment Reliability Data Guidelines. This data book built on the Chemical Process Quantitative Risk Assessment (CPQRA) Guidelines to show the process engineer how to make quantitative estimates of the risk of hazards identified, which can in turn identify the major contributors to risk. This project helped demonstrate the need for process industry data collection efforts like PERD and PSID.

3.7 EPRI/NRC-RES Fire Human Reliability Analysis (HRA) Guidelines

Although its focus is on human reliability, the Fire HRA Guidelines discusses the need to review past incidents experienced at a facility to understand whether the event was human-caused or how the workers' responded to the event as it evolved. These insights not only assist the HRA, but provide an understanding of how workers perceive their role during an accident and how they can influence accident progression, for good or for ill. These findings can then be fed back into the improvement of organizational and safety practices for accident prevention.

4 Considerations for Historical Data Analysis Application in the Chemical Process Industry

The Discussion Paper on Precursor Analysis for Offshore Oil and Gas Drilling [Ref. 3] identifies incident reports required under the Incident Reporting Rule (30 CFR 250.188) for oil, gas and sulphur operations in the outer continental shelf as an example of industry precursor data that would allow a system failure of concern [such as "significant uncontrolled escape of hydrocarbons" (SUEH)] to be evaluated through event trees based on generic commonality of events (such as those used in the US NRC ASP program).

Further, the Discussion Paper recommends an approach for the design of an ASP-enabled incident reporting system, which has been made more general here for application to a wider variety of chemical processes:

a) Define the class of events (such as SUEH) related to the system failure of concern.

b) On the basis of PRA models, design the set of generic event trees that have these events as possible endpoints.

c) Ensure that all nodes in these generic event trees are captured in an incident reporting system, such as PSID or under 30 CFR 250.188.

d) Design a control loop, possibly based on the Potential Incident of Non-Compliance (PINC) checklist inspection system, to ensure coverage of the incident reporting system, with appropriate sanctions for noncompliance.

e) Set up an analysis team to map the key even-related incidents onto generic event trees.

f) Set up a review board comprising industry and government agencies.

g) Design feedback mechanisms for involving industry in efforts to optimize industry-wide performance as measured by the ASP indicators.

h) Design an automation plan to streamline the ASP analysis.

The following additional tasks should be considered for translating these steps into industry guidance:

a) Documentation of insights and development of a draft guidelines for a selected process facility (to be defined) reflecting items a) through h) above.

b) Testing, validation and review of the guidelines. This may be done using one or more well documented industry events.

c) Final guideline for the selected process facility.

d) Adaptation of the guideline to other processes.

5 Conclusion

A recent article in CEP Magazine [Ref. 22] profiles five chemical process industry accidents that, while not as severe as Bhopal or Deepwater Horizon, still had serious results. The article states that "Accidents such as those discussed in this article are more likely to occur. These incidents demonstrate that even small mistakes can have disastrous consequences. Therefore it is imperative that engineers learn from previous incidents to reduce their likelihood of recurring." Near misses, like smaller scale events, are also instances that can provide valuable insights on lapses in process safety management.

As stated in EnergyWire [Ref. 2]. "ASP would use the data from these partial failures, including the failure of rig workers to respond appropriately, to compute the likelihood of a catastrophe. It would identify problems when they occur so regulators can stop them from happening again."

6 References

[1] "U.S. Chemical Safety Board Finds Multiple Safety Deficiencies Led to February 2015 Explosion and Serious Near Miss at the Exxon Mobil Refinery in Torrance, California," News Release, U.S. Chemical Safety and Hazard Investigation Board, Washington, DC, 16 January 2016.

[2] Vaidyanathan, G., "GULF SPILL: Could risk analysis prevent future deepwater disasters?" EnergyWire, Environment & Energy Publishing (E&E), http://www.eenews.net/stories/1059963200, 20 April 2012.

[3] Cooke, R.M., Ross, H.L. and Stern, A., "Discussion Paper - Precursor Analysis for Offshore Oil and Gas Drilling, From Prescriptive to Risk-Informed Regulation", RFF DP 10-61, Resources for the Future, Washington, DC, January 2011.

[4] Guidelines for Initiating Events and Independent Protection Layers in Layer of Protection Analysis, American Institute of Chemical Engineers (AIChE), Center for Chemical Process Safety (CCPS), New York, NY, 2015.

[5] "West Fertilizer Company Fire and Explosion", Final Investigation Report 2013-02-I-TX, U.S. Chemical Safety and Hazard Investigation Board, Washington, DC, January 2016.

[6] Presentation slides from CSB Public Meeting on ExxonMobil Refinery Explosion, U.S. Chemical Safety and Hazard Investigation Board, Washington, DC, 13 January 2016.

[7] Deep Water: The Gulf Oil Disaster and the Future of Offshore Drilling, Report to the President, National Commission on the BP Deepwater Horizon Oil Spill and Offshore Drilling, January 2011.

[8] The Bureau of Ocean Energy Management, Regulation and Enforcement, Office of Public Affairs, Fact Sheet, The Workplace Safety Rule on Safety and Environmental Management Systems (SEMS).

[9] Memorandum from Doug Morris, Chief, Office of Offshore Regulatory Programs to Brian Salerno, Director BSEE, SEMS Program Summary—First Audit Cycle (2011-2013), July 23, 2014.
http://www.bsee.gov/uploadedFiles/BSEE/Regulations_and_Guidance/Safety_and_Environmental_Management_Systems_-_SEMS/SEMS%20Program%20Summary%208132014.pdf

[10] Technical Basis and Implementation Guidelines for A Technique for Human Event Analysis (ATHEANA), NUREG-1624, Rev. 1, U.S. Nuclear Regulatory Commission, Washington, DC, May 2000.

[11] Final Report on the Investigation of the Macondo Well Blowout, Deepwater Horizon Study Group (DHSG), Center for Catastrophic Risk Management (CCRM), University of California – Berkeley, Berkeley, CA, March 2011.

[12] Minarick, J. W. and Kukielka, C. A., Precursors to Potential Severe Core Damage Accidents: 1969-1979, A Status Report, NUREG/CR-2497, U.S. Nuclear Regulatory Commission, Washington, DC, June 1982.

[13] Belles, R.J. et al., Precursors to Potential Severe Core Damage Accidents: 1998, A Status Report, NUREG/CR-4674, US Nuclear Regulatory Commission, Washington, DC, July 2000.

[14] Phimister, J.R., Bier, V.M., and Kunreuther, H.C. (Eds.), Accident Precursor Analysis and Management: Reducing Technological Risk Through Diligence, The National Academies Press, Washington, DC, 2004.

[15] Presentation slides, Overview: Process Safety Incident Database (PSID), American Institute of Chemical Engineers (AIChE), Center for Chemical Process Safety (CCPS)

http://www.aiche.org/ccps/resources/psid-process-safety-incident-database

[16] American Institute of Chemical Engineers (AIChE), website "About PERD"

http://www.aiche.org/process-equipment-reliability-database-perd/about-perd

[17] EPRI 1000894, Fire Events Database for U.S. Nuclear Power Plants: Update Through 1999, Electric Power Research Institute (EPRI), Palo Alto, CA, October 2000.

[18] EPRI 1025284, The Updated Fire Events Database: Description of Content and Fire Event Classification Guidance, Electric Power Research Institute (EPRI), Palo Alto, CA, July 2013.

[19] EPRI 3002000774, EPRI Guidelines for PRA Data Analysis, Electric Power Research Institute (EPRI), Palo Alto, CA, December 2013.

[20] Guidelines for Process Equipment Reliability Data with Data Tables, American Institute of Chemical Engineers (AIChE), Center for Chemical Process Safety (CCPS), New York, NY, 1989.

[21] EPRI 1023001/ NUREG-1921, EPRI/NRC-RES Fire HRA Guidelines, Electric Power Research Institute, Palo Alto, CA and U.S. Nuclear Regulatory Commission, Washington, DC, July 2012.

[22] A. Ness, "Lessons Learned from Recent Process Safety Incidents," CEP Magazine, American Institute of Chemical Engineers (AIChE), New York, NY, March 2015.

442384 Quantitative Risk Analysis of the Expansion of the Chlorine Room of the Industrial Water Plant of the Petrochemical Complex Jose Antonio Anzoategui of Petroquimica Venezuela Pequiven S.a

Abstract:

Quantitative risk analysis of the expansion of the chlorine room of the industrial water plant of the petrochemical complex Jose Antonio Anzoategui of Petroquimica venezuela PEQUIVEN S.A.

Venezuela Petrochemical PEQUIVEN SA has a management responsible for risk control, which makes this quantitative risk analysis to estimate the risk to employees and third parties for process safety events that may occur as a result of the loss of containment in the facilities and thus indicate recommendations aimed at reducing the level of risk associated with such events. Quantitative risk analyzes are made from the standard rules of Petroleos de Venezuela , PDVSA as IR-S-02: criteria for quantitative risk analysis and other international standards such as Center of Chemical Process Safety Guidelines for Chemical Process Quantitative Risk Analysis and the netherlands organization Yellow book: Methods for calculation of physical effects, through the computational tool PHAST RISK 6.7 . Developed by DNV Company.

PEQUIVEN has three complexes nationwide, in which fertilizers for the food industry, olefins and plastics for residential construction, among other products are produced. The Jose Antonio Anzoategui complex will make an expansion of the chlorine room of the industrial water plant and because of the dangers associated with chlorination process, a quantitative risk analysis has requested to engineering management risks.

The process of a quantitative risk analysis (QRA) started setting goals, objectives and depth of the study, once these are established, and in conjunction with materials, inventory, environmental conditions and the description of the process, one proceeds to identify scenarios. In this case, the HAZOP tool was applied, which gives the place a list of possible scenarios, of which those that could cause the worst consequences. With the previously selected scenarios, the model building on which the estimated frequency and consequence will be held is started. The frequency is estimated based on a review of probability theory study according to Chemical Process Safety Fundamentals with Applications Standard, through a theoretical data extracted from the norm Guidelines for Process Equipment Reliability Data, a probability of occurrence of the scenarios is established.

The estimation of Consequence answer a series of complex mathematical models that simulate the behavior of these events based on the data of the others that occurred already trough PHAST RISK software. Once the simulation is made, the software generates a series of results of contours risks in function of exposure to chlorine gas to the personnel working in this complex and surrounding facilities, allowing emit a series of observations than are aimed to decrease level of Risk associated with the expansion of the chlorine room.

12TH GLOBAL CONGRESS ON PROCESS SAFETY

Assessing Cognitive Human Errors for the Process Industries

Erin P. Collins
JENSEN HUGHES
333 E. 46th Street, #11L
New York, NY 10017
ecollins@jensenhughes.com

Bijan Najafi
JENSEN HUGHES
2105 S Bascom Ave, Suite 350
Campbell, CA 95008-3278
bnajafi@jensenhughes.com

Prepared for Presentation at
American Institute of Chemical Engineers
2016 Spring Meeting
12th Global Congress on Process Safety
Houston, Texas
April 11-23, 2016

Assessing Cognitive Human Errors for the Process Industries

Erin P. Collins
JENSEN HUGHES
333 E. 46th Street, #11L
New York, NY 10017
ecollins@jensenhughes.com

Bijan Najafi
JENSEN HUGHES

Keywords: cognitive, human error, human reliability analysis

Abstract

Through the analysis of actual incidents and the understanding of the underlying human errors that have contributed to them, it is apparent that 'first generation' HRA methods are appropriate for some instances but have their limitations. For example, failure to recognize the significance of high fluid returns during the negative pressure test and failure to perceive the importance of drill pipe pressure during well monitoring activities were key contributors to the Deepwater Horizon incident. These diagnosis and decision-making errors are not well addressed by first generation HRA methods and require particular timing, error mode and performance shaping factor considerations. This paper provides an introduction to HRA methods and then discusses those available for evaluating cognitive errors and their applicability to the chemical process industry.

1 Introduction

Layer of Protection Analysis (LOPA) provides a simplified risk analysis technique for a select set of postulated incident scenarios, evaluating an initiating event (IE) and subsequent independent protection layers (IPLs) as a series of probabilities of failure on demand (PFDs). Human error is assessed in LOPA as either an IE or IPL. Techniques that have been recommended for human reliability assessment (HRA) by the CCPS Guidelines for IEs and IPLs in LOPA [Ref. 1] are generally considered in other industries as 'first generation' HRA methods, meaning that they are appropriate for some instances but have their limitations. These limitations have become more apparent through the analysis of actual incidents and the understanding of the underlying human errors that have contributed to them. For example, failure to recognize the significance of high fluid returns during the negative pressure test and failure to perceive the importance of drill pipe pressure during well monitoring activities were key contributors to the Deepwater Horizon incident. These diagnosis and decision-making errors are not well addressed by first generation HRA methods and require particular timing, error mode and performance shaping factor (PSF) considerations. Several HRA quantification methods are available to assess the human error probability associated with these so-called "cognitive" aspects of a task, as well as the equipment manipulation.

While many HRA approaches had their genesis in the nuclear power industry, they have been successfully applied to other industries, such as chemical weapons destruction plants and nuclear fuel transport and storage, and their results have been accepted during peer and regulatory agency reviews. In each instance where HRA is applied to a new industry, however, an evaluation must be made regarding the degree of applicability of the standard error types, PSFs and quantification methods. Since the error types and PSFs have an underpinning in human factors, psychology and cognitive science, they generally apply regardless of the industry. However, it is important to evaluate the applicability of HRA tools and techniques to ensure that industry-specific issues are being appropriately addressed.

This paper provides an overview of HRA principles and methods, then discusses those available for evaluating cognitive errors and their applicability to the chemical process industry.

2 HRA Principles and Methods

HRA provides a framework to evaluate, prioritize, quantify and improve the human interface with complex systems. HRA tools and techniques were pioneered by the commercial nuclear power industry, but are based on fundamental precepts of human reliability and data collection and analysis that are not technology specific. They have been applied to an increasing number of facilities across various different industries, such as aerospace [Ref. 2], chemical weapons destruction [Ref. 3], HVAC system optimization, nuclear fuel disposal facility design risk evaluation [Ref. 4] and detection and suppression activities in response to fire events in the military [Ref. 5]. This experience and repetition have allowed lessons learned to be identified and fed back into safe practices, the development of a safety culture, and the ability to meet standards [Ref. 6] and regulatory goals (which are quite stringent in the nuclear power industry).

This history and associated lessons learned indicates that wider application of HRA in the chemical process industry could provide greater understanding of risks (both accident and economic), how to control them (by improving procedures, focusing maintenance and surveillance efforts on the actual errors), and how to develop a safety culture that focuses on human error prevention.

The steps of the HRA process discussed below are adapted from NUREG-1921, the EPRI/NRC-RES Fire Human Reliability Analysis Guidelines [Ref. 7]. The process begins with an identification and definition task to identify actions performed by various workers and to define the actions in terms of human failure events or HFEs. These actions may include both failure to perform a required task or undesired actions with deleterious impact on system performance and safety. For the chemical process industry, this identification and definition step could be expected to be very closely tied with the subsequent step of qualitative analysis because the same information gathering that helps define the events also feeds into the development of an "operational story" or HFE narrative that assists in further understanding the events.

A sound qualitative analysis allows the HRA to provide feedback to the facility operator on the factors contributing to both the success and failure of a worker action. This involves the evaluation of performance shaping factors (PSFs), which are situation-specific influences on human error probabilities (HEPs) and include:

- Cues and indications that indicate a system problem
- Timing of the task
- Procedures and training
- Complexity of the task
- Workload, pressure and stress
- Human-Machine Interface (HMI)
- Environmental factors, such as heat or confined spaces
- Special equipment, such as tools and protective gear

These PSFs are used to flesh out the HFE narrative and identify the key elements of the HFEs that require further understanding through the performance of subject matter expert interviews.

Finally, quantification or quantitative analysis is performed by applying the information gained from identification, definition and qualitative analysis of HFEs toward the estimation of HEPs. Quantification can be done at varying degrees of detail and rigor. The three levels of analysis performed in HRA are: Screening, Scoping and Detailed.

Screening analysis involves the assignment of a preliminary and conservative expert judgment value as a means for "screening" out the risk-significant HFEs from the non-contributing ones.

The Scoping approach, discussed in NUREG-1921, was developed to provide additional rigor to a semi-screening process, but still results in conservative values for initial quantification.

Detailed analysis is performed by evaluating the action steps, timing and PSFs in detail and assessing the probability of errors in the cognitive aspects, or diagnosis and decision-making process for the correct action to take, as well as the execution aspects, such as the equipment manipulation steps.

Several HRA quantification methods are available and it has been found to be most effective if the analyst evaluates the strengths and weaknesses of each method against the nature of the human error being evaluated to make the best match. Having a large HRA method toolkit to draw from allows the analyst the flexibility to assess the driving failure mechanism behind the error rather than simply filling out a checklist.

The following HRA quantification methods are most widely used in the nuclear power industry, and are contained in Electric Power Research Institute's (EPRI's) HRA Calculator software [Ref. 8]:

- Human Cognitive Reliability/Operator Reliability Experiment (HCR/ORE) [Ref. 9], which focuses on the cognitive or decision making errors that are constrained by time;
- Cause-Based Decision Tree Method (CBDTM) [Ref. 9] that addresses cognitive errors as well but from the standpoint of the error-causing mechanisms, such as unclear instrumentation or lack of procedural direction;
- The Technique for Human Error Rate Prediction (THERP) [Ref. 10], which has been used for many years for the evaluation of task performance failures (also called execution errors) that can occur either before or after an accident;
- The Accident Sequence Evaluation Program (ASEP) [Ref. 11] that provides a simplified version of THERP for human errors that contribute to but occur prior to the accident; and
- The SPAR-H method [Ref. 12], which only permits highly simplified assessments of cognitive and action errors (and has not been used in the latest generations of nuclear plant risk assessments since it does not allow the depth of analysis to support safety improvements).

For HRA applications other than nuclear power, such as chemical weapons destruction plants and nuclear fuel transport and storage, other quantification methods have been successfully applied and their results accepted during peer and regulatory agency reviews. These include the Cognitive Reliability Error Analysis Method (CREAM) [Ref. 13], and the Human Error Assessment and Reduction Technique (HEART) [Ref. 14]. As the name implies, CREAM focuses on the decision-making errors by providing HEPs for a standard set of task types, but also includes adjustment factors to evaluate facility and scenario-specific influences. HEART contains a list of generic task types and their HEP estimates that are modified using factors associated with 'error producing conditions'.

While the application of a method and generation of HEP values is important, the primary output of the HRA is the identification and prioritization of the human error issues of greatest risk significance and the development of risk insights. The benefit of conducting more detailed HRA is in gaining a better perspective on what influences human failure and therefore what can be done to improve human performance.

The decomposition of human error into cognitive vs. execution tasks and a detailed evaluation of these tasks has proven to be particularly important since the approaches to their improvement are different. For example, more instrumented processes attempt to provide information to the operator that allow him/her to diagnose the situation and take the appropriate action. In other cases, however, the process may not include such instrumentation or it may be not be available due to failure, so diagnosis becomes a critical issue. Even when there is no instrumentation or early warning system provided for an action, the HRA can provide an understanding of the cues required for diagnosis and allow these to be included in training and procedures/manuals. These cues may even be sensory in nature, such as the sound, smell or sight of a process that is going awry. This does not mean that the HRA will recommend instrumenting everything. If correctly done, the HRA instead allows the most critical information used by the worker to be identified and suggests how it should be presented or emphasized to optimize performance.

The reliability of execution tasks is generally improved by providing a formal set of procedures and training that are continually reviewed and refined. However there may be instances where procedures and training are not yet formalized, such as when the facility is a new design or is in the pre-operational testing phase, or because of historical precedent or preference. In these instances, the worker action is evaluated in HRA as a so-called 'recovery' action, where the worker utilizes previous experience with tasks they consider to be similar and applies them to a new situation. These insights are important to evaluating the reliability of task performance and can even be subsequently captured in procedure modifications and training enhancements. The risk benefit assessment might even recommend conducting a design change to the human-machine interface to improve the execution task reliability.

In summary, HRA is

- Widely used
- Applied across a variety of industries, which demonstrates that it provides value
- Allows the user to quantify and prioritize the human influence on safety
- Provides the means for improving process safety and reliability in the short term and a safety culture to reduce future errors over the longer term.

3 Cognitive Errors in the Process Industries

The Deepwater Presidential commission report [Ref. 15] identified several human error-related issues that led to the severity of the incident, including the following:

- Numerous changes to the temporary abandonment procedures for securing the well were made in the two weeks leading up to the operations plan for the worker activities, but "[t]here is no evidence that these changes went through any sort of formal risk assessment or management of change process"
- "The drilling crew and other individuals on the rig also missed critical signs that a kick [unplanned influx of gas or fluids] was occurring. The crew could have prevented the blowout—or at least significantly reduced its impact—if they had reacted in a timely and appropriate manner. What is not now clear is precisely why the crew missed these signals."
- "The crew should have diverted the flow overboard when mud started spewing from the rig floor. While that ultimately may not have prevented an explosion, diverting overboard would have reduced the risk of ignition of the rising gas. Considering the circumstances, the crew also should have activated the blind shear ram to close in the well. Diverting

the flow overboard and/or activating the blind shear ram may not have prevented the explosion, but likely could have given the crew more time and perhaps limited the impact of the explosion."

These issues can be considered as relating to cognitive human errors since they concern the lack of procedural guidance that assists with diagnosis and the decision-making on the part of the crew that appears to have been ill advised or ill informed. Hubbard and Embrey [Ref. 16] assessed these issues from a human factors standpoint and they raised the following concerns related to the cognitive errors during the incident:

- Why were procedures not used – is this typical?
- What was the effect on decision making and on practice of the pressure to get the well tested and capped?
- Why did the crew use the mud gas separator rather than pipework that would have discharged the mud more quickly?
- What emergency training did the team have? Did they have training or job support to help them identify and respond to escalating situations?

Questions such as these are key to the evaluation and assessment of cognitive errors.

A recent article by Ness in the CEP Magazine [Ref. 17] outlined lessons learned from several process accidents, many of which involved cognitive human errors:

- ARCO Channelview wastewater tank explosion, 1990 – "The wastewater tank was not considered part of the operating plant. Hence, the management and workers did not understand that a chemical reaction was taking place in the tank, generating oxygen. The lack of understanding enabled a series of poor decisions, such as discontinuing the nitrogen purge...no management of change (MOC) review of these decisions, and no pre-startup safety review."
- Terra Industries ammonium nitrate explosion, 1994 – "...the lack of procedures for shutdown and monitoring the equipment during shutdown led operators to perform actions that sensitized the AN solution and provided energy to the decomposition reaction."
- Formosa Plastics vinyl chloride release, 2004 - "The operator overrode an interlock, which led to a release of hot, pressurized VCM. Formosa Plastics did not have comprehensive written standards...for managing interlocks on the vessels. Employees were unprepared for a major accident at the facility."

These incidents demonstrate faulty understanding of the process and lack of procedural direction, which in turn led to improper decision-making and catastrophic consequences.

4 Treatment of Cognitive Human Error in Process Guidelines and Studies

The CCPS *Guidelines for Preventing Human Error in Process Safety* [Ref. 18] mentions cognitive failure assessment methods such as CBDTM, CREAM and HEART but does not provide much guidance to the user for implementing them. Furthermore, in the ten years since that guidance was issued, the Deepwater Horizon incident has occurred and much has been learned in various industries on the analysis and mitigation of human error using HRA.

The recently issued CCPS *Guidelines for Initiating Events and Independent Protection Layers in Layer of Protection Analysis* [Ref. 1] states that "developing PFD values for human IPLs is not straightforward." Note that PFDs as defined in these guidelines include not only the HEP, but the PFD of the cue for response, namely the "instrumentation or equipment used to detect the process deviation and/or take action to return the process to a safe state."

The IPLs for human actions that are cited in these guidelines are:

- Human response to an abnormal condition
- Human response to an abnormal condition with multiple sensors and indications and the operator has >24 hours to accomplish the required response action
- Adjustable movement-limiting device – this is actually equipment that is intended to prevent operator misoperation of equipment beyond a limit that would initiate a scenario of concern
- Personal Protective Equipment (PPE) – safety gear that acts as a barrier between a person and a hazard.

In an HRA context, the first two IPLs would be considered Human Failure Events (HFEs), but the specification of sensors/indications (cues) and time available for response would be evaluated in detail for both cases, in addition to other scenario-specific PSFs, such as procedures and training. Most importantly, the time available would be balanced against the time *required* for the action to assess the time margin and the HEP.

Note that the *nature* of the human response is not specified for these IPLs, so there is no differentiation between cognitive and execution errors, which have different aspects and methods for evaluation.

The adjustable movement-limiting device IPL would not itself considered to be a human failure, but would be considered as a mechanical form of interlock that would need to be defeated (either through another human action to leave it out of position after maintenance or through equipment failure) to allow the human error to occur. A Quantitative Risk Assessment (QRA) model allows such equipment and human failures to be functionally linked.

PPE is addressed in HRA not as an HFE but as a performance shaping factor on an existing human response action since it is considered to cause sight, sound and manipulation limitations on the human action, and can even lead to increased stress.

Papers presented in recent years discuss application of HRA to the petroleum industry through the use of the SPAR-H quantification method [Refs. 19 & 20]. Gould, Ringstad and Van De Merwe state that "The main advantages of SPAR-H is that it provides clear guidance on assigning numerical weights to the performance-shaping factors, is not very resource-demanding, and includes worksheets that promote analyst consistency." While SPAR-H provides a useful screening analysis of factors related to cognitive error, such as Available Time, Procedures and Ergonomics, it would seem to be best considered as the first step toward identifying risk-significant cognitive human errors, which would then need to be evaluated in greater detail. The benefit of detailed HRA lies in identifying the driving mechanisms behind the error that can then be applied to steps for error likelihood reduction.

Another current study [Ref. 21] has successfully applied the CREAM cognitive error evaluation technique to the cargo loading process of LPG tankers "to systematically predict human error potentials for designated tasks and to determine the required safety control levels on-board LPG ships." The human cognition and action context are determined in accordance with Common Performance Conditions (CPCs) which provide a well-structured and comprehensive basis for characterizing the circumstances and to define sets of possible error modes and error causes.

5 Information Processing Model for Cognitive Activities

Assessment of HFEs can be guided by a model of higher-level cognitive activities, such as an information processing model. The model that has been used by the authors is based on the discussion in Chapter 4 of the ATHEANA method [Ref. 22] and consists of the following elements:

- Monitoring and detection - Both of these activities are involved with extracting information from the environment. Also, both are influenced by the characteristics of the environment and the person's knowledge and expectations. Monitoring that is driven by the characteristics of the environment is called data-driven monitoring. Monitoring initiated by a person's knowledge or expectations is called knowledge-driven monitoring. Detection can be defined as the onset of realization by operators that an abnormal event is happening.

- Situation awareness - This term is defined as the process by which operators construct an explanation to account for their observations. The result of this process is a mental model, called a situation model that represents the operator's understanding of the present situation and their expectations for future conditions and consequences.

- Response planning - This term is defined as the process by which operators decide on a course of action, given their awareness of a particular situation. Often (but not always) these actions are specified in procedures.

- Response implementation - This term is defined as the activities involved with physically carrying out the actions identified in response planning.

The final bullet concerns the execution of the response, but is predicated upon the successful cognitive assessment of the course of action.

By assessing these features of a human action, the analyst can then select from the available HRA models to evaluate the HEP and moreover, identify the driving issues leading to error that can be factored into error prevention.

6 Conclusions

Recent studies are showing that existing HRA methods can be applied to assess cognitive human errors in the process industries, but these insights have yet to be incorporated into industry guidelines. While initial assessments can be made with simplified tools, the key to addressing and preventing major incidents is the identification of underlying human error mechanisms so that risk-mitigating solutions that truly address the diagnosis and decision-making process can be developed.

7 References

[1] *Guidelines for Initiating Events and Independent Protection Layers in Layer of Protection Analysis*, Center for Chemical Process Safety, American Institute of Chemical Engineers, New York, NY, 2015.

[2] Collins, E.P. and Fragola, J.R., "Advances in Human Error Evaluation", 1995 Reliability, Availability, Maintainability Symposium (RAMS), Washington, DC, January 1995.

[3] Umatilla Chemical Agent Disposal Facility Quantitative Risk Assessment, Report No. SAIC-00/2641. Volume I, Science Applications International Corporation (SAIC), Abingdon, MD, 2002.

[4] Yucca Mountain Project Canister Receipt and Closure Facility Reliability and Event Sequence Categorization Analysis, 060-PSA-CR00-00200-000-00B, Bechtel SAIC Company, LLC (BSC), November 2008.

[5] Garvey, M. (Dept. of the Navy), Joglar, F. and Collins, E.P. (JENSEN HUGHES, Inc.), "HRA for Detection and Suppression Activities in Response to Fire Events", 2014 Annual Reliability and Maintainability Symposium (RAMS), Colorado Springs, CO, January 2014.

[6] ASME/ANS RA-Sa-2009, Addenda to ASME/ANS RA-S-2008, "Standard for Level 1/Large Early Release Frequency Probabilistic Risk Assessment for Nuclear Power Plant Applications", The American Society of Mechanical Engineers (ASME), New York, New York, 2009.

[7] Electric Power Research Institute and U.S. Nuclear Regulatory Commission, EPRI/NRC-RES Fire HRA Guidelines, EPRI 1023001/ NUREG-1921, July 2012.

[8] Electric Power Research Institute, The EPRI HRA Calculator® Software Manual, EPRI 3002004030. Palo Alto, CA, May 2014.

[9] EPRI TR-100259, An Approach to the Analysis of Operator Actions in Probabilistic Risk Assessment, Electric Power Research Institute (EPRI), Palo Alto, CA, June 1992.

[10] Swain, A. D. and Guttman, H., Handbook of Human Reliability Analysis with Emphasis on Nuclear Power Plant Applications, NUREG/CR-1278, Washington, DC: U.S. Nuclear Regulatory Commission, August 1983.

[11] Swain, A., Accident Sequence Evaluation Program Human Reliability Analysis Procedure, NUREG/CR-4772, U.S. Nuclear Regulatory Commission, Washington, DC, February 1987.

[12] The SPAR-H Human Reliability Analysis Method, NUREG/CR-6883, Idaho National Engineering Laboratory for U.S. Nuclear Regulatory Commission, Washington, DC, August 2005.

[13] Hollnagel, E., *Cognitive Reliability and Error Analysis Method – CREAM*, Elsevier, New York, NY, 1998.

[14] Williams, J., HEART - A Proposed Method for Assessing and Reducing Human Error, 9th Advances in Reliability Technology Symposium, National Centre of Systems Reliability, University of Bradford, Bradford, UK, 1986.

[15] Deep Water: The Gulf Oil Disaster and the Future of Offshore Drilling, Report to the President, National Commission on the BP Deepwater Horizon Oil Spill and Offshore Drilling, January 2011.

[16] Hubbard, A. and Embrey, D., "Deepwater Horizon – Summary of Critical Events, Human Factors Issues and Implications," Human Reliability Associates, Ltd., 23 September 2010.

[17] Ness, A., "Lessons Learned from Recent Process Safety Incidents," CEP Magazine, American Institute of Chemical Engineers (AIChE), New York, NY, March 2015.

[18] *Guidelines for Preventing Human Error in Process Safety*, American Institute of Chemical Engineers (AIChE), Center for Chemical Process Safety (CCPS), August 2004.

[19] Gould, K.S. and Ringstad, A.J (StatOil) and Van De Merwe, K. (Det Norske Veritas), "Human Reliability Analysis in Major Accident Risk Analyses in the Norwegian Petroleum Industry", Proceedings of the Human Factors and Ergonomics Society 56th Annual Meeting, San Diego, CA, October 2012.

[20] Boring, R.L (Idaho National Laboratory) and Øien, K. (SINTEF), "Defining Human Failure Events for Petroleum Risk Analysis", Probabilistic Safety Assessment and Management PSAM 12, Honolulu, HI, June 2014.

[21] Akyuz, E. and Celik, M., Application of CREAM human reliability model to cargo loading process of LPG tankers, Journal of Loss Prevention in the Process Industries, Vol. 34, pgs. 39-48, Elsevier Ltd., 2015.

[22] Technical Basis and Implementation Guidelines for A Technique for Human Event Analysis (ATHEANA), NUREG-1624, Rev. 1, U.S. Nuclear Regulatory Commission, Washington, DC, May 2000.

442486 Critical Considerations in the Vacuum Protection of Butane Spheres

Abstract:

Vacuum protection of pressure vessels, including spheres, is not well defined in the ASME Boiler and Pressure Vessel Code (BPVC) Section VIII, Division I or API Std. 520. Vacuum protection of pressure vessels usually follows API Std. 2000 on low pressure tanks for guidance.

Vacuum can occur on butane storage spheres at temperatures of 25 to 30 F due to butane condensing. This paper addresses the challenges in vacuum protection of butane storage spheres, including both the positive pressure rating of vacuum devices and sizing of vacuum vents for a butane condensing scenairo.

442493 Assessing Risk in the Supply Chain Using the Bowtie Method

Abstract:

After a number of recent incidents involving the transportation and storage of hazardous materials, many industry end-users wish to understand their level of risk exposure during the supply chain. This exposure can either be via actual ownership or when the product can be associated with the concerned party. Ultimately, this may mean that an end-user's procurement group will want to see that their suppliers of products and services are performing as needed in the control of these risks.

A qualitative risk assessment is a technique by which these risks can be identified, understood, and evaluated. This paper will describe the application of one such method, where the bowtie barrier analysis is central to the means by which this assessment is completed. The barriers identified are those necessary to prevent and mitigate undesired events that could result in a fatality or a catastrophic incident to the environment, public, or a company's reputation.

The overall method employed in this process includes the following steps:

1. Hazard identification
2. Bowtie diagram development and barrier identification
3. Barrier screening to identify safety critical elements - those that provide a credible reduction in the risk of fatalities or other catastrophe
4. Performance standard development for the safety critical elements

A quantitative risk assessment can also be completed to provide additional risk knowledge to aid in prioritization and decision-making. The performance standards are then enforced appropriately throughout the supply chain. This will ensure that the transportation and handling of hazardous materials meet the purchasing company's expectations with regards to proper risk management.

442386 Reference Framework for the Application of Quantitative Risk Analysis for Hydrocarbon Pipelines, Coupled with Uncertainty Treatment Methods: Techniques for Analysis and Modelling Methods

Abstract:

Risk assessment is imperative to reduce consequences of accidents generated by leaks on pipelines. After a failure in an urban pipeline, the contents can be released rapidly forming a spreading pool in the surface or a vapor cloud that can lead to different events such as fires and explosions, affecting the population in the area. Generally, deterministic assessments are carried out where certain assumptions are made in the models used to calculate failure and event probabilities, event consequences and risk values. This causes the existence of uncertainty that is not considered and therefore it generates results only for "mean" or "worst-case" scenarios. Not taking uncertainty into account can lead to decisions with an unknown degree of conservatism: overestimate results could increase cost of decisions and underestimate results can lead to poor risk management. There is a lack of a clear directive to use methods of uncertainty assessment in risk analysis and for that reason, a framework is needed for carrying out quantitative risk assessment coupled with uncertainty treatment methods in order to handle, reduce, and properly inform uncertainty to the decision makers.

Uncertainty appears when we try to represent a system using different analysis techniques and therefore, depends on how accurately this process can be done. This in turn relies on how complex is the system under analysis and which techniques can be used. As part of the framework mentioned above, a classification is suggested to adequately select methods in probabilistic risk analysis to take into account different types of uncertainty in different kinds of representation and system analysis techniques. The classification is first based on the complexity of the system to be analyzed and then in the complexity of the modeling technique that is used to deal with system complexity. System complexity depends mainly on three variables: 1) understanding about the system (variables and its interactions) 2) the number of elements in the system 3) complexity of system interactions. Depending on system complexity, different techniques to represent it and analyze these systems in risk analysis are proposed and then, methods to propagate uncertainty are mentioned. A comparison of methodologies is presented to provide the engineer with tools to use adequately these methods in different situations and provide guidelines to deal with uncertainty in risk assessment.

Dynamics Operational Risk Management in Organizational Design, the Challenge for Sustainability

Salvador Ávila Filho
Universidade Federal da Bahia, UFBA
Rua Aristides Novis, Federação 02, Salvador Bahia
avilasalva@gmail.com

José Filipe Michel Gagliano Ferreira
Universidade Federal da Bahia, UFBA
Rua Aristides Novis, Federação 02, Salvador Bahia
gagliano.fm@gmail.com

Carlos Roberto de Oliveira Sousa
Universidade Federal da Bahia, UFBA
Rua Aristides Novis, Federação 02, Salvador Bahia
carlososousa.cs@gmail.com

Ricardo Araújo Kalid
Universidade Federal do Sul da Bahia, Itabuna
Ricardo.kalid@gmail.com

Prepared for Presentation at
American Institute of Chemical Engineers
2016 Spring Meeting
12th Global Congress on Process Safety
Houston, Texas
April 11-23, 2016

Dynamics Operational Risk Management in Organizational Design, the Challenge for Sustainability

Salvador Ávila Filho
Universidade Federal da Bahia, UFBA
Rua Aristides Novis, Federação 02, Salvador Bahia
avilasalva@gmail.com

José Filipe Michel Gagliano Ferreira
Universidade Federal da Bahia, UFBA
Rua Aristides Novis, Federação 02, Salvador Bahia
gagliano.fm@gmail.com

Carlos Roberto de Oliveira Sousa
Universidade Federal da Bahia, UFBA
Rua Aristides Novis, Federação 02, Salvador Bahia
carlososousa.cs@gmail.com

Ricardo Araujo Kalid
Universidade Federal do Sul da Bahia
Itabuna
ricardo.kalid@gmail.com

Keywords: business risk and corporative, human factors and organizational, tools, sustainability

Abstract

The maintenance of the organizational sustainability requires acknowledgment that business risks are dynamic and, therefore, depend on the balanced analysis of issues relating to product and process with people and culture. The analysis of the dynamic risk occurs in three steps as described as follows. The first analyzes the balance between the human and social factors with process and product factors, generating preliminary recommendations from probable scenarios through techniques such as LOPA and PRA with recommendations to avoid accidents and technical failures. Risks related to the environmental and human aspects are analyzed with PRA-SH to anticipate future events that may cause technical failure or evoke a stressor environment in the workplace. The second step requires knowledge of the organization in its aspects of management, technical and non-technical, inputs and productivity benchmark to define the assumptions and criteria used to project performance of operations, including the skills required by the business. The result of this study is the definition of technical, human, organizational and natural factors for the sustainable operation of production system. The techniques applied in this second stage are: HAZOP and HAZOP-Social; requirements for the task (GMTA, PADOP); events cause and consequence analysis; and analysis of barriers for events (BOWTIE). In the third stage of dynamic risk analysis rescues the criteria and assumptions established in the previous step by defining the management model and detailing the technology used. In this phase, it will be handled operational routines, asset control, control of failures and possible crisis. Thus, it will designed the systems of technical area (equipment, processes, products and control modes), interface (task, management, and human-machine interface), and social area (culture and biases, cognitive decision, stress analysis and behavior). The factors that make up the control of operations are analyzed in the standard procedure way and in the failure analysis using the following techniques: equipment (FMEA, FMEAH); product and process (monitoring and statistical control); task in complex mode (GMTA, CARMAN, PADOP); man-machine interface and control room (PRA-SH and Regional Criteria for control room design); culture (human reliability committee for organizational change and archetypes analysis); cognitive decision (study of cognitive failure); stress and decision (LOPA); and, stress and occupational health (LESHA). These techniques for analysis of operations in a dynamic way have metrics based on the operator's speech from the results of routine by reflecting the occupational consequences (accident), environmental (pollution and waste), climate (CO_2 emissions) and economic (competitiveness losses based in the rates of consumption and productivity). The result of these measurements allows the routine adjustments in the design of operations, the organization's project settings and business project. A deployment exercise will be proposed for the chemical industry. Diagram will be presented reflecting the dynamic in the management of operational risks.

Glossary

ANP - National Petroleum Agency
FMEA – Failure Mode and Effect Analysis
FMEAH – Human Failure Mode and Effect Analysis
FPSO – Floating Production Storage and Offloading
FTA – Fault Tree Analysis
GMTA – Goals-Means Task Analysis
HAZOP – Hazard and Operability Study
HAZOP-Social – Social Hazard and Operability Study
HSE – Health, Safety and Environment
LESHA – MultiLayer Stress Effect on Health & behavior Assessment
LODA – MultiLayer Operation Decision Assessment on Progressive Stress
LOPA – MultiLayer Operation Protective Assessment
MDA – Methyl Diphenyl Amine
PADOP – Patterns and Procedures in the Operation
PHR – Preliminary Hazard Analysis
PRA – Preliminary Risk Analysis
PRA-SH – Socio-Human Preliminary Hazard Analysis
THERP – Technique for Human Error Rate Prediction

1 A discussion on reliability and risk – an introduction

1.1 Industrial risk analysis

The initial thought on the issue of risk and subsequently his management was primarily driven by the financial sector, specifically post-World War II [1][2]. However, in parallel with accelerated development of industry and processes increasingly complex - which implies exposure to workers, society and the environment to hazards ever growing - this issue gained notoriety in the industrial environment.

The occurrence of major industrial accidents in Brazil, and around the world, became the subject Risk Analysis and Management more popular to meet the demands of society and governments about the safety of industrial processes. It is in this context [3] that is needed better management and mitigates the impacts of accidents. One reason for the importance of developing risk management was the change in constraints for operating license after the Seveso accident in Italy (1976).

Schneider and Martins [4] define risk as a result of the frequency of occurrence and the consequences generated by an unwanted event, so that risk reduction is by reducing the frequency of undesirable events and the mitigation of the consequences. The risks that can be quantified are risks of injury to the person, damage to property and environment, and risk of loss of reputation. Risk management depends on the level that is being analyzed: equipment, process unit, facilities, and other levels.

According to Bharatiya [5], an industrial hazard example is the handling of flammable products that can be released causing fire and explosion and resulting impacts on workers and third parties, as well as economic losses such as: profit loss caused by shutdown plant; damage to equipment; penalties for release of toxic material, etc. In this regard, it is necessary application of qualitative and quantitative techniques of risk analysis and its review to ensure a safe working environment for employees and neighborhood.

Qualitative techniques of risk analysis such as PHA/PRA (Preliminary Hazard Analysis / Risk) and HAZOP (Hazard and Operability Studies) allow identifying and assessing risks and hazards associated with industrial processes, and identify operational problems through scenarios identification. According to Fleming and Garcia [6] PHA/PRA technique performs a risk assessment for each identified scenario, develops mitigation measures and specific studies to calculate the probability and analyzes the effects through the vulnerability of these scenarios. The HAZOP allows analyzing risk scenarios using a procedure that generates questions and answers in a structured way by guiding words that enable the identification of possible deviations of the operation. The review team examines the links between the causes and consequences of these deviations.

Regarding the semi-quantitative techniques for risk assessment, there is a tool for layers of protective analysis (LOPA) that is used in the safety integrity level setting against unwanted events in safety instrumented system [7]. His concept was first treated by Center for Chemical Process Safety [8].

In quantitative analysis, the various methods are distributed from the analysis of consequences, which is measured the extent of damage caused by adverse events, to quantify the frequencies and probabilities of each event causing accidents [4]. Among the most widely used techniques are Vulnerability, Fault Tree Analysis (FTA) and LOPA. The techniques can be combined in a study of quantitative risk analysis.

The development of quantitative criteria can follow internationally recognized parameters. In general, the aim should be to develop criteria also based on the existing company culture and the most effective process in this direction would be a hybrid of internal and external internationally accepted criteria [9].

1.2 Risk analysis services

Brazil has been placing on a world stage major events that generate therefore great responsibilities. Mega-events such as road races, international shows, World Cup soccer (2014) and the Olympics (2016) demand new ways to manage the risks. Experiences with carnivals indicate the need for organization in public management to meet the security needs of the population, quality in the health area, mobility, and social life.

For each event has a positive impact on society and economy, it is important to have a control of risks in the event of scenarios. According to Correia [10], the factors that pose major difficulties in quality control and risks in the event derive from the fact of being subjective experiences and difficult to measure.

Each event is unique and presents its own security risks. Major events bring large crowds with factors to be analyzed for risk management [11]: demand complex planning with different stages in the design and operation; have different participating organizations that serving people of different customs; It works with many volunteers and inexperienced staff; It has national and international media projection; demands less than the time required for the complexity of the services; It has regional, state and national economic impact; and requires careful consideration for risk management.

For large events are ephemeral, not admit failures, delays, and its facilities and spaces are in most provisional, at the same time these experiences are characterized by subjectivity, It is more efficient to analyze the risk qualitatively [10]. In this way the analysis generates a more tangible and simple approach, easy to study and implementation by the organizing companies.

1.3 Integrating the reliabilities: equipment, operation and human

The first steps in the study of reliability were been conducted in the early twentieth century (the end of the World War I), but the concept of reliability as it is treated today only began to take shape during the World War II in an attempt to improve the reliability of missiles [12]. For this classic model of reliability, Kardec and Nascif [13] define reliability as "the probability of the system or device to perform the function successfully required, under specified conditions of operation and a certain time".

The development process of reliability engineering was important in product development and process improvements. In the latter case your application has a strong relationship with the evolution of maintenance management acting not only in the transition between corrective, preventive and predictive models, but above all, the definition of maintenance strategies where each model has a degree of importance that depends on the results of investigations into the system failure patterns. That is, even if the corrective, preventive and predictive models represent, respectively, the 1st, 2nd and 3rd phase of maintenance management, any one of them may be applicable depending on the operational context of the organization and its failure patterns. The concept of reliability then goes to lead the efforts for efficient maintenance.

To meet the demands of analyzing the nature of the failure and its consequences several tools were developed to improve the reliability of equipment. The most used are: Failure Mode and Effect Analysis (FMEA), and Root Cause Failure Analysis, Method of Analysis and Troubleshooting and Reliability-Centered Maintenance.

However, while many companies design the best systems based on this reliability model, taking into account the great difficulty in maintaining the availability and reliability of equipment and machinery at satisfactory levels and the cases of accidents with environmental impacts and fatalities, it is clear that such practices are not sufficient to maximize system reliability and/or guarantee the effectiveness of crash prevention barriers. Therefore, in order to fill this gap, many studies have been conducted to include the role of human in this process, what we call human reliability analysis. These studies began to gain strength from the sixth annual meeting of the Human Factors Society which had as its theme the quantification of human error. So was born the first formal method for human reliability analysis – the THERP technique [14].

According to Pallerosi *et al.* [15], human reliability can be defined as "the probability that a person does not fail to comply with a task in a given period of time, in appropriate environmental conditions and available resources to do it". The author also classified two types of procedures to quantify the human failures. One is called the prediction methods and includes tools as TESEO, HEART, SPAR-H, THERP, ASEP, MPPS and others. The other method is classified as evaluation methods and use theories and probability distributions as Gamma, Weibull, Exponential, Binomial and other distributions applied to each type of task analyzed.

The elements or factors that contribute to human error happen are related to the cognitive and operational processes. These elements of the cognitive process of decision-making are: perception, memory, abstraction, emotions, stress, fatigue, knowledge, skills, communication, tasks, procedures, organizational culture etc. Ávila [16] discusses these factors in the operator's performance in the control of socio-technical processes in task execution and learning stage. These factors can be measured by individual results in the task accomplishing, from the quantitative approach presented in verification methodology of learning parameterized (RGA - Reliability Growth Analyses).

The human factors analysis allows the projection and the calculation of risk and also the calculation of human reliability. Among the analytical techniques are: HAZOP SOCIAL, PRA-SH, FMEAH, LODA, etc. In the organizational dimension, the API 770 standard [17] aims to assess the company culture and the impact on human and operational reliability diagnosing aspects of policy, procedures, man-machine interface, and issues relating to the roles and task settings. The latter issues are present in David Embrey (2000), Frank Lees (1996), Rasmussen (1997) and Avila studies (2009).

The understanding of cognitive processes is essential to increase the human and operational reliability in organizations. Study the social, economic and emotional utility [16], cognitive function [18][19] and models for decision-making [20] are required to analyze the commitment and quality of the decision.

Therefore, approaches to socio-technical areas of reliability should not be an isolated study for each functional part of the industry [21]. The reliability of each part is interconnected with each other and their level of integration depends on the level of automation, the type of process, the manager's management style and the environment promoted by the organization. By understanding about the relationship between the functions of these systems makes it possible to study the integrated reliability. This knowledge enables the formatting of factors that increase the performance of staff and equipment leading to greater organizational efficiency.

1.4 Routine management and crisis management

The failure analysis and reliability project should be carried out at various stages of the project lifecycle - from conceptual design (detailing and assembly), through operation (control of operations and routine) to crisis management. The study of fault begins with the perception of signals from the process, people, equipment and tasks and advances in the analysis of incidents and / or accidents that cause financial loss (or image). These events, if not treated properly, can lead to loss of control in a chain reaction and It requires contingency actions not always known by the leaders, characterizing a crisis situation.

According Constantinides [22], crisis management is a systematic attempt to identify and detect potential crisis scenarios and take action to prevent them, contain its effects or interruption, and finally the recovery of the system normality. Muller [23] highlights the need for early warning systems projects of undesirable events and resolve them before they turn into a real crisis. He added that in case of failure of preventive measures is essential to take an immediate action.

2 Performance of GRODin and publications

The Group of Dynamics and Operational Risks (Grodin) located at the Federal University of Bahia intends to conduct diagnostic systems from applied research and works to develop solutions to complex problems in industrial sectors and services over the integration of academy, industry, and Public sector. Its constitution allows multidisciplinary making the investigative process dynamic starting in the subjective transformations of human and social behavior until the objective accomplishing of the task in industrial facilities with the search of metrics for decision making in organizations.

Currently (2015) the group consists of 25 masters, 5 doctoral and 10 graduate students with graduate completion forecast for masters and doctors around 10 per year. Large areas of concentration of Grodin's publications are: (a) Process Safety; (b) Human Factors; (c) Reliability and Risk analysis in socio-technical systems; (d) Energy Efficiency and Sustainability.

(a) Process safety

Publications performed in process safety seek always discuss the inclusion of human and organizational factors among the aspects considered in the accident process. Thus, several articles were discussed and presented at conferences as described: (a1) risk analysis of the factors that define human performance for the tasks in equipment and processes classified as critical [24]; (a2) the analysis of the performing task environment, the relationship between man and technology that result in failures and the discussion about leadership and the moment of decision [25]; (a3) analysis of underlying causes (human error) which are involved in the accident process (economic bias, risk management, conflicts between policy and practice, cultural traits, task planning, contracted management) [26]; (a4) analysis of decisions of the operation under progressive stress on the sinister situation [27]; (a5) analysis of high frequency and lower impact accidents considering the inclusion of human factors and complexity of the task [28]; (a6) analysis of social relations in oil platform and the implications for an enabling environment to an accident [29]; (a7) analysis of organizational change, drafting checklist, impact analysis of possible process safety events and the need for a more detailed discussion of the human factors [30]; and (a8) risk analysis of public services in cultural mega events (carnival) [31] and sporting events (World Cup) [32].

(b) Human factors

The publications performed on Human Factors always tries to discuss the analysis of the intrinsic variability of the man in an organizational and cultural environment and that brings consequences as gains or losses for the society and industry. Thus, several articles were discussed and presented at conferences as described: (b1) usability of procedures based in operators opinion of the fertilizer industry [33]; (b2) complex task evaluation taking into account the requirements of criticity [34]; (b3) cognitive deficit analysis during the performance of executive function in performing tasks [35]; (b4) social assumptions that change cognitive decision for the review of ergonomics standards in the control room [36]; (b5) analysis of archetypes and their derivatives to avoid improper behavior on safety of industrial plants [37]; (b6) Operator behavior analysis and leadership in emergency situations, high stress caused by chain reactions [27]; (b7) trace analysis of the Brazilian and global culture, biases, and bad habits that hinder the establishment of operational discipline [38]; (b8) stress analysis in industry, health care and diving and its consequences on the task, behavior and health [39]; (b9) changing operator behavior based on social dynamic culture, difficult to detect changes before impacting events happening like accidents with economic, human and environmental losses [40].

(c) Reliability and risk in socio-technical systems – complex processes

Publications performed in Reliability and Risk Analysis in complex processes discusses how to detect and calculate the risks and failures always considering the presence of human performance factors. Thus, several articles were discussed and presented at conferences as described: (c1) relationship between loss of industrial processes and connection to the cognitive processes with impacts measurement [24]; (c2) the operational failure exergy relating the energy loss in cognitive failure with the loss of energy in the industrial process heat systems [41]; (c3) evaluation of failure in complex processes, complex tasks and complex social relations in the chemical industry [34]; (c4) dynamic risk analysis for industrial sustainability [42]; (c5) human factors and equipment failures analysis using tools such as SPAR-H and Weibull ++ in the nuclear and chicken industries [43][44]; (c6) Analysis of socio-technical reliability in the refining industry [21]; (c7) risk analysis of the pre-salt considering: phenomenology, reliability model and risk factors [45][46].

(d) Energy efficiency and sustainability

Publications performed in energy efficiency and sustainability seek to discuss what are the opportunities to reduce consumption of energy and water and ensuring the sustainability of business. Thus, several articles were discussed and presented at conferences as described: (d1) mass balance and energy in cooling towers [24]; (d2) audit in organic effluent and rainwater to prevent process losses; (d3) audit of environmental deviations in industrial routine [47]; (d4) contamination control program at source by mapping effluent [48]; (d5) application of mass balance techniques and energy to identify opportunities to reduce energy and water losses in the industry; (d6) a discussion of systems and cooling towers, challenges to avoid energy losses and water; (d7) identification and calculation of thermal performance and energy efficiency in equipment and industrial processes [49]; (d8) Analysis of loss of condensate and steam heating systems and energy recovery [50].

3 A vision of sustainable organization based on risk

Maintaining organizational sustainability requires an understanding that the business risks are dynamic and the operational risks depend on a balanced analysis of issues related to product and process with the people and culture issues. So, to start the discussion on sustainability in high-risk industry a management model will be presented.

An efficient risk management of business develops premises to organizational project considering social-technical aspects. This risk management draws an industrial operation format including multidisciplinary analyzes of different levels of normality (default to failure and the crisis). It is intended with this model to measure the effects and impacts on society and then review the criteria and procedures adopted by project operations.

3.1 Inputs in the Risk Management and Organizational Design

Important inputs for the analysis of dynamic risk are: entrepreneur demand (vision and mission), technological conceptual design (product and process), analysis of the availability of resources and production factors (energy, materials and people), and external demands represented by the law and behavior of the stakeholders in the existence of this business. The risk analysis stage to address the dynamic risk intent to insert a discussion about the process and product taking into account the type of task (or the procedure), the cognitive models for decision (local or multicultural) and the social behavior of internal and external community of the organization.

3.2 Step one: Risk Management - socio-technical risk analysis

The first step of socio-technical risk analysis balances human-social factors with process and product factors generating preliminary recommendations. In the case of technical risks are evaluated the product, process and the likely scenarios through LOPA and PRA and getting as result recommendations to prevent accidents and technical failures. For non-technical risks such as environmental and human aspects is performed an PRA-SH to anticipate future events related to human errors that can cause technical failure or generate an stressor environment at the workplace. At this time are considered the risks of events that involving cognitive faults and failure on the task resulting from social processes.

3.3 Step two: organizational project - detailing of technical and non-technical project

The second step requires knowledge of the detailed organization design in the technical and non-technical aspects to define the assumptions and criteria used for the design of operations. Thus, the organizational project aims to open the product technology, process, service, and control to discuss which the necessary demands on leadership, behavior, attitude, and skills.

The intrinsic variability policies, humanities, social and variability of nature (climate, biota, etc.) are scored as important for maintaining the environment in which the industry is installed and intends to stay in a sustainable manner. The organizational design analysis requires detailed information of technology, knowledge in integrated management system, consumption analysis and energy benchmark, water, materials and people (productivity) to establish performance indicators. The result of these studies is the definition of technical, human, organizational, natural factors for the sustainable operation of the production system. The techniques applied in this second phase are: HAZOP, HAZOP-Social; requirements for Task (GMTA, PADOP); cause and effect analysis of the events; barriers analysis of events (BOWTIE).

PADOP is a method that starts with the analysis of the requirements for feeding the following phases: environment, planning and execution of the task. Others steps such as task control and conformity analysis under review are performed in operational routine (see item 3.4).

3.4 Step three: design and control of operations

In the third stage of dynamic risk assessment is intended to rescue the criteria and assumptions set out in the previous step, organizational design, where technology is detailed and an integrated management mode is defined. In the project of operations, it will be handled operational routines, asset control, and control of failures and possible crisis. So, it will be designed systems for operation in the technical area (equipment, processes, products and control modes), the interface (task management, and human-machine interface) up to non-technical area (culture and biases, cognitive decision and stress and behavior analysis). The results of these measurements allow adjustments on the routine of the design of operations and business and organization projects.

These factors and elements that make up the control of the operations are analyzed on standard procedure mode and failure analysis in the following areas and using the following techniques: equipment (FMEA, FMEAH); product and process (monitoring and statistical control); task in a complex way (GMTA, CARMAN, PADOP); man-machine interface and control room (PRA-SH and regional criteria for control room projects); culture (human reliability committee for organizational change and archetypes analysis); cognitive decision (study of cognitive failure); stress and decision (LOPA); and stress and worker health (LESHA). These techniques for analysis of dynamically operations present measurements and indicators that are based on operator's speech. This speech is announced from the routine results and reflects the occupational consequences (accident), environmental (pollution and waste), climate (CO2 emissions) and economic (loss of competitiveness based on consumption rates and productivity).

In this case, the PAPOD technique intent to compare the standard designed for task control rates with measurement of the process after the task performed. It is worth noting that these standards include variables of integrated management (safety, quality, environment, occupational health, etc) as well as socio-humans aspects (quality of communications). In case of deviations it will be realized preventive corrections with necessary revision to the task or written procedure.

4 Application of exercise

An implantation exercise is proposed based on the application of different technical and management systems in research in various chemical process industries. Activities and analyzed processes involve energy products (oil and LPG), chemical (sulfuric acid, methyl diphenyl amine – MDA, and explosives), petrochemicals (basic petrochemicals and fertilizers), polymers (polycarbonates) and plastic (plastic bottles).

These various research opportunities involve aspects of routine at the work place and characteristics of management that promote the company's results close to normal or fault situations. The experiences are gathered in an integrated way toward sustainability considering that risks are dynamic and that the systems are socio-technical.

Some of these techniques are applied with quantitative results indicating the intensity of the problem and the volume of investment to be taken and others indicate the existence of the failures and suggest the construction of safeguards, but in a qualitative way.

This exercise will only treat of techniques relating to the socio-environmental-human risk and it will quote only within the management system proposed and which evaluations (of process and product) should be interconnected.

4.1 Inputs in the risk management and organizational design

For this model that considers the dynamic risk, the entrepreneur discusses the social variability in the possibility of failure and practices a vision and mission more sustainable, which admits the existence of a multidisciplinary approach to the definition of the operating pattern. This discussion is part of the control of the operational routine. So, inputs to this management system from experienced cases it is suggested.

Entrepreneur demand

Case 1: the entrepreneur demand to identify the routine deviation is a requirement to draw up barriers that prevent systemic failure. To discuss the causes of failures that lead the operation to a mode of low productivity, an entrepreneur of the LPG industry recognizes the need to analyze the routine. The operator's speech feeds the construction of the fault tree for routine events and together with the task analysis identify the GAPs of competence that lead to failure [51].

Case 2: the perception of the manager must be attentive to behavioral deviations that can lead to accidents. In this case, the corporate HSE manager of the explosive industry believes that the study of human typology is necessary to change the criteria in risk management. This belief is based on the possible change of behavior, which caused two consecutive accidents not expected in different countries, after a history of 20 years without accident. This entrepreneur corroborates the following statement: "cultural changes alter the behavior of the workers and there is a need to study new criteria for dynamic risk management".

Analysis of the availability of resources and production factors (water, energy and people)

Production factor: energy. The water crisis in the state of São Paulo [52] and the reduction of energy quotas for petrochemicals due to its unavailability caused by climate change indicates that the analysis of availability is critical input in the organizational design to address the variability of nature.

Production factor: water. CEOs of Brazilian refineries are being instructed by National Agency of Water (for now voluntarily) for an actual 20% reduction in water consumption. This highlights the urgent need to change behavior and decisions in this organizational design.

Production factor: people. The opening speech of the Brazilian Association of Risk and Reliability (ABRISCO) indicates that the Brazilian engineer does not know and does not quantify industrial risks indicating the unavailability of appropriate skills to quantify investments in safeguards to prevent accidents.

External demands legislation for new behaviors of stakeholders

Stakeholder – National Petroleum Agency (ANP). After the FPSO production ship accident in the city of São Mateus, an investigation realized by the ANP [53] indicates the need to discuss the issues of contract management, omissions, commitment, crisis management and decision in stress. These issues are part of social- human variabilities.

Stakeholder – University which reflects demand for workers. In Risk Management is important to define the requirements necessary from technology and local culture for planning the task (PADOP). From the master's degree studies on the task analysis it was found that most of the 450 operators of a fertilizer industry accepted the importance of establishing social and human premises to carry out the critical task.

Stakeholder – Multinational oil companies. From international meetings with reliability consulting groups that serve the oil industry, it was found out the need to discuss regionalities to change the ergonomic criteria in control room. This indicates that the CEOs of major companies (Chevron, Statoil, BG) are reflecting on the question of risk review to include in their analysis the dynamic and social aspects (cultures and people).

4.2 Step one: risk and opportunity management - socio-technical risk analysis

After the hazard analysis and technical risks by conventional application of the techniques APP and APR (to business, processes and products) it is natural to use the vulnerability analysis in the most critical scenarios to indicate the level of impact on human and nature. Similarly to analyze the social and human risk seeks to detail the various performance shaping factors [54] that indicate the level of consequence in case of human errors.

Fault analysis (in 20 cases) in the chemical process industry indicates that elements such as organizational climate, leadership, skills, knowledge, and decision-making processes are related to human error in performing the task.

Several cases involving deviations, failures, incidents and accidents were analyzed and allowed the construction of the anatomy of failure in complex mode relating human and organizational factors [26]. Among the cases are: director's behavior in the plant starting management (polycarbonates); deviations in mitigation of waste generation reaction (MDA); operational control to avoid stoichiometric reaction stop (MDA); skill in the perception of vacuum loss in sulfuric acid unit; failures caused by improper turn change (MDA); investigation of effluent contamination source (MDA, sulfuric acid, lab, fertilizers); diagnosis to map the gaps in the process (MDA, polycarbonates, LPG); negotiation of new operating parameters without causing downtime (MDA); improper use of individual protection equipment and omission by fear of guilt (MDA); oring diameter change at startup leading to failures with solvent losses (polycarbonate); and others.

Based on the investigation of cases the initiators factors to human error are [26]: design criteria, risk management, economic bias, cultural traits, task planning, and conflict between policy and practice. It is also important to identify the social dangers to avoid accidents before assembling the installation and make the hires to operate the industrial site.

A specific case was applied at the stage of design of an oil exploration platform in northeastern Brazil during the graduate student research in the area of security and human factors analysis in 2013 [55]. After seven risk analysis sessions for human and social factors, the group of six experts (socio-technical) listed the following factors that can cause human errors in this installation that come into operation in 2015: difficulties in routine leadership; non-commitment to the safety result; low perception to failure; and improper internal communication.

It is worth listing the technique steps: identification of social and human dangers, social and human risk analysis, risk classification, interaction between risks (interaction matrix to correct risks), the final risk map, and treatment of social and human risk.

An item that is repeated in the workplace discussion refers to the effect of stress on the effectiveness of the task [56] and occupational health [57]. Thus, it interconnects the worker performance factors to the result of task with economic results for the industry. The causal relationship between work environment, cognitive fault and failure in the task are motivating ongoing research that result in techniques and applications.

Technique 1: analysis of social relations in the workplace. Application: oil platform where the rituals of loading and unloading, the distance and isolation are analyzed to list critical topics that can cause stress and human error [58] [29]. Technique 2: protocol for the analysis of stressor environment. Application: industry [56], nursing services [39] and diving services in deep waters. Technique 3: task analysis [33]. Application: chemical process industry. Technique 4: analysis of cognitive impairment [35]. Application: chemical industry.

4.3 Step two: organizational project - detailing of technical and non-technical project

Step 1 of this management model for sustainability is related to the fit project of product, process, and business, in a discussion of technology and economic viability. The analysis of social and human risk in the task is performed along with the APR technique of this future operation.

Step 2 is the organizational design that includes the adaptation of the technology for the future operation that includes product and process aspects, and non-technical issues related to teamwork in performing the task (culture and people). At this stage (detailing phase) are carried out studies of the conceptual design (made in step 1) and details the risks of failure to cause economic loss and / or the company's image.

Application of HAZOP-Social

The process deviation analysis during operation indicates the possibility of danger containment loss related to the industrial activities. The Social HAZOP tool works similarly to the conventional technique [16]. In its application the cognitive decision-making is tested through the deviations and recommendations are announced to prevent failure of attention, memory, competence, managerial decisions and others. Social HAZOP was applied to a draining operation of a decanter vessel in the refining industry as having maximum result the accident with fatality and with the critical lack of attention, centralized management profile and commitment to the task.

Diagnostic of skills gap

Identification of fault root cause in the chemical industry that has characteristics of processes and complex tasks [59] [34] requires the study of process variables and production beyond the qualitative and quantitative analysis of operator's speech [60][61]. This discourse analysis allows the organization to verify the main failure modes in socio-technical systems by locating the critical variables in the tasks, equipment, skills and operator behavior. When comparing the type of fault carried out (based on competence offered) with the demand for knowledge and skills in technology (competence demanded) have been related what are the gaps in competence.

This technique was applied in PhD research [51] of the gas compression unit (LPG) in northeastern Brazil. In this case, the database was collected from one year of statistical analysis of process and production variables, and four months of shift book analysis in low performance periods. Diagnosis of gaps indicates the lack of skills in the following areas [25][62]: Task control (abstraction and logical); concepts (thermal and material exchange); failure (rotary, motorized valve and burner); skills (thermal cycles, pressurized system and logistics); process control and safety (supervisory and risk in the routine).

The gap of skills already identified in the LPG plant may result from lack of commitment of the people, the team cooperation and inappropriate profile of leadership. A classification model of commitment, the leader's behavior and aggregation of the team is suggested by Avila [51][25][62].

Research on behavioral intrinsic variability

Demand for leaders in chemical industry and oil organizations during the GCPSs events indicates the importance of analyzing the changing worker behavior as a result of Dynamic Culture. A methodology is proposed to find the biases of global and regional culture that promote behavioral changes towards the accident [40]. This method was structured to identify potential changes in teams of explosives chemical company with unusual situations accident in facilities overseas. The suggested steps with the respective models are: (1) analysis of social culture; (2) analysis of the social and human types to work; (3) analysis of the dynamic factors of culture or affective socio-economic utility; and (4) evaluation of the result in the worker's behavior performing the critical task.

Archetypes and regionalities analysis

Entrepreneurs from different areas (chemical and oil) have discussed the ergonomic criteria for control rooms and raise the possibility that tacit aspects influenced by behaviors are not being considered. This hypothesis is due to continuity failures in critical operations such as plant startup and shutdown. These questions led the researchers to analyze the control room construction criteria based on the ergonomics requirements in the EEMUA and ISO standards [36]. The result of the research is the mapping of critical situations that can result in accidents.

In the case of a LNG industry control room were related aspects of local and global culture that impact on the cognitive processing leading to failure in decision making. This tool depends on the study of product risks and social risks (APR and APRSH), the analysis of the major failures and relations with human error in similar facilities, and brings results in a cross-matrix, which relates the social influences, cognitive functions with interfaces and control room devices.

The regional or global behavior that can turn into social risk (eg, default and inadequate leadership) may be due to cultural biases from primary or secondary archetypes. A case of industrial gases plant is analyzed and actions are suggested in the form of ritual to neutralize or deflect improper behavior [37].

Variability of climate change

The urgency of actions to reduce GHG emissions, fuel consumption and water motivate researchers in techniques seek to adjust the demand for water and energy in relation to confirmation of the offer. The demand adjustment actions involving routine management, knowledge of techniques for thermal performance analysis and new technologies such as process intensification or modularization of projects. The research conducted (LAB, Olefins, Urea) indicate that the equipment involved in research and routine adjustment are: cooling towers, boilers, furnaces, condensate systems and unit operations with their reflux [49][50].

4.4 Step three: design and control of operations

The Control of Operational Routine depends on the knowledge of appropriate standards for the maintenance of a stable process with the appropriate cost. It is increasingly announced by managers of high-risk companies that the economic result is guaranteed if the image is positive - that is, the non-occurrence of events related to accidents.

The definition of operating standard to routine management is part of a policy initiated by the management of technical and non-technical risks. So, the foreknowledge keeps the assets in normal mode. The occurrence of deviations, failures and incidents in the industrial operation can cause a crisis situation that requires a new leadership profile (crisis and contingency management).

Cases of task analysis

In LPG industry there are high-risk activities that can cause fire or explosion. Purification of LPG avoids failure through moisture removal. The moisture-removing columns have a saturation point and are removed from operation to be regenerated. This, in turn, requires the moisture expulsion using heated LPG itself resulting from passage through a furnace.

During the investigation concerning the study of root cause analysis in LPG process [51] was made the analysis of the operation in the task, startup and breakdown in regeneration furnaces using techniques such as GMTA[20], CARMAN[63] and PADOP[26][33]. These techniques used for fault analysis in the task involves the logic and the nature of the failure, the chronology of events, connectivity between factors beside the point where failure goes out of the social mode (eg. communication) to the physical mode in GLP [51] and Polycarbonate [60].

The task control actions include the procedure review (requirements, targets, or steps) with training. When the risk, complexity of processes, the complexity of social relationships and stress levels are high (in part or in full), the training actions has no effectiveness demanding action on culture [56][33] or to the redesign of equipment and processes.

Cases of failure in equipment and process

The benzene hydrogenation is an intermediate step in the production of caprolactam. The recycle gas is maintained at constant pressure to control the reaction through a centrifugal turbo-compressor of the brand Demag. Repeated interruptions on Monday (after the weekend) raised the possibility of having human error in the operation and/or maintenance which did not allow the effectiveness of technical actions.

A FMEAH study was performed to identify the failure mode which problems of perception, attention and expertise were occurring in routine production. The method resembles the traditional FMEA, but includes analysis of human error in the equipment failure mode [64].

The degradation systems, changes in technology, organizational changes [30] and climate change may cause variations in the process which results in the modification of the possible event scenario, which in turn increases the risk of accidents. In this regard, through the signals emitted in routine, seeks to review the product risk, process and operations through the application of new techniques such as APR, APRSH, HAZOP, Social HAZOP.

Human performance case in state of stress

A startup case of sulfuric acid purification discussed the influence of stressful environment. The elements that make up the work environment to achieve a state of stress that affects human performance are: task risk level; health impacts, the cost of business and social environment; function of the task in the process; and analysis of the psychological and organizational types. The protocol indicated that the level of stress was 4.8 exceeding the stress level manageable (under 4).

An accident analysis in progressive stress

An accident involving LPG furnace, as discussed in this paper, was used to build a fault tree of technical and non-technical events in different levels of stress.

The application of LODA technique [27] from gradually more dangerous events indicates that the decision of the shift varies according to the level of stress. The progressive stress is promoted by the need for decision-making by building possible increasingly serious consequences scenarios at the following levels: deviation from the routine, lack of process, lack of equipment and task, accident and sinister.

Unfortunately, the psychological unpreparedness of staff in dealing with contingencies, especially the leaderships, makes the human error probability goes from 10%, in steps of process deviations and lack of control of process, to 50% in sinister steps.

Cognitive failure cases and vices in the industry operators behavior

In a startup of a chemical industry reactor [35] are analyzed psychological functions in the executive work of the task. From social relationship models and the executive function analysis it was identified the following types of failure on the team with priority level 1: lack of knowledge and difficulties in memory recovery.

In the sulfuric acid industry [40] were analyzed behavioral addictions through fault trees that indicated the need for vaccines for cognitive diseases, among which: "frozen" perception, low motivation to cooperate, not enabled memory, commitment corrupted, communication and leadership inadequate, and very simple mental map. This finding allows the construction of behavior adjustment programs.

In industrial gases and chemical industries were found two types of behaviors in the control room which allowed the analysis of two different groups archetypes. Some archetypes favor a robust safety culture and others hinder their establishment [37].

Root-cause analysis in complex systems

The operator speech analysis is a tool that links events occurred in the shift such as: problems with equipment, product quality, double-shifts, missing and other facts. The form of communication, the statistics of events, the possibility of omission and team cooperation are some of the aspects included in this speech.

On several occasions in the chemical process industries [60] were applied root cause analysis techniques in complex systems. In the production of alkyl amines it was noted that the delays were the result of errors in the setup of the campaign. In the production of methyl diphenylamine it was observed that the interruptions were strongly influenced by the formation of groups that are not communicated with each other and by the lack of knowledge of the importance of recycle line for the quality in the process. In polycarbonates production sought to reduce loss of methylene chloride. In this case the operator's speech has enabled to review: procedures (from the task analysis) and manager behavior in the plant startup in control room.

In LPG industry in which the concern is the leakage events, fire or explosion, the operator's speech helped to identifying critical areas, gap of competence for decision-making, and team behavior in emergency situations [51].

5 Conclusion

The model of operational risk management requires technical and social aspects are correlated at different levels of normality / abnormality of the processes with the consequences and impacts that produce in their own business, on environment and society. By multidisciplinary involved in the analysis of these risks, the definition of management strategies should take into account the dynamics of the risks in business.

In some cases analyzed were used the operator' speech as a tool for building the event fault tree in the routine that jointly with task analysis identified the gaps of competence that lead to failure where the criteria and procedures adopted can be reviewed in the design of operations.

In another case (explosives factory) was found to be necessary to investigate the human type to change the criteria in risk management. In this case, the entrepreneur recorded accidents with serious injuries to the worker after a history of 20 years without accident (plants located in different countries) and admitted that cultural changes alter behavior which highlights the need to consider new criteria for dynamic risk management.

From the perspective of production factors such as water and energy, is also recorded vulnerability in the sustainability of the business in major change situations in the availability of these resources requiring that teams know how to prioritize investments to overcome such difficulties.

The applications of various techniques mentioned in this work are structured in a sustainable model of risk management. This model requires generation of multidisciplinary expertise (socio-technical aspects) to properly handle the dynamics of risk.

6 Annex

Figure 1 – Risk Management Model for Sustainable Organization

7 References

[1] G. Dionne Risk Management: history, definition, and critique. Risk Management and Insurance Review, 16 (2013), 147-166.
[2] J. L. Vesper. *Risk Assessment and Risk Management in the Pharmaceutical Industry: Clear and Simple*. Parenteral Drug Association, New Vernon, 2006.
[3] M. R. Martins *et al.* Undesirable events Consequences analysis during the transport, loading and unloading of Liquefied Natural Gas (LNG) on ships. Proceedings of Brazilian Association of Risk - ABRISCO. Rio de Janeino, 2013.
[4] A. M. Schleder and M. R Martins. Methodology for quantitative analysis of the consequence of flammable leaks and / or toxic environments with barriers or confined. Proceedings of Brazilian Association of Risk - ABRISCO. Rio de Janeino, 2013.
[5] B. K. Bharatiya. Projet Hazard Analysis In Hydrocarbon Processing Industry. Proceedings of 4th Latin American Conference on Process Safety - CCPS. Rio de Janeino, 2012.
[6] P. V. Fleming and C. B. Garcia. Industrial and Environmental Risk Assessment With Preliminary Hazard Analysis (PHA) and Fuzzy Logic. Proceedings of Brazilian Association of Production Engineering - ENEGEP. 1999.
[7] R. Freeman. Quantifying Lopa Uncertainty. Process Saf Prog 31(2012), 240-247.
[8] CCPS. *Guidelines for Safe Automation of Chemical Processes*. Center for Chemical Process Safety, American Institute of Chemical Engineers, New York, NY, 1993.
[9] J. N. Shan. Understanding and Developing Quantitative Risk Criteria. Proceedings of 8th Global Congress on Process Safety, Houston, 2012.
[10] A. H. L. Correia. Strategic Marketing of Sporting Event. International Seminar of Sports Event Management. Ministério da Juventude e do Desporto. Centro de estudos e formação desportiva. Lisboa, 2001.
[11] Worksafe Victoria. Advice for Managing Major Events Safely. 1st Ed. Austrália. 2006. Available at: https://www.worksafe.vic.gov.au/__data/assets/pdf_file/0010/12502/major_events_organisers_advice.pdf. Accessed on January 3, 2015.
[12] F. S. Flogiatto and J. L. D. Ribeiro. *Reliability and Industrial Maintenance*. Elsevier: ABEPRO, Rio de Janeiro, 2009.
[13] A. Kardec and J. Nascif. *Maintenance – Strategic Function*. 4th Ed. Qualitymark Editora, Rio de Janeiro, 2013.
[14] R. L. Boring. Fifty Years of THERP and Human Reliability Analysis. U.S. Department of Energy National Laboratory. U.S.A., 2012.
[15] C. A. Pallerosi; B. P. M. Mazzolini and L. R. Mazzolini. *Human reliability - Concepts, Analysis, Evaluation and Challenges*. AllPrint Editora, São Paulo, 2011.
[16] ÁVILA FILHO S.; PESSOA F.L.P. and ANDRADE J.C.S. Social HAZOP at an Oil Refinery. Process Saf Prog - 32 (2013), 17-21.
[17] D. K. Lorenzo. API770 – A manager's guide to reducing human errors, improving human in the process industries. API Publishing Services, Washington, 2001.
[18] J. Reason. *Human Error*. Cambridge University Press, New York, 2003.
[19] R. Stenberg Junior. *Cognitive Psychology*. 3rd Ed. Artmed, Porto Alegre, 2008.
[20] E. Hollnagel. *Human Reliability Analysis Context And Control - Computers And People Series*. Academic Press Inc, San Diego, 1993.
[21] S. Ávila Filho. Reliability analysis for socio-technical system, case propene pumping. Engineering Failure Analysis, 56 (2015), 177-184.

[22] P. Constatinides. The Failure of Foresight In Crisis Management: a Secondary Analysis of the Mari Disaster. Technological Forecasting & Social Change, 80 (2013), 1657–1673
[23] R. Muller. Corporate Crisis Management. Long Range Planning, 18 (1985), 38-48.
[24] S. Ávila Filho *et al*. Risk Management and Prevention Program to Avoid Loss of Process (LP3) based on Human Reliability, Practical Application and Measurement. Proceedings of 7th Global Congress on Process Safety – GCPS. Chicago, 2011.
[25] S. Ávila Filho and A. C. Carvalho. Human Factors Assessment to Avoid Accident at LPG Industry. Proceedings of 6th Latin American Conference on Process Safety – CCPS. Rio de Janeiro, 2012.
[26] S. Ávila Filho. Assessment of Human Elements to Avoid Accidents and Failures in Task Perform, Cognitive And Intuitive Schemes. Proceedings of 7th Global Congress on Process Safety – GCPS. Chicago, 2011.
[27] S. Ávila Filho. Dependent Layer of Operation decision analyzes (LODA) to calculate Human Factor, a simulated case with PLG event. Proceedings of 7th Global Congress on Process Safety – GCPS. Chicago, 2011.
[28] S. Ávila Filho. Review of Risk Analysis and Accident on the Routine Operations in the Oil Industry. Proceedings of 5th Latin American Conference on Process Safety – CCPS. Cartagena de Indias, 2013a.
[29] S. Ávila Filho. Risks in the Platform: Social and Human Guidelines Avoid Accidents and Alter Standards in Oil Production. Revista Proteção, 2013b.
[30] S. Ávila Filho; E. T. Dantas and J. Duarte. The Requirements & Tools to Treat Process Safety Risk as Result of Organizational Change. Proceedings of 9th Global Congress on Process Safety – GCPS. Austin, 2015a.
[31] S. Ávila Filho, *et al*. Analysis of risk scenarios and mapping of communication processes to avoid incidents in Salvador Carnival in 2015/16. Proceedings of Brazilian Association of Risk – ABRISCO/PSAM. Rio de Janeiro, 2015a.
[32] S. Ávila Filho, *et al*. Risk Analysis in Sports Mega-events, UFBA Project in the World Cup. Proceedings of 34th Brazilian Association of Production Engineering - ENEGEP. Curitiba, 2014.
[33] S. Ávila Filho; C. R. Sousa and A. C. Carvalho. Assessment of Complexity in the Task to Define Safeguards Against Dynamic Risks. Proceedings of 5th International Conference on Applied Human Factors and Ergonomics – AHFE. Las Vegas, 2015.
[34] S. Ávila Filho. Failure Analysis in Complex Processes. Proceedings of 19th Brazilian Chemical Engineering Congress – COBEQ. Búzios, 2012.
[35] S. Ávila Filho and C. Costa. Analysis of Cognitive Deficit in Routine Task: as a Strategy to Reduce Accidents and Increase Industrial Production. Proceedings of European Safety and Reliability Conference – ESREL. Zurich, 2015.
[36] S. Ávila Filho and F, L. P. Pessoa. Proposition of review in EEMUA 201 & ISO Standard 11064 Based on Cultural Aspects in Labor Team, LNG Case. JofPM, 3 (2015), 6101 – 6108.
[37] S. Ávila Filho, M.L.A. Menezes. Influence of Local Archetypes on the Operability & Usability of Instruments in Control Rooms. Industrial Production. Proceedings of European Safety and Reliability Conference – ESREL. Zurich, 2015.
[38] S. Ávila Filho, *et al*. Influence of Brazilian Culture in the Workers' Behavior and Operational Discipline. Proceedings of Brazilian Association of Risk – ABRISCO/PSAM. Rio de Janeiro, 2015b.

[39] K. R. V. Nunes, *et al.* Working Hours Excessive and its Influence on Accidents, Productivity and Decision Making of Nursing Technicians in an ICU. Proceedings of Brazilian Association of Risk – ABRISCO/PSAM. Rio de Janeiro, 2015b.
[40] S. Ávila Filho; C. C. Silva and C. R. O. Sousa. Worker Behavior Changing by Dynamic Culture. Proceedings of 9th Global Congress on Process Safety – GCPS. San Antonio, 2013.
[41] S. Ávila Filho. The Exergy of Operational Failure, a Philosophical Discussion with Basis Practices. Proceedings of 19th Brazilian Chemical Engineering Congress – COBEQ. Búzios, 2012.
[42] S. Ávila Filho, J. R. N. Lopes and C. R. O. Sousa. Strategies to Industry Sustainability. Annuals of 20th World Renewable Energy Forum – WREF. Denver, 2012.
[43] A. C. Alcoforado and S. Ávila Filho. SPAR-H Application in the Crushing Operation of Uranium Ore. Proceedings of Brazilian Association of Risk – ABRISCO/PSAM. Rio de Janeiro, 2015.
[44] S. S. Dórea and S. Ávila Filho. Failure Investigation in Operations of Chicken Industry with Based Human and Technical Factors. Proceedings of Brazilian Association of Risk – ABRISCO/PSAM. Rio de Janeino, 2015.
[45] M. N. G Abreu and S. Ávila Filho. Assessment of Risk in the Exploration and Production of Oil at Pre-Salt Layer – Brazil, a Discussion about Social-Technical Factors. Proceedings of Rio Pipeline Conference & exposition. Rio de Janeiro, 2015.
[46] M. N. G Abreu and S. Ávila Filho. Assessment of Reliability and Risk at Dynamic Systems: a Discussion of Social-Technical Factors in the Exploration and Production of Oil at Pre-Salt Layer – Brazil. Proceedings of Global Congress on Process Safety – GCPS. Austin, 2015.
[47] S. Ávila Filho, *et al.* Environmental Deviation Assessment, Record, Failure, Probability and Investment. Chemical Eng Trans, 26 (2012a), 441-446.
[48] S. Ávila Filho, *et al.* Program to Avoid Product Leakage at Source, Fertilizer Case. Chemical Eng Trans, 26 (2012b), 447-452.
[49] S. Ávila Filho, *et al.* Iniciatives of Energy Efficiency Toward Sustainability, Opportunities & Challenges In: Rio Oil Gas Conference & exposition. Rio de Janeiro, 2014.
[50] S. Ávila Filho, *et al.* Avoiding Loss of Energy in a Petrochemical Industry, Operation and Design. Proceedings of World Renewable Energy Congress – WREC. Linköping, 2011.
[51] S. Ávila Filho. Etiology of Operational Abnormalities in Industry: a Model for Learning. Doctoral thesis - Chemistry School. Federal University of Rio de Janeiro. Rio de Janeiro, 2010.
[52] ANP. Investigation Report of an Explosion Incident in FPSO Ship in São Mateus Town on November 02, 2015. National Agency of Petroleum, Natural Gas and Biofuels. Available at: http://anp.gov.br/?pg=78836&m=&t1=&t2=&t3=&t4=&ar=&ps=&1452081674560. Accessed on January 15, 2016.
[53] R. Bonalume Neto. 30% Loss of Water During Distribution Contributes to Crisis in São Paulo. Sustainability Forum. Folha de São Paulo. São Paulo, 2014.
[54] D. K Lorenzo. API770 – A manager's guide to reducing human errors, improving human in the process industries. API Publishing Services, Washington, 2001.
[55] M. P. Barroso. The influence of human factors in APP and APR risk analysis techniques: a study case of an oil drilling platform in the northeast. Master's Dissertation – Polytechnic School. Federal University of Bahia. Salvador, 2013.
[56] G. G. Gonçalves, *et al.* Protocol to evaluate the influence of stress on the task involving human types, behavior and health: An exercise proposal. ESREL 2015. Zurick. 2015.

[57] S. Ávila Filho. LESHA – Multi-Layer progressive stress & impact assessment on health & behavior. Proceedings of 9th Global Congress on Process Safety. San Antonio, 2013.
[58] S. Ávila Filho. Human Errors Assessment, Social Environment at Oil Platforms. Proceedings of 4th Latin America Conference on Process Safety. Rio de Janeiro. 2012.
[59] C. Perrow. *Normal accidents: living with high-risk technologies.* NY: Basic Books, 1984. 453 p.
[60] S. Ávila Filho. Methodology to Minimize Generation of Waste at Source from the Investigation of Operational Abnormalities. Master's Dissertation – Polytechnic School. Federal University of Bahia. Salvador, 2004.
[61] E.S. Drigo, S. Ávila Filho and C. A. O Sousa. Operator discourse analysis as a tool for risk management. Proceedings of European Safety and Reliability Conference – ESREL. Zurich, 2015.
[62] S. Ávila Filho. Diagnosis in Socio-Technical Systems Aiming to Increase Human Reliability, a LPG Case. Proceedings of International Applied Reliability Symposium – ARS. Fortaleza, 2011.
[63] D. Embrey. Preventing human error: developing a consensus led safety culture based on best practice. London: Human Reliability Associates Ltd. 2000.
[64] S. Ávila Filho, *et al.* Human Factors Analysis in Turbocharger Equipment Failure FMEAH in Chemical Plant. Proceedings of 27th Brazilian Maintenance Congress. Rio de Janeiro, 2012.

Water and Energy availability analysis, a proposal for management for chemical industry

Salvador Ávila Filho
Federal University of Bahia – UFBA
Salvador – Brazil
avilasalva@gmail.com

José Rafael Nascimento Lopes
National Service of Industrial Learning – SENAI
Salvador – Brazil
jrafael.lopes@gmail.com

Jean marcel Prazeres Silva
Federal University of Bahia – UFBA
Salvador – Brazil
jmarcel_7@yahoo.com.br

Prepared for Presentation at
American Institute of Chemical Engineers
2016 Spring Meeting
12th Global Congress on Process Safety
Houston, Texas
April 11-23, 2016

Water and Energy availability analysis, a proposal for management for chemical industry

Salvador Ávila Filho
Federal University of Bahia – UFBA
Salvador – Brazil
avilasalva@gmail.com

José Rafael Nascimento Lopes
National Service of Industrial Learning – SENAI

Jean Marcel Prazeres Silva
Federal University of Bahia – UFBA

Keywords: water availability, energy analysis, sustainability

Abstract

The evidences that the chemical industry is not ready to deal with interruption of operations resulting from climate change and the complexity of decisions is increasingly apparent. In the state of São Paulo / Brazil, petrochemical industries are using served water from sewage treatment plants to supply the cooling tower makeup water. However, industries are not striving to analyze the availability of water and energy to suit the demand to supply.

The setting of premises to build scenarios of water availability, energy and skills is required. This will allow to make strategic plans, control programs, techniques and methods of investigation that aspects of products and processes are considered as well as the characteristics of the culture and behavior of managers and employees.

The water supply scenario and its use in the agriculture, livestock and cities indicate that there will be reduced availability of water for industry. As these consumer sectors are disorganized and there are losses in the distribution of water in the order of 30%, it is up to industry to worry about meeting their demand while reducing your water consumption.

The aim of this paper is to present an analysis model of availability of water and energy in the chemical industry, along with a behavioral adjustment model to enhance a large reduction in water consumption and energy supply to meet the expectations of these. The reduction in water consumption and energy depends on: management actions; switching procedures; behavior changes in the short term; application reuse techniques and reprocessing wastewater and energy in the medium term; and development and installation of technologies to increase the supply of water and energy and to reduce demand for water and energy in the long run.

Initially normal trend of events without intervention in the provision of management or demand management is considered. If occur climate change altering the supply of water and energy, on the other is the population and industrial growth by increasing demand for water and energy. In this case the industry will become hostage to a situation of lack of water and energy when looking at demand in relation to supply. In our case this will occur in 2022.

The analysis of availability compensates for the reduced supply with even greater reduction in demand. The analysis method of availability is divided into the following steps considering the comparison between the supply of water and energy with climate change with the demand for water and energy with preventive actions. Should perform prediction of what should happen in the future to adjust the present with the routine programs. This methodology is divided into the following steps: (1) Definition of current and future supply scenarios based on IPCC; (2) Identification of current demand scenario and future to compare Offer; (3) Analysis of the availability of water, energy and competence; (4) Culture Analysis and People for behavioral change; (5) Analysis of Routine and Integrated Management; and (6) Compliance Analysis of consumption of water and energy. A case of the chemical fertilizer industry are tested.

1 Introduction: Context and Background

Water is essential for life and it is the base of almost all human activities. In this context, water is used in power generation processes, domestic use, agriculture, cattle raising, and industrial sector among others. Provide access to clear water in quantity and quality represent one of the major contemporary challenges [1].

According to the UN World Report, the demand for water is expected to grow by 55% worldwide by 2050 due to the fast urbanization mainly in developing countries. The report also states that the world's water supply will suffer a deficit by 40% until 2030 if the water management do not improve. The agricultural sector will produce about 60% more food, and the industry sector will increase water demand up to 400% until 2050. Developing countries such as Brazil, Russia, India, China and South Africa will drive this [2].

The concern about water scarcity increased in Brazil after the water crisis in the state of São Paulo the most populated area located in the southeast of the country. This region from Brazil instead of the Northwest was an area that there is not this kind of problems. The water crisis didn’t harm only the life of the local population, but also the industrial sector which account for 32% of the Brazilian GDP [3].

The industry needs to be alert and prepared to deal with this kind of event, because this issue affect the running and growth of production [4]. Moreover, scarcity and water pollution have amplified some planetary boundaries known by scientists as safe operating space for humanity. Be in the planetary boundaries is a precondition for sustainable development. According to Johan Rockstrom *et al.* freshwater problems is influencing and affecting some planetary boundaries as biodiversity, food security, population health, and the ecological functioning, such as climate regulation and carbon sequestration [5].

According to ANEEL (National Electric Energy Agency), approximately 75% of Brazilian power generation come from the use of renewable sources mainly hydropower with 62% [6]. Water is also a driving force in the generation of electricity in thermal power generation (steam generation), and necessary to chemical and petrochemical industries for heating (thermal energy) or cooling, among others services [7]. Thus, the availability of water is associated to the availability of energy in socio-economic activities in Brazil and the world.

The lack of structure industrial sector to deal with the interruption of operations resulting from non-availability of water and energy (effects of climate change) is increasingly evident. The structure is related to technological aspects (e.g., automation) and of complex decisions. Many decisions involve multiple factors such as climatic and behavioral aspects.

Although industries are trying to develop projects to reuse of water from sewage treatment plants to supply cooling tower makeup water as was done in the state of São Paulo during the water crises, it is not a tendency. The reuse of wastewater in the industrial sector has been considered synonymous with economic gains, in addition to the positive image transmitted. Nevertheless, the concern is the concentration of pollutants and the possibility of increased losses by corrosion. Although reuse of water from sewage treatment demonstrate that management procedures are related to the basic principles of preventing water shortages, energy and environmental impacts, it isn't clear a commitment to analyze the availability of water and energy in order to adequate demand over supply considering scenarios for 2035 [4].

2 Method

The method intend to compare the supply of water and energy to their demand in the present time and a projection for the future - forecasting. This future scenario permit to take place the necessary actions to maintain the water and energy availability over time – backcasting [8]. Furthermore, prediction scenarios of greenhouse will help adjust the current emissions through routine management programs and new technologies.

Thus, short and long-term actions can be preventive as reducing the emission of greenhouse gases (replacement of energy by renewable or decrease their use or reduction of fuel burning) or corrective as the reduction of water consumption.

The analysis of availability method (shown in Figure 1) is divided into the following steps. (1) Definition of current and future supply scenarios based on the IPCC. (2) Identification of the current demand scenario and future to compare with the supply. (3) Analysis of the availability of water, energy and competence. (4) Culture Analysis and People for behavioral change. (5) Analysis of Routine and Integrated Management. (6) Compliance Analysis of the consumption of water and energy.

2.1 Definition of current supply scenario and future based on IPCC

The scenarios of use and supply of water indicate further reductions in the availability of water for industry. Water users (agriculture, livestock and population) are messy and work with distribution losses that can reach up to 34% [9]. Thus, the industrial sector need to care about increasing the supply of water and reduce their consumption at the same time through demand management and local source control programs.

The discussion about the energy consumption is linked to the greenhouse emission. The rise of GHG into the atmosphere depends on the burning of nonrenewable fuels or the lower conversion efficiency producing CO. The environmental impacts result in climate change and consequently alterations in water availability for society and for the economy as well as impact on biodiversity.

According the annual report about the inventory of CO_2 in the atmosphere, the amount of emission should be minimal to reverse climate change. Thus, the planet earth must send less and less CO_2 into the atmosphere, gradually, even with the growth of economies. The value target is 14 GTY in 2050 [10] and [11]. The forecast on the growth of the economy, with no GHG emission reduction action is to increase the current 50 GTY to above 62 GTY in 2050 (value stated in 2009) [10] - indicating the irreversibility of the greenhouse effect.

This paper aims to demonstrate how an analysis model on the availability of water and energy in the chemical industry and a behavior adjustment model may cause a large reduction in water consumption and energy in order to meet the expectations mapped by future supply of these inputs. The reduction in water consumption and energy depends on: management decisions; switching procedures; behavior changes in the short term; application reuse techniques and reprocessing of waste water and energy in the medium term; and development and deployment of technologies to increase the supply of water and energy and to reduce demand for water and energy in the long run.

2.2 Identification of the actual and future demand scenario to compare with the offer

A analysis of production and shutdown reports allow to build, in a preliminary way, the knowledge about the cause of losses and the behavior type to deal with these losses mainly of time, materials water, and energy. A rapid assessment of water and energy restrictions can give an idea of the scenario in the company. The production growth rate indicates if the demand will increase with the view to forecast the water consumption in the future, without including the preventive and corrective actions to demand management. When the supply and demand are compared, we can estimate the gap and which information is required to improve the analyses.

Water is classified in industry in various levels to meet the demand of the production plant and the administration. Large volumes of water are used in operations such as washing, transportation, ion exchange and others. One of the largest industrial water uses are in the cooling towers that meets the warm currents of the separation unit operations in the chemical process industry. Clarified water is the type of water used in cooling towers and represent 70% of the raw water inlet to the industry [12].

The measurement of water consumption is a legal obligation in order to inform the government environmental agency who granted a license with a utilization limit. This grant lets to use a quantity of surface or underground raw water and returns it (in effluent format) a quality better than the shot. Thus, it is necessary knowing the use of water, qualities (restrictions of equipment and processes), disposal and the behavior of the team and the manager regarding this precious resource to reduce the current water consumption and future need.

Regarding to energy consumption, classifying the type of energy used by industry is the first step to establish a demand profile. The electrical or thermal energy are primary energy and they can be converted into chemical energy and work. Thus, the main inputs in chemical industry are electric power and fuel. The measurement of consumption allows calculating the use of energy per unit of product giving an idea of the efficiency of its use. When a baseline and tárget (benchmark) is established and these are compared with the requested reduction of emission of GHG it is possible to establish the programs and investments in medium and long term.

2.3 Analysis of the availability of water, energy and competence

The starting point of the availability analysis is building the graph indicating water consumption and how much in terms of grants is authorized consume (immediately). In the case of supply, it is necessary to build an algorithm to increase the supply due to new sources or the reduction of losses. Another algorithm must be built when the source of water decreases due to climate change, increasing urban population, rise of irrigation area to agriculture, and so on. However, demand for water could consider two possibilities: (1) any action to reduce demand during the time (2016-2035) indicating only increase in the forecast increase in production or (2) actions and investments in process design, control routine and team behavior to reduce water consumption.

When the annual supply is projected for the future (including economic changes in society and the climate) and compared to future demand without considering corrective action, there is an indication of interruption of industrial operations for lack of water. Thus, actions to reduce water demand is indicated.

The availability of energy analysis actually compares the possible amount of CO_2 emission (which changes with time due to more restrictive situations in the future) from estimates by the International Energy Agency, with the current and future amount of CO_2 emission in different situations as the action taken or not to prevent irreversible climate change.

The initial approach of water/energy supply and demand needs to be confirm by mapping the thermal systems, equipment controls, water and energy saving programs, and analysis of sources and water and energy uses. The demand mapping is confirmed and compared with the preliminary data, they are made of the demand adjustment exercises in the timeline considering new best-case scenarios or more critical every five years.

Thus, the master plan for water and energy is builds up with the demand adjustment regarding the supply by technologies, programs, and technologies to be installed gradually. It is considered to predict supply by 2035 criteria related to industry growth, population, issues related to agriculture and irrigation, rainfall forecast, and the probable faults keeping the same demand.

This initial analysis should be reviewed after a comparison with external benchmark and improve the quality of information from direct and not secondary measurements.

Among the technologies are: reverse osmosis, geothermal, flexible cooling towers for reuse, process intensification, and others. Among the techniques are: water balance, energy balance, pinch, reuse of wastewater, Exergy analysis, analysis of material losses and energy, and others.

2.4 Analysis of Culture and People for behavioral change

A series of procedures and attitudes in the levels of implementation, management and strategy are based on increasingly complex decisions [13] and [14]. It is necessary to know how these decisions occur based on local cultural characteristics [15] and likely rules that influence not reduce water and energy consumption. It is important to analyze the influence of equipment failure, processes and operations of the direct losses and indirect materials to the reprocessing [16]. The decision to adjust the quality of the product can impact on the increase in the consumption of water, energy and the rework of workers.

An analysis of people and cultures will be able to identify indicators that evidence the inappropriate behavior that enhance water loss and energy. This allows building a program to control losses from these resources. With this identification, it is possible setting a program of training or behavioral adjustment to be implemented in the company.

The cultural influence is discussed for building a robust safety culture and is based on the dynamic behavior of the staff. This behavior is the result of a field of cultural forces: organizational desire, local culture and features of global culture. Some trends should be strengthened and other inhibited and this action are based on archetypes analysis [17].

The human factors analysis allows to establish programs to increase human reliability. Human factors can be [18]: conflict between policies and practices, the conflict between production and safety priorities, control facilities, managerial aspects, skills analysis, and preparation procedures are part of human factors that should be analyzed and monitored.

The moment of decision is then influenced by biases of culture and human factors causing cognitive impairment [19] occurs following human errors mechanisms as the default [20]. In addition to the analysis of cognitive impairment seeks to reduce the risk of accident through the joint analysis of technical factors and not technical. This strategy allows us to discuss what are the aspects that can impact the security processes in an organizational change [21].

2.5 Analysis of Routine and Integrated Management

After identifying the behavioral constraints or characteristics that lead the employee and the manager to start faults that cause loss of energy and water, it is important to analyze the routine. Check the risks of system stop on a daily basis allows you to work in a review of equipment and critical tasks. The knowledge of water systems, heat exchangers, energy integration, wastewater allows for maximum use of energy by increasing the energy efficiency. The water and waste energy reuse will rationalize the consumption reducing dependence on bulk supply facilitating sustainability of the company.

2.6 *Analysis of the availability of water, energy and competence*

The establishment of comparison criteria for use of water, energy and behavior that avoids the technical and human error adjusts techniques, technologies, decisions, behaviors and knowledge. So this step enables the continuous improvement of the risk management program in Energy and Water. (ISO 50001) [22].

3 Demand management

Some global factors have to force the industry to make efforts to increase productivity and reduce the consumption of basic materials as raw materials, water and energy. The water is already being alert reason for society and for the industrial sector, where specific events of lack of water can turn into long periods of shortage impacting on the economy and the population.

Energy is a key input to ensure the economic and social development and the rationalization of its use is an essential strategy to avoid losses during periods of restrictions resulting from the emission of CO_2 and the scarcity of natural resources. Energy rationalization might be to change procedures, behavior change, control techniques and technologies to enhance the use of energy and thus improve efficiency and improved thermal performance of the equipment and systems.

The water consumption in the industry is directly linked with energy demand, for the use and distribution of water in a production process, the energy has a fundamental role to keep the equipment operating. Similarly, the electrical and thermal energy coming from a unit of basic inputs need the aid of water for its existence.

Industrial activities that involve the consumption of basic inputs has suffered deployments due to significant consumption and existing waste. Cycles of concentration in cooling towers, water use outside of specified boilers, mass loss and energy in steam distribution systems, heat losses in electric power generators, discarded wastewater without prior analysis, are some examples that show opportunities for deployment techniques or technologies that contribute to the reduction of utilities, waste and therefore environmental impact.

However, to reduce the consumption of water and energy are proposed a number of solutions management, various techniques used by the STAFF during routine operations and technologies.

3.1 *Management*

The management plays a key role in the implementation of programs to control routine related to the behavior of the staff and control of deviations. Each technique has suggested goal, steps and impacts are described below.

The control program at the source [23] is to identify equipment and pathways of contamination to the final disposal, liquid (effluent), solid (waste inventory) and gas (emissions or losses fugitive emissions). Understanding on which the carrier fluid and how it is possible to establish measures to avoid losses. The additional technical material balance loss is the mass calculation of losses and the impact on fixed costs, variable and billing.

Control of critical task, named as POS, Procedures and Operational Standards [24], aims to identify, in this case of water and energy availability, which tasks are that they can produce losses and analyze the environment, planning, execution, control and the failed task.

The analysis of environmental deviations [25] want based on the fault tree method and field observation, identify deviations that lead to contamination of everyday life. When this technique is well understood in the organization it enables a behavior change operation.

This research is in development and is about identifying what the proper conduct of the operation team to prevent energy loss. Follows the line of CCPS in Operational Discipline [26] and includes aspects of the company culture. It takes into account that energy is intangible and less noticeable to control losses.

The calculation of the reliability of socio-technical systems [27] depends not only on engineering and includes the social and managerial area of the company. For this reason it will be led by the manager. When the reliability of each functional part of the system (process, equipment, operation and human) is identified, after the process of the complexity of the classification, the calculus of the system reliability level is possible. The system reliability is then corrected through social attractiveness that indicates the strength of culture, the level of cooperation (communication) and adaptive leadership to new realities. Through this analysis it becomes possible to more policy action on production controls.

By calculating the reliability of systems and identified the impact of human reliability begins research on each performance factor indicated by SPARH [28] and discussed by API770 [18]. Thus, it becomes possible to define critical areas for action by the manager. These analyzes made by the management has focused on reducing energy losses!

3.2 Engineering

The engineering staff has the role to develop adaptations to current processes and equipment and provide new equipment to avoid or mitigate energy losses. These solutions require equipment yield calculation, energy efficiency and thermal performance.

Initially the operational control equipment and thermal systems adjusts the best operating point through procedures and control systems. It is therefore important that the process engineering and equipment to track and control the income and performance of furnaces, boilers, cooling towers, electric motors, power contracts, distribution of steam, condensate distribution, reflux in the distillation, and balance energy in the process equipment and utility in general.

This control can identify the need to revise the performance calculation methodology to be open systems, such as cooling towers. In furnaces and boilers may be a need to review the maintenance procedures (installation of refractory) or operating as purging. In collecting and recovering the condensate, on the other hand, several factors may impact on live steam loss to the atmosphere, it depends on the thermal load of the plant, steam quality, operation of steam traps, and quality of thermal insulation.

The effluent mapping to identify the loads of contaminants requires the establishment of networks of sources and uses of water by continuous or intermittent processes procedures. It is needed when discussing energy sources and uses, in this case, less visible, and thus requiring additional technical field, such as using pyrometer surface of hot equipment.

The mass balance and energy are more detailed technical and require methodologies for utilities and process equipment. simplifications may be needed, where process streams with low heat capacity variation are classified to 80% balance of the impact, reducing the number of streams.

For combustion systems are used to heat the lower and higher power. For open cooling tower systems we use the useful heat exchanged. In the case of energy balance that includes the axis of work it is important to check the income and losses of energy through heat. The energy balance is important to compare the fuel consumption with the issue of greenhouse gases.

3.3 Technology

In terms of technology we will discuss those involving new criteria for design of industrial plants and new equipment that reduce energy losses and water or increase its offer. Among the technologies involving new standards for process design and its control: modularity, intensification, integration and exergy. Among the technologies involving new equipment to reduce losses and increase in supply: desalination, heat pump, cogeneration, trigeneration and combustion bomb.

4 Application in the Chemical Industry

Applying this analysis methodology on the availability of water and energy it is based on the experiences in research projects to rationalize water use in the chemical industry of linear alkyl benzene (LAB) and the fertilizer industry (UREA) [29]. Other research has enabled the discussion of opportunities to reduce energy losses in the petrochemical industry from energy balances in processes and utilities [30] AND [31].

The experiences in the LPG industry and own fertilizer industry [24] allowed the present discussion of cultures and people. And finally, the argument based on ISO31000 and ISO50001 enabled compliance analysis.

The implementation of these steps during the period of two years will have as a result the establishment of control programs in source and strategic plans for water, energy and skills. Every five years is necessary to readjust the supply condition and analyze the indicators that demand occurred in accordance with the master plan. Organizational changes can alter the demand for water, energy and time.

The intended application for two years will be addressed through the following topics: contextualization, supply management, demand management, strategic analysis of the availability, culture and people analysis, routine management and compliance analysis.

4.1 Contextualization

Audits for use and water sources that used the effluent mapping technique and performance analysis in cooling towers indicated the high water consumption in the industry LAB mainly to feed the cooling towers [32]. It was noted also that occurs excessive contamination of waste water with several points of discharge of effluents and products. The LAB production process is complex, with the possibility of reprocessing with handling high toxicity products such as hydrofluoric acid, but a well-structured staff to decisions simply to achieve better control, access to appropriate techniques for controlling consumption energy and water.

The energy and water balance techniques, due to the complexity of petrochemical processes, indicates that more rigorous methods require mathematics reconciliation data. In other cases seeking the simplification based on the impact of the amount of energy involved, where the reduction of line processes is necessary, within the volume control for mass balance.

In the petrochemical industry of basic inputs where the type of naphtha defines differences in production costs, the equipment, the process parameters and the quantity to be produced of each product (olefins or BTX), it is important to review the energy balance. Flow rates in these processes are very high, so errors in the way of operating equipment can cause the plant to have high energy losses, indicating a need for high fuel consumption and high or make-up water for cooling towers.

The demand by up to 30% water cuts and agreements of pressure climates, and briefly change the intensifying industries in energy use. The lack of assertive decisions or appropriate behaviors have removed the goals in relation to the consumption of energy and water in the routine. This indicates that the transformation requires greater care in the industrial segment!

4.2 Supply management: control and planning for 2035

Petrochemical industry of cases and fertilizer supply of real data and water demand and actual data supply and energy demand are discussed.

4.2.1 Water supply management: premises for the analysis of supply management in the fertilizer industry in northeastern Brazil

Assumptions: (a) average climate change scenarios A1B with rainfall reduction impacts in Bahia (2070-2100) in about 60% on the coast and 35% in semi-arid with 2-5°C temperature increase [33]. (b) increase in supply due to improvements in the operation of river basins and due to climate change by 2018; (c) reduction due to climate change estimated at 28% to 60% of the current offer in surface and groundwater respectively, for 2035; (d) reduction of reliability in service of the Paraguaçu River Basin [34] for 2011 to 2040, mainly for irrigation, industry and power generation; (e) Increase in the number of months during which there were failures in care:

the Savior supply (1% to 5%) and industrial demand (1% to 9%) with reduced power generation by 66% in Dam of Horse Stone [35].

Supply: the water supply to the industrial sector in the basin of Paraguacu and Salvador city between 2012 and 2034. **Offer 2012:** Surface water = 20.9 m^3/s, groundwater = 3.9 m^3/s, total = 24,8 m^3/s. **Offer 2018:** Climate change with increased rainfall: 24,8 to 78,8 m^3/s (3.2 x). **Offer 2034:** Reduced rainfall = 25% in 2018, 19.7 m^3/s.

Fertilizer unit: in the specific case of fertilizers unit [29] is considered that the initial flow supplied (2012) is equal to the demanded flow rate, that is, 700 m^3/h. **The Offer in 2018:** since part of the overall offering for the industry shifts to supply the city (an increase of 2 X instead of 3.2 X), that is, 1400 m^3 / h. **Offer 2034:** decrease due to climate change by 33%, that is, 462 m^3/h. The demand for water in the fertilizer unit should be less than this flow.

4.2.2 Power supply management: premises for the analysis of supply management in the petrochemical industry in northeastern Brazil

Considers that energy supply is limited by the capacity of the atmosphere of the planet receive GHG to prevent climate change. Thus, the supply limit is based on climate studies by the IPCC. On the other hand, the demand for energy in industry depends on the activities carried out by thermal and electrical systems. Thus, the balance between environmental restriction is sought (supply) and the actual greenhouse gas emissions by industry (demand).

The power availability analysis will be handled by the comparison between supply and demand. Supply management will have an upper limit maximum environmental capacity to receive GHG emissions to prevent climate change (14 GTY in 2050), this is the blue scenario. In this scenario emissions are monitored since 2005 with 28 GTY emissions and tend to grow up to the value of 62 GTY when the climate change process becomes irreversible. This emissions growth occurs if no action is taken towards sustainable economic growth of BRICS and developed countries [10].

Global targets need to be resized to establish regional or industrial sector or by enhancing industry in energy consumption targets. This schedule allows you to define local action strategy seeking to guarantee the reduction of GHG emissions.

In the case of a petrochemical in Camaçari (Brazil): since 1000 there are on planet earth petrochemical same capacity and the global industry emits 100 times the amount of CO_2 (based on fuel consumption) of the global petrochemical segment. The proportion of the global emission reduction for local equivalent to a reduction of 10-5 Industrial overall goal [10] for this local industrial. In the case of switching to renewable matrix, the overall goal is 9 GTY after the local design amounts to 0.09 MTY for this petrochemical industry.

Offer in 2050: (a) replacement by renewable energy → 10 gigatons year of global emissions or 0.1 mega ton year local issue; (b) reduction in CO_2 consumption by effective use of thermal energy resulting from the burning → 11,5 gigatons year of global emissions or 0.115 mega ton year local issue [10].

4.3 *Demand Management: control and planning for 2035*

Petrochemical industry of cases and fertilizer supply of real data and water demand and actual data supply and energy demand are analyzed.

4.3.1 Water demand management: assumptions for demand management analysis on the fertilizer industry in northeastern Brazil

Assumptions: demand management analysis in a fertilizer industry in the Northeast of Brazil: (a) average growth of 2,5% per year of urban demands of the cities of the RMS (increased population projection and reduction of per capita consumption) [36] and (b) growth in consumption by industrial expansion and irrigation.

Urban and irrigation demands will reflect the reduced availability of water for industry.

Demand: the demand for water for industrial thread in the basin of Paraguaçu and nearby cities including Salvador city between 2012 and 2034 considering that there is no demand tuning activities. **Demand in 2012:** 15,5 m^3/s city and 6,2 m^3/s for industry, total = 21,7. **Demand in 2034:** 40% of consumption by the population increase and reduction of losses of approximately 25% due to investments in the operation of water supply equivalent to 5% increase in the city, 16,3 m^3/h. New industry and new units with 15% increase indicating 7,2 m^3/s for the industry. Total 23,5 m^3/s. Indicating that the offer will be lower (19,7) than the demand (23,5 m^3/s). That is, if no action is taken to adjust to demand, the industry will be at high risk of interruption.

Fertilizer unit: in the specific case of fertilizers unit [29] is considered that the initial flow supplied (2012) is equal to the demanded flow rate, that is, 700 m^3/h. **Demand by 2034 without setting activities,** will increase by 15% in water consumption = 805 m^3/h. This unit of fertilizer also have problems which need to consume less than 462 and will be consuming 805 m^3/h.

4.3.2 Energy demand management: assumptions for demand management analysis in the petrochemical industry in northeastern Brazil

The suggestion by the International Energy Agency (IEA) is that there are different ways to reduce GHG emissions: energy efficiency (electrical and thermal), renewable energy (wind, solar, biomass), nuclear energy and carbon capture. The IEA also shows that the main actors for this reduction are: processing industry, power industry, transport and buildings.

Avila *et al.*, 2011, showed that a petrochemical site:

(a) replacing 20% of energy consumption based on fuel oil with biomass energy. Following the assumption that this biomass, which compose a forest previously cut off from the atmosphere by photosynthesis, the same amount of CO^2 emitted in their use. This petrochemical be taking 20% of CO^2 emissions initially issued.
(b) 50% reduces the loss of live steam (estimated at 30%) in the condensing system, steam systems, boilers and furnaces.
(c) recover 50% of losses due to non-use of cooling water recirculation flow in heat exchange in cooling towers (ideal thermal performance (dti) = 95%; dti real = 75%; 20% difference). Thus, an original temperature difference of 10°C, with higher

energy efficiency reaches 11°C, a reduction of 1°C in cold water → bringing as positive impact improving the profile of the separation columns and reduction of fuel demand thereof magnitude and CO^2 emissions as a result, 10%.

Note that:

1. The amount of biomass used (petrochemical-based local → global petrochemical → global industry) inhibits the emission of CO^2 with the change of 20% for renewable matrix 8,45 GTY, compatible with the demands of the International Energy Agency.

2. Increased energy efficiency in thermal systems (petrochemical-based local → global petrochemical → global industry) reduces burn and CO_2 emissions by 12,5 GTY, compatible with the demands of the International Energy Agency.

4.4 Analysis of availability: strategic plan

Exercises for analysis of the availability of water, energy and skills

The program to control wastewater at the source and master plan provided operational activities and new projects to achieve the demand reduction of 750 to 310 m3/h, so below the 462 m3/h.

Scheduled actions are [29]: (2013-2018) (1) pattern of change in the final product plant shutdown; (2) stability in the operation of purification and reuse; (3) reprocessing the waste water treatment, reuse in cooling tower; (4) increase in operational reliability; (5) reuse of disposal with a high level of contaminant; (5.1) purifier contamination reduction - reduction of product load; (6) condensate polishing removal of organics. (2019-2020) (7) energy conservation project, condensate (I); (8) control work in the steam generation (II); (9) condensate return for demineralized water; (10) the entry of new plants with improved design pattern. (2020-2024) (11) of rainwater use project; (12) improvements in the use of cooling towers; (13) Water reuse for fire and utilities; (14) improvements in water treatment plant; (15) rationalize water use in the area and the administrative use. (2025 and 2028) (18) effluent reuse in cooling towers; (19) advanced technologies to reduce vapor loss - cogeneration and heat pumps. (2029 and 2034) (20) new techniques for heating and cooling: intensification; (21) new cooling technology projects: geothermal energy and application of the products.

The availability of energy and competence [8] are discussed from Figures 1 and 2. Figure 1 is considered the emission of CO2 in a temporal analysis which compares the issue (with and without adjustment of shares) with the offer is the maximum capacity of the atmosphere to process GHGs without climate change. In Figure 2 we analyze the defendant competence for technology with the expertise offered by people with and without adjustments regarding the offer.

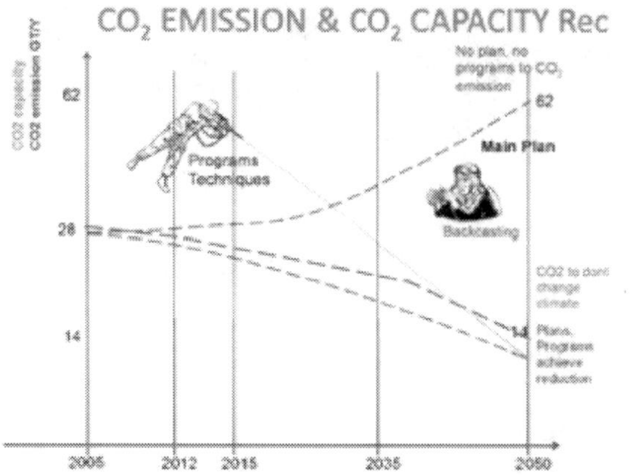

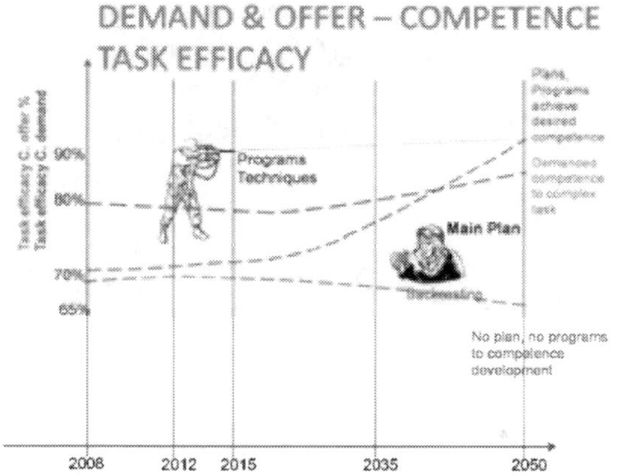

Figure 1 - Supply and demand GHG [8] Figure 2 - Supply and demand Competence [8]

Master Plan for Water and Energy [8]

The Master Plan is a strategic tool that enables meet the challenges of climate change and its consequences for the industry. Planning for adjustment and availability of water & energy for the industry is complemented with technical and management tools for the implementation of the Master Plan.

The Master Plan discusses the legal and institutional aspects of the laws dealing with water, energy and environmental impacts. supply management depends on legal issues that allow water and energy management in industry. The current and future availability is performed by comparing supply of water resources in local basins with the demand of these resources to the industry seeking the necessary adjustments. This discussion should also is made to address the constraints of the Brazilian energy matrix with the conditions for water use and CO2 emissions.

Analysis of water and energy classes and specifications of equipment and processes enables us to diagnose the possibility of water and energy reuse. The identification of classes used in the industry indicates their effective use to have qualities in different sources. The water treatment and effluent depends on the water balance enabling reuse projects.

Concepts and applications are detailed for adaptive management of supply and demand and which proposals for long-term strategic adjustments as credit demand reduction, industrial symbiosis concepts and opportunities, the importance of monitoring indicators, and the risk of lack of water for 10 and 40 years.

The ways to manage supply take into account: availability of water sources, climate change, hydro-geological perspective, costs and energy content. A calculation procedure is suggested for the adjustment of supply relative to the projected demand. In case of power must calculate the CO2 emission target for the specific case of the installation. The management of future demand includes the control program in the source material balance losses and others.

In the Annex of this plan are presented technical and managerial and operational methods. The intention is to keep demand and controlled supply. The main technologies to maintain demand and controlled supply for Industry are classified and detailed.

Control Equipment and Thermal Systems

The thermal control equipment and systems, process control and reliability of the systems are extremely important to regulate the demand for water and energy. The demand adjustment requires lower value to water supply (based on state water policy) and energy (based on local and global constraints of GHG emissions).

The equipment that has high impact on fuel consumption and the emission of greenhouse gases are: (a) cooling towers; (B) heat exchangers that utilize utilities; (C) generating steam boiler; (D) heating stoves or pyrolysis reaction; (E) steam distribution and condensate energy recovery (including traps); (F) current of high enthalpy process; (G) operating point and balance of separations such as distillation; and (h) electrical systems in general.

At this point will be discussed energy efficiency in some of these systems: (a) Deten-water research project in 2003, Ávila noted the importance of working with clean cooling water circuit to avoid blockages in the upper pools and highlighted the potential losses if maintaining the cooling tower structure (and water quality) is poorly made, with losses that may cause heating of water up to 1 ° C. By maintaining one should prevent the cooling water flow paths follow without heat and mass transfer with the dry air. (b) This project was highlighted the importance of working with cycle highest concentration in the towers to reduce water consumption. This control increases the risk of corrosion in order to be header exchangers and hinders water reuse in cooling towers [37] [38].

(c) There are several opportunities to recover waste heat in the boilers that generate steam in enhancing industry in energy consumption [30] [31] [38]. Since preheating of the water, control of stoichiometric combustion, the specification and maintenance of refractories, the utilization of heat of flue gas (new boiler, flue pump). (e) the design assumptions for steam distribution, steam traps installation and condensate energy recovery (including traps) should be based on the expansion possibilities of production and operating limits of the systems [30] [31] [38] . Thus, due to the limitations of operating with traps and thermal systems, search is the best point and with new needs, install new production modules.

Water balances, effluent, fuel, GHG and energy

In research projects to rationalize water and energy were used mathematical techniques mass balance (FAFEN-WATER, DETEN-WATER) and enthalpy (Ecobraskem) with multiplier weights varying according to quality information and then closing balance by data reconciliation. This technique was used for LAB production plants, Olefins, urea and petroleum [30] [31] [38]. As for the balance of utilities in equipment such as cooling towers, boiler and furnace energy balance it was based on ICPs and the formulas simplified the treaters water that is based on the wet bulb temperature, the flow and the temperature difference [38].

Sustainable projects: new techniques and technologies

The need to reduce energy consumption requires changes in equipment design and process. The main reason for these changes is to reduce the gaps, allowing a reduction in the consumption of utilities through direct heat transfer without the need for thermal fluid design adjustments considering the external environmental changes and changes in thermal load.

The spare capacity of operations and equipment is adjusted to increase the capacity of industrial plants. In industrial processes it is possible to carry out this adjustment due conservative design factors that enhance the capacity of the equipment, but in the case of utilities is not always possible. Thus, you can use the spare capacity of reactors, pumps, but it is not possible to trap and electric motors. With restrictions on the use of energy and knowledge in human reliability, it is intended to modularize the projects through more accurate system design [38].

Another option in the project area is to reduce the use of utilities and size of equipment by intensifying operations and processes where the heat necessary to the achievement of operations is transferred directly by the equipment wall. Where the heat is released from certain operation and received in another operation without the need for use of thermal fluids. If energy balance in petrochemical is noted that the use of exchangers with hot thermal fluids (steam) and cold (cooling water) consumes 4000 GJ / h compared effectively with the necessary heat (80 GJ / h) to unit operations work without utilities. With intensive use of energy can significantly reduce power consumption and size of equipment [38].

The rationalization of energy and water can be conducted using techniques which integrate the process with the utility and waste. The identification of water users with the material restrictions for equipment allows the water and energy savings through its reuse in equipment. This technique is applied in sewage networks and networks of heat exchangers.

In the case of wastewater networks (pinch), the contamination levels of waste water streams are analyzed to compose a quality goal that meets the requirements of equipment that can use this water. If exchanger network (pinch), the residual heat through the cooling water reuse allows a better use to reduce energy losses.

The main equipment that consumes water and has high potential for use of waste water is the cooling tower [32]. Thus, it is the integration of processes (sources) with the effluent (waste water standard), and returns to the utilities (water cooling).

Energy recovery from the best thermal performance in cooling tower and application of process integration (pinch) has as a result the reduction of water loss and the reduction of CO2 emissions. To account for this loss reduction in CO2 emission area is seeks to study the integration of the cooling tower, heat exchangers, separation operation (reflux), the heat demand (steam oven), burning fuel and CO2 emissions.

The decrease in water supply from climate change requires research on the feasibility of industrial desalination including comparing other solutions involving energy supply and water (eg lifting deep brackish water). The current technology of reverse osmosis membrane desalination is still not profitable for water treatment industries in the private sector.

4.5 Culture analysis and people

Private and public institutions recognize the importance of investigating the matter culture and people for sustainability, but do not yet know what priority to this issue. To establish an organizational and strong safety culture is required: to know the tasks and risks; analyze the behavior through the routine signals; and establish metrics to identify the stage of safety culture.

An onshore oil industry indicated that the image directly affects the profitability of the company, so often the actions prioritize avoid independent accident effect on production and on revenues. A proposed action was suggested for intervention in three years: (1) Task: To establish indicators of efficiency, formalizing forms of communication in organizational levels to review the inadequate standards establishing best practices; (2) Behavior: analyze routine deviations, establish a causal link with the process losses, see how minor emergencies are handled (decisions) and indicate gaps in relation to skills, knowledge and standard of acceptance; (3) Culture: Establish indicators for technical culture (bottom-up) and measure the difference in perception between the entrepreneurial vision and practices. Set organizational practices (rituals) to neutralize or mitigate biases culture.

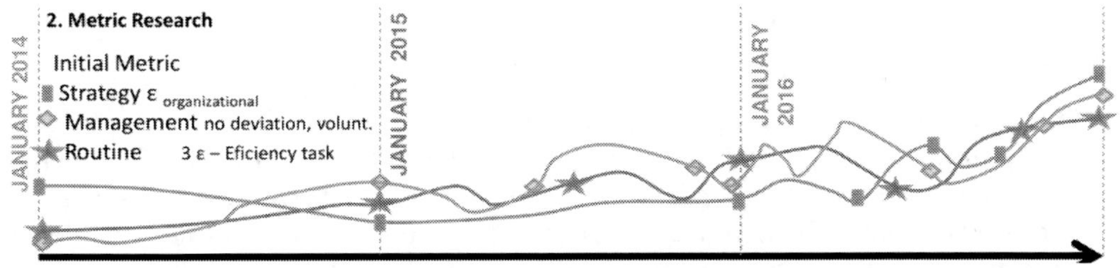

This initiative for the onshore site has been replicated through intervention in the fertilizer industry to analyze events in the routine that involves the effluent, where old vices of operation contribute to the process losses. The suggestion to change behavior and neutralization of decisions that led to contamination of effluent was through the educational program "Friends of emergency pond." The tools used in this program were: lectures in the control room and field, event analysis and hypothesis of the root cause, exchange of information through informative texts and making changes in the decisions of supervision and management.

The process that involves loss materials are more easily perceived as contamination of wastewater, fugitive emissions or the generation of solid waste. Large losses that affect the company's image as an accident or leaks has immediate impact on the stock market. Since energy losses are less visibility difficult behavior change to achieve targets in reducing GHG emissions. So it is important to consider cultural aspects and behavior to take more assertive decisions.

Analysis of elements or human factors [24]

In this GCPS organizational factors related to regional-global culture that are entered in 20 events in the chemical industry were discussed. These factors can initiate failure mode with final impact on energy losses. Thus, conflicts between policies and practices, economic difficulties, poor planning of the task, cultural traits that induce inappropriate decisions, risk management that does not include human factors are important factors to analyze the operational routine which leads to energy losses. The routine in deviant behavior difficult to establish operational discipline to avoid accidents and loss of energy [26].

Cultural analysis based on Archetypes [17]

Cultural phenomena do not exactly follow a logical order and are based on ancient customs rooted through tribal and new rituals introduced by the media. In an attempt to develop cultural barriers to the accident and GHG emissions seeks to analyze the phenomena resulting from the action of primary and derivative archetypes in the organization's tasks. This kind of technique indicates which new rituals should be deployed to vaccinate cognitive impairment.

Analysis of Cognitive Failure [19]

The cultural environment induces undue behaviors in individual or group during the task. Understanding the type of cognitive failure depends on the analysis of executive function in the various stages of cognitive processing: memory operations, environmental awareness for the construction of knowledge and action, attention calibration, job planning, job control and socioeconomic utility affective. It is important to identify the type of human error in the energy losses to prevent the ineffectiveness of action taken.

Reliability Analysis Sociotechnical [27]

The integration of production technology (product, process and control) with its operations on critical equipment indicates the importance of joint analysis of reliabilities. The operation achieves success if the understanding of the task objectives and knowledge of the process conditions (standard and fault). Thus, reliability of operations, processes and equipment depends on human reliability that receives the direct influence of cultural environments.

A propylene pumping operation in the refining industry was analyzed and it was concluded that the redundancy pumps is not enough when cognitive functions are weakened in conflict between policies and practices environment.

4.6 Routine management

The quality brought about the revolution of the administration [39] but did not ensure the sustainability of their results. The complex decisions and the investigation of systems failure mode indicate that to avoid losses should measure the efficiency and effectiveness of the task should be controlled thermal systems, critical procedures and also analyze the level of material losses and energy.

In the area of utilities and processes which circulates high amount of energy it is important to perform measurements and balances to prevent high losses of energy and water industry. The control of thermal systems involves critical steps such as maintaining water quality, purge tower and boiler, the level of control and the use of cells in cooling towers [31].

The production control by type of raw material defines the economic quantities of schedules based on variable cost of energy. In the petrochemical industry this issue has great importance in the definition of production programs knowing input naphtha type (weights and average).

Depending on the input naphtha occur demand for changes in processes, for example, the recycle gas will have another quality and will need to review the burners in the process gas boilers reuse [38]. To carry out the control of these tasks with high impact for the loss of energy is important to apply task analysis techniques in complex processes and have control over the realization schedule these tasks [40].

The mass balance and energy (water, materials and supplies, steam, fuel) enables the detection of losses and control at source [35] [23]. The balance of goods and inputs is closed with the quantification of losses: solid waste; liquid contaminants to the effluent; and fixed and fugitive emissions to the atmosphere.

4.7 Compliance Analysis of water and energy consumption

ISO 50001 [22] are described important procedures for the management and control of energy. It is important to identify sources (equipment and systems) lose water and energy in the production process. It is important to define if the losses are isolated or continuous, which the physical state and properties, the transport fluid and the impact receiving body. The output of thermal systems is measured by measuring the consumed energy and the comparison with expectations. Part of the energy loss and water is recoverable and the cost outweighs the benefit. compliance analysis indicates where the main losses and what the distance from the expected default behavior.

The fixtures and fittings that consume fuel or indirectly affect this consumption are extremely important: boiler, furnaces, heat exchangers, steam traps, cooling towers, water and steam distribution pipes, flash vessels, safety valves, and distillation columns .

It is important to define the consumption profile (baseline), the expected pattern of consumption, benchmark-based target, the difference between the measurement and the expected pattern (loss) and the percentage of impairment. It is also important to describe the operational procedures that can impact on the loss of water and energy consumption, expected standard and goal. These losses may have been caused by misconduct or lack of knowledge.

The criteria for equipment performance analysis are based on rules established by institutions such as ASME (boilers and furnaces) and CTI (cooling tower). Other criteria related to behavior and procedures may be based on API770 and operational discipline guidelines CCPS.

5 Conclusion

Initially considered the normal trend of events without intervention in the provision of management or demand management. On the one hand occur climate change altering the supply of water and energy, on the other hand is the population and industrial growth by increasing demand for water and energy. In this case the industry will be trapped in a situation of lack of water and energy across the demand relative to supply. In our case this will occur in 2022 (for water).

According to the problems already experienced shortages, so there is a balance between supply and water demand and energy are already demonstrated the means and methods that need to be taken into consideration and applied in the productive sectors and society. A case of the fertilizer and petrochemical industry have been discussed.

Already have the knowledge that technologies and techniques have been implemented in the industrial sector considerably reducing the waste of raw materials and fuel and thus environmental damage, but their implementations are not prioritized the way it should, as a general in contrast, there is a cultural influence, behavioral, economic and management that becomes increasingly slow the process of improvement.

With the projection of a future scenario and possible interventions mainly developing countries GHG emissions planned for the year 2050 is approximately 14GTano, very difficult task to achieve.

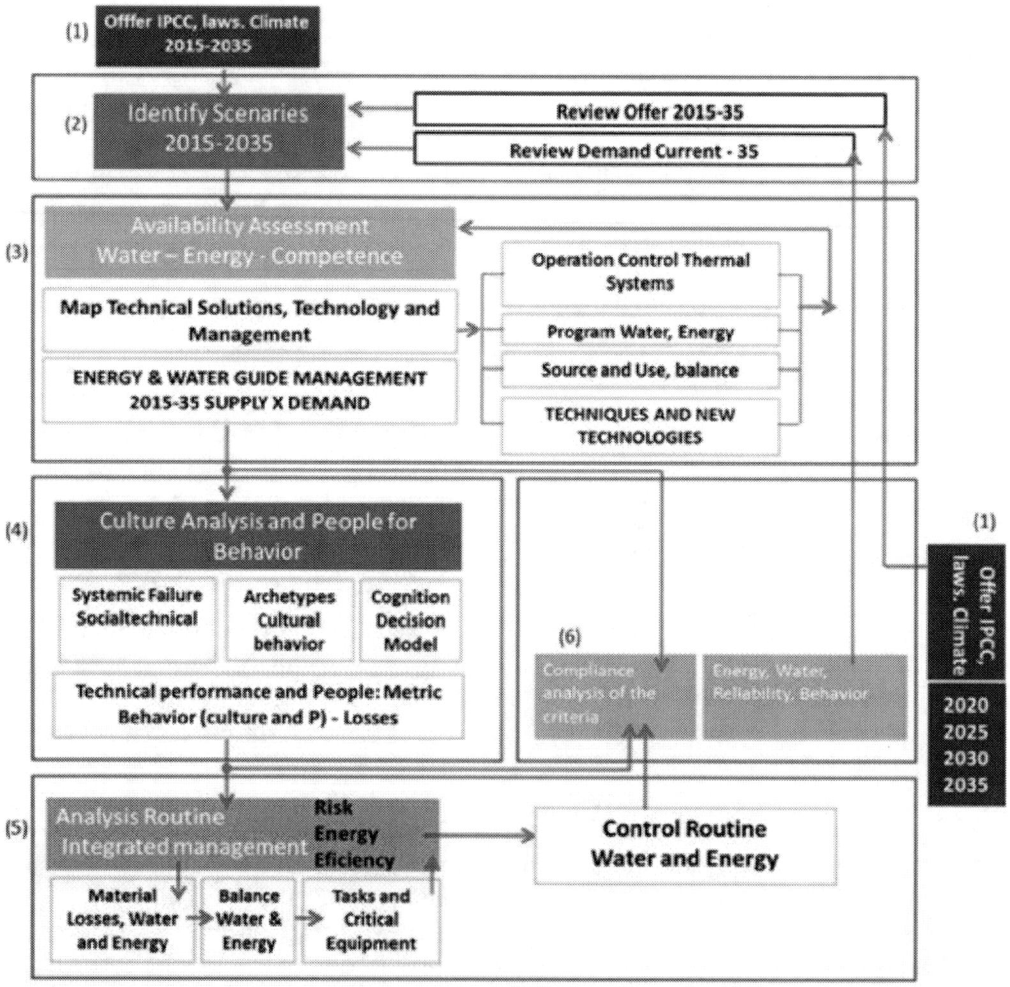

Figure 1. Analysis of Water Availability, Energy and behavior.

6 Bibliography

[1] CARMO, R. L.; DAGNINO, R. S.; FEITOSA, F. F.; JOHANSEN, I. C.; CRAICE, C. População e consumo urbano de água no Brasil: interfaces e desafios. In: SIMPÓSIO BRASILEIRO DE RECURSOS HÍDRICOS, **Anais...** Bento Gonçalves, 2013.

[2] WWDR – The United Nations World Water Development Report 2015: Water for a Sustainable World. Paris, UNESCO, 2015.

[3] FUNDAÇÃO SISTEMA ESTADUAL DE ANÁLISE DE DADOS – SEADE. Disponível em: < http://www.seade.gov.br/pibtrimestral/analise/>. Acessado em: 06 fev. 2016.

[4] PQU. SABESP. Sabesp e Polo Petroquímico lançam o maior projeto de Água de Reuso do País. 2008. Disponível em: <http://www.agenciagabc.com.br/noticias.asp?id=1045>. Acesso em: 06 fev. 2016.

[5] ROCKSTRÖM, J. *et al*. Planetary Boundaries: Exploring the Safe Operating Space for Humanity. Ecology and Society 14(2): 32. 2009. Disponível em: <http://www.ecologyandsociety.org/vol14/iss2/art32/>. Acessado em: 06 fev. 2016.

[6] ANEEL – Agência Nacional de Energia Elétrica. Matriz de Energia Elétrica. 2016. Disponível em: <http://www.aneel.gov.br/aplicacoes/capacidadebrasil/OperacaoCapacidadeBrasil.cfm>. Acesso em: 06 fev. 2016.

[7] MIERZWA, J. C.; HESPANHOL, I. Água na indústria: uso racional e reuso. São Paulo: Oficina de Textos. 2005.

[8] ÁVILA, S. F.; LOPES, J. R. N.; SOUZA, C. R. O. Strategies for industry sustainability. In: WREF 2012 - WORLD RENEWABLE ENERGY FORUM. 2012, Denver.

[9] BONALUME NETO, R. Perda de 30% da água durante distribuição contribui para crise em SP. **Folha de São Paulo**, São Paulo, 05 de jun. 2014. Fórum de Sustentabilidade.

[10] DOORNBOSH, R.; GIELEN, D.; KOUTSTAAL, P. 2008. Mobilising Investments in Low-Emission Energy Technologies on the Scale Needed to reduce the Risks of Climate Change, OECD SG/SD/RT (2008) 1, Paris, France.

[11] IPCC (2007), Metz, B.; Davidson, O. R.; Bosch, P. R.; Dave, R.; Meyer, L. A. Technical Summary in: Climate Change 2007: Mitigation. Contribution of Working Group III to the Fourth Assessment Report of the Intergovernmental Panel on Climate Change, Cambridge University Press, Cambridge, United Kingdom and New York, NY, USA.

[12] FURTADO, M. Água – Clientes exigem resfriamento de baixo custo. Revista Química e Derivados. São Paulo, pp. 32 – 34., 2009.

[13] PARRY, M. L. et al. (Ed.). Climate change 2007: impacts, adaptation and vulnerability; contribution of Working Group II to the Fourth Assessment Report of the Intergovernmental Panel on Climate Change. Cambridge (UK): Cambridge University Press, 2007f. 976p.

[14] ÁVILA, S. F. Análise de falha em processos complexos. In: CONGRESSO BRASILEIRO DE ENGENHARIA QUÍMICA, 19,, 2012. **Anais...** Búzios, Rio de Janeiro.

[15] ÁVILA, S. F.; PESSOA, F. L. P. **Proposition of review in EEMUA 201 & ISO Standard 11064 based on cultural aspects in labor team, LNG case.** Procedia Manufacturing, vol. 3, pp. 6101 – 6108, 2015.

[16] ÁVILA, S. F. **Reliability analysis for socio-technical system, case propene pumping**. Engineering Failure Analysis, vol. 56, pp. 177 – 184, 2015.

[17] ÁVILA, S. F.; MENEZES, M. L. A. Influence of local archetypes on the operability & usability of instruments in control rooms. In: EUROPEAN SAFETY AND RELIABILITY CONFERENCE – ESREL 2015. 25., Zürich, Switzerland.

[18] AMERICAN PETROLEUM INSTITUTE. API 770.2001: A manager´s guide to reducing human errors. Improving Human Performance in the Process Industries. Washington, DC., 2001.

[19] ÁVILA, S. F.; COSTA, C. Cognitive deficit analysis in routine tasks, as a strategy to reduce accidents and increase of industrial production. In: EUROPEAN SAFETY AND RELIABILITY CONFERENCE – ESREL 2015. 25., Zürich, Switzerland.

[20] REASON. J. Human error. New York: Cambridge University Press, 2003. 302p.

[21] ÁVILA, S. F.; DANTAS, E. T.; DUARTE, J. The Requirements & tools to treat process safety risk as result of Organizational Change. In: GLOBAL CONGRESS ON PROCESS SAFETY. 11., 2015. **Anais…** Austin, Texas.

[22] ABNT, Associação Brasileira de Normas Técnicas. **NBR ISO 50001:2011: Sistemas de Gestão de Energia: Requisitos com orientações de uso.** Rio de Janeiro: ABNT, 2011.

[23] ÁVILA, S. F.; KALID, R. A.; KIPERSTOK, A.; CARVALHO, V. S. Program to Avoid Product Leakage at Source, Fertilizer Case. Chemical Engineering Transactions. vol. 26, 2012.

[24] ÁVILA, S. F. Assessment of Human elements to avoid accidents and failures in task perform, cognitive and intuitive schemes. In: GLOBAL CONGRESS ON PROCESS SAFETY. 7., 2011. **Anais…**Chicago, Illinois.

[25] ÁVILA, S. F.; KIPERSTOK, A.; KALID, R. A.; SANTOS, A. C.; CARVALHO, V. S. Environmental Deviation Assessment, Record, Failure, Probability and Investment. Chemical Engineering Transactions. vol. 26, 2012.

[26] CCPS (Center for Chemical Process Safety), Conduct of Operations and Operational Discipline: For Improving Process Safety in Industry, American Institute of Chemical Engineers, New York, NY, 1985.

[27] ÁVILA, S. F. Reliability analysis for socio-technical system, case propene pumping. Engineering Failure Analysis, vol. 56, pp. 177 – 184, 2015.

[28] GERTMAN, D.; BLACKMAN, H.; MARBLE, J.; SMITH, B. C. The SPAR-H Human Reliability Analysis Method. Idaho National Laboratory, Washington DC, 2005.

[29] ÁVILA, S. F.; LOPES, J. R.N.; KALID, R. A.; KIPERSTOK, A. F. Gestão da disponibilidade de água e energia através do Plano Diretor. In: RIO OIL & GAS EXPO AND CONFERENCE 2014. Rio de Janeiro.

[30] ÁVILA, S.; KIPERSTOK, A.; BRAGA, B.; KALID, R. Avoiding loss of energy in a petrochemical industry, operation and design Congress 2011. In: WORLD RENEWABLE ENERGY CONGRESS 2011. Linköping, Sweden.

[31] ÁVILA, S. F.; COIMBRA, H.; SOUZA, C.; GOLÇALVES, A. Iniciativas de eficiência energética para sustentabilidade, oportunidades & desafios. In: RIO OIL & GAS EXPO AND CONFERENCE 2014. Rio de Janeiro.

[32] ÁVILA, S. F.; JESUS, Z. M. R.; KALID, R. A.; NASCIMENTO, V. Good practices in Cooling Towers: Comparison between industrial audits. In: Cooling Technology Institute Annual Conference. Corpus Christi, Texas, 2003.

[33] GENZ, F., TANAJURA, C. A. S., DE ARAÚJO, H. A. Impacto das Mudanças Climáticas nas Vazões do Rio Pojuca sob o Cenário A2 do IPCC - 2070 a 2100. In.: I CONGRESSO BAIANO DE ENGENHARIA SANITÁRIA E AMBIENTAL. Salvador. 2010.

[34] SILVA, S. F. Análise da disponibilidade e demanda para o sistema de abastecimento de água de salvador frente a cenário de mudanças climáticas. Dissertação (mestrado). Programa de pós-graduação em Meio Ambiente, Águas e Saneamento. UFBA. Salvador BA. 2012.

[35] KIPERSTOK, A.; BRAGA, E. A.; CESAR, M.; ÁVILA, S. F.; SAMARA.; KALID, R. A. Eds: 2012. Base de Informações para Plano Diretor de Água Fafen-BA, TECLIM RESEARCH FAFEN – UFBA. Camaçari, Brasil.

[36] SEDUR, SECRETARIA DE DESENVOLVIMENTO URBANO DA BAHIA. Estudos de Aproveitamento dos Mananciais da RMS – Revisão e Atualização das Proposições do Plano Diretor de Abastecimento de Água, Incluindo o rio Pojuca, e Visando o Uso Racional dos Recursos Hídricos Regionais - Estudo de Demandas. Relatório 1100.11-REL. EST. DEMANDAS-00. 73p. Salvador. 2010.

[37] SANTIAGO, P. S., & SANTOS, R. P. Major Improvements in cooling system performance at an ammonia/urea plant. IWC – International Water Conference, Proceedings: Pensylvania: 2000.

[38] ÁVILA FILHO, S., NASCIMENTO LOPES, J.R. Cooling Water Systems: An Overview, in: Z. Amjad, K.D. Demadis (Eds.), Mineral Scales and Deposits: Scientific and Technological Approaches, 2015, pp. 499–531.

[39] DEMING, W. E. Qualidade: A revolução da Administração. Rio de Janeiro: Editora Marques-Saraiva. 1990. 367 páginas.

[40] ÁVILA FILHO S., SOUSA C. R., CARVALHO A. C. Assessment of complexity in the task to define safeguards against dynamic risks. 6th International Conference on Applied Human Factors and Ergonomics (AHFE 2015) and the Affiliated Conferences, AHFE 2015. Procedia Manufacturing 3 (2015) 1772 – 1779.

442403 A Novel Open Source CFD Code for Explosion Modelling

Abstract:

The porosity parametrization of complex geometry is very important in explosion simulation. This project proposes a novel approach to calculate the mesh porosity (volume and area) based on the Gilbert–Johnson–Keerthi (GJK) distance algorithm. The GJK algorithm can be used to check the collision between convex objects. This concept is used in the current work to check the collision between an element of the mesh and an object of the geometrical model. The parametrized geometry is combined with the combustion model based on flamelet formulation in accordance with BML (Bray-Moss-Libby) model. The typical chemical libraries are replaced by the integral length scale of wrinkling based on semi-empirical formulation. One simple step reaction mechanism is considered. The turbulence closure problem is addressed via Boussinesq formulation using two equations model for turbulent kinetic energy and the rate of dissipation of turbulent kinetic energy.

The CFD (Computational Fluid Dynamics) solver relies on an explicit finite-volume code based on a structured Cartesian grid. The classical fourth order Runge-Kutta time marching approach is considered. The PDR (porosity distributed resistance) approach is used to model all geometry under analysis (including the large-scale objects and the small ones). The central differencing scheme is applied to discretize the governing equations and artificial viscosity is used to smooth out numerical wiggling. Additional resistance terms are considered in the momentum equation. Additional generation source term for turbulence is considered in the turbulent kinetic energy equation. The simulation of standard flow across different geometries presented consistency with the literature. Good agreement is found for similar simulations using a commercial CFD software. The developed open CFD code seems to be promising as an alternative to solve complex explosion problems from the engineering view point.

442442 Compliance with Process Safety Management Elements

Abstract:

OSHA promulgated 29 C.F.R. § 1910.119 regulation in 1992 and the fourteen Process Safety Management (PSM) elements. The main objective of defining these elements is to prevent unwanted discharge of highly hazardous chemicals exposing personnel to serious hazards. The elements are linked to each other and have intricate interdependencies. Every element provides information to other element(s) or uses information from other element(s). Given the interdependencies of these elements, an effective Process Safety Management Program (PSM) has to be in place which considers process design, technology, procedures (both normal and emergency), emergency response plans, training programs etc. The presentation talks about all these fourteen elements in details with main emphasis on the relief systems design.

442454 A Fuzzy Logic and Probabilistic Hybrid Approach to Quantify the Uncertainty in Layer of Protection Analysis

Abstract:

Layer of Protection Analysis (LOPA) is a widely used semi-quantitative risk assessment method. It provides a simplified but less precise method to assess the effectiveness of protection layers and the risk of an incident scenario. The outcome frequency and consequence are intended to be conservative by prudently selecting input data, given that design specification and component manufacturer's data are often overly optimistic. There are many influences, including design deficiencies, lack of layer independence, availability, human factors and wear in testing and maintenance, which are not quantified and are dependent on type of process and location. This makes the risk usually overestimated. So, there are different sources and types of uncertainty in the LOPA model that need to be identified and quantified. In this study, a fuzzy logic and probabilistic hybrid approach was developed to quantify the uncertainty of frequency of initiating event and the probabilities of failure on demand (PDF) of Independent Protection Layers (IPLs) based on the available data and expert experience. The method was applied to a distillation system with a capacity to distill 40 tons of flammable n-hexane. The outcome risk of the new method was proven to be more precise comparing to the results of the conventional LOPA approach.

442474 Quantitative Risk Analysis of Nitrogen Fertilizer Plant, of Moron Petrochemical Complex

Abstract:

Quantitative Risk Analysis of Nitrogen Fertilizer Plant, of Moron Petrochemical Complex

Considering the possible risk scenarios associated with new Nitrogen Fertilizer Plants (Ammonia, Urea and Utilities) of the Moron Petrochemical Complex, in general, are common to most petrochemical complexes in the world and are a result of materials hazardous that are manufactured, processed, handled and stored at the facility; the various processing systems and conditions; the different procedures used for the operation and maintenance of the facility; as well as the response of the detection and mitigation of hazards. Previous experiences in similar plants indicate that the most likely hazards are those who identify with the physical-chemical properties of the materials used and the process conditions.

For petrochemical facilities, the most common consequences are:

- Toxic Gas Effects
- Jet Fire
- Flash Fire
- Pool Fire
- Bleve
- Vapor Cloud Explosions.

Based on the premise of "Only what is known can be controlled," This paper aims to develop the Quantitative Risk Analysis (QRA) for new Nitrogen Fertilizer plants (Ammonia, Urea and Utilities) of Moron Petrochemical Complex, which it is located in the jurisdiction of the Municipality Juan Jose Mora, Carabobo State from Bolivarian Republic of Venezuela, considering for it the most severe scenarios identified previously by HAZOP analysis, SIL studies and previous plannings studies for the control and emergency response and contingency, then determining the levels of social and individual risk based on the probability of occurrence of the events of greater impact and its consequences and impact on people, equipment and buildings, comparing the risk levels obtained with tolerance criteria established by the corporation.

The Quantitative Risk Analysis (QRA) are developed from the standard and guidelines of Petroleos de Venezuela, (PDVSA); Center of Chemical Process Safety (CCPS), and The Netherlands Organization (TNO), through the computational tool PHAST RISK 6.7 . from Det Norske Veritas (DNV).

Management of Plant IPL's

Richard Stanley
Tehama Control Systems Consulting Services
Consultant to Mangan Software Solutions
26741 E Clifton Drive
Aurora, CO 80016
rick@tehamacontrolservices.com

Prepared for Presentation at
American Institute of Chemical Engineers
2016 Spring Meeting
12th Global Congress on Process Safety
Houston, Texas
April 11-23, 2016

Management of Plant IPL's

Richard Stanley
Tehama Control Systems Consulting Services
Consultant to Mangan Software Solutions
26741 E Clifton Drive
Aurora, CO 80016
rick@tehamacontrolservices.com

Keywords: IPL, LOPA, Independent Protection Layer, Safety Life Cycle Management, Safety Instrumented System, Safety Instrumented Function

Abstract

Independent Protection Layers (IPL) are usually identified in a Layers of Protection Analysis (LOPA) during a Process Hazards Analysis. The IPL portion of a LOPA identifies Functions or Design Features within a processing facility that serve to prevent or minimize the consequences of a deviation from the expected processing conditions. Many times a LOPA may identify various IPL's, but the facility then has difficulty in clearly associating the functions identified by an IPL with real plant assets. They then have difficulty in assuring that those assets have the necessary operational and maintenance visibility, and are operated and maintained to assure their availability. This paper discusses these issues and describes the characteristics of an effective Management Program for IPL's.

1 Introduction

During Process Hazard Analysis (PHA) for processing facilities, a number of Independent Protection Layers (IPL) may be identified as having the capability to reduce the risk of a given consequence occurring. A given scenario may require that one or more of these IPL's be in place to reduce risk to tolerable levels. Typically, these IPL's are identified during a Layers of Protection Analysis (LOPA), but other methods can be used.

Once these IPL's have been identified, the process of managing them has just started. Designation of an IPL adds Operational and Maintenance requirements that often exceed typical practices. This presents an operating facility with the challenge of adapting Operation and Maintenance practices to assure that the IPL's that are credited are well understood, that their protective functions remain available and meet the performance requirements that were defined for them.

This paper discusses the various type of IPL's and the requirements and methods for assuring their availability. The basic ownership requirements are described along with observations relative to the difficulty that real facilities have had in addressing these requirements. This particularly applies to the less visible IPL types such as those IPL's that are not associated with Safety Instrumented Functions (SIF).

After discussion of the ownership requirements and obstacles to meeting those requirements, the use of a Safety Life Cycle database to overcome these obstacles is discussed. The effective application of a Safety Life Cycle database can enable an operating organization to manage those requirements with maximum effectiveness and value with minimum impact of existing organizational structures and accountabilities.

2 Independent Protection Layers

A key concept in managing Process Hazards is that of an Independent Protection Layer. The formalized concept of an IPL was developed along with the LOPA process. During a PHA the potential consequences of various deviations in plant operation are evaluated, and the potential consequences of those deviations if no preventative measures are taken are defined in qualitative terms.

When an unmitigated risk identified by the PHA is higher than an organization's tolerance, the probability of the event occurring may be further evaluated to better understand the risk and the effectiveness of the various potential safeguards that could prevent consequence from occurring.

The most common technique for performing this analysis is the Layers of Protection Analysis (LOPA). In a LOPA, functions or design characteristics that are credited as being able to prevent the consequence are called Independent Protection Layers (IPL).

2.1 IPL Identification

The measure of effectiveness of identified IPL's is generally assessed against an organization's risk policies using a Target Mitigated Event Likelihood (TMEL). TMEL is usually based upon the severity of the LOPA scenario unmitigated consequences. For example, a severity level may have a required TMEL of 1e-04, or one occurrence of the unmitigated consequence in every 10,000 years.

During the LOPA, various Initiating Causes for a Deviation scenario are identified and each is assigned an expected frequency of occurrence. Some organizations may choose to apply additional frequency mitigation factors, such as the probability of personnel being in the area at the time of the Initiating Cause. Some organizations will choose to sum Initiating Causes, while others do not.

The product of expected frequency of the initiating causes, any additional frequency mitigation and Probability of Failure Upon Demands (PFD's) of the credited IPL's is used to calculate the value of the Total Mitigated Event Likelihood ($TMEL_{act}$). If the $TMEL_{act}$ is below the threshold specified for the consequence level associated with the event, the number of IPL's is considered to be sufficient. If the $TMEL_{act}$ exceeds the threshold the LOPA team recommends corrective actions, often recommending that additional IPL's be added. However, other recommendations such as process modifications to reduce the consequences or reduce the initiating causes may also be recommended. An example of a LOPA TMEL analysis is shown in Figure 1.

Figure 1 – Sample LOPA TMEL Analysis

LOPA TMEL Calculation Example

Consequence TMEL = 1.0E-05 (1 consequence Event per 100,000 years)

IC1 - 0.1 (once per 10 years)
IC2 - 0.1 (once per 10 years)
MF - 0.1 (Personnel present 10% of the time)

IPL1 PFD = 0.1
IPL2 PFD = 0.01

TMELact = (IC1+IC2) * MF * IPL1 * IPL2
TMELact = (0.1+0.1) * 0.1 * 0.1 * 0.01 = 2.0E-06

Actual TMEL (TMELact) is less than TMEL, so the IPL protection is acceptable. No additional protection is required

2.2 What is an IPL?

An Independent Protection Layer is an instrumented or physical function which, if effective, would prevent the unmitigated consequences of a scenario identified in a PHA or LOPA from occurring.

The primary characteristics of an IPL are:

- The proper functioning of the IPL would prevent the consequence associated with a scenario from occurring
- The IPL is independent of any of the scenario initiating causes and is independent of other IPL's cited for the scenario.
- The IPL is reliable enough to perform the expected function with the probability of failure being with the range with which the IPL is credited.

- The physical and functional design of the IPL meets the required independence and availability requirements.
- The IPL is Operated and Maintained in a manner that assures the credited reliability is achieved during plant operation.

CCPS [1] provides guidelines as to what functional, design or physical aspects of a process installation can qualify for being credited as an IPL, and describes the limitations for where multiple IPL credits may be taken. However, it is up to the facility owner or operator to determine specific local requirements. It should be noted however, that a 2015 US OSHA interpretation [2] appears to extend the concept of Recognized and Accepted Good Engineering Practices (REGAGEP) to publications that are not necessarily consensus documents. CCPS publications, such as the LOPA publication cited above, fall into this category. Furthermore, the OSHA interpretation notes that internal standards must represent REGAGEP and will reject standards that do not comply.

2.3 IPL types

Some of the more common IPL types are:

- Operator response to an Alarm
- A Basic Process Control System Automatic Action
- A Safety Instrumented Function (SIF) implemented in a Safety Instrumented System (SIS)
- A non-SIF interlock
- Pressure relief devices
- A physical barrier that would contain process fluids after a loss of primary containment. An example is a tank dike or a drainage system
- Physical equipment that would prevent a scenario from causing the unmitigated consequence associated with it. For example, this could be a valve limit stop, a restriction orifice or other flow restriction that prevents excessive inflow.
- Inherently Safe Design

2.4 IPL's and the Safety Life Cycle

The concept of the Safety Life Cycle is often considered to be associated with Safety Instrumented Systems (SIS) because its prominence in IEC 61508 and IEC 61511. However, the concepts are applicable to any protective functions or systems and are used in other contexts.

An important factor to recognize is that the risk reduction provided by IPL's is the product of their RRF's, and that if any IPL fails to meet the performance credited by a LOPA, the associated risk reduction of that IPL is lost. CCPS recognizes this by requiring that any IPL be Operated and Maintained in manner that assures that the credited Risk Reduction is effective.

The IEC 61511 and ISA 84.00.01-2004 Standards provide a framework for managing of Safety Instrumented Systems (SIS) and the Safety Instrumented Functions (SIF) that were identified by the Hazard Analysis process. However, these Standards are silent on managing of other non-SIF IPL types.

This has led some operating organizations to conclude that there are no corresponding requirements relative to the non-SIF IPL's. This is not the case: non SIF IPL's require sufficient care to assure their availability and cannot just be run to failure or be treated like non-IPL functions. The sections in the remainder of this paper discuss the practicalities of IPL implementation and ownership, the risk gaps that have been observed, and suggested management practices to prevent these gaps from developing.

3 What Should an IPL Management System Look Like

One of the fundamental gaps observed in practice is that there is often a perception that once an IPL is credited in a LOPA, and something exists in the facility that appears to perform the function defined by the LOPA, no additional action is required. This results in a system in which the effectiveness of an IPL is left to chance and the assumed risk reduction becomes an illusion.

An effective IPL Management System must be able to assure that IPL's are visible, that their design meets the functional and independence requirements for an IPL, and that the IPL's are Operated and Maintained to assure the required availability.

IPL's types and their implementation can vary considerably. While SIF IPL's may be paid more attention because of standards such as IEC 61511 and ISA 84.00.01-2004, non-SIF IPL's are usually shared with other basic plant systems such as the Basic Process Control System (BPCS) or blend in with other mechanical systems. Without an effective IPL Management System, non-SIF IPL functions often are either not effective to begin with, or atrophy to the point were they do not deliver the risk reduction with which they were credited.

Table 1 below identifies some of the gaps in IPL Management that have been observed in real operations. Unfortunately, these gaps are more common than not.

Table 1 – IPL Management Gaps

IPL Management Gap	Cause	IPL Management Requirements
Improper IPL credits are taken	Personnel are not properly trained on what the requirements and limitations for IPL qualification are, or local standards do not comply with REGAGEP	LOPA facilitators must be well trained on IPL requirements and advise participants on IPL requirements including IPL Types allowed and restrictions taking multiple credits. LOPA's should be reviewed by qualified and empowered individuals to assure validity.
IPL's as implemented do not meet the effectiveness, independence and reliability requirements defined in the LOPA's	Plant assets are identified of being IPL's, but are not evaluated to verify that they meet the IPL requirements.	Checklists for assessment of IPL qualifications are available and all IPL's are assessed against those requirements.
IPL functions are not understood by Operations and Maintenance personnel	IPL's are identified in LOPA documents and are not clearly identified in documents that are used by Operations and Maintenance personnel. LOPA documents are not readily available to non-PSM personnel and then are difficult to correlate to actual plant installations.	A system is needed that clearly identifies plant assets that are used to implement IPL's. The data should identify real plant assets that perform IPL functions, the Operational and Maintenance expectations for the IPL's, including operator actions, restrictions on disabling or bypassing the IP and Maintenance and Testing Requirements.
IPL's are bypassed or disabled	Assets credited as IPL's are not clearly identified. Operating and Maintenance procedures for IPL's are not prepared, available or enforced	Operating procedures need to exist that define criteria for bypassing or disabling of IPL's and requirements for management or operations while bypassed or disabled.
IPL's are modified or removed without proper MOC	The IPL list is not available, or is out of date. Projects making modifications do not recognize devices which are in IPL service and treat their modification like modification of non-IPL services	A continuously up to date IPL list is provided and linked to all LOPA, IPL design and operations and maintenance data. Projects evaluating proposed modifications have all the data needed to perform a proper MOC available in one place.
IPL's are not tested at the intervals required to maintain performance	IPL's are not identified in maintenance systems or do not have test procedures available. When procedures are available, management is ineffective because of many documents in many places, many of which may not be consistent or up to date	Test intervals, maintenance procedures and testing procedures are integrated in one database. Test records are captured in the Operations and Maintenance Module to allow for identification of next required test interval.

3.1 Why Is Compliance So Hard?

Many organizations have demonstrated difficulty in incorporating the requirements of IPL ownership and operation into their day to day practices. The basic reasons for this are that the importance of IPL's is not always recognized, and that effective management of IPL's generally does not fit within traditional plant management structures.

Effective IPL management requires that information cross traditional organizational boundaries. In traditional operating facilities, these boundaries usually result in many parallel information and documentation systems that have limited accessibility or provide only the functionality that a particular group requires and does not necessarily support the needs of other groups. Some examples of this are:

- Process Safety Groups have proprietary software that often has highly restrictive incensing terms that effectively lock up HAZOP and LOPA data access to just a few people.
- Maintenance Management Systems have historically been developed for management of major rotating equipment by and for rotating equipment personnel. Other functions are focused upon parts inventory management and work order tracking. These systems perform poorly when other systems are involved. They are particularly bad at handling Instrumented Systems. Resistance to making them effective for Instrumented Systems, or handling safety related non-rotating equipment has typically been quite high.
- Much of other traditional information management is done with discrete paper files or electronic files that are often not coordinated or readily available. These records are held by various organizations and are difficult coordinate and keep up to date. Often these records do not survive changes in personnel.

Management of IPL related data across these multiple platforms and formats is extremely inefficient and easily broken. Most facilities do not have the personnel or skill sets required to manage these problems. The Systems, even when well intentioned, tend to fail under the burden the highly labor intensive requirements of maintaining this data.

3.2 Safety Life Cycle Management

Effective management of IPL's across the entire Safety Life Cycle requires Safety Life Cycle Management systems that address the cross discipline nature of the data required and supports the following types of information:

- Defining relationships between Hazard Identification, Functional Requirements and Design of Protective Functions and Equipment, and Operation and Maintenance requirements.
- Capturing the Operational and Maintenance Events that are related to the IPL functions and equipment and using that data to provide Operation and Maintenance Management a clear picture of compliance and performance.

- Allowing users from various organizations to readily access IPL related data including the Hazard associated with IPL Assets, IPL functional requirements and equipment specifications and IPL Asset events and performance

Figure 2 illustrates a high level view of the relationships that should be defined and maintained and their general relationship to the Safety Life Cycle.

- HAZOP scenarios that have a high consequence are related to LOPA's
- LOPA's and the IPL's that were credited by the LOPA are identified
- Plant Assets that are used to perform the functions identified by a LOPA IPL are identified and their relationships to to the LOPA IPL's are shown. Note that in some cases, multiple LOPA IPL's may refer the the same Plant Asset.
- The Plant Asset data contains definition of the functions required, performance requirements, testing requirements, expected operator responses and physical equipment used to implement the functions.
- Operation and Maintenance Events are recorded and associated with the Plant Assets or Plant Equipment.
- KPI's are derived from the Operation and Maintenance Events and used to manage such things as period testing and assess performance such as demands, failures, bypasses.

Figure 2 – IPL Asset Data and Relationships

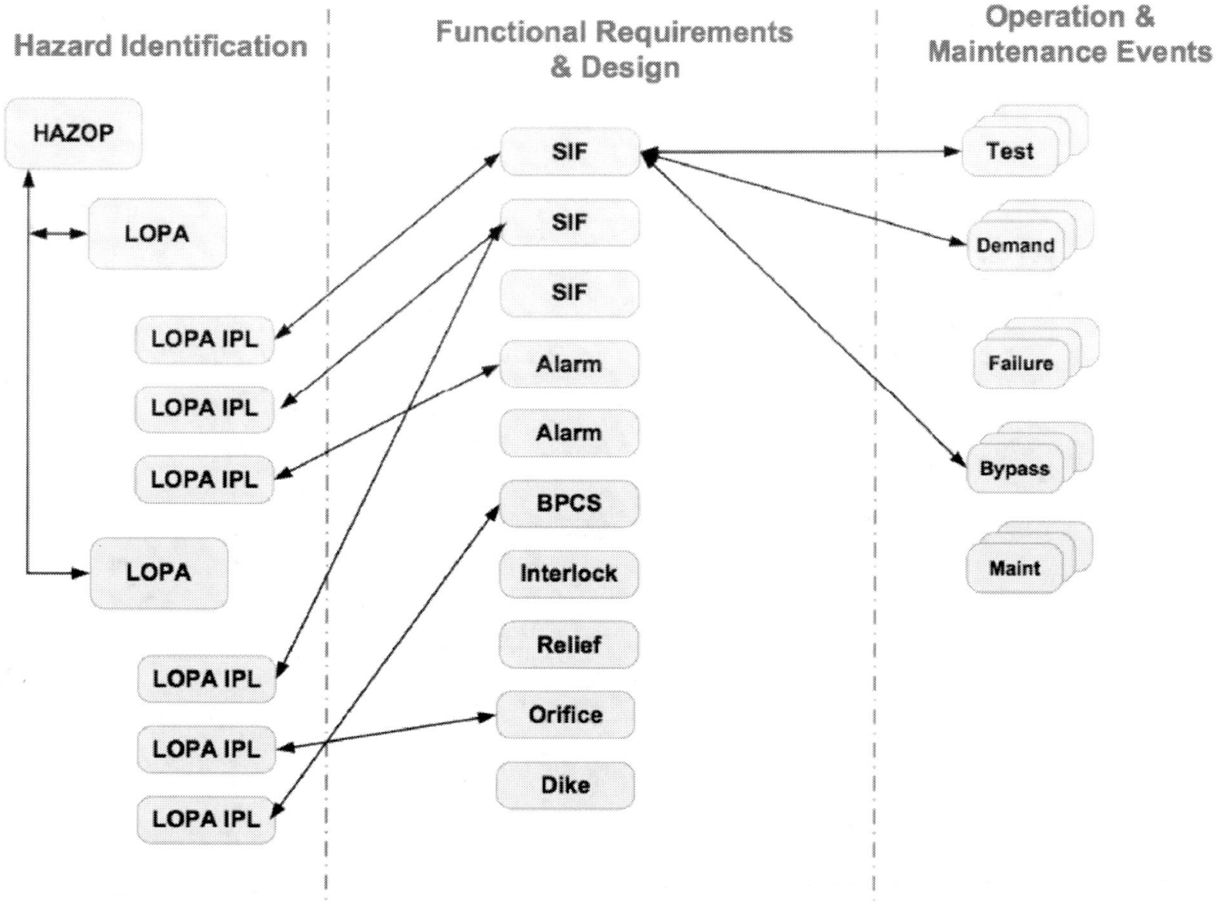

Figure 3 shows a more detailed set of data and relationships that focuses on one Plant Asset that is used to perform IPL functions.

- The Asset is related to one or more LOPA IPL's and to the equipment used to perform the Asset functions
- Plant Equipment that is used to implement an IPL Asset, such as instrumentation or other physical equipment is associated with the IPL Asset
- Assessment checklists and results are associated with the LOPA IPL's. Design Assessment checklists are associated with each Plant Asset and its associated equipment.
- Important documents such as Operating and Maintenance procedures or design documents are associated with the Asset for quick access.
- Operating and Maintenance Events are recorded and associated with their related equipment or Plant Assets

Figure 3 – IPL Asset Data and Relationship Details

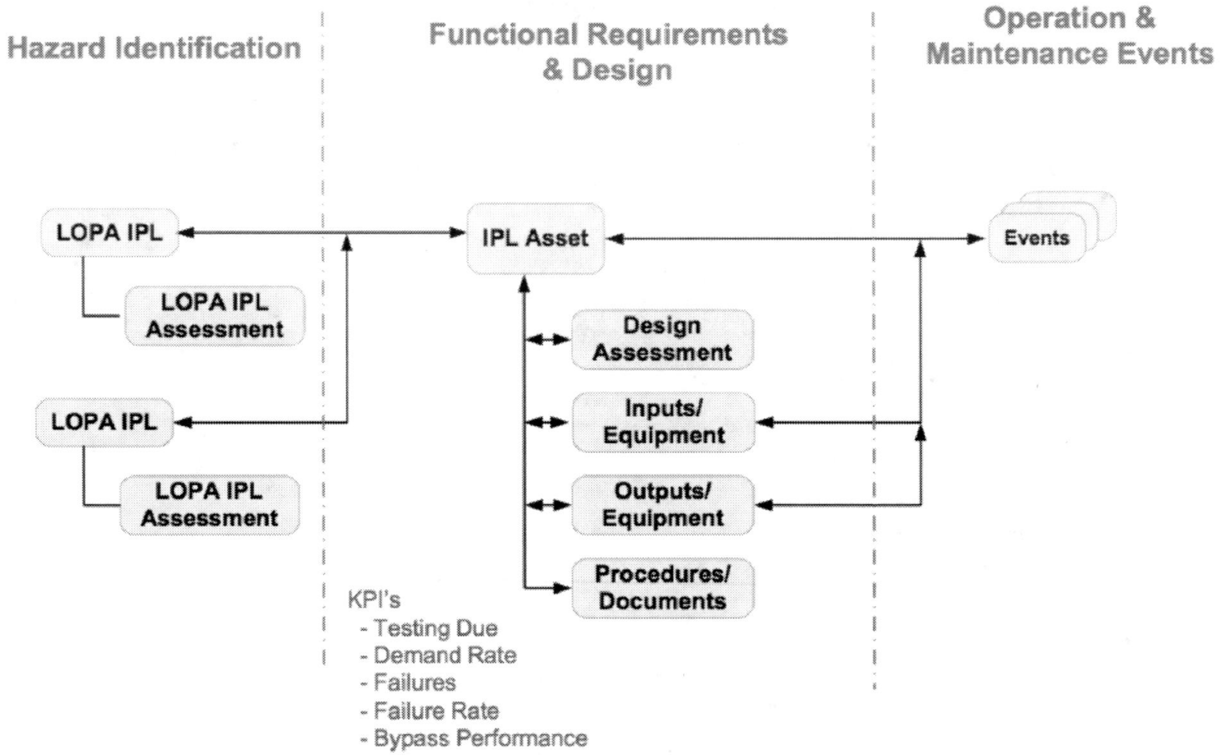

4 Using a Safety Life Cycle Management Data Base

A Safety Life Cycle Management System is best implemented in a database that is constructed specifically to manage the data and events that apply to the Safety Life Cycle. The data should be highly linked, so that a User can readily access related data. For example, an operator would be able to view the functional requirements for an IPL Asset, using a tag or other ID that is used within the Unit, and then directly access the Hazard data associated with the IPL. Appropriate Operations, Maintenance or Engineering personnel should then be able to associate Events with the IPL Assets and view KPI's associated with those Assets.

4.1 Data Relationships

A Safety Management Life Cycle Management Data Base should contain data objects that provide relationships among the various items shown in Figure 3. Among these are

- LOPA IPL's that define basic definitions of the IPL function and performance requirements and relate the LOPA IPL to the appropriate LOPA and HAZOP scenarios

- IPL Assets that define the actual plant assets that are used to implement the IPL functions. These should be identified using the same identifiers used in other plant documentation and by which plant operators, engineers and maintenance personnel know the asset. The IPL Asset data should define the design requirements such as function, expected response, accuracy and test and maintenance requirements
- Assessment Check Lists that define the items relative to performance, design, documentation or testing that is required of the IPL
- Input and Output Asset and Device objects that define the functional requirements for the Asset, Inputs or Outputs that are used to implement the IPL function and the specification data for the Devices that are installed in those Asset locations
- If IPL Performance is to be tracked, Event data objects that allow for tracking of events such as Demands, Failures, Testing, Bypasses, etc.
- Reference Documents such as procedures or design or specification data

4.2 Data Content

The data for each of the objects shown in Figures 2 and 3 should capture information necessary to support IPL Management. Some examples are:

- LOPA IPL's define the type of IPL, the performance requirements - typically in terms of a Risk Reduction Factor (RRF), and a functional description of the IPL. If operator response to an alarm is involved, the expected operator action and response time should be specified
- SIF Asset data should define the functional requirements, performance expectations and testing and maintenance requirements. In some implementations, data for a SIF could include all data necessary for definition of the Safety Requirements Specification (SRS) as defined in IEC 61511 and ISA 84.00.01-2004.
- Non-SIF IPL Asset data should define define the functional requirements, performance expectations and testing and maintenance requirements appropriate to the type of IPL.
- A LOPA IPL Assessment is a check list of items and their assessment results necessary to validate that the IPL will be effective and independent of other IPL's credited for the LOPA scenario
- An IPL Design Assessment is a check list of items and assessment results necessary to verify that the IPL Asset as designed meets the functional, performance and independence requirements for the IPL.
- Input/Equipment and Outputs/Equipment should define the functional requirements for Inputs and Outputs used by the IPL. The Equipment sections should contain or be linked to specification data for the specific equipment performing the functions.
- Procedure/Documents data should identify the specific Operating Procedures, Maintenance and Test Procedures or other Design Documents that apply to the IPL.

- Events – Events are used to describe the date, time and results of various types of Operation and Maintenance Events that occur during the IPL Life Cycle. The data for each event should be appropriate to the Event Type, such as as-found and as-left data for Functions and Equipment, start and end of bypasses, etc.

5 AN EXAMPLE IPL

The Figures below illustrate how a Safety Life Cycle data base application can be used to manage an IPL. An Alarm IPL is shown as an example.

5.1 IPL Functional Requirements

Figure 4 shows an example of an IPL Functional Specification for an Alarm with Operator Action IPL. This allows a User to define the functional requirements for the Alarm including such data as

- The Alarm ID – Typically this is the Tag Number of the Alarm as it appears on P&ID's, Operator Interfaces and other documentation
- Alarm functions, set points and basic implementation requirements
- Expected operator response and the time within which the response is expected
- Alarm operating restrictions such as when the alarm may be disabled or restrictions on changing set points or other configuration data
- Alarm Testing Requirements – type of testing and frequency of testing.
- LOPA Associations including the LOPA scenario for which the Alarm was cited as an IPL, the LOPA IPL and other IPL's that were cited by same LOPA scenario
- Associated Documentation such as Test Procedures, Operating Procedures

Figure 4 – Alarm IPL Functional Requirements Data

Alarm O M Data | Alarm Functional | Alarm Devices | Revisions

Print | 1 Document | Attach Tasklist | Subscription | Bookmark | Admin Tools

Alarm Functional Requirements - FCC-FAL-138, Feed to East Riser

Alarm Requirements

Alarm ID:	FCC-FAL-138
Alarm Description:	Feed to East Riser Low Flow
Functional Description:	Feed flow to the East Riser is indicated on the BPCS by the analog input poing FCC-FI-138. This point has a low flow alarm configued to alert the operator of low riser feed flow
Asset Protection Category:	IPL
Operator Repsonse Time:	10 Min
Operator Repsonse Time Notes:	Nominal time to responsd to a high priorty alarm
Defined Corrective Action:	The operator shall place the East Riser Assist Steam flow controller FCC-FC-282 into automatic with a set point of 1100 PPH. Verfiy that the controller is properly proving steam to the riser

Protective Category:

Health and Safety	Environmental	Commercial	Reputation
Yes	No	No	No

Performance Requirements:	RRF 10 PFD 1.0E-01
Alarm Operating Requirements:	Alarm must be enabled at all times when feed is in the unit. If the alarm must be bypassed or disabled prepare Safety Bypass Authorization. Set point is not to be changed without an MOC being performed
Alarm Response Time:	5 sec
Alarm Response Time Notes:	BPCS must activate alarm within 5 seconds of when the transmitter senses low flow
Alarm Location:	FCC Operator's Console in Main Control Room
Alarm Implementation:	Soft alarm on analog input tag
Alarm Priority:	High
Alarm Priority Rationale:	The risler low feed flow alarm indicates that conditios may exist the contribute to a riser flow upset but the situation is no yet crtical. Additional SIF layers of protection exist
Alarm Testing Requirements	Test transmitter, calibrate and test alarm activation and verify set points at turnarounds
On-Line Testing Interval	N/A
Off-Line Testing Interval	60 Months

Alarm Set Point (Software Based)

Set Point ID	Value	Type	Tolerance	Time Delay	Basis
FCC-138-SPL	115	Decreasing	5%	1 sec	Minimum HC flow w/o assist steam

Alarm Set Point (Device Settings)

Set Point ID	Value	Type	Tolerance	Time Delay	Basis
N/A	N/A	N/A	N/A	N/A	N/A

LOPA References

LOPA IPL ID	LOPA ID	LOPA IPL Description	LOPA Hazard	LOPA Consequnces	Required RRF
FCC-LOPA-15-1	FCC-LOPA-15	Low Feed Flow Alarm	Increase in riser catalyst density and loss of flow stability. Potential loss of flow and flow reversal	Hydrocarbon in regenerator. Very high regenerator temperataures and potenial hydrocarbon in precipitator. Potential explosion	10

Associated IPL's

ID	Type	Description
FCC-SIF-02	SIF	East Riser Emergency Steam
FCC-SIF-05	SIF	FCC Catalyst Reversal Protection System

Alarm Testing Procedures

Document ID	Title		Description
CE-TEST-7764	Alarm Testing Procedure		General Refinery Alarm Testing Procedures

5.2 IPL Operating and Maintenance History

Figure 5 shows an example of a view of the Alarm IPL's Operating and Maintenance history for the same Alarm as illustrated above. This view repeats the basic Alarm Functional Requirements data and the provides data specific to the Alarm operating status and the history of the events which have been entered into the database for the Alarm. This view provides the following types of data.

- Basic Alarm Function Requirements
- Alarm Testing and Performance Data – this section lists things such as the testing status for the alarm, the last test date and next test data, commissioning data, hours in service and a summary of performance statistics such as the number of demands and failures and the in-service failure rate
- Lists of the events that have been entered into the database for the alarm listed by type of event. In the example, this includes the Alarm Service Status history, Test Events, Demand Events, Bypass Events, etc.

In this view, many data fields are linked so the User can click on things such as Event ID's to go directly to the event detail view.

Figure 5 – Alarm Operation and Maintenance Data

Alarm O-M Data | Alarm Functional | Alarm Devices | Revisions

Print | 1 Document | Attach Tasklist | Subscription | Bookmark | Admin Tools

Alarm O-M View - FCC-FAL-138, Feed to East Riser

Alarm Data

Alarm ID: FCC-FAL-138

Alarm Description: Feed to East Riser Low Flow

Functional Description:

Feed flow to the East Riser is indicated on the BPCS by the analog input poing FCC-FI-138. This point has a low flow alarm confirgued to alert the operator of low riser feed flow

Operator Repsonse Time: 10 Min

Defined Corrective Action: The operator shall place the East Riser Assist Steam flow controller FCC-FC-282 into automatic with a set point of 1100 PPH. Verify that the controller is properly proving steam to the riser

Testing and Performance Data

Alarm Testing Requirements: Test transmitter, calibrate and test alarm activation and verify set points at turnarounds

On-Line Testing Interval: N/A

Off-Line Testing interval: 60 months

Test Status: Current

Last Test Date: 3/26/12

Next Test Due Date: 3/26/17

Due Date Based Upon: Turnaround Schedule

Determining Devices	ID	Type	Description
	FCC-FT-138	Flow Transmitter	Feed to East Riser

Alarm Service Status: Commissioned

Service Hours: 108,624 hrs

Alarm Commission Date: 4/28/08

Alarm Commission Notes: None

Alarm Decommission Date: N/A

Alarm Decommission Notes: N/A

Alarm Performance	Life of Plant	5 Years	1 Year	Year to Date
No. of Failures	1	1	0	0
No. of Demands	6	6	0	0

In Service Failure Rate: 7.0 E-02

Alarm Service Status History

Event ID	Date	From	To
FCC-FAL-138-Serv-001	4/28/2008	Created	Commissed

Test Groups

Test Group ID	Test Group Description	Date Last Run	Test Event ID
FCC-Alarms-001	FCC Safety Related Alarm Testing at Turnaround	4/4/2008	FCC-Alarms-001-Test-002
FCC-Alarms-002	FCC Safety Related Alarm Testing at Turnaround	3/24/2012	FCC-Alarms-001-Test-001

Functional Test Events

Event Date	Event ID	Parent Event ID	As Found	As Left	Event Notes	Event Status
4/4/2008	FCC-FAL-138-Test-001	FCC-Alarms-001	Failed	Passed	Pre-Commissioning Test	Approved
3/4/2012	FCC-FAL-138-Test-002	FCC-Alarms-002	Passed	Passed	Routine Turnaournd Tst	Approved

Demand Events

Demand Date	Demand Event ID	As Found	As Left	Event Notes	Event Status
4/17/2008	FCC-FAL-138-Demand-001	Passed	Passed	Startup	Approved
9/8/2008	FCC-FAL-138-Demand-002	Passed	Passed	Feed Pump Trip and Spare Start	Approved
12/17/2008	FCC-FAL-138-Demand-003	Passed	Passed	Feed Pump Trip and Spare Start	Approved
7/22/2011	FCC-FAL-138-Demand-004	Passed	Passed	Feed control valve sticking	Approved
10/9/2012	FCC-FAL-138-Demand-005	Passed	Passed	Feed Pump Trip and Spare Start	Approved
11/30/2013	FCC-FAL-138-Demand-006	Passed	Passed	Feed Pump Trip and Spare Start	Approved

5.3 IPL Assessment

As mentioned above, a critical part of the Safety Life Cycle is verification of an IPL's qualification to perform as an IPL. Check Lists can be created and tracked to document this process. Examples for an IPL Independence Check List and an Alarm Design IPL Check List are shown below.

In Figure 6 illustrates a Check List that would be completed near the time an IPL is cited in a LOPA is shown. This task list is intended to verify that the basic requirements to qualify the function as an IPL have been satisfied.

Figure 6 – LOPA IPL Check List

Item No.	Description	Finding	Comments
1.1	Verify that the IPL function, if effective, will prevent the consequence of the initiating cause from occurring	Passed	
1.2	Verify that the IPL is independent of the LOPA scenario imitating cause	Failed	The Alarm appears to use the same transmitter whose failure was the initiating cause.
1.3	Verify that the IPL is independent of any other IPL's that have been credited for the LOPA scenario	Passed	

Figure 7 shows a Check List summary that appears in the LOPA IPL view. This provides the user with a quick overview of the verification status and typically appears on the IPL Functional Requirements view.

Figure 7 – LOPA IPL Check List Status

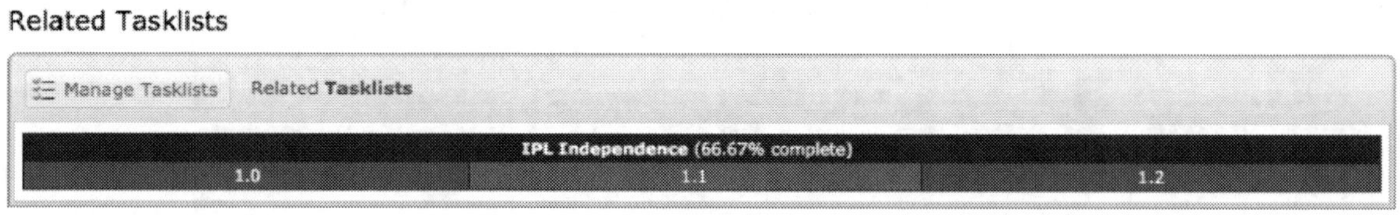

Figure 8 illustrates an IPL Design Check List for an Alarm. This check list addresses design requirements and operational preparedness items such as operation and maintenance preparedness items.

Figure 8 – Alarm IPL Checklist

Item No.	Description	Finding	Comments
3.1	Verify that the Alarm and its associated operator actions are capable of preventing the consequence for which the alarm has been cited as an IPL	Passed	
3.2	Verify that the alarm location is continuously attended	Passed	
3.3	Verify that the initiating or sensing device is independent of the initiating cause of the LOPA scenario	Failed	This alarm is derived from FCC-FT-138 whose failure is the LOPA scenario IC. Need an independent transmitter
3.4	Verify that the initiating device or sensor is independent of other protection layers claimed as IPL credits for the same LOPA scenario	Passed	
3.5	Verify that an operator action has been defined for this alarm, that the action can be accomplished and the result effective within the time between the initiating cause and the consequence occurring	Passed	
3.6	Verify that a written procedure for the operator response to this alarm has been prepared, that operators have been trained to locate the procedure and that the procedure is readily available to the operator	Failed	Operating procedures require updating.
3.7	Verify that this alarm has a suitable priority and it is identified as an IPL Alarm on Operator Graphics and Alarm views	Failed	Alarm needs to be added to operating graphics and IPL Alarm graphic

Item No.	Description	Finding	Comments
3.8	Verify that the settings for this alarm and disabling, inhibiting or bypassing of the alarm is under control of a Bypass Control Procedure and suitable approvals are required for the above actions and that return to operation is managed by this procedure.	Failed	The alarm needs to be added to the list of critical items that require Safety Bypass Authorization.
3.9	Verify that modifications to this alarm are under Management of Change Controls	Passed	Site Standard MOC Procedure
3.10	Verify that Maintenance and Testing Procedures have been prepared which are applicable to this alarm, and that the Procedures include capture of all testing records	Passed	Site Standard Maintenance procedure for testing of alarms
3.11	Verify that the P&ID; correctly shows this alarm	Failed	P&ID has not yet been updated to show this alarm as an IPL
3.12	Verify that the Alarm has been included in the Unit IPL List	Passed	

5.4 KPI's from Data Base

Once IPL data is captured in a data base, KPI's associated with the IPL Operation and Maintenance can be readily created and displayed. Some of the KPI's and summary data that can be developed are

- Identification of IPL Testing that is upcoming or overdue
- Identification on IPL's that have had high Demand or Failure rates
- Calculation of In-Service IPL Failure rates
- Reporting on IPL Assessment status and and completion of open items
- IPL quantities for Units, Sites or Enterprise-wide
- Identification of LOPA IPL's that are not linked to a Plant IPL Asset
- Identification of Plant Assets that are associated with multiple LOPA IPL's

Figure 9 shows a few examples of the types of IPL KPI data that can be provided to Users

Figure 9 – Typical IPL KPI's

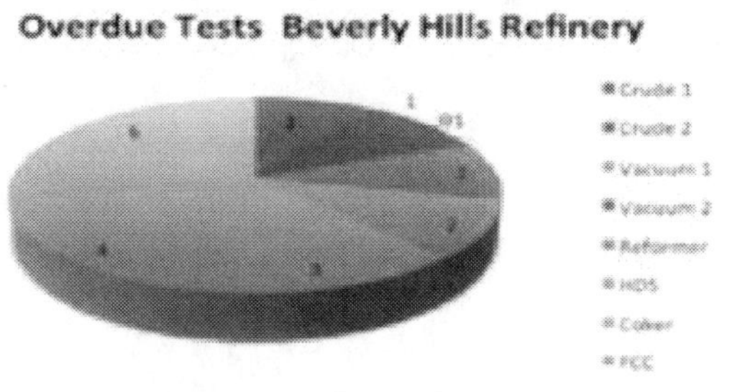

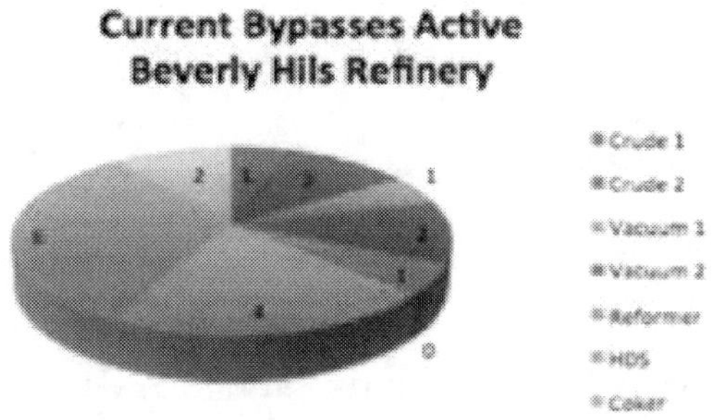

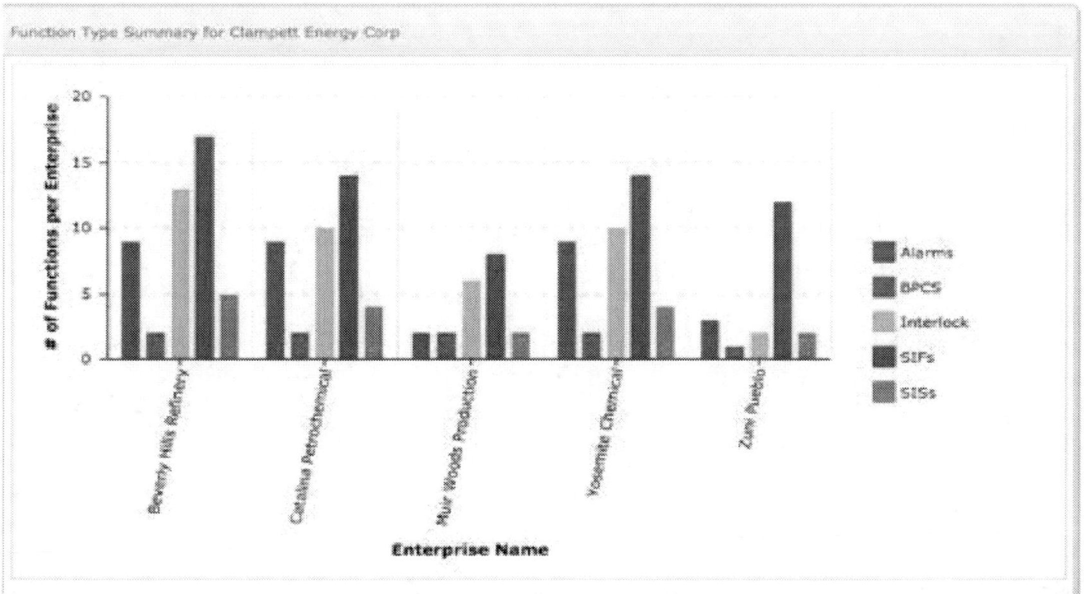

Enterprise Name	Alarms	BPCS	Interlock	SIFs	SISs
Beverly Hills Refinery	9	2	13	17	5
Catalina Petrochemical	9	2	10	14	4
Muir Woods Production	2	2	6	8	2
Yosemite Chemical	9	2	10	14	4
Zuni Pueblo	3	1	2	12	2

6 Conclusions

The Management of IPL's within an Operating Facility presents distinct challenges to Facility Management and trickles down to many organizations within the Facility. Many organizations are having signification problems in managing these requirements due to lack of the clarity of requirements and the cross discipline aspects that do not fit within the practical restrictions of management tools traditionally used. Force fitting IPL Safety Life Cycle management requirements as an overlay to these tools has generally been highly inefficient and generally ineffective.

The compliance gaps that have been observed as operating facilities struggle with management of IPL's and the Safety Life Cycle are a result of these problems. A practical and effective means of compliance is to apply a Management Process that is designed specifically for support of the Safety Life Cycle. This Management Process can maximize effectiveness of compliance and assure that the risk reduction that has been assumed for a facility is actually realized.

7 References

[1] – Layer of Protection Analysis, Center for Chemical Process Safety, ISBN 0-8169-0811-1, 2001

[2] – OSHA Memorandum – June 5, 2015 – Thomas Galassi, RAGAGEP in Process Safety Management Enforcement.

Rick Stanley has over 40 years' experience in Process Control Systems and Process Safety Systems with 32 years spent at ARCO and BP in execution of major projects, corporate standards and plant operation and maintenance. Since retiring from BP in 2010, Rick has consulted with Mangan Software Solutions (MSS) on the development and use of MSS' ProSys SLM Safety Lifecycle Management software and has performed numerous Functional Safety Assessments for both existing and new SISs.

Rick has a BS in Chemical Engineering from the University of California, Santa Barbara and is a registered Professional Control Systems Engineer in California and Colorado. Rick has served as a member and chairman of both the API Subcommittee for Pressure Relieving Systems and the API Subcommittee on Instrumentation and Control Systems.

442704 Jack Rabbit II: Experimental Large-Scale Chlorine Release Trials

Abstract:

In August and September of 2015, the U.S. Department of Homeland Security (DHS) Science & Technology (S&T) Chemical Security Analysis Center (CSAC) conducted the Phase 1 field trials of the Jack Rabbit II (JR II) program, a series of 5 large-scale outdoor chlorine release experiments with a collaborative team of partners from government, industry, and academia. Sponsored by the CSAC, the Defense Threat Reduction Agency (DTRA), and Transport Canada, these field trials were performed to fill critical knowledge, data, and capability gaps to support and improve chemical release modeling, industrial safety, and emergency response procedures. This paper summarizes the execution of the 5 successful chlorine release trials ranging from 5 to 9 tons, the instrumentation deployed, the data collected, and preliminary analyses of the high-definition video footage, cloud movement, and behavior and concentration/time profiles as a function of distance from the release site. This paper will also present plans for the JR II Phase 2 field trials scheduled for late 2016 which will include additional chlorine releases up to 16 tons.

445083 Novel Methodology for Inherent Safety Assessment in the Process Design Stage

Abstract:

Various physical and chemical hazards are intrinsic to the material itself or its manner of storage and use. Inherently safer design of the chemical process and equipment has been increasingly used by the design engineers to minimize or eliminate these hazards. This approach aims to mitigate hazards by design instead of add-on measures. This presentation describes a new design approach, introducing comprehensive inherent safety index (CISI), for use in the early stages of process development. Chemical, process and connectivity scores are calculated for individual units in the process, based on which the CISI assigns safety scores for the equipment. Each chemical in the unit has a weighted severity score. Reactivity score is calculated for mixture of chemicals in that unit. Connectivity score in introduced because the units can be highly interconnected, leading to compounding of hazards. Biodiesel and methyl methacrylate processes were studied to demonstrate the new safety assessment methodology. The results of the assessment are used to clearly identify key areas for improvement in a root-cause analysis.

Creating a Culture of Operational Discipline within a Management of Change Process

Stacey Elizabeth Jackson Griffith
Eastman Chemical Company
100 Lincoln Street
Kingsport, TN 37660
segrif@eastman.com

Prepared for Presentation at
American Institute of Chemical Engineers
2016 Spring Meeting
12th Global Congress on Process Safety
Houston, Texas
April 11-23, 2016

Creating a Culture of Operational Discipline within a Management of Change Process

Stacey Elizabeth Jackson Griffith
Eastman Chemical Company
100 Lincoln Street
Kingsport, TN 37660
segrif@eastman.com

Keywords: Management of Change, MOC, Operational Discipline, Six Sigma

Abstract

In 2010, Eastman Chemical Company ("Eastman") chartered an "All In for Safety" team to lead the company on a transformational safety journey. Eastman's goal is to achieve a sustained world-wide result of zero injuries and process safety incidents. To this end, the All In for Safety team coordinated efforts in many areas of personal and process safety. Numerous individuals and teams participated in this effort. This paper highlights the work of the Management of Change (MOC) Process Improvement team.

Eastman management chartered the MOC Process Improvement team to institute changes to the company's MOC process. As part of the All In for Safety effort, principles from the CCPS Book, *Conduct of Operations and Operational Discipline*, were to be incorporated. Conduct of Operations and Operational Discipline (COO/OD), put simply, requires organizations to establish standards and structure processes in such a way that these standards become the framework for performance. "A good COO/OD system can be viewed as a type of 'control system' for human performance. It permits appropriate variability of intelligent human control yet achieves desired performance within established constraints."[1]

As the MOC Process Improvement team applied this principle to Eastman's MOC process, we realized the necessity to drive a paradigm shift. Users often treated the MOC process as the compliance mechanism itself, rather than as a framework that empowered them to make safe decisions. Using Eastman's Business Design for Six Sigma methodology, the team made changes to enhance the productivity and accuracy of the electronic process. At the same time, the team improved the users' ability to make decisions and control outcomes when addressing the hazards associated with MOC changes. The team also sought to create a process and accompanying tool robust enough to provide COO/OD structure for the changing company profile in terms of site complexity and regions of operation.

The project objectives were achieved by implementing the following changes:

1. Standardizing job expectations and training for those individuals involved in the risk identification and mitigation of changes;
2. Instituting screening tools that help the users tailor the level of review with the risk profile of the change;

3. Improving the electronic tool to leverage modern collaboration tools;
4. Improving the electronic tool to allow for management of routine and non-routine tasks associated with MOCs;
5. Improving documentation of risks and their associated mitigation plans; and
6. Providing clear, consistent, and visible measures that are standard across the company.

1 Introduction

In 2010, Eastman Chemical Company ("Eastman") chartered an "All In for Safety" team to lead the company on a transformational safety journey. This team of senior managers worked to create step changes in all areas of personal and process safety. The long range goal for this effort is that Eastman will achieve a sustained world-wide result of zero injuries and process safety incidents. Many individuals and teams participated in this effort. This paper highlights the work of the Management of Change (MOC) Process Improvement team.

Eastman management chartered the MOC Process Improvement team to institute changes to the company's MOC process. As part of the All In for Safety effort, principles from the CCPS Book, Conduct of Operations and Operational Discipline, were to be incorporated. Conduct of Operations and Operational Discipline (COO/OD), put simply, requires organizations to establish standards and then structure processes in such a way that these standards become the framework for performance. "A good COO/OD system can be viewed as a type of 'control system' for human performance. It permits appropriate variability of intelligent human control yet achieves desired performance within established constraints." [1]

As the MOC Improvement team began scoping the work to be accomplished through the lens of operational discipline, we identified a number of improvement opportunities. In terms of the mission of MOC, the team found a need to refocus the effort. Eastman's MOC process has been existence for more than twenty years. After such a long period of time, "process creep," had occurred, creating bureaucracy and inefficiency. Users often felt punished by the amount of bureaucracy and did not view the process as a helpful means to get work done but as a regulatory driven impediment.

The core mission of management of change is to protect employees and equipment from safety incidents due to changes made to manufacturing processes. Eastman expanded this mission to include both environmental and quality considerations. However, other business processes had worked their way into MOC. All of these processes were worthy, but most did not align with the core vision of MOC. Examples of these processes include reliability efforts and non-manufacturing information systems.

The team also identified the need to drive a paradigm shift for users of the MOC process. The Eastman process, while thorough, was rigid with screening questions that were "Yes/No" type questions. This rigidity limited the user's ability to document judgment calls and concerns which were evaluated and determined to be adequately controlled without further mitigation. It limited the user's ability to document collaboration on risks. It also made it possible for a user to rely on the process as the compliance mechanism, since he was not required to make text input on risk screening questions.

Finally, during the scoping phase, Eastman began to rapidly expand through acquisitions. The team recognized the need to provide a consistent framework of performance for all sites related to MOC. This framework needed to be immediately available to new acquisitions. The structure had to provide mechanisms for MOC compliance that ensured operational discipline with company principles, while at the same time, allowing for the varying site profiles due to size, manufacturing processes, and regional regulations. This framework also needed to be independent of the eMOC tool so that sites could maintain compliance regardless of the integration timeline for use of the electronic tool.

2 Applying Business Design for Six Sigma to Management of Change

Once scoping work was complete, the team used Eastman's Business Design for Six Sigma methodology to execute improvements to Eastman's Management of Change process. Composition of the team included a Six Sigma Black Belt to lead and facilitate, an IT resource to advise on the electronic tool, and representatives from Manufacturing, Technology, and Engineering. These individuals were high enough in organization management to make decision and drive change, but close enough to the actual users of the MOC process to have an accurate perspective. The team followed the five steps of Business Design for Six Sigma: Define, Conceptualize, Design, Optimize, and Control. The sections below will detail the results.

2.1 Define

During the define phase, the team utilized Voice of Customer interviews, external partner benchmarking, and an extensive literature review. The team evaluated the Eastman MOC process compared to other industry MOC processes, commercial processes based on software evaluation, and technical literature resources including the CCPS Books *Conduct of Operations and Operational Discipline* and *Guidelines for the Management of Change for Process Safety*.

The evaluation sought to identify:

1. Opportunities to improve the productivity of the process by removing bureaucracy and inefficiency.
2. Opportunities to better engage users to collaboration and document risk mitigation evaluation
3. Opportunities to improve task management within the MOC process
4. Adjustments needed in light of the changing company profile; and
5. Recommendations for leading indicator measures.

Working with our senior management steering team, the MOC Improvement team determined that the project would include all Eastman sites, regardless of size and region. Per Eastman policy, all manufacturing processes and pilot plant operations were already included in the scope of the project, not just those subject to OSHA 1910 regulation. Lastly, Eastman has always included Quality and Environmental risks as part of management of change.

2.2 Conceptualize

Using the principles outlined in the Define phase of the project, the team employed failure mode effects analysis (FMEA), Lean techniques[1], and other project methodologies to identify the following conceptual improvements:

1. Standardizing job expectations and training for those individuals involved in the risk identification and mitigation of changes;
2. Instituting screening tools that help the users tailor the level of review with the risk profile of the change;
3. Improving the electronic tool to leverage modern collaboration tools;
4. Improving the electronic tool to allow for management of routine and non-routine tasks associated with MOCs;
5. Improving documentation of risks and their associated mitigation plans; and
6. Providing clear, consistent, and visible measures that are standard across the company.

2.3 Design

The design phase involved many different parallel efforts. Figure 1 illustrates the model used to identify the work to be done and also to communicate the process to the users. As seen in this figure, this phase addressed many aspects of the MOC process. The center of the process is the tool used at the manufacturing site. The company provided tool is an electronic system called MOC Global, developed by Eastman. The framework pieces of Procedures, Training, Skilled Personnel, Audits, and Measures surround the tool and provide the structure for appropriate use of the tool. Finally, the operational discipline safety belt holds all the pieces together so that they work as designed. Subsequent paragraphs identify design information for the six improvements listed in the Conceptualize section, above.

[1] LEAN is a project methodology used in Six Sigma. LEAN methodology is well described in literature. For example, additional information may be found on the web at The Lean Enterprise Institute (http://www.lean.org)

Operational Discipline is the safety belt for our success.

Figure 1. Eastman MOC Process Logo

2.3.1 Standardizing job expectations and training for those individuals involved in the risk identification and mitigation of changes

With regard to skilled personnel, Eastman's MOC process defines six required roles that are responsible for risk identification and mitigation. Table 1 describes these six roles.

Table 1. Six Key MOC Roles Standardized by Eastman

Position	General Description of Role
Originator	Individual who is responsible for the MOC Change.
MOC Coordinator	Individual responsible for day to day operations of the MOC process. The MOC Coordinator reviews all MOCs for quality control and site specific practices not customizable in the software.
Subject Matter Experts	Safety, Quality, and Environmental personnel who assist the originator in identifying risk and developing risk mitigation plans.
Engineering Subject Matter Expert	Engineering staff person who is responsible for detailed engineering design.
Primary Approver	Manager who is responsible for the manufacturing process where the MOC Change is being made.
Facility Upper Manager	Manager who is responsible for ensuring that the site complies with the Eastman MOC process, and that the roles specified above are staffed appropriately. This is typically the site manager.

Eastman's increase in total number of sites introduced variability with regard to staffing these required roles. In order to ensure consistent performance across the company, the corporate MOC document defines minimum responsibilities and training[2] for individuals filling the six key roles. The corporate standard focuses on creating a consistent structure for performance while allowing sites to select personnel for these positions based on the management structure of the site. As shown in Table 1, the SOP gives the Facility Upper Manager the responsibility to ensure that its requirements are met.

2.3.2 Instituting screening tools that help the users tailor the level of review with the risk profile of the change

Eastman processes more than 7000 MOCs each year in the company. The improvement team identified the ability to tailor the level of review for individual MOC changes as important to improving productivity, safety, and discipline. By putting thousands of MOCs through the exact same review, regardless of size and potential impact, the team believed that we could, in fact, be diluting focus on those changes with the largest potential impact. Though some amount of "right sizing" was already present in the existing process, the team sought to make this decision clearer for users. The team selected a set of screening criteria to guide the users to determine the appropriate amount of review and the risk content areas that could be impacted by the change. The initial set of criteria was put through ten table top pilots in various sizes and types of operating areas. Existing MOCs were put through the criteria list. The recommended amount of review was compared to the outcome the team would expect. The criteria were refined as needed. This process was continued until four consecutive tabletop exercises were held that resulted in no changes to the screening criteria. As a result, non-value-adding reviews have been reduced by 15 to 20%. In terms of quality of reviews, other changes were made to ensure that review quality was increased. The section on improving the documentation of risks, below, discusses these changes.

2.3.3 Improving the electronic tool to leverage modern collaboration tools

Previous versions of Eastman's electronic tool were very linear in nature and were written prior to recent collaboration software. The improvement team sought to improve both efficiency and communication by leveraging modern collaboration tools. The following tools were incorporated:

- Skype for Business®, allowing users to contact each other directly from any individual MOC Dashboard;
- Online status updates, so that any user can see the approval status and identified risks for the MOC in real time;
- Tracking tools, showing collaborators the relevant changes made by each user;
- A discussion board for communications that do not need to be archived with the MOC; and
- An electronic log display that allows users to see every action taken on an MOC by user and date.

---

[2] Training requirements are primarily Eastman specific and will not be mentioned. However, certain user roles in the Eastman process are required to complete Process Hazard Analysis (PHA) training. Content requirements for PHA training are documented in the corporate procedure.

2.3.4 Improving the electronic tool to allow for management of routine and non-routine tasks associated with MOCs

For any given type of work done on an MOC, there exist certain activities that are routinely done. The Eastman MOC system calls these activities "tasks." Tasks are included in the list of activities to be executed with every MOC, though the MOC Global system allows customizing of tasks based on the work to be done. Examples of tasks are detailed engineering design, construction scheduling, construction execution, and field validation steps for facility change MOCs.

The system calls specific activities that must be completed on individual MOCs "action items." Action items are not standard enough to be included on every MOC. As an example, most facility change MOCs require no change to the site air emissions permit. Therefore, it is not productive to include a task for permit update on every MOC. However, a small subset of facility changes requires Title V minor modifications, which must be obtained before the standard "Start Up" task is executed. An even smaller subset requires Title V construction permits, which must be obtained before the "Construction" task begins.

Previous versions of Eastman's electronic system did not aid the users in identifying action items. Action item tracking also had to occur outside of the software program, which introduced inefficient record keeping and chasing of information. It inherently contained an opportunity for execution errors by individuals because non-routine activities could not be tied directly to the MOC execution software.

In the MOC Global software, users have the ability to create action items. Most action items are the result of risk mitigation plans developed as part of the risk screening. The section on improved documentation of risks, below, contains more information about these action items. Action items may also be identified independently of the risk mitigation plans during the collaboration process and even during MOC execution.

Because of the fluid nature of MOC changes, action items do not key off of dates. The software ties action items to one of three MOC process milestones: construction, startup, or closeout. When the execution software for the tasks (routine items) reaches a milestone, it checks for the status of action items tied to that milestone. If any milestone action items are incomplete, execution of tasks will stop until completion of all the appropriate activities. Eastman conservatively estimates a productivity savings of 30 minutes for each action item tracked inside the MOC system versus the former external method.

2.3.5 Improving documentation of risks and their associated mitigation plans

Previous versions of the electronic MOC software had a linear hazard review process. The originator would use the screening tools to indicate the presence of risks and could make comments on the main MOC form regarding these risks, but only in a single comment box for the entire form. There were no comment sections for individual screening questions. When the subject matter expert (SME) reviewed this material, the SME could make changes and approve or reject the MOC. If the MOC was rejected, the approval process was required to restart fully. In addition, documentation of risk mitigation plans could only be done via the general comments box, or in external documents attached to the MOC.

The team identified a second opportunity regarding the nomenclature used during risk screening. When preparing an MOC proposal, an originator entered all the information he knew about the change, including answering a series of Yes/No questions regarding potential risks. When an originator was uncertain regarding a particular hazard, the Y/N nomenclature created ambiguity in terms of passing on information to the SME. It was also difficult for the users to document situations where a concern was identified about a risk, but resulting investigation showed that the current controls for this risk were adequate. Again, the Y/N nomenclature, with no comment field, did not lend itself to clear communication for these situations.

Figure 2 provides an illustration of how this collaboration opportunity was addressed in the MOC Global software.

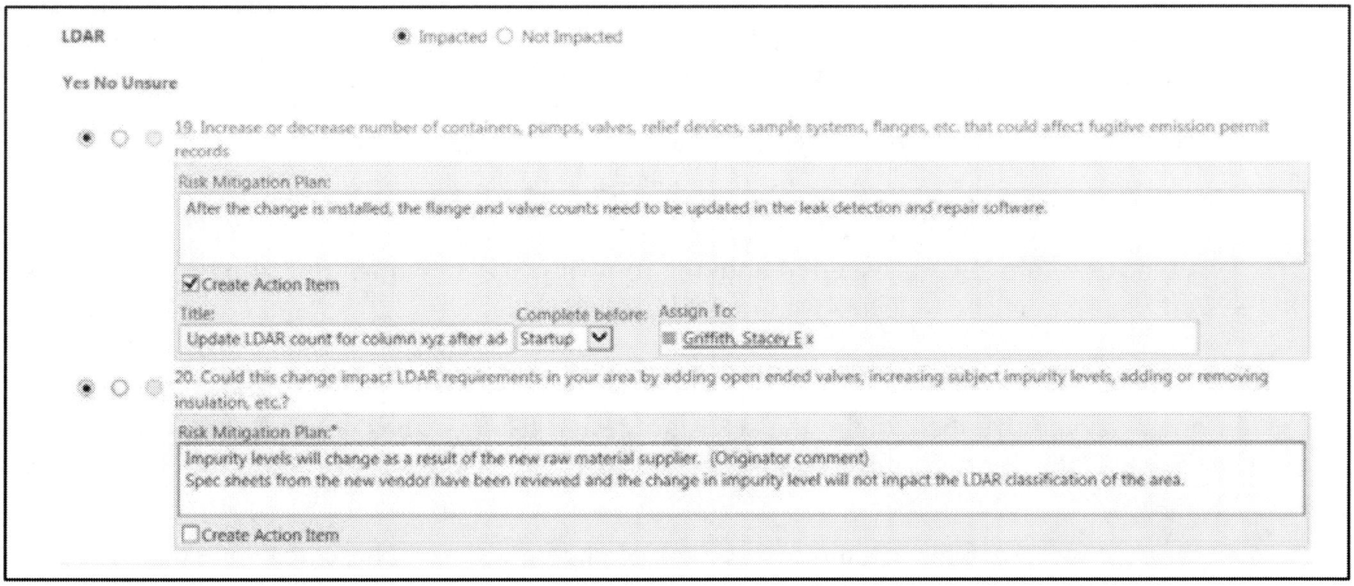

Figure 2. Example risk screening questions with risk mitigation documentation

When an originator or subject matter expert (SME) indicates yes to a risk screening question, a mandatory risk mitigation plan field appears. This field must be filled out with information about what needs to be done to mitigate the risk. If current controls are adequate, this information would be documented instead. This allows risk screeners to document that a particular hazard was a concern, that they evaluated it, and that they determined it to be adequately controlled. The added detail is beneficial in terms of compliance documentation and for PHA revalidations. If risk mitigation is needed, the users may select to create an action item for activities required, when appropriate.

The originator also has use of an alternate selection, which is "unsure." An "unsure" answer indicates to the SME that the originator is concerned about a particular hazard, but does not have enough knowledge to make a firm determination. The SME must change unsure answers to "yes" or "no."

Finally, there is no rejection button for SMEs to use when they receive the MOC. Because they can make changes and document information pertinent to the risk, there is no need to reject the MOC. When the SME makes changes and approves the risk area checklist, MOC Global notifies the originator. The originator can then review the SME's changes which are highlighted in yellow for clarity. This process will continue until both the originator and SME are satisfied that all risks have been fully identified and mitigated.

2.3.6 Providing clear, consistent, and visible measures that are standard across the company

Lastly, any operational discipline related process requires clear measures to indicate performance. Eastman instituted a measures system that tracks key company measures; however, the software utilized allows users at the individual site and operating area level to customize the view to the subset of MOCs within their control. This ability ensures that all sites using the MOC Global software are evaluating their MOC process against the same set of criteria and data. It also represents a productivity improvement because measures are automatically generated without specific sites having to submit data. Examples of measures that are kept:

- Numbers and types of MOCs processed in a given time period;
- Average number of days for different tasks to be accomplished; and
- Action Item usage.

This type of information is useful to both the company and individual operating areas to keep the system transparent and to identify potential improvements.

2.4 Optimize

In order to optimize the system, the software design team established a user feedback process. Feedback from users is processed monthly and divided into one of four categories. These categories are:

- FAQs – user feedback indicates a lack of understanding. These items are answered and posted as frequently asked questions.
- Backlog – These are items that are suggestions for future process enhancements. Items that do go into the backlog are rated by a criteria system to generate a prioritized list for continual improvements.
- Resolved Items.
- Items requiring no action.

The backlog list generates continuous improvement opportunities for the system. In February 2016, the Company implemented a number of enhancements based on Q3/4 2014 and Q1/2 2015 feedback.

2.5 Control

Eastman had a number of thorough documents describing the management of change process. In order to obtain a complete understanding of the MOC process; however, users had to review multiple documents in detail. In some cases, users also had to filter out additional PSM related content such as PHA process information that was not pertinent to the MOC process.

The team also identified a need related to acquisition sites. A simple, clear, and easily accessible document on management of change would facilitate faster and easier compliance for new sites trying to integrate many Eastman processes at one time.

As a result, the team generated a single corporate standard operating procedure (SOP) containing all MOC related content. Any other PSM policies or guiding documents refer to this MOC procedure rather than containing MOC content. This procedure centralizes information in one location, facilitates training, and provides reference for users. The corporate SOP outlines the appropriate structure and basic process for MOC. It is sufficient as a stand-alone compliance mechanism for smaller sites. Larger sites, with additional requirements, processing profiles, and more complex organization structures have developed site specific companion documents to further document their practices.

3 Conclusion

In conclusion, Eastman's MOC Process Improvement team utilized business design for Six Sigma to strengthen the Operational Discipline associated with management of change. The changes made to the system were designed to integrate human judgment and collaboration into a solid support framework of company procedure and tools. As shown in Figure 1, above, the team strengthened existing framework. The process is facilitated by an enhanced electronic tool. Measures, audits and an MOC Process Leadership Council work to ensure that operational discipline is being followed. As is stated in our training and in Figure 1, "Operational Discipline is the safety belt for our success."

Reference in order of first citation

[1] Center for Chemical Process Safety. (2011). *Conduct of Operations and Operational Discipline.* New York, New York: John Wiley & Sons, Inc.

Managing Alarms to Support Operational Discipline

Denise Chastain-Knight. PE, CFSE
exida Consulting LLC
64 North Main Street
Sellersville, PA, USA
dchastainknight@exida.com

Todd Stauffer, PE
exida Consulting LLC
64 North Main Street
Sellersville, PA, USA
tstauffer@exida.com

Prepared for Presentation at
American Institute of Chemical Engineers
2016 Spring Meeting
12th Global Congress on Process Safety
Houston, Texas
April 11-23, 2016

Managing Alarms to Support Operational Discipline

Denise Chastain-Knight. PE, CFSE
exida Consulting LLC
64 North Main Street
Sellersville, PA, USA
dchastainknight@exida.com

Todd Stauffer, PE
exida Consulting LLC

Keywords: Process Safety, Alarm Management, Rationalization, Prioritization, Independent Protection Layers, Nuisance Alarms

Abstract

Process alarms, coupled with operator action, are frequently cited as a safeguard in a Process Hazard Analysis (PHA) and an Independent Protection Layers (IPL) in a Layer of Protection Analysis (LOPA), but does the alarm management system really support the safeguard/IPL?

According to ISA-18.2 / IEC 62682 an alarm must indicate an equipment malfunction, process deviation, or abnormal condition that requires a timely operator action. If no action is taken, then the alarm is either invalid or the operator is not doing their job. Both scenarios represent a breakdown in operational discipline for alarm management as does the presence of nuisance alarms and alarm floods. This breakdown in operational discipline for alarms has been cited as a contributing factor in many significant safety incidents, some of which will be analyzed in this paper. If operational discipline for alarms is lacking, then it is very possible that the desired risk reduction for a process alarm used as an IPL will not be achieved and the probability of an ineffective operator response will increase.

As systems have evolved from hardwire to computer control, alarms have become easier and less expensive to implement leading to more and less purposeful alarms. Operators must contend with multiple alarms at one time with only their experience to determine priority. Alarms may be added to or removed from a control system without proper management of change. Systems may include alarms for which there is no possible action, or inadequate action time. What can an organization do to take control of their process alarms and improve operational discipline?

1 Introduction

Operational discipline is the practice of doing the right things the right way every time. Success is predicated on having a well-defined process and trained personnel. The process is usually described through operating procedures, work instructions and tools. Personnel are trained to follow the procedures, understand the instruction and use the tools. An organization can become very experienced and repeatable in executing the well understood normal processes. By definition, alarms are triggered when there is an abnormal condition in the process. The situation may or may not be addressed in the procedures, instructions, tools and operator experience. In many control rooms, alarms sound so frequently that operators miss important alarms in the constant flow across the alarm banner. Worse yet, they become accepting of the condition as 'normal' and no longer recognize alarms as a warning of potential trouble. This paper describes techniques to establish an alarm management program that will equip operators to take the correct action every time they are presented with an alarm.

2 Notation

BDI	Burst Disk Indication
DCS	Distributed Control System
ISA	International Society of Automation
I/O	Input/Output
IPL	Independent Protection Layers
KPI	Key Performance Indicator
LOPA	Layer of Protection Analysis
MOC	Management of Change
OD	Operational Discipline
OSHA	Occupational Safety & Health Administration
PHA	Process Hazard Analysis
PLC	Programmable Logic Controller

3 Role of Alarm Management in Operational Discipline

3.1 Background

Prior to the advent of DCS and PLCs in the 50s, process control systems consisted of a few wires and mechanical devices. Each device required a capital investment so theoretically, every control loop and interlock were well thought out and justified in the design process. Process alarms were so rare that each one was taken very seriously by operators. As DCS/PLC capabilities have evolved, it has become cheaper and easier to implement alarms. Control system designers will often enable multiple alarms (3, 5 or more) by default per analog I/O point without thinking about whether they are really needed. Many are configured as a matter of routine. PHA and quality control interest may dictate additional alarms to make the operator aware of a condition trending toward abnormal. The result is a situation where operators are presented with so much information that they cannot effectively manage the influx of data and are likely to miss critical situations. An operator may correctly identify an alarm, but the time between alarm and event is inadequate for an operator to effect corrective action. Frequent alarms with no corrective action may desensitize operators so they become slow to act in response to a truly abnormal event. Alarm overload, inadequate response time and nuisance alarms have been contributing factors in industrial accidents. Specific examples are described in the sections below.

3.2 The Human Element

Operational Discipline (OD) is defined as the deeply rooted dedication and commitment by every member of an organization to carry out each task the right way every time [1]. For effective operational discipline, procedures are well documented and codified into best practices. Three personal characteristics are required of employees in an organization with effective OD [2]:

- Awareness – the ability to anticipate potential problems and recognize unusual situations;
- Knowledge – an understanding of how to do a task correctly and safely;
- Commitment – dedication to doing the task the right way, every time.

Unfortunately many organizations use alarm management practices that diminish operator effectiveness and undermine operational discipline. When alarm management procedures are not defined and the expected operator response to alarms is not documented, how do operators gain knowledge to act correctly? A process plagued by nuisance alarms reinforces negative behaviors and breeds indifference to certain alarms. Alarm floods mask real problems within a deluge of alarm signals making it difficult for the operator to be aware of what is occurring within the process. If metrics are not established to monitor alarm system performance, issues that negatively impact OD go unidentified. An undefined alarm management program cannot be audited and improved to achieve best practice.

The operator is charged with the responsibility to <u>detect</u> abnormal situations (awareness), <u>diagnose</u> the situation (knowledge) and <u>respond</u> to correct the condition (commitment). D.G. Dunn, et al. [3] performed a study of eleven incidents to analyze the mechanisms where operational discipline broke down and the alarm protection layer failed. The mechanisms studied in the review are summarized in Figure 1.

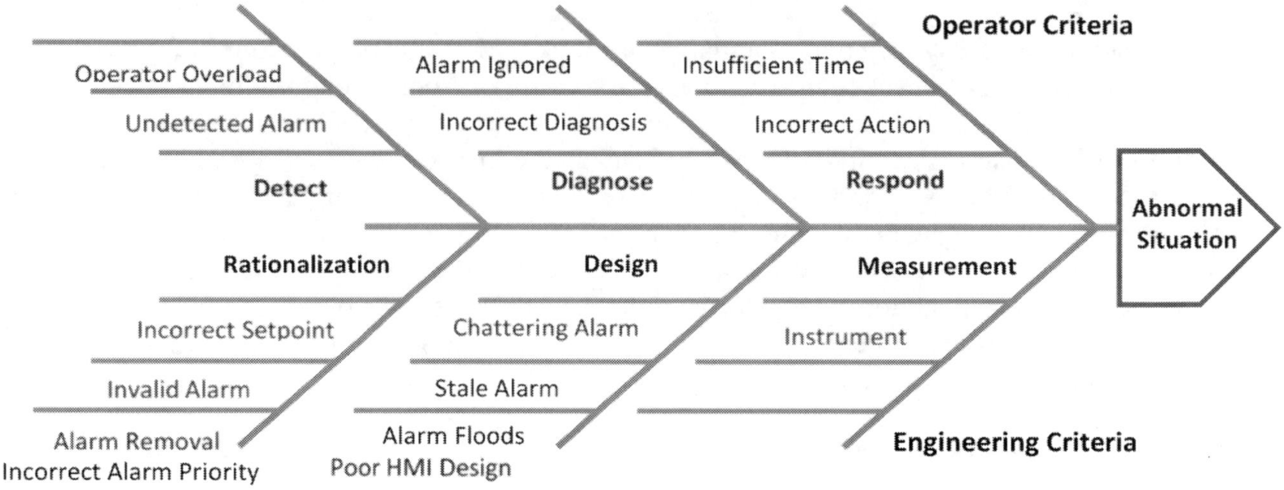

Figure 1. Failure Mechanisms (Fishbone)[a]

Poorly designed alarms (chattering, stale, flood) lead to operator overload and undetected alarms. Absence of a rationalization process can lead to alarms with incorrect priority or presence of invalid alarms drawing operator attention away from a more important situation, hinders diagnosis and produces a culture of ignoring alarms. Poorly defined alarms and incorrect set points lead to incorrect response and inadequate operator response time. A well-defined and executed alarm management program includes OD principals to eliminate alarm failure mechanisms.

4 Metrics

For a brownfield facility the best place to start development of an Alarm Management program is to understand the types of existing operational issues. Some performance indicators can be extracted from existing operating logs, while other information may be gleaned from antecedent evidence. Though a formal process is necessary long term, useful insights can be gained by performing a clipboard test and informally documenting alarm activity for a period of 20 minutes [4]. Information such as the number of new alarms, operator actions taken, and standing alarms should be noted. The highest priority alarm and alarms remaining active for the test period should be captured. Operator response to alarms should be noted. Do the operator(s) show signs of stress, or desensitization? Impressions from the informal test should be considered alongside statistics collected from the alarm history.

Analytical information is an important tool to assess the health of the alarm management system. Initial data should be analyzed to identify the types of alarm issues that are present in order to set priorities for improvement, and identify gaps in other parts of the overall process safety program (e.g. design, training). A successful alarm management

[a] Reference [3], pg. 4.

program includes routine monitoring and periodic auditing. Metrics considered in the preliminary analysis will form the basis of long term metrics. Within the Alarm Philosophy, a set of Key Performance Indicators (KPIs) should be established and a process for routine monitoring of the KPIs established. Goals should be determine for system performance, and threshold limits for performance tolerance. Figure 2 is an example set of KPIs.

Metric	Target	Action Limit
Average alarm rate per operator, alarms/day	<300	>600
Average alarm rate per operator, alarms/10 minutes, %	1-2	>4
Time alarm system is in flood, i.e., >10 alarms/10 minutes, %	<1	>5
Hours with >30 alarms, %	<1	>5
Average number of alarms out of service, %	<1	>5
Low priority alarms in total alarms, %	~ 80	<50
Medium priority alarms in total alarms, %	~ 15	>25
High priority alarms in total alarms, %	~ 5	>15
Top ten most frequent alarms' contribution to total alarms, %	<1 - ~ 5	>20
Number of stale alarms, i.e., active for > 24 hours, on any day	<5	>5
Number of chattering and fleeting alarms	0	>5

Figure 2. Monthly Performance Metric Analysis

A key OD paradigm is that there is "no improvement without action". Routine monitoring will identify alarms where initial rationalization failed to achieve desired results. These should be re-rationalized challenging original assumptions and evaluating supplementary information. In addition to the routine monitoring, periodic audits must be performed. The audit should look at the entire process comparing actual implementation against the processes defined in the philosophy. A comparison of the DCS configuration against the alarm rationalization tool is useful to identify gaps in the management of change process. Action items from the audit need to be tracked and closed out in a timely manner.

5 Establishing Best Practice

5.1 Anatomy of an Alarm Management Program

Operator action in response to alarms is an important safeguard that is frequently credited as an Independent Protection Layer (IPL). Safety incidents do occur when the alarm response layer fails. Mistakes happen when operators must filter through overwhelming or confusing alarm information to determine what situations need action, which is most important to resolve, or what action should be taken. A robust alarm management system will reduce alarms to a manageable level, establish clear priorities, and provide direction to the operators. The lifecycle approach described in ISA 18.2 [5] provides a best practice approach to defining alarms, monitoring performance, managing change and Auditing progress. The lifecycle is given in Figure 3.

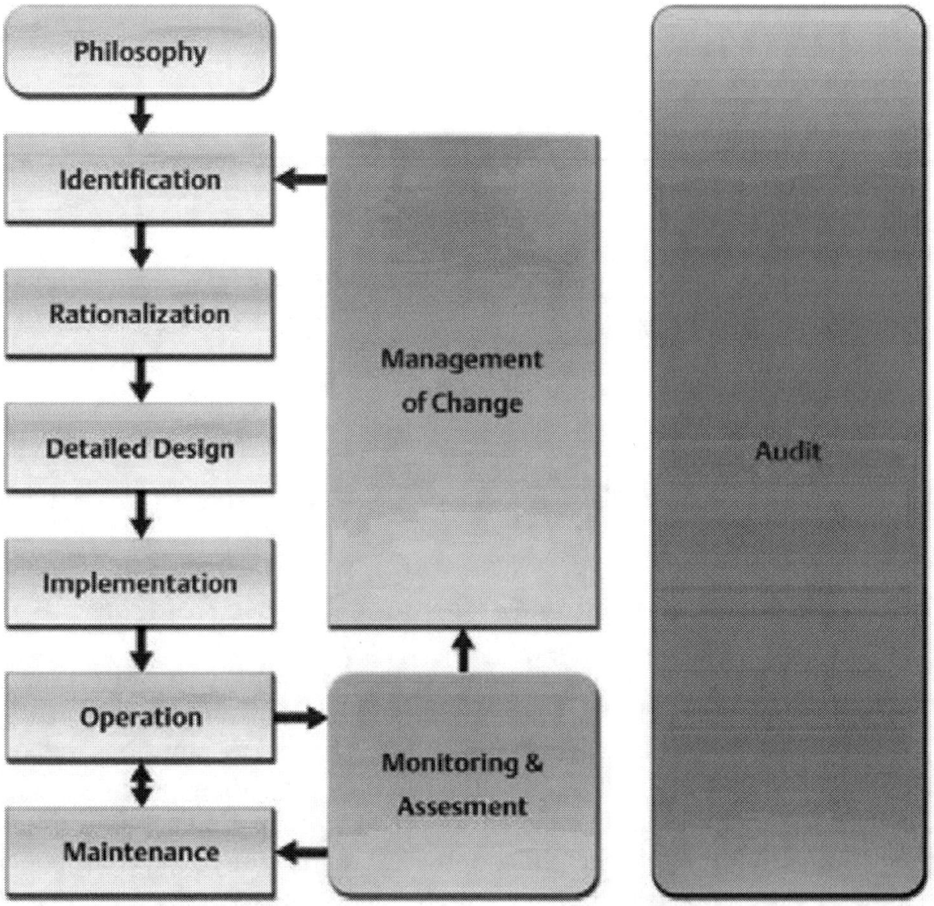

Figure 3. ANSI/ISA-18.2 Alarm Management Lifecycle

Implementation of an alarm management system is a journey that requires an ongoing commitment. A detailed discussion on how to implement a program is available in the July 2012 issue of Chemical Engineering Progress (CEP) [6]. Key elements of an alarm management program are summarized here as background to illustrate how specific OD issues cited within this paper can be addressed.

5.2 Alarm Philosophy

The first step in developing a robust alarm management system is to document the philosophy. The philosophy is a statement of best practices and establishes lifecycle activities, roles and responsibilities, procedures and processes, and monitoring metrics. It will define the alarm and non-alarm notification categories, provide rules for prioritization and disposition of alarms and non-alarm notifications, and document the alarm rationalization procedure, Guidance should be included on how to establish alarm set points to allow an operator adequate time to respond to alarms. Techniques to handle fleeting, chattering and other nuisance alarms, as well as advanced alarm management should be included. Finally the procedure for monitoring alarm performance, metrics to measure success, management of change requirements, personnel training and auditing intent is included.

5.3 Defining Alarms

When faced with an alarm overload situation an operator only has a short time to process and address each alarm signal they receive. The situation is greatly improved by sorting out what is truly an alarm from other notifications, and redirecting the notifications away from the alarm banner. ISA 18.2 defines an alarm as:

> An audible and/or visible means of indicating to the operator an equipment malfunction, process deviation or abnormal condition requiring a (timely) response.

Within a DCS, the modules for process variables include a default set of configuration points (high, high-high, low, etc.) that offer a means to identify a potentially abnormal process condition and communicate when the condition is met. These are often referred to as alarm points. In practice people refer to events generated from these configuration points as alarms. In reality, a large percentage of these communications do not meet the ISA definition of alarm and should actually be considered as different types of notifications. When alarm points are rationalized, they should first be sorted into alarm and non-alarm categories. Figure 4 illustrated four major categories and the rationale for each.

	Operator Must Act	FYI to the Operator
Abnormal	Alarm	Alert
Normal	Prompt	Message

Figure 4. Notification and Alarm Categories

The categories may be further subdivided based on the responsibility for and method of disposition. Some characteristics include:

- An alarm must be due to an abnormal condition and have a defined operator response. There must be adequate time for the operator to carry out the response. Each alarm presented to the operator should be useful, relevant and unique. The alarm will alert the operator of the condition, inform them of the proper response, and guide operator action. Examples: High level alarm on a storage tank; high rate of temperature increase in a reactor.
- A prompt is a normal condition that requires an operator response. Prompts are usually associated with a routine batch type operation where operator intervention is required to proceed to the next stage. Examples: Signal for operator to perform manual charge in a batch reaction; indication that truck unloading is complete.
- An alert communicates to the operator that an abnormal situation has occurred, but there is no action to be performed by the operator or the timeframe for response is long (e.g., extends beyond a single shift). They are typically displayed to the operator on a separate screen from the alarm list. Alerts may require action by others and may also be communicated directly to the responsible parties. Examples: A bad measurement signal indicating that a transmitter requires maintenance attention; an indication that an interlock has tripped.
- A message includes things that are both normal and informative. It includes communication that requires no action but may be relevant in a diagnostic or reporting perspective. Examples include: indication of an automated pump run status, indication that an automated regeneration cycle has completed a particular step.

Although a prompt, alert and message are not truly alarms, they are often assigned an alarm priority if the DCS alarm system is to be used to communicate all types of operator notifications. The alarm philosophy document will define the various categories and establish priority and communication criteria for each. This may be presented in the form of a guidance table. Figure 5 is an example:

Alarm Category	**Alarm Priority**	**Alarm Summary** (Where the alarm appears)
Safety shower and eye wash alarm	High	Alarm List
Gas Monitors (low O2, NO2,NOx N2)	High	Alarm List
Process interlock trip with no corrective action required	Alert	Alert List
Process interlock trip with corrective action required	Per Alarm Prioritization matrix (Refer to section 5.4)	Alarm List

Bad process variable of a BPCS sensor	See note 1	Alarm List, Maintenance Alarm List
Power System Diagnostic alarms	Medium	Alarm List , Maintenance Alarm List
Note 1: The alarm priority should be set equal to the highest priority alarm condition of the tag. If no alarms are configured for the tag, the priority of a BAD_PV alarm should be set equal to Alert.		

Figure 5. Example Alarm Priority by Category

5.4 Alarm Rationalization and Prioritization

Since an alarm's priority should be indicative of its relative importance, when presented with multiple alarms, the operator diagnose and respond process may be confounded by alarm priorities assigned in a subjective fashion rather than by rationalization. Many alarm systems are configured based on a simple perception of the importance of an alarm, leading to a near even distribution between low, medium and high priority alarms, or a disproportionally high percentage of the alarms assigned to the highest priority. In a stressful situation, an operator faces additional decision challenges when the most critical priority level includes alarms the operator believes to be of lower significance. ISA 18.2 recommends the distribution for annunciated alarm priorities given in Figure 6.

Priority	3 Priority Distribution	4 Priority Distribution
Highest	N/A	~1%
High	~5%	~5%
Medium	~15%	~15%
Low	~80%	~80%

Figure 6. ISA 18.2 Annunciated Alarm Priority Distribution

The alarm philosophy will establish prioritization levels for alarms and the criteria for each level. Three and four level schemes are most common. Priority is established based on the immediate consequence of operator inaction, and the time requirement for operator response. An example of consequence decision criteria is given in Figure 7.

Type of Consequence (Impact Area)	Direct Consequence of Inaction or Incorrect Action			
	NEGLIGIBLE	MINOR	MAJOR	SEVERE
Personnel	Injury requiring medical treatment with no lost time.	Injury requiring medical treatment, time off work and rehabilitation.	Permanent disabling injury and/or long term time off work.	Fatality
Public or Environment	Negligible impact	Onsite impact	Short-term, offsite impact	Long term impact.
Financial	Event or lost production costing <$100K.	Event or lost production costing $100K to $1M	Event or lost production costing $1M to $10M	Event or lost production costing >$10M

Figure 7. Example Alarm Consequence Description

The consequence of operator inaction is assessed based on the next immediate (direct) consequence in the chain of events, not the ultimate consequence in the scenario. For example, in a tank overflow scenario, the immediate consequence of operator failure to act on the high alarm may be the activation of a high-high interlock stopping feed to the tank. If tripping the interlock would cause a process upset costing the organization more than $100K in lost production, the consequence would be defined as minor per the consequence criteria in Figure 7.

Operator response time is the second factor used to set priority. The operator response time is the maximum time between the annunciation of the alarm and the time the operator must take corrective action to avoid the consequence. Priority is typically determined by considering both response time and severity of the consequences. An example of a priority matrix is given in Figure 8.

Time Available to Respond	Alarm Priority			
	Event has Negligible Consequence	Event has Minor Consequence	Event has Major Consequence	Event has Severe Consequence
Not Urgent (>30 Min)	No Alarm	Re-engineer the alarm for urgency (Invalid)	Low/ Re-engineer the alarm for urgency	Medium/ Re-engineer the alarm for urgency
Prompt (<30 Min)	No Alarm	Low	Low	Medium
Rapid (<15 min)	No Alarm	Low	Medium	High
Immediate (<5 min)	No Alarm	Medium	High	Invalid

Figure 8. Example Alarm Prioritization Matrix

The operator action timeline is important to the decision criteria. In the minor consequence example above, if the relative set points of the high alarm and high-high interlock were defined so that at the fill rate an operator had less than 5 minutes response time to avoid interlock trip, the priority would be set to Medium per Figure 8. Available response time of significantly more than 5 minutes would deliver a priority of Low.

The success of alarm prioritization is predicated on the assumption that the priority matrix accurately reflects company business / risk criteria. At the end of the day when multiple alarms are presented to the operator at the same time, the priority should indicate which one is more critical to safety and the business.

Operator training is also important. Operators must be trained to immediately respond to every alarm and use alarm priority to determine relative importance when multiple alarms are active at the same time. For example, a board operator receiving a high and medium priority alarm at the same time would assign a field operator to the high alarm first, then seek assistance from a different field operator to address the medium alarm.

Figure 8 includes notations in the chart that indicate different disposition is required.

- "No Alarm" is assigned whenever the consequences are negligible. These may be handled as messages or alerts if the rationalization team feels there is value in having a notification, or they can be disabled.

- "Re-engineer the alarm for urgency": These alarms may be set to the indicated priority or should be considered for re-engineering to ensure that there is a level of urgency to the response (for consistency with other types of alarms).
- "Invalid" indicates a design scenario that should be avoided (< 5 mins to prevent significant consequences such as a fatality). Alarms falling into this band should be re-engineered to automate the response, incorporate additional safeguards, or utilize inherently safer design.

6 Detect (Awareness)

A nuisance alarm is one that annunciates excessively, unnecessarily or does not return to a normal state after correct response has been taken. It will lead to ignoring of alarms and failure to detect an abnormal condition. Too often organizations respond to the nuisance condition by failure of operational discipline. Reflexive silencing, removing annunciation, ignoring visual indication or inhibiting/disabling the alarm are common responses.

The January 2010 Methyl Chloride release at the DuPont Belle, WV facility [7] occurred when a nuisance rupture disk burst indication alarm went un-addressed for an extended period of time. The rupture disk is located on a common vent line between two reactors and a scrubber. A 0.5 inch weep hole (source of leak in the incident) was installed in the piping just downstream of the rupture disk inside the process building, and the main discharge line vented to the roof. The rupture disk burst indication (BDI) had a long history of nuisance alarms, mostly due to a low battery signal on the BDI sensor. Operators became accustomed to frequent alarms but experienced only a few true activations, so they had become desensitized to the alarm. In the four-month turnaround just prior to the event, the rupture disk style had been changed to one with a more reliable indication, and the indicator power source was converted from a battery (monthly replacement) to a continuous power supply. The disk is believed to have ruptured during outage and the alarm had been ignored (became stale) because there were no process hazards present at the time of the alarm. When chemicals were reintroduced into the process several weeks later, operators did not recognize the stale alarm as a concern because of the nuisance history. An estimated 2,000 pounds of methyl chloride was released over 5 days until a process area gas detection alarm brought the situation to the attention of operators and the process was shut down.

Nuisance alarms undermine operational discipline. They distract and desensitize the operator leading them to ignore alarms. Frequent, chattering or fleeting alarms increase operator stress and can lead to missing other alarms.

Frequently occurring alarms, or bad actors, annunciate excessively but rarely indicate a true problem. The DuPont rupture disk burst indication event is a classic example. The alarm was originally configured so that the operators received the same signal for a low battery warning as they did for a burst disk indication. Operators had become so desensitized to the alarm being active from battery warnings, that it became a stale alarm and was not cleared prior to start-up following the outage. Because the operators were accustomed to seeing a false alarm, they failed to act in the correct way and resolve the alarm before introducing hazards into the process. Prioritizing the low battery indication as an alert separately from the burst indication alarm would have reduced the number of alarms and would have retained the importance of the BDI indication.

7 Diagnose (Knowledge)

The DuPont Belle, WV incident [7] is an example where a nuisance alarm was tolerated and fortunately, the consequences were minor. The general (evacuation) alarm on the Deepwater Horizon [8, 9] is an example of where inhibiting an alarm so as to avoid a potential nuisance alarm can have fatal consequences. It also illustrates how operational discipline can crumble in a high stress situation when operators are uncertain of the proper response to a situation.

The April 2010 Macondo well blowout on the Deepwater Horizon Rig provides a lesson on the importance of well-considered alarm management practices. The event occurred during temporary abandonment[b] operations of the well. A kick[c] escalated to a blowout[d], then to an explosion and fire that continued two days until the rig sank [8, 9]. Eleven people were killed, 17 critically injured and almost 5 million barrels[e] of oil discharged into the Gulf of Mexico. The general alarm system on the rig was operating in "inhibited" mode[f] so the general alarm would not automatically trip on actuation of multiple area gas detection alarms in different areas of the rig. According to testimony of the chief electronics technician, the system was configured in inhibit mode because "they did not want people woke up at 3 o'clock in the morning due to false alarms" [10]. According to the event times reported in the US Department of the Interior report[g], multiple local gas alarms sounded 1-2 minutes before the first explosion, but the general alarm was not sounded until about 10 minutes after the explosion. Witness testimony suggests there was some uncertainty as to when personnel should manually trigger the general alarm. The investigation Panel[h] determined:

> The *Deepwater Horizon* operated a manually-functioned general alarm system. If the general alarm of the *Deepwater Horizon* had been set to automatically sound when "high-high" gas alarms sounded in multiple compartments of the rig, personnel in the pump room likely could have moved to a location where their chances of survival were greater. **The "inhibited" general alarm system was a possible contributing cause of the response failure.**[i]

[b] The temporary closure of a well so that the drill rig can be moved off and a production rig can be put in place.
[c] An unplanned flow of fluids into the wellbore.
[d] An uncontrolled hydrocarbon release into the environment.
[e] Reference [9], pg. 31.
[f] Reference [8], pg. 4.
[g] Reference [8]
[h] The Joint Investigation Team of the Bureau of Ocean Energy Management, Regulation and Enforcement ("BOEMRE") and the United States Coast Guard.
[i] Reference [8], pg. 114.

Assurance that the operator is equipped to 'do the right thing every time' requires an assessment of the hazards, documented required operator action, operator training, and tools to assist operators. The alarm rationalization process will identify and document likely cause(s), consequence, confirmation and corrective action related to an alarm. Figure 9 is an example of SILAlarm rationalization documentation.

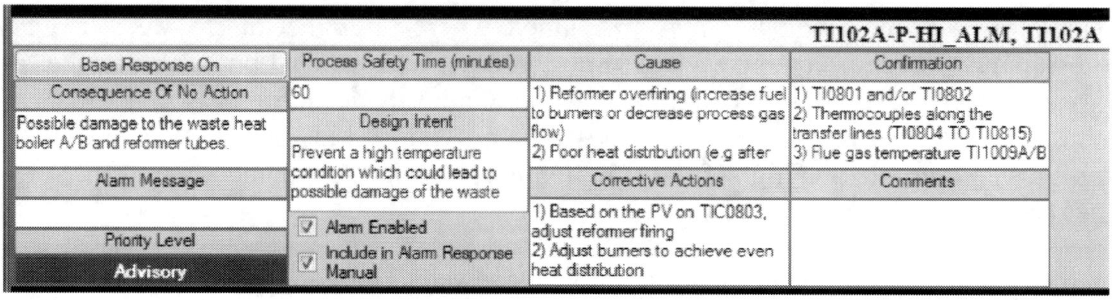

Figure 9. Example Alarm Response Procedure

In the rationalization example, the potential consequences of inaction, causes for the alarm, means to confirm that the alarm is real, and corrective action for each cause is identified. The information clarifies operator response action and timeline to the rationalization team for priority ranking, and provides important information for operator training and operating procedure content. The information captured during the rationalization exercise can be displayed to the operator in the form of an interactive help file (in some systems) or from the alarm response procedure (manual) shown in Figure 10.

Alarm Response Procedure Report

Tag ID	TI102A				
Tag Description	Compressor Bearing Temp A				
Location	REACTORS_AREA/ UM_REACTOR_1 /				
Range	0	To	100		no units
Alarm Name	TI102A-P-HI_ALM				
Alarm Message					
Alarm Type	High				
Priority	Advisory				
Limit	95				
Time to Respond	15-30 Minutes				
Classifications	• 4 - HAZOP Safeguard				
Cause	1) Reformer overfiring (increase fuel to burners or decrease process gas flow) 2) Poor heat distribution (e.g after burning and tubes with high DPs)				
Confirmation	1) TI0801 and/or TI0802 2) Thermocouples along the transfer lines (TI0804 TO TI0815) 3) Flue gas temperature TI1009A/B				
Consequence Of No Action	Possible damage to the waste heat boiler A/B and reformer tubes.				
Corrective Actions	1) Based on the PV on TIC0803, adjust reformer firing 2) Adjust burners to achieve even heat distribution				

Figure 10. Example SILAlarm Alarm Response Manual

Well trained operators should possess knowledge necessary to diagnose an alarm; however, as the Deepwater Horizon incident illustrates, in times of stress, operator decision processes can be slow and/or incorrect. Making this type procedure available to the operator will support operational discipline by providing information that can improve the speed and accuracy of diagnose, especially in times of stress.

8 Respond (Commitment)

Operator response time is dependent on the process dynamics (including rate of change) and alarm set point. In a properly rationalized system, the operator will view alarm priority as an indicator of which alarm needs attention first. In a system where alarms are not rationalized, the multiple alarms may sound with the same priority, or with priorities that are inconsistent with operator experience. The operator may not know where to respond first, and may not have adequate response time available to be effective. Alarms are cited as a protection layer in a PHA and/or LOPA; however, the review teams don't always accurately consider the operator's ability to respond in a given timeframe. This can lead to a situation where the operator can do the right thing and still have an undesirable outcome.

The 2008 Bayer CropScience runaway (decomposition) reaction in Institute, West Virginia [11] is an example of the operator doing the right thing without sufficient response time built into the design of the alarm. The event occurred during restart of the methomyl unit following a control system and residue treater replacement. Operational discipline failed in several ways enabling the event conditions:

- During start-up, operators failed to complete steps that would pre-fill the residue treater with solvent prior to introducing methomyl containing flasher bottoms feed.
- The start-up was complicated because some systems were still being modified and the operators were not completely familiar with the new control system workstations and their function.
- Issues with centrifuge operation resulted in high methomyl centrate eventually reaching residue treater. This coupled with the solvent pre-fill error led to a much more reactive composition than normal in the residue treater prior to startup of the vessel.
- On start-up of the residue treater, recirculation and minimum temperatures interlocks were bypassed.

The vessel was heated and recirculation was established. As the contents began to react, the temperature rate of increase moved from 0.6°C/min to 2°C/min while pressure held steady, until the critical decomposition temperature was reached and pressure began increase. By the time the high pressure alarm sounded, vessel pressure was climbing rapidly. The lead operator dispatched two operators to the area for troubleshooting and manually switched to full cooling. The vessel exploded 8 minutes after the high pressure alarm sounded. The explosion, fire and toxic chemical release resulted in the death of two operators, eight injured and shelter in place ordered for 40,000 people.

Thoughtful determination of alarm and set point is critical to assure that operators have adequate time to address the alarm to avoid the immediate consequences. The 2008 Bayer CropScience runaway reaction illustrates the consequence of improper alarm selection and set point determination. The multiple failures leading to the runaway scenario had not been identified in a PHA, therefore was not considered in alarm design. However, an analysis of the chemical reaction behavior may have identified an improved alarm design providing operators more response time to resolve the situation. Figure 12 is a chart of process indication leading up to the vessel rupture.

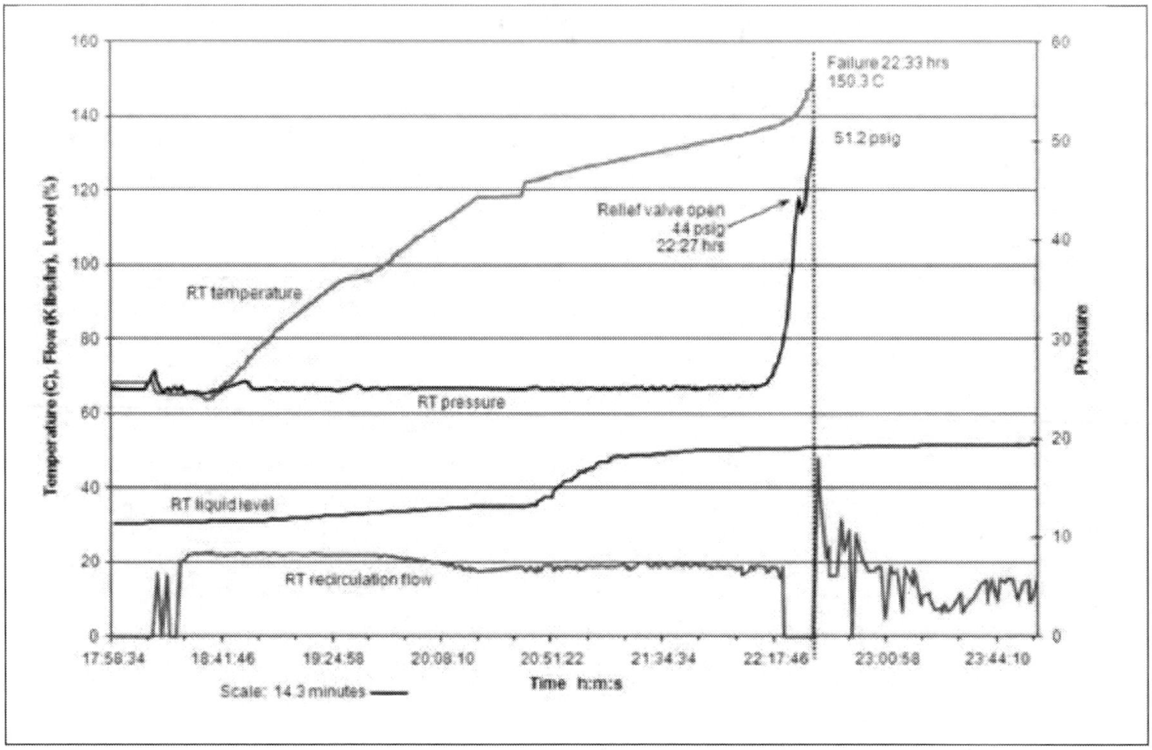

Figure 11. Residue Treater Process Variables before the Explosion[j]

Runaway and decomposition reactions often show a predictable temperature/pressure relationship. Laboratory testing can be utilized to determine the nature of this relationship. The information is then used to select the best alarm indicators for a potential event, and establish set points for those alarms. In a decomposition runaway, the temperature will increase at constant pressure until the reaction initiates, then pressure increases rapidly. The process data in Figure 11 illustrates classic decomposition reaction behavior. There is no indication in the CSB report that the operator received any temperature alarms prior to the high pressure alarm. Since the pressure alarm set point had been established with the expectation that the methomyl concentration would be

[j] Reference [11], Figure 11, pg. 39.

much lower, and rate or pressure rise much slower, there was insufficient response time available to the operator and inadequate mechanical relief capacity.

If the chart given in Figure 11 had been available to an alarm rationalization team, they might have determined that absolute temperature or rate of temperature rise was a better indication of potential decomposition reaction and could have established a new alarm more indicative of the situation. The set point for the new temperature could be determined based on the process safety time (time from alarm set point to consequence threshold), operator response time (time for operator to detect, diagnose and respond), and process dynamics (time for process to respond to the operator action).

9 Design

Operators can possess an awareness about the operation, be highly knowledgeable about the process, and be committed to doing the right thing all the time, but a poorly designed alarm system is an enabling condition for several failure mechanisms. The incidents above illustrate what can happen if an operator is de-sensitized by nuisance alarms, has insufficient time and guidance to diagnose a problem and respond properly, or when the alarm is not the best indicator of process dynamics. This section discusses some alarm design issues that may undermined OD.

9.1 The silent alarm

The INDSPEC Chemical Corporation oleum release [12] demonstrates the consequences of a missed alarm. The incident occurred in an oleum storage building normally staffed 7-3 weekdays and as necessary on weekends. The building contained three pressure vessels used for railcar unloading, and two process tanks that fed the continuous resorcinol process in another building.

On Saturday, October 11, 2008, an operator was called into the facility to perform a periodic but routine weekend operation to prepare soda ash for the process. While on site, the operator would also transfer oleum from three railcar unloading pressure vessels to the two continuous use process tanks to ensure that the pressure vessels are empty, or nearly empty, on Monday mornings.

The three pressure vessel discharge pumps were powered by plugging individual power cords into the "Normal" or "Emergency" power supply receptacles. The process tanks were equipped with alarms that displayed locally and in the oleum storage DCS for high level (audible and amber visual indication) and high-high level (audible and red visual indication). The high-high alarm was also interlocked to shut down the transfer pump for the "Normal" power supply but not the "Emergency" power supply. During weekend operations, the operator would utilize both power supplies to speed the transfer of oleum, and would shut down operations and leave the site when tasks were complete.

On the day of the event, the operator performed operations as usual, shut down the pump connected to the "Normal" power supply, and left the facility. The operator failed to shut down the pump connected to the "Emergency" power supply, so oleum continued to transfer from the pressure vessel to the process tank. High level alarm in the process tank triggered after the operator had left the process area; however it only sounded in the oleum storage building and unoccupied control room. The alarms were not routed to the control room occupied 24/7; therefore, they went unnoticed by personnel remaining on site. The High-high alarm was not connected to emergency power supply, so transfer continued. About two hours after the initial alarm, an employee noticed a white mist escaping the oleum storage building and an emergency was declared, triggering an evacuation or shelter-in-place affecting 2500 residents of Petrolia, Bruin, and Fairview Pennsylvania. Fortunately, there were no fatalities and only one responder suffered a minor injury. INDSPEC faced over $120,000 in OSHA violations and unreported damages.

In this example, the alarm system was ineffective because it sounded in an unoccupied space and there was no operator present to detect the alarm, thus no corrective action taken. A rationalization review of the high level alarm might have detected this design flaw and resolved by relaying an alarm to the continuously staffed control room. The high-high level alarm was designed to interlock with the normal power and shut down the pump, but there was no similar interlock for the emergency power. The situation was known to operators, but since the unwritten work practice had been used for more than 25 years, it was not recognized as being a failure in operational discipline. PHAs conducted on the system failed to identify the potential hazards. The alarm rationalization process focuses specifically on identifying and documenting the cause, consequence and corrective action related to a single alarm. This level of detail often identifies potential gaps the PHA may have overlooked, and existing flaws in operational discipline.

9.2 Alarm Overload

ISA 18.2 [5] recommends an alarm annunciation rate of 1-2 alarms per operator per 10 minutes. This is based on providing sufficient time for the operator to detect the alarm, diagnose the problem (retrieve relevant data from the control system, analyze the situation) and respond (perform corrective action(s)). The standard recommends that the average daily rate does not exceed 150-300 alarms in order to be manageable. Undocumented reports suggest that industry groups experience 900-2000 alarms per day or 5-8 alarms per 10 minute interval. In *exida*'s experience, more than 1000 a day is common, particularly for systems that have not undergone rationalization. The alarm overload inhibits the operator's availability to address any one alarm, and increases the probability that an important alarm will be missed. The following paragraphs discuss a few types of nuisance alarms that may be addressed with alarm design techinques.

Stale Alarm

A stale alarm is one that remains in the alarm state for an extended period of time. They can distract the operator by cluttering the alarm summary screen and may interfere with identification of new alarms. The DuPont Belle rupture disk burst indicator alarm is one example of a stale alarm. OD broke down when the operators became so accustomed to false alarms that they silenced the alarm but allowed it to remain active for an extended period of time. Another example is an alarm that indicates a sump pump has stopped after pumping out the sump (a normal operation). Operators may grow tired of the recurrent annunciation and allow it to become stale to stop the repeated cycling. The rationalization process will help determine if the indication should even be an alarm or whether it belongs in one of the other categories. If it is determined to be an alarm, the analysis will identify opportunities to make the alarm more relevant.

9.3 *Chattering and Fleeting Alarms*

Chattering and fleeting alarms can be annoying and difficult to diagnose. A chattering alarm will repeatedly transition between alarm and normal states often without operator interaction to clear. Fleeting alarms transition between the alarm and normal state, but do not immediately repeat. This type of alarm distracts an operator like flies at a picnic. The alarms seem to appear and clear at random. Operators spend time chasing an elusive cause, grow frustrated, and eventually fail to respond to the alarm at all. These alarms are easily identified by analyzing data from the historian, specifically alarm frequency and duration. Many of these can be addressed by analyzing the deadband (hysteresis), filters and on/off delays as part of the rationalization review.

9.4 *Redundant Alarms*

Redundant, or cascading, alarms are those that consistently occur within a short period of time of other alarms. They are generally associated with the same event. For example, when a pump is stopped, there may be an alarms from the motor run indication, low pressure at the pump discharge, and low flow from the downstream flow meter. This type of alarm can compromise operational discipline by presenting an alarm (low pump discharge pressure) which may or may not be relevant and which relies on the operator to assess its validity in real time.

Redundant alarms may be addressed by using advanced alarming techniques such as conditional or designed suppression. Conditional suppression will prevent certain alarms from alarming when a specified normal condition is present. For the pump example above, low flow and low pressure alarms would be suppressed when the pump is in a STOP status. Suppression by design would be used to automatically suppress the consequential alarms associated with a planned event, such as equipment shut down (preventing stale alarms) or an unplanned event such as a compressor trip (preventing an alarm flood). To prevent an alarm flood from resulting after a compressor trip, the alarms (temperature, flow, pressure, vibration etc.) associated with the trip are included in a flood group to be suppressed, with latching enabled to identify first out, and a common trouble alarm indicating there was a trip. Advanced alarm techniques can be used to safely reduce the number of alarms sent to the operator so they are able to more quickly identify and diagnose a situation.

10 Conclusion

Process alarms and operator action is heavily relied upon as an independent protection layer in the process safety model. Four events have been analyzed demonstrating how this safety layer can be rendered ineffective when an operator is faced with too many alarms, unclear priorities, and nuisance alarms. Alarms where operator action is unclear, or there is inadequate time to affect response can lead to deadly consequences. A successful alarm management system can filter out useless alarms, direct notifications to responsible parties, and address design flaws that generate nuisance alarms. It supports documentation of process knowledge that can be used in operator training and become the basis for operator help. Discipline is established in alarm management giving the operator the tools needed to do the right thing, the right way every time.

11 References

[1] B. Rains, Operational Discipline: Does Your Organization Do the Job Right Every Time?, DuPont 2010.

[2] J.A. Klein, and B.K.Vaughen, Implementing an Operational Discipline Program to Improve Plant Process Safety, Chemical Engineering Progress, pp. 48-52 June 2011.

[3] D.G. Dunn, N.P. Sands, and T. Stauffer, When Good Alarms go Bad: Learning from Incidents, 70th Annual Instrumentation and Automation Symposium – Texas A&M University, January 2015.

[4] T. Stauffer, P.E. and K. VanCamp, Making Some Alarming Moves, Chemical Processing, http://www.chemicalprocessing.com/articles/2012/make-some-alarming-moves/, April 2012.

[5] ANSI/ISA–18.2–2009. Management of Alarm Systems for the Process Industries. June 2009.

[6] T. Stauffer, P.E., Implement an Effective Alarm Management Program, Chemical Engineering Progress, pp. 19-27, July 2012.

[7] U.S. Chemical Safety and Hazard Investigation Board, http://www.csb.gov/, Investigation Report: E.I. DuPont DE Nemours & Co., Inc. Methyl Chloride Release, Oleum Release and Phosgene Release. Report No. 2010-6-I-WV, September 2011

[8] The Bureau of Ocean Energy Management, Regulation and Enforcement, US Department of the Interior, Investigation Report Regarding the Causes of the April 20, 2010 Macondo Well Blowout. September 14, 2011

[9] U.S. Chemical Safety and Hazard Investigation Board, http://www.csb.gov/, Investigation Report Volume 1: Explosion and Fire at the Macondo Well. Report No. 2010-10-I-OS, June 5, 2014

[10] Testimony of Mike Williams, Joint Investigation Hearing, July 23, 2010, at 34-35, C-Span, http://www.c-span.org/video/?294728-1/investigation-deepwater-horizon-explosion-mike-williams, Accessed February 9, 2016.

[11] U.S. Chemical Safety and Hazard Investigation Board, http://www.csb.gov/, Investigation Report: Pesticide Chemical Runaway Reaction Pressure Vessel Explosion. Report No. 2008-08-I-WV, January 2011

[12] U.S. Chemical Safety and Hazard Investigation Board, http://www.csb.gov/, Case Study: Uncontrolled Oleum Release, Petrolia, Pennsylvania. 2009-01-I-PA. September 2009

12TH GLOBAL CONGRESS ON PROCESS SAFETY

Achieving operational excellence through a sustainable approach within process safety

Kumar (Chris) Israni
Principal Consultant - ERM
840 West Sam Houston Parkway North, Suite 600
Houston, TX 77024
Chris.Israni@erm.com

Steve Hawkins
Partner - ERM
9825 Kenwood Road, Suite 100
Cincinnati, OH 45242
Steve.hawkins@erm.com

Prepared for Presentation at
American Institute of Chemical Engineers
2016 Spring Meeting
12th Global Congress on Process Safety
Houston, Texas
April 11-13, 2016

Achieving operational excellence through a sustainable approach within process safety

Kumar (Chris) Israni
Principal Consultant - ERM
840 West Sam Houston Parkway North, Suite 600
Houston, TX 77024
Chris.Israni@erm.com

Steve Hawkins
Partner - ERM
9825 Kenwood Road, Suite 100
Cincinnati, OH 45242
Steve.hawkins@erm.com

Keywords: process safety, sustainable safety, risk management,

Abstract

In the last four decades, we have seen a lot of changes within the various industries on recognizing and acknowledging risks. Learnings from catastrophic incidents in the industry led to evolution and enhancements to risk techniques as well as regulatory changes globally. Also, the last decade has seen the evolution of the concept of sustainability. One of the key elements for a company in evaluating their sustainability performance is safety that ties into company's goal to strive for operation excellence.

Investigations and awareness during recent major accident events have indicated that when assessing both the short term and long term consequences of such events, the impacts seen are economic, environment and social in nature. This affects the organization's governance processes. A robust proactive risk management exercise when initiated during the concept of facility, and then carried out effectively and continuously until the demolition and decommissioning of the facility, renders success in the excellence program.

Several operating companies have driven forward in this direction with the use of the either the safety case regime or various international onshore and offshore regulations. The safety program would be considered to be a sustainable program if it covers the entire process safety management lifecycle for the facility using a risk based approach, rather than a prescriptive approach. The various techniques involved would be an array of qualitative and quantitative studies. Each of these studies brings about a plethora of

actions. The actions need to be addressed using a risk based approach rather than a cost based approach.

This paper illustrates the authors' experience on the sustainable and sequential application of process safety methods in a risk based manner. That assures longevity for the facilities and thus operational excellence.

1. Introduction

The term sustainability comes with a lot of interpretations [1]. Irrespective of the source, for an industry to remain sustainable, the key is endurance and long term retention of resources for future generations. In order to endure and prosper, it is important to uphold the facets of safety, environment and the business which typical receptors when we conduct risk assessments.

It is critical that we have a clear understanding of these interpretations. This gives rise to the concept of sustainable safety.

2. Elements of Sustainable Safety Operational Excellence

Any kind of manufacturing facility is comprised of –

- The plant
- The people and
- The process.

In order to optimize operations and excel,, it is important the interactions between these three elements be well managed. For a safe running facility, it is necessary that there be a complete and coherent coordination between the plant equipment and their application in the processes involved. This requires implementation of various technical studies, in the various stages of the evolution for this facility. At the same time, for people to work on the processes safely, there must be an environment that promotes continuous learning and knowledge exchange. Hence training and data management become strategically important. At the same time, the human and machine interface (i.e. human factors) cannot be overlooked, or undervalued.

It is paramount to have the right mix of technical safety, knowledge transfer / management and human-machine interface, to optimize facility operations and obtain operational excellence.

We have termed this concept "sustainable safety" and it is illustrated in Figure 1.

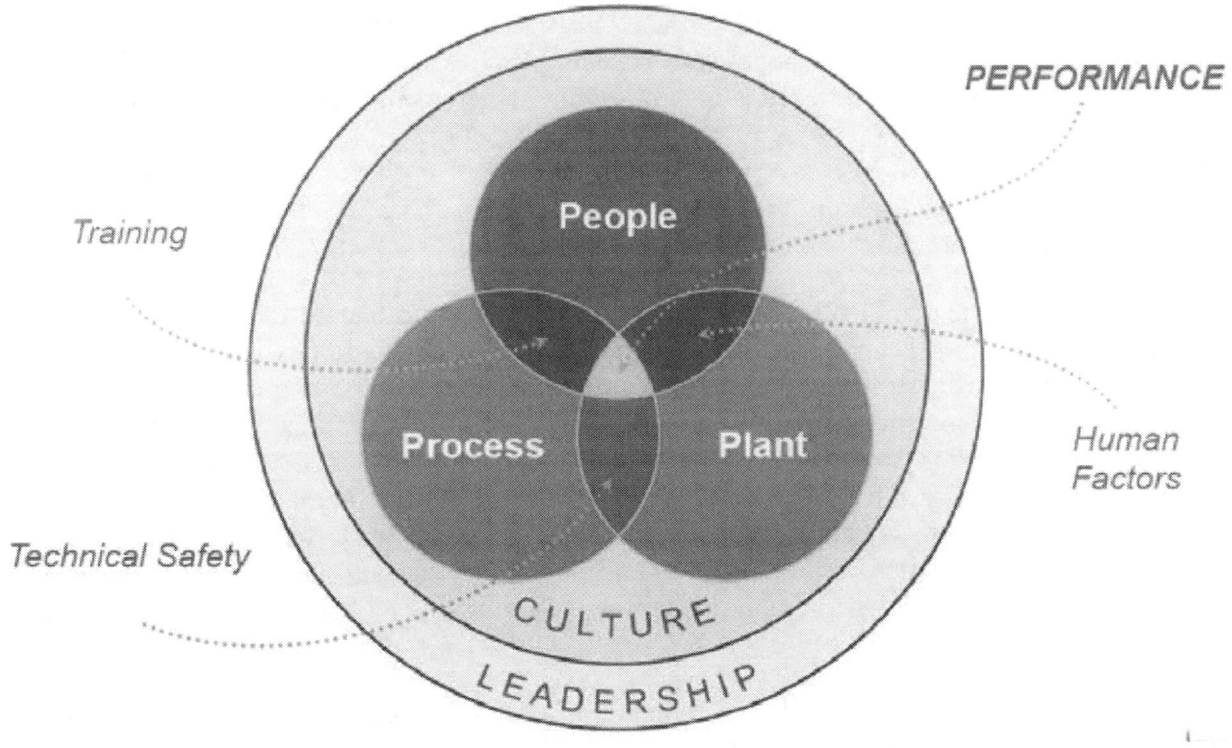

Figure 1. Explanation for Sustainable Safety

Effective and sustainable process safety programs are vital to overall company health and viability from a business continuity perspective (we have all seen the devastating business impacts of catastrophic accidents), but these same programs also drive operational excellence in several important ways. For example, implementation of robust asset integrity programs increases equipment "up-time", extends equipment life, increases overall productivity rates, and consequently reduces overall capital expenditure. We have recently implemented an asset integrity and reliability program for an oil and gas operating company which will drive an approximate 18 % increase in equipment "up time" and a corresponding increase in productivity. This implementation included development of an enhanced mechanical integrity program to align with recognized and generally accepted good engineering practices, taking into account the existing gaps as compared to compliance requirements and industry best practices.

Likewise, we have seen that by developing and implementing detailed Standard Operating Procedures (SOP) in conjunction with structured training programs, companies can reduce operational errors and rework, while improving productivity and product quality. An example of this includes a recent project for an injection molding manufacturing process in which we overhauled the facility SOPs and developed a structured On the Job training program to qualify operators. The facility's management has estimated an approximate 7% to 10% increase in process productivity and significant waste reduction based on the elimination of pre-existing operator errors and rework.

In our experience, it is not uncommon to see 3% to 10% overall productivity increases simply from stronger technical procedures and structured training of employees. Robust Management of Change programs can also provide direct "up-front" capital cost savings by eliminating subsequent equipment retro fit changes and reducing process downtime, often with significant monetary savings. In addition, there is also strong evidence to indicate that sustainable process safety programs increase the overall employee satisfaction and moral, which can reduce employee turnover rates, hence saving expenditures to hire and train new employees, increase overall productivity, and lessen the likelihood of catastrophic events from human factors considerations and operator error.

Operational excellence can also be driven by examining existing hazard identification processes and evaluating human-machine interfaces. For example, a non-regulated OSHA PSM cotton processing pilot plant facility in the eastern United States had recently experienced multiple fire events. The site had completed traditional HAZOPs for multiple processes, but was missing a complete comprehension of their existing human-machine interface. A critical control study resulted in the identification of existing gaps and disconnections amongst their human factor dependent processes. Associated corrective actions resulted in a more robust pilot plant processing operation, while minimizing fires and potentially catastrophic impacts. The analysis proved so successful that it was implemented at the company's r commercial production site, located in Asia.

These are just a few examples of how sustainable process safety programs can have a profound, positive impacts on a company's bottom line

3. Process Safety Basic Principles

We have seen that the best way to promote sustainable safety, from a process, or "technical safety" standpoint, is through the development and implementation of focused, risk-based approaches. As outlined by the Center for Chemical Process Safety (CCPS) Risk Based Process Safety (RBPS) program, there are four primary accident prevention pillars to sustainable process safety programs: Commitment to Process Safety (the foundation for the reminder of the program elements); Understanding the Process Hazards and Associated Risks; Managing the Risk; and Learning and Continuous Improvement.

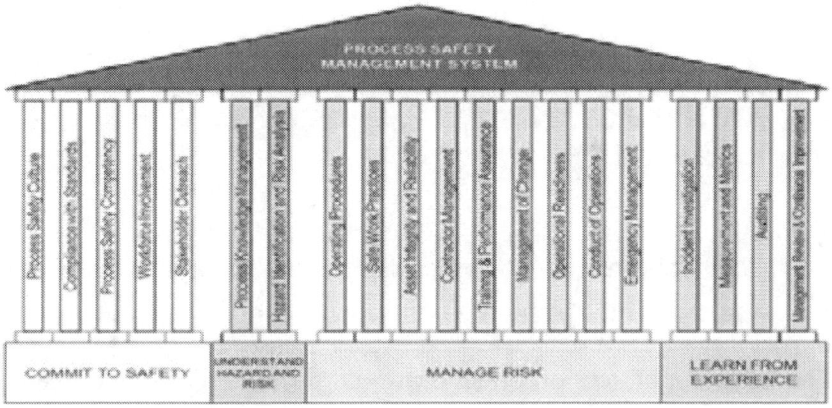

Figure 1. Illustration of Process Safety [3]

An authentic and real commitment to process safety throughout all levels of the organization is foundational to any process safety program and of paramount importance in building and implementing sustainable process safety management programs. Frontline workers tend to gravitate towards the value systems that are held important by their supervisors and the organization's top management. As such, management commitment to process safety, when clearly demonstrated on a daily basis cascades throughout the entire organizational structure and promotes an engaged and proactive process safety culture which ultimately leads to sustainable process safety performance.

4. Understanding Process Hazards

At the core of sustainable process safety programs is the mandate to ensure that resources and programs are focused on the areas that will provide the highest level of protection from a risk based standpoint in preventing catastrophic events. As such, it is extremely important that the organization clearly and quantitatively understand the hazards and corresponding risks associated with various process operations. Detailed hazard analysis which clearly document the hazards of the process and provide a meaningful way to quantify the risks associated with these hazards are crucial in allocating limited resources and effort to the areas that provide maximum, sustainable effectiveness.

Having such programs or techniques implemented in a consistent manner, is what generates are Industry experience shows that organizations that utilize proven hazard analysis methodologies and risk mitigation techniques to design and implement lower-risk operations enjoy long term success and sustainable results not only in decreasing catastrophic accidents, but improving quality and operational excellence.

5. Managing Risk

In accordance with the CCPS guidelines [3], managing process safety risks essentially falls into one of three primary categories:

1. Prudently operating and maintaining processes that pose the risk;
2. Managing changes to those processes to ensure that the risk remains tolerable; and
3. Preparing for, responding to, and managing incidents that do occur.

As previously discussed in our section on Operational Excellence, the following individual elements that form this pillar can have a profound positive effect on not only preventing catastrophic accidents, but on a company's bottom financial line:

- Operating Procedures;
- Safe work Practices;
- Asset Integrity and Reliability;
- Contractor Management;
- Training and Performance Assurance;
- Management of Change;
- Operational Readiness;
- Conduct of Operations; and
- Emergency Management.

Development of specific systems and tools within each of these areas helps company's experience long-term and sustainable success in an incident free and profitable manner.

6. Learn from experience and continual improvement

The ability to reflect on one's mistakes and change future behavior is a key concept in human growth and development, as well as in the viability of most companies. At the heart of sustainable process safety is the tenant that despite best efforts to manage process safety risks, that things will at some point not go as planned. Mature organizations turn their mistakes, and the mistakes of others, into opportunities to improve upon their overall process safety performance and establish more sustainable and effective programs going forward. Additionally, by developing meaningful and measurable Key Performance Indicators (KPIs) or process safety metrics, companies can honestly assess how well the individual elements of a sustainable program are functioning, and make real-time adjustments to drive continuous improvement.

Development and implementation of systems and tools in alignment with these four foundational pillars should enable organizations to sustain ongoing safe and profitable operations.

A recent project with a leading global chemical manufacturer that started off as a PHA program development, resulted into an overall process safety and risk management system enhancement. The initial deployment was limited to just providing hazard identification guidance. But this escalated quickly into the need of having a consistent hazard analysis program, tying into follow up of the action items and the subsequent management of change (MOC) needs. The various sites belonging to this chemical manufacturer decided to eventually adopt the PSM 1910.119 approach for all their locations worldwide, as the bare minimum accepted tolerance criteria.

7. Industrial tools to represent process safety and risk management [4- 10]

There are various tools in the industry such as safety case regime [4,5], SEMS [7,8], PSM [9] or any other safety management system. Such systems must be in place from the time that the facility is being designed and kept current once the facility is operational. Table 1 [11] shows the various elements of each safety regulations and the similarities between the various methods. The three regulations have the same goal to educate the operators about the procedures in place, the importance of preventive and mitigation mechanisms and to proactively address hazards. They stress the need of the implementation of these systems that are essential for the safe operation without affecting the people or the environment. It may seem that there may be multiple formats for safety case, but they all have same objective to enable a goal based safety approach that can be easily adopted **Error! Reference source not found.**

Table 1. Comparison of Safety Case Versus SEMS and PSM [11]

Concept	Safety Case	SEMS	PSM
Introduction	General content	General content	
Overview of the facility	Facility description	Safety and environmental information	Process Description (part of PSI)
Safety assessments	Hazard identification, control measures	Hazard analysis	Process Hazard Analysis
Analysis	Safety analysis, evaluation of residual risks	Pre-startup review	Pre-startup reviews
Emergency response	Emergency response plan	Emergency response and control	Emergency Response and Planning, Incident Investigation
Training	Training	Training	Training

Concept	Safety Case	SEMS	PSM
Communications	Communications	Employee participation	Employee Participation, Trade Secrets
Management of change	Management of change	Management of change	Management of Change
Safety Management Systems	Inspection, testing and maintenance	Mechanical integrity, safe work practices	Mechanical Integrity
Work Control Systems	Permit to work, etc.	Stop work authority, ultimate work authority, reporting unsafe work conditions	Hot work permits, Contractors
Performance standards	Audits	Audits	Compliance Audits
Performance standards	Records and documentation	Records and documentation	Process Safety Information, Operating procedures

8. Summary

The various sections above introduced the concept of sustainable safety highlighting its various components, of people, process and plant. One of the key interactions is the process and plant interphase, which is technical safety. Technical safety based on the region the facility exists, may have different terminology such as safety case regime, SEMS, PSM, etc. A robust technical safety program assists in minimizing undesired consequences that impact people, environment and the business. Hence this ensures longevity in the program, leading towards greater performance or operational excellence.

Additional interphases of training and human factors also contribute towards operational excellence. These will be discussed in future publications.

9. References

[1] Definition of Sustainability, https://www.epa.gov/sustainability/learn-about-sustainability#what, accessed on 24 February, 2016 "
[2] Guidelines for Risk Based Process Safety, CCPS Publication, published March 2007.
[3] S.P. Wilson, T.P. Kelly, and J.A. McDermott, Safety Case Development: Current Practice, Future Prospects, Department of Computer Science, University of York, England

[4] Worksafe Group Victoria, Guidance Note – Overview of a Safety Case Regime for a Major Hazard Facility, April, 2011.

[5] "Safety case versus SEMS, are they really that different", Kumar (Chris) Israni, Dennis Butts and Radhika Bompelly, presented at OTC 2015, Houston, TX.

[6] Department of Mines and Petroleum, Submission of a Petroleum Pipeline Safety Case – Guidelines: Resource Safety, Department of Mines and Petroleum, 2011.

[7] The Bureau of Ocean Energy and Management, Regulation and Enforcement: The Work Place Safety Rule on Safety and Environmental Management Systems (SEMS), October, 2010

[8] The Bureau of Ocean Energy Management, Regulation and Enforcement: Interior. 30 CFR 250 Oil and Gas SEMS (Proposed Changes), 14 September 2011.

[9] OSHA, Process Safety Management of Highly Hazardous Chemicals, 29 CFR 1910.119, 1 July 2012.

[10] Leidos, Elements of PSM, Available at https://www.leidos.com/engineering/psm/elements#ep, Accessed on 11 February, 2015.

[11] "Achieving the safety goal: Safety case versus SEMS and PSM", , Kumar (Chris) Israni, Dennis Butts and Joshua Yang, presented at GCPS 2015, Austin, TX.

445436 Getting The Most From Your PHA – Lessons Learned From A Seasoned Facilitator

Abstract:

This presentation will include activities and techniques which can be used to improve efficiency and outcomes of PHAs. Topics covered will include:

- Deciding on PHA methodologies
- Preparation activities
- Making the most of team time
- Techniques for effective participation
- Execution management

445437 LOPA Boot Camp – Basic Training with Practical Examples

Abstract:

Layer of Protection Analysis has become a cornerstone risk evaluation tool in the chemical industry since the original CCPS text on the topic was published in 2001. Now, 15 years later, the method is used by many companies as the primary means of determining the adequacy of controls for hazard scenarios as well as the required safety integrity level for recommended safety instrumented systems. This presentation will present the basics of the LOPA methodology including the critical evaluation of the hazard scenario, establishing the initiating event and its frequency, investigation of safeguards and selection of independent protection layers, a brief overview of enabling conditions and conditional modifiers, and comparison of the resulting calculated risk to established targets. Practical examples of the application of LOPA to several scenarios will be used to illustrate the use of the technique.

Hazards Inherent to Control Systems: Case Studies and Lessons Learned

Brenton L. Cox
Exponent, Inc.
4580 Weaver Parkway, Suite 100, Warrenville, IL 60555
bcox@exponent.com

Stephen W. Garner, Andrew R. Carpenter, and Mark T. Fecke
Exponent, Inc.
4580 Weaver Parkway, Suite 100, Warrenville, IL 60555
sgarner@exponent.com, acarpenter@exponent.com, mfecke@exponent.com

Prepared for Presentation at
American Institute of Chemical Engineers
2016 Spring Meeting
12th Global Congress on Process Safety
Houston, Texas
April 11-23, 2016

Hazards Inherent to Control Systems: Case Studies and Lessons Learned

Brenton L. Cox
Exponent, Inc.
4580 Weaver Parkway, Suite 100, Warrenville, IL 60555
bcox@exponent.com

Stephen W. Garner, Andrew R. Carpenter, and Mark T. Fecke
Exponent, Inc.

Keywords: Process control; Case studies; Lessons learned

Abstract

Control systems are an integral part of almost all chemical processes, regardless of size or complexity. As the demands for performance and safety increase, control systems will only proliferate. With automation comes the ability to develop more complex and interrelated processes, and to respond more rapidly to disturbances. However, increased reliance on control systems, both discrete and continuous, requires increased diligence with respect to their development and testing, as well as the implementation of independent layers of protection.

A control system that provides optimal product for typical feed conditions may behave poorly during abnormal conditions, such as idling or turnarounds. Similarly, independent control systems charged with safety instrumented functions may not be called upon to act until they become necessary to prevent or mitigate an emergency situation. In order to recognize and mitigate hazards associated with control systems, process design and process hazard assessment teams may need to consider these and other potential issues.

Presented here is a brief overview of the hazards associated with control systems. Finally, case studies are used to highlight the need for considering the potential hazards associated with control systems, from the design stage, through startup, operation and turnarounds.

1 Introduction

In 1993, the Center for Chemical Process Safety (CCPS) published *Guidelines for Safe Automation of Chemical Processes*, the seminal text on the subject of process safety with respect to process control systems.[1] However, in the intervening 23 years since its publication, incidents and near-misses related to process control systems continue to occur, and in some cases, a better understanding of the hazards inherent to these control systems by designers, builders, operators and other personnel may have allowed prevention or mitigation of these issues. With the advent of self-driving vehicles and the Internet-of-Things, there can be no doubt that automation and interconnectivity in the chemical process industry will only increase.

Here, we aim to highlight some of the hazards described in reference texts, e.g., those by the CCPS, and through the lessons learned from case studies.

2 Background

In 1993, when the *Guidelines for Safe Automation of Chemical Processes* was published, the microprocessor was about twenty years old, and had already begun to see widespread use within the chemical process industry, through microprocessor-based sensors, operator workstations, and distributed control systems. At the time of its publication, the prevalence and complexity of microprocessors was still in its relative adolescence. For comparison, in 1993, only about 23 percent of American households had computers.[2] In 2016, the average operator has more computational power in their pocket when they arrive at the plant than was available to operate many of the early process control systems.

Similarly, automated process control began with single-loop controllers, primarily subject to single event failures. [3] Today, more advanced control algorithms allow the operator to execute commands directly targeting production specifications, rather than single set-point values such as pressure and flowrate. However, as a result of the increased complexity of the underlying process control system, plants may be more susceptible to equally complex failure modes, including multiple event failures which could include operator error.[4] Nevertheless, the guidance provided within this text remains valid, especially the framework for chemical process safety and the role process control systems and safety instrument systems play within this framework.

Building on *Guidelines of Safe Automation of Chemical Processes*, in 2007, the CCPS published *Guidelines for Safe and Reliable Instrumented Protective Systems*.[5] The latter text guides the reader through the seven core attributes of independent protection layers (IPL) as they apply to the life-cycle of a plant's instrumented protective system (IPS). These attributes are: independence, functionality, integrity, reliability, auditability, access security, and management of change.

Guidelines for Safe and Reliable Instrumented Protective Systems also discusses the application of recognized and generally accepted good engineering practices (RAGAGEPs) that exist within the field of process control and instrumented protective systems. Of particular relevance are the International Electrotechnical Commission's (IEC's) international standard IEC 61511, *Functional Safety: Safety Instrumented Systems for the Process Industry Sector*, and its close counterpart in the United States, ANSI/ISA 84.01, *Application of Safety Instrumented Systems (SIS) for the Process Industry*.

Process control textbooks, e.g., *Process Dynamics and Control* by Seborg, Edgar and Mellichamp, also deal with issues related to process safety.[6, pp. 248-255] However, in the cited textbook, only a cursory discussion of issues is included, and the reader is referred to the other sources for more detail. This background section roughly follows the Seborg, Edgar, and Mellichamp discussion of the components of an automated process control system, with emphasis on some of the specific safety aspects not discussed in the textbook.

2.1 Basic Process Control Systems

The basic process control system (BCPS) is defined as the control equipment installed to support normal production functions. This excludes specialized safety instrumentation and the safety interlock system (SIS). The lifecycle of the BCPS mirrors that of the plant: design, selection, installation, operation, and maintenance. CCPS provides guidance on safety considerations in the selection and design of BCPSs in Chapter 4 of *Guidelines of Safe Automation of Chemical Processes*.[1]

The reliance upon components to fail safely is a common safeguard cited during a process hazard analysis (PHA). However, microprocessor-based process control systems involve complex interactions of software and hardware that may exhibit unanticipated failure modes, and the assurance that all of these failure modes end in a fail-safe state may be challenging. It is important that knowledgeable team members weigh in on the ability of a particular component of the BPCS to fail safely if credit is to be taken during a PHA. For example, some process controllers allow an engineer to specify a fail-safe value for controller outputs.[1, p. 121] In the event of an unexpected controller restart, the outputs will reset to these fail-safe. However, the system should be carefully reviewed for circumstances where this could create an unsafe condition.

An additional consideration is the performance of the BCPS during abnormal conditions. Operators may require additional training or guidance for circumstances such as loss of instrument air, loss of cooling water, a fire, or other out-of-the-ordinary situations.[1, p. 261] Procedures and training may prevent escalation of an incident due to an improper response by the operators.

The BPCS may be used for protective functions, but there are limits to the level of separation that can be achieved between safety and control functions of the same system.[5, p. 275] CCPS guidance suggests that "the risk reduction allocated to a safety function in the BPCS must be less than 10." They also recommend failure modes with a hazard rate greater than 10^{-6}/hr not rely on a BPCS serving both control and protective functions, even if using separate sensors, controllers and elements.

2.2 Sensors and Alarms

Sensors and alarms are typically configured to monitor the process during steady state, and alert an operator when a material deviates from anticipated steady state conditions. However, during shutdowns or idle periods, the sensors and alarms may not function as expected, if at all, and the onset of hazardous conditions may not be recognized by plant personnel. If the sensors and alarms for the system during normal process conditions are to be relied upon to alert operators or maintenance staff of unexpected behavior during a shutdown or turnaround, it is important to consider these non-routine operating conditions.

For example, a recent article in Chemical Engineering Progress discussed an explosion that occurred during a plant shutdown.[6] An EPA investigation into the incident concluded that the incident was at least in part due to a lack of procedures for putting the vessels into a safe state at shutdown, or for monitoring the vessel during shutdown. A key lesson cited by the article was that a hazard assessment of the shutdown step would have revealed that key sensors could not function during shutdown. In order to ensure this is the case, *Guidelines for Safe Automation of Chemical Processes* advises that risk experts consult with process design and instrumentation specialists to "facilitate the understanding of interactions between different portions of the total system and processes."[1, p. 40]

Another concern associated with alarms is the presence of too many alarms with too tight of limits, referred to as "nuisance alarms." In this case, operators may become less responsive to alarms, regardless of their importance.[8] Additionally, during process upsets, many alarms are not critical, and provide the operator with excessive information at a time when quick action is critical.[9] The CCPS guidance suggests considering importance, grouping (priority) and context as criteria in managing alarms.[1, p. 139] It is also important to prove plant operators with sufficient data to take appropriate action when alarms occur, and to log all alarm and event data.

2.3 Safety Interlock Systems

As described by the *Guidelines for Safe Automation of Chemical Processes*, a Safety Interlock System (SIS) "is the backup device that intervenes and takes a system to a safe state when the BPCS fails to do so."[1, p. 140] In order to ensure a failure of the BPCS does not interfere with the SIS, the SIS should be physically distinct from the BPCS, with its own sensors and actuators.[6, p. 252] The alarm panel for the SIS and the BCPS may also need to be physically separated to reduce the possibility that both fail simultaneously.

Components of the SIS are expected to provide accurate process surveillance and timely and effective automated corrective actions. However, many of these systems charged with safety instrumented functions may never be called upon to act until they become necessary to prevent or mitigate an emergency situation. Thus, testing of the SIS is essential to ensure its protective function. The frequency at which SIS components should be tested depends on the safety protection provided, and the safety integrity level availability requirements. *Guidelines for Safe Automation of Chemical Processes* provides detailed guidance on testing procedures and frequency.[1, p. 278-283]

2.4 Interlocks

Interlocks, much like process alarms, should be carefully chosen to not result in unnecessary plant shutdowns and startups.[6, p. 254] The result of an excessive number of interlocks or too sensitive of trip conditions is not only financial. Periods of plant shutdown and startup are often considerably more hazard-prone than steady-state operation. Guidance on the design and selection of safety interlocks is provided in *Guidelines for Safe Automation of Chemical Processes.*

Another aspect to consider is the fail-safe characteristics of the interlock, such as whether the interlock operates by being energized or deenergized. If plant power or other utilities are lost, the safety interlocks should fail to a safe condition.

3 Case Studies

3.1 Furnace Explosion

3.1.1 System Background

The incident occurred at grain and corn processing complex. As part of the corn-to-ethanol process, corn-based material (fiber) was dried in an industrial dryer. The dryer use gas burners which co-currently dried the wet fiber. The wet fiber was fed into the dryer just beyond the combustion zone of the gas fired burners. A process flow diagram (PFD) for the Fiber Dryer #1 process is provided in Figure 1. For reference, the rotating drum component of Fiber Dryer #1 is approximately 12 feet in diameter and 60 feet long.

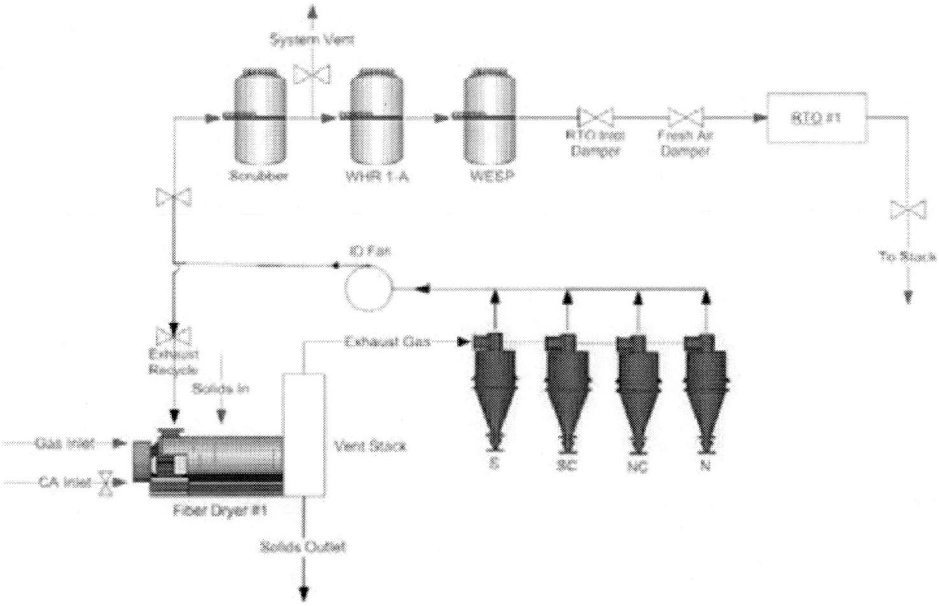

Figure 1. Process flow diagram for the Fiber Dryer #1 process.

The dryer consisted of a burner assembly connected to a large rotating drum which has internal baffles that mix the wet material with the burner exhaust and carried it down the length of the dryer. The dried fiber was discharged from the exit of the dryer via conveyor and then travels to other process equipment within the complex. During normal operation, the exhaust gases from the dryer were fed through several stages of emissions control prior to being vented to the atmosphere. The downstream path from the dryer to the atmosphere included:

1. four cyclone dust collectors (labeled South, South-Center, North-Center, and North),
2. an induction fan (driving force for moving the exhaust gases through the process),
3. a sulfur scrubber (referred to as scrubber),
4. a waste heat recovery unit (WHR 1-A),
5. a wet electro-static precipitator (WESP),
6. a regenerative thermal oxidizer (RTO), and
7. an atmospheric vent stack.

During startups and atypical operations, such as what occurred on the day of the explosion, the exhaust gases may be vented directly to the atmosphere following the scrubber, indicated in Figure 1 as the "System Vent."

Fiber Dryer #1 utilized a 90 million BTU burner that operated using a flammable gas as the fuel source. A portion of the facility complex consists of a biorefinery, which digests organic material and produces a flammable biogas that can be used as a natural gas supplement. This biogas is typically blended with natural gas and fed as the flammable gas fuel to the Fiber Dryer #1 burner.

A simplified process flow diagram for the dryer-only portion of the Fiber Dryer #1 process is shown in Figure 2. The burner portion of the dryer combines combustion air (CA) with flammable gas in a combustion chamber and ignites the flammable mixture. The combustion air and flammable gas flows are modulated via flow control valves (FCVs). The pilot air and pilot gas are only utilized during startup; following successful ignition of the main burners, the pilot lines are deactivated. The main burner consists of two parts: a radial burner assembly and primary burner assembly.

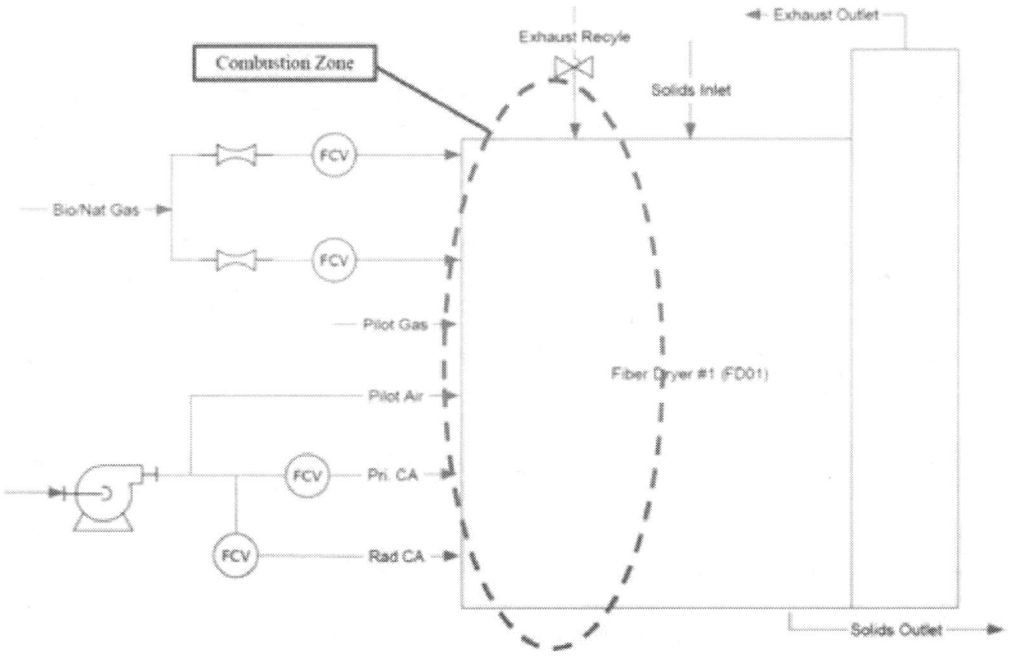

Figure 2. Process flow diagram for Fiber Dryer #1 portion of dryer process.

To control temperatures of the combustion zone of the dryer (indicated by a red circle in Figure 2), a portion of the exhaust gases could be recycled to the burner via louvers at the top of the combustion zone. The Fiber Dryer #1 process was controlled by a distributed control system (DCS). The burner was also controlled by a burner management system (BMS), which provided safety functions and combustion controls and included associated wiring, sensors, controllers, and actuators; the BMS also communicated with the Distributed Control System (DCS). The control system was monitored by plant operators who interfaced with the system from the control room.

3.1.2 Incident Summary

Fiber Dryer #1 was shut down for routine maintenance. The work order describing the maintenance activity for this shutdown was titled "#1 Fiber Dryer Quarterly Inspection." This work order included nine major tasks; one of the major task's description stated "Inspect #1 Fiber Dryer Gas Line Flow Meter and Clean." The primary gas flow rate flow meter was used to measure the gas flow being introduced into Fiber Dryer #1's primary burner. Following completion of the work order, the dryer was restarted.

To restart the Fiber Dryer #1 process, the operators first purged the system with steam. Following the steam purge, the operators initiated a restart on the console and the system performed a pre-programmed restart. The system starts by purging the process using combustion air at 100% flow for 5 minutes. Following the purge, the pilot gas and pilot air are introduced into the combustion chamber and the flammable mixture is ignited. The valves that block the radial and primary gas lines (one set of valves are shared between both) are opened and the radial gas and radial combustion air are introduced into the combustion chamber where it is ignited by the pilot. If flame is confirmed (via a UV flame scanner) within a certain time, the system is considered to have lightoff and the pilot is turned off. If the flame is not confirmed, then the startup process is terminated and must be reinitiated (starting with another steam purge). Immediately after lightoff, the system goes into a standby mode during which the radial flame is at minimum fire position and there is no primary flame. This standby mode will persist until the operator releases the system to "autofire." This is the first operator interaction with the process since initiating a restart. Once in autofire, the primary flame is at minimum fire position and the operator manually then brings the system up to the desired operating temperature. Once the dryer outlet temperature is near the desired condition, the operator then releases the system to modulate, or automatic control. The system remains in automatic until a system trip occurs, an operator switches out of automatic, or a process upset/alarm returns the system to the standby mode (where radial flame is at minimum fire and there is no primary flame).

During the startup process on the day of the incident, three restarts were attempted before lightoff was achieved (referred to as 1st lightoff). Operators and personnel indicated that unsuccessful lightoff during the first several restart attempts is typical for restarts following maintenance. A description of the first three restart attempts follows:

- The unit tripped during the first restart attempt due to a flame safety alarm.
- The unit tripped during the second restart attempt during the purge due to a loss of the purge permissive.
- Successful lightoff was achieved during the third restart attempt.

After radial flame lightoff, the operator initiated lightoff of the primary burner flame. The burner flames were on for approximately 2 minutes before a system alarm brought the burners down to radial burner only standby mode (primary burner off) due to a lid limit switch on the wet fiber feed screw. Approximately 30 seconds later, a BMS permissive caused the burner system to trip due to an interlock.

Following the 1st lightoff and subsequent shutdown, there were three more restart attempts before lightoff was achieved again (referred to as 2nd lightoff). A description of the second set of restart attempts follows:

- The unit tripped during the fourth restart attempt during the purge due to a loss of the run permissive; several other alarms occurred.
- The unit tripped during the fifth restart due to a flame safety alarm.
- Successful lightoff was achieved during the sixth restart attempt.

After the radial flame lit off the second time, the operator initiated lightoff of the primary burner flame. The primary burner was lit and the unit was operating for approximately 5 minutes and 30 seconds before the explosion.

Following a detailed root cause analysis (RCA) it was determined that following the removal and cleaning of the gas flow meter for the burner, the subsequent reinstallation of the gas flow meter was performed incorrectly; specifically, the sensing lines had been reconnected in reverse. As a result of the reversed sensing lines on the primary gas flow rate meter, the primary gas flow rate signal provided to the controller reached its low-saturation setting. When evaluated in the context of the other control signals and system parameters, this indicated an abnormal operating condition, called a process upset. At least three opportunities to identify the existence of a process upset and make timely corrections were possible, including:

1. Detecting a signal under-range (low-saturation) condition on the gas flow rate meter;
2. Detecting a primary gas low flow deviation for the primary gas flow rate control loop (non-responsive control loop);
3. Detecting an air-to-fuel deviation (out-of-range) condition.

The control system was not configured to detect the first of the three conditions, the control system provided no alarms or automatic trips for it, and the operator did not respond to the condition. The control system was configured to detect the last two of these three conditions, and, as the alarm log indicates, did provide alarms for them. However, there were no automatic trips in response to these two alarms, there was no operator response to alarms that were triggered, and there was no operator response to the abnormal operating condition. Neither the operator nor the control system performed any corrective action in response to the abnormal operating conditions (summarized in Table 1). Had the control system or the operator taken timely corrective action, the explosion may have been prevented or its severity reduced.

Table 1. Causal factors 1 through 3: Alarms, Trips, and Operator responses.

Causal Factor	Alarm	Trip	Operator Response
1 - Sensor Under-Range	None	None	None
2 - Non-Responsive Control Loop	Yes	None	None
3 - Air/Fuel Ratio Deviation	Yes	None	None

3.2 Fire at a Specialty Chemical Manufacturer

A specialty chemical manufacturer of a very high purity chemical used in the semi-conductor materials industry suffered a fire which resulted in the total destruction of the facility and an environmental release of chlorine gas. The specialty chemical, which they produce at a purity levels as high as 99.99%, is both pyrophoric and water reactive, and its reaction with the moisture in air can form a solid compound that is both shock and friction sensitive. The chemical is produced and purified in batches, including a multi-day controlled batch distillation step at a very high reflux ratio.

At the time of the fire and explosion, three employees of the facility were in the process of cleaning up a spill of the high purity chemical which had overflowed receiving vessels in the reactor room. Fortunately all three of the employees were able to quickly exit the facility at the onset of the event and were not injured.

3.2.1 System Background

The chemical produced at the facility is first generated in a dedicated reactor vessel and subsequently refined to very high purity via batch distillation with a very high reflux ration. The incident occurred in the distillation room which the chemical was being distilled to a purity level approaching 99.99% and as such, the reactor and other proceeding processes including the first stage of purification will not be discussed further in the paper.

During the batch distillation phase, the crude product is distilled in glass distillation columns operating at a very high reflux ratio. The distillation room included four batch distillation systems which were all completely independent from each other with the exception of sharing a common control system and instrument air supply. Figure 3 below is a schematic of a batch distillation system utilized at the facility:

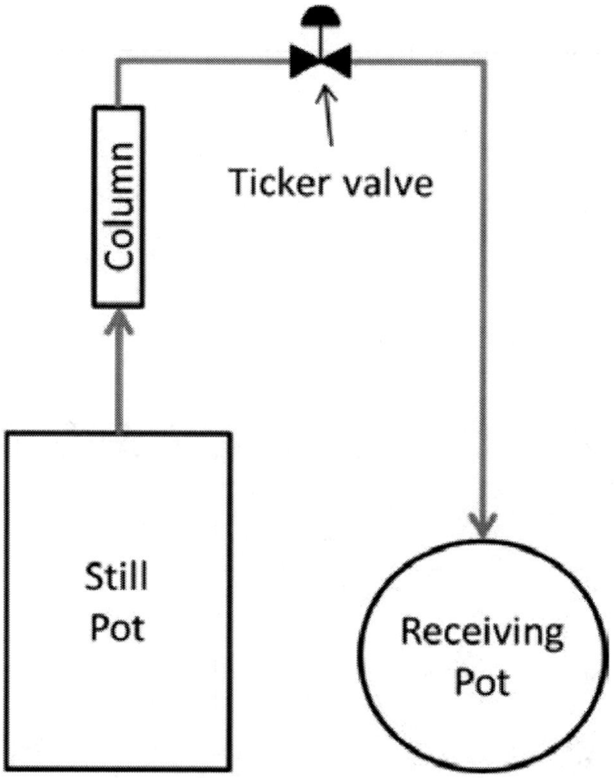

Figure 3. Approximate sketch of batch distillation system.

In simple terms, the crude product is transferred into the still pot and the distillation is accomplished by heating the still pot and pulling off very small amounts of high purity product from the top of the column. The high purity product would be transferred from the column to a receiving vessel by the opening of a solenoid valve. This valve would be opened very briefly allowing the transfer of the distilled chemical. To obtain the desired purity, the reflux ration was set to pull only a few ml of product from the column at a time. The opening of the solenoid valves, as well as the heating of the still pots for all four of the distillation systems was controlled the process control system which had been custom designed and programmed for this application.

3.2.2. Incident Summary

The facility operated the distillation columns in batch mode with each batch requiring several days of processing to complete. The facility was typically only staffed for a single shift during week days, so while the facility was unoccupied, the distillation systems were in operation solely under the control of the control system without any operator oversite. The control system was configured to send alerts and alarms via email and text message to several employees while the facility was unoccupied. During the day of the incident, two of the batch distillation processes were in operation. The incident occurred on the weekend when the facility would not be normally staffed, with the exception of a planned weekend walk through by one of the operators.

The first signs of any problems with the process occurred at approximately 6:00 am on the morning of the incident when the control system sent an email alert that one of the operating still pots had alarmed on a high temperature. The control system was intended to shut down power to the still pot heaters and isolate the distillation system; in essence putting the distillation system into a safe state. It was reported that the emailed alert was first noticed by one of the operators approximately two hours later at approximately 8:00. The responding operator was planning on visiting the facility later in the day for a walk through, but he decided to arrive earlier than he planned to check on the alarm and if possible, return the distillation system to service.

When the responding operator arrived, he observed that there was a white fog (consistent with the presence of spilled high purity product) in the distillation room, and that two of the distillation systems receiving vessels were completely liquid full. One of the overfilled receiving vessels was for the distillation system that was in alarm, while the other overfilled receiving vessel was from the other operating distillation system from which no alarms had been reported. The responding operator called in additional support from the facility to assist in the cleanup and investigation of the alarm. While waiting for additional support, the responding operator noticed that the alarmed distillation system still pot was still in the heating mode, which was a deviation from the intended safe state.

When additional personnel arrived, the responders donned PPE and began the process of cleaning up the spill. During the cleanup activities, a small fire occurred which was attributed to the ignition of an accumulation of dried high purity product from frictional heating while some of the equipment was moved to allow for cleanup of the spill. All three operators safely exited the distillation room. It was quickly determined by the responding operators that this initial, small fire had self-extinguished, and they made the decision to reenter the distillation room and resume cleanup activities, however they made the decision to also stage hand held firefighting equipment in the distillation room as a precaution should another small fire occur. While staging the firefighting equipment inside the reactor room, another fire occurred which very quickly flashed to engulf the entire distillation room. At the time of this ignition, all three responding employees were located just inside the doorway to the distillation room and were able to make an immediate exit from the room. One of the operators was taken by ambulance to a local hospital for medical observation, but was later determined to have not been injured during the incident, and he was subsequently released.

3.2.3 Control system failure

The fire which initiated in the distillation room quickly spread to engulf the rest of the facility, and the responding fire department acted to set up a perimeter around the facility, but due to the concern of the water reactive materials stored inside the facility, allowed the fire to burn itself to competition. As a result, the entire facility was completely destroyed, and no elements of the control system could be recovered for examination. It was determined however that the control system took two unintended actions and failed to take one intended action while responding to the initial high temperature alarm. The control system failed to shut down heat to the still pot as was the intent. The control system also called for the opening of the valve to the receiving vessel (a normally closed valve, and a valve intended to remain closed during the alarm state) as well as call for the opening of the valve to the receiving vessel for the other distillation system in operation which was not reported or found to be in the alarm state. Unfortunately the extent of damage to the facility precluded a thorough investigation into the control system, so the exact nature of the control system failure, including equipment failure or programing failure could not be determined.

4 Conclusions

As the breadth and depth of process control systems increase, the importance of understanding the failure modes and hazards associated with these systems becomes increasingly important. In depth review of guidance from the CCPS and other sources, and sharing lessons learned from process safety incidents and near misses both help foster continued improvement, and help stakeholders grow along with the technology available.

Of particular importance are the instruments and control systems that are relied upon as safeguards. Careful consideration may be necessary to determine circumstances (e.g., specific operating modes, idling, or turnarounds) where instrumentation and controls do not function as intended and thus may not be reliable. Also, for instrumentation or controls not frequently used or monitored during operation, testing may be necessary to demonstrate operability on demand. Lastly, it is important to determine whether instrumented and/or automated safeguards are independent, or may be affected by a common failure mode.

5 References

[1] CCPS. *Guidelines for Safe Automation of Chemical Processes*. Center for Chemical Process Safety, American Institute of Chemical Engineers, New York, NY, 1993.

[2] "Computer Use in the United States." P20-522, United States Census Bureau, October 1997.

[3] "Overcoming the Hurdles in Chemical Process Automation." *Chemical Processing*, November 15, 2002.

[4] R.A. Ogle, D.M Morrison III, A.R. Carpenter. "The relationship between automation complexity and operator error" J. Haz Mat, 159, pp. 135-141, 2008.

[5] CCPS. *Guidelines for Safe and Reliable Instrumented Protective Systems*. Center for Chemical Process Safety, American Institute of Chemical Engineers, New York, NY, 2007.

[6] D.E. Seborg, T.F. Edgar, and D.A. Mellichamp. "Process Dynamics and Control." 2nd Ed., John Wiley & Sons, Hoboken, NJ, 2004.

[7] Ness, A. "Lessons Learned from Recent Process Safety Incidents." *Chemical Engineering Progress*, March 2015, pp. 23-29.

[8] C.S. Connely. "Lack of planning in alarm system configuration is, in essence, planning to fail." *ISA Trans*. Volume 36, Issue 219, 1997.

[9] J.A. Shaw. "Smart Alarm Systems: Where Do We Go From Here?" *InTech*, December 1985.

Human Reliability and Process Safety Metrics: Considerations from an Abnormal Situation Management Perspective

Dr. Peter T. Bullemer
Human Centered Solutions
Independence, MN
pbullemer@applyhcs.com

Dr. Dal Vernon C. Reising
Human Centered Solutions
Canton, MI
dreising@applyhcs.com

Prepared for Presentation at
American Institute of Chemical Engineers
2016 Spring Meeting
12th Global Congress on Process Safety
Houston, Texas
April 11-23, 2016

Human Reliability and Process Safety Metrics: Considerations from an Abnormal Situation Management Perspective

Dr. Peter T. Bullemer
Human Centered Solutions
Independence, MN
pbullemer@applyhcs.com

Dr. Dal Vernon C. Reising
Human Centered Solutions

Keywords: Human Reliability, Abnormal Situation Management, Root Cause Analysis, Process Safety Metrics, Operations Discipline

Abstract

The ASM Consortium conducted a series of studies investigating the role of operations practice failures in major process safety incidents. The results of the studies improved our understanding of human reliability factors contributing to major process safety incidents. The paper discusses the implications of these operations practice failure modes for process safety metrics. The paper includes recommendations on how process safety metrics could be enhanced to measure the effectiveness of operations practices and their potential impact on process safety performance.

1 Introduction

In 1992, Honeywell formed an alarm management task force with representatives from several of their customer organizations who had and were making the transition from pneumatic panel control boards to distributed control systems (DCSs). These individuals discussed alarm management problems with Honeywell and suggested potential improvements to the alarm system functionality of the TDC3000 DCS. During the discussions, customer representatives told Honeywell that some of the alarm management problems were a manifestation of a larger problem. They referred to the larger problem as the abnormal situation management challenge. This problem was characterized as a human reliability challenge wherein there is a disturbance in the process and the control system cannot cope with the disturbance requiring human operators to intervene to bring the process back to the normal operating regime and preventing an escalation to a process safety incident.

A characterization of the problem was that during normal, steady-state operations the available control system functionality was quite good but once the process became abnormal, operations team members had significant challenges in detecting the abnormal condition, understanding the source and the implications, and then taking appropriate actions in a timely manner. In some of these situations, the significant number of alarms (i.e., alarm flooding) contributed to failures to detect and understand the nature of the abnormal situation.

The industry tendency was to try to solve this human reliability issue with more automation. However, the increase in automation led to more sophisticated processes that were more complex for human operators to understand. Consequently, the more sophisticated processes had a tendency to reduce human reliability. This phenomenon became known as the paradox of automation [1, 2, 3].

At the conclusion of the alarm management task force activity, Honeywell and a few of the customer representatives agreed to establish a research project to better understand the abnormal situation management challenge. In 1993, a study team was formed to conduct visits to plant facilities to better understand the abnormal situation management problem and identify the requirements to improve operations team's ability to prevent and respond to abnormal situations.

The 1993 project team's findings were powerful enough to motivate several industry competitors to form a research and development consortium focused on developing solutions to improve plant safety. The key findings of this foundational 1993 study led to the founding of the Abnormal Situation Management (ASM) Joint Research and Development Consortium, which recently celebrated its 20th year working together.

In the remainder of this paper, we present some key findings from the 1993 study [4] and examine the ASM Consortium research progress in addressing the ASM requirements identified in that foundational study in terms of (i) the industry's understanding of the contribution of human reliability to major process safety incidents and (ii) the implications for operations practices to effectively manage human reliability in reducing risks that may contribute to process safety incidents.

Abnormal situations need to be managed through Prevention, Early Detection, and Mitigation, in order to reduce unplanned outages and process variability that increase production, safety and environmental risk to plant employees and local communities. Historically, mechanical integrity is important to process safety, however the ASM Consortium focus on the operations team has led to an increased emphasis on operational integrity to reduce the likelihood of abnormal situations and ultimately improve process safety and plant performance. Hence, the perspective of the ASM Consortium on process safety metrics has focused on the impact of operations practices on plant process safety.

2 1993 Foundational Study

Peter Bullemer led the Honeywell Abnormal Situation Management (ASM) Study Team on five plants visits in North America and Europe between September, 1993 and February, 1994. The team interviewed plant personnel to understand their concerns regarding current limitations facing industrial plant operations during abnormal conditions. The team spent five days at each site conducting structured interviews of plant personnel including operators, supervisors, engineers, managers, maintenance specialists, and automation specialists. These plant visits were conducted with the following objectives:

- Identify problems by interviewing site personnel and observing operations personnel in their work environment
- Identify potential changes in methodologies, practices, and operations and the technical solutions to achieve best practices

Abnormal situations were found to occur almost on a daily basis in most large process industrial plants. Most abnormal situations were relatively small and adeptly handled by the operating, maintenance and engineering staff in the plants with perhaps minor impact on the unit. An example of a small event would be a controller being pushed beyond its limits and placed in manual to stop cycling, or an operator entering the wrong set point value for a controller. In some situations, the disturbance would result in poor quality product or reduced production. A smaller percentage would result in situations that mandate a process shutdown. An even smaller fraction would cause significant equipment damage, release of undesirable materials into the environment, and even human injury or death.

Although individual perceptions of abnormal situation management varied, there was consensus that normal and abnormal represent two distinct modes of operation. Furthermore, abnormal operations were more likely during transition events such as startup, feed change, product grade change, and shutdown. Errors in situation assessment could be a source of abnormal situations because personnel did not understand the plant or the software system and therefor made erroneous interventions. Furthermore, the current interface design of the DCS at the time led to predominately a reactive rather than a proactive mode of intervention.

Based on these findings, it was clear that a common definition of the nature and sources of abnormal situations and the problems associated with their management was the critical first phase in developing solutions. The next section presents a general overview of sources of abnormal situations and their average rate of incidence in the plant visits.

2.1 Sources of Abnormal Situations

The study team examined the plant incident reports to identify sources of abnormal situations. Root and contributing causes may have appeared in isolation or in combination with each other. However, in most cases, the abnormal situation was a result of the interaction among multiple sources. For example, a frequent plant practice was to extend the process to its limits to maximize production. In purposely pushing to the limits of the process, the probability of

- the equipment failing
- the process becoming less resilient
- and/or personnel making errors

greatly increases. As a result, the equipment's original design limits were challenged, the process was operating very near or at its original design constraints, and personnel were being asked to monitor and interact with a process that was complex and reaching the limits of their cognitive and physical capabilities. At any point in time, one or more of these factors may contribute to the onset and escalation of an abnormal state. The resulting abnormal situations vary in their complexity and effect on the process.

The study team characterized the sources of abnormal situations in terms of three basic types of sources or causes of abnormal situations. Figure 1 shows the average percentages for each type of source.

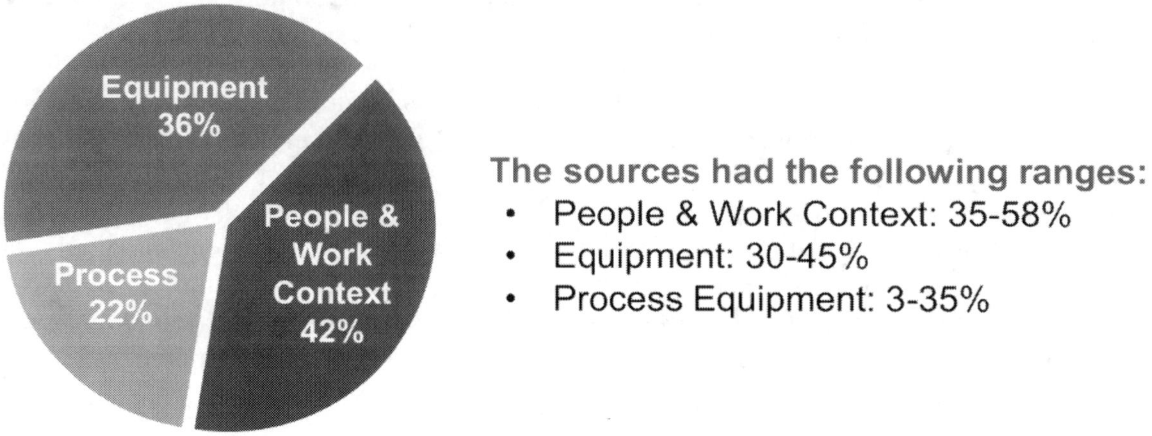

Figure 1. Average percentages and ranges for the 3 types of sources of abnormal situations identified in plant incident reports

The data was obtained from 1992-3 incident reports at each of the five sites. Of all the incidents reports, the study team selected only those events that had an impact on process operations, i.e., did not include slip, trips and fall reports. These findings differed from a contemporary industry report at the time from the Chemical Manufacturers Association [5] that indicated up to 80-85% of human errors contributed to process safety incidents. The study team identified a general reluctance of plant personnel to report people as the cause of an incident. Hence, the study team concluded that the available data was probably biased towards reporting equipment and/or process sources rather than people or work context.

For the incidents involving People & Work Context factors, the following root causes were reported:

- Inadequate or no procedure (27%)
- Fail to follow procedure (24%)
- Inadequate or incorrect action (24%)
- Inadequate work practices (15%)
- Fail to recognize problem (5%)
- Defective installation (5%)

Hence, the reported root causes indicated a significant contribution of ineffective procedural operations to the occurrence of the reported process safety incidents.

2.2 Human Reliability Influences

Based on the interviews with site personnel, the study team identified the following human reliability influences that contributed to the perceived abnormal situation management challenge:

- Minimal understanding of the source and impact of abnormal situations

- Operating close to design limits
- Ineffective use of procedures
- Inadequate communication
- Inadequate training methods
- Ineffective access and presentation of information

Additional detail on these human reliability influences can be found in a recent conference paper by the authors [6].

2.3 Not by Technology Alone

A significant conclusion from this initial study of the abnormal situation management challenge was that the effective human-machine system solution involved more than developing the right technologies [7]. While the capabilities of technology can impact human performance, it is not the only factor that needs to be considered when organizations seek to improve human reliability. Effective solutions involve addressing the work culture, organizational structure, work processes, user acceptance and adoption, as well as appropriately designed technology.

The key to successful technology deployment requires understanding the problems that need to be solved and designing solutions that fit the work context. This means that organizations need to refrain from *only* identifying applications of the latest technological advances just because the technology is available.

3 Human Reliability and Process Safety Incidents

Historically, the reporting of failures has tended to emphasize root causes associated with equipment reliability and less so on human reliability root causes [8]. Consequently, there is limited information available on the frequency and nature of operations failures pertaining to human reliability. This tendency has limited the ability of process industry operations organizations to identify improvement opportunities associated with their management systems and operations practices.

In an effort to improve on the understanding of the impact of ineffective operations practices and management systems on safe plant operations, the ASM Consortium conducted a series of studies between 2008 and 2012 involving the root cause analysis of existing major incident reports. An initial study examined fourteen public and private incident reports to determine the impact of communication and coordination practice failures [9]. Based on the results of the initial study, a follow-on study was conducted to analyze an expanded set of incidents reports for failure modes of operations practices in general [10, 11]. Additional studies focused on particular problem areas, a third study used the same methodology to examine the impact of procedure execution failures during abnormal situations [12] and a fourth study examined incident reports involving heater operations [13].

This section presents the results of the root cause analysis across a total of 42 major process safety incidents reports. The methodology used to analyze these incident reports was developed by the ASM Consortium research team to investigate the impact of operations practice failures in major process safety incidents (see [10] for detailed description of the methodology).

3.1 Incident Analysis Methodology

The purpose of the incident analysis technique is to generate information to enable an understanding of why the incidents occurred and develop improvement programs and corrective actions to address weaknesses in operations practices or management systems—thereby addressing human reliability supported by them. The methodological emphasis is on the elimination of common and systemic problems.

The project team conducted a search to identify potential public incident reports from sources world-wide and private incident reports from ASM Consortium member companies. The details of the method for identifying and selecting incident reports are available in the initial study report [9]. In the selection process, priority was given to recent refining and chemical incident reports with severe consequences (where recent was within the previous 10 years) and the reports had sufficient detail to conduct a root cause analysis. In addition, the ASM Consortium wanted the analysis to represent operations practice failures from a global perspective so there was an attempt to obtain a global distribution. Table 1 shows the selection distribution results in terms USA versus non-USA incident reports as well as type of industry.

The project team used only the documentation that was made available through the public domain or ASM member companies, such as the formal incident reports and supporting analysis documents. The project team did not conduct any interviews or additional incident investigation as part of the incident analysis methodology.

Table 1. Location, source and industry type for the 42 incident reports

Location	Public	Site		Total
USA	21	8		29
Non-USA	6	7		13
Total	**27**	**15**		**42**

Industry	Public	Site	Total
Refining	9	10	19
Chemical	16	4	20
Oil & Gas	2	1	3
Total	**27**	**15**	**42**

Figure 2 illustrates the seven steps in the work process used to analyze the incident reports.

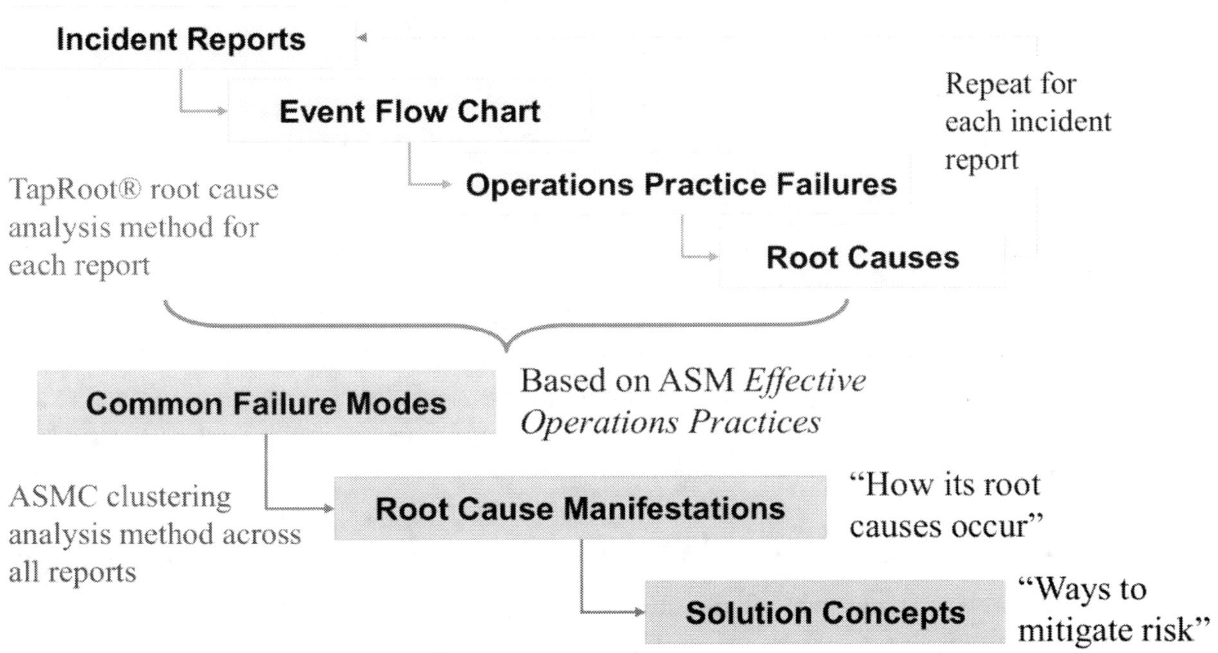

Figure 2. Work flow for the incident analysis methodology

The output of the initial three steps was a list of operations practice failures in the language of respective the incident investigation teams and the associated root causes per the TapRoot classification scheme [14]:

- **Operations practice failure** (see upper half of Figure 2) is any failure that, if corrected, could have prevented the incident from occurring or would have significantly mitigated its consequences. An operations failure describes 'What went wrong' in the specific incidents and is typically in the investigation team's own language/terms. An example of an operations failure is *Ineffective supervision of procedure execution*.
- **Root cause** (see upper half of Figure 2) is the most basic cause (or causes) that can reasonably be identified that management has control to fix and, when fixed, will prevent (or significantly reduce the likelihood of) the failure's (or factor's) recurrence [14]. A root cause describes 'Why a failure occurred.' In the research project, the team used the root cause tree available in the TapRoot methodology. Two root cause examples are *No Supervision* and *No communication* which can both result in the *Ineffective first line leadership* common failure mode.

In the previous ASM research study of 32 major process safety incidents [10], the project team found that additional analysis was valuable to identify the systemic failures across incidents. Identifying the systematic failures provided a better understanding of how to address the operational risk indicated by its operations failures and root causes with an ASM solution (or solutions) that would have the broadest impact. This additional, second analysis clustered the initial findings into common operations practice failures and common root cause manifestations to indicate where ASM solutions may reduce the apparent operational risks identified in the incident analysis. These two clusters are defined as:

- **Common operations practice failure** (lower half of Figure 2) is a description of multiple operational failures that appeared across incidents. A common failure mode represents a common problem across industry sites. The project team characterized these common failures using language from their *Effective Operations Practices* [15]. If a common failure mode did not map to one of the Effective Operations Practices, the project team created a new failure mode description. An example of a failure mode is *Ineffective first-line leadership roles*.
- **Common root cause manifestation** (lower half of Figure 2) is the specific expression or indication of a root cause in an incident. The root cause manifestations describe 'How' operational failure modes are expressed in real operations settings. The root cause manifestation characterizes the specific weakness of an operations practice failure mode. *Supervisor not in accessible to control room to discuss problems* is an example manifestation for the *No Supervision* common root cause and the *Ineffective First Line Leadership Role* common failure mode.

3.2 Common Operations Practice Failures

Table 2 below shows the number of operations practice failures identified as a function of the seven ASM Operations Practice categories. The most common operations practice failure category is Organizational Roles, Responsibilities, and Work Processes (46%) accounts for almost half of all of the failures indicating management systems is a significant source of human reliability failures. Communications is the next most common category (16%) followed by Process Monitoring, Control & Support Applications (9%), Procedures (8%) and Knowledge and Skill Development (7%).

Table 2. Operations practice failures as a function of the seven categories of ASM operations practices as defined in the Effective Operations Practices guideline document [15]

Rank	Operations Practice Category	Total Failures	% of Failures
1	Organizational Roles, Responsibilities, and Work Processes	288	46%
2	Communications	101	16%
3	Process Monitoring, Control, and Support Applications	54	9%
4	Procedures	53	8%
5	Knowledge and Skill Development	43	7%
6	Understanding Abnormal Situations	21	3%
7	Work Environment	5	1%
	Other	59	9%
	Total	**624**	**100%**

Table 2 shows the top common operations practice failures in rank order. The top ten operations practice failures account for 71% of the operations practice failures.

Table 3. Top Common Operations Practice Failures shown in rank order of most frequent to least frequent

Rank	Top Common Failure Modes	Total Failures	% of Failures
1	Lack of a comprehensive hazard analysis and communication program.	92	15%
2	Ineffective first line leadership roles	74	12%
3	Ineffective continuous improvement program	72	11%
4	Poor safety culture	37	7%
5	Ineffective task communications	35	6%
6	Lack of a comprehensive Management of Change (MOC)	34	5%
7	Failure to establish initial and refresher training based on competency models for all modes of operation	32	5%
8	Ineffective cross-functional communications	24	4%
9	Failure to ensure compliance with an explicit policy on the use of procedures	23	3%
10	Lack of effective design guidelines and standards	17	3%
	All other failure modes	184	29%

The earlier ASM Consortium research project referenced above investigated operations practice failures in 32 major process safety incidents [10]. Interestingly, a comparison to the top ten operations practice failures in that previous study found the same list of failure modes. Hence, the addition of ten additional incident reports [11] did not change the overall finding from those reported in the initial study [10].

3.3 Common Root Cause Manifestations

To better understand the nature of the operations practice failures, the common root cause manifestations were analyzed for each of the operations practice failures. Table 4 shows the common root cause categories associated with the operations practice failures. Eighty-two percent of the common root causes were human performance-related with the top category indicating ineffective management systems. Eighteen percent of the common root causes were associated with equipment related sources:

- Design review (11%)
- Design specification (5%)
- Preventative maintenance (2%)

Table 4. Common root cause categories associated with the operations practice failures

Root Cause Categories	#	%
Management System	279	29%
Work Direction	159	16%
Communications	131	14%
Procedures	82	9%
Training	64	7%
Human Engineering	56	6%
Quality Control	27	3%
All other failure modes	166	18%
Total	**964**	**100%**

Hence, the common root cause findings are consistent with the 1990 CMA report indicating 80-85% of the contributing causes of major incidents is related to human reliability. Additional detail on the common root causes is available in a recent publication by the authors [6].

3.3 Implications for Operations Practices

The information obtained through the analysis of 42 major process safety incidents provides a much richer source of information than was obtained in the 1993 foundational study of the abnormal situation management challenge. The foundational study was limited both by the comprehensiveness of the reported human-related contributions to incidents and the reluctance of individuals to report human reliability sources.

This most recent analysis indicates many of the practice areas identified in the 1993 foundational study of abnormal situation management challenges still persist today and are contributing to major process safety incidents. The good news is that in many of these practice areas there are organizations in the industry that have established effective practices to address these process safety risks. The challenge is getting the entire industry to improve the quality of their abnormal situation management practices to reduce the overall industry risk to process safety incidents (PSIs).

4 Reducing Process Safety Risk with Human Reliability Metrics

The ASM Consortium has identified a gap in the capabilities of plant incident reporting systems [8]. Designed to support Process Safety Management (PSM) programs, current incident reporting systems do not support the plant's need to understand and control the bottom layers of the Safety Pyramid – those dealing with near misses, unsafe behaviors, and insufficient operating discipline. The ASM Consortium believes that changes are needed in the process industry's capability to identify, analyze and measure sources and impacts of abnormal situations in the plant. Moreover, advances in current practices are necessary for the successful integration of metric usage with daily work processes within the operations group to enhance operating team performance in anticipating and minimizing the impact of abnormal situations.

Because of this concern, the ASM Consortium has taken a keen interest in the recent developments of the CCPS to improve the industry's metrics on process safety [16]. The ASM Consortium sees an opportunity to improve the process industries alignment on metrics for process safety and abnormal situation management, particularly in the area of near-miss indicators and leading indicators.

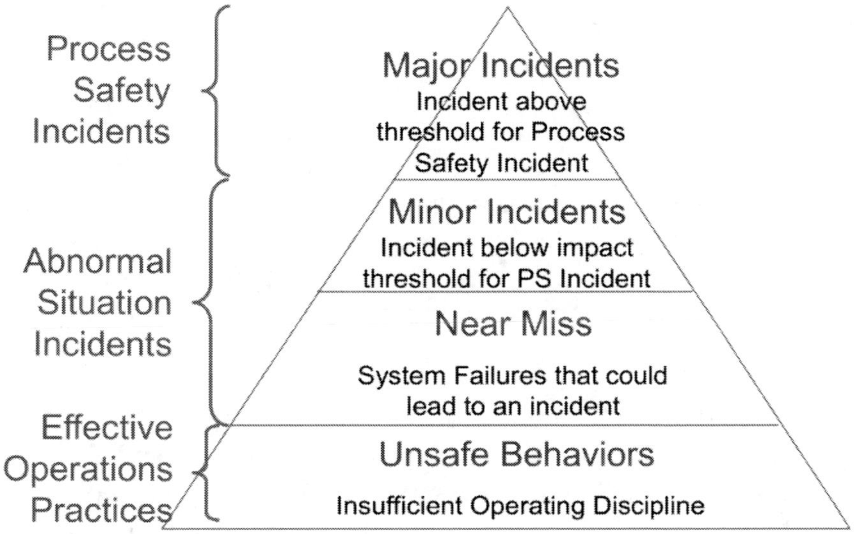

Figure 3. Relation of Abnormal Situation Management to Process Safety Management illustrated with the Process Safety Pyramid [16]

Figure 3 illustrates the relation between abnormal situations and process safety incidents with the Process Safety Pyramid [16]. At the bottom of the pyramid is unsafe behaviors characterized as insufficient operating discipline. The ASM Consortium's *Effective Operations Practices* [15] specifically address operating discipline that is influential in preventing and responding to abnormal situations. At the second and third level of the pyramid, there are near miss events and minor incidents representing system failures that can lead to a PSI or have an impact below the threshold criteria for a PSI. Abnormal situations represent near-miss or minor events that may have the potential to escalate into a major incidents.

5 Near-Miss Indicators

The conclusion of the CCPS committee was that a near-miss metric must have three essential elements based on the review of industry trends in near-miss event reporting [16]:

- an event occurs, or the discovery of a potential unsafe situation
- the event or unsafe situation had a reasonable potential to escalate
- the potential escalation would have led to adverse impacts

Hence, the following definition is offered: "A near-miss is an undesired event that under slightly different circumstances could have resulted in harm to people, damage to property, equipment or environment or loss of process." [16, Page 36]

While this definition potentially applies to personnel and process safety, a process safety specific definition is proposed by the CCPS committee as the following: "A process safety near miss is (A) any significant release of hazardous substance that does not meet the threshold for a PSI lagging metric, or (B) a challenge to a safety system, where challenges to safety systems can be divided into one of three categories:

1. Pressure Relief Device (PRD) challenge,
2. Safety Instrumented System (SIS) challenge, or
3. Process deviation or excursion."

The first component of the definition is fairly obvious because it is equivalent in nature to the PSI only at a lower threshold of impact. The second component defined as safety system challenges is quite logical and appropriate since the PRDs and SIS are critical elements of the mechanical barrier established to prevent process safety incidents. The third category may not be as obvious and potentially challenging for organizations to implement. However, from an ASM perspective, this category is equally important and comprises a key link between abnormal situation management and process safety management.

As stated above and defined as part of 1993 foundational study, an abnormal situation is a disturbance or series of disturbances in a process that cause plant processes to deviate from their normal operating state and operations intervention is necessary to return to normal operating state. The abnormal situation management objective is to return the plant to normal before the process safety systems are challenged. Once these systems are challenged the operations shifts from abnormal situation management to emergency response management. Consequently, the ASM Consortium perspective on process deviations is that if abnormal situation management practices are ineffective the situation has the potential to progress into a process safety incident.

It is true that not all process deviations or excursions from normal operating limits will lead to a PSI. However, it can be argued that any deviation or excursion that is classified as an alarm in an effectively rationalized alarm system and that is brought back within normal operating limits could be counted as a near miss. Figure 4 below illustrates the role of an effectively design alarm system relative to an ineffectively designed system.

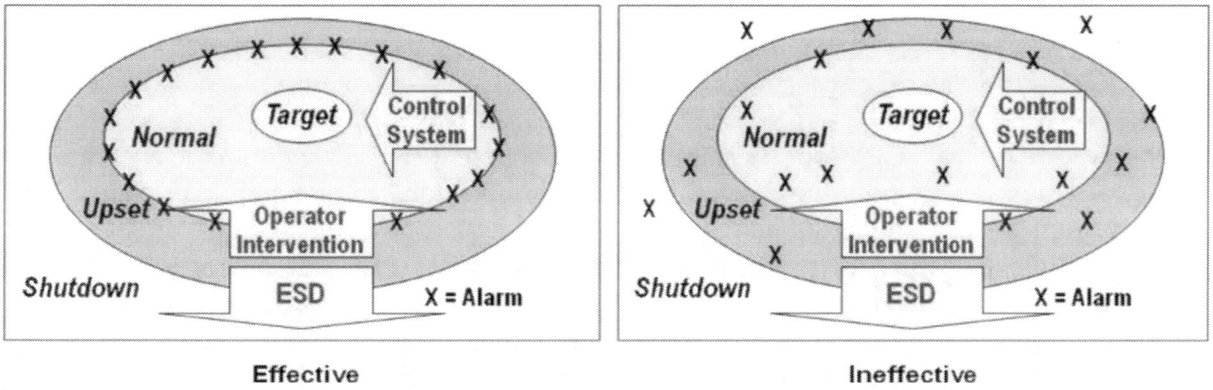

Figure 4. Illustration of the difference between an effectively designed alarm system and an ineffectively designed alarm system (from EEMUA [17])

The "normal" region in Figure 4 represents process deviations that the control system can manage. The "upset" region represents process deviations that require operator action to bring the process back to normal. It has been established in alarm management circles such as EEMUA and the ISA SP18 Committee that a situation calls for an alarm only if specific operator action is required as a response. If timely operator action is not required, other notification mechanisms may be used for operator communication (such as operator alerts and messages) within the "normal" region; but such alerts or messages should not be classified as alarms, in order to avoid operator confusion on urgency. Hence, in an effectively rationalized alarm system, the count of alarms represents the count of transitions into the "upset" region.

It is true that alarms can be configured to protect the process from potential problems other than safety, such as (non-safety) environmental exceedance, product quality, equipment life, and economic objectives. However, the mechanisms of control, loss of control, and recovery are essentially the same for all these potential problems. Hence it can be argued that any loss of control detected "under slightly different circumstances" could result in a safety incident, and thus should be categorized as a near miss.

As with other metrics, alarm rate as a near-miss metric should be normalized in some way to enable comparison across control rooms and across plants. Our recommendation is to calculate the average alarm rate using the number of alarms associated with a console position over a specific time period such as 10 minutes. ASM research has established that a console operator can effectively respond to a peak rate of about 10 alarms in a 10-minute period [18]. The current industry defacto standard on alarm system configuration published by the Engineering Equipment and Materials Users Association [17] asserts that an average alarm rate of more than 1 alarm per minute is "very likely to be unacceptable", that 1 alarm per 5 minutes is "manageable", and that less than 1 alarm in 10 minutes is "very likely to be acceptable". Because operators have other tasks besides responding to alarms, and since operating philosophies differ between sites, the most important strategy for a site is to establish reasonable goals that are consistent with their operating philosophy and to track to those goals on a regular basis.

For this approach of using alarm rates as a near-miss metric to be effective, the production organization needs to have an alarm management system that includes an alarm rationalization methodology capable of producing an "effective" as opposed to an "ineffective" alarm system as shown in the Figure 4. In fact, the absence of an effective alarm system can be taken as a leading indicator of process safety risk; and perhaps should be considered in the Safety Culture evaluation under Leading Metrics.

An observation on the CCPS process safety metric document [16] is that with the exception of the "Errors of Omission/Commission" category, the examples tend to be equipment-centric, in that, (1) the examples tend to point out human error in the installation or maintenance of equipment and (2) the examples describe the problem in terms of the state of the equipment but not the work processes or operation of the equipment. Whereas, the CCPS focuses on the Process Safety Management System, the ASM Consortium perspective provides a slightly broader consideration of all Operations Management Systems that can impact operations ability to prevent and respond to abnormal situations. Thus, from a human reliability perspective, consideration should be given to failures of the procedure and training management systems as well.

6 Leading Metrics

The objective of leading metrics is to provide an early indication of any deterioration in the effectiveness of the safety systems or Safety Management Systems and enable remedial actions to restore safeguards to prevent the occurrence of PSIs. The CCPS leading metrics proposed are based on the assessment of hazards inherent in operations, critical causal factors from major incidents [16]:

1. Maintenance of Mechanical Integrity
2. Action Items Follow-up
3. Management of Change (MOC)
4. Process Safety Training & Competency (& training competency)
5. Process Safety Culture

From an ASM perspective, there is some overlap with the ASM Consortium recommendations for effective operations practices. As reported above (Table 3), the ASM Consortium research studies provide findings on common operations practice failure modes associated with major process safety incidents. A comparison between the recommended leading metrics to the top 10 operations practice failure modes indicates some additional areas for leading metrics consideration.

In the last column of Table 5, we present our assessment of the quality of coverage by the leading metrics defined in the CCPS pamphlet relative to the top 10 failure modes from the ASM Consortium root cause analysis study. Since we did not have a chance to review the Safety Culture survey, we were unable to characterize the coverage of issues identified in our study. The assessment looked at four different aspects of the proposed metrics to characterize coverage of relevant aspects of the management system:

1. Scope of coverage
2. Timeliness and efficiency
3. Effectiveness
4. Compliance with stated policy/practice

Table 5. The top 10 failure modes in rank order of impact in relation to coverage by the CCPS five critical causal factors

ASM Operations Failure Modes	CCPS Critical Causal Factors	Coverage Completeness
1. Lack of Comprehensive hazard analysis & communications	2. Action item follow-up	Partial
2. Ineffective first line leadership Roles	5. Safety culture	Unknown
3. Ineffective continuous improvement program	2. Action item follow-up	Partial
4. Poor safety culture	5. Safety culture	Unknown
5. Ineffective task communications	None	None
6. Lack of comprehensive MOC	3. MOC	Good
7. Failure to establish training for all modes of operation	4. Process safety training	Partial
8. Ineffective cross-functional communications	None	None
9. Failure to ensure compliance on use of procedures	Process safety competency	Partial
10. Lack of effective design guidelines and standards	None	None
N/A	Maintenance of Mechanical Integrity	Good

A key finding was that there is some overlap between the top 10 failure modes identified in the ASM Consortium study with the five leading metric categories proposed in the CCPS pamphlet. The one area not identified in our top ten was mechanical integrity. However, the ASM Consortium would agree with the value of including the mechanical integrity metric category.

One significant category not addressed in the leading metrics is communications as illustrated by three of the top 10 failure modes: #1. Lack of comprehensive hazard analysis & communications program, #5. Ineffective task communications, and #8. Ineffective cross-functional communications. These three failure modes reflect breakdowns in communications between shift teams, within shift teams, and between operations, management and maintenance. Communications between plant personnel is fundamental to effective operations under normal, abnormal, and emergency situations. It is often taken for granted that people know what, when, and how to communicate effectively. In fact, as abnormal plant conditions evolve effective communication often makes the difference between success and failure in the intervention.

Successful communication involves the transmission of a message between at least two parties, where the receiver of the message correctly comprehends the intended meaning of the sender and the sender knows that the message was received and understood. In critical situations, this requires that those involved share the same assumptions about the plant conditions, their goals, and their options. A failed communication can contribute directly to the initiation of an abnormal condition as well as to an unsuccessful intervention via an inappropriate response, slowed response, or no response.

One other operations practice category not addressed in the leading metrics is Process Control, Monitoring and Support Applications. In this category, the common operations failure mode of #10. Lack of effective design guidelines and standards was found. Process monitoring, control and support applications comprise the software and hardware platforms used by the operations team such as operator graphic displays, alarm system configuration, regulatory control systems, advanced control, safety-instrumented systems, programmable logic controllers, SCADA, communications, equipment diagnostic and monitoring applications, procedure, and logbook applications. Process monitoring and control applications provide console operators the primary means of maintaining situation awareness through visualizing and interacting with the plant process. Operations support applications are becoming more pervasive as many paper-based functions are migrating to electronic-based solutions such as training, lab reporting, logging, procedures, communications, and diagnostics. As the number, complexity and importance of monitoring, control and support applications increase, there is a corresponding need for more formal, consistent work processes for engineering and IT functions to ensure that these applications deliver the intended benefits and do not make the operations task more difficult.

7 Conclusions

Abnormal situations are managed by prevention, early detection, and mitigation, in order to reduce unplanned outages and process variability that increase production, safety and environmental risk to plant employees and local communities. Historically, mechanical integrity is important to process safety, however the ASM Consortium's focus on the operations team has led to an increased emphasis on operational integrity to reduce the likelihood of abnormal situations and ultimately improve process safety and plant performance. Hence, the perspective of the ASM Consortium on process safety metrics has focused on the impact of operations practices on plant process safety.

The ASM Consortium has a strong interest in the recent developments of the CCPS to improve the industry's metrics on process safety. The ASM Consortium sees an opportunity to work together with CCPS to improve the process industries alignment on metrics for process safety and abnormal situation management, particularly in the area of leading indicators and near-miss lagging indicators.

The CCPS-proposed near-miss process safety category of process deviation or excursion comprises the key linkage between abnormal situation management and process safety management. In fact, we propose that the plant alarm system may be an effective basis for capturing metrics related to critical process deviations or excursions.

The ASM Consortium has been studying operations management systems and practices to identify key challenges to effective prevention and response to abnormal situations – some of which may have the potential to escalate into process safety incidents. The Consortium's understanding enables it to offer some potential enhancements to the leading metrics that are currently based on process safety management systems.

Finally, from an abnormal situation management perspective, the proposed lagging metrics are appropriate and adequate for tracking process safety impacts. While it is important to have good metrics for measuring process safety incidents, the categories for causes are equally important to establish a common understanding of where continuous improvement is required. This is an area where we have witnessed significant variability between organizations. Moreover, there should be some relationship between the causal categories and the leading metrics, if in fact; the leading metrics are indicating opportunities to reduce the risk of process safety incidents. Consequently, we suggest the CCPS also considers addressing some key root cause categories to help gain some consistency in understanding contributors to PSI occurrences.

8 Acknowledgements

This paper was funded by the ASM® Consortium, a Honeywell-led research and development consortium. The Abnormal Situation Management® (ASM®) Consortium (www.asmconsortium.com) is a long-running and active research and development consortium of 16 companies and universities concerned about the negative effects of industrial plant incidents. The consortium identifies problems facing plant operations during abnormal conditions, and develops solutions. Deliverables from the collaboration among member companies include products and services, guideline and other documents, and information-sharing workshops; all incorporating ASM knowledge. Also, the authors acknowledge the technical contribution of Doug Metzger in the ASMC analysis of the relation of process system alarms and the CCPS process safety metrics.

9 References

[1] E.L. Wiener, *Human factors of advanced technology ("glass cockpit") transport aircraft*, NASA Contractor Report 177528, Ames Research Center, Moffett Field, CA, 1989.

[2] E.L. Wiener, and R.E. Curry, Flight-deck automation: Problems and promises, *Ergonomics*, *23(10)*, (1980), 995-1011.

[3] L. Bainbridge, Ironies of automation, *Automatica*, *19(6)*, (1983), 775-779.

[4] P. T. Bullemer, *Towards an understanding of abnormal situation management*, Honeywell Technical Report, Honeywell International, Minneapolis, MN, 1994.

[5] D.K. Lorenzo, *A manager's guide to reducing human error*, Chemical Manufacturers Association, Washington, DC, 1990.

[6] P. T. Bullemer, and D. C. Reising, Managing human reliability: An Abnormal Situation Management historical perspective, *Proceedings of the Mary Kay O'Conner Process Safety Center International Symposium*, College Station, TX, 2015.

[7] E. Cochran and P. Bullemer, Abnormal Situation Management: NOT by new technology ALONE, *Proceedings of the 1996 AICHE Conference on Process Plant Safety*, Houston, TX, 1996.

[8] P. T. Bullemer, Better metrics for improving human reliability in process safety, *Proceedings of the 11th Process Safety Symposium at the 5th Global Congress on Process Safety*, Tampa, FL, 2009.

[9] J.C. Laberge, P.T. Bullemer and D. Whitlow, Communication and coordination failures in the process industries, *Proceedings of the 52nd Annual Meeting of the Human Factors and Ergonomics Society*, New York, NY, 2008.

[10] P. T. Bullemer and J.C. Laberge, Common operations failure modes in the process industries, *Journal of Loss Prevention in the Process Industries*, *23*(*6*), (2010), 928-935.

[11] P.T. Bullemer and D.C. Reising, *Extend root cause analysis*, ASM Consortium Research Report, ASM Consortium, Minneapolis, MN, November 12th, 2012.

[12] P.T. Bullemer, L. Kiff and A. Tharanathan, Common procedural execution failure modes during abnormal situations, *Journal of Loss Prevention in the Process Industries*, *24* (*6*), (2011), 715-916.

[13] P.T. Bullemer and D.C. Reising, *Addressing the process safety challenges associated with heater operations in the process industry*, Presented at the 19th Annual Meeting of the European Refining Technology Conference, Lisbon, PT, 2014.

[14] M. Paradies and L. Unger, *TapRoot®: The system for root cause analysis, problem investigation, and proactive improvement*, System Improvement, Inc., Knoxville, TN, 2000.

[15] P.T. Bullemer, *Effective operations practices*, ASM Consortium Guideline Document, ASM Consortium, Minneapolis, MN, 2014.

[16] CCPS *Process Safety Leading and Lagging Metrics: You don't improve what you don't measure*, American Institute of Chemical Engineers, New York, NY, 2011.

[17] EEMUA Publication No. 191, *Alarm systems: A guide to design, management and procurement* (3rd Ed.), Retrieved from http://www.eemua.org, 2013.

[18] D.C. Reising, J. Downs and D. Bayn, Human performance models for response to alarm notifications in the process industries: An industrial case study, *Proceedings of the Human Factors and Ergonomics Society 48th Annual Meeting*, Human Factors and Ergonomics Society, Monterey, CA, 2004.

12TH GLOBAL CONGRESS ON PROCESS SAFETY

Security and Risk Mitigation: Practical and Cost-Preferred Solutions by the U. S. Nuclear Industry

Gary Hayner
Manager, Safety and Reliability
Jensen Hughes
2015 S. Bascom Ave, Suite 350
Campbell, CA 95008 USA
ghayner@jensenhughes.com

Prepared for Presentation at
American Institute of Chemical Engineers
2016 Spring Meeting
12th Global Congress on Process Safety
Houston, Texas
April 11-23, 2016

Security and Risk Mitigation: Practical and Cost-Preferred Solutions by the U. S. Nuclear Industry

Gary Hayner
Manager, Safety and Reliability
Jensen Hughes
2015 S. Bascom Ave, Suite 350
Campbell, CA 95008 USA
ghayner@jensenhughes.com

Keywords: nuclear, security, emergency preparedness

Abstract

This paper will examine and discuss the challenges faced by the U.S. nuclear industry after 9/11 and the Fukushima accident, the process to determine the appropriate path to success, and the solutions themselves. It will show that the solutions were effective, useful, and cost-effective. It is possible the fuel and petrochemical industry may glean valuable lessons from the experience of the nuclear industry regarding security and risk mitigation.

1 Introduction

The highly regulated United States nuclear industry must balance safety, performance, and cost. Safety to the public, competition with other energy producing industries, and the large capital and O&M costs of nuclear are exaggerated compared to other industries. Among these clashing constraints is the intersection of plant safety and security. Many view security as taking away from the bottom line of a plant's performance; however, the current global climate dictates the need to examine critical infrastructure security in a dynamic and effective manner while minimizing the cost and operational impacts to highly complex systems.

The nuclear industry has successfully navigated this challenge in the last 15 years. While the almost 100 operating nuclear plants in the U.S have been spent collectively billions on security, emergency response, and beyond design basis events, the process to determine the requirements and the results can be labeled as success. There have also been residual benefits that the industry is now realizing.

A combination of several factors, such as early industry involvement, engaged stakeholders, and regulatory negotiation, have allowed the nuclear plant operators and

U.S. nuclear regulator to conclude in a mutually-agreeable, reasonable, and effective response to the need for safety and security change.

While the petrochemical industry is notably less regulated in this area than the nuclear industry, the lessons learned by the nuclear industry throughout this process may benefit the petrochemical industry in establishing more consistency and effectiveness in the safety and security area in a cost-effective manner. The massive number of petrochemical facilities in the U.S. and across the globe dwarfs the number of nuclear plants in the same locations, and face similar threats.

The challenges faced by the U.S. nuclear industry after 9/11 and the Fukushima accident, the process to determine the appropriate path to success, and the solutions themselves may provide useful insights and benchmarking to the petrochemical industry for safety, security, and emergency response, as well as regulatory interface and management.

2 Background

Two major events have occurred in the last 20 years which have profoundly affected the U.S. nuclear industry – the terrorist attacks of 9/11/2001 and the nuclear disaster at Fukushima, Japan in 2011. These events have caused waves within the industry that are still felt today through regulatory, operational, and design changes. The uncertainty surrounding the events, particularly in the periods just after the events, led to a variety of responses across the spectrum. Some believed nothing needed to be done. Others believed that a massive overhaul of the regulatory process and physical plants were required. Between these extremes were most opinions which desired a balance between effectiveness, usefulness, and near and far term costs.

The Nuclear Regulatory Commission (NRC) is the federal regulator of the U.S. nuclear industry. It consists of five Commissioners appointed by the President and confirmed by the U.S. Senate for five year terms. The primary decision making tool is written correspondence from the NRC staff to the Commission. This includes policy, security, rulemaking, and adjudicatory matters. The Staff will typically engage the industry in some degree during this process, if only to get feedback on issues and potential recommendations. The independence between regulatory and industry may be categorized as high, but with a respectful consideration of one another.

The NRC develops the regulations with which the nuclear licensees must comply. They also develop some guidance documents and standards, provide oversight, identify and resolve generic issues, and perform support activities for decisions (e.g., research, risk assessment). They typically perform a regulatory analysis of cost and benefit for regulatory positions.

3 Discussion

Challenges

There are approximately 100 nuclear operating plants in the U.S. The energy generation industry is extremely competitive, operating in both regulated and deregulated environments. Nuclear plants have a very low profit margin compared to other power producers, such as many fossil plants. Nuclear plants have very high initial construction costs capital, often amortized over the life of the plant, as well as high operation and maintenance costs. Plant modification costs are consequently also high, particularly when considering the stringent design and construction controls and reviews for safety-related systems. Plant owners and utility shareholders desire to avoid additional costs as much as possible if they do not add to increased output or reduced cost. The types of regulations imposed by Fukushima and 9/11 are generally "overhead" costs which detract from the bottom line.

Both Fukushima and 9/11 made international headlines and echoes of those events are present in everyday life to this day. The public and political response, often knee-jerk, demanded a quick response, especially for critical infrastructure. It is no secret that the general perception of nuclear power is "scary" with high consequences. Nuclear plants are considered a high value target to terrorists and with each unit averaging almost 1500 MWe, they are considered critical infrastructure. A quick response must be balanced with a reasonable, effective, and useful solution to avoid losing time, cost, or stakeholder confidence.

As 9/11 and Fukushima were examined in further detail with respect to the nuclear industry, it was found that the potential threats, either man made (i.e., terrorism/sabotage) or natural (i.e., tsunami, earthquake) have a very large set of possible circumstances and definition of specific threats is not realistic. For example, the initial approach for 9/11 response was to defend against an aircraft attack similar to that used for 9/11. Further examination of the problem statement concluded that the size of the aircraft may vary, the speed of the aircraft may vary, or the weight or approach characteristics may also change. Then terrorist tactics were investigated, and the permutations associated with this were endless. The use of ground attacks, the number of adversaries, the adversary characteristics, and other elements were so varied that it became clear that a design solution to mitigate all possible threats was not realistic or feasible. The "what if's" for defining initial conditions or causes of events becomes endless. In order to solve the issue, a reasonable box must be drawn around the problem.

The NRC is a regulator independent of the industry, but still evaluates the cost effectiveness, benefit, and necessity of regulatory policies. The NRC must follow its own rules when developing new requirements.

Process

The NRC and the Nuclear Energy Institute (NEI), representatives of the nuclear utilities, perform independent research on issues. This includes threats, technical solutions, bounding conditions, and impacts to the population of affected plants. There are often both formal and informal discussions between NRC and NEI during this period, with both groups normally sharing research to reach a realistic basis for decisions. NEI may provide feedback on impact to the licensees, yet ultimately the NRC will decide on its approach independent of the industry. NRC may provide some guidance for implementation of orders, or may rely on NEI to write such guidance and then endorse that guidance.

Whatever the NRC's position may be, an implementation timeframe accompanies any orders. Plants may have 30 days or 24 months to implement requirements. During this period, plans may be submitted to NRC for verification. Subsequent to implementation, NRC may perform inspections or audits. The NRC may accept exceptions to any requirements but normally requires extensive technical supporting information.

Solutions

Through extensive research, technical studies, and discussions, the industry and the regulatory body came to consensus on an approach. The use of realistic, best-estimate calculations, unlike the conservative, specific design-basis calculations typically performed in the past, resulted in a reasonable picture of consequences and expected actions. A measure of relative risk and risk improvement for such changes can also be quantified and compared.

The defense-in-depth concept was appealing and ultimately implemented. This approach minimized the impact to existing plant configuration. Flexibility was also desired to avoid limiting the assessments to specific scenarios. It was desired to have the solutions be as independent from the initiating event as possible to maximize its usefulness.

From 9/11 the industry response evolved into what was called "B.5.b." A quick response was needed due to the sudden new threat vectors and public and political pressure. There were many revisions to physical security regulations with commensurate spending, but there was also a need for operational and emergency response revisions to meet these new threat scenarios. It couldn't be said that a security event would not occur, and the plant had to have an integrated response plan. B.5.b was the operational and emergency preparedness response and had a "broader" application than just a security-related incident. Initially, it was defined as a loss of large area of the plant due to fire or explosion (i.e., aircraft crash). This later evolved into an undefined, nonspecific threat. The industry responded to the short and general orders with sets of 10 to 15 plant procedures and limited new equipment (portable pumps, hoses) along with minor modifications (connection points to existing systems). Diversity and mitigating strategy entry points are not specific.

From Fukushima the industry response was called FLEX (Flexible and Diverse). A task force was established in response to Commission direction to conduct a systematic and methodical review of U.S. Nuclear Regulatory Commission processes and regulations to determine whether the agency should make additional improvements to its regulatory system and to make recommendations to the Commission for its policy direction, in light of the accident at Fukushima. This task force concluded that, among other things, the sequence of events like the Fukushima accident is unlikely to occur in the United States and some appropriate measures have been implemented. The task force then concluded that that a more balanced application of the Commission's defense-in-depth philosophy using risk insights would provide an enhanced regulatory framework that is logical, systematic, coherent, and better understood. The NRC recognized that there must be a balance between voluntary industry initiatives and regulatory actions. Some key actions included re-examination of the seismic and flooding studies for all plants, as well as developing revised strategies for extended losses of AC power to prevent or mitigate the consequences of such events. The industry pushed back on some recommendations, such as installing filtered containment vents. These were viewed as minimally effective with a high cost. Eventually, the NRC concurred on this issue. Nonetheless, the industry has recognized the importance of such advanced planning for such low likelihood, high consequence events. The use of procedural revisions, portable equipment, and relatively minor plant modifications resulted in a diverse, flexible mitigation strategy that extended well beyond the intended scenarios.

Results, Effects, and Benefits

Although each nuclear plant has spent tens of millions of dollars on the response, the use of portable equipment and regional response centers still helped to mitigate the costs of implementing these solutions. For example, the installation of a permanent water injection source for core cooling would include design, procurement and fabrication of components, installation, testing, procedure development, and training. All of this would be performed on safety-related systems under rigid quality assurance and control programs. A plant modification of this type may cost upward of $25 million and take several years to plan and implement. The flexible approach using portable equipment utilizes a simpler design which reduces design costs. The portable pumps are typically off the shelf purchases which result in lower costs than custom design components. The physical modifications to plant systems are also less. A permanent system would require complete piping systems, tie in to existing systems, connections to power and control, and all while working in a contaminated area. The portable equipment requires only a connection point to the existing plant system. This requires a much simpler plant modification of installing a blank flange or small piping run with isolation capability. The procedures used for the portable systems are higher level, and due to the components, are inherently simpler and shorter. This results in cheaper implementation and training costs. Long term inspection and maintenance of the portable equipment is also less due to less stringent quality requirements on the equipment. Overall cost for the flexible designs are estimated at $5 to $10 million per plant, which is a significant savings.

Nuclear plants rely on risk assessments and risk can be greatly impacted by defense in depth. Daily plant operations rely on risk insights to support safe maintenance planning. NRC performs risk-informed decision making and risk-informed enforcement. A benefit of the FLEX and B.5.b initiatives has been reduced risk profiles for plants. The results include more flexible operation, more confidence in operating principles, and cost reduction due to improved risk margins for systems and components.

While the B.5.b and FLEX initiatives were directed at specific scenarios, additional benefit to the plants was later realized. For example, it was found that the B.5.b pumps and equipment provided a means to add water to cool the reactor core or spent fuel pool in events other than security-related events. This additional defense in depth extended beyond the intended scenario, and the benefit of having an additional plant system to provide makeup water to such vital plant components reduced the overall risk of the primary systems as well as the cumulative risk of the plant based on detailed probabilistic risk analyses. The FLEX equipment provided similar benefits. This additional risk margin is beneficial to the plant for providing margin for maintenance windows, outage times, inspection and surveillance frequencies, and system management.

The application of these new capabilities was further realized during emergency preparedness drills. These drills often simulate loss of many primary systems, and exercise the operator and support personnel response. It was found that these beyond design basis procedures provided additional resources and methods to perform key safety functions. Classroom training on these procedures is required, but they are not normally used during live simulator exercises or full scale drills. They have been incorporated more into such training elements because of the realized benefits.

Full scale drills, including offsite responders such as local law enforcement, FBI, DHS, and others, are now required for all plants. The command and control enhancements included in the B.5.b efforts allow effective management of numerous organizations during such emergency events.

Comparison with Petrochemical Industry

The petroleum and petrochemical industry are less regulated in the security and risk area than the nuclear industry. For most petroleum facilities, the risk and security assessment is developed and implemented by the respective facility owner. Although API has issued several guidance documents with best practices for the industry for performing such assessments, there is still variability within the industry.

Additional regulation upon the petroleum industry in this area would not be well received, and is an impractical approach. The threat matrix for such facilities, given the diverse locations across the globe, require a more flexible and realistic approach where a wide variation of capabilities can be applied. This minimizes cost and still provides reasonable assurances of risk mitigation and public safety. Sabotage or attack on a single facility does not have the degree of consequence of the same on a nuclear plant. Also, the public perception of such an event is much less extreme for a petroleum facility. A consistent approach to risk mitigation within the petroleum may also provide a solid contingency plan for the industry in response to any future events to manage public, media, and political responses. The image that the industry is united and in control during and after such events is a powerful message. This is a cornerstone of NEI for the nuclear industry.

4 Conclusion

The defense in depth philosophy of the NRC and the nuclear industry is a beneficial approach to risk and security management and can be applied to the petroleum industry. The nuclear industry solutions were effective, useful, and cost-effective. The use of flexible, portable equipment saved tens of millions of dollars for each site compared to the installation of permanent plant systems. It was also found that the flexible solutions provided additional benefits in improving risk margins which allows reduced plant maintenance and inspection requirements, as well as additional plant emergency response layers of defense which allows the plant to better respond to severe accidents or plant events. It is possible the petroleum and petrochemical industry may glean valuable lessons from the experience of the nuclear industry regarding security and risk mitigation.

5 References

[1] NEI 06-12, "B.5.b Phase 2 & 3 Submittal Guideline," Revision 3, July 2009.

[2] NEI 12-06, "Diverse and Flexible Coping Strategies (FLEX) Implementation Guide," Revision 0, August 2012.

[3] "Recommendations for Enhancing Reactor Safety in the 21st Century – The Near-Term Task Force Review of Insights from the Fukushima Dai-Ichi Accident," U.S. NRC, July 12, 2011.

[4] American Petroleum Institute, "Security Guidelines for the Petroleum Industry," April 2005.

[5] American Petroleum Institute, "Security Risk Assessment Methodology for the Petroleum and Petrochemical Industries," ANSI/API Standard 780, First Edition, May 2013.

[6] American Petroleum Institute, "Security Vulnerability Assessment Methodology for the Petroleum and Petrochemical Industries," May 2003.

Compressor System Performance at Abnormal or Non-Design Conditions

Presenting Author:

Dylan Grosscup
Consultant
dylangrosscup@gmail.com

Co-Authors:

Carl D Ramlakhan
Director – Engineering
Atlantic LNG Company of Trinidad and Tobago
cramlakhan@atlanticlng.com

Christian Jarvis
Operations Specialist-Project Engineering and Optimization
Atlantic LNG Company of Trinidad and Tobago
cjarvis@atlanticlng.com

Charles Lea
Senior Process Safety and Risk Management Consultant
ioMosaic Corporation
lea.c.mn@iomosaic.com

Prepared for Presentation at
American Institute of Chemical Engineers
2016 Spring Meeting
12th Global Congress on Process Safety
Houston, Texas
April 10-13, 2016

Abstract

When attempting to optimize facility performance or determine the effectiveness of process safety systems, accurate evaluation of compressor performance at conditions deviating from design is critical. This paper presents a joint study involving the Atlantic LNG Company of Trinidad and Tobago, the ioMosaic Corporation, and an independent contractor. At the Atlantic LNG facility, a large feed gas compressor was tested at varying inlet conditions and the operating data were analyzed to determine the impact on compressor performance. A comprehensive data set spanning 2 years and more than 400,000 measured conditions were used to develop a method combining empirical data and design calculations to optimize compressor performance and provide a framework for evaluating compressor safety.

A validation of 3 methods of predicting compressor performance (the Isentropic Energy Balance Model, the Ideal Polytropic Model, and the Real Polytropic Model) was performed to determine which method/s are most suitable for process safety/optimization work. The analysis considered both, the fundamental and empirical bases. A mathematical model was derived for projecting the behavior of the polytropic head and efficiency curves, provided by the compressor manufacturer, based on numerical integration and data residualization techniques. Then, the data set were analyzed to find behavioral patterns which may indicate potential hazardous conditions or deviations from efficient operation. Additionally, a detailed list of hazardous scenarios, design considerations, and operating practices that have resulted in safety incidents in current industry applications are given for consideration during compressor system design and evaluation.

It was found that, under optimal conditions, the Isentropic Energy Balance Model and the Ideal Polytropic Model predict exactly the same results for compressor performance. It was found that the polytropic head and efficiency curves furnished by the vendor are only accurate at certain, optimal, suction pressures and that these optimal pressures are only consistent with the reference pressure on the vendor supplied curves at very specific operating points. Furthermore, it was found that deviations from optimal suction pressure result in a reduced head and flow capacity of the compressor. The pattern of inefficiency is consistent with operating conditions that have been shown to lead to equipment failure and loss of containment from vibration induced fatigue in existing compressor system applications. The final result of the study found that it is possible to use a combination of operating data and design calculations to generate suction pressure curves that will improve safe operations, optimal efficiency, and can be used by operators while the system is online.

1. Introduction and Previous Work

Since 1698, when British inventor Thomas Savery patented the first steam engine, compressors have been integral components of virtually every process facility on the planet. As such, compressor performance has been widely studied from both the theoretical and measured operational data aspects. The foundation of modern compressor assessment techniques was published in 1850, in the paper, "On the Moving Force of Heat" by Rudolf Julius Emanuel Clausius. Today Clausius' theories are known as the First Law of Thermodynamics and the Second Law of Thermodynamics and they drive more than just compressor system analysis. A few decades later, in 1869, J. Homer Lane, credited with first publishing the polytropic model in his paper, "On the Theoretical Temperature of the Sun Under the Hypothesis of a gaseous mass Maintaining it's Volume by it's Internal Heat and Depending on the Laws of Gases Known To Terrestrial Experiment." The Polytropic Model, with Ideal Gas assumptions, is the method most commonly used today to assess compressor performance. In 1962, John Shulz wrote, "The Polytropic Analysis of Centrifugal Compressors" in which he derives a modification of the Ideal Polytropic equation, the polytropic work factor, intended to correct for deviations due to non-ideal conditions. Additionally, he develops two compressibility factors, X and Y, to supplement the traditional Z factor, and applies the factors using a method known as the Shulz method. ASME PTC 10-1997, "Performance Test Code on Compressors and Exhausters", first published in 1965, gives guidance on how to determine appropriate compressor testing methods as well as several dimensionless parameters for using mechanical design criteria to predict compressor performance. Methods for designing compressors to meet specific performance criteria have been well explored in the current body of work; however, there is little direction on safe design criteria, and almost nothing related to a more practical holistic approach for performance analysis of an entire system.

This paper presents information, tools, and methods for design/validation of compressor safety systems as well as methods for optimizing performance using a combination of empirical data and design calculations.

2. Compressor Performance Prediction Models

Before abnormal operating conditions can be effectively, and quantitatively, assessed, it must be shown that there exists a mathematical model capable of predicting meaningful and sensible values for compressor performance across a wide range of conditions.

Three calculation methods were analyzed to determine their suitability for predicting compressor system performance at abnormal conditions:

1. Isentropic Energy Balance Model
2. Ideal Gas Polytropic Model
3. Real Gas Polytropic Model

The Ideal Gas Polytropic Model and the Real Gas Polytropic Model are outlined in ASME PTC-10[1]. This section provides a fundamental description of each model as well as the applied calculation method used for this analysis. The Isentropic Energy Balance Model approach defines the system boundaries using and analyzes the system based on the 1st and 2nd laws of thermodynamics[2]. Simplifying assumptions are made and a method is produced to generate predictions of the compressor discharge pressure for given suction conditions, composition, and compressor speed.

2.1 Stagnation Pressure and Temperature Translation

In order to apply the calculation methods mentioned within this paper to operational data it is necessary to convert measured pressure and temperature values to stagnation pressures and temperatures[1]. Per Bernoulli's equation[2]:

$$P_{stagnation} = P_{static} + P_{velocity} + P_{gravitational\ potential} \tag{2.1-1}$$

Pressure sensors typically measure static pressure. Static pressure does not account for velocity pressure from kinetic energy of a flowing fluid, nor the impact of differences in height along the cross sectional area of the pipe (gravitational potential pressure is normally assumed to be radially constant).

Temperature sensors typically measure a temperature that is somewhere between the static temperature and the stagnation temperature. The difference between the measured temperature and the static temperature is defined by a recovery factor (r_f)[1] Manufacturers often publish these values for specific sensors. If the r_f is unknown a value of 0.65[1] can be used based on most sensor configurations.

$$r_f = \frac{H_{measured} - H_{static}}{H_{stagnation} - H_{static}} \tag{2.1-2}$$

2.1.1 Calculation Methodology

1. Find $ke_{measured}$ using $P_{measured}$ and $T_{measured}$.

$$ke = \frac{1}{2}mv^2 \quad (2.1.1\text{-}1)$$

2. Find H_{static} using:

$$H_{static} = H_{measured} - r_f ke_{static} \quad (2.1.1\text{-}2)$$

3. Using a fixed pressure (P_{static}), vary T until the enthalpy satisfies the condition in step 2
4. Now that you have P_{static} and T_{static}, find, $P_{stagnation}$ and $T_{stagnation}$ by performing isentropic translations (it is assumed that $S_{stagnation}$ is constant) from P_{static} and T_{static} until the following condition is satisfied:

$$H_{stagnation} = H_{static} + ke_{stagnation} \quad (2.1.1\text{-}3)$$

2.2 Isentropic Energy Balance Model

2.2.1 Derivation

General Energy Balance[2]

$$\frac{d(mU)_{cv}}{dt} = -\Delta[\left(U + \frac{1}{2}v^2 + gz\right) * m]_{fs} + Q + W \quad (2.2.1\text{-}1)$$

cv = control volume fs = flowing stream

Steady State Energy Balance:*

$$0 = -\Delta[\left(U + \frac{1}{2}v^2\right) * m]_{fs} + Q + W \quad (2.2.1\text{-}2)$$

*assumes the following:

1. Changes in potential energy due to changes in elevation are negligible such that gz = 0.
2. Accumulation of energy within the system is negligible (all the energy moving into the system is moving out of the system) such that $d(mU)_{cv} = 0$.

Definition of Enthalpy[2]

$$H \equiv U + PV \quad (2.2.1\text{-}3)$$

Isentropic Manipulation of Steady State Energy Balance:*

$$0 = -\Delta[\left(H_{is} + \frac{1}{2}v^2\right) * m]_{fs} + W_{is} \quad (2.2.1\text{-}4)$$

is = isentropic

*isentropic manipulation assumes the following:

1. The work is reversible (i.e. no work is lost to inefficiency) such that in $H \equiv U + PV$, PV = 0 therefore U = H
2. The system is adiabatic (i.e. no heat transfer into or out of the compressor system) such that Q = 0
3. Kinetic energy effects should not be assumed to be negligible per ASME PTC 10-1997.E.3.1 which states "The isentropic work for the purposes of this Code is the work done in an isentropic process between the inlet stagnation state and the discharge stagnation state."

Energy Balance Base Equation:*

$$W_{is} = [\left(H_2 + \frac{1}{2}v_2{}^2\right) * m_2 - \left(H_1 + \frac{1}{2}v_1{}^2\right)] * m_1 \quad (2.2.1\text{-}5)$$

*due to leakage rates, often $m_1 \neq m_2$

Leakage Rate[1]

Compressor systems exhibit leakage such that $m_1 \neq m_2$. The compressor test code requires estimation of the leakage ratio[1]. If the design/tested leakage rates are known this should be accounted for in the calculations and, if possible, an estimation of a potentially increased leakage rate at abnormal conditions should be considered as this may lead to increased pressure. If the calculations are being done to assess an existing system for which there is no information about the design/tested leakage rates, and no leakage information is available it may be necessary to assume the leakage rate is 0.

Simplified Energy Balance Equation:*

$$H_2 = H_1 + W_{is} + \Delta ke \quad (2.2.1\text{-}6)$$

*a value from the polytropic head curves may be used here if W_{is} is multiplied by the mass flow rate m. Validity of this assumption is shown in section 3.1.

2.2.2 Calculation Methodology

1. Convert measured/static conditions to stagnation conditions using 1.2
2. Perform isentropic translations from P_1 and T_1 to P_2 and T_2' until the condition in 1.3-6 is satisfied. (T_2' is the temperature that is coincident to the isentropic flash from suction conditions to discharge pressure, P_2.)

2.3 Ideal Gas Polytropic Model

2.3.1 Derivation

The ideal gas polytropic method is based on, as the name implies, approximating the polytropic path of a reversible compression process. A polytropic process must satisfy the following equation[3]:

$$W - \Delta ke = \Delta H \quad (2.3.1\text{-}1)$$

*ke = kinetic energy

The polytropic path describes the relationship between Pressure (P) and Specific Volume (υ) for a polytropic process and can be characterized by the following function[3] (n is defined as the polytropic exponent):

$$-\frac{1}{n} = \frac{\mathrm{P}}{\upsilon}\left(\frac{\partial P}{\partial \upsilon}\right)_S \quad (2.3.1\text{-}2)$$

Rearranging 1.4-2 and integrating (assuming n is constant) gives:

$$n\int\frac{\partial \upsilon}{\upsilon}+\int\frac{\partial P}{P}=\int 0$$

$$n\ln\upsilon+\ln P=const$$

$$P\upsilon^{n}=const \qquad (2.3.1\text{-}3)$$

The polytropic work equation[3] is an approximate solution of the work of compression integral for a reversible (or ideal) and adiabatic work process:

$$W=\int_{P_1}^{P_2}VdP=\left(\frac{n}{n-1}\right)P_1\upsilon_1\left[\left(\frac{P_2}{P_1}\right)^{\frac{n-1}{n}}-1\right] \qquad (2.3.1\text{-}4)$$

Alternatively[1]:

$$W=\left(\frac{n}{n-1}\right)Z_1\mathrm{RT}_1\left[\left(\frac{P_2}{P_1}\right)^{\frac{n-1}{n}}-1\right] \qquad (2.3.1\text{-}5)$$

*R = universal gas constant divided by molecular weight

Rearranging 1.4-3, the polytropic exponent (n) can be found using the following equation[3]:

$$n=\frac{\ln\frac{P_2}{P_1}}{\ln\frac{\upsilon_2}{\upsilon_1}} \qquad (2.3.1\text{-}6)$$

for calculational ease use:

$$\upsilon=\frac{1}{\rho} \qquad (2.3.1\text{-}7)$$

$$n=\frac{\ln\frac{P_2}{P_1}}{\ln\frac{\rho_1}{\rho_2}} \qquad (2.3.1\text{-}8)$$

Solving equation 1.4-5 for P_2 gives the following equation:

$$P_2=P_1\left[\left(\frac{n}{n-1}\right)\frac{W}{Z_1\mathrm{RT}_1}+1\right]^{\frac{n}{n-1}} \qquad (2.3.1\text{-}9)$$

2.3.2 Calculation Methodology:

1. Convert measured/static conditions to stagnation conditions using 1.2
2. Find Z_1, ke_1, and ρ_1 using P_1, T_1, MW_1

3. Perform an isentropic translation from P_1 and T_1 to P_2 (= P_1 * 1.1 for the initial calculation) to find ρ_2
4. Calculate $n_{calculated}$ using equation 1.4-5 with $P_{2initial}$ and ρ_2
5. Find $P_{2calculated}$ using equation 1.4-8, work, W_p - Δke (W_p is from the compressor design curves), and set $n_{guessed} = n_{calculated}$
6. Continue replacing $n_{guessed}$ with $n_{calculated}$ until $n_{calculated} = n_{guessed}$ meaning $P_{2calculated}$ converges to a constant value
7. Perform a new isentropic translation from P_1 and T_1 to $P_{2calculated}$ and redo steps 4-7 until $P_{2calculated}$ converges on a constant value

(this method involves iteratively solving steps 5-6 in an inner calculation and then iteratively solving 4-7)

2.4 Real Polytropic Model[3]

2.4.1 Derivation:

In order to compensate for potential deviation of the polytropic exponent (n) from a perfect isentropic path, the polytropic equation can be modified with a polytropic work factor (f) as follows:

$$W = f\left(\frac{n}{n-1}\right)P_1\upsilon_1\left[\left(\frac{P_2}{P_1}\right)^{\frac{n-1}{n}} - 1\right] \qquad (2.4.1\text{-}1)$$

$$f = \frac{H_2' - H_1}{\frac{n}{n-1}(P_2\upsilon_2' - P_1\upsilon_1)} \qquad (2.4.1\text{-}2)$$

Applying this to the polytropic equation results in the following:

$$W = (H_2' - H_1)\frac{P_1\upsilon_1}{P_2\upsilon_2' - P_1\upsilon_1}\left[\left(\frac{P_2}{P_1}\right)^{\frac{n-1}{n}} - 1\right] \qquad (2.4.1\text{-}3)$$

Solving equation 1.5-3 for P_2 results in the following:

$$P_2 = P_1[\left(\frac{P_2\upsilon_2' - P_1\upsilon_1}{P_1\upsilon_1}\right)\frac{W}{(H_2' - H_1)} + 1]^{\frac{n}{n-1}} \qquad (2.4.1\text{-}4)$$

2.4.2 Calculation Methodology:

1. Convert measured/static conditions to stagnation conditions using 1.2
2. Find H_1, ke_1, and ρ_1 using P_1, T_1
3. Perform an isentropic translation from P_1 and T_1 to $P_{2guessed}$ (= P_1 * 1.1 for the initial calculation) to find ρ_2 and H_2

4. Find $P_{2calculated}$ using equation 1.4-8 and work, W_p - Δke (W_p is from the compressor design curves)
5. Continue guessing a new $P_{2guessed}$ and performing steps 3 and 4 until $P_{2calculated} = P_{2guessed}$

3. Numerical Integration Method of Compressor Mapping

Projection of the compressor polytropic head and efficiency curves is critical in accurately predicting abnormal compressor performance. Due to the irregular nature of polytropic head and efficiency curves, it is difficult to accurately interpolate between speeds. Extrapolation of system behavior to speeds above or below the normal system boundaries is even more difficult. It is often necessary to analyze power requirements during start up conditions as well as safety system design requirements at over-speed conditions. Traditional methods of analyzing the behavior of compressor curves look at changes in the X and Y axis to project or interpolate polytropic values. The traditional methods are time and labor intensive to develop and the predicted values are generally inconsistent with operational data.

The Numerical Integration method of compressor mapping is based on mathematical analysis and can be automated and give stable values at speeds well above the mechanically achievable range.

3.1 Derivation

To assess the mathematical behavior of a system of curves it is necessary to determine both the direction and magnitude of a change in the system. A compressor map (as shown in Fig. 3-1) appears to be a traditional X-Y plot of inlet actual volumetric flow rate and polytropic head or polytropic efficiency. Similar to a topographical map, changes in the X-Y directions of the compressor map are related to 3 variables: inlet actual volumetric flow rate (X), polytropic head/efficiency (Y), and compressor speed (shown in terms of X and Y).

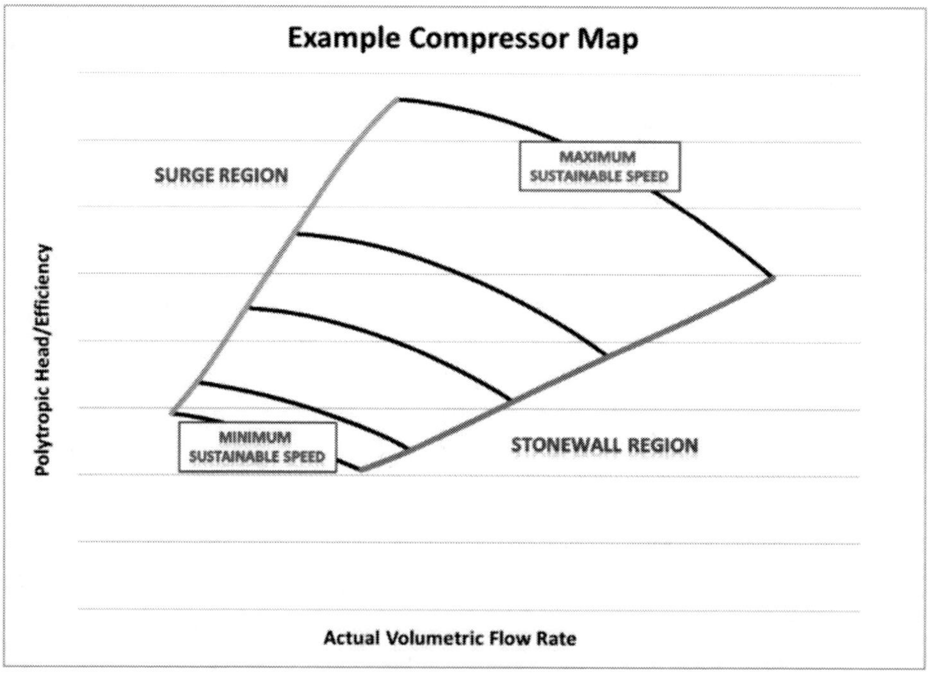

Figure 3-1. An example compressor map depicting the typical regions that define rotating multi-speed compressor performance behavior.

The fundamental concept of this analysis method is rooted in the idea that the shape and length of the individual curves are defined by continuous functions bounded by the same limits. These limits are the surge line and the stonewall line. The surge line is defined as the inlet volumetric flow rate at which the back pressure in the machine exceeds the developed head and a cyclic reversal in flow occurs (this coincides with the point of minimum inlet volumetric flow at which the compressor develops the maximum head for a specific speed). The stonewall point is defined as the point at which the compressor develops a choked or critical flow region at the flow limiting diameter of the machine. At the stonewall point there is a pressure discontinuity on the compressor discharge that acts as a physical barrier to flow based on the sonic velocity of the fluid at the conditions in question (this coincides with the maximum inlet actual volumetric flow rate at a specified speed). It is assumed that limitations based on thermodynamic properties will consistently define the relationship between inlet volumetric flow rate and polytropic head at constant speed.

The functions governing the mathematical tendencies of a set of irregular can be expressed as:

$$f(polytropic\ head, inlet\ volumetric\ flow\ rate, compressor\ speed) = \frac{\mathrm{W}_p}{\mathrm{V}}\left(\frac{\partial \mathrm{W}_p}{\partial V}\right)_{Speed} \qquad (3.1\text{-}1)$$

Integration of 3.1-1 yields:

$$\int f(inlet\ volumetric\ flow\ rate, compressor\ speed) = \int_{V_{surge}}^{V_{stonewall}} \frac{\mathrm{W}_p}{\mathrm{V}}\left(\frac{\partial \mathrm{W}_p}{\partial V}\right)_{Speed} \qquad (3.1\text{-}2)$$

The Numerical Integration method of compressor mapping assumes 3.1-2 can be solved numerically by the common distance equation at very small intervals to find the length of a fixed speed best fit line on the compressor map.

$$length = \sum_{V_{surge}}^{V_{stonewall}} \sqrt{(V_2 - V_1)^2 + (W_{p2} - W_{p1})^2} \qquad (3.1\text{-}3)$$

Using equation 3.1-3 to estimate the direction of the curve assumes that any point on curve A has a corresponding behavior point on curve B that occurs at the same length ratio at every speed. More basically stated, a point on curve A that occurs at 10% of the total length of curve A is equivalent to the point on curve B that occurs at 10% of the total length of curve B. This relationship becomes more accurate at predicting the direction axis as the length of the segments calculated in equation 3.1-3 decreases, so it is essential to use a large number of points (on the order of several thousand) to assess system behavior.

By creating a series of best fit line functions, from the numerically integrated, direction axis, a speed length axis is generated. These unusual axes comprise a unique coordinate system for each compressor map. The created coordinate system is functionally similar to the radial-angular axes in the radial coordinate system (r-θ). One of the axes used to project the compressor maps in the Numerical Integration method is magnitude of speed. The other axis is a combination of polytropic head/efficiency and actual volumetric flow in one parameter that indicates direction.

Often it is convenient to use the polynomial model for the best fit lines as well as the contours that characterize curve behavior. When doing so it is necessary to limit the order of the polynomial by the number of data points minus one. For example, a 3rd order polynomial cannot be characterized by less than 4 data points, and a 6th order polynomial cannot be characterized by less than 7 data points. This becomes relevant when considering the number of speed curves available.

As stated earlier, this method can be automated in spreadsheet software, and performs high/low speed projections and interpolations that are consistent with historical data. Figure 3.2 depicts the application of this method used in the Atlantic LNG Feed Gas Compressor analysis, projected to 130% of the design speed. Even at 30% above the design speed, the projection is extremely stable. The contour lines shown "perpendicular" to the vendor supplied curves are used to characterize the mathematical behavior of the entire curve set. These curves are generated from points at a specific distance ratio along the integrated curve. The green line represents a projection of the polytropic head values at 6071 rpm, which is 130% of the design speed.

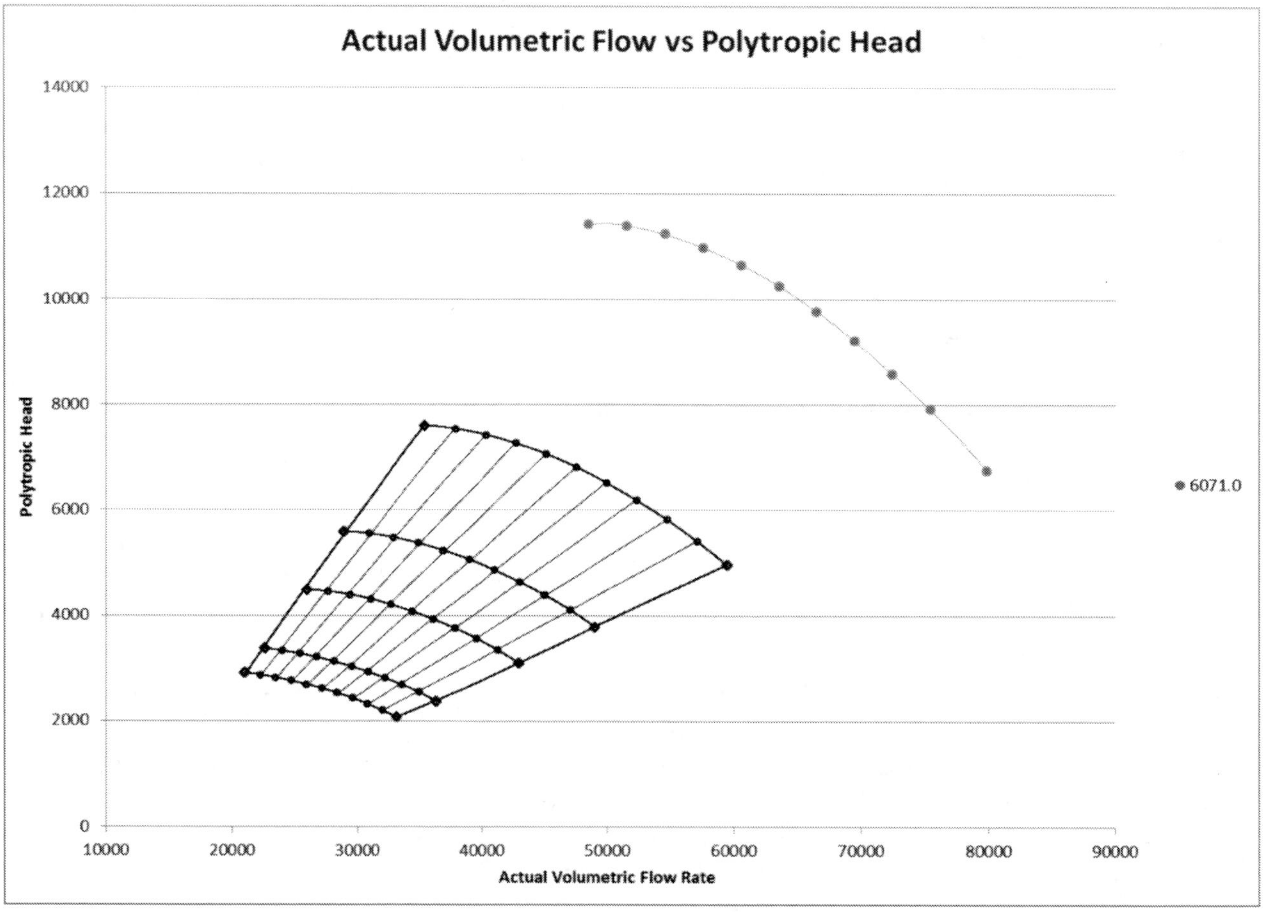

Figure 3-2. Polytropic head curves for Atlantic LNG feed gas compressor.

With curve sets that behave in a similar manner at each speed, using the correct best fit model is less important as the contour lines used to characterize the polytropic head/actual volumetric flow ratio exhibit a less dynamic behavior. In this case the contours can be characterized by the same model and a 2nd or 3rd order polynomial is often used used on a consistent basis. When the curves become more complex, as in the polytropic efficiency curves shown in Figure 3-3 below, it is important to characterize each curve and contour on an individual basis. This becomes especially significant when projecting values above or below the limits of the vendor supplied curves. Best fit models that are extremely accurate within the supplied boundaries may change dramatically when even slightly above or below so it is necessary to project at 10-15% above the desired speed when choosing the appropriate method. The below figure displays the polytropic efficiency values, as given by the compressor vendor, as well as the contour lines that were generated to predict the mathematical behavior of the curve system. Polytropic efficiency maps are often more complex in behavior and require more complex contour lines. The green line represents a projection of the behavior to 5700 pm, which is over 120% of the design speed.

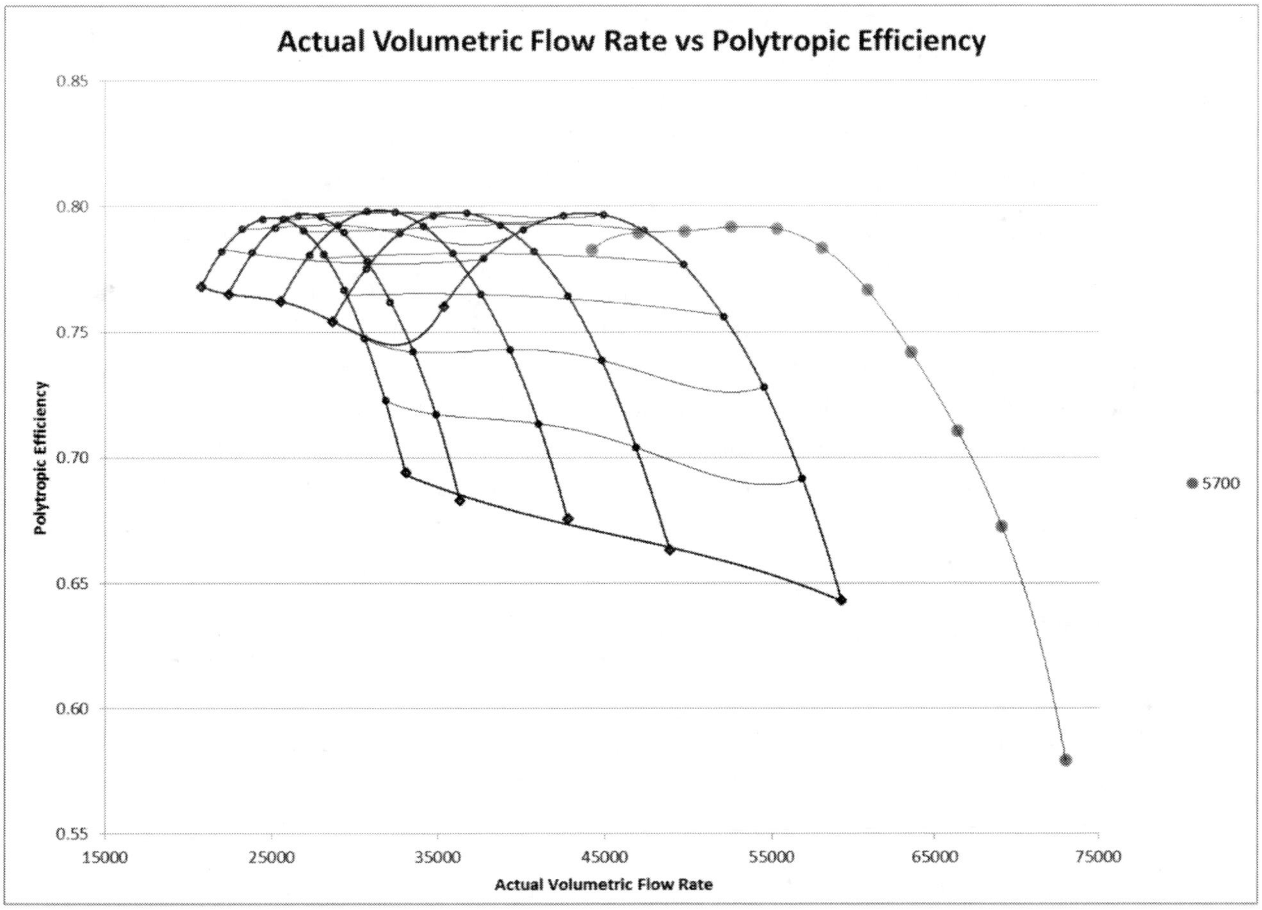

Figure 3-3. Actual Volumetric Flow Rate vs Polytropic Efficiency.

3.2 Calculation Method

1. Gather numerical data points for each set of curves from the Polytropic Head or Polytropic Efficiency maps. Normally it takes 10-15 data points per curve to create a curve with adequate accuracy. Plot the points on a chart.
2. Develop best fit lines for each compressor speed. Normally 5-8 compressor speeds should be used for increased accuracy. If you do not have at least two speeds, use one speed and the origin for the most conservative assumption.
3. Use the best fit lines from step 2 to calculate 5000-10000 individual data points for each curve in order to maintain the inaccuracy below 1%.
4. Find the surge point, 10%-90% length lines, and the stonewall point for each speed.
5. Create a best fit line equation from all of the surge points on each curve. Do the same for the 10%-90% points and then create a best fit line from the stonewall points on each curve. This results in best fit curves based on speed.
6. Chose a speed, solve each best fit line from step 5 for the chosen speed.

7. Create a best fit line from the points generated in step 6.

4. Comparison of Calculation Methods

To assess the accuracy of the calculation methods in Section 2, in predicting compressor behavior at abnormal conditions, an array of data points was chosen for analysis. More than 7,000,000 data points, at more than 415,000 conditions, across two years of operating data, including shut downs, startups, instrumented trips, and normal operating conditions were sorted through to find the most stable/repetitive conditions experienced as they are expected to be fully representative of the compressor system. Discharge pressure was chosen as the basis for this analysis since increasing pressure is the fundamental purpose of a compressor.

Data points at compressor speeds 3040 rpm, 3965 rpm, 4685 rpm, and 4904 rpm were analyzed, grouping the data to 5 rpm increments (or 0.1% of the design speed). Within each speed, representative points were taken in a range from the highest suction pressure to the lowest suction pressure experienced by the compressor within the span of the data. The data points at each suction pressure were grouped by inlet pressure, inlet temperature, discharge pressure, discharge temperature, and recycle opening. Then the largest grouping of points at each suction pressure was chosen in order to ensure the stability of the data. To determine the accuracy of the calculations, the predicted discharge pressure and the predicted compressor differential pressure were compared to the operating data values.

Suction Pressure:	27.1 - 54.2 barg
Suction Temperature:	18.1 – 29.3 C
Mass Flow Rate:	784,717 – 1,315,437 kg/hr
Inlet Volumetric Flow Rate:	24,272 – 41,177 m3/hr
Compressor Speed:	3040 – 4904 rpm (65 – 105%)
Recycle Valve Opening:	5 – 100%

The reference conditions set on the as tested design compressor curves are as follows:

Suction Pressure:	37.2 barg
Suction Temperature:	15.9 C
Compressor Speed:	3770 rpm (80%)
Recycle Valve Opening:	0%

Table 4-1, shows 19 conditions supported by approximately 14,000 data points, upon which each of the 3 calculational methods were performed. The data were chosen to represent the entire range of inlet pressures, inlet temperatures, inlet densities and mass flow rates experienced by Atlantic LNG's feed gas compressor during a period of two years for the speeds considered in the analysis.

Table 4-1. Calculation Method Comparison.

Calculation Method Comparison									
Speed	Measured Pressure			Isentropic Energy Balance Model		Ideal Gas Polytropic Model		Real Polytropic Model	
	Inlet Pressure	Discharge Pressure	Delta P	Discharge Pressure	Delta P	Discharge Pressure	Delta P	Discharge Pressure	Delta P
RPM	barg	barg	bar	barg	bar	barg	bar	barg	bar
3043	43.4	54.0	10.7	54.2	10.8	54.2	10.8	53.0	9.6
3043	43.9	54.3	10.4	54.6	10.7	54.6	10.7	53.3	9.4
3043	46.1	55.7	9.6	56.8	10.7	56.9	10.8	55.6	9.5
3043	47.3	56.3	9.0	58.5	11.1	58.4	11.1	57.2	9.9
3043	50.7	58.2	7.5	60.5	9.8	60.5	9.8	59.3	8.6
3043	50.8	57.9	7.1	60.8	10.0	60.8	10.0	59.6	8.8
3043	51.8	59.7	7.9	62.2	10.3	62.2	10.3	61.0	9.2
3043	52.7	60.8	8.1	63.0	10.3	63.0	10.3	61.8	9.1
3043	54.2	62.8	8.7	65.1	11.0	65.1	10.9	63.8	9.6
3964	37.2	53.3	16.0	53.7	16.5	53.7	16.5	52.2	15.0
3964	38.8	55.4	16.6	56.3	17.5	56.3	17.5	54.9	16.0
3964	41.3	55.6	14.3	58.1	16.8	58.1	16.8	56.7	15.4
4683	33.5	54.4	20.9	55.2	21.7	55.2	21.8	53.6	20.1
4683	34.4	54.9	20.6	56.8	22.4	56.8	22.4	55.1	20.7
4683	35.7	56.1	20.4	59.4	23.7	59.4	23.7	57.7	22.0
4904	27.1	45.7	18.5	46.3	19.1	46.3	19.2	44.6	17.5
4904	32.3	53.9	21.5	55.9	23.5	55.9	23.5	54.2	21.8
4904	33.7	55.0	21.3	57.9	24.2	57.9	24.2	56.2	22.5
4904	34.0	55.6	21.7	58.6	24.6	58.5	24.6	56.8	22.9
	Average Deviation			3.0%	12.4%	3.0%	12.4%	0.5%	3.1%
	Max Deviation			5.5%	28.9%	5.5%	28.9%	2.8%	19.2%
	Min Deviation			0.3%	1.5%	0.3%	1.4%	-2.4%	-11.5%
	Range of Deviation			5.2%	27.5%	5.2%	27.4%	5.3%	30.6%

4.1 Discharge Pressure Prediction

As shown in Figure 4-1., provides the comparative information regarding predicted discharge pressure for each point in Table 4-1., for each calculation method and the measured discharge pressure for reference. The Isentropic Energy Balance model and the Ideal Gas Polytropic Model unexpectedly return the same values for every point. The largest deviation calculated was less than 0.2% across the data set and the majority of the time the agreement is on the order of 99.9%. For this reason these two methods are considered to be equivalent for the remainder of this paper. The highest deviation from predicted discharge pressure by the energy balance method is +5.5% while the lowest deviation is +0.3% giving a range of 5.2%. These methods produce results that are close to, but consistently above, the measured discharge pressure.

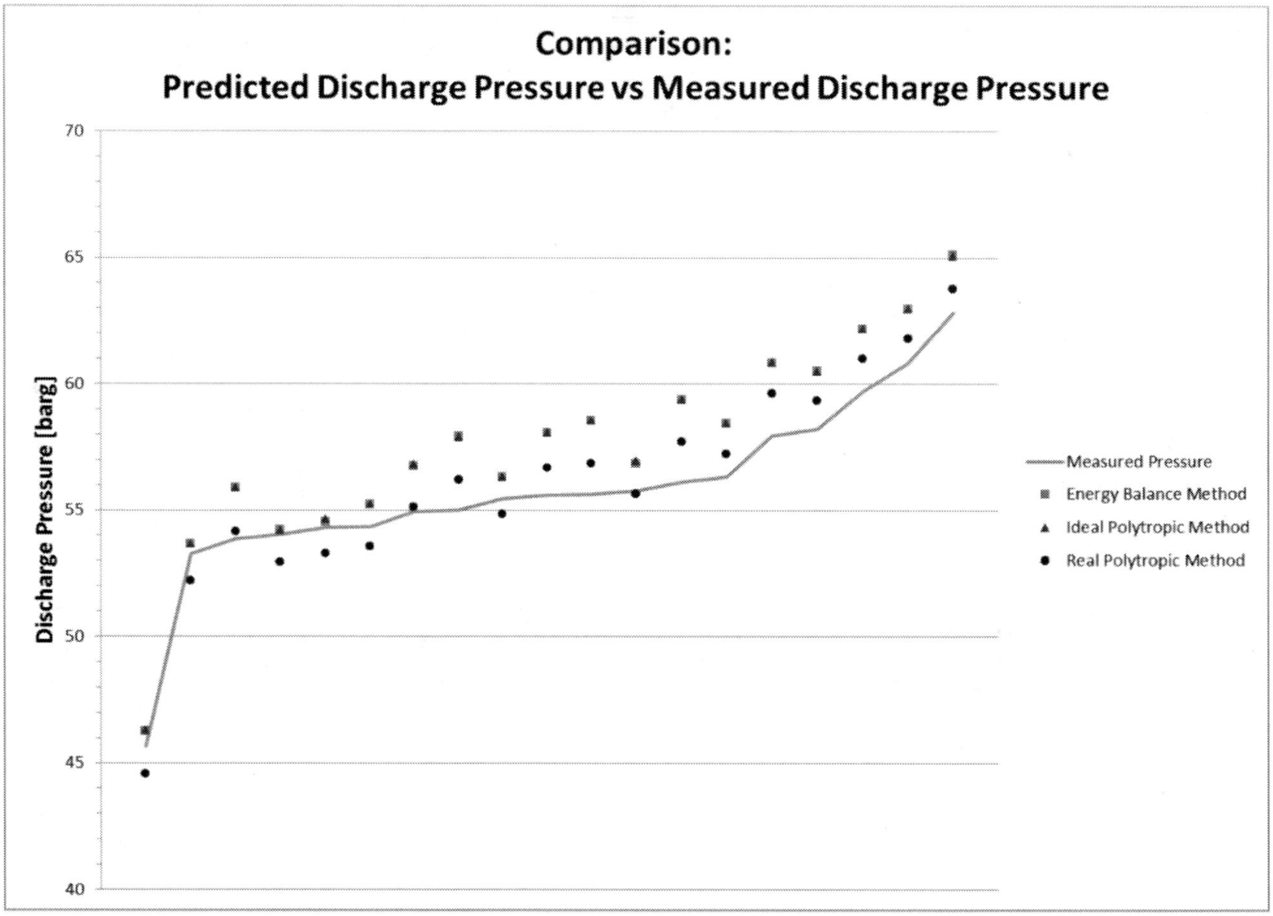

Figure 4-1. Comparison: Predicted Discharge Pressure vs Measured Discharge Pressure.

The real polytropic method returns results that are more closely grouped around the measured pressure such that the average deviation is smaller; however, the range of deviation is larger. The highest deviation from the measured discharge pressure is +2.8% while the lowest deviation is -2.4% giving a range of 5.3%.

4.2 Pressure Differential Prediction

Since differential pressure is a function of the work applied to the fluid by the machine, shown in Figure 4-2., deviations in the prediction of this value more accurately predict the behavior of the compressor system. The figure shows the graphical comparison of the pressure differential predicted for each point in Table 4-1, for each calculation method and the measured pressure differential for reference.

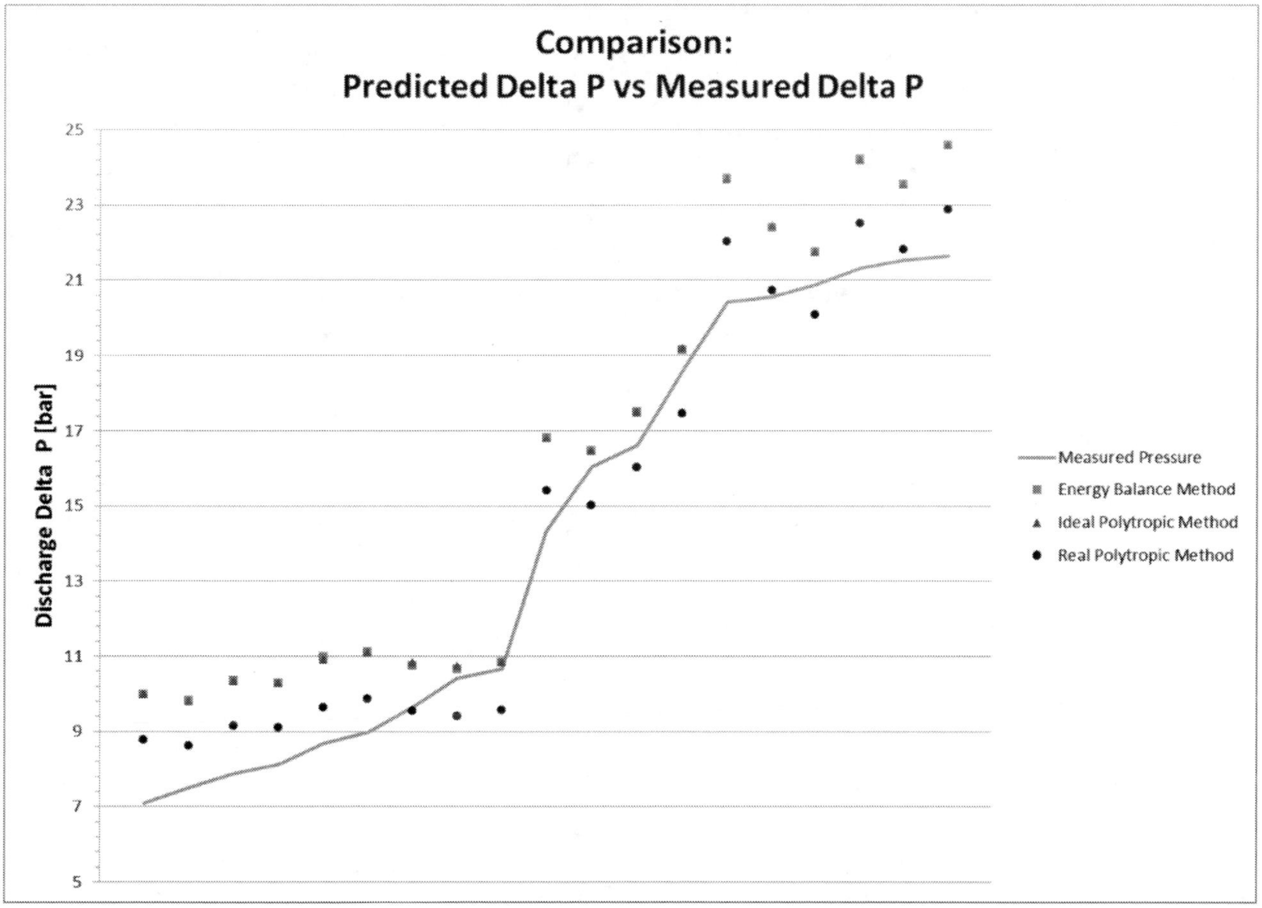

Figure 4-2. Comparison: Predicted Delta P vs Measured Delta P.

The highest deviation from predicted discharge pressure by the energy balance method is +28.9% while the lowest deviation is +1.5% giving a range of 27.5%. The deviations here appear to be much more significant even though the deviation in discharge pressure is only +3.3 barg.

The real polytropic method returns results that, again, appear to be more closely grouped around the measured pressure, such that the average deviation is smaller; however, visual inspection of the results seems to show the difference between the ideal gas polytropic equation and the real gas polytropic equation is merely a reduction in value instead of a modification of behavior[3]. The highest deviation from the measured discharge pressure is +19.2% while the lowest deviation is -11.5% giving a range of 30.6%.

4.3 Reference Suction Pressure

Analysis of the deviations in pressure differential from the measured values for both the Energy Balance method and the Real Gas Polytropic Equation method shows no direct correlation between varying values of suction pressure, suction temperature, suction density, mass flow rate, volumetric flow rate, recycle valve opening, nor compressor speed.

All dual combinations of the aforementioned variables were considered showing no indirect correlation with the exception of one combination. Considering the combination of compressor speed and suction pressure, at each compressor speed, there is a pressure below which the deviation in predicted differential pressure levels off. This relationship is demonstrated in the Figure 4-3 below which is the same as Figure 2-1 above, with the higher pressure points removed at each speed. In order to differentiate between inefficiency and inaccuracy, the high suction data points were removed.

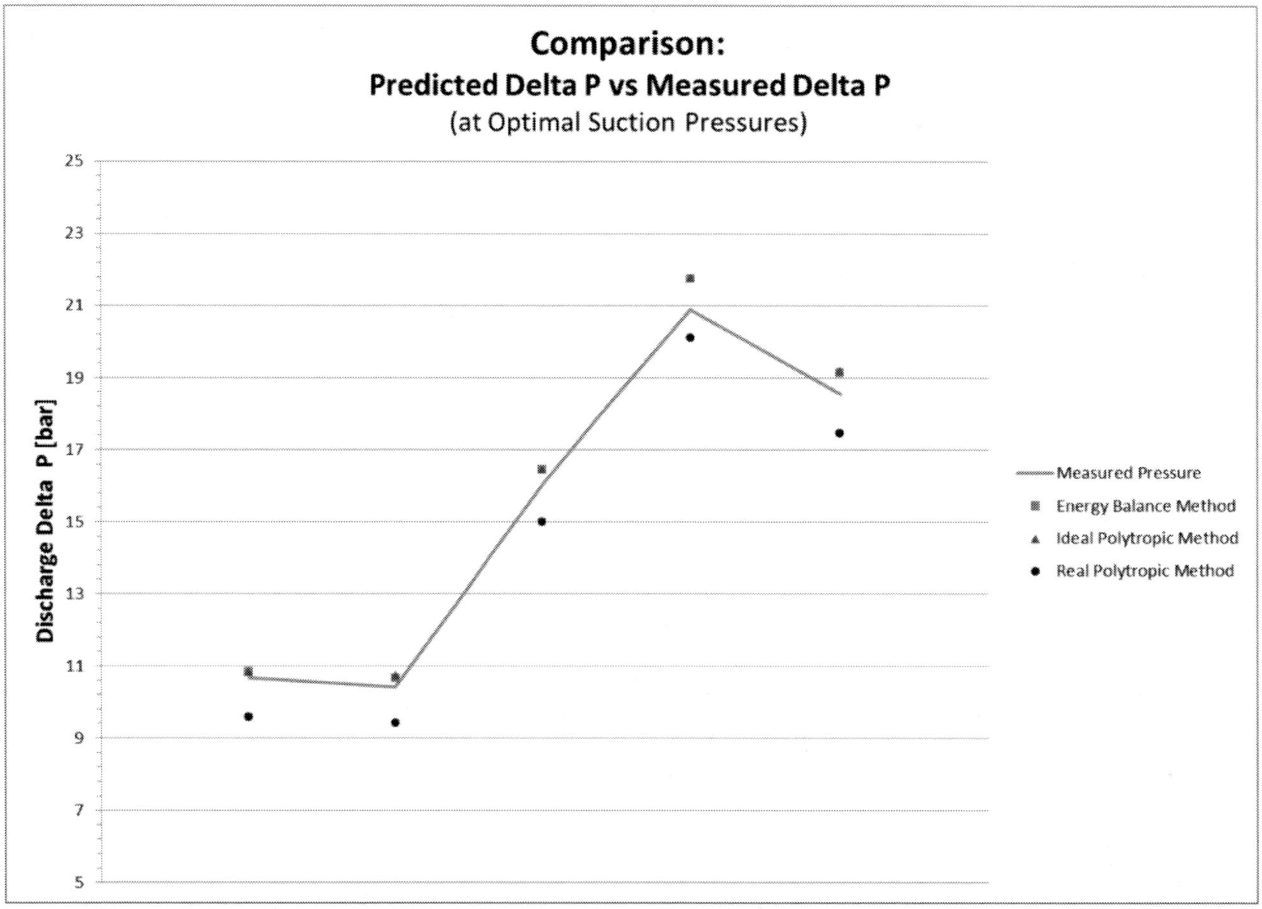

Figure 4-3. Comparison: Predicted Polytropic Work vs Measured Polytropic Work (Low Pressure).

At suction pressures on the lower end of the range of pressures considered for each speed, the Ideal Polytropic Equation method is extremely accurate. At these conditions the predicted discharge pressure deviates from the measured values by 0.3-1.6% and the predicted differential pressure deviates by 1.4-4.0%. In this range the Ideal Polytropic equation yields results that are far more accurate than those from the Real Polytropic equation. There is a range for which this same phenomenon is true for the Real Polytropic method; however, since this behavior implies a machine efficiency related to a speed sensitive reference pressure, and the Real Polytropic method predicts negative efficiency values for work in reference to the ideal system work, this would violate the 2nd Law of Thermodynamics[2], and thus does not appear to be a reliable method for determining compressor performance at abnormal operating conditions.

4.4 Compressor Efficiency vs Suction Pressure

In order to determine if efficiency based on deviation from a speed sensitive reference pressure is a valid concept, a more comprehensive data set, covering 64 conditions supported by more than 50,000 data points, was analyzed using the Energy Balance method and the results were plotted on a chart of polytropic work vs actual volumetric flow rate. Both, the polytropic work predicted by the Isentropic Energy Balance method and the polytropic work as tested compressor design curves were placed are shown in Figure 4-4. This is a representation of the ideal work that should be developed per vendor documentation to the actual work developed per the measured operating data. For each speed the darker color represents the stream work predicted by the vendor supplied polytropic head curves based on inlet volumetric flow rate. The lighter color represents the work that is obtained from isentropically flashing from the measured inlet pressure and temperature to the measured discharge pressure.

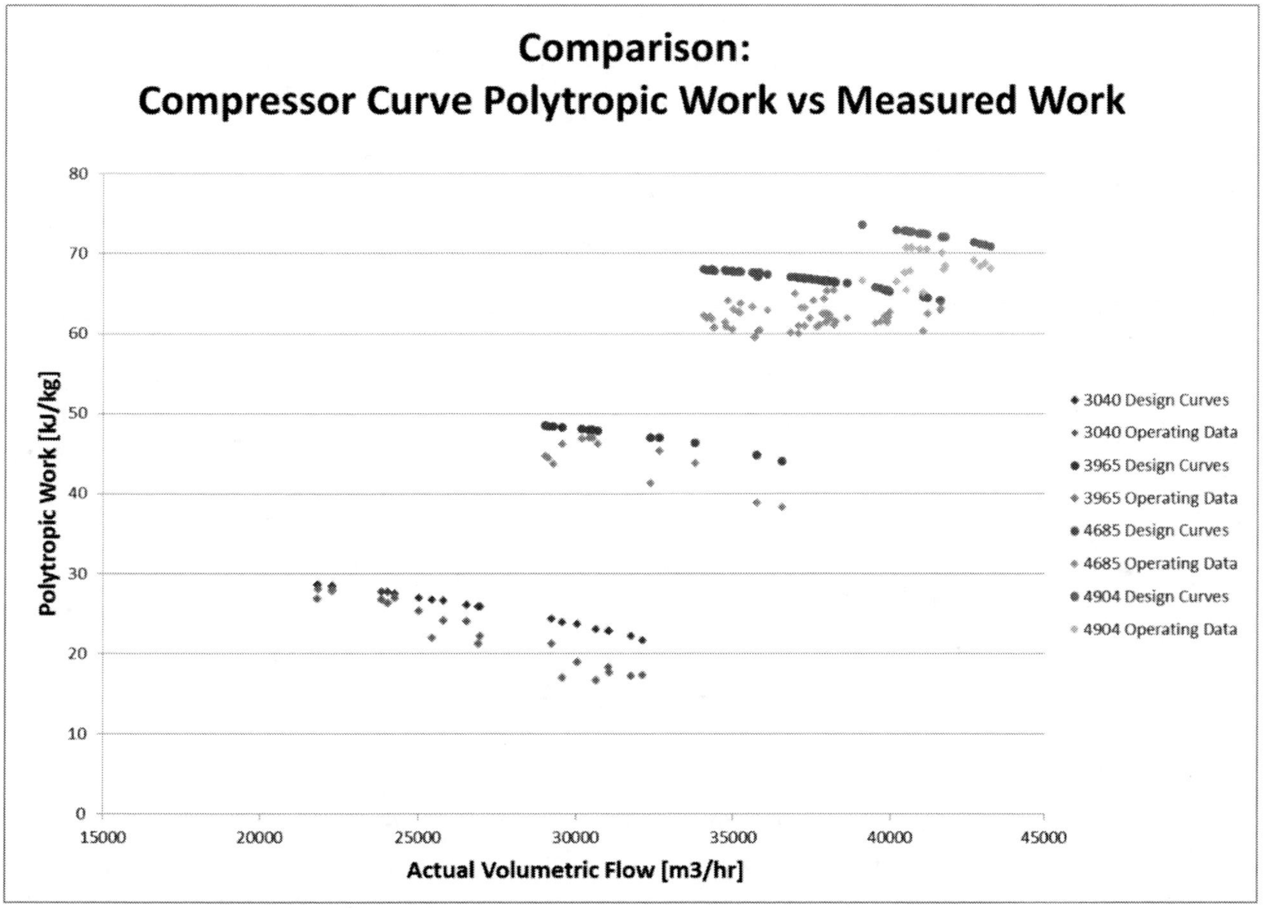

Figure 4-4. Comparison: Compressor Curve Polytropic Work vs Measured Work.

The data points shown at lower operating speeds (3040 rpm and 3965 rpm) indicate a reduction in machine efficiency based on increased volumetric flow rate; however, data points at the higher speeds

(4685 rpm and 4904 rpm) contradict this concept as they are predicting polytropic work rates very close to the as tested curves across the entire range of volumetric flow rates analyzed. Consistent and smooth predictions of a reduction in the capacity to develop head at lower speeds indicate the deviation may be a characteristic of the system rather than a lack of accuracy in the calculations. The pattern potentially indicates a discrepancy between the available power supplied by the turbine and the power required for compression. It is likely the data trends are influenced by both of these phenomena.

Both polytropic work and volumetric capacity have been shown to be relatively independent of pressure in Figure 4-4. As shown in Figure 4-5 below, the data points at each speed were sorted by increasing suction pressure and the higher pressure data points were removed from the analysis. The data points consistent with the highest 75-80% of the suction pressures at each analyzed speed were removed to demonstrate if there is a comprehensive relationship between elevated suction pressure and the ability of the compressor to generate head.

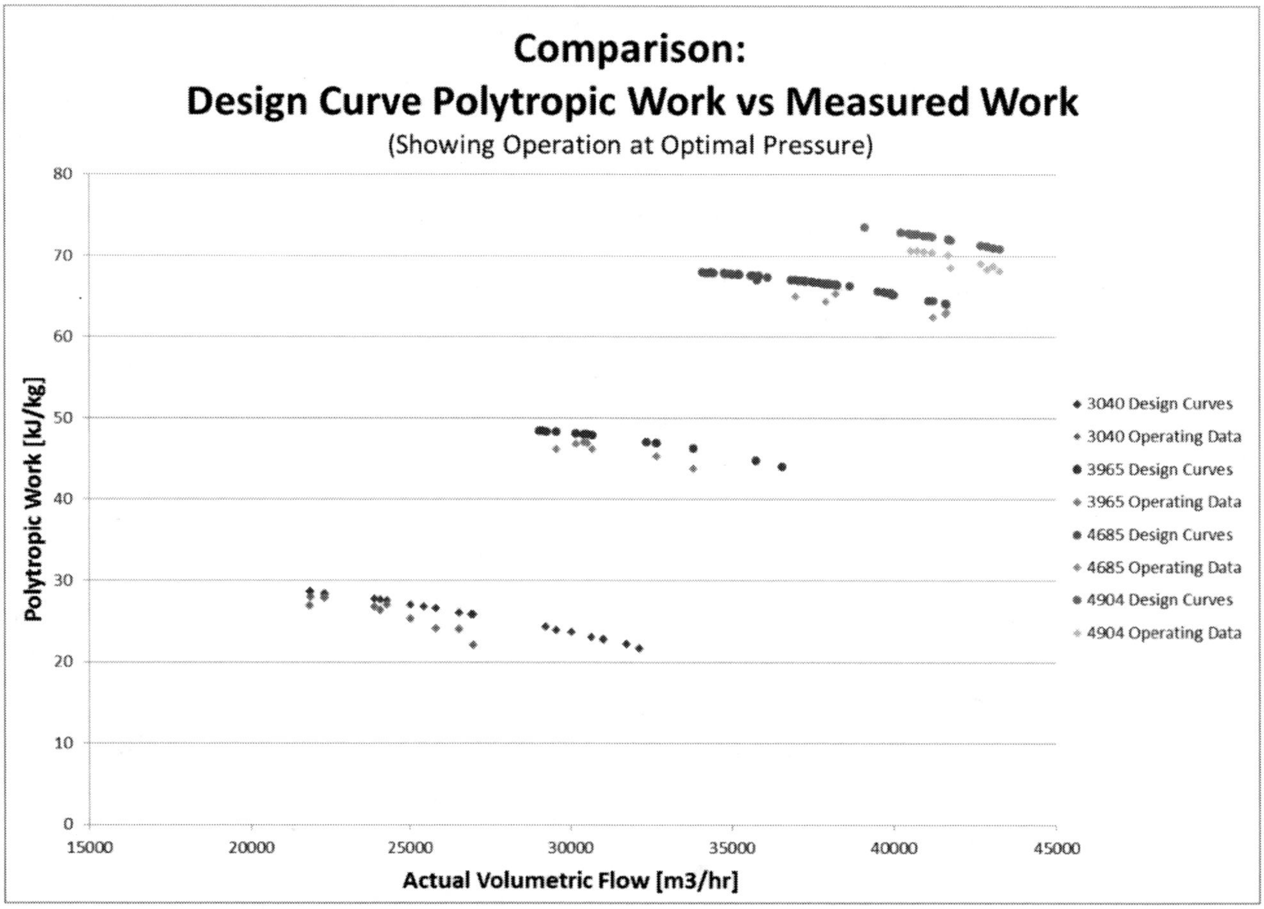

Figure 4-5. Comparison: Design Curve Polytropic Work vs Measured Work.

This manipulation of the data shows a direct correlation of polytropic efficiency with a binary relationship between suction pressure and speed. The loss of high volumetric flow data points at some compressor speeds while maintaining flow capacity at other speeds shows the compressor will operate stably at conditions where the supplied driver power deviates from the from the polytropic head maps. This second point is extremely important when assessing the impact of abnormal operating conditions on safe compressor operation.

The above analysis supports the following statements to be true:

1. There is a speed sensitive reference pressure, unrelated to the reference conditions reported on the as tested compressor curves, which defines the machine efficiency across a large range of operating conditions.
2. The optimal suction pressure can vary by a large amount, on a percentage basis, from the reference suction pressure from the vendor.
3. A compressor system can operate stably across a wide range of conditions, even when the supplied driver power is lower than the power required to move the flow across the compressor, as predicted by the polytropic compressor maps.
4. The deviation in efficiency is not related to deviations in compressor mass, or volumetric, flow rate.
5. The Isentropic Energy Balance Model and the Ideal Gas Polytropic Model are essentially equivalent and accurate.
6. For the purposes of process safety assessment, the Isentropic Energy Balance Model and the Ideal Gas Polytropic Model are well suited to performing process safety design calculations and optimization calculations.

Since the data used in this analysis covers the operation of only one compressor system, it was not considered within the scope of this paper to determine a system independent correlation between the design suction pressure and the speed sensitive, optimal suction pressure. Additionally, it was not considered within the scope of this project to determine a system independent correlation between the decay of machine efficiency at abnormal suction pressures and the polytropic work curves presented by the vendor.

5. Safe Compressor System Operation and Design

The complex nature of system design, combined with the highly variable nature of compressible fluid properties presents a variety of safety considerations that, while not inherently different than the normal range of process safety considerations, do require a specific perspective. From the perspective of hazardous scenario identification, compressors are unusual in that the majority of the scenarios needing to be analyzed are more closely related to failures in the systems feeding the machine and not related to any failure in the compressor system itself. From the design perspective, abnormal operating conditions often require more careful and varied analysis even than what is required for normal machine operation. Additionally, there are many safety concerns that actually arise when the operators are not given enough system specific information to accurately characterize compressor behavior at abnormal operating conditions such as start-up, shut down, and coping with upstream and downstream system changes as well as deviations from facility design rates and conditions due to plant upgrade and optimization projects.

5.1 Hazardous Scenario Identification

Compressor systems can offer unusual challenges when it comes to identifying and quantifying safety hazards. It is important to remember that, when considering overpressure, no allowance for beneficial control action should be taken unless it tends to increase the relief load[4]. This same concept can be reasonably applied to other potentially hazardous conditions such as under pressure, high/low temperature excursions, and vibration induced fatigue. Many times, instrument initiated shut downs due to pressure/temperature/liquid level deviations are used to rule out hazardous scenarios even though the SIL rating of the compressor trip does not meet the minimum requirements for a High Integrity Protection System[4]. It is vital to evaluate these systems considering the assumptions implicit to the step in the overall process hazard analysis method being used. When calculating overpressure relief load, or other safety hazards, conservative assumptions should be used and instrumentation or operations based intervention should not be considered in this step in order to properly perform the risk assessment. When performing the risk assessment and/or LOPA step in the process hazard analysis, the perspective assumptions need to be modified such that both operator and instrumentation intervention need to be assessed based on the order of magnitude of the potential safety hazard from the calculation phase. It is critical to avoid making non-conservative or "realistic" assumptions during relief load calculation in order to maintain the integrity of the analysis.

The following scenarios should be evaluated to determine safe compressor operation:

5.1.1 Blocked Vapor Outlet

Failure closed of automated control valves, inadvertent closure of manual isolation valves on the compressor discharge line, failure closed of a bleed or vent line on the compressor discharge, or loss of cooling/condensing service on the compressor discharge may lead to overpressure or surge conditions.

The relief load for this scenario can be determined using the maximum normal operating mass flow at the lowest system maximum allowable accumulation pressure, as defined by the appropriate reference code or standard for the piece of equipment or piping that defines the limiting pressure.

5.1.2 Elevated Suction Pressure

It is critical to understand that elevated suction pressure may lead to over pressure in the compressor discharge even though no failure has occurred in the compressor system itself. This should not be considered double jeopardy as the fundamental purpose of a compressor is to increase the pressure in a system and no credit for positive instrumentation action should be taken at this stage of the analysis[4]. Many times the upstream pressure is limited to pressures that are safe for the suction of the machine, but may lead to unsafe conditions in the discharge side of a compressor due to the increased pressure differential. Most compressors have instrument initiated shut downs based on high suction/discharge pressures; however, if these shut downs do not meet the requirements of a High Integrity Protection System[4], they should be considered as possible latent failure points and therefore should not be considered as removing elevated suction pressure from analysis. The following are potential sources of elevated suction pressure:

1. Failure of Automated Controls – failure open of control valves supplying normal process flow to a compressor may lead to elevated suction pressures
2. Inadvertent Manual Valve Opening – inadvertently opening a manual bypass valve on a line that normally supplies process flow, or, inadvertently opening a manual valve connected to a high pressure source that is normally used during a different operating mode or configuration may lead to elevated suction pressures
3. Abnormal Vapor Generation – when a compressor is being fed by a flashing liquid system, loss of cooling, loss of heating, abnormal heat input, and mixing of hot and volatile fluids scenarios may potentially result in an elevated suction pressure due to increased vapor generation or the presence of light components flashing across pressure let downs.

To determine if elevated suction pressure is a concern, use the calculations from Section 2 at the maximum normal operating speed. Credit can be taken for normal volumetric vapor outflow if there is not a condenser on the discharge. If there is a condenser on the discharge then either take credit for the normal liquid outflow or perform iterative calculations considering the condensing capacity of the heat exchanger at relieving conditions to determine how much vapor will drop out to control the system pressure. If there is outflow capacity at relief conditions, either from credit for outflow or from relief valves on the compressor discharge, then determine the head by moving down the polytropic head curve until the corresponding flow point is reached. Remember the curves work on inlet volumetric flow but the relief valves are moving the discharge flow. If there is no outlet capacity or the outlet flow is less than the minimum flow at the surge point on the maximum speed curve, then the surge head should be used in conjunction with the maximum suction pressure to determine the maximum discharge pressure.

Elevated suction pressures will result in high discharge pressures within the compressor casing independent of the flow capacity downstream of the discharge. Running the system to stonewall will generate a pressure discontinuity downstream of the compressor discharge which may protect downstream equipment. This should only be considered if the maximum allowable working pressure of the compressor casing is higher than the maximum pressure developed by the compressor at elevated suction conditions.

If the maximum allowable working pressure of the compressor casing is lower than the calculated discharge pressure, considering the maximum suction pressure and discharge flow capacity, it is necessary to limit the suction pressure the machine can see. This can be accomplished by two methods. The first method, limit the set pressure of pressure relief valves upstream of the compressor suction. These pressure relief valves are normally set to the design pressure of the system in order to reduce the size/cost of the relief system; however, it is often more cost effective to reduce the set pressure of upstream valves to the maximum safe suction pressure. The maximum safe suction pressure is defined as the compressor suction pressure at which the discharge pressure is equal to the minimum maximum allowable working pressure of the system being considered. At or below the maximum safe suction pressure it is not physically possible for the compressor to overpressure the discharge. If this approach is used, some assessment of the operating range should be performed to ensure the system will function properly at the lower set pressures. The second method is to use an appropriately SIL rated trip at the suction pressure that will prevent overpressure in the discharge using the same concept of maximum safe suction pressure. Most applications require a SIL 2 or SIL 3 to be considered equivalent to a pressure relief valve. In large capacity systems a High Integrity Protection Systems is often less costly than several large, modulating pressure, relief valves, the associated piping, and the potential implications for flare sizing.

5.1.3 Abnormal Suction Composition

Since the thermodynamic properties of a fluid vary a great deal with changes in composition, a significant change in the composition of the suction gas can produce discharge pressures higher than the maximum allowable working pressure even at the normal suction pressure. Any potential sources of significant variation in inlet composition should be evaluated for their impact on discharge pressure. The following events may lead to large changes in inlet gas composition:

1. Fire on an upstream vessel - vessels under fire can produce dramatic changes in vapor composition in downstream equipment
2. Control Valve Failure – when the source of flow to a compressor is from a mixture of flow sources, failure open of a control valve on the compressor suction may lead to changes in the inlet composition
3. Inadvertent Manual Valve Opening – inadvertently opening a manual valve from a source of flow that does not normally contribute to the process flow may result in changes to the inlet composition
4. Tube/Core Rupture – often, flow through the other side of a heat exchanger has a significantly different composition such that introduction of this flow into the normal process fluid may result in significant changes in the inlet composition

Abnormal suction composition is often overlooked because it does not involve a direct failure in the compressor system. The calculation method and mitigation considerations from Section 5.1.2 should be applied for abnormal suction composition. It should be considered that elevated suction pressures and abnormal suction compositions will occur simultaneously. Each source of abnormal suction composition should be considered independently. Additionally, special consideration should be made for fire on upstream liquid handling systems that supply flashed vapor since the maximum suction pressure will be 121% of the relief valve set pressure and the composition can vary significantly. Fire on upstream liquid handling equipment can result in extremely high discharge pressures and relief loads so using fire proof insulation such that time to reach relieving conditions is longer than the length of time considered for fire scenarios (usually 2-4 hours) is often a cost effective mitigation even when factoring in the cost of maintenance needed to prevent corrosion.

5.1.4 Reverse Flow

When multiple compressors, with different drivers, are operating together in a system, inevitably, there are times when one machine will be in operation while the other machine is down. When this occurs there are often several paths across which flow may move from one machine to the other. Since the system is not moving flow during this mode, leaks that would normally not impact the safety of the system may result in overpressure to isolated piping or equipment in the compressor system which is down.

5.2 Safe Compressor System Design

There are a variety of hazardous operating conditions that may arise more from insufficiently detailed design rather than variations in operating conditions.

5.2.1 Compressor Inlet Isolation Valve Placement

Compressor recycle lines need to tie in downstream of automated isolation valves in order to reduce the likelihood of under pressure due to mechanical failure of the inlet isolation valve to the closed position. If the entire suction system piping and equipment are rated for full vacuum this is not a concern.

5.2.2 Pocketing in Compressor Recycle Lines

Pocketing in recycle lines and inappropriate failure position of quench valves can lead to accumulation of liquid. Actuation of recycle/anti-surge valves with accumulated liquid may result in significant damage to the piping or an unexpected source of liquid to the compressor inlet. To reduce the risk of hazardous operation, the following configurations should be considered:

1. There should be a knock out drum with a demisting pad to remove liquid formed in suction lines due to pressure variation. A compressor knock out drum should not have a liquid supply during normal operation.
2. For applications with quench fluid injection, the appropriate failure position needs to be chosen such that liquid will not feed in upstream of the anti-surge valve and lead to slug flow concerns upon anti-surge valve actuation.

5.2.3 Pressure Relief Valve Selection

Pressure relief valves on compressor suction and discharge lines need to exhibit modulating behavior across the entire range of actuation. Compressor surge is a direct function of deviation in mass balance between the compressor suction volume and the compressor discharge volume. Anti-surge valves work to maintain the appropriate mass balance across the system. Spring Loaded and Pop Action relief valves, sized to remove the normal operating mass flow of the compressor, will cause a sudden imbalance in the discharge side of the system which will interfere with the ability of the anti-surge system to operate properly. More simply, the dynamic nature of the initial actuation of spring-loaded and pop action pressure relief valves will cause the system to go into surge at the moment of actuation. This is not a concern for pressure relief valves whose capacity is small in relation to the design mass flow capacity of the system.

5.2.4 Settle Out Pressure

The equipment on the suction side of the compressor should be designed to handle the maximum settle out pressure that can be developed by the system. Often settle out pressure is only calculated at normal operating conditions; however, since settle out pressure is based on total system mass abnormal operating conditions result in varying settle out pressures. Additionally, changes in piping layout during design or upgrade projects can have a significant impact on settle out pressure. Ratio in suction side volume to discharge side volume is a key indicator in determining if this may be a problem assuming a fixed maximum pressure. If the ratio goes down then the settle out pressure needs to be recalculated.

If, while the system is isolated, there are liquids at equilibrium contained within the system, the settle out pressure should be calculated considering the system at thermodynamic equilibrium based on initial shut down conditions. Additionally, the settle out pressure should be calculated with the system warmed up to ambient temperature. If the pressure at this condition exceeds the rated pressure of the system, some assessment should be done to determine how long it will take for the settle out pressure to accumulate from initial settle out to relief pressure due to ambient warming. This information should be used in determining the feasibility of safely deinventorying the system, either to liquid storage, or to flare. Additionally, the results should be furnished to the operations department for use in prioritizing safe facility depressurization and deinventory requirements during emergency situations.

5.2.5 Pilot Tubes on Pilot Valves

Vibration analysis should be performed on pilot tubes attached to the diaphragm on pilot operated relief valves. These tubes are often very thin walled. Pilot tube failure is not uncommon, and failure on the process pressure side of the pilot diaphragm results in a complete failure of the pressure relief valve.

5.3 Risk Factors Inherent in Compressor Operation

There are certain risks that stem from the manner in which a compressor system is operated. Many times, automated control systems are tuned to satisfy theoretical design limitations and do not take into consideration any increased risk to the system based on differences between static system design and the actual performance characteristics. Due to the inherently dynamic behavior of a compressor system and the large number of automated and manual controls, certain, potentially unaccounted for, risks to safe operation should to be assessed.

5.3.1 Oversized Recycle/Anti-surge Valves

Data analysis presented in this paper, consistent with incidents experienced by operating facilities, shows a direct correlation between loss of effective energy translation between the compressor and the process fluid at high suction pressures. Efficiency losses are shown consistently to be higher than 30% at high suction pressures and low operating speeds. Operating facilities have experienced significant increases in non-AIV related vibration in the compressor system piping. During start-up and full recycle mode operation, many anti-surge valves will operate in the 100% open position. Since the suction pressure is proportional to the magnitude of recycle valve opening for both events, oversized recycle valves, at non-design conditions, may lead to vibration induced fatigue that is not shown when considering AIV or mechanical vibration alone.

For systems in the design phase it is often more favorable from a cost and safety standpoint, to use two valves for anti-surge conditions. There are two classes of considerations when considering anti-surge valve sizing, control flow capacity and trip flow capacity. The control flow capacity requirement is based on the range of flow needed to keep the compressor out of surge during expected operating conditions. The trip flow capacity requirement is determined by the flow needed to keep the system out of a surge condition during an instrument initiated trip. The trip flow requirement is often several times the control flow requirement. In design this indicates an opportunity to improve safety and simultaneously decrease cost. During automated anti-surge control, it is necessary to make fine adjustments. This requires more sophisticated, more expensive valves with a higher flow resistance characteristic than needed for trip flow considerations. A higher flow resistance characteristic leads to the need for larger anti-surge line sizes. Instead of using one valve that can both, control the flow during the automated flow phase, and pass adequate flow at trip conditions, two valves can be used. One, a control valve in a "bypass" piping configuration, since it experiences lower flow rates, and the other valve in the main anti-surge line, that exhibits more favorable flow characteristic and can open at a much faster rate at lower expense. This will reduce vibration concerns since the trip flow valve will be closed during start-up and recycle operation.

5.3.2 Excessive Recycle/Anti-Surge Valve Actuation

Liquid droplet formation in compressor suction lines as a result of retrograde condensation is not an uncommon occurrence when the inlet flow is at, or near, thermodynamic equilibrium. Normally, these droplets are small and disperse in the turbulent region at the compressor inlet; however, pressure waves in the suction line, caused by aggressive control valve action, will propagate faster than the temperature can adjust and large liquid droplets will form. Larger liquid droplets, which do not disperse at the compressor suction, will lead to excessive erosion of the compressor rotor. This phenomenon is difficult to diagnose without the aid of dynamic simulation and is often misidentified as carry over from the knock out drum. The rate of Anti-Surge control action should be optimized, based on the specific installation configuration, in order to reduce the propagation of pressure waves in the suction line.

5.3.3 Local vs Remote Controls and Instrumentation

Stability of compressor operation and ability for instrumented controls to maintain system parameters are essential for remote operation of systems. Instrumented systems, when in automatic control, reduce the requirement for operator intervention and, hence, human error.

Automatic compressor control systems are sometimes required to handle a wide range of operating scenarios: start-up, compressor loading, various levels of turndown and trips to name a few. Often, it is the case that no single or even multiple control loops are capable of achieving sufficiently stable operation. As a result, the Operator may employ the manual mode of operation. Increased risk to the facility engendered by taking the system out of automated control, to place it in manual mode, introduces another failure point of the instrumented systems. Additionally, the risk register needs to be modified based on the need for people to go to the compressor building to change things.

6. Performance Optimization of Compressor Systems

Similar to distillation columns and reactive systems, compressor systems exhibit complex behaviors which are very sensitive to changes in upstream as well as downstream conditions. Recycle streams and multi stage compressors complicate things further. Focusing on the improvement of one performance aspect, to the detriment of another, can lead to unsafe system behavior or intended improvements becoming unintended reductions. Any time mechanical or operational changes to any part of a compressor system are proposed, both, the increased risk, and the sustainability of those changes, needs to be assessed.

6.1 Dangers of Optimization

It is important to recognize that facility performance optimization, while desirable from a financial standpoint can have an impact on every aspect of process safety. Many times the impact of facility upgrades on a single aspect of safety is assessed and the results are assumed to be the same for the other systems.

For example, the impact of changes may be assessed as far as the pressure relief valves are concerned and the changes may be acceptable from that perspective; however, the changes may impact the ability of the Emergency Depressurization systems to depressurize in the required time. In another case, the changes in pressure and flow rate may not be a problem for the relief systems, but, since the heat and material balance simulations are often never updated across the entire system, the performance of the safeguards on peripheral systems may no longer be adequate due to variance in composition, heat transfer rate in integrated systems, or flare depressurization capacity.

The performance of most safety systems is rooted in design conditions (pressures, temperatures, compositions, flow rates) and operational ranges (vessel liquid levels, distillation column flow rate splits, etc.). Changes to either the design conditions or operational ranges will render invalid many aspects of previous safety system analysis.

Since most hazardous operating scenarios that apply to compressor systems actually involve failures in peripheral equipment that cascade down to the compressor, comprehensive system analysis needs to be performed in order to get a realistic picture of the potential dangers of system optimization.

6.2 Sustainable Changes

In addition to impacting facility safeguards, deviations from design conditions may have a significant impact on equipment reliability and availability. Unexpected shut downs, even though short term, have a large cost when looked at cumulatively. Additionally, turnarounds for large compressor systems are costly and time consuming so any changes that may impact the recommended maintenance schedule should be considered carefully. Evaluation of increases in compressor capacity should involve detailed analysis of system reliability and availability to ensure they are not increasing instantaneous production by decreasing long term capacity.

6.3 Example: Optimal Suction Pressure

In an effort to maximize throughput during periods of lower suction supply, focus is generally placed on maximizing compressor speed by exhausting the available horsepower of the driver. Operating at lower suction pressures to deliver maximum flow to the downstream equipment is the current general approach taken by operations. These conditions of highest speed and lower suction pressures often lead to increased recycle rates on the compression systems and negatively impact overall system performance. Negative impacts may include higher discharge temperatures and reduced flow to the downstream equipment (products of elevated compressor recycle). During some periods of 'maximization' it is not uncommon that some compressor controls be placed in manual mode to take advantage of the 'assumed' stable point of operation at maximum output.

Controls are preferred to be in auto to reduce human error, but this is not always the most stable option. Maximization efforts by current industry methods may not be delivering the best overall performance. Performance, in this case, meaning optimization of mass flow and temperature to downstream equipment with compressor controls in automatic mode and negligible operator intervention. Performance may be enhanced if compressor operation is optimized.

In the case of an LNG facility, whose main purpose is to reject heat to the atmosphere, inefficient compressor operation can degrade the facility performance by more than a loss of effective compression power. Some of the power that is lost to compression goes into the process fluid as enthalpy. Obviously, since the increase in inefficiency is non-linear, the increase in energy input into the stream is non-linear and, since the energy of inefficiency manifests as enthalpy, it can be very difficult to see without performing any calculations. Many times the energy difference is fairly significant and the total impact of compressor inefficiency can only be measured in terms of total system performance which, in the case of an LNG facility, can only really be seen in the mass that comes into or goes out of the facility.

The calculation methods in sections 2.1, 2.2, and 4.2 were used, along with data analysis at the speeds from section 3 to determine if the current operation methods will effectively optimize compressor performance on both, an individual equipment, as well as a holistic system basis. It would be preferable to show a relationship between LNG production and compressor performance; however, changes in the front of the plant can sometimes take several hours to manifest fully in the back of the plant so mass flow into the plant was analyzed as an indicative basis. A visual analysis of 2 years of operating data, shown in Figure 6-1 below, indicates is a very clear relationship between the feed gas compressor suction pressure and the facility inlet mass flow from the pipeline. Validating the hypothesis that increased inefficiency leads to unexpected negative performance, the data in Figure 6-1 shows, if the suction pressure is low enough, the facility inlet mass flow rate is even higher at design speed (4685 rpm) than at the maximum operating speed (4904 rpm). All of the pipeline flow moves through the compressor; however, taking considering recycle flow the compressor flow and the facility flow can vary a great deal. The data from Figure 6-1 were analyzed using the triangles shown. This was done to indicate a directional shift in performance as opposed to concrete performance deficiency.

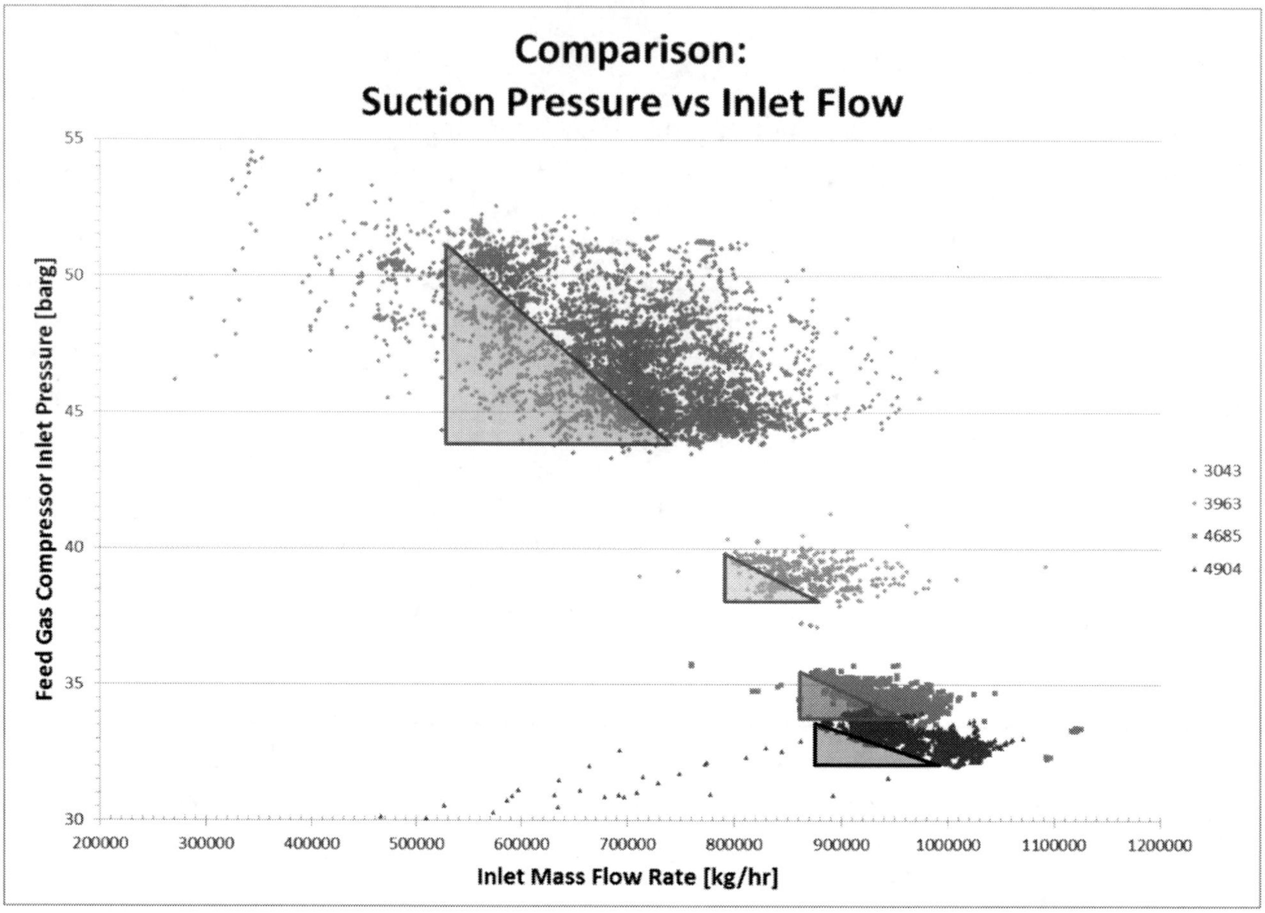

Figure 6-1. Comparison of Feed Gas Compressor suction pressure to the amount of mass flowing into the facility from the pipeline.

The data in Table 6-1 below shows the potential relationship between decreasing suction pressures and increasing mass flow rates as the speed of the machine increases.

The initial assessment, shown in Table 6-1 below, indicates as much as a 7% loss of efficiency, which is manifesting as a reduction in flow capacity, at sub-optimal conditions.

Table 6-1. Visual Analysis of Suction Pressure vs Inlet Flow data.

Speed	Low Pressure	High Pressure	Low Flow Rate	High Flow Rate	Flow Inc vs Press Decay
rpm	barg	barg	kg/hr	kg/hr	kg/barg
3040	38.8	51.2	527000	740000	17177
3963	37	39.8	790000	880000	32143
4685	33.7	35.5	860000	960000	55556
4904	32	33.5	875000	980000	70000

Figure 6.2, below, was generated to determine if there is a regular and usable pattern in the relationship between enthalpy inefficiency and compressor suction pressure.

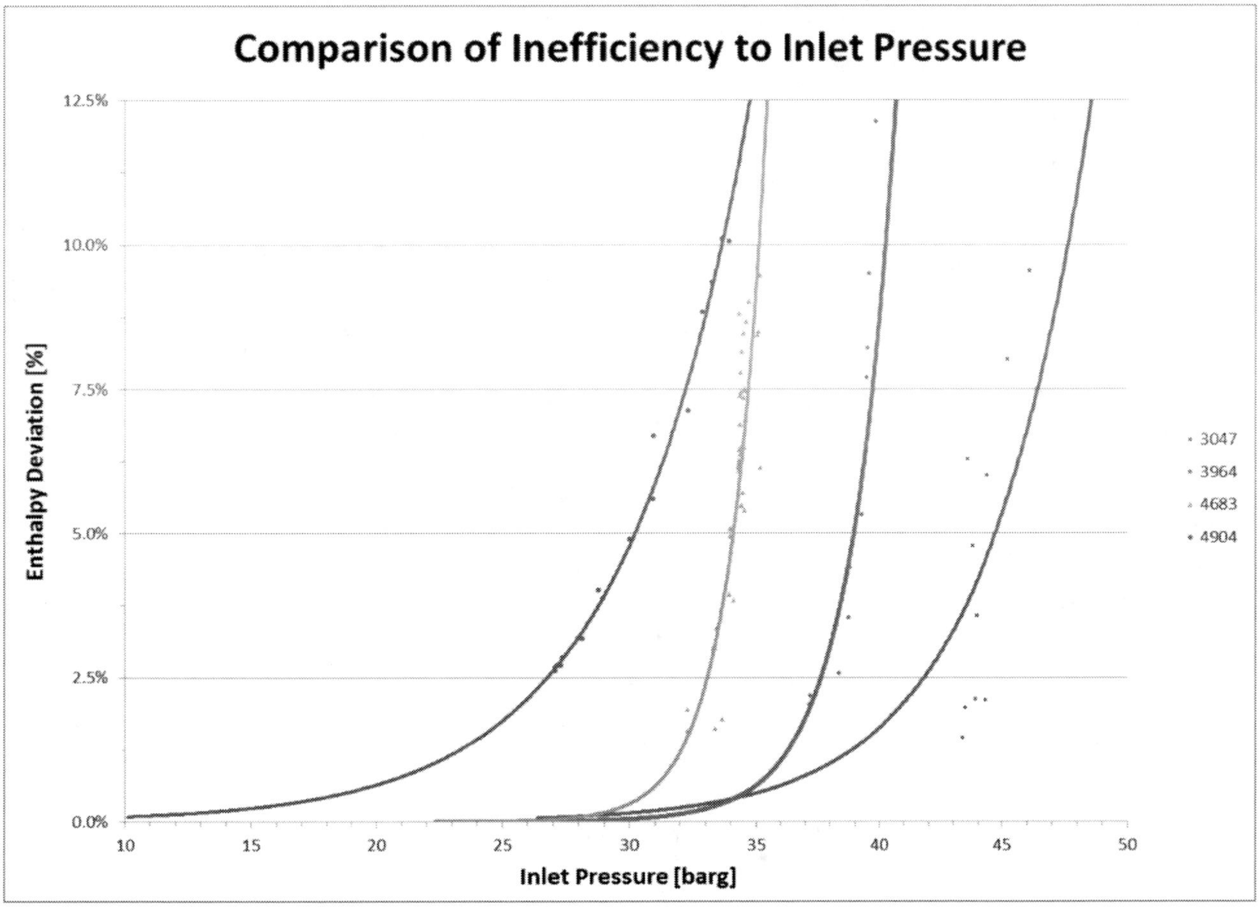

Figure 6-2 Comparison of Inefficiency to Inlet Pressure. Enthalpy deviation was calculated by comparing the polytropic work from the vendor supplied polytropic head curves to the ideal work transferred to the stream, calculated from an isentropic flash from measured inlet pressure and temperature to measured discharge pressure.

Enthalpy deviation exhibits an exponential behavior, at each speed, which approaches 0% inefficiency at 0 barg. The results here further support the idea that deviations between measured values and calculated values are a function of machine inefficiency. It is also encouraging that the behavior is consistent across all of the projected curves.

In order to develop a tool that may be used by the operations personnel to improve safety considerations and improve overall facility performance, best fit lines were generated from the lines in Figure 6.2. The best fit lines generated in Figure 6-3., below, defines a usable relationship between compressor inlet pressure and compressor speed. This chart indicates a range of suction pressures, for each operating speed of the evaluated compressor, that correspond to levels of compressor inefficiency. Once an acceptable degree of inefficiency is chosen the operator should maintain the suction pressure below pressure indicated on the curve.

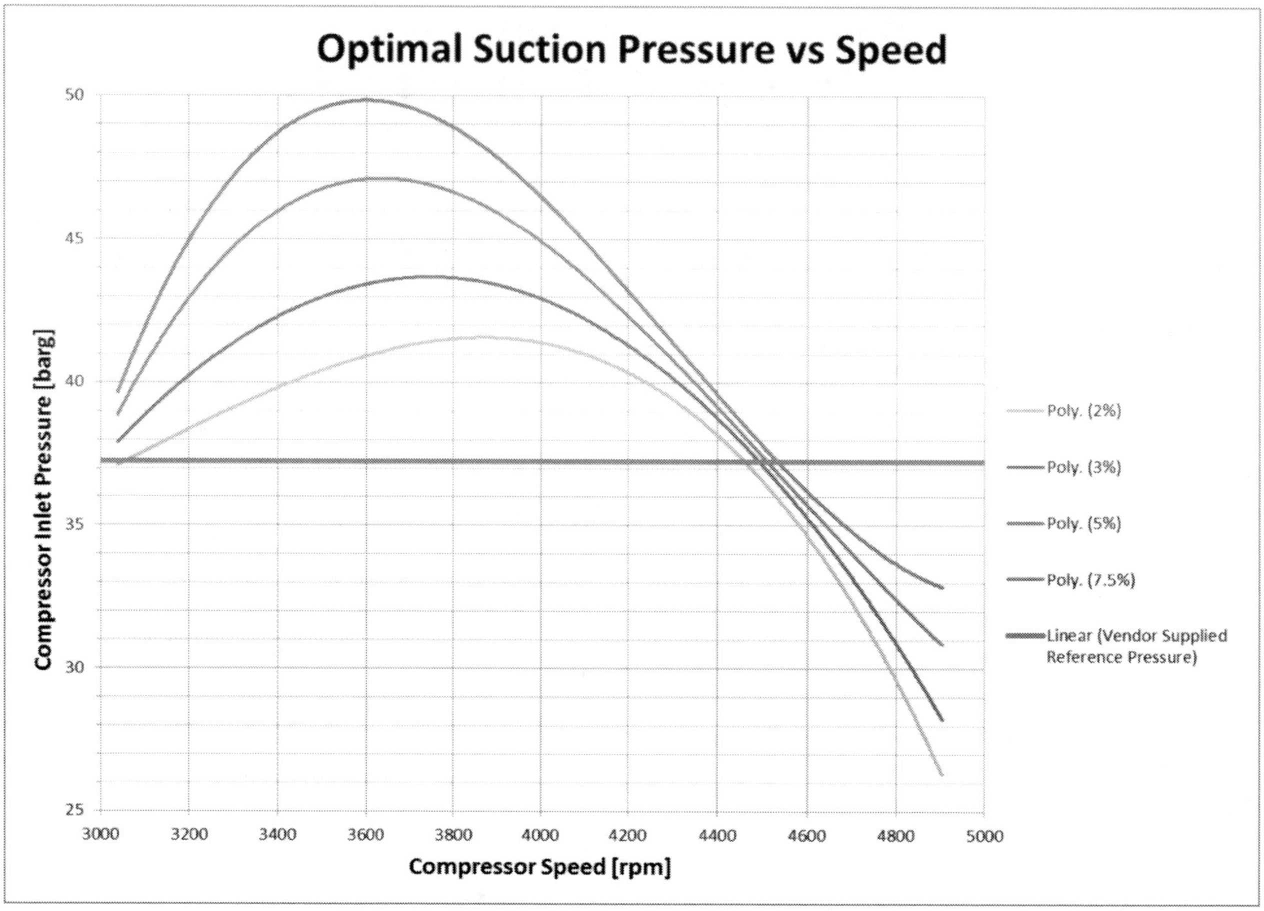

Figure 6-3. Optimal Suction Pressure vs Compressor Speed.

It is of some benefit to the operating team to receive guidance on potential for optimization of compressor operation. As compressor speed varies so does the suction pressure at which it gives best efficiency and recycle rate can be minimized. If this variation is known and defined as outlined in this paper, using Figure 6-3., the operator may adjust controls to minimize recycle rate, thus improving efficiency, increasing compressor availability, and achieving more stable compressor operation.

The example exercise above shows some potentially large improvements in facility performance while simultaneously improving process safety concerns. For the intended application to the Atlantic LNG facility, adjusting their operating procedures to use this chart is expected to result in a reduction in fuel gas expenditure of 3-7%, with a corresponding increase in LNG production. In many facilities, especially where the process is not driven by compressor performance, an increase in capacity is not beneficial; however, long term reductions in fuel cost can be significant. This style of analysis may be used to optimize virtually all aspects of compressor performance.

7. References

[1] ASME, "ASME PTC-10 – Performance Test Code on Compressors and Exhausters", 1997.

[2] Smith, J.M., Van Ness, H.C., Abbott, M.M., "Introduction to Chemical Engineering Thermodynamics", Seventh Edition, New York, NY, 2005.

[3] Schulz, J.M., "The Polytropic Analysis of Centrifugal Compressors", Transactions of the ASME, Series A. Vol. 84, Journal of Engineering and Power. January 1962, p. 69-82.

[4] API Standard 521, "Pressure-relieving and Depressuring Systems", 2014.

APPENDIX A – Definition and Description of Basic Terms

A_c	Cross sectional area – the cross sectional area of the piping through which the fluid being considered is flowing (around 8-10 diameters from the compressor)
cv	Control volume – denotes a fixed volume which is being analyzed
Z_1	Suction compressibility factor
ρ	Density
ρ_1	Suction density
ρ_2	Discharge density
$\rho_{measured}$	Measured density – the density of a fluid that is calculated from the fluid conditions measured by the operating facility
H	Enthalpy
H_1	Suction enthalpy
H_2	Discharge enthalpy
H_2'	Discharge isentropic enthalpy – enthalpy of a fluid based on isentropic flash to discharge pressure using "ideal" work
H_{is}	Isentropic enthalpy – enthalpy calculated using an isentropic flash
$H_{measured}$	Measured enthalpy – the enthalpy calculated from the fluid conditions measured by the operating facility
$H_{stagnation}$	Stagnation enthalpy – the enthalpy of a fluid considering the static, velocity, and gravitational contributions to pressure
H_{static}	Static enthalpy – the enthalpy of a fluid at static conditions not considering velocity or gravitational acceleration
fs	Flowing stream – denotes properties related to a mass moving through a fixed system
g	Gravitational acceleration
Q	Heat
z	Height
U	Internal energy
is	Isentropic – describes a process that has the same entropy value on the inlet as the outlet of a process, otherwise described as process which is reversible (no pressure volume work) and adiabatic (no heat input across system boundaries)
ke_1	Suction kinetic energy
ke_2	Discharge kinetic energy
$ke_{measured}$	Measured kinetic energy - the kinetic energy of a fluid that is calculated from the fluid conditions measured by the operating facility
ke_{static}	Static kinetic energy – the kinetic energy of a fluid calculated based on the static pressure and temperature (not considering velocity of gravitational acceleration)
$ke_{stagnation}$	Stagnation kinetic energy – the kinetic energy of a fluid calculated based on stagnation conditions
m	Mass flow
m_1	Suction mass flow
m_2	Discharge mass flow
n	Polytropic exponent – a condition specific exponent which describes the slope of a polytropic process on a pressure specific volume diagram
$n_{guessed}$	Polytropic exponent guessed for the purposes of calculation only
$n_{calculated}$	Polytropic exponent calculated based on a guessed or assumed polytropic exponent value
f	Polytropic work factor – a factor applied to the Real Gas Polytropic model intended to correct for the deviations of real gases from ideal behavior

P	Pressure
P_1	Suction pressure
P_2	Discharge pressure
$P_{2calculated}$	Discharge pressure calculated by the polytropic equation within an iterative calculation loop
$P_{2initial}$	Discharge pressure based on a guessed value for the purposes of calculation only
$P_{gravitational\ potential}$	Gravitational potential pressure – the part of stagnation or total pressure that manifests as a function of the acceleration of gravity acting on a fluid
$P_{stagnation}$	Stagnation pressure – the total pressure exerted by a fluid
P_{static}	Static pressure – the part of stagnation or total pressure that manifests as a function of the volume expansivity of a fluid
$P_{velocity}$	Velocity pressure – the part of stagnation or total pressure that manifests as a function of the velocity of the fluid
r_f	Recovery factor – a factor of the fluid enthalpy that represents the difference between the enthalpy at the measured temperature and the enthalpy at the static temperature
T_1	Suction temperature
t	Time
R	Universal gas constant – used here as the universal gas constant divided by the molecular weight of the fluid
v	Velocity
v_1	Suction velocity
v_2	Discharge velocity
V	System Volume
υ	Specific volume – reciprocal of density
υ_1	Suction specific volume
υ_2	Discharge specific volume
υ_2'	Isentropic specific volume – the specific volume that is coincident to the discharge isentropic enthalpy
W	Work
W_{is}	Isentropic Work – the "ideal" amount of work that is represented by a process which has the same entropy/disorder value at the inlet to a process as it does at the outlet of a process
W_p	Polytropic work – generally given by the polytropic head on the vendor supplied compressor curves

Risk Criteria Selection and the Impacts on LOPA Results:

To Sum or Not to Sum, That is the Question

Aaron Huberman
aeSolutions
3800 Centerpoint Drive, Suite 620
Anchorage, AK 99503
aaron.huberman@aesolns.com

Prepared for Presentation at
American Institute of Chemical Engineers
2016 Spring Meeting
12th Global Congress on Process Safety
Houston, Texas
April 11-13, 2016

Risk Criteria Selection and the Impacts on LOPA Results:

To Sum or Not to Sum, That is the Question

Aaron Huberman
aeSolutions
3800 Centerpoint Drive, Suite 620
Anchorage, AK 99503
aaron.huberman@aesolns.com

Keywords: Layer of Protection Analysis (LOPA), Target Mitigated Event Likelihood (TMEL), Target Frequency (TF), Risk Summation, SIL Selection, Risk Criteria, Risk Tolerance

Abstract

In the CCPS book *Layer of Protection Analysis – Simplified Process Risk Assessment*, Layers of Protection Analysis (LOPA) is initially described as the analysis of a single cause-consequence pairing. However, later in the book, there is the discussion of summing risk for multiple scenarios. In practice, several companies prefer to sum the frequencies of multiple causes leading to a single consequence when conducting LOPA. Summing the causes can be a useful tool in that it will ensure proper integrity of a safety function to address all of the causes for a single consequence and assist in the reduction of the numbers of Independent Protection Layers (IPLs) necessary at a facility. However, caution must be taken in using this method, as there may be unrealized effects on LOPA results, and therefore unintended impacts to the entire safety lifecycle. In contrast, evaluating only a single cause-consequence pair also poses different concerns when relating the results to selected risk criteria.

This paper will seek to provide insight into the effects of each choice, including the pros and cons of each method. Deeper examinations into the definitions of risk criteria and consequence will be explored.

1 Introduction

Layers of Protection Analysis (LOPA) is a simplified risk analysis that uses order of magnitude approximations to provide more precise estimations of likelihood for hazardous scenarios. LOPA is intended to be a semi-quantitative middle ground between the qualitative hazard identification techniques, such as Hazard and Operability Study (HAZOP), and full Quantitative Risk Assessment (QRA). The methodology allows a process to be analyzed in finer detail than can be achieved with HAZOP, without requiring the extensive time and effort of QRA. The LOPA methodology has been more extensively used in industry since publication of the CCPS book *Layer of Protection Analysis: Simplified Process Risk Assessment* [1]. Further guidance on the application of LOPA has been provided through the recent publication of the CCPS *Guidelines for Enabling Conditions and Conditional Modifiers in Layer of Protection Analysis* [2] and *Guidelines for Initiating Events and Independent Protection Layers in Layer of Protection Analysis* [3].

Despite the guidance in these publications, there are still aspects of LOPA for which guidance either does not exist or is very broad, as LOPA is still an evolving methodology. In *Layer of Protection Analysis: Simplified Process Risk Assessment* [1], as well as IEC 61511 [4], there is a brief discussion on the topic of summation of risk for multiple scenarios. However, in application in a wide range of industries, companies have chosen to sum the likelihoods of causes before comparison to risk tolerance criteria (referred to as "summing causes"). Although this can be a very useful variation of the methodology, it does create additional potential for inaccuracy of results, if not properly addressed.

2 Risk Tolerance Criteria

In order to discuss the idea of risk tolerance criteria, it is first necessary to define risk. In CCPS *Guidelines for Developing Quantitative Safety Risk Criteria* [5], the definition of risk is provided as follows:

> A measure of human injury, environmental damage, or economic loss in terms of both the incident likelihood and the magnitude of the loss or injury. A simplified version of this relationship expresses risk as the product of the likelihood and the consequences (i.e., Risk = Likelihood x Consequence) of an incident.

This definition, while effective, leaves much to the interpretation of those making risk-based decisions. Many factors must go into the proper evaluation of risk. One of the major questions regarding risk tolerance criteria is how to categorize risk to varying receptors (e.g., cumulative risk to personnel, acute risk to personnel, risk to offsite persons, risk to environmental receptors, etc.). LOPA is conducted on the basis of a single consequence and therefore often does not necessarily match the ideas of acceptable risk at a facility, unless this is taken into account when developing risk tolerance criteria.

The idea of risk tolerance is the notion that a certain level of risk may be endured in order to gain the benefits of the activity creating the risk. For example, a petroleum refinery operation creates significant hazards due to heat, pressure, flammable materials, and other sources. However, in order to create useful products out of crude oil, the refining process is necessary. The benefit of gasoline, diesel, or other refining products in terms of the potential for profit due to societal demand provides sufficient reason to continue the refining operation. Another example is the operation of chlorination facilities for water treatment. Although there is the potential for a chlorine release, which could have acute toxic effects, the overall benefit of providing disinfected drinking water necessitates the need for the inherent risk in the disinfection process.

"Tolerable" is a term often preferred over "acceptable" when discussing risk, as hazards are not generally viewed as bearable, but are necessary to gain benefits. In either case, the goal is to reduce the risk to a level where the benefits outweigh the potential hazards. All hazardous operations aim to avoid consequences, but only if prevention of the consequences does not exceed reasonable costs.

The LOPA process is designed to estimate the likelihood portion when evaluating risk. In the LOPA process, risk tolerance criteria are usually defined in terms of a Target Mitigated Event Likelihood (TMEL) or Target Frequency (TF). This is basically a goal of how often the consequence of interest will occur for the process being evaluated (e.g., the likelihood that a fatality will occur). Since this is defined as a frequency or likelihood (often used interchangeably, despite potential differences between the two), it is usually expressed on a per year basis (1/year).

The LOPA methodology is intended for use with a single consequence or hazardous event (to be discussed later), usually taken directly from the HAZOP. Part of the reason for this is that LOPA is intended as a hazard assessment technique, unlike HAZOP which is intended as a hazard identification methodology and inherently identifies the consequence as part of scenario development. For example, a HAZOP would be used to determine the ways a process upset could lead to a fatality, but a LOPA would be used to more accurately define the likelihood of the same scenario.

LOPA is known to be effective for only a limited range of scenario types. Most companies develop criteria to determine which scenarios are further reviewed using LOPA. The most common approach is to apply LOPA to high consequence severity scenarios where an instrumented solution may be applied to prevent the scenario. Using this method, choosing the severity of the consequence in the HAZOP also assigns the TMEL for the subsequent LOPA. Figure 1 provides an example of an onsite safety risk matrix.

For simplicity, this risk matrix is only intended to address the risk estimation for workers onsite. For a complete analysis, the risk matrix would need to be expanded upon, or additional risk matrices would need to be created to address other categories of risk, such as offsite impacts, environmental impacts, and commercial impacts. Many companies will choose to provide descriptions for different consequence types while using the same general matrix for simplicity. In addition, other companies will never define any risk as intolerable, but instead will define the level of approval required to continue the operation (i.e., the higher the risk, the higher position, potentially up to president of the company, required to approve the risk).

In Figure 1, locations in the matrix that represent tolerable levels of risk are green. Therefore, when evaluating a scenario, the severity is a fixed row on the matrix, so the goal is to reduce the likelihood in order to move left in the matrix until tolerable risk is reached. For example, if a scenario is being analyzed that may result in a single fatality, additional mitigations are required until the frequency of the scenario is less than one in 10,000 per year (i.e., TMEL = 10^{-4}/year). It should be noted that this implies the target for the single scenario being evaluated on the risk matrix.

Severity								
	A	3+ Fatalities	L	M	M	M	H	H
	B	1-2 Fatalities	L	L	M	M	M	H
	C	Severe Injury	L	L	L	M	M	M
	D	Recordable Injury	L	L	L	L	M	M
	E	First Aid Injury	L	L	L	L	L	M
			$<10^{-5}$/year	10^{-4}/year - 10^{-5}/year	10^{-3}/year - 10^{-4}/year	10^{-2}/year - 10^{-3}/year	10^{-1}/year - 10^{-2}/year	$>10^{-1}$/year
			1	2	3	4	5	6
			Likelihood					

H	**High Risk** – Immediate mitigation or shutdown necessary
M	**Medium Risk** – Mitigation should be implemented as soon as possible
M	**Marginal Risk** – Mitigation should be implemented, if reasonable
L	**Low Risk** – Tolerable as is, no further mitigation required

Figure 1 – Example Onsite Safety Risk Matrix

By having tolerable risk defined by a risk matrix used on a per scenario basis, it must be understood that the cumulative risk at the facility is actually higher. At a facility, there may be a large number of different scenarios that could result in the same severity which would have to be aggregated to determine the overall risk to personnel. This must be considered when developing risk criteria. For example, using the risk matrix above, if there were 100 possible scenarios that result in a fatality, the overall potential for fatality at the facility would actually be one in 100 per year (10^{-2}/year). While this may still be a reasonable target for overall risk at a facility, this must be considered against the appropriate risk tolerance. If a company had determined a fatality one in 10,000 per year (10^{-4}/year) was the overall requirement, the risk target would not be met in this case.

Another consideration in defining the risk tolerance criteria is the application of the criteria. If corporate standards define the risk tolerance for all locations, the cumulative risk at each facility must be taken in to account. For instance, an oil and gas company may own multiple refineries in different locations. Due to advances in technology, varying dates of construction of each facility, and desired products from the facility, the complexity and design of each facility may vary greatly. However, if the corporate risk criteria are set on a per scenario basis, without taking the potential variations into account, the overall risk at each facility may be very different, as the more complex or older facilities may have a significant number of additional scenarios, potentially reaching an order of magnitude difference. This variation may be acceptable based upon greater production and therefore greater benefit of operation of a larger plant, but it must be taken into consideration at the time of risk criteria development. In turn, a plant with less production that uses old technology may have a greater overall need for risk reduction due to the lack of improvements. Once again, the key is to ensure these factors are taken into account on the front end.

3 Consequences and Hazardous Events

As mentioned previously, the LOPA process is designed to estimate the likelihood portion when evaluating risk, but it depends on proper definition of consequence before conducting the LOPA. Therefore, the definition of consequence defined in *Guidelines for Developing Quantitative Safety Risk Criteria* [5] should be examined:

> The undesirable result of an incident, usually measured in health and safety effects, environmental impacts, loss of property, and business interruption costs.

Although this definition provides a starting point for estimating the consequence of an event, due to the potential impacts of interpretation of the consequence, there can be significant impacts on the results of a LOPA, especially when summing the frequencies of multiple causes leading to the same consequence.

When conducting a LOPA using a single cause-consequence pairing, the precise definition of consequence becomes less important. If the cause is pressure control valve failure, it is simple to define the potential for overpressure, rupture, and potential fatality. The sequence of events from the cause to the consequence can be clearly laid out. The primary concern is proper selection of the appropriate severity ranking, and therefore, usually, the appropriate target mitigated event likelihood (TMEL) or target frequency (TF). The target (i.e., TMEL or TF) is automatically determined by selecting the severity according to the risk matrix being used in the analysis.

It should be noted that in some cases a company will set TMEL based on a combination of the severity and likelihood chosen as part of the risk ranking portion of the hazard identification effort (often with the use of the HAZOP methodology). While this method may be applied, it adds subjectivity to the likelihood ranking in the HAZOP, which may therefore add inaccuracy in the analysis. If this method is used, proper definition of the TMEL is further complicated, both in the case of single cause-consequence LOPA and summed-cause LOPA.

However, when summing causes (i.e., aggregating the mitigated event likelihoods of multiple causes for comparison to a single target), the consequences must be more clearly defined to determine which items should actually be summed. Depending on the definition of consequence, the LOPA results could range by an order of magnitude or more, which can result in unnecessary costs and/or meaningful impacts to the actual risk in terms of the level of protection provided.

Often used interchangeably with the term "consequence" is the term "hazardous event." However, the differentiation between the two terms becomes important in the context of summing causes in LOPA. IEC 61511 Edition 2 [4] defines a "hazardous event" as an event that can cause harm. Since this does not provide clear guidance for the term "hazardous event," in the context of risk analysis, it could be thought of as a specific event leading to the outcome of concern. In other words, the hazardous event requires a specific chain of events to lead to the consequence. For example, while a fatality may be defined as a consequence, a hazardous event may be defined as overpressure of a vessel with release and personnel exposure to a toxic material, leading to a fatality.

4 Summing Causes in LOPA

The concept of summing causes in LOPA has been mentioned previously, but has not been defined to this point. In a single cause-consequence pair LOPA, the basic calculation performed is as follows:

$$RRF = \frac{IEF \times \prod_{j=1}^{J} PFD_j \times \prod_{k=1}^{K} FM_k}{TMEL}$$

Where:

RRF = Risk Reduction Factor, a unitless measure of the additional protection necessary, this is the inverse of the probability of failure on demand (PFD)

IEF = Initiating Event Frequency (often referred to as Initiating Event Likelihood), the order of magnitude approximation of the frequency of occurrence of the initiating event per year (1/year)

$\prod_{j=1}^{J} PFD_j$ = Product of the probabilities of failure on demand of each independent protection layer (IPL), a unitless measure of the probability that an IPL will fail when needed to protect against a consequence

$\prod_{k=1}^{K} FM_k$ = Product of the Frequency Modifiers (often referred to as Conditional Modifiers) applicable for the scenario, a unitless adjustment factor to take into account items such as Probability of Ignition, Occupancy Factor, Time at Risk Factor, Enabling Conditions, etc.

When summing causes, the formula modifies slightly to become the following:

$$RRF = \frac{\sum_{i=1}^{I}(IEF_i \times \prod_{j=1}^{J} PFD_{ij} \times \prod_{k=1}^{K} FM_{ik})}{TMEL}$$

It may be noted that, in this version of the equation, different protection layers and/or frequency modifiers may be applicable for different causes of the same consequence. Depending on the way the methodology is applied at a specific company, this equation may have to be slightly modified.

Now that the equation has been defined, what does it really mean to sum causes? What LOPA really does is estimate a frequency of reaching the final consequence and compare that frequency to a target. If causes are summed and compared to the same target that is applied to a single cause-consequence pair, the precision of the results may be skewed.

4.1 What to Sum?

The basic concept behind summing causes seems simple; however, selecting the appropriate causes to sum is where the difficulty lies. Some companies will choose to sum the causes that lead to a specific consequence, which raises the question of how to properly define a consequence. For the purpose of severity determination in HAZOP, the consequence may be looked at as the impact to persons or the environment, but not both simultaneously. Using that definition of consequence, any fatality that occurred due to any process upset could possibly be summed. This would prove ineffective in application to the LOPA process. There would be no reasonable method to reach the risk target in LOPA, as Independent Protection Layers (IPLs) in general would not be able to address all of the reasons for fatality at once.

Another summing methodology often applied in industry is to sum for all causes that lead to the same hazardous event. However, defining which hazardous events are actually equivalent is difficult. For example, in a fired heater, there are multiple reasons for loss of flame, which could lead to deflagration and fatality. Results of the LOPA may vary depending on the definition of the hazardous event. If flameout due to high pressure of the fuel gas is considered the hazardous event, there may only be a small number of causes that are summed. However, for the same heater, if flameout by itself is considered the hazardous event, the number of potentially summed causes, and therefore requirements for any installed SIF, will increase.

Yet another example of when to sum is to sum all causes that lead to the same hazardous event, and will share the same SIF. Using this methodology, it can be argued that the results of the LOPA will then reflect the overall demand rate for the SIF being evaluated. However, care must be taken to ensure the SIF definition and application are appropriate, as loss of precision of the results can easily occur if improperly defined.

4.2 Summing Example

Below is a simplified drawing of a forced draft fired heater. The primary method to protect the heater is to shut off the flow of fuel gas to the heater by closing the Shut Down Valves (SDVs). Several different inputs may result in shutdown of the heater. In general, it could be assumed that low fuel gas pressure on PT-1, high fuel gas pressure on PT-1, low air flow on FT-1, high air flow on FT-1, and loss of flame on BE-1 would all cause the SDVs to close.

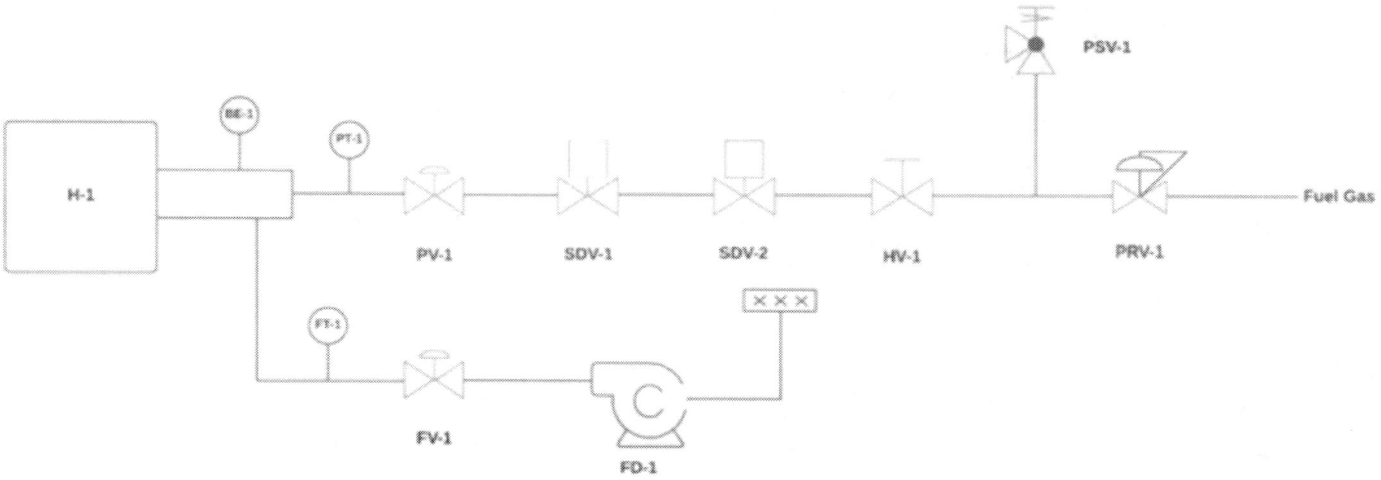

Figure 2 – Example Heater

It is assumed that any upset in the balance of air and fuel may lead to flameout of the burner (i.e., more or less flow or pressure of air or fuel would result in the mixture in the heater not being appropriate to sustain combustion), which could lead to accumulation of fuel gas in the heater and subsequent deflagration. Examples of causes of loss of flame include failure of the combustion air blower, or an upset in any of control loops, such as PV-1 or FV-1. In this example, it is assumed that the heater is located in an area where personnel are not constantly present, allowing the use of an occupancy factor frequency modifier of 0.1 (i.e., it is assumed that only 10% of the time will there be a person present in the event of deflagration). Because the heater would be assumed to be in operation, the residual heat is assumed to be an ignition source, and therefore the probability of ignition is conservatively taken as 1 (meaning 100% chance of providing an ignition source). For simplicity it is assumed that the initiating event frequency (IEF) for each cause is one in ten per year (0.1/year). Also for simplicity, there are assumed to be no enabling conditions for any of the causes. Deflagration of the heater could lead to one or two fatalities if the area was occupied. Using the risk matrix provided earlier in Figure 1, the TMEL for the event is therefore one in 10,000 per year (10^{-4}/year).

When examining a single cause-consequence pair with LOPA, assuming no credit for any IPLs other than the proposed SIF, the calculation would look as follows:

$$RRF = \frac{IEF \times P_O \times P_i}{TMEL} = \frac{0.1 \times 0.1 \times 1}{10^{-4}} = 100$$

Where:

RRF = Risk Reduction Factor

IEF = Initiating Event Frequency

P_o = Occupancy Factor

$P_i =$ Probability of Ignition

Note that for the single consequence pair method for this example, the specific cause being examined does not matter, as it was assumed that each cause has the same IEF. By not applying any credit for the SIF in the scenario, it helps determine the total RRF that will be required by a SIF to provide the necessary Safety Integrity Level (SIL)[1] to protect against the consequence. With each different SIL level, there are different requirements for the SIF. IEC 61511 [4] relates SIL Levels to the RRF according to the following table:

Safety Integrity Level (SIL)	**Risk Reduction Factor (RRF)**
4	>10,000 to ≤100,000
3	>1,000 to ≤10,000
2	>100 to ≤1,000
1	>10 to ≤100

Now, if it was assumed that summing was used for LOPA for all causes that lead to flameout of the heater, either ignoring which SIFs were related or assuming that the only SIF was a shutdown on loss of flame, the initiating events that would have to be considered would be pressure valve PV-1 closing, PV-1 opening, manual valve HV-1 being closed, pressure regulating valve PRV-1 closing, PSV-1 opening, PRV-1 opening, flow valve FV-1 closing, FV-1 opening, forced draft fan FD-1 failing, and filter F-1 plugging.[2] All of these causes may lead to an imbalance in the fuel to air ratio in the heater leading to flameout. Note that for causes that would lead to reduced fuel input, it is assumed that the cause could be either partial closure, providing insufficient fuel to sustain combustion or closing allowing loss of flame, then reopening (e.g., instrument air system issue causing a momentary loss of air which could close the valve, then reopen upon air being reestablished). Being that there are ten causes, assuming each at 0.1/year, as before, the calculation would look as follows:

$$RRF = \frac{10 \times (0.1 \times 0.1 \times 1)}{10^{-4}} = 1{,}000$$

Note that although it is shown as multiplication, the calculation is actually the addition of ten similar causes, and shown as multiplication for simplicity. This assumes the same risk tolerance criteria as the single cause-consequence calculation. This results in an entire order of magnitude increase in the necessary risk reduction. Note that if the only SIF on the heater that existed was

---

1 Safety Integrity Level is a term used to describe the reliability of a safety instrumented function. Industry standards such as IEC 61511 [4] are used to define requirements for SIF design based on the SIL.

2 It should be noted that for certain control loops, failures resulting in a valve going both open and closed could be counted as one cause. In this case, they are assumed as separate causes to support the subsequent discussions. In general, this level of detail may be more appropriate for a more rigorous methodology.

shutdown based on loss of flame, and there were no other IPLs, the last calculation would provide the target for the SIF.

For the next calculation, it is assumed that summing is based on a combination of hazardous event and SIF for the case of high pressure causing flameout, which will be protected by a single SIF that will close SDV-1 and SDV-2 based on loss of flame on BE-1 or low air flow on FT-1 (i.e., either loss of flame or low flow alone would close the valves, creating a 1oo2 voted input to the SIF). Once again, the same risk tolerance criteria will be applied. In this case, there are three causes that would have to be considered: flow valve FV-1 closing, forced draft fan FD-1 failing, and filter F-1 plugging. Therefore, the calculation would look as follows:

$$RRF = \frac{(0.1 \times 0.1 \times 1) + (0.1 \times 0.1 \times 1) + (0.1 \times 0.1 \times 1)}{10^{-4}} = 300$$

4.3 What Do the Results Mean?

In the previous three calculations, three different results were achieved when analyzing what can be considered to be the same process, depending on the summing method applied. The question then becomes: Which answer is correct? In reality, there is no wrong answer. In any of the cases above, protection layers would be added to ensure safety of personnel based upon the above described risk tolerance criteria.

However, in the first case where only a single cause-consequence pair was examined, the SIF built to address the scenario would only require a SIL1 system to be put in place (although a SIL2 would likely be more appropriate). This means that there would be lower lifecycle costs for the system than the other results. However, the fact that the SIF would be less robust would result in slightly higher risk at the facility. In either case, it may meet the risk tolerance criteria set by the company. On the other hand, in one of the subsequent cases, the SIF must meet SIL2 requirements which will cause increased lifecycle costs. In order to increase the integrity of a SIF from SIL1 to SIL2, hardware redundancy would be required which would increase the capital costs of SIF. For example, a simple SIL1 SIF may only be comprised of a single transmitter, single SDV, and simple Programmable Logic Controller (PLC), whereas a SIL2 SIF may require two transmitters in a one out of two (1oo2) voting scheme, with two SDVs in series (meaning either closing may prevent the hazard), and an upgraded PLC. In addition, the varying requirements may require more frequent testing, which equates to additional operational costs throughout the life of the process. While creating a SIF that is more robust than necessary may be a positive in terms of the safety of persons at risk, the added safety may not be worth the increased lifecycle costs. In addition, the more robust system could increase the potential for spurious trips, which could negatively impact the company commercially. While a redundant architecture, such as 1oo2, is less likely to fail to trip when needed, there are also more devices that could fail which could lead to a spurious trip, resulting in lost production and less revenue.

4.4 To Sum or Not to Sum?

All of this variation brings up the question, is it better to sum causes in LOPA or to evaluate on a single cause-consequence basis? Either methodology will result in evaluation of the hazards at facility. Conducting a LOPA on a single cause-consequence basis may result in additional time and/or paperwork for analysis because each one needs to be documented separately. In addition, the results may be less conservative, as they do not look at combined demand rates on a specific protection. On the other hand, summing causes introduces some level of subjectivity into the analysis by choosing what items to sum. The results may be either excessively conservative, or not conservative enough based on what items are chosen to sum. In the previous examples, the resultant RRF could range from 100 to 1,000 depending on what items were summed. However, once again, there is no correct answer about whether to sum or not. Some companies may choose to apply multiple levels of risk criteria to address both single cause and summed causes separately. Some companies may choose to perform the analysis on a single cause basis, but adjust the target results if it is noted that there are a significant number of causes for a consequence (e.g., more than four causes leading to the same consequence results in adjusting the SIL target by a level from SIL1 to SIL2 to account for the multiple causes). Any methodology can provide an effective analysis if applied properly. The major caveat is that whichever method is chosen, the risk tolerance criteria need to be designed appropriately. The lack of precision between the results from summing or not is less important than providing accurate results by defining the risk tolerance criteria appropriately for the chosen methodology.

5 Conclusion

Developing risk tolerance criteria for LOPA is an extremely important activity as it sets the basis for a potentially large amount of work that may follow as part of the safety lifecycle. While there are many examples and much guidance available of the topic, there is no definitive answer given by any source. In general, there is no "correct" answer for the results of a LOPA, but how the risk tolerance criteria are defined may skew the results which could lead to a process being safer or costing more to operate than intended. The goal of developing risk tolerance criteria is to find the appropriate middle ground that will allow for worker safety while minimizing negative impacts on the economics of the business. There are a variety of intricacies of LOPA that may differ between industries, companies, and even facilities. These differences are tolerable, as long as the differences are understood and taken into account before the LOPA begins. Regardless of which option is chosen, it is important to understand the strengths and weaknesses of each option, to ensure proper application. Also, it is important that those responsible for selection of the risk criteria have an understanding of the available options for the range of quantitative assessment techniques.

6 References

[1] CCPS. *Layer of Protection Analysis: Simplified Process Risk Assessment.* Center for Chemical Process Safety, American Institute of Chemical Engineers, New York, NY, 2001.

[2] CCPS. *Guidelines for Enabling Conditions and Conditional Modifiers in Layer of Protection Analysis.* Center for Chemical Process Safety, American Institute of Chemical Engineers, New York, NY, 2013.

[3] CCPS. *Guidelines for Initiating Events and Independent Protection Layers in Layer of Protection Analysis.* Center for Chemical Process Safety, American Institute of Chemical Engineers, New York, NY, 2015.

[4] Committee Draft IEC 61511, Edition 2. *Functional safety – Safety instrumented systems for the process industry sector, Parts 1–3*. Geneva: International Electrotechnical Commission. 2015.

[5] CCPS. *Guidelines for Developing Quantitative Safety Risk Criteria.* Center for Chemical Process Safety, American Institute of Chemical Engineers, New York, NY, 2009.

A Call to Action: Understanding the Key Driver for Continual Process Safety Performance Improvement

David A. Jones
JCL Risk Services LLC
16225 Park Ten Place, Suite 500
Houston, Texas 77084
djones@jclsc.com

Kay A. Modi
JCL Risk Services LLC
16225 Park Ten Place, Suite 500
Houston, Texas 77084
kmodi@jclsc.com

Prepared for Presentation at
American Institute of Chemical Engineers
2016 Spring Meeting
12th Global Congress on Process Safety
Houston, Texas
April 11-13, 2016

A Call to Action: Understanding the Key Driver for Continual Process Safety Performance Improvement

David A. Jones
JCL Risk Services LLC
16225 Park Ten Place, Suite 500
Houston, Texas 77084
djones@jclsc.com

Kay A. Modi
JCL Risk Services LLC
16225 Park Ten Place, Suite 500
Houston, Texas 77084
kmodi@jclsc.com

Keywords: process safety management, action system, continual performance improvement, human factors

Abstract

A well-established design concept for management systems is the continual improvement life cycle involving the steps: Plan-Do-Check-Act (PDCA). Similar to standards for managing quality, environment, occupational health and safety, and security, design models for effective process safety management (PSM) systems also utilize the PDCA cycle.

The activities within each element of an integrated PSM system include the identification of deficiencies and closure of these gaps in order to improve overall system performance in reducing and/or controlling process safety risks. Root cause analysis is a valuable learning tool for sustained improvement. Without gaining a deep understanding of the cause(s) for why a deficiency exists or why an incident occurred, there can be no meaningful corrective or preventive action for improvement. Instead, only a correction as a short-term "quick fix" is possible, which results in a higher likelihood (and hence risk) that the same or similar deficiency (or incident) will be repeated. As a result, some improvement opportunities can be missed and many others only offer partial (or at best temporary) solutions to identified deficiencies within the PSM system itself or in the engineered process system to which PSM is being applied to reduce and/or control risks.

Yet successfully executing the "Plan", "Do", and "Check" steps for the identification and analysis of the causes of deficiencies, but failing during the "Act" step to properly implement and close out action items (from e.g., process hazard analyses, equipment inspections and tests, incident investigations, compliance audits, etc.) on time or, even worse, failure to close them at all, defeats the whole purpose. Besides lost opportunities for risk reduction, open actions can result in legal "smoking guns", as well-intended but unfulfilled actions that could be viewed as willful inactions.

1. Introduction

A well-established design concept for management systems is the continual improvement life cycle involving the steps: Plan-Do-Check-Act (PDCA). Similar to standards for managing quality, environment, occupational health and safety, and security [2,9,11,12], design models for effective process safety management (PSM) systems also utilize the PDCA cycle [13,14].

At a higher level, the PSM system elements address all four of the PDCA steps:

- **Plan** – Process Safety Information, Process Hazard Analysis (PHA), Operating Procedures, Operator Training, and Contractors address planning activities.
- **Do** – Employee Participation, Management of Change, and Emergency Planning/Response address operational activities.
- **Check** – Pre-Startup Safety Review, Mechanical Integrity, Incident Investigation, and Compliance Audits address verification activities.
- **Act** – Most of the PSM elements address activities or systems for managing actions.

In a well-design PSM system, each PSM element represents an individual management process which is aimed at contributing to an integrated, comprehensive, and risk-based approach for controlling risks and improving process safety performance. At a lower level, the PDCA life cycle model for each of these management processes assures a robust treatment of the process resources and activities required to achieve effective results for continual improvement.

Any deficiencies within the design or the implementation of these management processes will have a cumulative degrading effect on the results. For example, if one does not even plan to manage at the "Plan" step, there is no hope of ever reaching the actions needed for improvement. Similarly, planned activities that are poorly executed at the "Do" step will not achieve the desired results. Further, without proper monitoring and measuring at the "Check" step to verify whether results have actually been achieved, the likelihood of identifying system deficiencies is reduced. And finally, failure to effectively act on the findings or recommended actions in the "Act" step results in missed opportunities for continual improvement.

Yet successfully executing the "Plan", "Do", and "Check" steps for the identification and analysis of the causes of deficiencies, but failing during the "Act" step to properly implement and close out action items (from e.g., process hazard analyses, equipment inspections and tests, incident investigations, compliance audits, etc.) on time or, even worse, failure to close them at all, defeats the whole purpose. Besides lost opportunities for risk reduction, open actions can result in legal "smoking guns", as well-intended, but unfulfilled actions that could be viewed as willful inactions.

Thus, managing process safety performance for continual risk reduction and/or risk control requires robust management processes that are effectively managed throughout the PDCA life cycle of key entities. These entities include critical inputs, resources, activities, and outputs for the overall PSM system, as well as the engineered process system for which the risks are being managed. For example, processes for managing leadership and employee participation are key inputs aimed at engaging and involving the entire workforce. Managing personnel and contractor competence, information quality, and asset integrity are all aimed at providing quality human, informational, and physical resources, respectively, which are needed to design, construct, operate, and maintain the engineered process system. Processes for managing the hazards and risks before an incident, managing emergencies during an incident, and managing lessons learned after an incident, are the three critically opportune times for reducing and/or controlling process safety risks. Finally, processes for managing plans (policies, programs, procedures), managing changes (organizational, procedural, technical), managing performance (conformance, compliance, results) and managing actions (corrections, corrective actions, preventive actions), are each aimed at the critical outputs of the risk-based process safety management system processes for Plan, Do, Check, and Act.

This paper stresses the integration of the "Check" and "Act" steps for several of the PSM elements and their management processes. Specifically, verification results of the "Check" steps for each of the management processes includes, for example, employee survey results, equipment tests and inspections, PHA findings, emergency response drill evaluation findings, incident investigation findings, audit findings, etc. Root cause analysis is a valuable learning tool for sustained improvement. Without gaining a deep understanding of the cause(s) for why a deficiency exists or why an incident occurred, there can be no meaningful corrective or preventive action for improvement. Instead, only a correction as a short-term "quick fix" is possible, which results in a higher likelihood (hence risk) that the same or similar deficiency (or incident) will be repeated.

Further, the action results of the "Act" step include the resolution and treatment of the recommended actions to correct deficiencies, address the findings, or improve other processes that could impact process safety performance. In order to achieve effective and lasting results from the action system, it is crucial to understand the differences between: (1) correction to eliminate deficiencies; (2) corrective action to eliminate the cause of an identified deficiency in order to prevent recurrence; and (3) preventive action to eliminate the cause of potential deficiencies in order to prevent future occurrences.

Ten real-life examples will be presented to demonstrate several elements of a PSM system where improperly managed deficiencies and actions have led to actual process safety incidents. Further, these examples will show how human factors are most often the prevalent root cause(s). That is, human errors or decisions causing deficiencies associated with designing and/or implementing either the management system or the engineered process system, if not properly identified and eliminated, represent a potential multi-fold increase in the risk of process safety incidents. This paper will draw upon the authors' extensive experience in conducting PSM system audits, incident investigations, and process hazard analyses, as well as expertise in designing robust systems for managing process safety, including corrective and preventive actions.

2. Repeat Performance – The Need for Action

Changes, both planned (e.g., design changes, operational improvements) and unplanned (e.g., equipment deficiencies, human error) create a continual challenge to the management of process safety risks. If not properly managed, these changes can lead to deficiencies within the PSM system as well as the engineered process system. Identifying system deficiencies, including the causes of process safety incidents, is the first step in treating them. If not properly managed through actions, these deficiencies can repeat themselves or manifest as a systemic problem for similar situations.

2.1 System Deficiencies – The Opportunity for Improvement

Existing deficiencies with any system, whether a management system or the engineered process system, create process safety risks associated with ineffective risk controls required to prevent or mitigate process hazards. Each management process that employs a PDCA model presents opportunities to identify system deficiencies in the "Check" step, while certain processes (e.g., Process Hazard Analysis, Incident Investigation, and Compliance Audits) are actually intended as verification processes and therefore generate findings of deficiencies. Example deficiencies derived from verification activities for typical PSM system elements are shown in Table 1.

Table 1: Example PSM Deficiencies

PSM Element	Verification Activities	Example Deficiencies
Employee Participation	Culture survey	Inadequate communications
Process Safety Information	Design reviews P&ID walk-downs	Inaccurate piping & instrument diagrams (P&IDs), missing data
Process Hazard Analysis	Evaluation of process hazards and safeguards	Inadequate safeguards, incomplete design basis, design deficiencies
Operating Procedures	Periodic review for currency and accuracy	Outdated operating procedures
Operator Training	Testing / re-testing, on-the-job evaluation	Inadequately trained operators
Contractors	Pre-job and periodic evaluation of performance	Unqualified contractors or contractors not performing to expectations
Pre-startup Safety Review (PSSR)	Safety review of design, operating procedures, training, etc.	Construction and equipment not in accordance with design specifications
Mechanical Integrity	Equipment tests and inspections	Equipment deficiencies outside of acceptable limits
Management of Change	Technical review for proposed change	Impact of change on safety and health not assessed or not acceptable
Incident Investigation	Causal factor analysis and root cause analysis	Various management system deficiencies as findings
Emergency Planning / Response	Post-drill or post-incident response evaluation	Inadequate planning, lack of resources, inadequate emergency equipment
Compliance Audits	Interviews, records reviews, physical inspections/observations	Various management system deficiencies as findings

Note: **Bolded** PSM elements above require an action system per the OSHA PSM Standard [17].

Effective PSM systems that rigorously identify system deficiencies present opportunities for reducing and/or controlling risks at three stages of the incident life cycle [13,14]:

- ***Before* an incident** – Prevention through design and operational discipline in order to identify, avoid, eliminate, or minimize risks to prevent occurrence (e.g., Process Hazard Analysis, Compliance Audits, and most other PSM elements);
- ***During* an incident** – Emergency response in order to prepare for and respond to an incident in order to mitigate consequences during occurrences (e.g. Emergency Planning and Response); and
- ***After* an incident** – Incident investigation with root cause analysis and lessons learned in order to prevent recurrence (e.g., Incident Investigation).

However, missed opportunities for actions to improve risk reduction and/or risk control can occur at each step of the PDCA improvement life cycle for any of the management system processes:

- **Plan** – No plan or other mechanism to identify deficiencies; i.e., you didn't look for possible problems or opportunities for improvement.
- **Do** – Deficiencies not adequately identified; i.e., you didn't find existing problems.
- **Check** – Identified deficiency not adequately analyzed; i.e., you didn't determine root causes and settled for only corrections, instead of corrective and/or preventive actions.
- **Act** – Actions not adequately developed; i.e., you didn't develop proper solutions (poorly worded, wrong solutions) or any at all (ruled out based on risk, time, cost, etc.).

2.2 Correction, Corrective & Preventive Action – The Drivers for Improvement

Effectively integrating identified deficiencies with actions in a formal system allows us to gain the most out of managing the opportunities for improvement. For each of the management processes for the PSM system elements, the aim is to take the appropriate actions on a continual basis to improve the quality of the PSM system and its effectiveness in reducing and/or controlling process safety risks. The design intents of each PSM system element support the overall performance improvement objective of the PSM system, as shown in Table 2.

Table 2: Example PSM Action Objectives

PSM Element	Performance Objectives to Reduce/Control Risks
Employee Participation	To improve participation and ownership
Process Safety Information	To improve information quality
Process Hazard Analysis	To improve the process design & safeguard effectiveness
Operating Procedures	To improve operational control
Operator and Maintenance Training	To improve competence and performance
Contractors	To improve qualifications and performance
Safe Work Practices (Hot Work Permit)	To improve control of work
Pre-startup Safety Review	To improve the adequacy of PSM elements before startup
Mechanical Integrity	To improve the asset integrity
Management of Change	To improve the control of changes
Incident Investigation	To improve the system and prevent incident recurrence
Emergency Planning & Response	To improve response capability and mitigate consequences
Compliance Audits	To improve the effectiveness of the PSM system

In the context of management, actions are things we as humans do with intent. We often use the term "action plan" to define what and how we will carry out an action. For people to effectively carry out actions, management needs to define who (person responsible), what (the task), when (the schedule), where (the application), why (the intent or objective), and how (the method).

Types of Actions with Examples

Both industry and regulators use terms for "actions" that appear to be synonymous. To many, the terms *recommendation*, *resolution*, *action*, *correction*, *corrective action*, and *preventive action* all refer to the same thing: to fix a deficiency or address a finding or nonconformity. In fact, there are significant differences in these terms and their understanding can offer a better path toward performance improvement.

First, *recommendation* typically refers to a recommended action to address a finding, or deficiency.

Next, a common misunderstanding is with the term *resolution* (of recommendations from PHAs, incident investigations, etc.) as used by the OSHA PSM Standard. Resolution does not mean to close the action, but instead refers to a decision by management to either adopt the recommendations or to decline to do so. Since the promulgation of the PSM Standard in the early 1990s, OSHA has clarified their intent within their Compliance Directive for PSM, as follows:

> "OSHA considers an employer to have 'resolved' the team's findings and recommendations when the employer either has adopted the recommendations, or has justifiably declined to do so. Where a recommendation is rejected, the employer must communicate this to the team, and expeditiously resolve any subsequent recommendations of the team. An employer can justifiably decline to adopt a recommendation where the employer can document, in writing and based upon adequate evidence, that one or more of the following conditions is true:
>
> 1. The analysis upon which the recommendation is based contains material factual errors;
> 2. The recommendation is not necessary to protect the health and safety of the employer's own employees, or the employees of contractors;
> 3. An alternative measure would provide a sufficient level of protection; or
> 4. The recommendation is infeasible." [18]

Such a clarification, originally provided as direction to OSHA Inspectors, also provides industry with a better understanding of the steps to resolve, then take, appropriate action.

Finally, there are three different types of *action* intended to address an identified deficiency, each with a specific purpose. As a minimum action, we generally make a *correction* to eliminate an identified deficiency. However, *corrective and preventive actions*, both of which address the root causes of deficiencies, offer longer-lasting and farther-reaching solutions, respectively. These three types of actions are commonly used in international standards for quality, environment, occupational health and safety, and security management systems [2,9,11,12].

For consistency, the following definitions based on quality management terms from ISO 9000:2015 are used throughout this paper [19]:

- **Nonconformity** – Non-fulfilment of a requirement. Note, the term "deficiency" is a type of nonconformity.
- **Correction** – Action to eliminate a detected nonconformity (or deficiency).
- **Corrective action** – Action to eliminate the cause of a nonconformity (or deficiency) in order to prevent recurrence.
- **Preventive action** – Action to eliminate the cause of a potential nonconformity (or deficiency) or other undesirable potential situation in order to prevent occurrence.

The important distinction here is that a correction only addresses eliminating the immediate deficiency, whereas a corrective action goes further to eliminate the cause(s) of that deficiency in order to prevent recurrence. Even more importantly, a preventive action extends the corrective action to other potential deficiencies, perhaps in other management system elements, other equipment or process units, or even other facilities.

The risk reduction opportunity of actions is increased dramatically by extending a correction to a corrective action, and adding preventive actions, to address the elimination of causes of deficiencies. Thus, lessons learned from eliminating causes can reap a larger return on investment. To illustrate this concept with a simple example, consider the possible actions and improvement opportunities for the following hazard scenario:

> During a walkthrough inspection of a manufacturing plant, lubrication oil was observed on the floor next to operating machinery.

Possible actions to treat this hazardous condition are postulated in Table 3:

Table 3: Actions to Treat an Oil Spill

Action Type	Possible Action	Benefit
Correction	Eliminate the detected deficiency by cleaning up the oil spilled on the floor.	Prevents someone from slipping on the spilt oil (in the short term).
Corrective Action	Eliminate the cause of the detected deficiency by replacing a defective hose to prevent recurrence of a spill.	Prevents someone from slipping on the spilt oil (in the long term) and reduces repeated cleanup efforts.
Preventive Action	Eliminate the cause of a potential deficiency by replacing similarly defective hoses on other machinery at this and other plants to prevent occurrence before they leak.	Prevents several people from possibly slipping on spilt oil at several locations and eliminates future cleanup efforts.

The important implication here is that many process safety incidents occur as a result of ignoring corrections or not properly implementing the corrections. Further, other incidents have occurred despite correcting the deficiencies, by not analyzing and eliminating the causes of the deficiencies that could have prevented their recurrence. Of even more significance, some incidents could have had a more profound effect on the prevention of similar incidents within the industry had published recommendations been more effectively shared and more rigorously applied as preventive actions.

The "Check" steps for PSM system elements should aim to identify the system deficiencies and to analyze their causes so that the appropriate actions can be developed and taken to correct the deficiencies and eliminate their causes. Thus, actions should be aimed at answering both "what exists/happened" and "why it exists/happened".

Causes can be analyzed through a simple or formal root cause analysis process, depending on the complexity of the deficiency. Causal factor analysis and root cause analysis are quite common in incident investigations to determine the causes of process safety incidents [4,7]. Similarly, failure analysis and trending is often used for engineered process systems to determine the root causes of process equipment deficiencies to assure mechanical integrity [5,7].

While inspections, tests, and other verification methods may be used to identify engineered process system deficiencies, auditing techniques are aimed at identifying the systemic causes of management system deficiencies, such as a weak safety culture and other organizational and human factors [3]. But without the proper analysis of the causes of such deficiencies, the actions which are generated tend to address only corrections and not corrective and preventive actions.

To further illustrate the points discussed above, three example process safety incidents with tragic outcomes from published investigations are summarized in Table 4 for the cases of (1) no corrections, (2) no corrective actions, and (3) no preventive actions, covering three different life cycle phases for the engineered process system: design, operation, and maintenance.

Table 4: Example Incidents Involving Failure to Carry Out Three Types of Actions

Actions Not Taken	Incident Summary	Deficiencies and Impacts
Corrections (deficiencies not identified)	• D.D. Williamson & Co. (DDW) facility in Louisville, Kentucky. • Caramel Coloring Feed Tank Overpressure and Explosion, 2003 • Life cycle phase – Design CSB Investigation Report No. 2003-11-I-KY [28]	• Deficient Management Process – Hazard Evaluation System (i.e., Process Hazard Analysis) • 1 fatality, community evacuation, and shelter-in-place
Corrective Actions (corrections made, but causes of deficiencies not identified)	• Giant Industries' Ciniza refinery in Jamestown, New Mexico. • Hydrofluoric Acid Alkylation Unit Pump Fire and Explosion, 2004 • Life cycle phase – Maintenance CSB Case Study Report No. 2004-08-I-NM [26]	• Deficient Management Process – Mechanical Integrity program • 6 employee injuries
Preventive Actions (potential deficiencies not identified for previous incidents)	• First Chemical Corporation (FCC) facility in Pascagoula, Mississippi • Mononitrotoluene (MNT) Distillation Column Runaway Reaction, Explosion and Fire, 2002 • Life cycle phase – Operations CSB Investigation Report No. 2003-01-I-MS [29]	• Deficient Management Process – Incident Investigation program • 3 employee injuries, offsite consequences

Example Incident #1 – No Corrections

Lesson to be Learned – *Without a verification process to identify design deficiencies and evaluate the associated process hazards, corrections to reduce risks cannot be made.*

The CSB found:

> "D. D. Williamson did not have effective programs in place to determine if equipment and processes met basic process and plant engineering requirements.
> - There was *no program to evaluate* [emphasis added] necessary layers of protection on the spray dryer feed tanks. Likewise, there was no recognition of the need to provide process control and alarm instrumentation on the two feed tanks.
> - The feed tanks were installed for use in the spray dryer process *without a review* [emphasis added] of their design versus system requirements.
> - Safety valves on the spray dryer feed tanks had been removed to transport the tanks to Louisville and were never reinstalled. There was *no* evidence that DDW conducted an *engineering evaluation* [emphasis added] to determine the hazards of this change." [28]

The CSB concluded:

> "D. D. Williamson did *not have adequate hazard analysis systems* [emphasis added] to identify feed tank hazards, nor did it effectively use contractors and consultants to evaluate and respond to associated risks.
> - Neither DDW nor its contractors and consultants recognized the need for overpressure protection for the two feed tanks used in the spray dryer process." [28]

Example Incident #2 – No Corrective Actions

Lesson to be Learned – *Unless the causes of deficiencies are identified and eliminated, these same deficiencies can recur and corrections to the same problem may be repeated multiple times, thus increasing the risk that their potential consequences will be realized in future incidents.*

The CSB found:

> "A review of repair work prior to the incident revealed a history of repeated pump failures. The primary, electric, and steam-driven spare isostripper recirculation pumps had 23 work orders submitted for repair of seal-related problems or pump seizures in the one-year period prior to the incident" [26].

The CSB concluded:

> "Giant's mechanical integrity program did not effectively prevent these repeated failures of the pump seals… Giant's approach to these frequent pump seal problems was an example of breakdown maintenance. In other words, pump failures were addressed when the equipment finally broke down, instead of identifying causes of breakdowns and preventing them before they occurred again" [26].

Example Incident #3 – No Preventive Actions

Lesson to be Learned – *When instances of deficiency identification and the actions for their solution are treated in isolation and not shared, valuable lessons learned cannot be effectively applied to other potential deficiencies to prevent similar incidents.*

The CSB noted from *Essential Practices for Managing Chemical Reactivity Hazards* by CCPS: "Multiple facilities in an organization may have similar chemical reactivity hazards . . . or use similar technology to control the associated hazards. If so, it may be more efficient for a corporate office or personnel to assume responsibility for some improvement activities… This can also facilitate communication of incidents and best practices between facilities" [29].

The CSB revealed the following root cause:

> "The FCC Pascagoula facility did not have an adequate system for evaluating the hazards of processing mononitrotoluene (MNT) in its continuous process and *did not apply lessons learned from hazard analyses conducted on similar processes in the plant* [emphasis added] … The facility became aware of the hazards of allowing MNT to be exposed to elevated temperatures for an extended time during a batch project in 1996, but the lessons learned (including operating considerations and the addition of safety interlocks) were not applied to the existing MNT columns… there was *no system to apply evaluation results* [emphasis added] from the batch process to continuous processing equipment" [29].

The CSB also found:

> "FCC experienced an explosion and fire in a batch process under development for a third party in 1986… The incident involved a runaway reaction and overpressurization of equipment in a column that had no provisions to mitigate a thermal runaway… One of the recommendations was to perform hazard analyses of existing processes. FCC *did not apply lessons learned from this event* [emphasis added] to the MNT distillation system. OSHA made the following determinations: The PHA was deficient because it did not identify 'any previous incident which had a likely potential for catastrophic consequences'" [29].

Action Systems

A means for effectively managing corrections, corrective actions, and preventive actions is an action system, often narrowly referred to as a corrective action system. Requirements for these systems have been defined by international standards and guidelines for quality, environment, occupational health and safety, and security management systems, as well as by regulations [2,9,11,12].

Good practice beyond the minimum stated compliance requirements would be to incorporate these features into a common action system and apply them to all PSM system processes requiring the management of actions throughout their life cycle.

Table 5 shows "what good looks like" for an action system using a PDCA model, with references to explicit requirements in the OSHA PSM Standard [17] for certain aspects of such a system. An interesting observation from Table 5 is how the wording of the requirements for similar action systems can vary, even from the same source.

Table 5: Action System Model

Step	Task	Example OSHA PSM Requirements [17]
Plan	• Address findings (deficiencies) and recommendations. • Resolve findings (deficiencies, contributing factors) and recommendations; document resolution. • Determine causes of deficiencies. • Document actions (corrections, corrective and preventive actions) to be taken. • Assign responsibilities for actions. • Develop written schedule to complete actions.	**Process Hazard Analysis** – "The employer shall establish a system to promptly address the team's findings and recommendations; assure that the recommendations are resolved in a timely manner and that the resolution is documented; document what actions are to be taken; develop a written schedule of when these actions are to be completed." **Incident Investigation** – "The employer shall establish a system to promptly address and resolve the incident report findings (i.e., factors that contributed to the incident) and recommendations. Resolutions and corrective actions shall be documented." **Compliance Audits** – "The employer shall promptly determine and document an appropriate response to each of the findings of the compliance audit."
Do	• Complete actions (correct deficiencies, eliminate causes of deficiencies). • Communicate actions and lessons learned to those affected.	**Process Hazard Analysis** – "Complete actions as soon as possible; communicate the actions to operating, maintenance and other employees whose work assignments are in the process and who may be affected by the recommendations or actions." **Incident Investigation** – "The report shall be reviewed with all affected personnel whose job tasks are relevant to the incident findings including contract employees where applicable."
Check	• Monitor timeliness of closeouts. • Review actions taken. • Review effectiveness of actions. • Monitor trends for common deficiencies.	
Act	• Follow up on overdue actions. • Document closeout.	**Compliance Audits** – "Document that deficiencies have been corrected."

Properly designed and implemented action systems, whether manual or electronic, would incorporate a robust process for managing the aforementioned life cycle of actions. The next section addresses examples of process safety incidents that have occurred as a result of inadequately designed or ineffectively implemented PSM system management processes, including the action system.

3. Example Process Safety Incidents – Lessons to be Learned

Industry is replete with examples of process safety incident investigation reports and case histories from which we can learn lessons and apply to the improvement of PSM systems [8,15,16,19,20].

Ten real-life examples are presented in this section to demonstrate several elements of a PSM system where improperly managing deficiencies and actions have led to actual process safety incidents. These examples show a wide range of issues related to deficiencies within both the PSM system and the engineered process system for a cross-section of industries, types of hazards, and geographic regions.

These examples also demonstrate how human factors are often the prevalent root cause(s) [6]. That is, human errors or decisions causing deficiencies associated with designing and/or implementing either the management system or the engineered process system, if not properly identified and eliminated, represent a potential multi-fold increase in the risk of process safety incidents. Organizational and human factors contributing to incidents beyond human errors and poor decisions may include a lack of commitment and recognition of the value of safety, an inadequate safety culture, misperceptions of risk, a higher appetite for risk taking, and simply poor leadership and management of the business.

For each of the PSM system elements and corresponding example process safety incidents presented in Table 6, a synopsis from the respective incident investigations reports by the U.S. Chemical Safety and Hazard Investigation Board (CSB) is presented in the following sub-sections.

These briefs synopses demonstrate how PSM system deficiencies within PSM system elements (individually or in combination) contributed to the ultimate failure of management to take appropriate action(s) that could have prevented such incidents.

From the example incidents selected, we see that there are different ways to break the PDCA continual improvement life cycle in such a way that actions were not able to be taken (see Table 7). Such gaps prevent achieving the goal of continually improving process safety performance.

For instance, some incidents involved missing or ineffectively implemented policies and programs that would have required in the "Plan" step the need for identifying deficiencies, and therefore prevented any progress in the subsequent "Do", "Check", and "Act" steps.

Next, some incidents which may have had successful planning also inadequately identified deficiencies in the "Do" step, which did not allow for meaningful causes to be determined in the subsequent "Check" step.

Further, other incidents included no root causes or improperly determined causes of the identified deficiencies in the "Check" step, which did not allow for suitable actions to be proposed for the "Act" step.

Finally, one incident demonstrated that despite apparent success in the "Plan", "Do", and "Check" steps, ineffectively managed action within the action system prevented improvement and actually contributed to the incident.

Table 6: Example Incidents from Ineffective Management of Deficiencies & Actions

PSM Element	Incident Reference	Location, Impacts & Process Deficiencies
Example Incident 1: Employee Participation	Refinery Crude Unit Pipe Rupture and Fire, 2012, CSB Investigation Report No. 2012-03-I-CA [27]	• Refinery in Richmond, California. • 6 minor employee injuries. • Stop Work Authority program; Safety Culture Assessment process.
Example Incident 2: Process Hazard Analysis	Refinery Propane Deasphalting Unit Fire, 2007, CSB Investigation Report No. 2007-05-I-TX [30]	• Refinery in Sunray, Texas. • 3 employee injuries; 1 contractor injury. • Process Hazard Analysis program.
Example Incident 3: Operating Procedures	Refinery Isomerization Unit Overfill, Fire and Explosion, 2005, CSB Investigation Report No. 2005-04-I-TX [31]	• Refinery in Texas City, Texas. • 15 contractor fatalities; 180 employee and contractor injuries. • Operating Procedures process; Management of Change (MOC) process; Pre-Startup Safety Review (PSSR) process.
Example Incident 4: Operator Training	Ink Dust Collection System Explosion and Flash Fire, 2012, CSB Case Study No. 2013-01-I-NJ [24]	• Manufacturing facility in East Rutherford, New Jersey. • 7 employee injuries. • Operator Training Program.
Example Incident 5: Contractors & Hot Work Permit	Oilfield Production Tank Explosion, 2006, CSB Investigation Report No. 2006-07-I-MS [23]	• Oilfield in Raleigh, Mississippi. • 3 contractor fatalities; 1 serious contractor injury. • Contractor Safety Program; Hot Work Program.
Example Incident 6: Mechanical Integrity	Refinery Sulfuric Acid Storage Tank Explosion, 2001, CSB Investigation Report No. 2001-05-I-DE [32]	• Refinery in Delaware City, Delaware. • 1 employee fatality; 8 employee injuries. • Mechanical Integrity management system.
Example Incident 7: Management of Change (MOC) & Pre-Startup Safety Review (PSSR)	Peroxide/Alcohol Piping Explosion and Flash Fire, 2004, CSB Safety Bulletin No. 2004-03-B [21]	• Petrochemical facility in Port Neches, Texas. • 2 serious employee injuries. • Management of Change process; Pre-Startup Safety Review process.
Example Incident 8: Incident Investigation	Thermal Decomposition in Polymer Catch Tank, 2001, CSB Investigation Report No. 2001-03-I-GA [33]	• Plastics plant in Augusta, Georgia. • 3 employee fatalities. • Incident Investigation system.
Example Incident 9: Emergency Planning / Response	Mixing Tank Heptane Vapor Cloud Explosion, 2006, CSB Case Study No. 2006-08-I-IL [25]	• Concrete chemicals facility in Bellwood, Illinois. • 1 contractor fatality; 1 serious employee injury; 1 minor employee injury. • Emergency Planning & Response process.
Example Incident 10: Compliance Audits	Vinyl Chloride Tank Welding Flammable Vapor Explosion, 2010, CSB Case Study No. 2011-01-I-NY [22]	• Chemical plant in Buffalo, New York. • 1 contractor fatality; 1 serious contractor injury. • Compliance Audit process.

Table 7: Summary of PSM System Deficiencies for Ten Example Incidents

PSM Element	Plan	Do	Check	Act
	Policies/Programs developed?	Deficiencies identified?	Causes determined?	Actions taken?
Example Incident 1: Employee Participation	✓	✓	✓	✗
Example Incident 2: Process Hazard Analysis	✓	✓	✗	✗
Example Incident 3: Operating Procedures	✓	✓	✗	✗
Example Incident 4: Operator Training	✓	✗	✗	✗
Example Incident 5: Contractors & Hot Work Permit	✗	✗	✗	✗
Example Incident 6: Mechanical Integrity	✓	✓	✗	✗
Example Incident 7: Management of Change & PSSR	✓	✗	✗	✗
Example Incident 8: Incident Investigation	✓	✓	✗	✗
Example Incident 9: Emergency Planning / Response	✗	✗	✗	✗
Example Incident 10: Compliance Audits	✓	✗	✗	✗

Note: ✓ indicates a successful PDCA step; ✗ indicates an unsuccessful step.

3.1 Example Incident 1 – Employee Participation

Incident Reference	Plan	Do	Check	Act
	Policies/Programs developed?	Deficiencies identified?	Causes determined?	Actions taken?
Refinery Crude Unit Pipe Rupture and Fire, 2012, CSB Investigation Report No. 2012-03-I-CA [27]	✓	✓	✓	✗

Lesson to be Learned – A Stop Work Authority (SWA) program at the Chevron Refinery in Richmond, California was in place and periodic safety culture assessments required by the local regulations provided a means of verification to identify deficiencies (findings), develop an improvement plan with a list of actions and milestones, and provide a rationale for prioritizing actions and their justification [27]. *However, actions were not effectively taken to improve identified weaknesses in the SWA program and therefore contributed to the incident in 2012.*

Finding – The CSB concluded:

> "In the years leading to the August 6, 2012, incident, the Chevron Richmond Refinery identified weaknesses in its Stop Work Authority program due to employee hesitation to use Stop Work Authority when witnessing an unsafe act. However, the regulator did not require the Chevron Richmond Refinery to take quality, constructive steps to improve these areas." [27]

Human Factors Issue – The CSB noted:

> "No one formally invoked their Stop Work Authority. In addition, Chevron safety culture surveys indicate that between 2008 and 2010, personnel had become less willing to use their Stop Work Authority. Regardless of how a Stop Work program is portrayed, there are a number of reasons why such a program may fail related to the 'human factors' issue of decision-making; these reasons include belief that the Stop Work decision should be made by someone else higher in the organizational hierarchy, reluctance to speak up and delay work progress, and fear of reprisal for stopping the job." [27]

3.2 Example Incident 2 – Process Hazard Analysis

Incident Reference	Plan	Do	Check	Act
	Policies/Programs developed?	Deficiencies identified?	Causes determined?	Actions taken?
Refinery Propane Deasphalting Unit Fire, 2007, CSB Investigation Report No. 2007-05-I-TX [30]	✓	✓	✗	✗

Lesson to be Learned – A Process Hazard Analysis (PHA) program at the Valero McKee Refinery in Sunray, Texas was in place and periodic PHA revalidations should have provided a means of verification to identify safeguard deficiencies, generate and resolve recommendations, and follow up on their implementation [30]. *However, a recommended action for installing remotely-operated shut-off valves (ROSOVs) made during the initial PHA eleven years prior was never implemented, was subsequently inaccurately closed out as "complete", and was not discovered during a 5-year PHA revalidation.*

Finding – The CSB concluded:

> "The McKee Refinery did not apply Valero's mandatory *Emergency Isolation Valve* procedure when evaluating risks in the PDA unit to ensure that the large quantities of flammable materials in the unit could be rapidly isolated in an emergency" [30].

Human Factors Issue – In addition to the abovementioned action being incorrectly closed out as being complete as a result of human error, the verification processes failed on multiple occasions from an organizational perspective to detect this critical error.

3.3 Example Incident 3 – Operating Procedures

Incident Reference	**Plan**	**Do**	**Check**	**Act**
	Policies/Programs developed?	Deficiencies identified?	Causes determined?	Actions taken?
Refinery Isomerization Unit Overfill, Fire and Explosion, 2005, CSB Investigation Report No. 2005-04-I-TX [31]	✓	✓	✗	✗

Lesson to be Learned – A management process for developing and implementing written Operating Procedures was in place at the BP Texas City Refinery, with the intent to provide clear instructions for safely conducting activities involved in each covered process (as required by the OSHA PSM Standard) [31]. Furthermore, BP guidelines requiring that procedures be reviewed as often as necessary and certified annually as being current and accurate, including changes managed under the MOC policy, as well as PSSR reviews, should have provided a means of verification to assure that they reflected current operating practices (as required by the OSHA PSM Standard) [31]. *However, the operating procedures did not reflect actual practice, were not corrected over time, were not reviewed for changes as required before multiple start-ups, and were routinely deviated from during operations.*

Finding – The CSB found in the two previous PSM audits prior to the incident (another verification means for identifying management system deficiencies), that:

> "a number of operating procedures were not current and did not accurately reflect practices on particular units" and that "process safety action item resolution was still a problem for the refinery (20 percent of open action items were overdue), and that changes were still being made before MOC sign-offs and action items had been resolved" [31].

Human Factors Issue – The CSB noted:

> "Addressed here are the human factors that explain why feed was added to the tower for three hours without liquid being removed. While recognizing that human errors were made in the raffinate startup, this investigation goes beyond individual failures to gain a deeper understanding of why the incident occurred, which is more useful in major accident prevention... The broader aspects of this investigation revealed serious management safety system deficiencies that allowed the operators and supervisors to fail. The following underlying latent conditions contributed to the unsafe start up:
>
> - A work environment that encouraged operations personnel to deviate from procedure.
> - Lack of a BP policy or emphasis on effective communication for shift change and hazardous operations (such as unit startup).
> - Ineffective supervisory oversight and technical assistance during unit startup.
> - Insufficient staffing to handle board operator workload during the high-risk time of unit startup.
> - Lack of a human fatigue-prevention policy.
> - Inadequate operator training for abnormal and startup conditions.
> - Failure to establish effective safe operating limits." [31]

3.4 Example Incident 4 – Operator Training

Incident Reference	Plan	Do	Check	Act
	Policies/Programs developed?	Deficiencies identified?	Causes determined?	Actions taken?
Ink Dust Collection System Explosion and Flash Fire, 2012, CSB Case Study No. 2013-01-I-NJ [24]	✓	✗	✗	✗

Lesson to be Learned – Although combustible dust is not covered under the OSHA PSM Standard and at the time OSHA had not promulgated a combustible dust standard, certain management system processes were in place at the U.S. Ink/Sun Chemical black ink manufacturing facility in East Rutherford, New Jersey, including operator training and related verification processes for managing risks, including PHA and MOC [24]. *However, these processes were not effectively implemented to prevent injuries from an ink dust explosion and flash fire in a new dust collection.*

Finding – Further, a Capital Appropriations/Asset Request (CAR) process was in place, but the CSB found:

> "In the CAR environmental health and safety section, a checkbox indicating the need for a process hazard analysis (PHA) or management of change (MOC) was not checked, indicating that neither a PHA nor a MOC was necessary for the dust collection system" [24]. This precluded an important means for verifying that procedures and training were in place, based on an understanding of the combustible dust hazards.

Human Factors Issue – The CSB found that:

> "The lack of adequate oversight by Sun Chemical Corporation management personnel in the planning, design, installation, and commissioning of the dust collection system likely contributed to the October 9, 2012, incident. The CSB identified significant management issues, including inadequate project oversight, ineffective employee training on the dust collection mechanism, and failure to develop and implement corrective actions from a previous incident" [24].

3.5 Example Incident 5 – Contractors & Hot Work Permit

Incident Reference	**Plan**	**Do**	**Check**	**Act**
	Policies/Programs developed?	Deficiencies identified?	Causes determined?	Actions taken?
Oilfield Production Tank Explosion, 2006, CSB Investigation Report No. 2006-07-I-MS [23]	✘	✘	✘	✘

Lesson to be Learned – Neither a Contractor Safety Program nor a Hot Work Program were in place by Partridge-Raleigh (oilfield owner) or Stringer's Oilfield Services (contractor) at an oilfield in Raleigh, Mississippi when an explosion incident occurred on June 5, 2006 as flammable vapor inside two tanks was ignited by welding activities on a nearby tank [23]. *Such programs could have provided for safe work practices and emergency procedures, as well as a means for verification of the welding task (e.g., hot work permit) and verification of contractor safety management (e.g., evaluating contractor safety performance when selecting contractors, informing contract employers of the known potential fire, explosion, or toxic release hazards related to the contractor's work, and periodically evaluating on-the-job performance).*

Finding – The CSB found [23]:

- Partridge-Raleigh and Stringer's did not use available industry guidelines for hot work safety such as that provided in NFPA 326 and API 2009.
- A flammable gas detector was not used prior to welding activities. "Instead, workers used an open flame, or tank 'flashing' to verify that flammable vapor was not present in the tank."
- Partridge-Raleigh did not have established safety requirements for personnel at the oilfield. Stringer's had not established a formal safety program for its employees, and Partridge-Raleigh did not require Stringer's to have one.
- "Stringer's and Partridge-Raleigh did not adhere to OSHA requirements and precautions for burning and welding. OSHA Standard 29 CFR 1910.252 contains requirements in a number of areas including the use of guards to confine heat, sparks and slag generated during welding; special precautions when welding in the presence of explosive atmospheres; and requirements for cleaning used drums, barrels, tanks or other containers prior to welding".

Human Factors Issue – Neither a Contractor Safety Program nor a Hot Work Program at the facility were found to be in place. Without a culture of safety and the organizational commitments by both parties to require the most basic safety programs and safe work practices, this incident was not able to be prevented.

3.6 Example Incident 6 – Mechanical Integrity

Incident Reference	**Plan**	**Do**	**Check**	**Act**
	Policies/Programs developed?	Deficiencies identified?	Causes determined?	Actions taken?
Refinery Sulfuric Acid Storage Tank Explosion, 2001, CSB Investigation Report No. 2001-05-I-DE [32]	✓	✓	✗	✗

Lesson to be Learned – A Mechanical Integrity management system for the Motiva Enterprises Delaware City Refinery was in place and periodic inspections should have provided a means of verification to identify and correct equipment deficiencies that are outside acceptable limits in a safe and timely manner (as required by the OSHA PSM Standard) [32]. *However, this program was not effectively implemented to determine causes and to correct deficiencies before a process safety incident occurred on July 17, 2001.*

Finding – The CSB concluded as a root cause:

> "Motiva did not have an adequate mechanical integrity management system to prevent and address safety and environmental hazards from the deterioration of H_2SO_4 storage tanks.
> - The repeated recommendations of the tank inspectors that tank 393 be taken out of service 'as soon as possible' for an internal inspection were unheeded.
> - A leak in the shell of tank 393, observed in May 2001, was not repaired. Instead, the tank liquid level was lowered below the leak point and the tank remained in service.
> - Management failed to recognize the imminent hazard posed by the holes in tank 393 and did not promptly initiate repairs or take the tank out of service" [32].

The CSB also concluded as a contributing factor:

> "The Motiva refinery system for investigating Unsafe Condition Reports, informing workers about such reports, and tracking the satisfactory resolution of issues was inadequate.
> - In the 3 weeks between submittal of the Unsafe Condition Report on June 27 and the day of the incident, management did not correct the reported deficiencies or implement temporary safeguards" [32].

Human Factors Issue – The CSB noted a lack of management accountability:

> "The Motiva Enterprises LLC management oversight system failed to detect and hold Motiva refinery management accountable for deficiencies in the refinery's mechanical integrity, engineering management, and MOC systems" [32].

3.7 *Example Incident 7 – Management of Change*

Incident Reference	Plan	Do	Check	Act
	Policies/Programs developed?	Deficiencies identified?	Causes determined?	Actions taken?
Peroxide/Alcohol Piping Explosion and Flash Fire, 2004, CSB Safety Bulletin No. 2004-03-B [21]	✓	✗	✗	✗

Lesson to be Learned – MOC and PSSR processes at the Huntsman Petrochemical Corporation facility in Port Neches, Texas, were in place and should have provided a means of verification to manage changes to assure that considerations for changes to process chemicals, piping, and procedures (including safe work practices such as opening process equipment or piping), as well as the impact on safety and health, are addressed prior to any change, and also to confirm that prior to start-up that safety procedures are accurate (as required by the OSHA PSM Standard) [21]. *However, these processes did not adequately address the deficiencies associated with safety effects of a procedural change from water flushing to nitrogen purging for removal of hazardous chemicals as a precursor step to steam purging, nor the identification of low-point traps from a previous piping change, by the time a flash fire incident occurred in January 2004 [21].*

Finding – The CSB found:

> "To reduce waste volume, Huntsman revised the procedures to substitute inert gas purging for water flushing. However, purging with inert gas or steam does not necessarily remove trapped liquid. The revised procedure:
>
> - Failed to address the importance of identifying low points in the piping.
> - Failed to require the use of low-point drains to remove trapped hazardous liquids" [21].

Human Factors Issue – Based on a lack of oversight for managing technical changes in existing processes, the CSB concluded:

> "A comprehensive review of the as-built drawings, combined with a walkdown of the entire peroxide/alcohol transfer pipe – if required in the management of change (MOC) process – would have likely identified the low-point trap" [21].

3.8 *Example Incident 8 – Incident Investigation*

Incident Reference	Plan	Do	Check	Act
	Policies/Programs developed?	Deficiencies identified?	Causes determined?	Actions taken?
Thermal Decomposition in Polymer Catch Tank, 2001, CSB Investigation Report No. 2001-03-I-GA [33]	✓	✓	✗	✗

A site Incident Investigation system was in place at the BP Amoco Polymers, Inc. plant in Augusta, Georgia and should have provided a means of verification to thoroughly investigate and analyze process safety incidents and near misses by identifying and determining the contributing factors (as required by the OSHA PSM Standard) [33]. *However, the system did not adequately identify causes or related hazards for previous incidents by the time a process safety incident occurred on March 13, 2001. As a result, no effective measures had been developed to prevent recurrence of numerous fires or tank overfills and no review system was in place to detect trends.*

Finding – The CSB found:

> "Since 1993, there had been several near-miss incidents involving both the polymer catch tank and the waste plastic. Had these incidents been more thoroughly investigated, they could have provided insight into the hazards associated with the operation" [33].

The CSB also concluded as a root cause:

> "The Augusta site system for investigating incidents and near miss incidents did not adequately identify causes or related hazards. This information was needed to correct the design and operating deficiencies that led to the recurrence of incidents.
>
> - Sound technical theories were not developed to explain the spontaneous ignition of waste plastic or the phenomenon whereby lumps of waste plastic burst.
> - Incidents and near misses tended to be treated as isolated events. Management did not have a review system to detect trends and patterns among incidents.
> - The polymer catch tank had been overfilled and the vent lines plugged on other occasions. No effective measures were developed to prevent recurrence.
> - Fires occurred at the extruder on numerous occasions. No effective countermeasures were developed" [33].

Human Factors Issue – Based on weaknesses with management oversight of the incident investigation system, the CSB recommended:

> "Implement a program to conduct periodic management reviews of incidents and near-miss incidents. Look for trends and patterns among incidents. Address root causes and implement and track corrective measures" [33].

3.9 Example Incident 9 – Emergency Planning & Response

Incident Reference	Plan	Do	Check	Act
	Policies/Programs developed?	Deficiencies identified?	Causes determined?	Actions taken?
Mixing Tank Heptane Vapor Cloud Explosion, 2006, CSB Case Study No. 2006-08-I-IL [25]	✗	✗	✗	✗

Lesson to be Learned – An Emergency Planning & Response management process at the Universal Form Clamp (UFC) facility in Bellwood, Illinois, was not in place. *However, had this process been in place, it would have provided employees a means for planning, training, and verification to prepare for emergencies such as the release of heptane vapor from a heated mixing tank which exploded on June 14, 2006 [25].*

Finding – The CSB found:

> "UFC had no emergency action plan, employees had not received any emergency action training or conducted an evacuation drill, and the facility was not equipped with an employee alarm system." The CSB also found: "…there was no procedure or system to initiate a facility-wide evacuation" [25].

Human Factors Issue – The CSB noted:

> "The Process Safety Management (PSM) standard provides a structured program for a systematic approach to chemical process safety and the prevention of catastrophic incidents… However, at the time of the incident, UFC had not implemented a program to comply with this standard" [25]. A lack of management commitment to comply with existing regulations resulted in a missed opportunity for the facility's employees to prepare for emergencies.

3.10 Example Incident 10 – Compliance Audits

Incident Reference	**Plan**	**Do**	**Check**	**Act**
	Policies/Programs developed?	Deficiencies identified?	Causes determined?	Actions taken?
Vinyl Chloride Tank Welding Flammable Vapor Explosion, 2010, CSB Case Study No. 2011-01-I-NY [22]	✓	✗	✗	✗

Lesson to be Learned – A Compliance Audit process was in place and used by DuPont corporate to conduct PSM audits in 2006 and 2009 at the DuPont facility in Buffalo, New York in order to evaluate compliance (as required by the OSHA PSM Standard), and should have provided an independent means of verification that the procedures and practices developed under the OSHA PSM Standard were adequate and being followed [22]. *However, the CSB concluded: "These latest audits carried out by DuPont at the DuPont Buffalo facility missed many deficiencies that became apparent as a result of the November 9, 2010 incident" [22].*

Finding – Specifically, the CSB noted:

> "In November 2006, DuPont corporate performed a PSM audit of the DuPont Buffalo facility. DuPont awarded DuPont Buffalo a score of 99% on this audit, the highest score the auditor team had ever awarded a facility. The audit commended the Buffalo facility on their operating procedures and safe work practices and made no recommendations to improve these programs. No recommendations were made for the Buffalo facility's MOC procedure for subtle changes, stating that 'all systems comply'. Another DuPont audit was carried out in October 2009. This document commended DuPont Buffalo's PHA process. It stated that the PHA's were 'very well managed and executed' and are of 'consistently high quality'" [22].

Human Factors Issue – An apparent discrepancy between corporate audit scores and actual facility PSM system compliance may have resulted in missed opportunities for process safety performance improvements.

4. Conclusions

In its independent safety review report of five BP refineries following the explosion at Texas City, Texas in 2005, the Baker Panel commented [1]:

> "BP had not instituted effective root cause analysis procedures to identify systemic causal factors that may contribute to future accidents. When true root or system causes are not identified, corrective actions may address immediate or superficial causes, but not the true root causes."

> "BP, however, has sometime failed to address promptly and track to completion process safety deficiencies identified during hazard assessments, audits, inspections, and incident investigations. The Panel's review, for example, found repeat audit findings at BP's U.S. refineries, suggesting that true root causes were not being identified and corrected."

Despite the significant progress for managing process safety within the chemical industry over the past 25 years, more recent experience of process safety incidents has shown us that there is much room for improvement and lessons still to be learned. This is especially true for: (1) how we can better identify deficiencies with both the PSM system and the engineered process system; (2) how we can better analyze the causes of these deficiencies; and (3) how we can better resolve and make commitments to rigorously carry out corrections, corrective actions, and preventive actions to closure, in order to reduce and/or control risks.

A call to action for leadership to drive such continual improvement of process safety performance includes the following challenges:

1. **Plan** – Instill a continual improvement culture of learning and commit the necessary resources to seek out deficiencies within both the engineered process system and each management process of the PSM system.
2. **Do** – Demonstrate leadership by proactively identifying deficiencies through the verification mechanisms designed within each element of the PSM system.
3. **Check** – Adopt and apply a rigorous, analytical approach to understanding the causes of both existing and potential deficiencies in order to generate effective actions.
4. **Act** – Implement a system to manage the life cycle of actions in order to correct identified deficiencies, eliminate the causes of existing deficiencies through corrective actions in order to prevent their recurrence, and eliminate the causes of potential deficiencies through preventive actions in order to prevent their occurrence in the future.

This paper has shown that we have at our disposal a multitude of lessons learned from industrial process safety incidents through published investigation reports and other sources. We also have at our fingertips industry-standard methods and technology to enable us to manage deficiencies through meaningful actions to continually reduce and control process safety risks.

Towards this end and with these means, we can live up to our policies to protect people's safety and health, the environment, and our communities.

5. References

[1] Baker III, James A., et al. *The Report of the BP U.S. Refineries Independent Safety Review Panel*, January 2007.

[2] BSI. "Occupational health and safety management systems – requirements", OHSAS 18001:2007, BSI (British Standards Institution) OHSAS 18001:2007. London: BSI, 2007.

[3] CCPS. *Guidelines for Auditing Process Safety Management Systems*, 2nd Edition, ISBN: 978-0-470-28235-9, Center for Chemical Process Safety (CCPS) of the American Institute of Chemical Engineers (AIChE) and John Wiley & Sons, New York, New York, 2011.

[4] CCPS. *Guidelines for Investigating Chemical Process Incidents*, 2nd Edition, ISBN: 978-0-8169-0897-4, Center for Chemical Process Safety (CCPS) of the American Institute of Chemical Engineers (AIChE) and John Wiley & Sons, New York, New York, 2003.

[5] CCPS. *Guidelines for Mechanical Integrity Systems*, 1st Edition, ISBN: 978-0-8169-0952-0, Center for Chemical Process Safety (CCPS) of the American Institute of Chemical Engineers (AIChE) and John Wiley & Sons, New York, New York, 2006.

[6] CCPS. *Guidelines for Preventing Human Error in Process Safety*, 1st Edition, ISBN: 978-0-8169-0461-7, Center for Chemical Process Safety (CCPS) of the American Institute of Chemical Engineers (AIChE) and John Wiley & Sons, New York, New York, 2004.

[7] CCPS. *Guidelines for Risk Based Process Safety*, 1st Edition, ISBN: 978-0-470-16569-0, Center for Chemical Process Safety (CCPS) of the American Institute of Chemical Engineers (AIChE) and John Wiley & Sons, New York, New York, 2008.

[8] CCPS. *Incidents That Define Process Safety*, 1st Edition, ISBN-13: 978-0470122044, ISBN-10: 0470122048, CCPS (Center for Chemical Process Safety) of the American Institute of Chemical Engineers (AIChE) and John Wiley & Sons, New York, New York, 2013.

[9] ISO. "Environmental management systems – Requirements with guidance for use", ISO 14001:2015, Geneva: International Organization for Standardization (ISO), 2015.

[10] ISO. "Quality management systems –Fundamentals and vocabulary", ISO 9000:2000, Geneva: International Organization for Standardization (ISO), 2000.

[11] ISO. "Quality management systems – Requirements", ISO 9001:2015, Geneva: International Organization for Standardization (ISO), 2015.

[12] ISO. "Specification for security management systems for the supply chain", ISO 28000:2007, Geneva: International Organization for Standardization (ISO), 2007.

[13] Jones, David A. "How to Design Effective EH&S Management Systems for Continual Improvement of Risk Reduction Performance", ASSE Safety 2002, American Society of Safety Engineers (ASSE), Nashville, Tennessee, June 2002.

[14] Jones, David A. "ISO 9000: A Framework for Integrating Safety, Health & Environmental Management Systems," 28th Annual Loss Prevention Symposium, AIChE, Atlanta, Georgia, April 1994.

[15] Kletz, Trevor. *What Went Wrong? Case Histories of Process Plant Disasters and How They Could Have Been Avoided*, 5th Edition, ISBN-13: 978-1856175319, ISBN-10: 1856175316, Butterworth-Heinemann/IChemE, 2009.

[16] Mannan, Sam (Editor). *Lee's Loss Prevention in the Process Industries: Hazard Identification, Assessment and Control*, Fourth Edition, ISBN: 9780123971890, Elsevier Butterworth-Heinemann, Burlington, MA, 2012.

[17] OSHA. "Process safety management of highly hazardous chemicals", 29 CFR 1910.119, (e), Process Hazard Analysis, Occupational Safety and Health Administration (OSHA), *Federal Register*, February 24, 1992 (57 FR 6356).

[18] OSHA. "Process Safety Management of Highly Hazardous Chemicals – Compliance Guidelines and Enforcement Procedures.", Directive Number CPL 02-02-045 (REVISED) [formerly Directive Number CPL 2-2.45A (REVISED), CPL 2-2.45A, CPL 2-2.45A CH-1], 09/13/1994.

[19] Reason, James. *Managing the Risks of Organizational Accidents*, 1st Edition, ISBN-13: 978-1840141054, ISBN-10: 1840141050, Ashgate Publishing, 1997.

[20] Sanders, Roy E. *Management of change in chemical plants: learning from case histories,* Fourth Edition, ISBN 0750611359, 9780750611350, Oxford, Butterworth-Heinemann, 2015.

[21] U.S. Chemical Safety and Hazard Identification Board. "Case Study – Confined Space Entry – Worker and Would-be Rescuer Asphyxiated" (Huntsman Petrochemical Corporation facility in Port Neches, Texas, January 2004), No. 2004-03-B, July 2004.

[22] U.S. Chemical Safety and Hazard Identification Board. "Case Study – Flammable Vapor Explosion, Welding on Tank Containing Vinyl Chloride" (E.I. DuPont de Nemours & Co Inc. facility in Buffalo, New York, November 9, 2010), No. 2011-01-I-NY, 2011.

[23] U.S. Chemical Safety and Hazard Identification Board. "Case Study – Hot Work Control and Safe Work Practices at Oil and Gas Production Wells" (Partridge-Raleigh Smith County Oilfield, Raleigh, Mississippi, June 5, 2006), No. 2006-07-I-MS, June 2007.

[24] U.S. Chemical Safety and Hazard Identification Board. "Case Study – Ink Dust Explosion and Flash Fire" (U.S. Ink/Sun Chemical, East Rutherford, New Jersey, October 9, 2012), No. 2013-01-I-NJ, January 2015.

[25] U.S. Chemical Safety and Hazard Identification Board. "Case Study – Mixing and Heating a Flammable Liquid in an Open Tank" (Universal Form Clamp, Inc., Bellwood, Illinois, June 14, 2006), No. 2006-08-I-IL, April 2007.

[26] U.S. Chemical Safety and Hazard Identification Board. "Case Study – Oil Refinery Fire and Explosion" (Giant Industries' Ciniza Oil Refinery in Jamestown, New Mexico, April 8, 2004), No. 2004-08-I-NM, October 2005.

[27] U.S. Chemical Safety and Hazard Identification Board. "Final Investigation Report – Chevron Richmond Refinery Pipe Rupture and Fire" (Chevron Richmond Refinery #4 Crude Unit, Richmond, California, August 6, 2012), Report No. 2012-03-I-CA, January 2015.

[28] U.S. Chemical Safety and Hazard Identification Board. "Investigation Report – Catastrophic Vessel Failure" (D.D. Williamson & Co., Inc. facility in Louisville, Kentucky, April 11, 2003), Report No. 2003-11-BK-Y, March 2004.

[29] U.S. Chemical Safety and Hazard Identification Board. "Investigation Report – Explosion and Fire" (First Chemical Corporation facility in Pascagoula, Mississippi, October 13, 2002), Report No. 2003-01-I-MS, October 2003.

[30] U.S. Chemical Safety and Hazard Identification Board. "Investigation Report – LPG Fire at Valero – McKee Refinery" (Valero Energy Corporation, Sunray, Texas, February 16, 2007), Report No. 2007-05-I-TX, July 2008.

[31] U.S. Chemical Safety and Hazard Identification Board. "Investigation Report – Refinery Explosion and Fire" (BP, Texas City, Texas, March 23, 2005), No. 2005-04-I-TX, March 2007.

[32] U.S. Chemical Safety and Hazard Identification Board. "Investigation Report – Refinery Incident" (Motiva Enterprises LLC, Delaware City Refinery, Delaware City, Delaware, July 17, 2001), Report No. 2001-05-I-DE, October 2002.

[33] U.S. Chemical Safety and Hazard Identification Board. "Investigation Report – Thermal Decomposition Incident" (BP Amoco Polymers, Inc., Augusta, Georgia, March 13, 2001), Report No. 2001-03-I-GA, July 2002.

12TH GLOBAL CONGRESS ON PROCESS SAFETY

Risk Based Process Safety Program in R&D Facility

Monica Prasad
Huntsman International (India) Limited
India
monica.kamta@gmail.com
monica_prasad@huntsman.com

Praveen Kumar Bommaganti
Indian Synthetic Rubber Private Limited
India
loginpraveen@gmail.com
praveen.bommanganti@isrpl.co.in

Prepared for Presentation at
American Institute of Chemical Engineers
2016 Spring Meeting
12th Global Congress on Process Safety
Houston, Texas
April 11-23, 2016

Keywords:

CCPS	Center for Chemical Process Safety
DSR	Design Safety Review
HIA	Hazard Identification and Assessment
HIDA	Hazard Identification and Design Assessment
HOTO	Hand Over/Take Over
IH	Industrial Hygiene
JCC	Job Cycle Check
JSA	Job Safety Analysis
KPI	Key Performance Indicators
LEL	Lower Explosive Limits
LFI	Learning From Incident
MOC	Management of Change
NFPA	National Fire Protection Association
PPE HA	Personal Protective Equipment Hazard Assessment
PSI	Process Safety Information
PSSR	Pre Start-up Safety Review
RBPS	Risk-Based Process Safety
SOP	Standard Operating Procedure
SLP	Standard Laboratory Practices
SWP	Safe Work Practices
TNI	Training Needs Identification
WI	Work Instructions

Abstract

Research and Development facilities are comprised of pilot plants and laboratories which are designed and constructed to perform new experiments and develop advancements in processes. These facilities investigate the critical aspects of a process, with the goal of replicating it in a commercial facility. There may be a general perception that R&D facilities are less hazardous, safer than commercial plants, and don't require the same level of precautions or safety management systems as commercial plants. However, these facilities are complicated by a range of factors, including new, unknown and changing technology, numerous unknown chemicals, less information about the materials and processes, frequent experimental modifications, lack of waste characterization and limited automatic safety interlocks. This complexity creates significant hazard potential that can lead to injuries, fatalities, property damage and environmental releases. Therefore, it is a very challenging task for R&D organizations to develop a comprehensive and standardized process safety program capable of addressing and mitigating all potential scenarios that may lead to process safety incidents.

This paper attempts to provide a risk-based process safety program for R&D facilities based on four pillars, all of which can help us to understand potential hazards and their identification, assessment and mitigation pathways in R&D facilities. These four pillars [1] are:

- Understanding of hazard and risk
- Managing risk
- Learning from past incidents
- Commitment to process safety

Risk-based process safety programs provide the structured, strategic, functional and technical baseline that helps R&D facilities achieve sustainable safety.

1. Introduction

The purpose of Risk-Based Process Safety (RBPS) is to provide guidelines for designing systems which are not only useful to big manufacturing units, but equally applicable to R&D facilities. The risk-based approach primarily focuses on identifying hazards and their mitigation methods based on the risk involved. In R&D facilities, the risk-based approach is more appropriate than the standard-based approach or compliance-based approach as the operations involved, chemicals used, and process variables are not always the same. This paper will provide some insight into the potential safety issues encountered when handling lower quantities of chemicals with flexible operations and unknown situations in R&D. RBPS in R&D is primarily based on three things which need to be developed in order for pilot plants and laboratories to maintain a high level of process safety; they are as follows:

-Understanding of risk associated with activities and ways to manage those hazards.

-Development of procedures and checklists based on the risk involved.

-Development of review processes and process safety culture within the R&D facility.

2. Overview of Risk-Based Process Safety in R&D:

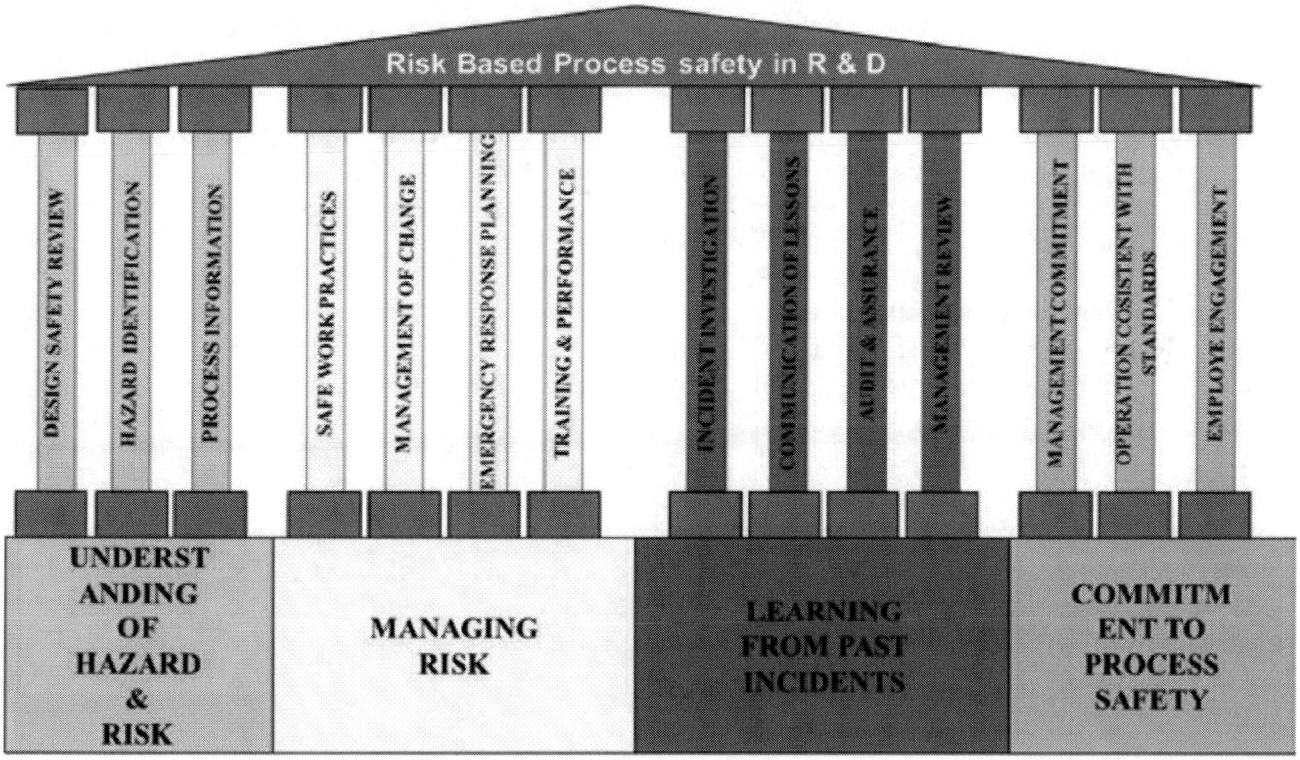

Figure 1: Overview of RBPS in R&D

2.1 Understanding of Hazard and Risk:

The core of this step is to identify hazards associated with activities involved in laboratories and pilot plants. For pilot plants, a detailed design safety review is followed by hazard identification and assessment, whereas for laboratory operations, the first step is to conduct an authorized process safety review and use it as a basis for further analysis. Simple laboratory processes (e.g. simple analytical laboratories) can be operated with a combination of authorized process safety reviews, standard laboratory practices, and PPE hazard assessments. For complicated laboratory operations, authorized safety reviews are followed by hazard identification and design assessments, as well as standard laboratory practices and PPE hazard assessments.

2.1.1 Pilot Plant*:*

In order to understand the hazards involved with design, operations, and maintenance of a pilot plant, a detailed review of Piping and Instrumentation Diagrams (P&IDs) is followed by an evaluation of other aspects of the operation, including chemicals, critical activities, and waste handling. The two steps of the identification and assessment process are:

- Design Safety Review
- Hazard Identification and Assessment

Design Safety Review: The objective of the design safety review process is to verify the adherence to engineering standards, as well as good design and engineering practices. This also certifies that the engineering design meets the operational objectives under both normal and reasonably anticipated upset conditions. A comprehensive review of the engineering design package consists of process flow diagrams (PFDs) and P&IDs. Set process operation envelopes are validated during this step to ensure the mechanical integrity of the equipment. Alarm levels and safety interlocks are assessed to make sure that they are well within the operational envelope of the equipment. A design safety review must be done by a multi-disciplinary team consisting of a process expert, a design engineer, instrumentation and electrical experts, operations staff, and Environment, Health, and Safety (EHS) personnel.

Hazard Identification and Assessment: The objective of this study is to identify the physical and chemical hazards associated with chemical operations. Physical hazards associated with chemicals include high pressure, highly exothermic reactions, etc. Chemical hazards are hazards associated with properties like corrosiveness, toxicity, chronic and acute exposure hazards, etc. Once hazards are identified, a control plan specifically designed to eliminate or minimize these hazards needs to be made. In addition to developing a hazard control plan, the hazard identification and assessment process is used to develop the PPE hazard assessment matrix and personal monitoring plan for chemical exposure assessment. The hazard assessment concludes with a set of action items that need to be completed, which should then be reviewed during the startup of the process.

2.1.2 Laboratory:

In laboratories, the first step is to identify risks through an authorized process safety review, after which resources can be allocated accordingly. Every early stage of the laboratory process must be reviewed by the authorized safety reviewer in order to select appropriate hazard evaluation methodologies and determine the level of process safety requirements. In this section we discuss the hazard identification and design assessment methodology, which is used for the most complex laboratory process. Simpler processes are covered by simple methodologies like standard safety information, standard laboratory practices, work instructions and informal inspections etc. these methodologies will be covered under section 2.2.1. However, this section will cover:

- Authorized process safety review
- Hazard identification and design assessment

Authorized process safety review: An authorized process safety review is the first step in identifying the level of risk involved in the design and modification of a new laboratory process or equipment, which helps to allocate resources as needed. Without this, the effectiveness of every subsequent step in the process will be compromised. Authorized process safety reviews must be done with the participation of various stakeholders, such as operations/laboratory process experts, authorized reviewers from outside of the reviewed process, EHS experts, and researchers. Depending upon the complexity and hazard level of the laboratory process, methodologies and steps are defined to identify the hazard and mitigation.
Authorized reviewers are independent (i.e., not accountable for the design or working directly on the process) of the operations they review. The authorized process safety reviewer is an individual selected by EHS department and management who is given the authority to classify operations, approve Safe Laboratory Procedures (SLP), and conduct management of change reviews and HIDA, as well as perform PSSRs and informal inspections of laboratory operations. A typical checklist of an authorized process safety review is shown below.

AUTHORIZED PROCESS SAFETY REVIEW FORM-LABORATORY

Unit Number	Process Name	Date

NOTE: THIS TABLE IS COMPLETED BY AUTHORIZED PROCESS SAFETY REVIEWER ONLY –The reviewer is responsible for classifying the operation in the following table and determining any required follow-up actions. When important aspects of the operation are outside the reviewer's area of knowledge, the reviewer will call in additional expertise to assist in making the classification and identifying action items.

Check all that apply	Classification of Operation
☐	**Standard Safety Information only-** This is a low hazard operation and only requires a "Personal Protective Equipment Hazard Assessment", "Personal Protective Equipment matrix and an emergency procedure"
☐	**Safe Laboratory Procedures-** This is a lower hazard operation. Some written Safe Laboratory Procedures (SLPs) are required in addition to the standard safety information to adequately address the hazards of the operation. The reviewer requires those documents checked below:
	☐ **Job Safety Analysis** ☐ **Startup, Operating, and Shutdown Procedures** ☐ **Chemical Handling Procedure** ☐ **Flow/Device Diagram** ☐ **Emergency Procedure** ☐ **Waste Disposal Procedure**
	☐ **Other** (please explain)
☐	**Hazard identification and Design Assessment (HIDA) -** This operation having high volume operation, high flow rates at high temperature and pressure (not including gas cylinders), has chemistry that can produce deflagrations, has a fairly complex equipment and control scheme or the reviewer decided it needs further review.

Person Responsible for the Operation:	Sign:	Date:
Authorized Process Safety Reviewer	Sign:	Date:
Technical Expert Reviewer	Sign:	Date:
Additional Reviewers:	Sign:	Date:
The level of check before startup of the operation		

☐	**PSSR** (mandatory whenever the operation has an HIDA)	Please include the date the informal inspection was completed and the signature of the reviewer who performed it.	Sign:
☐	**Informal Inspection**		Date:

Action items from the Authorized Safety Review should be documented and tracked for closure before start up the operation.

	Action Item	Description of Resolution	Responsibility	Date Completed
1.				
2.				

Hazard Identification and Design Assessment (HIDA): Complex laboratory operations are further assessed through HIDA, which is shown on the authorized process safety review form above. HIDA is a combination of DSR and HIA; however, the detailed P&ID review of a full DSR is not required here. Instead, HIDA requires hazard identification and assessment of the operation and the chemicals used, as well as leak-proof system design. Table 1[2] provides a list of the hazards often encountered in laboratory research activities.

Table 1: Typical hazards in laboratory

Type of Hazard	Examples
Chemical Hazard	Carcinogenic, corrosive, pyrophoric, toxic, mutagenic, reproductive hazard, explosive, nonionizing radiation, biological, hazard/pathogenic, flammable, oxidizing, self-reactive or unstable, potentially explosive, reducing, water-reactive, sensitizing, peroxide forming catalytic, or chemical asphyxiate etc.
Physical Hazard	High pressure, low pressure, electrical, uneven surfaces, pinch points, suspended weight, hot surfaces, extreme cold, steam, noise, simple asphyxiant, oxygen-deficient spaces, ultraviolent radiation, or laser light etc.
Activity based Hazard	Creation of secondary products, lifting, chemical mixing, long-term, use of dry boxes, repetitive pipetting, scale up, handling waste, transportation of hazardous materials, handling glassware and other sharp objects, heating chemicals, recrystallizations, extractions

The output of a hazard evaluation should be a qualitative, and sometimes quantitative, understanding of a hazard. A risk assessment of the hazards of a given experiment will guide the selection of risk management techniques [3] and tools such as:

- Intensification: Using a smaller quantity of hazardous material.
- Substitution: Replacing more hazardous material with less hazardous material.
- Attenuation: Using less hazardous process conditions.
- Simplification: Using a process/equipment that will make operating errors less likely and will be more forgiving of errors made.
- Engineering Controls: Fume hoods, local ventilation systems, gas cylinder and flame proof cabinets for storage, restrictive toxic gas flow, etc.
- Personal Protective Equipment (PPE) Assessment: Using PPE (safety goggles, respirators, dust mask, chemical suits, etc.) to avoid chemical exposure.

2.1.3 Process safety information (PSI):

DSR and HIA provide a detailed compilation of PSI, as well as a documented review of process hazards in pilot plants. PSI in laboratories includes standard laboratory practices, work instruction, machine operating procedures etc. The major contributors to PSI for laboratories and pilot plants are shown in Fig. 2.

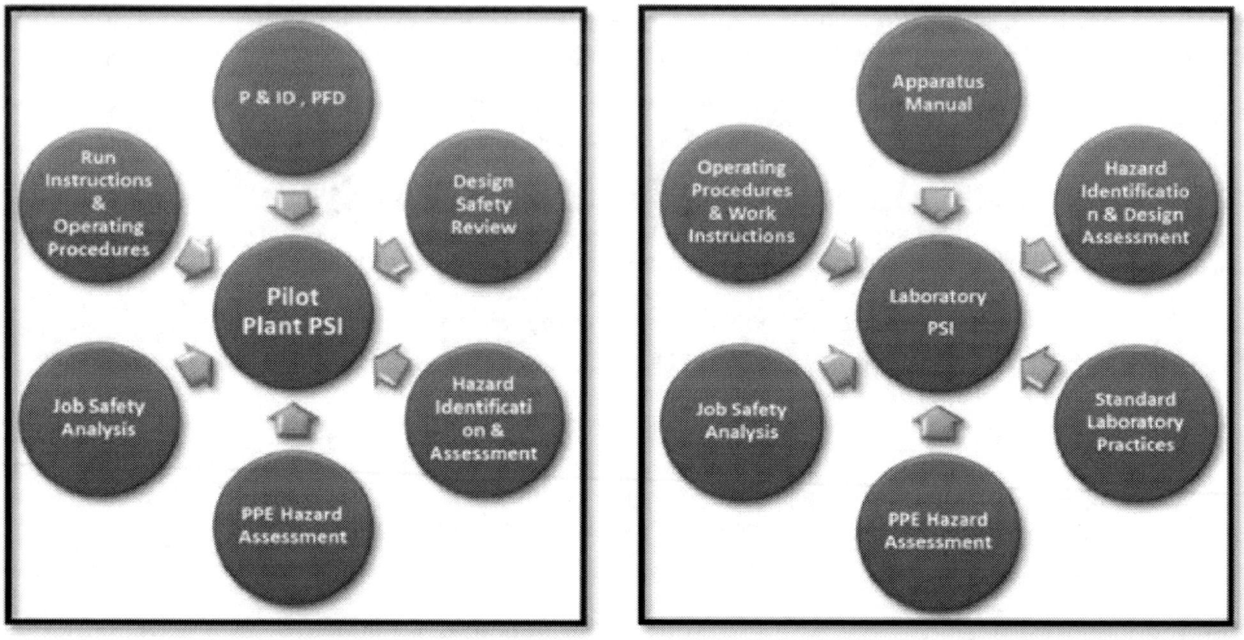

Figure 2: Sample process safety information in pilot plant and laboratory

2.2 Managing Risk:

The hazards identification and evaluation process will be ineffective if the results of the hazard analysis are not applied. Once an evaluation is complete and the necessary hazard controls have been identified, it is imperative that operational personnel/researchers understand the hazard analysis information and that they are committed to following the agreed-upon controls.

2.2.1 Safe Work Practices (SWPs):

The key principle of managing risk in R&D through different SWPs is managing the risks associated with routine and non-routine activities. Pilot plants and laboratories must have written documents defining all of the operation guides like normal startup/shut down, emergency shut downs, chemicals and hazards and their controls, safety systems, consequences and controls, operation actions, etc. General SWPs like lock out tag out, line breaking, energy isolation, standard maintenance procedures, leak-proof design, well-planned storage, and well-designed ventilation are applicable to both as required. The content and scope of Standard Operating Procedures (SOPs), Work Instructions (WI), and Personal Protective Equipment Hazard

Assessments (PPE HAs) are same for pilot plant and laboratories. Figures 3 and 4 illustrate SWPs in laboratories and pilot plants, respectively. Various elements which constitute the SWPs for pilot plants and laboratories are discussed below in detail.

Job Safety Analysis (JSA): All routine and non-routine activities must be reviewed for potential hazards and suitable controls.

Job Cycle Checks (JCC): This is an advanced method used to verify the effectiveness of SOPs and JSAs. With the help of this methodology, senior line managers physically review and validate the actual implementation of defined SOP/JSA steps.

Work Instructions (WI): Work instructions are derived from the operating procedures or the equipment run manual. Key points which are required to perform any activity can be summarized as work instruction. WI should not be more than one or two pages and consists of steps to perform particular small steps or activities from SOPs.

PPE Hazard Assessment (PPE HA): As a part of an industrial hygiene (IH) program, PPE requirements for a particular area according to the hazards present is listed in a PPE HA. These requirements must be displayed at the strategic entry location of any hazardous area to control any unwanted risk inside the area.

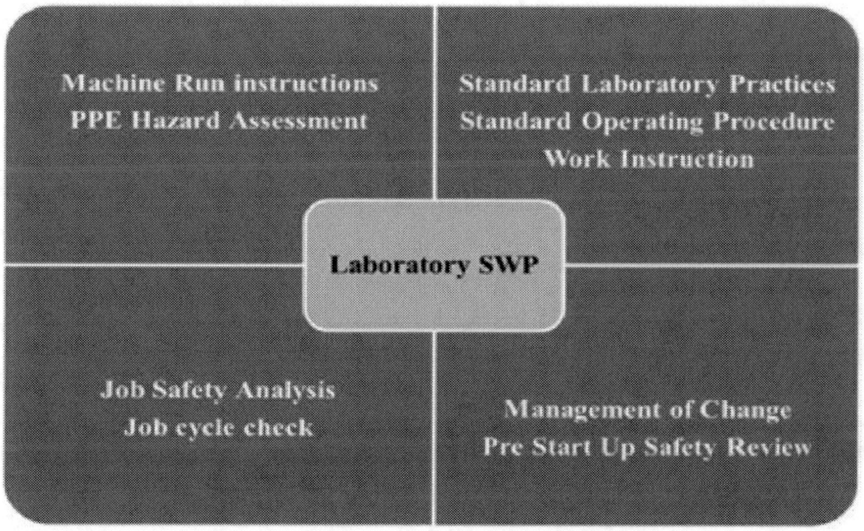

Figure 3: Safe Work Practices in a Laboratory

Standard Laboratory Practices (SLP): SLP provides the framework within which laboratories studies are planned, performed, monitored, reported, and archived.

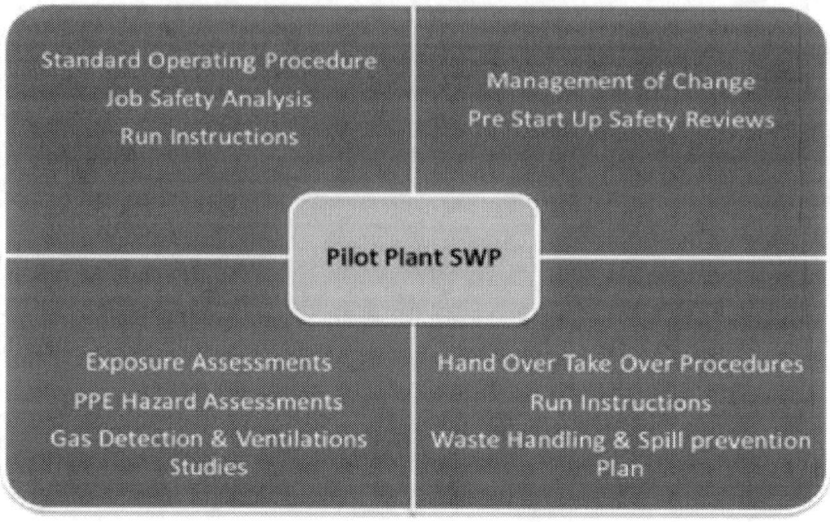

Figure 4: Safe Work Practices in Pilot Plant

Run Instructions: Considering the complexity and flexible operations of pilot plants, pilot plant associates are trained to follow step-by-step for how to run the particular pilot plant. Run instructions are aligned with operating procedures but are somewhat more detailed and contain the operating envelopes under which an experimenter wants to run the process to get the desired results.

Hand Over/Take Over (HOTO): A written hand over/take over process must be defined to capture critical points of shift communication to the next group in the required amount of time.

Gas Detection Studies: A gas detection system is required for the inclusion of any hazardous chemicals in the pilot plant. Under the restricted flow of the gases, alarms and safety interlocks must be set for different TLV, STEL, and IDLH values (for toxic gases) and LEL values (for flammable gases).

Waste Handling Plan: This program includes characterization, monitoring, treatment and handling of waste generated by pilot plant and laboratory.

Spill Prevention Plan: Spill prevention plan includes plan for prevention of chemical spillage, preparedness for spill response. This is the detail plan which includes the containment requirement for every operation in R&D.

2.2.2 Management of Change (MOC):

As pilot plants and laboratories are made to cater wide range of processes and chemicals, a modification review is necessary to ensure that:

- Unacceptable risks are not introduced
- Proposed modifications fall under the ambit of the existing design envelope
- Communication is maintained with operating personnel

- Relevant documentation is always updated

The MOC process is completed for all changes which are not in the scope of original HIDA, HIA, and DSR. Figure 5 is a decision tree that shows the categorization of the modification processes which can be implemented to create a strong MOC system. Based on the category of the modification, appropriate checks are completed.

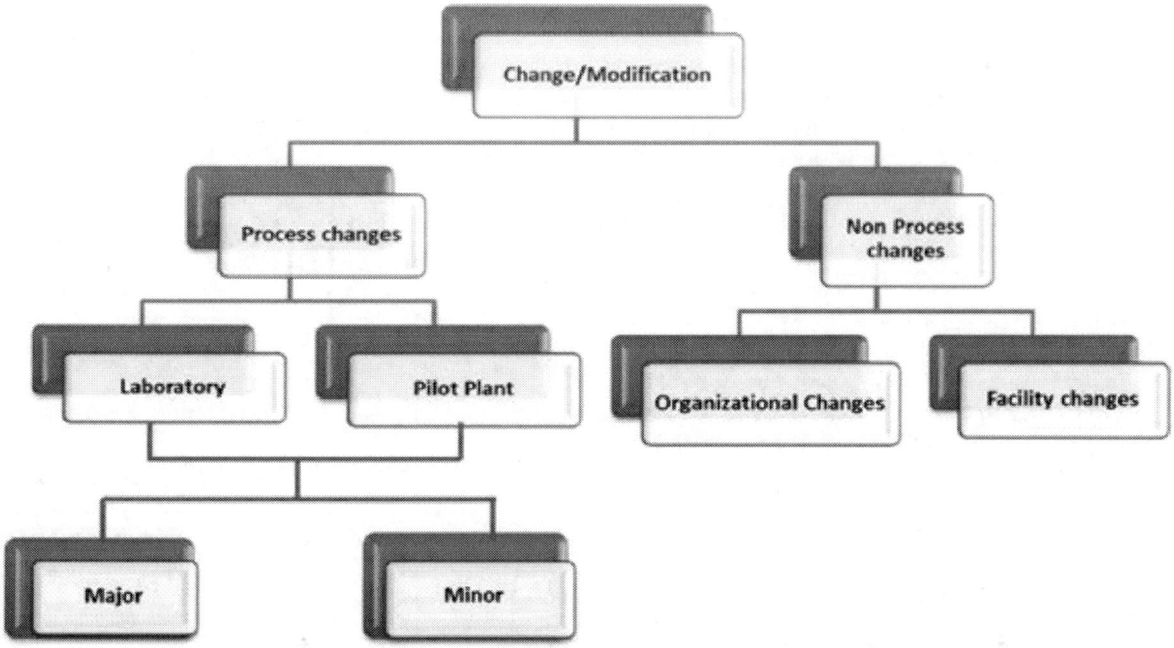

Figure 5: Sample MOC modification categorization in R&D

Whenever changes are initiated or conceptualized, the category of modification is decided based on the decision tree. The decision tree gives guidelines which helps categorize the changes as major or minor based on the chemical hazards involved, the quantity of chemicals, used, and original design parameters. These major and/or minor modifications can be performed in the form of a checklist which will result in action items and their closure before start up. PSSR must still be done before the startup of the process. A typical decision tree for the categorization of MOC modifications as major or minor is shown below:

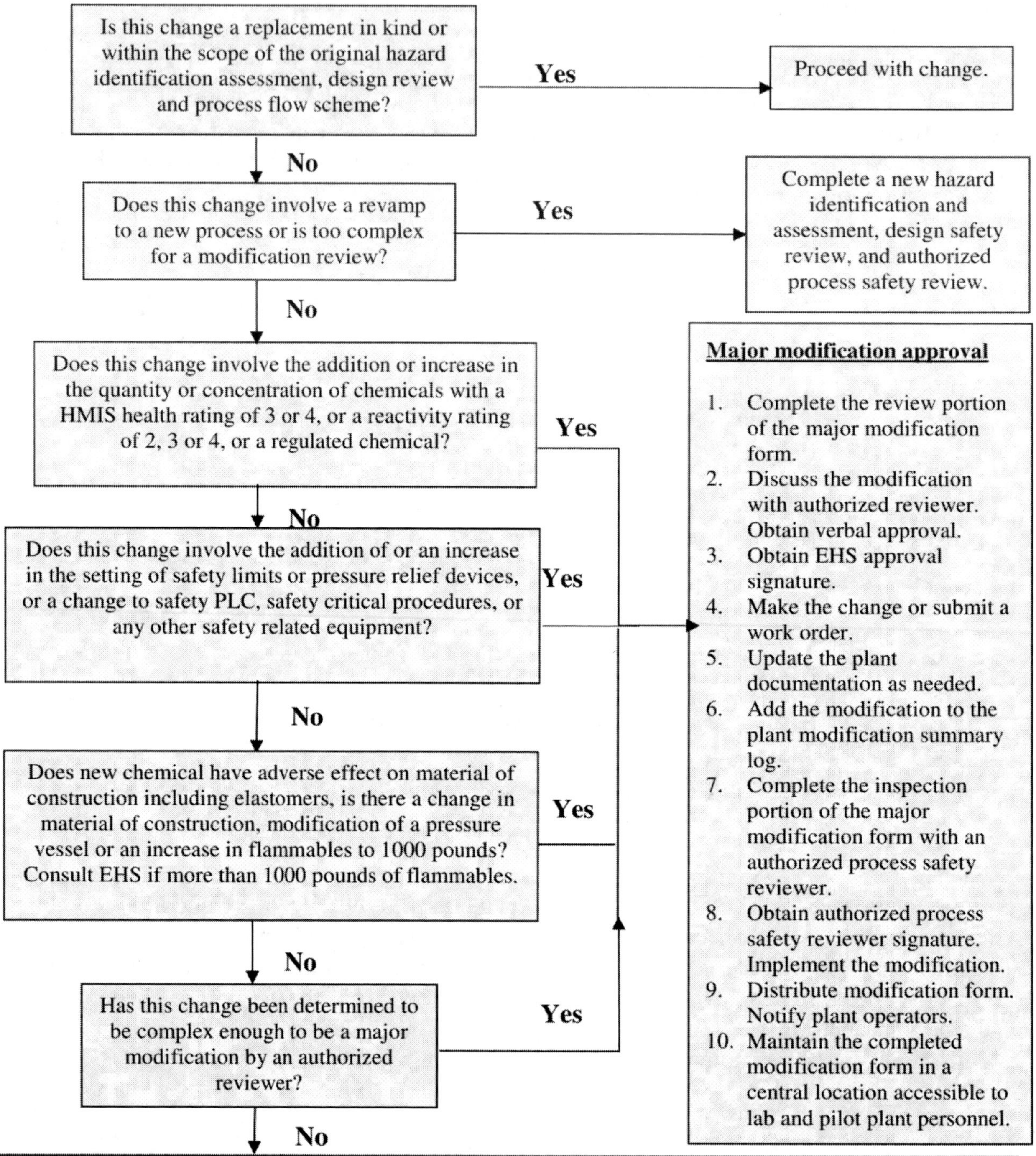

Minor modification

1. Complete the review portion of the Minor modification form.
2. Discuss the modification with authorized reviewer. Get verbal approval before starting work.
3. Provide before (P&ID) and after markup sketches of modification.
4. Make the change or submit the work order.
5. Add the modification to the plant modification log.
6. Update plant documentation and procedure as needed (HIA/DSR or HIDA, P&ID, Revision Log, etc.)
7. Notify concerned personnel.
8. Distribute updated modification log to reviewer plant process safety area manager, area supervisor, operations staff, and automation (if applicable).
9. Obtain authorized process safety reviewer signature.
10. Scan completed modification form into process safety folder.

2.2.3Emergency Response Planning(ERP): Emergency management activities are generally triggered by an incident or an event. ERP is used as a tool to handle emergency situations in an orderly and effective manner. In addition to the standard emergency response procedures such as evacuation, safe assembly, head count, emergency response team, and pre-incident planning, some additional requirements for laboratories and pilot plants include unattended operation safety (required during off hours when researchers are not available on site) and emergency controls in case of single occupancy in the laboratory.

2.2.4 Training and Performance Assurance:

The most important element of implementing process safety behavior is training. Training can be divided into three levels: awareness, skill, and expertise. All elements of process safety and EHS must be covered under this element. For new employees, comprehensive training should be administered according to the process safety criticality of the employee's job role. Pilot plant personnel and laboratory researchers must be informed of the risks and hazards associated with flexible and complex operations and the large number of chemicals being handled. Developing and maintaining process safety competency of employees encompasses three interdependent elements:

- Right person - Right information philosophy.
- Continual improvement of knowledge and skill.
- Learning application.

For complete assurance for the competency of employees, detail learning plan and Training Needs Identification (TNI) matrix must be prepared, communicated and followed. A TNI matrix specifies the training requirements based on the employee's job role. Training efficacy can be verified by written or verbal assessments, depending upon the job requirement. Procedural refresher trainings are to be scheduled for once every three years.

2.3 Learning from Past Incidents:

Pilot plants and laboratories are always associated with new research, flexible operations, and different levels of control at each stage. Every time a process is undertaken, there is a possibility of encountering different events and incidents which can leave a different set of footprints on future operations. Therefore, it is very important to follow a structured method to capture root causes and lessons from past incidents. Methods of documenting process safety incidents are the same for both pilot plants and laboratories. All process safety incidents and near misses are to be investigated and the recommendations are to be communicated and implemented. Apart from this, auditing helps to reduce the risk by proactively identifying and correcting weaknesses in management system design and implementation.

2.3.1 Incident Investigation:

An unplanned or uncontrolled release of any material from primary containment, including non-toxic and nonflammable materials (e.g., steam, compressed air) is known as Loss of Primary Containment (LOPC). An LOPC is also known as a Process Safety Event (PSE), a term which also includes near misses, which are undesired events or conditions that could have potentially resulted in a LOPC had they occurred under slightly different circumstances. When a PSE happens at a site, the essential parts of an incident investigation include finding the root cause of the incident, prior identification of process safety events for similar events in future, reporting the process safety event, sharing learnings from incident (LFI) in order to prevent reoccurrence, and updating existing processes and procedures. Methodology for the risk ranking of the PSE may differ from company to company. The simpler way to identify the risk ranking of the PSE is with a 3x3 (or 5x5) risk matrix as shown in Fig. 6. Based on the risk ranking of the event, reporting and investigation methodology/effort changes. This risk ranking matrix can also be used for risk assessment quantification in the hazard identification stage.

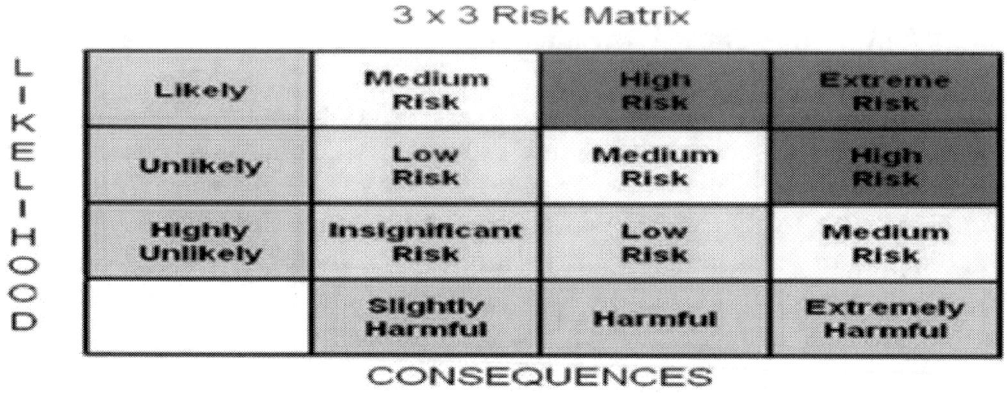

Figure 6: Sample 3x3 risk matrix for risk ranking

2.3.2 Communication of Lessons:

Communicating investigation recommendations can be accomplished by (1) incorporating results into formal presentations that are specific to an incident, (2) presenting the lessons learned during safety meetings, (3) distributing them through email, and (4) posting the investigation results on a facility bulletin board. These lessons need to be imparted to concerned persons and operating staff in order to avoid reoccurrence.

2.3.3 Auditing and Assurance:

An RBPS management system audit is a systematic review used to verify the suitability of the system's elements and ensure their effective, consistent implementation. Used in both pilot plants and laboratories, these types of audits use a methodical, typically team-based approach to confirm that the process and its safety components are operating within the applicable procedures and standards. Data are gathered through the review of program documentation and implementation records, direct observations of conditions and activities, and interviews with individuals responsible for implementation or oversight of the element(s) or who might be affected by the RBPS management system. The data are analyzed to assess compliance with requirements, and the conclusions are documented in a written report. Recommendations for addressing any performance gaps identified by the audit are proposed, responsibilities and schedules for addressing the recommendations are assigned, and recommendations are tracked to their resolution. The audits can also be used to enhance the effectiveness of RBPS by monitoring

RBPS performance over time, identifying the management system's weaknesses, and implementing enhancements to address these weaknesses.

2.3.4 Management Review:

Management review is the routine evaluation of whether management systems are performing as intended and producing the desired results as efficiently as possible. In R&D facilities where pilot plants and laboratories work separately with different sets of people, it is necessary to have a tiered management review process. For effective implementation of process safety standards, clear responsibility and accountability needs to be defined at all levels. Periodic management review of set process safety Key Performance Indicators (KPIs) will help the facility establish a safety culture. Definition and investigation of near miss events is also important factors. In management reviews, in addition to leading and lagging factors, recommendations from audits and their compliance evidence must be reviewed.

2.4 Commitment to Process Safety:

2.4.1 Management Commitment Towards Process Safety and Employee Involvement:

Commitment to any methodology involves two different groups, both of which are responsible for making any changes in the system: **management and employees**. To build up the process safety culture within an organization, there must be a combination of management values and employee behavior. In R&D facilities, where there are multiple groups working, the biggest challenge is to synchronize the groups to achieve the process safety goal. Here a tiered level approach is used through which one can develop the process safety culture in R&D facilities effectively. With this method, different tiers are established in order to simplify the tracking of actual RBPS implementation. The structure of a tiered level management system will be the same for laboratories and pilot plants, but with different leading and lagging KPIs. The flow of this tiered level system is from bottom to top and vice versa, where policies, guidelines, standards and procedures are made and approved by middle and top level management and are implemented on the ground by middle and bottom line management as depicted in Fig. 7. Tiered level meetings are designed to ensure full employee engagement at all levels. Different KPIs, action items, and escalations are defined according to personnel roles in organization. In tiered level meetings, all critical communications and learnings are shared among all levels of employees.

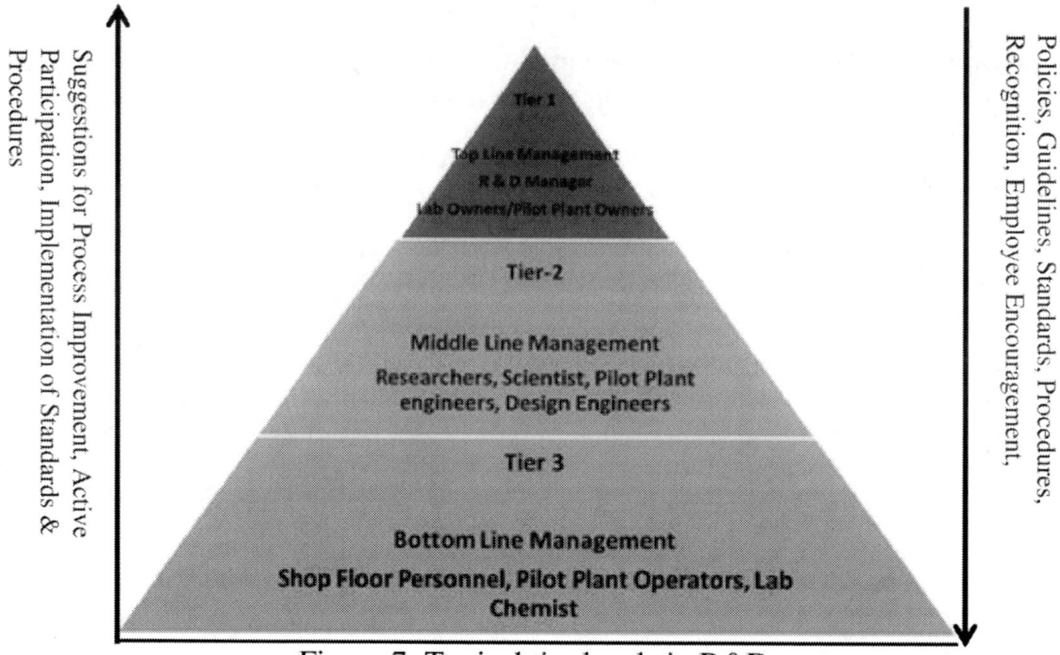

Figure 7: Typical tier levels in R&D

Two levels of responsibilities are defined to establish a process safety management system in R&D: champions and procedure owners. Procedure owners are defined for each RBPS element at Tiers 1 and 2, and champions are defined Tier 3. Champions are accountable for implementation of respective elements in their areas. Management must define the roles and responsibilities through a written matrix and procedure to ensure clear understanding of roles and expectations. This is closely related to the training and performance assurance elements discussed in Section 2.2.4. Assurance of implementation and scope of improvements are identified by a periodical audit program, which is discussed in Section 2.3.3. Group activities like mock drills, safety trainings, quizzes, and celebrations are other examples of employee engagement programs.

Documentation management is the parent of the RBPS management system where all relevant procedures, policies, forms, checklists, training records and permits, etc. are maintained according to their period of storage. DSRs, P&IDs, HIAs, and HIDAs are maintained for the life of the asset or process and are modified as required or whenever changes happened. Companies can adopt a variety of storage methods, including company intranet and hard copy files.

2.4.2 Operation Consistent with Standards:

A R&D facility handling varying quantities of a large number of chemicals may or may not fall under the defined threshold quantity limit for identifying the facility as an OSHA/PSM regulated site. In this case, OSHA PSM or EPA RMP may not be applicable, but many standards and guidelines are useful. For example, NFPA, IS, CCPS guidelines, OSHA standards (like 29 CFR 1910.1450[Occupational Exposure to the Hazardous Chemical in Laboratories Standard] or 29 CFR 1910.1200 [The Hazard Communication Standard]), are applicable and followed to an extent in laboratories and pilot plants.

NFPA 45-11[4] applies to laboratory buildings, units, and work areas where hazardous chemicals are handled and stored. For example, pilot plant buildings are constructed to class A (high fire hazard), whereas laboratories are considered hazard type I and type II. Steel structures, civil structures, doors, floors and roofs must have a 2- to 3-hour fire rating. All pilot plant areas and laboratory units must comply with maximum permissible quantity of flammable and combustible hydrocarbons.

For storage of hydrocarbons in non-laboratory units, NFPA 30-19[5] (Flammable and Combustible Liquids Code) guidelines are followed. The standard defines the limits of flammable chemical storage. For example, the design of an inside room is limited to a maximum of 500 square feet. The room must have at least a 2-hour fire rating and must be electrically classified as class I division II if class I flammable hydrocarbons are present. Ventilation as per the room capacity and chemical storage quantity must be designed.

All laboratory units and pilot plant areas are classified according to the NFPA 70: National Electrical Code (NEC) [6] as general purpose. The general purpose classification is applied based on the very high ventilation rates which dilute the concentration of hydrocarbons exceeding the lower explosive limits.

3. Conclusion:

A structured methodology for identifying and managing hazards though a formal, systematic and comprehensive framework of hazard identification assessments, design safety review, safe work practices and management of change can put an organization on the right path towards excellence in R&D operations and process safety. The RBPS strategic approach has been built on the principle that appropriate levels of detail and rigor in process safety practices are dependent on three factors:

- A sufficient understanding of the risk associated with the processes.
- The level of demand for process safety activities based on risk associated.
- The process safety culture within which the process safety practices will be implemented.

As L.H. Baekeland suggests pilot plants and laboratories operate on the principle

"Make your mistakes on a small scale and your profits on a larger scale".

4. References

[1] Center for Chemical Process Safety. "Guidelines for Risk-Based Process Safety," John Wiley and Sons: Hoboken, NJ, pp. 37-66, 125-135, 245-253, 255-259, 2007.

[2] OSHA 3404-11R "Laboratory Safety Guidance," 2011.

[3] National Research Council, Prudent Practices in the Laboratory: Handling and Management of Chemical Hazards, National Academies Press: Washington, DC, pp. 15,25, 2011.

[4] NFPA 45-11 Standard on Fire Protection for Laboratories using chemicals, pp. 14, 2011.

[5] NFPA 30-19: Flammable and Combustible Liquids code, pp. 22-39, 2012.

[6] NFPA 70 Recommended Practice on Static Electricity.

12TH GLOBAL CONGRESS ON PROCESS SAFETY

Restructuring of Process Safety at Air Products

Presenter:
Brian E. Farrell, Air Products and Chemicals, Inc., Allentown, PA
farrelbe@airproducts.com

Prepared for Presentation at
American Institute of Chemical Engineers
2016 Spring Meeting
12th Global Congress on Process Safety
Houston, Texas
April 10-14, 2016

Restructuring of Process Safety at Air Products

Presenter:
Brian E. Farrell, Air Products and Chemicals, Inc., Allentown, PA
farrelbe@airproducts.com

Keywords: *Process Safety, restructuring*

Abstract

Air Products restructured its Process Safety organization in January 2015 as part of an overall corporate restructuring. Process Safety was moved from a centralized function to a dispersed capability located within the operating businesses. The reorganization offered the potential for simplicity and speed within Process Safety if executed well, or a loss of focus, competency and institutional memory if executed poorly. Twelve months later, the reorganized Process Safety teams have delivered on commitments and the changes can be characterized as successful. This Case Study will discuss the planning process for the transition, the outcomes of the reorganization, and how the following factors were managed during the transition to enable success.

- Process Safety Culture
- Executive Leadership Commitment
- Management of Change for Process Safety Leadership Positions
- Standards, Tools, and Management Systems
- Metrics

The paper will also examine the strengths and weaknesses of centralized vs. decentralized Process Safety teams, the sustainability of Process Safety culture, and the ongoing challenges to retain institutional memory and improve competency of Process Safety professionals in a decentralized organization. As with any reorganization of this size, all aspects did not proceed according to plan or go as smoothly as anticipated. The Lessons Learned from the reorganization will be shared.

1 Introduction and Background

Air Products (NYSE:APD) is a world-leading Industrial Gases company celebrating 75 years of operation. The company's core Industrial Gases business provides atmospheric and process gases and related equipment to manufacturing markets, including refining and petrochemical, metals, electronics, and food and beverage. Air Products is also the world's leading supplier of liquefied natural gas process technology and equipment. The company's Materials Technologies business, serves the semiconductor, polyurethanes, cleaning and coatings, and adhesives industries.

Air Products has developed and maintained a strong Process Safety culture, starting in the 1970s, based on strong leadership from our founder Leonard Pool and his successor Ed Donley. This culture is seen today in our Corporate Goal and Total Safety Philosophy.

The Corporate Goal is to be the **safest** and the **most profitable** industrial gas company in the world, providing excellent service to our customers. Our corporate culture is built upon a Total Safety Philosophy that is at the foundation of all of our business activities.

We believe that:
- Nothing is more important than safety - not production, not sales, not profits
- All accidents and injuries are preventable - not inevitable
- Safety is a management responsibility - and safety can be managed
- Safety is an individual responsibility - and a condition of employment
- Safety is a way of life - around the clock
- Every task must be performed with a concern for safety - for ourselves, our fellow employees our contractors, our visitors, our customers and the communities in which we operate

The company had fiscal 2015 sales of $9.9 billion. Approximately 20,000 employees in 50 countries strive to make Air Products the world's safest and best performing Industrial Gases company, providing sustainable offerings and excellent service to all customers.

Air Products operates approximately 750 production facilities and more than 1800 miles of industrial gas pipelines in 50 countries. The vast majority of the production facilities and pipelines were also designed and engineered by Air Products. For these reasons, the Process Safety function at Air Products includes the following design and operational activities.

<u>Engineering</u>
Building siting analysis
Risk and consequence analysis (LOPA, Fault Tree, QRA)
Hazard and Operability Reviews
Relief System Design
SIL Assessment and Verification
Operational Readiness Inspection

<u>Plant Operations</u>
OSHA PSM and EPA RMP
OPHR Revalidations
Transportation / Pipeline Risk Models
MOC to existing operations
Seveso Compliance

In addition to the plant design and operational activities, the Process Safety function is also responsible for functional development activities.

Functional Development
- Process Safety training across the company
- Process Safety standards
- Competence within the Process Safety function
- Process Safety tools
  - Development and maintenance
  - Internally developed and 3rd-party
- Process Safety metrics and reporting

In September 2014, Air Products' chairman, president and chief executive officer, Seifi Ghasemi, announced a major company restructuring, with the goal of regaining industry leadership by providing excellent service to customers as the safest and most profitable industrial gas company in the world. A key feature of the reorganization was the move to a decentralized, simpler, and more efficient structure to create true profit and loss (P&L) accountability at many levels of the organization. To enable true P&L accountability, operations, distribution, and portions of other enabling functions were integrated within the Industrial Gases and Materials Technologies businesses. One of the enabling functions that was integrated into the businesses was Process Safety.

2 Process Safety Prior to Restructuring

The Process Safety function consisted of a central team of 32 individuals under the direction of a Global Process Safety Director who reported to the Vice President of EHS, and a group of 8 individuals who were imbedded into the Engineering teams, reporting to the Process Technology Managers of those teams. The Engineering Team members had a secondary reporting relationship to the Global Process Safety Director. It is important to note that none of the Process Safety team members were directly assigned to our Plant Operations group.

The Central Process Safety Team had primary responsibility for Process Safety in Plant Operations and for Functional Development. The majority of the company's most senior Process Safety Engineers resided in the Central Team. The Engineering-based individuals were responsible for Process Safety activities on new capital projects. The Central Team was often called upon to supplement the Engineering team, or to lead Process Safety activities in times of high project workload.

The Central Team members were located primarily in our three regional headquarters locations in Allentown, PA, in Hersham, England (a suburb of London) and in Shanghai, China. To support Plant Operations across Asia, several Central Team members were located in Korea, Taiwan, and Singapore. The Central Team members would support Plant Operations activities through frequent travel and teleconferences.

Engineering Team members were located in Allentown and Hersham, with none in Asia, and supported Industrial Gases capital projects. Process Safety work on all capital projects executed in Asia was led by Central Process Safety Team members. The Process Safety work on air separation projects was often led by members of the Central Process Safety Team due to the small number of Process Safety Engineers assigned to the Air Separation Engineering Team. There were no Engineering Team members assigned to support the Materials Technology business, and as a result, Materials Technology Process Safety activities were led by the Central Team.

3 Transition Planning and Execution

Shortly after Air Products announced plans for a corporate restructuring in September 2014 and just before the restructuring of the Process Safety function began in January 2015, the incumbent Global Process Safety Director (Shakeel Kadri) retired and accepted a new position as the Executive Director of CCPS. While this was an outstanding opportunity for Shakeel and fully supported by his management team and his Process Safety team members, the timing could not have come at a less opportune time in terms of the restructuring.

The author was appointed as Global Process Safety Director in December 2014 to replace Shakeel Kadri. Simultaneously, it was announced that the Process Safety reorganization needed to be complete by 31 March 2015. The goals of the reorganization were:

1. To transform from a centrally-based Process Safety organization to a set of business-aligned teams
2. To reduce costs (as is true with most reorganizations)
3. To provide clear accountabilities for all Process Safety activities
4. To maintain functional excellence in Process Safety
5. To deliver an improvement on Process Safety metrics through clearer alignment of Process Safety and business/operating activities

Prior to the reorganization, Process Safety was not broken at Air Products. Just the opposite – it was recognized as one of our strongest core competencies and an area in which the company was a leader in the chemical and processing industries. The reorganization was not undertaken to 'fix' Process Safety, but rather, to align it more closely with the operating businesses.

The team leading the corporate reorganization completed the new corporate structure in early December 2014, and this work included determining where Process Safety teams would be located in the new organization, as listed below.

Industrial Gases Operations – Americas
Industrial Gases Operations – Europe, Middle East, Africa, India (EMEA)
Industrial Gases Operations – Asia
Materials Technology
Engineering – Air Separation
Engineering – Hydrogen

Central Process Safety

A Process Safety Leadership Team was formed in December 2014 by the author to facilitate the reorganization. Representatives were selected from each of the organizations in Operations, Engineering, and Central Process Safety that would receive a Process Safety Team. With only three months to transform the Process Safety function, the Leadership Team met weekly to track progress and make decisions. The Leadership Team decided to follow a collaborative decision-making model as most of the decisions would impact more than one of the new Process Safety teams. In the event that a collaborative decision could not be made, the Leadership Team agreed that the new Global Process Safety Director would make the decision. The Leadership Team reported on progress to the EHS Vice President on a biweekly basis or as needed.

Following formation of the Leadership Team, the first step was to develop a complete list of Process Safety responsibilities and activities, and assign them to the different teams.

Team	**Manager**	**Responsibility**
IG Operations - Americas	Regional Plant Process Technology Manager	OSHA PSM and EPA RMP OPHR Revalidations MOC to existing operations Process Safety metrics and reporting
IG Operations - EMEA	Regional Plant Process Technology Manager	Seveso Compliance OPHR Revalidations MOC to existing operations Process Safety metrics and reporting
IG Operations - Asia	Regional Plant Process Technology Manager	Country and local compliance OPHR Revalidations MOC to existing operations Process Safety metrics and reporting
Materials Technology	Plant Engineering Director	OSHA PSM and EPA RMP Seveso Compliance OPHR Revalidations MOC to existing operations Process Safety metrics and reporting
Engineering - Hydrogen	Process Technology Manager	Building siting analysis Risk and consequence analysis (LOPA, Fault Tree, QRA) Hazard and Operability Reviews Relief System Design SIL Assessment and Verification

		Operational Readiness Inspection Corporate Expertise in H2 / Syngas Safety
Engineering – ASU	Process Technology Manager	Same as Engineering – Hydrogen Corporate Expertise in Air Separation Safety
Central Process Safety	Global Process Safety Director	Process Safety training across the company Process Safety Policy, Procedures, and Standards Competence within the Process Safety function Process Safety tools Development and maintenance Internally developed and 3rd-party tools Corporate Expertise in all PS Tools Transportation / Pipeline Risk Models

The next step in the process was to assign the Process Safety engineers to the new teams and simultaneously to determine which positions to eliminate to meet the required cost reduction target. To meet the target necessitated elimination of approximately 20-25% of the existing Process Safety positions. This was accomplished by cutting a layer of management positions within Process Safety and slotting the new teams under existing technical managers, through retirements, and finally through position eliminations.

Position eliminations do not succeed without a corresponding elimination of work. The Leadership Team chartered a sub-team of experienced Process Safety Engineers to examine our work processes, our reporting procedures, and other activities to identify areas where work could be eliminated. As an example, one area identified for simplification was in monthly Process Safety KPI reporting. At Air Products, all Process Safety and other operating incidents are reported using a single global reporting tool. In the old organization, each Process Safety team separately reviewed the entire reporting database monthly and picked out the Process Incidents relevant to their team. Each team then determined whether the events were Tier 1, 2, or 3 and assigned severity points according to the facts of each incident. The representatives from each team then met to compile a corporate report. In addition to the Process Safety events, the teams also compiled leading and lagging indicators such as audit findings, overdue relief device inspections, overdue critical safety circuit revalidations, overdue OPHR Action Items, and expired temporary MOCs. This manual method was time-consuming and unsustainable.

In the new organization, the IT department was commissioned to automate the Process Safety metrics reporting procedure. Now instead of a manually intensive process in which multiple Process Safety engineers were combing the corporate database each month, each team receives a customized report for their business at the end of each month. The Central Process Safety Team receives a similar report for the entire company. The main manual task remaining is to assign Tier Level and Severity to Process Safety events. The new method has cut the time expended to compile the monthly reports by 75% and improved data quality at the same time. In the time since the reorganization was started through today, our Process Safety metrics have actually improved. This improvement is the result of an already strong foundation being in place at the time of the reorganization, and increased focus on Plant Operations Process Safety resulting from moving Process Safety engineers into the operating organizations.

The largest reduction of effort was observed from better alignment of Process Safety resources with Process Safety tasks and activities. One example is that Process Safety work required for new capital projects is now 100% under the direction of the Engineering Teams that are building these projects. In the past, Process Safety work for capital projects was supported by Process Safety engineers from the Engineering Teams, from Central Process Safety, or from Operations, depending on the technology involved and the geographical location of the project.

Another example of better alignment of resources and work is in the area of OPHRs. Air Products has a corporate policy that requires OPHR revalidation of all operating plants at regular intervals. For plants that are covered by OSHA PSM or that would fall under OSHA PSM requirements if they were located in the US, the revalidations are held every 5 years, or more frequently if required by local codes. All of Air Products hydrogen plants and Materials Technologies plants fit under this umbrella. For plants that are not subject to OSHA PSM requirements, the Air Products policy is to complete an OPHR revalidation every 10 years. All of the air separation plants are in this category. With 750 operating plants, and either a 5-year or 10-year revalidation interval, this results in about 80-90 OPHR revalidations every year.

In the old organization, OPHR revalidations were led primarily by the Central Process Safety Team, as there were no Process Safety Engineers assigned to the Operating groups. Timely closeout of OPHR Action Items was a major issue. The Accountability for closing out the Action Items rested with plant management. Many of the Action Items required Process Safety work to close, yet the plant managers did not have direct access to the Process Safety team and had no control over the priorities of the Process Safety team members. In the new organization, the operating groups each have Process Safety team members who report up through line management. These Operations-based Process Safety engineers support Plant Operations 100% of the time and therefore are able to assign a higher priority to OPHR activities than in the past when this work was handled by the Central Process Safety Team (whose competing priorities included capital project support, standards writing, and functional development).

Credit must be given to the new teams themselves, as the majority of work simplification efforts have been organically generated and executed within the teams. The Central Process Safety Team serves as referee in these cases, to certify that the simplifications do not undermine our Process Safety policies and standards.

The Central Process Safety Team went from 32 individuals to 6 individuals following the reorganization. The new Central Team has primary accountability for sustaining Process Safety excellence at Air Products through a variety of activities.

- Ownership of Policies and Standards
- Development of in-house Process Safety tools
- Expertise in third-party and internally developed Process Safety tools
- Development of Process Safety science such as
    - Improved methodologies for quantitative risk assessment
    - New models of Process Safety phenomena such as explosions and gas dispersion

    - Oxygen Safety and Flammability
- Process Safety training and functional excellence for Process Safety engineers
- Process Safety training for non-PS personnel
- Primary contact with third party PS firms such as Baker Risk and DNVGL
- Resolution of Process Safety items that are elevated from the teams

One of the major concerns in moving to a decentralized Process Safety organization was the potential loss of the strong Process Safety voice at Air Products. Prior to the reorganization, Air Products had a well-developed and practiced risk escalation process, whereby individual Process Safety engineers could raise an item for review to increasingly higher levels in the company if certain risk criteria were not satisfied. The process was successful in no small part due to the strong presence of the Global Process Safety Director to whom the majority of the Process Safety engineers reported. The risk elevation process was maintained in the new organizational structure, and training was provided to the new Process Safety management team to make them aware of the process and when to use it. In the new organizational structure, the Global Process Safety Director maintains independence from the business teams and serves as the conduit for risk elevation and the risk elevation process owner. The process has continued to function well in the new structure.

4 Strengths and Weaknesses of the Organizational Models

No organizational construct is perfect. Every organizational construct has strengths and weaknesses. The Process Safety reorganization at Air Products was completed approximately one year ago, and with the passage of time, the strengths and weaknesses of the central Process Safety organization and the decentralized Process Safety organization have come into sharper focus.

	Centralized	Decentralized
Strengths	• Single leader to drive Process Safety culture across the company and serve as strong voice for Process Safety • Facilitates development of PS Engineers via variety in assignments and technologies supported without having to change jobs • Independence of PS from line management • Easier to develop long-tenured Process Safety engineers	• Greater focus and accountability for Process Safety work on capital projects and in Plant Operations • Simplified work processes • Increased PS expertise in Engineering and Plant Operations • PS is closer to 'where the action is' and more effective in making positive changes • Easier to move people into and out of Process Safety assignments

Weaknesses	• Competing priorities and no direct control of resources by the user groups (Engineering, Plant Operations) • Inefficient work processes	• PS excellence harder to develop and sustain in dispersed teams • PS teams not independent from line management

5 Sustainability and Process Safety Competency

Twelve months after completing the reorganization it is apparent that the largest challenge in the new structure is developing and maintaining functional excellence in Process Safety. With one large Central team of co-located individuals, Process Safety training and professional development was easier than in the new decentralized structure, if for no other reason than the logistics were more amenable for collaborative learning in the old structure. We realized that our approach to training and professional development needed to change to be successful in the new organization. While we are still moving towards this new state, the main elements are:

Community – The entire Process Safety community meets bi-monthly to discuss events that have cross-organizational implication, to roll out new tools and methodologies, and to review standards updates.

Mentoring – New or inexperienced Process Safety engineers are paired with a senior Process Safety engineer. The senior person provides mentoring, explains work processes and serves as a general resource for the less experienced individual. The mentoring relationship will last for several years.

Leadership – The Process Safety Leadership Team continues to meet monthly to review matters of common interest and to assess the health of the Process Safety community. Typical agenda items include manpower sharing across teams, professional development and training, review of significant Events, and related items. The Global Process Safety Team continues to independently review all Tier 1 and Tier 2 Events to ensure that the investigators arrived at the correct root causes and remediations.

Training Curriculum – Air Products has an internally developed Process Safety training curriculum that is delivered by experienced Process Safety engineers to other Process Safety team members and the general employee base. We have developed over 50 training classes that cover topics including Hazard Identification, SIS lifecycle, SIL Assessment and Verification, Consequence Analysis, Dispersion Modeling, Radiation Modeling, Mechanical Integrity, MOC, OSHA PSM, Seveso, HAZOP, LOPA, Fault Tree Analysis, Relief System Design and Analysis, Vent Analysis, Oxidizer Safety, and H2/Syngas Safety. In the old organization, the training philosophy was that all Process Safety Engineers should complete all of the training classes, with the order of training dependent upon the current or next work assignment. In the new organization, we are arranging the curriculum around the needs in each group – Engineering, Operations, and Central. For example, The Engineering team members will receive HAZOP, Quantification, and SIS/SIL training before receiving OSHA PSM or OPHR training. The opposite is true for team members in Operations. The new philosophy is to structure the training so that the Process Safety engineer receives training in an order that corresponds with the major responsibilities of the team in which they reside.

6 Lessons Learned

The Lessons Learned are intended as general guidelines to consider if/when your company proposes a reorganization of the Process Safety function, and may also be helpful if you are a new Process Safety Manager and want to overhaul the organization that you inherited.

1. Reorganization is not a substitute for Leadership Commitment or Process Safety Culture. If these factors are not strong before the reorganization, they are not going to become stronger because of a reorganization (unless they are specifically addressed as part of the reorganization).
2. Have a clear goal that explains the desired outcomes. For Air Products, the reorganization was designed to deliver Speed and Simplicity in the Process Safety arena by more closely aligning resources with the businesses they served.
3. Faster is better. Only having three months to complete the reorganization forced us to make decisions faster than if we had a longer schedule. It also minimized the time period during which there was disruption.
4. Use the reorganization as an opportunity to eliminate work and to streamline work processes.
5. Ensure that all team members continue to do their 'day job' during the reorganization. The reorganization cannot be used as an excuse to slowdown or stop critical Process Safety activities.
6. Monitor the Metrics to ensure that Process Safety performance is not slipping.
7. Meet regularly with the new Leadership Team. Use these meetings to advance the Process Safety Culture and to ensure that the new leaders are effectively working through Management of Change for their new responsibilities and their personnel.
8. Recognize that Standards, Tools and Management Systems are company-wide enterprises and need to continue no matter how the new organization is configured. In our case, we located these activities in the Central Team and they continue to remain under the direction of a Global Process Safety Director.
9. Communicate with stakeholders as often as possible, preferably face-to-face or in one-to-one teleconferences. Listen to their concerns and put plans in place to address them.

7 Conclusion

A reorganization of the Process Safety function will deliver tangible improvements if done well, but also has the potential for disastrous results if not done well. Strong leadership and a clear vision of the 'after' state are necessary for a successful transition. It has been twelve months since we completed the reorganization at Air Products, and it is apparent that the transition has been successful, with a continued strong Process Safety Culture and Leadership Commitment.

LIST OF ACRONYMS

EHS: Environmental Health and Safety function or department

EPA RMP: A Risk Management Program administered by the Environmental Protection Administration in the United States

HAZOP: Hazard and Operability Review

LOPA: Layer of Protection Analysis

MOC: Management of Change

P&ID: Process and Instrumentation Diagram, also called a flowsheet

OPHR: Operating Plant Hazard Review

OSHA PSM: A Process Safety Management program administered by the Occupational Safety and Health Administration in the United States

QRA: Quantitative Risk Assessment

SIL: Safety Integrity Layer

12TH GLOBAL CONGRESS ON PROCESS SAFETY

Becoming "Wiser" in Management of Change

Sean J. Dee
Exponent Failure Analysis Associates
4580 Weaver Parkway, Suite 100, Warrenville, IL 60555
sdee@exponent.com

Mark Fecke, Trey Morrison, Russell Ogle
Exponent Failure Analysis Associates
4580 Weaver Parkway, Suite 100, Warrenville, IL 60555
mfecke@exponent.com, tmorrison@exponent.com, rogle@exponent.com

Prepared for Presentation at
American Institute of Chemical Engineers
2016 Spring Meeting
12th Global Congress on Process Safety
Houston, Texas
April 10-13, 2016

Becoming "Wiser" in Management of Change

Sean J. Dee
Exponent Failure Analysis Associates
4580 Weaver Parkway, Suite 100, Warrenville, IL 60555
sdee@exponent.com

Mark Fecke, Trey Morrison, Russell Ogle
Exponent Failure Analysis Associates

Keywords: Management of Change, Process Safety Management, Case Studies

Abstract

An Arabian Proverb identifies four types of people based on their knowledge level and self-awareness: the fool who is to be shunned, the simple who is to be taught, the sleeping who is to be awoken, and the wise who is to be followed. In a Management of Change (MOC) program, it is important to establish the known (and unknown) impact of a proposed modification. When effectively utilized, MOC programs evaluate and control proposed modifications to existing processes, ensuring that new hazards and existing hazards are effectively managed. Unfortunately, the existence of an MOC program alone does not guarantee that process incidents related to facility modifications will not occur. When post incident investigations find root causes related to process changes, a question commonly raised is: "Why wasn't this incident's consequence identified and evaluated through the site's MOC program?" This paper presents several case studies where process incidents occurred as a result of a modification in a facility with an existing MOC program. Cases will highlight critical points within an MOC program where oversight can lead to an unintentional increase in hazards and risks. Corrective actions and lessons learned will be discussed in an effort to improve the overall design, development, implementation, operation and monitoring of a facility's existing MOC program.

1 Introduction

Management of change (MOC) is recognized as a critical pillar to a process safety management program. [1] An MOC program establishes the process for systematically evaluating facility modifications to ensure that hazards and risks are adequately controlled. An MOC program can help a facility maintain safe operation while adapting to changing environmental, financial, or technological constraints. When effectively implemented, an MOC program not only assists in the safe execution of process modifications, but it can also be a key aspect of a facility's process safety culture.

In many instances, the change to the process that precedes a process safety incident can be readily recognized as a significant departure from a prior safe design. But, process safety incidents sometimes occur due to unintended consequences of seemingly insignificant modifications to equipment, controls or operating methods. These types of incidents can provide the impetus for a continual improvement in a company's implementation of their MOC programs. In the following manuscript, several case studies will be presented that highlight areas for improvement related to an existing MOC program. Several additional resources are available for information related to establishing, operating, evaluating and improving an MOC program. [1-3]

2 The Four Types

Take the following version of an old Arabian proverb –

> *Men are of four types:*
> *He who knows not, and knows not he knows not: he is a fool – shun him.*
> *He who knows not, and knows he knows not: he is simple – teach him.*
> *He who knows, and knows not he knows: he is asleep – wake him.*
> *He who knows, and knows he knows: he is wise – follow him.*

Consideration of the proverb reveals key insights into human interaction, which can be applied in life…and in the chemical process safety field. The proverb speaks of four specific types of people; the fool, the simple, the asleep, and the wise. These four types are grouped according to two key criteria, their knowledge level, and their self-awareness of their knowledge level. The fool not only lacks relevant knowledge, but also the self-awareness of the lack of knowledge. The simple recognize their own knowledge gaps, and thus are open to teaching and training to address them. The asleep have the necessary tools, but lack recognition and must be awoken. The wise are both knowledgeable and self-aware and are followed as leaders. In its simplest sense, one would expect that everyone would strive to be categorized as wise according to the proverb. However, an individual can find value in understanding which of the four types applies to himself or herself, and identify the necessary steps to become wise. There is also additional benefit in being able to understand how others fit into the four types, to push them in a collaborative environment to develop toward the wise. In this way, the proverb speaks not only to *inter*-personal relationships but also *intra*-personal qualities.

On its surface, the proverb may seem more relevant to a high school English student than a process safety manager, but several aspects of the passage are relevant to developing a robust MOC program. In an MOC program, several different departments are involved in the evaluation process (i.e. management, safety, engineering, maintenance, operations, etc.). In these types of interactions, it can be beneficial to understand and identify the four types of people described in the proverb. Then supplemental resources (i.e. additional people, training, testing, documentation, etc.) can be utilized to address specific limitations. Departments can also understand their value, as well as knowledge gaps, to effectively contribute to the MOC process. Several of the shortcomings of the four types can be found at the root of some of the following commonly identified challenges with MOC programs.

- Changes slip through the MOC system without being evaluated
- There is confusion/indecision regarding the level of detail for a hazard review
- The reviews of proposed changes are not thorough enough

- The quality of reviews is inconsistent
- Personnel turnover leads to:
    - Lack of follow-through on unfinished reviews of a predecessor
    - Loss of appropriate process knowledge

The relevance of the four types of individuals as related to the proverb is underscored in the Flixborough case study discussed in the CCPS book on MOC. [1] In the 1974 accident, a temporary reactor bypass piping loop failed, and the release created a flammable vapor cloud that ignited leading to a catastrophic explosion. As stated in the CCPS reference:

No engineering support was available in the plant at the time of the accident. The temporary modification was constructed by people who did not know how to design large pipes equipped with bellows. As stated in the official report: **"... they did not know that they did not know."** An effective MOC system should have discovered the design flaw before the change was implemented, thus averting the disaster. [1]

Herein we present case studies that highlight the relevance of the Arabian proverb to an MOC program. In the following examples, key learnings, insights, and recommendations will be discussed related to the continual improvement in the design, evaluation, and execution of a facility's MOC program.

3 Case Studies

The following case studies are used to illustrate the important of understanding an MOC evaluation team's knowledge, and their self-awareness of their knowledge. The case studies describe real industrial incidents, but some details have been changed to protect the confidentiality of the parties in the matters. In the following cases, oversight related to what people knew/didn't know, or information that was incorrectly assumed to be known, led to process safety incidents. In all the cases presented, understanding the limitations of the knowledge of the team members, and addressing them during the MOC program could have potentially prevented incidents that resulted in unscheduled downtime, lost raw materials and/or personnel injury.

3.1 Batch Processing of Combustible Dust and a Flammable Liquid

A facility operated a batch mixing process, where a combustible dust and a flammable liquid were mixed, heated, and then pumped to the next stage of the production process. An overall diagram of the equipment involved in the batch mixer is shown in Figure 1. When an order was placed for product, an operator would add powdered catalyst, which was a combustible dust, through an overhead chute into the tank. The contents of the chute would then be added to the tank when the operator started the rotary airlock at the base of the addition chute. A typical batch required the addition of several drums of the powdered catalyst to the mixing tank.

Facility personnel recognized the hazard associated with adding combustible dust to a tank containing a high vapor pressure flammable liquid. The site used nitrogen gas for purging and inert blanketing of the mixer vessel and associated additive chute to limit the potential for fires or explosions within the system. Several valves including a rotary airlock, and multiple slide gates served as isolation points between the tank headspace and the addition chute. The tank headspace was vented to an emission control system (flare). A controller was configured such that when combustible dust was being added to the tank, the tank vent valve to the flare closed, and a nitrogen blanket opened and applied an inert blanket to the tank. The N_2 blanket valve was interlocked, such that the valve would not open unless the rotary airlock motor was operating. When the rotary lock was not running, the controller would close the N_2 blanket valve, and open the tank's vent valve to the flare. A flow-through nitrogen purge system was also installed on the addition chute. This system gave operators the ability to purge the addition chute with nitrogen prior to, during, or after the addition of combustible dust.

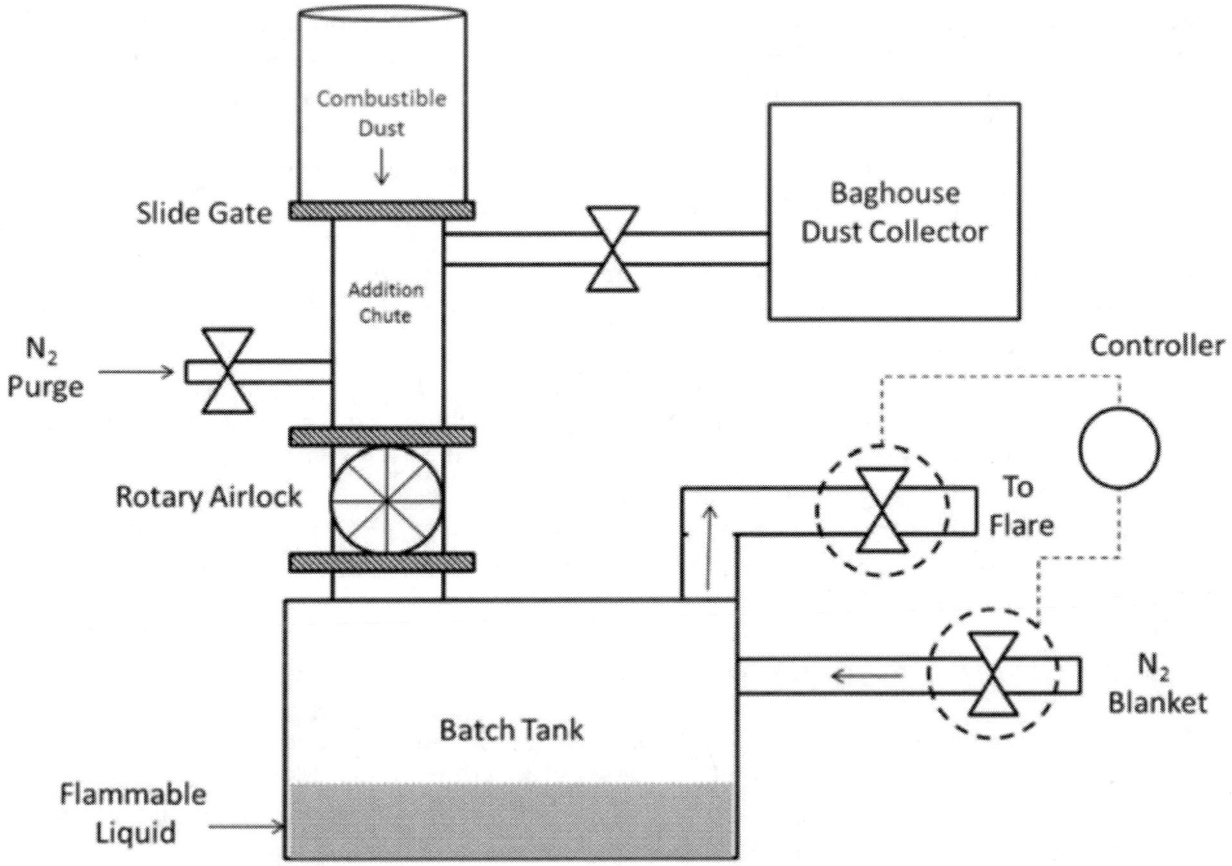

Figure 1. Schematic for addition of combustible dust to a flammable liquid

The facility operated the batch mixing system without incident for several years before several flash fires occurred in the addition chute. The incident investigations discovered that there was operator-to-operator variation in the combustible dust addition steps and manual purge steps used for the system. One of the noted differences was whether an operator would add the contents of the addition ductwork as separate one-drum additions, or one addition that contained all drums. As a corrective action, the site standardized the procedure, and ultimately decided that the addition procedure should consist of separate one-drum additions, with a purging step performed before and after each addition.

After standardizing the procedure the mixing procedure was performed without incident for several months, before another flash fire occurred in the addition chute. Post-incident investigation found that flammable vapors were leaking past the rotary airlock and slide gates into the addition chute system. The valves were repaired, but the site experienced a third flash fire during the next mixing operation. The facility ceased all mixing operations until the root causes of the recurring flash fire incidents could be determined.

After the third incident the engineering staff was puzzled due to the fact that flash fires were occurring in a system that was under an inert atmosphere and isolated from the tank's flammable vapors. However, upon further examination several deficiencies were identified. First, while the purge system established an inert atmosphere prior to catalyst addition, air would infiltrate the system at the drum/addition chute interface. Second, it was determined that the split level controller took approximately 5-7 minutes to transition from venting the mixing tank to applying the nitrogen blanket. As a result, an inert blanket was not established and maintained in the tank during catalyst addition. Additionally, the rotary lock used in the system required solid material above it to maintain its seal. When the addition chute emptied the rotary lock would spin for several minutes allowing infiltration of flammable vapor into the addition ductwork.

In this particular case, the facility had several areas to potentially improve its MOC process. The facility felt an MOC was not necessary because there was no modification to existing equipment. They also felt the standardization of the procedure did not require an MOC because some of the operators had been utilizing the standardized version of the procedure. However, the standardization of the procedure was a significant change. Interfacing with the various departments involved in an MOC process could have potentially revealed the deficiencies of the nitrogen inert blanket and purging system. The facility also made several assumptions related to the establishment and maintenance of the inert atmosphere with the system. When testing revealed infiltration of both oxygen and flammable vapors into the addition chute, the assumptions were shown to be invalid.

3.2 Tank Failure Due to Reverse Flow Condition

A facility processed a highly viscous liquid product with solid sediment. In the finishing portion of the facility, large sedimentation tanks were used to allow the solids to settle, so pure, high viscosity liquid could be withdrawn and packaged. The facility went through an upgrade, where a large storage tank (approximately 100 feet tall, 30 feet in diameter) was converted into a new sedimentation tank. An MOC was completed for the conversion of the storage tank to a sedimentation tank. The newly converted storage tank operated for several years, when an operating modification was proposed. The facility had conducted a study that determined if the tank was bottom filled, liquid product could be drawn off the top of the tank at a rate that would double the throughput of the sedimentation tank. The facility performed a MOC for the modification to the flow condition, and implemented the new design. An overview of this modified design is shown in Figure 2. The facility operated with the bottom fill sedimentation orientation for approximately 10 months, when suddenly the tank wall failed at the 70 foot elevation level, causing the upper portion of the tank to sink approximately seven feet. The facility was shut down, and a root cause investigation was undertaken.

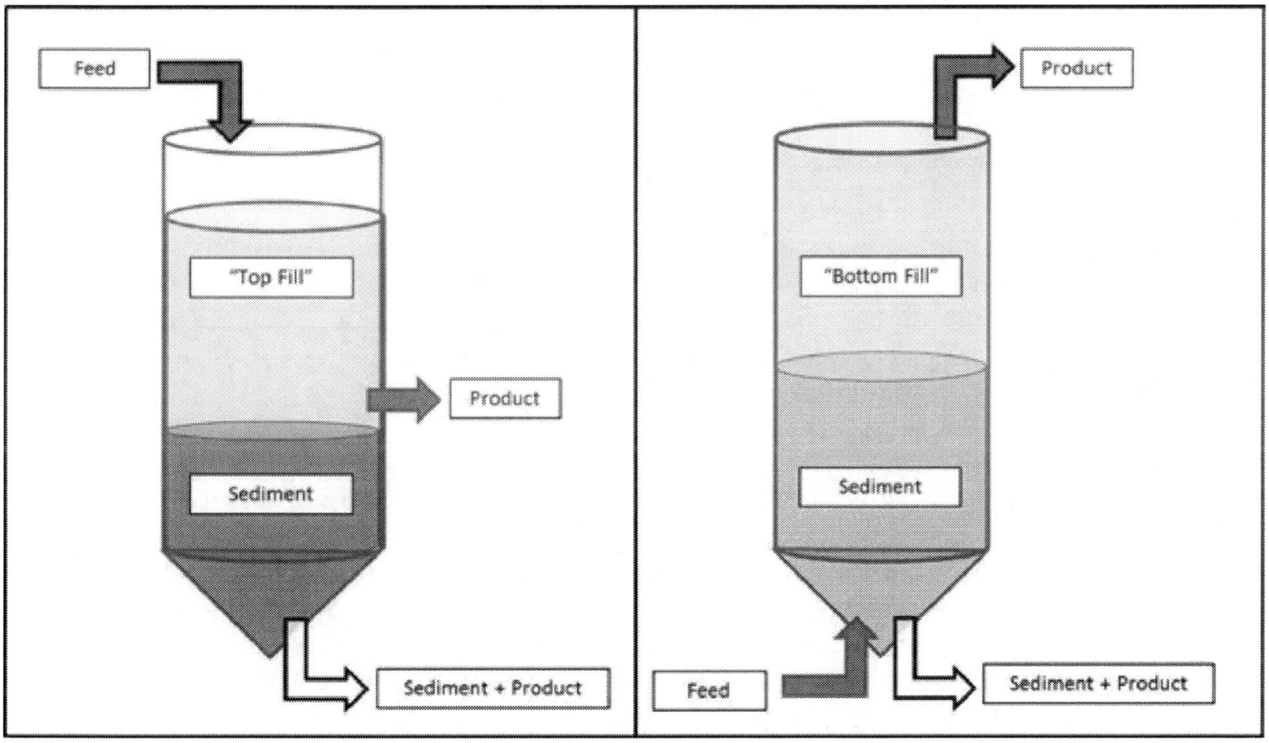

Figure 2. Comparison of Bottom Fill and Top Fill arrangements for a settling tank

Post incident investigation determined the root cause of the failure as excessive stress loading on the tanks internal wall. In certain storage tank designs (for example tanks that are much taller than they are wide), special design considerations are necessary to maintain the vessel's structural integrity as the weight of the wall increases. The failed tank wall consisted of several different thicknesses of metal, gradually thinning as the tank height increased. The thinner wall material at the top of the tank results in less weight that has to be supported by the lower courses. The post incident investigation determined that the tank wall failed at an elevation corresponding to a transition between two different course thicknesses.

During the root cause investigation, the MOC process and paperwork for both modifications (storage tank conversion and reverse flow configuration) were reviewed. The evaluation team from the conversion project included a representative from the storage tank manufacturer, because the site wanted to ensure that the tank was structurally capable of handling the loads associated with its new service. The facility recognized the necessity to involve the tank manufacturer when the tank was converted from static storage, to a flow through tank. The impact of transitioning a tank from static loading to continuous flow loading was outside the expertise of the facility, so the manufacturer was needed to approve the tanks structural integrity. The tank manufacturer was comfortable with the top filled flow configuration because the interstitial flow velocity was low, and the areas of the largest localized stresses were at the bottom of the tank and corresponded to the elevations with the thickest wall courses. Given the proposed design, the tank manufacturer approved the MOC paperwork.

In the several years between the approval of the storage tank conversion MOC, and the proposed flow modification, the facility experienced a high degree of personnel turnover. The MOC evaluation team for the flow modification consisted of completely new members, none of whom were involved in the MOC for the storage tank conversion. The proposed flow modification was considered an operating condition change, and the tank manufacturer did not have a representative on the MOC evaluation team. As a result, when the evaluation was finalized, the impact of increasing the throughput flow rate (thereby increasing the interstitial flow velocity and shear stress loading on the walls) was not evaluated for impact on structural integrity. The impact of having these larger stresses located at the top of the tank, where thinner courses of wall material were located, was also not considered in the MOC evaluation. The MOC was approved, and the potential issues associated with the structural integrity of the vessel in the new flow condition went unnoticed until the structural failure occurred.

The facility originally recognized their knowledge gaps, related to converting a large static storage tank to a dynamic flow through vessel. As a result, the facility brought in the necessary expertise of a firm that designs both static and dynamic vessels to assist with their MOC evaluation process. However, when the facility experienced turnover, they lost the insight and knowledge that they had gained through the initial MOC. The newly completed MOC did not include a thorough evaluation of the critical parameters to operate the system safely. This is one reason why personnel turnover and capturing knowledge that could be lost during turnover is a situation that can be addressed exclusively through a specific MOC.

3.3 How Reducing Air Emissions Created Explosive Wastewater

A toll manufacturer produced an organic intermediate using a nitration process. The process unit, regulated under OSHA's Process Safety Management regulation, was a batch operation with numerous reactors, separation units, and storage tanks. The process unit generated a number of waste streams including hazardous organic wastes (both liquid and solid), wastewater, and a nitrogen exhaust stream. Over a period of a few years, the manufacturer implemented a number of undocumented process improvements to improve their compliance with environmental regulations. One of these improvements was to reduce their air emissions from the nitrogen exhaust stream by passing it through a condenser. The condensed organic liquid was stored in a storage tank (called the condensate tank) and on occasion the liquid organic waste stream was recycled back into the front of the process.

During a maintenance turnaround, it was deemed necessary to drain the contents of the condensate tank. The organic liquid was transferred to a wastewater receiving basin by pumping it through a heat-traced pipeline. During the transfer there was an explosion that injured several workers and demolished half of the process unit. The explosion was later determined to be a condensed phase detonation of the organic liquid condensate.

Besides producing the desired molecule, the nitration of organics typically produces a number of undesired byproducts that are themselves nitrated organics. It was well known to the facility management that nitrated organics can often pose serious chemical reactivity hazards. When the condenser was added to the process, the proper MOC paperwork was completed, but the potential for the condensed liquid to pose a chemical reactivity hazard was not considered. During the investigation of the accident it was determined that the organic liquid condensate was a mixture of mostly trinitromethane and tetranitromethane. Both of these nitrated organics are shock-sensitive and thermally unstable compounds.

In the toll manufacturer example, the MOC was completed, but a specific hazard scenario that was later experienced was not considered as part of the MOC process. It is important for a facility to gather relevant information necessary to characterize a change's impact across the various aspects of the processing equipment. In this particular case, an unidentified hazard was not evaluated and controlled, because the facility personnel involved in the MOC did not know that the organic liquid condensate contained unidentified hazardous materials. Instead, they assumed that the stream had a composition that was consistent with a waste they could handle safely. In an MOC process, questioning the assumptions is critical to determine the specific facts that have been characterized and measured as part of the preparation to justify a particular change. A waste stream characterization study could have been requested as part of the MOC evaluation, which would have revealed the presence of trinitromethane and tetranitromethane. When personnel are asked to make decisions with incomplete information, it is possible questionable assumptions will be proven incorrect with future incidents.

3.4 Changes to the Valve Configuration for a Legacy Process

A beverage processing facility vaporized pressurized liquefied carbon dioxide (cryogenic LCO_2) in a semi-continuous process in order to add carbonation to drink products. A new process line was being added downstream of this system. Temporary changes to the upstream LCO_2 system led to accumulation of trapped cryogenic liquid. The cryogenic liquid warmed then eventually caused catastrophic explosion of vessels in the system. Fortunately, no employees or members of the public were injured.

The LCO_2 plant worked on a simple concept of flowing LCO_2 through a vaporizer heated by a closed loop water heating system. LCO_2 flow was driven by the pressure differential between the source bulk LCO_2 tank and the downstream carbonation process. Operators typically only started or stopped the LCO2 system and did not attempt to manually regulate LCO_2 flow. The vaporizer was equipped with a flow-restricting orifice and automatic fail-closed shut-off valves with a manual bypass around the vaporizer. From examination of the system, the uninsulated LCO_2 supply piping likely allowed the liquid to vaporize due to ambient temperatures prior to reaching the vaporizer under most operating conditions.

A series of high-pressure surge vessels was used to accommodate fluctuations in CO_2 gas pressure with changes in the downstream demand. The high-pressure surge vessels were connected overhead to the vapor space of the bulk liquid storage tank by an equalization line. The vessels manifolded on the inlet side and on the outlet side. The surge vessels were all equalized at the same pressure due to the interconnected manifolds. The operating pressure of the vessels was normally about 250 psig or less, the same as the bulk storage vessel due to the gas equalization line.

Prior to the incident, operators adjusted the positions of manual valves in the LCO_2 vaporizer system in an attempt to give the downstream system a boost of gas pressure. Most operators had worked in the beverage processing plant for several years, but few understood the operation of the LCO_2 plant. No MOC process was conducted as part of this temporary, abnormal operating state. As a result, the system was unintentionally allowed to bypass a significant quantity of LCO_2 into the downstream processes and pressure vessels. The operators subsequently isolated the system and attempted to drain the LCO_2 to atmosphere. Unfortunately, several vessels plugged with solid CO_2. After warming for several hours, some of the vessels catastrophically ruptured causing significant damage to the facility.

The facility staff did not employ an MOC process to identify potential hazard of a change in operating state. Instead, the experience of the operators was implicitly relied upon to perform the unique operation safely. Without an engineering analysis of the change and a plan for monitoring the status of process during the changes, the process was allowed to go out of control. An adequate MOC would have identified the hazards. The MOC could also have provided a plan for preventing the trapping of cryogenic LCO_2 in the system and a response plan in case such an event did occur.

4 Conclusions

Several case studies were examined with respect to process incidents related to insufficient MOC programs. The particular framework with which the cases were evaluated focused on an evaluation of the information/assumptions that a facility made as part of the MOC process, and the degree to which the facility was aware of their knowledge (or lack of knowledge) regarding a proposed change. Among the key concepts discussed were:

- Small, seemingly insignificant changes can have large consequences on the performance and control of other related systems. Therefore, it is important that when a decision is made to forgo an MOC for a proposed change, that the decision to forgo is approached with the same rigorous evaluation as the actual MOC process. The assumption that a change is not important enough to perform an MOC, without a thorough analysis, can lead to increased risk.
- When performing an MOC evaluation, it is important to distinguish between assumptions versus facts regarding the impact of a proposed change on a system's performance. Evaluations should be cautious in using pre-change performance as an indication of the proposed continued success after a change is implemented.
- MOC processes should aim to be broad in their evaluation scope and involve members with varied, diverse backgrounds to evaluate the breadth of potential consequences of a proposed change. Facilities should consider bringing in external resources and expertise when it is not available within the facility staff.
- Facilities should be mindful of previous MOCs that have been performed on equipment that is under evaluation for a new proposed change. Previous MOCs could contain relevant knowledge, learnings, and documentation that is relevant to a new MOC. A new MOC should not neglect a consideration that was noted in a previous MOC.
- Facility staff turnover can impact a team's ability to effectively conduct an MOC evaluation. When turnover occurs, knowledge should be captured and documented, such that it is not lost after the transition.

In process safety management, and life in general, one should strive to be characterized as "wise" according to the definition in the referenced Arabian proverb. This wisdom is rooted in two key factors, one's knowledge and their self-awareness of their knowledge. It is not enough to possess knowledge – one must also be aware of the knowledge they possess, and any gaps in the knowledge they possess. MOC programs can benefit from approaching their evaluations by asking the following key questions: "What is known?", "What is unknown?", and "How confident are we regarding what is known and unknown?"

5 References

[1] CCPS. *Guidelines for Management of Change for Process Safety*. Center for Chemical Process Safety, American Institute of Chemical Engineers, New York, NY, 2008.

[2] CCPS. *Guidelines for Managing Process Safety Risks During Organizational Change*. Center for Chemical Process Safety, American Institute of Chemical Engineers, New York, NY, 2013.

[3] CCPS. *Guidelines for Auditing Process Safety Management Systems*. Center for Chemical Process Safety, American Institute of Chemical Engineers, New York, NY, 2011.

12TH GLOBAL CONGRESS ON PROCESS SAFETY

Successfully Managing a Changing Workforce – Application to Process Safety Management

Charles A. Soczek
Principal Consultant
DuPont Sustainable solutions
Wilmington Delaware

Prepared for Presentation at
American Institute of Chemical Engineers
2016 Spring Meeting
12th Global Congress on Process Safety
Houston, Texas
April 11-23, 2016

Introduction

Many companies around the globe are faced with the difficult challenge of managing workforce change in a time when budgets are tight, pressures to decrease costs and improving productivity are high, and there is still an expectation to maintain safe operations. This is not a new challenge, but workforce changes are even more complex than in the past.

The Channel Industries Mutual Aid (CIMA) is a cooperative organization of over 100 emergency response agencies in the Texas counties of Harris, Chambers, and Brazoria and industrial fire departments from the petrochemical, chemical, and refining operations. They were formed in 1955 under the name of Ship Channel Industries Disaster Aid Organization. They have 60 years' worth of serious industrial accident records. The head of CIMA believes the single most significant factor in incident increases is when the workforce changes as a result of new processes and major shutdowns.

In today's environment, companies have much more sophisticated technologies and streamlined supply chains, and mistakes can have severe consequences. The problems are compounded by the fact that many companies have a large number of retirees, a large number of new employees entering the workforce for the first time, and few in the middle.

New engineers entering the workforce today have an extraordinary broad range of skills. Successful companies will take advantage of what these new employees bring by changing work processes to be increasingly efficient and productive, while still utilizing the experiences of the more senior workforce.

DuPont Sustainable Solutions (DSS) has identified a multi-part process to help companies thrive in this time of workforce change.

This paper will explore key characteristics of new employees entering the workforce today and will outline the steps developed at DSS to meet the needs of this new type of employee. The goal of these changes is to ultimately provide a suitable working environment and improve employee retention which will have the added benefit of improving process safety performance.

Managing Highly Hazardous Operations

Successful safe operations in high hazards process industries require talented engineers with broad ranges of skills and experience. This includes chemical, mechanical, instrument, and electrical engineers who are involved in conception, design, construction, and operations. They build and modify facilities, and keep the plants running efficiently and safely. They work on longer range problems to implement safety, cost, and capacity improvements. Most operations run around the clock, 24/7, 365 days per year, occasionally interrupted by breakdowns in equipment. Because of this, engineers that support operations need to be available at all times to assist in the resolution of problems. Engineers plan and execute the maintenance and inspection programs. They are doing this today with fewer people, tighter budgets, and greater demands for production. Workforce changes, whether a result of retirements, transfers, leaving for other opportunities, or other reasons can put a great strain on the remaining workforce, and on the new people on the job.

Manufacturing operations have always required personnel with skills, experience, and the much more elusive and difficult to define characteristic, wisdom, which couples skill and experience with good judgement. Engineers learn their basic skills in college, refine and expand them in industry, and add new experiences along the way. Experience and wisdom can reside anywhere in the organization, in practicing engineers, line management, and also in the operators and their supervision. Experience is gained not only from direct operational experience, but also from incidents that have occurred in related companies and industries. People in the workforce who combine their experiences with a sense that hazardous events can happen in their operations bring sound judgement. They continuously seek to avoid events that can have bad consequences.

The role of immediate supervisor to new engineers is critical. Much has been made in recent years of the need to achieve a work-life balance. Many millennials, or generation Y (those born from the mid-1980s to the mid-1990s), achieve a work-life blend, where one is not entirely separate from the other. Technology allows employees to work anywhere, anytime. The impact of this is that people will need to be judged by performance, not presence.

Skills and Qualities of Engineers who have Recently Entered the Workforce

There is a lot of information on the generational traits of millennials. However it is wise to not assume that all entering engineers have these qualities, as there are significant differences between individuals. With that caveat, let's looks at some of the more important traits that can impact a millennial's process safety management (PSM) performance. Some of these may be considered positive aspects, and some negative.

General Traits of Millennials, from Sally Kane on Generation Y[1], (millennials, born in the mid-80s or later)

Tech Savvy

Smart phones, tablets, email, text messaging, Twitter, Facebook, LinkedIn and whatever new comes along is candy for engineers. They use these devices and tools to stay in continuous communication with each other with little distinction between personal and work life. They prefer these forms of electronic communication versus phone calls and face-to-face. Printed materials are avoided wherever possible. While previous generations may feel something is lost with mostly electronic interaction, millennials have been educated using electronic tools, and can use them while maintaining a personal touch. With these tools, they can work anywhere, anytime. For example, a parent needs to leave at 3 to pick a child up at day care, but after 8, when children are in bed, the parent works until 11, and communicates with others who are connected electronically.

Family Centric

"Generation Y have a different vision of workplace expectations and prioritize family over work." This can create some difficulties for an engineer in manufacturing operations. Shift cover for start-ups and other operating problems can greatly interfere with family life. Many married engineers from two-working-parent families or who are single parents demand flexible work schedules. Many companies have adjusted to this by incorporating core work hours, such as 9 to 3, where all employees are present.

Achievement Oriented"Generation Y wants meaningful work and a solid learning curve." They want these good assignments to gain experiences as well as a reputation for getting the job done. In particular, the first assignment is critical if a company hopes to retain talent. There is also an expectation of promotions.

Team Oriented

"They value teamwork and seek input and affirmation of others." They have been involved in team activities their whole lives, from sports to an educational process that involves working together to solve complex engineering problems.

Attention Craving

"Generation Y craves attention in forms of feedback and guidance. They appreciate being kept in the loop and seek frequent praise and reassurance." This can be a problem for supervisors heading an organization with a very large number of engineers versus previous generations.

Some misconceptions about millennials are that they are spoiled, immature, and lazy. These are character traits of individuals, not a generation. Many millennials are highly motivated and expect to move up rapidly in an organization.

Generation Z

The millennials, or Generation Y, are generally thought of as being born between the mid-80s to the mid-90s. After the millennials comes Generation Z, those born after 1995. They are only now beginning to enter the workforce, and will have some general characteristics that distinguish them from the millennials. Laura Montini (2) gave some characteristics of generation Z. They prefer person to person interaction and they come to the workforce well prepared. Both generation Y and Z say honesty is the most important quality of a good leader.

Integrating New Engineers into High Hazards Operations

The large number of new engineers entering the workforce makes it a challenge to continue safe operations. Faced with this challenge, a small cross-generational group in DSS was identified with the purpose of defining the keys for a successful integration of new employees into high-hazard operations. The objective was to develop a process that could be used with minimal changes for engineers entering the workforce. The process starts with understanding what is important to the engineers and the company, then moves to identifying action items, and finally developing and implementing plans.

A six sigma process to identify key areas, develop the CTQ (items critical to quality) and the CTXs (a factor that impacts the CTQ) was used. Initially six key areas were identified and developed. This was later expanded to eight. Most of these areas were drawn from practices at DuPont sites.

1. Critical functions and skills need to be identified

This step identifies not only the PSM functions and skills for today, but seeks those required in the future to assure competitiveness. There is clarity in what is done internally, and what is outsourced. This also includes the identification of "Technology Guardians." Technology Guardians are senior technical employees who oversee a wide range of activities over a single or group of processes for a long period of time. They help assure continued safe operation. For example, this includes review of all proposed changes within a given process. Career development of a guardian is critical. In some cases this is accomplished by providing gradually expanding responsibilities to cover multiple facilities or additional processes.

2. Individual Competency Development Plans

These plans help assure growth of all personnel, but can go a long way in retaining individuals that are necessary for future success. It is important that development plans meet the need of the company as well as individuals. They are developed jointly with the manager and include training and development assignments. The company has a process to identify high performers and talent for leadership positions.

3. Succession Planning for all Critical Positions

Identification of critical positions is the first step in process to manage a changing workforce. Once the critical positions are identified, up to five-year projections are made for new hires, promotions, reassignments, and retirements. These plans need to be periodically reviewed with senior management. Leadership assures resources are made available. Plans are put in place to overcome gaps identified.

4. Knowledge Transfer Management

Leveraging of experienced personnel is necessary through assignment, coaching, and mentoring, along with the development of a searchable technical database. All process safety information must be entered into the system and there must be a process to keep the information up-to-date.

5. Use of Blended Learning Approaches

Individual training plans are developed, and personnel are encouraged to participate in internal and external opportunities. Modern training methods include less lecturing, and more doing. The effectiveness of learning is regularly evaluated.

6. Cultivate Learning Communities

Leadership is strong on driving involvement of engineers in internal and external networks. Team and individual accomplishments are identified and rewarded.

7. On-Boarding Process for New Employees

The on-boarding process is critical to the success of new employees. Two key factors in this are assigning challenging and meaningful initial assignments, and using a mentorship program.

8. Cultural Environment

This is a multi-faceted subject that can require a company to evolve to take advantage of the workforce skills and expectations of employees entering the workforce for the first time. Examples include an environment that is sensitive to the needs of women, promoting teamwork for problem solving, and valuing speed in implementation. For instance, a company has an older phone system that doesn't pick up a woman's higher pitched voice, or the company does not allow employees to work from home when needed, such as caring for a sick child. Maternity leave also continues to be an important subject.

Below is an example of some of the CTXs for the initial on-boarding of a new employee. These are given in the form of questions that an organization can use to gain feedback from their new employees and their supervisor.

On-Boarding
Questions for the new employee:
What are your thoughts on your first assignment? Is it an important assignment? Is having an important assignment necessary or critical to you?
Have others in the organization been identified to help you on your assignments? Does this include both senior and junior people other than your direct supervision?
Do you feel there is there a high probability of success?
Questions for the supervisor:
Will the new employee be given access to management to present findings?
Will the first assignment give the opportunity to advance to more meaningful assignments?
Will reward and recognition be provided?

Conclusions

With ongoing changes and the loss of process safety experience, organizations can be put under great stress. In particular, the direct supervisors of new employees will face the greatest challenges. Typically start-ups of new facilities often require junior engineers to work shifts. This can range from a few days to weeks or even months, greatly disrupting the lives of engineers who may have more family obligations than previous generations. New engineers entering the workforce have less loyalty to an organization they hire into as a result of company policy changes, the need for rapid promotion, and pay increases that may not meet their expectations. Increased span of control for a supervisor means less time with employees. These and other issues mean companies must develop good processes to first understand their specific problems, and then implement changes to sustain safe operations.

Sources

1. Generation Y. (2014, December 16). Retrieved from http://legalcareers.about.com/od/practicetips/a/GenerationY.htm

2. Montini, L. (2014, September 2) Generation Z: Meeting your future hires http://www.inc.com/laura-montini/meet-you

16AIChE
Spring Meeting, Houston, TX

Conducting a Human Reliability Assessment to support PHA and LOPA

Dave Grattan, P.E., CFSE
aeSolutions
10375 Richmond Ave. Suite 800
Houston, TX 77042
dave.grattan@aesolns.com

Prepared for Presentation at
American Institute of Chemical Engineers
2016 Spring Meeting
11th Global Congress on Process Safety
Houston, Texas
April 10-14, 2016

UNPUBLISHED

Conducting a Human Reliability Assessment to support Process Hazard Analyses (PHA) and Layer of Protection Analyses (LOPA)

Dave Grattan, P.E., CFSE
aeSolutions
10375 Richmond Ave. Suite 800
Houston, TX 77042
dave.grattan@aesolns.com

Keywords: Ergonomics, Human Factors, Human Reliability, Task Analysis

Abstract
PHA and LOPA facilitators may struggle with Human Factors and Administrative controls in their team sessions for a variety of reasons. These could include: complaints from the PHA team that Operators are not given enough credit; scenarios where an Operator is the "cause" and the "cure" (the same or different Operator); scenarios that involve high hazards that are purely human factor and administrative control related (i.e. not related to equipment malfunction or external events); and scenarios that involve Operator tasks performed on a high frequency basis (i.e. one or more times per day).

The PHA team may get "stuck" and be tempted to deem these scenarios as "non-LOPAable", which is correct, but then fail to follow-up with an appropriate analysis.

Other tools such as Human Factor Checklists and Procedural PHA do not directly address these issues in a meaningful or quantifiable way.

A better methodology is needed to handle human factors and administrative controls when quantifying initiating cause frequencies and Independent Protection Layer (IPL) credits in PHA and LOPA, and is the topic of this paper. The methodology is aligned with the work of Swain and Guttmann (1983) Handbook of Human Reliability Analysis (NUREG/CR-1278). This paper will describe how the method can be applied to the semi-quantitative needs of PHA and LOPA. The results may also be used as an input to further QRA (Quantitative Risk Assessment).

The method begins with a well conducted PHA that will uncover the high hazards associated with mis-application of a Standard Operating Procedure (SOP), or other critical Operations based human error. From here, these scenarios become candidates for a Human Reliability Assessment (HRA).

The HRA is conducted as a "talk-through" consisting of two parts: [1] a qualitative task analysis of the specific steps of an SOP that, if omitted could lead to the scenario of concern from the PHA (errors of commission may be treated as well), or a task analysis of the Operator response

IPL credited in the PHA; and [2] a quantitative judgment of the key PSFs (Performance Shaping Factors) related to operator performance as initiating event or IPL.

The first part of HRA (task analysis) provides a qualitative benefit for understanding the Operator's relationship to the SOP, or IPL, in question. Is the procedure followed, is it reviewed before the task is performed, is it carried as the task is performed, does it utilize a check-list format, and is the procedure performed so often it becomes a skill-based task? Task analysis also scores the quality of a procedure in terms of complexity and specificity. Task analysis addresses dependencies within an individual (awareness of one's own errors) and between individuals (second operator). Task analysis identifies recovery factors from human error.

The output of task analysis includes drawing a Human Reliability (HR) Event Tree. The (HR) Event tree is fit for purpose. It is constructed step-wise from the critical steps in the SOP and looks different than traditional event trees. When the HR Event Tree is solved the results can be fed back into PHA and LOPA as a frequency or probability.

The second part of HRA (look at key PSFs) provides adjustment factors for the basic human error probabilities (BHEP) found in the Swain and Guttmann (1983) Handbook. The adjusted BHEP are used to quantify the HR Event Tree from the task analysis. The selection of the adjustment factors is relative to the best available technology at the time the Handbook was published. There are literally 1000's of PSFs in addition to human psychology (fears, motivations, etc.) that make quantification of human performance problematic. The method presented will focus on those PSFs that are measurable and have the greatest impact on human performance as it relates to initiating cause frequency or IPL.

A follow-up process walk-through, or model review (if the facility is under construction), is performed as well to validate the findings of the "talk-through".

This paper will present an overview of the HRA methodology, worksheets used to develop and document the HRA, examples of HR Event Trees, a method to incorporate the results back into PHA and LOPA, and lessons learned from conducting HRAs.

1. Introduction to Human Reliability Analysis

1.1 Overview

Human Reliability Analysis (HRA) is a tool used to evaluate human performance in the context of:

- human error as the initiating cause for hazard event scenarios
- human response to abnormal events as a safeguard or independent protection layer (IPL)
- failure to restore safety critical equipment following ITPM (inspection, test, and preventive maintenance) activities

HRA is a method under the umbrella of Quantitative Risk Analysis (QRA). This paper will use the term human reliability study interchangeably with HRA. A significant component of HRA is "task analysis" including human factors. HRA includes both qualitative and quantitative analysis.

The discipline associated with the study of human reliability is called "Human Factors Engineering" or equivalently the term "Ergonomics" has historically been used as well. Practitioners of this craft are called "human factors specialists" or "engineering psychologists" [1]. In modern parlance, the term human factors is used to mean anything that affects human performance of a task (e.g., labelling of field equipment). Ergonomics in contemporary usage means designing the environment to fit human capabilities (e.g., using a 3-D model review to locate manual valves so they are accessible).

HRA is especially useful when handling hazard scenarios that involve only human action, i.e., human error is the cause and human response is the solution. HRA may be used to "sharpen the pencil" in these cases to provide more realistic estimates of unrecovered error probability. The methodology can handle the same operator making and recovering from their own error, as well as handling human redundancy. The method is particularly well suited to account for dependencies within and between individuals.

The output of the HRA is a probability or frequency of human error, to use if needed as an IPL or initiating cause frequency, respectively.

To conduct a human reliability study one does not necessarily have to be a human factors specialist. A practitioner of traditional safety studies such as HAZOP, What-if, FMEDA, Fault Tree, or LOPA will have the most relevant competencies to perform HRA. The activity of task analysis which includes human factors is the most unfamiliar part of HRA. Gaining competence in Human Factors work will include formal education, training, and experience.

1.1. The essential steps to conducting a human reliability study are:

1. The PHA/ LOPA has identified a good candidate for HRA.
2. Determine if you will be solving for a probability (as an independent protection layer) or frequency (as an initiating cause). This will affect how you set up the human factors worksheets.
3. Gather process safety information (applicable procedure, P&IDs, etc.).
4. Perform the task analysis focusing on human factors

   — talk-through of task with Operations and Engineering

   — walk-through with Operations demonstrating task
5. Create task analysis Event Tree
6. Quantify Event Tree
7. Report results to Team

1.2. Benefits of HRA

1.2.1. Qualitative

A human reliability study provides many qualitative benefits apart from quantification. The task analysis provides the mechanism to obtain the qualitative insights. Every procedure can be improved. Procedures where errors of omission or commission could lead to catastrophic results need additional review. Fundamental improvements in procedure layout or content are likely to be recommended. Evaluating if procedures are used correctly (if not, why not?), if there is a better way to do the task, and human factors issues associated with the task (e.g., sending Operators up and down stairs to perform the task) may result in other qualitative improvements including design changes.

1.2.2. Quantitative

Quantitative HRA allows an estimation of human error probability (or frequency). Why is this important? Many tools in use today including fault tree, event tree, and Bow-tie provide quantitative estimation of hardware and process failure frequency or probability as an input to methods such as LOPA, SIL verification, or QRA. Ignoring the human component part of the equation is not a sustainable position any longer.

A benefit of quantitative HRA is that it provides a rigorous way to treat human error scenarios outside of LOPA. The results of which can then be used similarly to LOPA for making risk based decisions or fed into further QRA. It can also provide validation of the associated LOPA calculation for human error cause frequency or human IPL credit, and resolution of any gap closure recommendations.

1.3. Limitations of HRA

Limitation in this context primarily refers to aspects of the quantitative treatment part of HRA. The benefits gained from the qualitative application of HRA will outweigh any limitations found in the quantitative analysis.

"How can one quantify human error performance – humans are unpredictable." This is a common criticism. Humans are unpredictable, this is true, however the tasks that HRA investigates do not involve uncertain or capricious situations involving human behavior.

HRA task analysis is performed in a well-defined context of procedure, equipment, and plant surroundings. To be sure, there may be deficiencies in, for example, the procedures used (or not used) for a task, however the task itself (and the goal of the task) is defined. Any deficiencies in procedures, equipment, plant or people, are considered performance shaping factors that will impact the human error probabilities. These can be accounted for. The method is vetted even for the treatment of operator response to an abnormal event where significant cognition may be involved.

"Where does the human error data come from? It can't be reliable." This is another common criticism. The "problem" of human reliability data is not trivial. Primarily because there is not much of it. However, the data source referenced in this paper, has been rigorously researched and vetted [1]. In addition, follow-on work has validated the methodology and data set used for HRA [2].

Another limitation is the selection and more importantly the interpretation of human factors when used to modify basic human error probabilities. There may be subjective interpretations between analysts making results difficult to compare.

Kletz [3] suggests to use quantitative HRA for the purpose of risk based decisions. He acknowledges that because there is so much judgement involved with human data, there may be a temptation to "jiggle" the numbers to get the "right" answer, but also states that data on equipment failure is really data on people, so that judgment must also be exercised even when using equipment failure databases.

The limitations listed in this section, although valid and worthy of discussion, should not be used as a reason to exclude qualitative or quantitative HRA from use.

2. When to use HRA

2.1. What is a task?

HRA is a method to analyze and improve tasks. What is a task? Tasks will cover routine (production related) and non-routine (non-production related) activities. A task is performed to achieve a goal.

A task may involve,

2.1.1. Physical action

E.g., performing a critical operations procedure.
E.g., restoring safety critical equipment following ITPM.

2.1.2. Checking or inspection

E.g., Quarterly walk-around to verify chain locked valves are locked and in proper position.

2.1.3. Communications

E.g., shift hand-over in the middle of a critical procedure

2.1.4. Diagnosis or decision making (higher dependence on cognition)

E.g., response to abnormal situation (critical alarm)

Not all tasks are considered to be safety critical tasks. Critical in this context means an error made during the task may lead to or make more likely a major accident. Each organization will have to define what a critical task is.

2.2. *What is a "task analysis"?*

The task analysis is the most important activity of an HRA study. Task analysis involves dissecting the specific step(s) in the procedure that have been identified from the PHA as safety critical. Task analysis will look at all the ways the step could be performed incorrectly (errors of commission, including doing what is not in the procedure), or omitted entirely. Task analysis is performed in a team context. Most importantly, task analysis, will evaluate the relevant human factors (as performance shaping factors (PSF) or performance influencing factors) for that step. The relevant PSFs will then be used to modify the base human error probabilities for those errors. The pool of potential performance shaping factors to choose from is in the hundreds [8].

At a minimum, the task analysis should consider the following human factors:

- Training
- Experience
- Procedure
- HMI/ information
- Time

In addition, the following human factors are often accessible in the task analysis:

- Fatigue
- Workload
- Task complexity
- Intentional non-compliance
- Ergonomics
- Communications (e.g., across shift change)

The following human factors are less accessible to obtain (harder to estimate):

- Staffing needs
- Decision making criteria
- Organization factors
- Safety culture

There are many human factors related to human psychology:

- Intelligence
- Self-esteem
- Sociability
- Personality
- Norms and beliefs
- Risk taking
- Attitude

However, there is no need to investigate or quantify psychological factors. The goal of human factors is to change the environment, not the person (i.e., to match the environment to the abilities of the person).

In quantifying human error via task analysis, we assume that the pool of Operations and Maintenance personnel employed in the facility have the minimum intelligence required to do their assigned work (validated by documentation showing successful completion of relevant education and training). We also assume any error prone individuals will be re-assigned or replaced over time.

2.2.1. Using increased competency to improve human factors

"Competency" is a buzzword. An entire CCPS book is devoted to it [4]. What is competency? It's something that is hard to define, but you know it when you see it.

Competency is related to training and instruction, as well as KSA (knowledge, skills, and ability.)

The tools of competency produce skills and ability which leads to competency. The outcome of competency is "understanding" [5].

Tools:
Knowledge (formal education, book learning)..................→
Training (hands-on, vocational, workshops, class-room).......→...Skills/ ability →...Competence = Understanding
Experience (practice, simulation, OJT) ....→
Instructions (written and verbal)..................................→

What human factors does competency affect? Better competency will impact all human factors in a positive way. A few are listed here.

- Human: cognitive states (response to abnormal event), motivation (day-to-day tasks)

Motivation is an elusive human factor. What motivates us? What causes individuals to like their work, pay more attention, and improve task quality? Understanding via competency does.

- Task: better procedures
- System: better design

2.3. "European" approach to HRA

HRA appears to be more widely used in Europe as there is more awareness of this technique than in the United States. European regulators are expecting to see human factors task analysis included in safety cases and safety reports. HRA in Europe is primarily qualitative based, not quantitative. The Energy Institute (EI) has published guidance [6] which is necessary reading if planning to practice HRA, qualitative or quantitative.

2.4. Other HRA methods.

"There are so many human reliability methods to select from, which should I choose and why?" This is a common question. It is true there is an alphabet soup worth of acronyms related to human reliability methods. A few being: THERP, HEART, SLIM, CREAM, IDEAS, and SPAR-H. Dozens more methods could be referenced. THERP may be considered as the original method for manipulating PIF (performance influencing factors). The point to keep in mind is that PSF (performance shaping factor) or PIF (performance influencing factor) based methods such as those listed share common features, primarily task analysis and modifying base HEP (human error probability) by applying PSFs or PIFs. Most of the newer methods try to improve on THERP or other early techniques [7]. The THERP method [1] is appealing because it comes with its own data set.

There is so much low hanging fruit it is currently a moot point to argue methodology when any of the methods will yield benefits.

3. Types of human error accounted for in HRA

The HRA will quantify errors of omission and errors of commission, whether they occur due to slip or lapse (unavoidable), or mistake (improved through better training and instructions).

Omission is forgetting to do something (e.g., skip a step in a procedure). An error of commission involves performing the wrong action. Commission is more complex because there are usually hundreds of wrong actions a person could make while carrying out a task (e.g., instead of opening valve 'A', the operator closes valve 'A'. This is referred to as a reversal error). The end result of an error of commission is an error of omission – the desired action is skipped. The error of commission may produce other undesirable effects. Because the pool of potential errors of commission is so large, it is most practical to identify only the most credible errors of commission.

A third type of error is referred to as a "violation" (a.k.a., non-compliance). These are deliberate errors made with good intentions. The best that HRA can do is treat these in a qualitative way by discovery during task analysis.

The task analysis will uncover all of the credible ways human error can happen.

In all cases the Operator is assumed to be acting with good intentions. HRA does not model the effects of disgruntled employees.

4. How good is my data (human error probabilities)

A typical generic value for human error probability that originated from study of human interaction with weapons systems during WWII is 1/100. This value, considered conservative, was used for many years until better estimates were needed [1].

Many of us don't walk in the shoes of an Operator, so it is hard for us to gauge how our performance would degrade under certain conditions.

The following example may be more familiar to some of us. If we are in an elevator with someone rushing to catch it before the door closes, we may quickly diagnose the situation and respond by reaching for the "door open" button on the elevator panel. The "door open" push button and the "door close" push button are adjacent to each other. Further they look nearly alike and use cryptic symbols (<> or ><). There are also multiple other similar buttons that could be pressed. Add to that we are in a rush to press the button as the door is about to close or is closing. Under these conditions the likelihood of failure is high, approximately 0.5 probability. The value of 0.5 may be regarded as an upper bound for human error failure probability [3].

A lower bound for human error failure probability may be taken as 1/1,000 [3]. This is per procedural step. Plant tasks that may be performed weekly or daily— regeneration; catalyst handling; taking samples; starting rotating machinery; preparing equipment for maintenance, all performed very reliably. These tasks are performed very reliably because they involve "well-practiced associations". They become "second nature". However, an occasional slip or lapse is to be expected. This is referred to in HRA as "skill-of-the-craft." These tasks are typically proceduralized, and experienced operators typically keep the procedure in their back pocket during the task.

A task that is not "practiced" frequently may be reinforced with procedures so the performance becomes good enough to justify a 1/1,000 human error probability (e.g., omitting a step in a procedure) or even better, depending on other human factors. The caveat here is that the procedure must be used correctly (e.g., checklists should not be pencil-whipped, the operator must not rely on their memory, etc.).

The human error probabilities discussed above are considered "base" probabilities. In practice, we ask if an error has the potential to be recovered. Recovered errors will achieve lower probabilities than the base error rates. Recovery factors include awareness of one's own error (through feedback indication), or human redundancy (checking and inspection).

There is an optimum between how often a task is performed and resulting complacency. Motivation is an important (yet elusive) performance shaping factor that can help prevent complacency for frequently performed tasks.

5. What makes a good procedure?

Procedures should be written by Operations. They should be accessible. They should be periodically reviewed and updated. At a minimum, any critical procedure should include the following:

- One action verb per step
- Feedback that the desired action has taken place
- Relevant notes or safety issues, per step
- Not rely on memory (e.g., if the Operator is having to scribble notes on scrap paper to carry for which valves to turn, expect selection errors)

Procedures should be written in a column/ table format, not as a narrative.

Procedures need to state what to do (but not in a verbose manner). It would be impractical to list everything the Operator is *not* to do (there could be hundreds of possibilities). An Operator may think, "The procedure didn't say I *couldn't* do it." Training on the procedure should address the issue of using discretion.

If procedures are lacking the Operator will "fill in the gaps." Operators will perform a task the best way (usually the simplest way—which may be fine, except the simplest way doesn't match the procedure).

A task analysis that includes the critical procedure will discover the issues discussed above.

6. Achieving low human error probability

Human error may creep in many ways, and at any time during the life of a process facility. The operations, maintenance, and modifications phase of a plant represents the largest time fraction of its life. To be sure, design, fabrication, and installation provide opportunity for human error. Many of these errors will be discovered by testing, checking, inspection, or premature or sudden failure in the early life of the plant, hopefully without catastrophic consequences.

The concepts in this section applies to the operations, maintenance, and modifications phases.

6.1. Control human factors

The following human factors if controlled well will promote low human error probability:

- Procedures (quality)
- Safety communications (across shift change)
- Task Practice (minimize cognitive demand)
- Non-Reliance on memory (between procedural steps)
- Motivation (competency = understanding)
- Fatigue. 12 hr shifts (manage per API-755)
- Administrative control systems for the above

6.2. Provide task "recovery factors" for human error

- Managing "restoration" of safety equipment following ITPM
- Ensure proper restoration following ITPM
- Second check following ITPM
- Periodic check between ITPM
- Chain lock open root valves to SIS/ IPLs

6.3. Reduce task dependencies

1. Awareness of one's own errors
    - Provide feedback
        — Statuses, positions, indications
2. Checking someone else's work
    - Active inspections (with check-list)
    - Passive inspections (walk-around, scan the console, etc.)

- PSF (performance shaping factors)
  — It is better not to know whose work is being checked

7. Applying learning in HRA to PHA/ LOPA

7.1. Basic LOPA assumptions are conservative

In the real world, Operators may be able to recover from an error they make. In the LOPA world, they cannot. This sometimes leads a LOPA team to ask if a second operator is available to respond and cover for the first operator, for credit. If so, a different issue arises: are we sending the second operator into a potentially dangerous situation? There is also a human factors issue: how should the procedure be written that incorporates the response of the second operator? For example, say the console operator is monitoring the first outside operator perform a critical task. The practiced response of the console operator is to radio the first operator if he sees a problem. It would not be correct to train contrary to this well practiced association by writing the procedure that the console operator contact the second available operator. Why is this incorrect to do? Because the procedure will not be followed.

Procedures should be written, and work situations should be designed, such that an Operator has indication that his action was performed successfully (or unsuccessfully). This will make the recovery from one's own error more likely.

For example, say an operator locally starts a pump into a dead-headed line. This will be obvious to the operator and he will shut-off the pump immediately. Let's assume a base error rate of 1/100 per opportunity. The pump is started following maintenance approximately once per year, giving a 0.01 per year error rate. The recovery factor is awareness by the Operator that he has dead-headed the pump and will shut it down prior to pump damage, a credit of 0.15 assuming medium dependence with himself. The unrecovered error frequency is then 0.0015 per year, a more realistic value.

Drill-down, are procedures followed? At a minimum, procedure numbers should be referenced in the PHA study where relevant. The procedure should be reviewed by at least one team member during the PHA to verify the desired instruction is included. Recommendations should be made to improve a procedure if needed. Questions related to operator response to an abnormal condition (response time, etc.) need to be asked. This is the minimum a PHA team can do. Any further analysis would have to be performed outside the PHA.

7.2. Listen to the team

If Operations states that selection errors are common then there are likely human factors issues to solve. This could include labelling or physical equipment arrangement. If mistakes are made, then procedures are likely deficient (missing, not used, etc.) or training is lacking.

7.3. Using time-at-risk to reduce human error frequency

It is stated (e.g., CCPS) not to apply time-at-risk to human error initiated scenarios.

The explanation usually given is: Since the failure (human error) is not random in time, it is not appropriate to take credit for time-at-risk for human error. The failure (of interest) can only occur while the activity is being performed. The failure is not independent of the time-at-risk window (it can only occur during the time-at-risk). Therefore, it would not be proper to multiply the frequency of the error by the fraction of time-at-risk, i.e., P(time-at-risk | failure) = 1.

Looking at it another way: Over a 12-hour (or 8-hour) shift, the Operator is engaged in (nearly) continuous tasks. Of course, not every task an Operator performs represents the same level of risk (this is the concept of peak risk). However, all tasks are at some level of risk to human error. Therefore, applying time-at-risk to any one Operator action is misguided.

Compare this to a field device that must perform a safety critical control function once per month for 8 hours. The balance of the time the device is either dormant or performing low-level control. Time-at risk for this device would be appropriate to apply to the initiating cause frequency as long as certain conditions are met: As long as failures are random in time, and the probability of failure of the device is independent of its use during the time-at-risk, and any latent fault is detected prior to the time-at-risk use, a credit may be taken. The failure data selected for the device should also be based on continuous service time.

7.4. How do I better incorporate human factors issues into my PHA?

The answer does not involve a silver bullet node, guideword, or deviation.

The answer is surprisingly straightforward: Get facilitators trained in human factors and practicing in human reliability studies. A large facility will have several or more scenarios that are obvious and good candidates for a human reliability study.

For example, offloading or loading a highly hazardous chemical; manual dryer regeneration; manual catalyst handling. These are examples of good candidates for a human reliability study. There are many more. These are rich proving grounds for gaining experience in human factors.

7.4.1. Using human factors checklists during PHA

The generic human factors checklists that are used in a typical PHA are not the correct tool to uncover and analyze critical or safety related tasks. A PHA team can't successfully apply a checklist across all nodes and hope to capture anything meaningful.

The second issue associated with the human factors checklist is that in any well-designed and maintained facility, the big human factors issues have already been caught. The checklist can only scratch at the surface of these issues. A critical task analysis is needed to uncover obscure but potentially hazardous human factors issues associated with human initiating causes or human response to abnormal events.

7.4.2. Performing a procedural PHA

A procedural PHA will closely parallel a task analysis with two big exceptions. A procedural PHA will not drill down into human factors, and will not quantify results.

7.5. Performing HRA for abnormal events (alarm response IPLs)

This is the most complex situation to analyze because typically responding to an abnormal situation will require cognition (higher order thinking). Even if a critical alarm is programmed to generate the step-by-step response on the alarm display or console, for low-probability events (such as those catastrophic events identified in PHA/ LOPA) the initial response will most likely be influenced by skepticism and disbelief. This is called the "incredulity response" in the literature [1]. Diagnosis will be involved to figure out why the critical alarm sounded (is the alarm "real"?). The only way to reduce the "incredulity response" is to practice (talk-throughs, table-top exercises, what-ifs?, drills, etc).

Fortunately HRA can model this cognitive response primarily as a function of allowable response time in addition to alarm flood. Even in an era of alarm rationalization and management, it is wise to model some degree of alarm flood for these types of rare catastrophic events.

It is very important that the PHA/ LOPA identify at a minimum the procedure associated with the operator response to critical alarm. Ideally the team would open up the procedure to identify the specific steps that can be credited as operator response to an abnormal event.

7.6. Performing HRA for human error that results in catastrophic events (initiating events)

This type of analysis is typically less complex than analyzing abnormal response. The reason is the level of cognition is significantly less. Initiating events caused by lapses or mistakes are primarily procedural or skill-of-the-craft (i.e., second nature) based.

It is very important the PHA/ LOPA identify at a minimum the procedure associated with the operator error, and ideally the team would open up the procedure to identify the specific steps that would lead to the catastrophic event if an error were made.

If a procedure is not used, the PHA/ LOPA should note this (call it a "practice" or "skill-based" activity) and then recommend to investigate if the activity should be proceduralized.

7.7. Is a procedure needed to get credit in LOPA?

The more a procedure is used (practiced) the more it becomes a skill (second-nature). At that point the operator may only print and review the procedure prior to the task, and carry it in his back pocket, versus continually referring to the procedure during the task. This is acceptable for an experienced operator. A new operator would have the procedure to follow as the task is performed.

Checklists may suffer the same fate. To be sure, a properly used checklist will yield the lowest error probabilities. An improperly used checklist (not used or pencil-whipped) offers no additional benefit and is a negative contributor to safety culture.

It is therefore important to identify only those tasks that will benefit from procedures, or procedures with checklists.

It is true that as a procedure is used more often, an operator increases his skill in using it (has lower error probability), but it is equally true that a task that is performed more often will have a higher error rate (more opportunities). The increase in operator skill does not cancel out the increase in opportunities for error.

7.8. Performing a dedicated HFTA (Human Factors Task Analysis) in addition to PHA/ LOPA

All human error initiating events potentially leading to catastrophic events, and all operator response to an abnormal situation with potential for catastrophic harm should undergo a human factors task analysis with human error quantification, even if generic credits were assigned in LOPA.

The opportunity for improvement is large compared to the cost of the study.

8. Examples of documenting an HRA with event trees

8.1. Finding initiating cause frequency (ICF)

A human error (e.g., caused by an operator) may be the initiating cause for a hazard scenario. The following are examples of real case studies of human reliability event trees depicting operator error as an initiating event (and the same or different operator as the safeguard). An equivalent fault tree may be constructed. However, for HRA, the presentation below is more efficient and easier to read than fault tree.

"S" represents a success path. "F" represents a failure path (this is the path of interest). Dashed lines "------------" represent recovery paths back to a success path. The following event trees may be quantified by applying human error probability at each discrete sub-task (A,B,C,…) and recovery factor, and then using probability math to solve for 'F'. Human error data is found in reference [1].

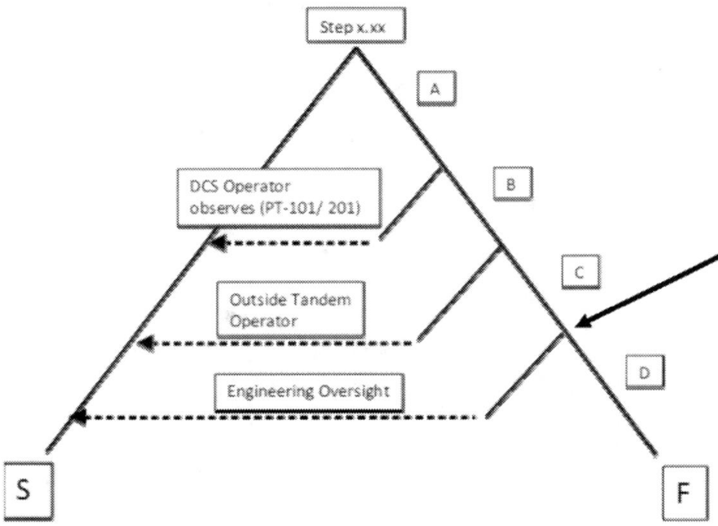

Credit for active walk-around inspection with low dependence between Operator and engineer. (Active inspections are those in which a person is directed to inspect specific items of equipment.) Low dependence assumes status, training, responsibility, and other social and psychological factors are different between the two. Per talk-through with the PHA team, the engineer is in the field with Operations as this procedure is executed. This procedure is owned by Engineering.

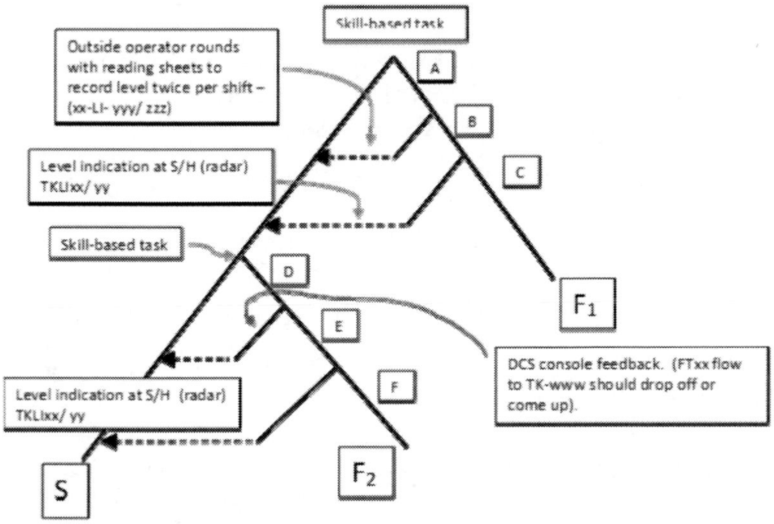

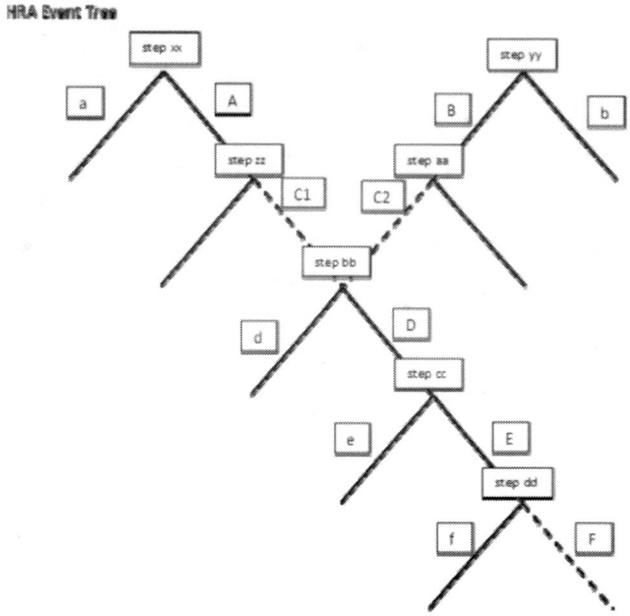

8.2. Finding IPL probability of failure to mitigate Abnormal Event

A human response (e.g., from an operator) may prevent a hazard scenario from propagating. The following example is a real case study of human reliability event trees depicting operator response to abnormal situation. An equivalent fault tree may be constructed. However, for HRA, the presentation below is more efficient and easier to read than fault tree.

"S" represents a success path. "F" represents a failure path (this is the path of interest). Dashed lines "-------------" represent recovery paths back to a success path (not applicable in this example). The following event trees may be quantified by applying human error probability at each discrete sub-task (A,B,C,...) and recovery factor, and then using probability math to solve for 'F'. Human error data is found in reference [1].

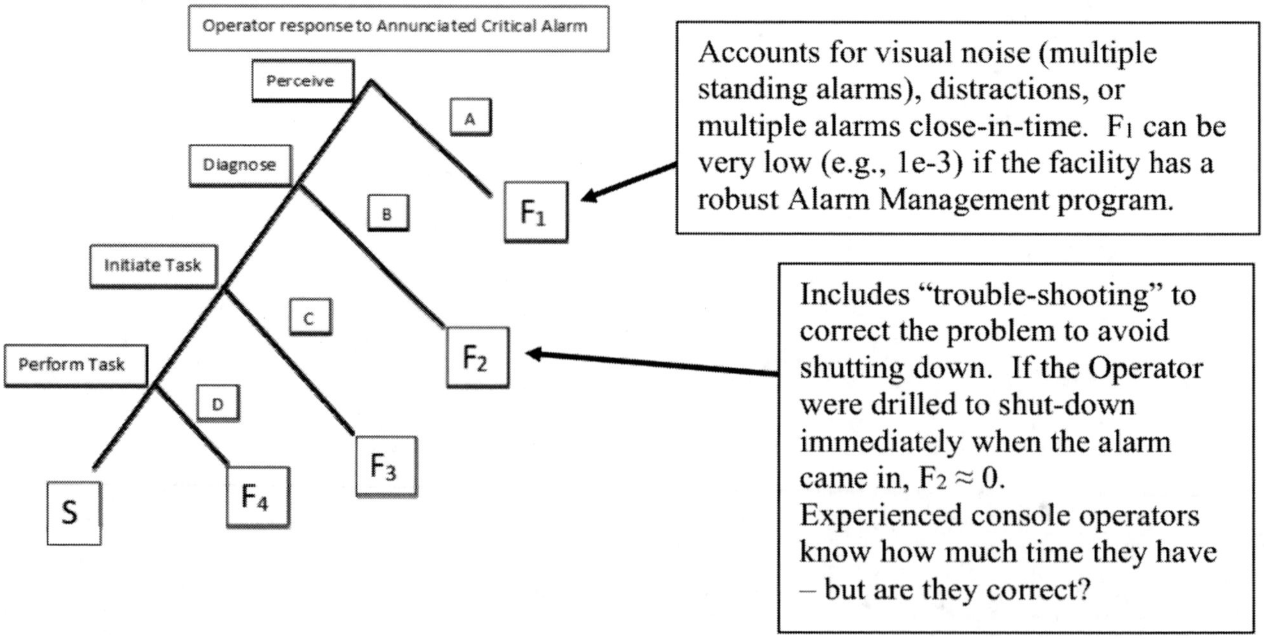

9. Summary

Human reliability analysis is a qualitative and quantitative tool the benefit of which is unrealized yet in the process industries. There is much low hanging fruit to be grabbed, the benefits to process safety are real, and the cost to conduct an HRA for potential significant human error related scenarios is small compared to the improvements gained.

Process safety personnel that facilitate PHA and LOPA studies, and also have used quantitative tools such as fault tree and event tree, possess the minimum qualifications to perform HRA. Human factors skills will be acquired through practice and training. Begin with an obvious scenario to start HRA and build from there.

10. References

[1] Handbook of Human Reliability Analysis with Emphasis on Nuclear Power Plant Applications, Final Report, A. D. Swain, H. E. Guttmann, NUREG/CR- 1278. August 1983.

[2] The validation of three Human Reliability Quantification techniques — THERP, HEART and JHEDI: Part II — Results of validation exercise. Barry Kirwan, Richard Kennedy, Sally Taylor-Adams, Barry Lambert. Applied Ergonomics, Elsevier, Volume 28, Issue 1, February 1997, Pages 17–25

[3] An Engineer's View of Human Error, Trevor Kletz, Taylor & Francis, 3rd Edition, 2001.

[4] Guidelines for Defining Process Safety Competency Requirements (CCPS) 2015.

[5] Guidelines for Risk Based Process Safety (CCPS) 2007.

[6] Guidance on human factors safety critical task analysis. Energy Institute, London. 1st edition March 2011.

[7] Human Factors Methods for Improving Performance in the Process Industries (CCPS) 2006.

[8] A Taxonomy of Performance Influencing Factors for Human Reliability Analysis of Emergency Tasks. Jae W. Kim and Wondea Jung. Journal of Loss Prevention in the Process Industries. 2003.

12TH GLOBAL CONGRESS ON PROCESS SAFETY

Quantitative risk analysis for decision support in process plant operations

Professor Stein Haugen
Norwegian University of Science and Technology
N-7491 Trondheim, Norway
stein.haugen@ntnu.no

Nathaniel John Edwin
Safetec – an ABS group company
N-7037 Trondheim, Norway
nathaniel.john.edwin@safetec.no

Olav Brautaset
Safetec – an ABS group company
N-7037 Trondheim, Norway
olav.brautaset@safetec.no

Jan Erik Vinnem
Norwegian University of Science and Technology
N-7491 Trondheim, Norway
Jan.erik.vinnem@ntnu.no

Prepared for Presentation at
American Institute of Chemical Engineers
2016 Spring Meeting
12th Global Congress on Process Safety
Houston, Texas
April 11-23, 2016

Quantitative risk analysis for decision support in process plant operations

Professor Stein Haugen
Norwegian University of Science and Technology
N-7491 Trondheim, Norway
stein.haugen@ntnu.no

Nathaniel John Edwin[2], Olav Brautaset[2], Jan Erik Vinnem[1]

[1]Norwegian University of Science and Technology
[2]Safetec – an ABS group company

Keywords: activity-based risk, oil and gas, QRA

Abstract

Design of complex technological systems with a potential for major accidents such as nuclear power plants, offshore installations and high hazard process facilities is often supported by quantitative risk analysis (QRA). The QRA has been developed and improved over several decades for this purpose. In recent years, the offshore oil and gas industry in Norway has put significant effort into application of QRA also for supporting decisions in day-to-day operations. An important objective is to ensure that major accident risk is taken into account in an adequate manner. The MIRMAP project (Modelling Instantaneous Risk for Major Accident Prevention) attempts to address this issue in a systematic manner. The starting point has been the types of decisions made in daily operation, compared to design decisions. Based on this, risk models have been developed which are intended to give up-to-date risk information with limited effort and sufficiently quickly to be available when the decisions are being made. In traditional QRA, focus in the modelling is on technical systems and layout, and how the systems may fail; i.e. aspects that are relevant in design. In operation, however, the design is fixed and changes in the risk level are driven primarily by the activities taking place and how they may interact or even interfere with the technical systems. The risk analysis should therefore also focus on activities. This is a different approach compared to traditional QRAs. The paper describes the background to the project and the principles of the modelling of risk. Practical implications and feasibility of the approach is also discussed.

1 Introduction

Design of complex technological systems with potential for major accidents, such as nuclear power plants, offshore installations and high hazard process facilities, is often supported by quantitative risk analysis (QRA). This contributes to a safe design and more efficient use of resources for controlling risk. Within the nuclear industry, this started in the 1970s, with the publication of the WASH-1400 report [1]. This formed the basis for later Probabilistic Risk Assessments (PRA) for nuclear power plants. In fact, in the years that followed, the use of risk information for decision making was encouraged by the United States Nuclear Regulatory Commission (USNRC) resulting in effective risk management and much improved plant performance.

The oil and gas industry in Norway was also inspired by the nuclear industry when the first quantitative risk analyses were performed around 1980. The first studies and requirements were firmly aimed at supporting the design process. The first guideline from the Norwegian Petroleum Directorate (NPD), now the Petroleum Safety Authority, Norway (PSAN), were called "Guidelines for safety evaluation of platform conceptual design" [2]. This specifically pointed to the early design phases of new installations. Over the 35 years since then, the Quantitative Risk Assessments (QRA) have been extensively developed as a tool for design, with detailed requirements being laid down in the Norsok Z-013 standard for Risk and Emergency Preparedness Assessment [3].

In recent years, the offshore oil and gas industry in Norway has put significant effort into the use of QRA also for supporting decisions in day-to-day operations of the oil and gas installations. An important objective is to ensure that major accident risk is accounted for in an adequate manner. Success in meeting this has however been somewhat limited so far.

A key reason for this is the nature of the risk models used in QRAs. The QRA models the relevant identified hazardous events and accident scenarios. Any risk model will however be a simplification of the complete set of accident scenarios that can occur. Only a representative set of scenarios are modelled and normally only factors that are considered important for the result are modelled explicitly. This is where the basis of the QRAs as a tool for supporting design, becomes an important limitation. In a QRA, the layout of the installation, the equipment and all technical systems that influence risk significantly are usually modelled explicitly. Thus, the QRA can guide the design process by pointing out what contributes most to risk, how changes in systems can reduce risk and how we can introduce improvements. However, in the design phase, operational details are not yet fixed and it is typically assumed that they will be some sort of "industry average" and operations are therefore modelled explicitly only to a limited degree.

An example to illustrate this can be gas leaks, where experience has shown that on the Norwegian Continental Shelf, more than 50% of the leaks are related to intervention in the process plant by operations or maintenance personnel [4], [5]. However, in the QRA, the frequency of gas leaks is dependent only on the quantity of equipment, not on the number of operations [3]. The most important factor influencing the frequency of leaks is thus not modelled explicitly, but only taken into account implicitly through the use of historical leak frequencies. We can thus not see the effect on the risk level of making changes in the number of operations.

In the design phase, we can argue that the purpose of the QRA is to enable improvements in the design and technical systems, and that the operations is a more or less fixed "framework" for the risk calculation. However, in the operational phase, the situation is the other way round; the operations vary from day to day, while the technical systems and design is a more or less fixed context for the operations. This will also require a different approach to QRA, where the activities taking place in the plant are at the center of the risk models.

The MIRMAP (Modelling Instantaneous Risk for Major Accident Prevention) project attempts to address this issue in a systematic manner. As indicated in the title of the project, the focus is on major accidents. The starting point has been the types of decisions taken in daily operation, compared to the types of decisions in a design project. Based on this, risk models have been developed which are intended to give up-to-date risk information with limited effort and sufficiently quickly to be available when the decisions are being made.

Another aspect that also is relevant for the applicability of the QRA is that risk in operation will vary from day to day and even from hour to hour. However, the QRA models average risk over a long period of time, usually for a "typical" year in the operation of the plant. As a basis for developing a design that can be safe "on average" over the lifetime of the installation, this is adequate. However, it will not support operational decision-making. In the nuclear industry, the concept of living risk analysis has been applied for some time, denoting an analysis that is updated regularly/daily, taking into account the status of the plant at a specific point in time.

The main objective of this paper is to describe the method. The rest of the paper is structured as follows. In the next section, a brief overview over relevant literature is provided. This is followed by a description of the method in Section 3, and a discussion of some key aspects in Section 4. In Section 5, conclusions are presented.

2 Background

A structured literature review has been performed of similar methods and applications of real-time risk analysis for decision support in different industries and settings. The concept of a real-time or living risk analysis has its beginnings in the nuclear sector. One such example is Risk Spectrum RiskWatcher [6] that is a software which uses event and fault trees to model systems and reflect the point-in-time risk based on the current system configuration, ongoing activities and other relevant environmental factors. The availability of detailed event and fault tree models from the nuclear Probabilistic Safety Assessments (PSA) makes these models realistic to implement in practice. Another example is RiskVu from Isograph that falls into this class of risk monitors. The main focus of these tools are plant and barrier conditions and not the effect of activities and combinations of them [7].

Several efforts have been made in the oil and gas industry to extend existing QRAs in an attempt to operationalize them. A few of these efforts include the Barrier and Operational Risk Analysis (BORA) project [8] that focused on barrier and operative issues to setup a basic risk influence modelling framework and the Operational Condition Safety (OCS) method [9] that looked at human and organizational factors and their influence on barrier performance. The RiskOMT method [10] suggests an approach to risk influence modelling for updating the hydrocarbon leak frequency based on technical, operational and organizational factors. A significant limitation of all of these methods was the complexity and associated work to extend existing QRAs to fit with these approaches and make them relevant for real-time decision support.

Parallel to the BORA, OCS and RiskOMT work, a lot of research effort has also gone into developing indicator sets for major accidents to aid regular monitoring of trends within a system with major accident potential. However, a review of literature on indicators [9] shows that there is very limited work on defined approaches to establish and validate these large indicator sets. Research presented [11] builds on this to provide a generic and systematic method to develop these indicator sets wherein a risk model links technical, operational and organizational factors and their underlying aspects (such as root causes, background factors etc.). These factors are measured through indicators. Although not quantitative, this method provides a valuable foundation for how to systematically approach identification of relevant influencing factors and indicators.

In other related developments, the MARI method [13], also based on traditional risk models, uses causal chains to illustrate the influence relations between factors and a major accident event. The Hybrid Causal Logic (HCL) [14] combines traditional risk analysis tools with Bayesian belief networks. This quantitative method is tested on the aviation industry as a case and the relevance of the method for offshore risk analysis is also discussed [15]. The HCL approach is a very flexible risk-modelling framework that in fact forms the basis for the activity-based risk modelling approach discussed in this paper.

In the aviation sector, the Aviation System Risk Model (ASRM) [16] combines the use of a human error taxonomy, probabilistic Bayesian Belief Networks and case-based scenarios to assess a relative risk intensity and develop a graphical profile of the updated risk measure in a risk matrix. Scenarios that provide the best risk reduction are selected and subjected to further drill down diagnostics to identify the critical causal factors that define the model performance. In the process industry, Storybuilder [17] is a tool that uses bow tie structures to graphically depict failures in barrier elements from prior occupational incidents and accidents. This is primarily intended as a systematic instrument to record historical barrier performance and not specifically for proactive use.

Shifting focus back to the oil and gas industry, a number of offshore operator companies have recently developed their own proprietary software-based tools for better status and monitoring across their facilities. These tools promote integrated management of safety critical information (status of barriers, deviations, ongoing activities etc.) for better visualization, data management and reporting. Two examples are iSee from ConocoPhilips [18] and Total Risk from Shell UK [19]. These tools provide an overview of the safety critical activities and deviations at a facility and thereby provide a better decision basis, improved data accessibility and a better learning from experience.

3 Method Description

The development of a practical model to measure changing risk levels in operations comprises of three main phases and is illustrated in Figure 1. Phase 1 involves defining the scope of the risk analysis and therein developing a generic risk model for the process plant or facility concerned. Phase 2 covers calibrating the generic model from Phase 1 and scaling the model for actual use. Important aspects in this phase include identification of data sources and model quantification. Finally, Phase 3 looks at providing suitable risk results from the risk model that are relevant for decision support. This paper mainly covers Phase 1 and provides only a brief reflection on some aspects from Phase 2. Phase 3 – risk reporting and decision support – is not covered in this paper but is part of ongoing work. The examples in this section relate to a typical oil and gas process facility as this has been the focal point of the project and overall method development.

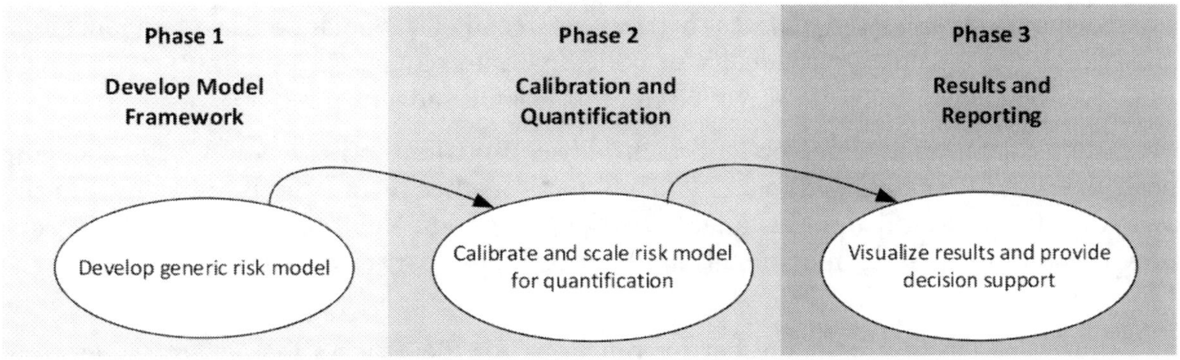

Figure 1 Outline of the three main phases of the method

3.1 Phase 1 - Develop model framework

This phase involves a systematic seven-step process that first defines the scope of the activity-based risk analysis and lays the foundation for the development of a generic risk model for a process plant. Figure 2 illustrates these seven steps. The final result from Phase one is a generic activity-based risk model that provides a detailed description of the risk by modelling the most significant process hazards, the means of protection and the various relevant factors that can influence or affect these means of protection.

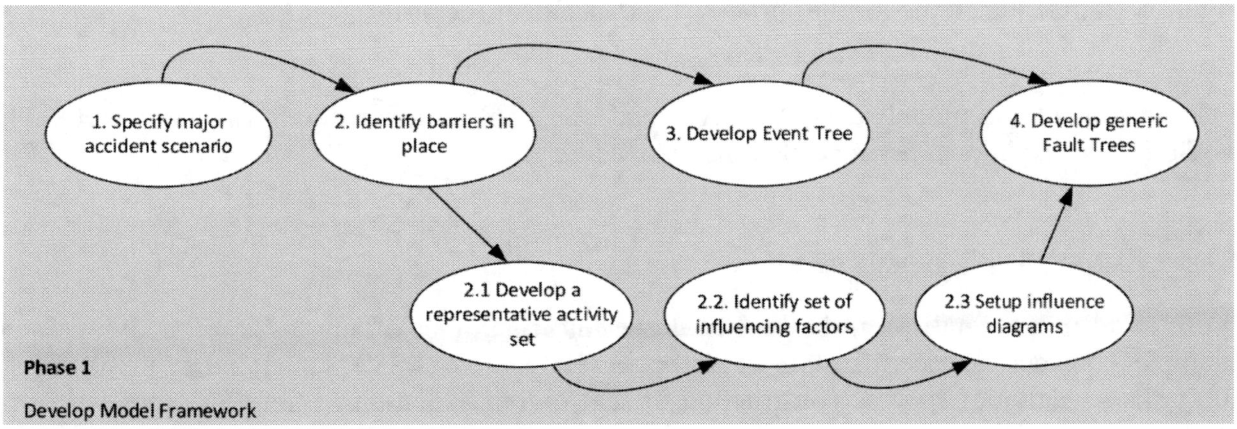

Figure 2 Illustration of Phase 1 - develop model framework

3.1.1 Major accident scenarios and barriers

Steps 1 and 2 define the context and scope for the risk model. The existing design QRA is a valuable input to identify the main risk contributors for a facility. For instance, for offshore oil and gas facilities, process leaks, blowouts and ship collisions are the highest contributors to risk [20]. These scenarios are further analysed to study how daily operational activity affects this risk level. Only risks that vary dynamically with changing operational conditions will require real-time monitoring. Here, process leaks followed by ignition and escalation are the major accident scenarios.

Several pre-designed layers of protection or barriers prevent the uncontrolled progression of an accident scenario. PSAN [21] defines barriers as technical, operational and organizational elements which are intended individually or collectively to reduce the possibility for a specific effect, hazard or accident to occur, or which limit its harm/disadvantages.

Barrier systems form an integral part of each barrier function and are comprised of concrete technical, operational or organizational elements that together realize the barrier function in its entirety. All relevant installed and available barriers functions, their constituent barrier systems and corresponding technical, operational and organizational barrier elements are identified and listed.

Figure 3 illustrates the four essential barrier functions for the process leak event sequence and the associated accident scenario. For example, BF3 "Prevent Ignition" includes barrier systems such as ignition source isolation, control of hot work and ventilation systems. Similarly, all the other barrier functions and their related barrier systems are identified for each major accident scenario shortlisted from Step 1.

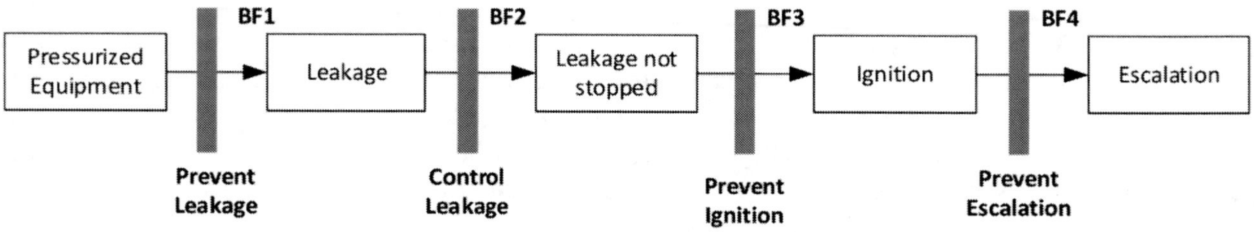

Figure 3 Barrier Functions for the "process leak" accident scenario

This version of the model does not cover end consequences, i.e. fatalities, completely. This would require additional barrier functions to be added.

3.1.2 Representative activity sets

Work at a process facility can be broken down into a number of sub-tasks or "activities". For example, consider the recertification of a process safety valve (PSV). This job may involve many sub-tasks or activities such as construction of scaffolding, isolation of the PSV, removal of the PSV, removal of isolation after completion etc. The purpose of Step 2.1 (see Figure 2) is to identify a representative set of sub-tasks or activities that are typical and relevant for the process plant at hand.

Only activities influencing major accident risk are relevant to include, and we categorize activities into two groups – "Activity A1" that introduces a hazard that may affect the integrity of a barrier, or "Activity A2" which may directly impair/weaken a barrier system/element. For instance, "disconnection of a gas detector" is an example of a category A2 activity that directly weakens/impairs gas detection. "Critical lifting" is an example of a category A1 activity that introduces a hazard that can lead to a process leak due to dropped or swinging loads, an impairment of BF1.

In a similar fashion, through a thorough review of each barrier function and its associated systems and elements, all possible activities or conditions that may directly or indirectly represent or lead to an impairment or deviation in the barrier is identified. This exhaustive activity list may also correspond to different types of operational deviations for a process facility, for instance functional tests, ongoing maintenance work, alarms from condition monitoring services and possible notifications from the control system amongst many others.

3.1.3 Influence factors and influence diagrams

Steps 2.2 and 2.3 involve identifying influence factors and understanding the influence relation between the factors and the activity concerned. The term Risk Influencing Factor (RIF) is commonly used in literature related to risk influence modelling. In the current context, a RIF is defined as an aspect of an activity that influences the changing risk level [22]. These factors describe the nature and degree to which the RIFs affect the safe execution of the task with respect to major accident risk and thereby the risk level. Examples of RIFs for a type A1 activity, "Work on hydrocarbon systems – isolation", may be Competence (Mechanical crew group), Competence (Process crew group), Time pressure, Design/HMI, Documentation and Isolation plan.

3.1.4 Event Trees

Step 3 builds on the accident scenarios specified earlier in Step 2 to develop event trees for each scenario. The main objective of the event trees is to model the barriers that were identified in Step 1. The pivotal events are ordered based on where they influence the accident sequence.

3.1.5 Fault Trees

Step 4 covers development of fault trees for each pivotal event from the event trees. The basic events applied in the fault trees can be divided into four categories:

A1. Activities that introduce a hazard

A2. Activities that represent a direct barrier deviation

B. Technical integrity issues, e.g. equipment degradation due to age, environmental conditions and other factors

C. Inherent design characteristics or deficiencies, e.g. limitations in firewater capacity, limitations in coverage of gas/fire detection etc.

Inclusion of all relevant A1, A2, B and C factors as a part of the barrier fault tree models result in detailed and comprehensive activity-based fault tree structures that also cover relevant design and technical integrity issues.

3.1.6 Generic risk model

A run-through of all steps from Phase 1 produces a generic risk model as seen in Figure 4. In this model, event trees describe the selected major accident scenarios and activity-based fault trees model every pivotal event from the event trees. The lowest failure level in the fault trees are the individual activities and these link to influence factors via influence diagrams.

The various end consequences or sequences marked ES in Figure 4 are weighted together to define the risk level. The weights of the ES are determined based on the potential for fatalities due to heat exposure, toxic gas, inability to escape etc. Therefore, the number of personnel or manning levels is an important parameter for the risk measure. This is a rather simple interpretation of the consequence aspect of the risk measure. Further discussion on this is provided in Section 4.

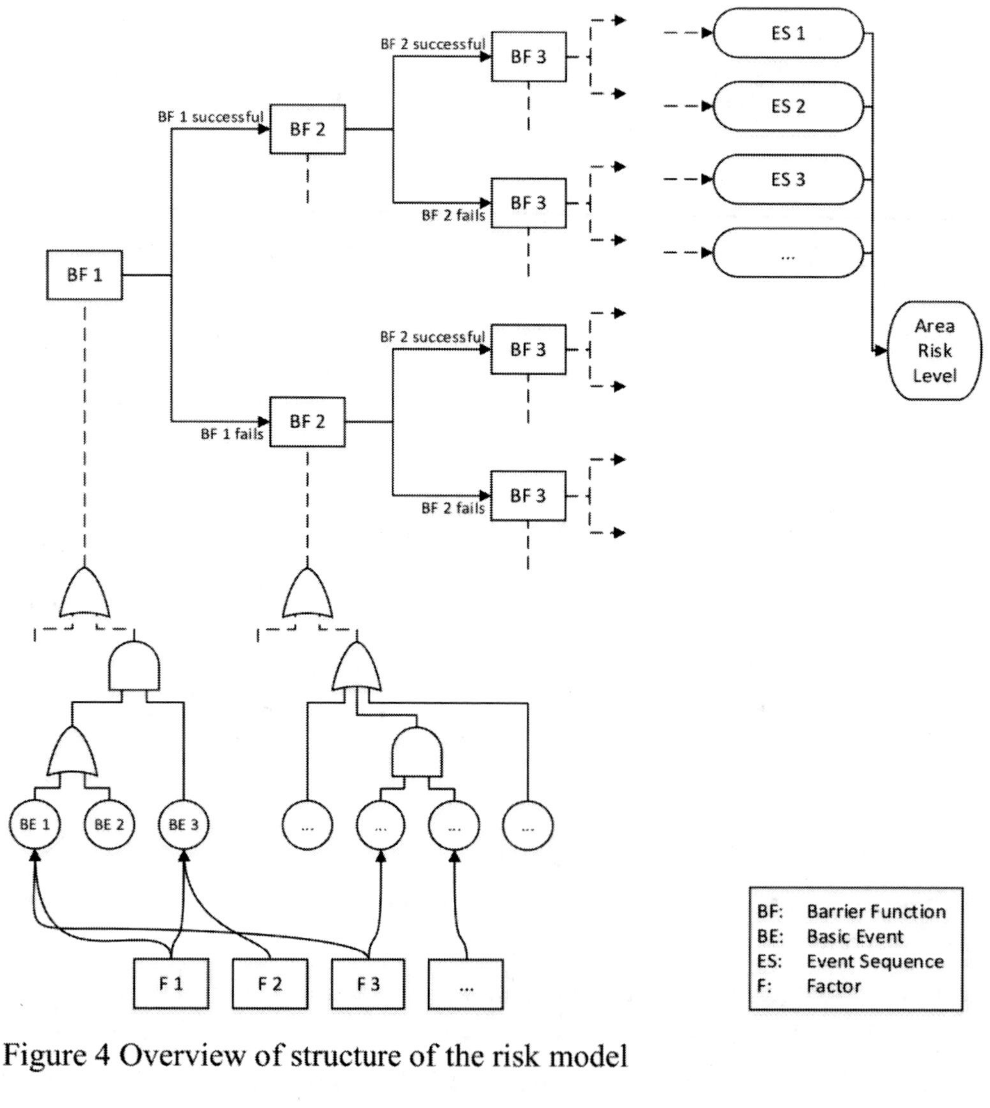

Figure 4 Overview of structure of the risk model

3.2 Phase 2 – Model calibration and quantification

This phase requires necessary modifications to the model – structural and numerical, to prepare the model for quantification of risk at the process facility.

It is common practice for hazardous process plants to be divided into different areas – for example, analysis areas, main areas, fire areas, construction areas etc. The need for these divisions arise due to various reasons – for instance, the analysis areas are created to support the risk analysis, the construction areas are used for physical reference to locations across a facility etc. This step selects one such area classification for the risk model. Model configuration, further calibration and final quantification is performed based on the selected area classification.

Scaling the risk model involves selecting an area subdivision and making the risk model specific per area for the process facility. For example, while one area might have automatic firefighting, another area might have only manual firefighting. In such a manner, calibration involves removal of redundant or irrelevant lower level events or basic events from the fault trees based on the actual conditions in the given area.

Calibrating the risk model per area requires identifying relevant data sources and setting the probabilities for the basic events for the fault trees. The widespread use of computerized solutions for monitoring and reporting in the process industries means that the amount of available data often is huge and available in a variety of formats. This is an advantage as well as a challenge at the same time. This is discussed more in Section 4.

Model quantification is inspired from the Hybrid Causal Logic [15] and RiskOMT [10] methods wherein Bayesian belief networks integrate with the traditional event and fault tree quantification. The use of Bayesian belief networks provide an elegant way to model uncertainty in information by specifying probability distributions at the influencing factor level. These aspects are not detailed further in this paper.

The resulting risk model produces risk results per area of the installation, which together provides an overall risk picture at the entire facility. This may in principle be updated on a daily basis.

4 Discussion

4.1 More focus on probability than risk

In risk analysis, risk is usually calculated as a combination of probability and consequence, giving a statistically expected consequence, normally over one year (see further discussion in Section 4.6 in relation to the PSAN revised definition of risk). Common risk metrics, such as PLL (Probable Loss of Life), is an expression of an average or statistically expected consequence. This is a useful way of comparing alternatives when we want to minimize future, average losses.

In an operational setting, our primary concern is to avoid accidents. The goal each day is to complete operations without anyone being injured or killed. Minimizing statistically expected losses will contribute to this, but is not necessarily the only or even the best measure we need. This means that in addition to risk, the probability of accidents and the performance of individual barriers to help avoid these accidents are important parameters.

This is of relevance in the approach to the modelling of risk as this implies that most of the effort can be placed on modelling the probability. The probability will also tend to vary more from day to day than the consequence. It is not always like this, but in many cases this is likely to hold true. The advantage of more detailed consequence modelling also needs to be seen in relation to the potential effect on computation time. A significant increase of computational time would most likely affect the usability of the analysis drastically, because the effect of changes to input parameters should be seen almost instantly.

4.2 Work required to do risk analysis

The method description indicates that the effort required to develop risk models like this will be quite extensive. Three different modelling techniques are used and each method has to be applied to all types of activities in a plant where there is a major accident risk. This requires a considerable amount of work and it may be questioned whether the effort is cost-effective. We will not explicitly attempt to discuss the question about cost-effectiveness, but may point out a couple of issues that are relevant.

Practice in the Norwegian offshore industry has been to update the installation QRAs regularly, and the risk assessment standard NORSOK Z-013 describes criteria for updating [3]. This implies extensive work, in particular detailed consequence modelling using CFD (Computational Fluid Dynamics) tools has been resource demanding. The purpose is to ensure that up-to-date risk models are available for decision making in operations. In our opinion, the need for this is significantly reduced if risk models like described in this paper are developed. Considerable savings in cost can thus be achieved.

Another aspect is that it may be possible to develop risk model "templates" for a variety of activities. Many activities will be similar, if not identical, from one plant to another, and many influencing factors and other related aspects will also be similar. Thus, with a library of models covering the most common activities as a starting point, developing a risk model for a specific plant will require picking relevant models, modifying these, and putting this together. Clearly, there will be aspects that cannot be handled in this way, but we do not foresee that all details of all models will require development from scratch each time.

4.3 Availability of data

Availability of data is always an issue in risk analysis. Within the offshore industry, data collection has been ongoing for many years. A well-known example is the OREDA database [23]. We are therefore in a relatively fortunate situation with access to recognized databases to support risk quantification.

Some input from the QRA will be applied. As long as a QRA is available, extracting this data should be relatively easy. This information is also more or less static, implying that it can be gathered from the QRA once (or at least with long time intervals) and applied without changes for a long period. This may be seen as part of an initial QRA adaptation phase, to prepare the QRA for use in operational risk models.

However, information about types of activities, number of activities, where they are taking place, how many people are involved, who is involved etc. must be updated daily. In addition, status of safety critical systems is required. This includes information about systems/components that have failed, maintenance status, systems out of operation etc. This will not necessarily change every day, but sufficiently often to require frequent updating. The implication of this is that the data collection must be automatic to make this feasible and cost-effective in practice.

A lot of the information that is required is typically available in the maintenance management/-planning system and the work permit system (in many cases these may also be integrated). Provided automatic transfer of information from these systems to a computerized risk tool is possible, it is realistic to get updated information on a frequent basis.

The availability of data is a key issue for successful implementation, but as pointed out above; our experience is that a lot of the information is available and can be used. It is more a matter of adapting to this way of thinking and making an effort to make the data available in a suitable manner. This is a one-time effort provided automatic data transfer could be arranged and in our view this is possible to achieve.

4.4 Practical application

We have not gone deep into practical use of the risk model, but our conclusion thus far is that this is a method that can be used to support primarily the planning process. We can rank individual activities, identify high risk activities that need additional attention and suggest possible risk reducing measures. This is useful throughout the planning process until a final decision is made to approve execution of an activity. This could thus contribute to reduce the need to postpone or even stop activities late in the planning process.

The tool can also evaluate the effect of combining activities, and highlight possible interaction effects that contribute to increased risk (e.g. performing work that can lead to leaks and hot work in the same area and at the same time). These can also be identified early in the planning process and potential problems can be addressed and fixed early.

Both of these uses will contribute not just to reduce risk but potentially also to improve the planning process, reducing the need for late rescheduling and thus potential inefficiencies in the use of resources.

We have not so far considered potential uses during work execution. During the execution phase itself, it is hard to foresee that formal risk assessment will have a significant role in decision-making. Here, it is more a question of the workers experience, their knowledge of the systems and activities and of course, their knowledge of what the risks are. This then forms their basis for decision-making.

A more general aspect of practical application of this model is the balance between complexity (and completeness) of the model versus the ability to produce results in time and the trust that the personnel involved in the planning process will have in the model. It is likely that there will be a lot of compromises necessary to develop a model that works in practice. This needs to be carefully weighed against what we lose through simplifications and whether we reach a point where the model is no longer useful.

4.5 *Presentation of results*

Part of the work has also been to look into how the resulting presentation can be improved, to support better decisions. This is still ongoing, but some important principles have been laid down as a basis for the model development.

We have used inspiration from the description of fast and slow thinking by Kahneman [24] as a "model" for how we want the risk model to work. Fast thinking relies very much on simple decision rules and heuristics, enabling us to make quick decisions. On the other hand, slow thinking implies a more comprehensive deliberation of key factors that influence our decision. For low risk activities, fast thinking can be applied, in the sense that standardized risk reducing measures are applied and simple decision-criteria are applied. There may even be a possibility that these decisions can be more or less automated. However, for high-risk activities, the aim is to "push" the decision-maker into slow-thinking mode by presenting more information about why the risk is high (the "risk driving factors"), not just the risk level as such. Of course, there is still a possibility that the decision-makers will focus only on a few of these factors and over time develop heuristics that help them stay in fast thinking mode. The idea is however to try to support the process of staying in slow thinking mode.

How the risk should be presented is also an issue that requires careful consideration. The traditional risk metrics that we use, such as PLL and FAR are all representations of statistically expected consequences. As pointed out earlier, this is not necessarily the most useful information for operational decision-making.

4.6 *Uncertainty*

PSAN defines risk as "the consequences of the activities, with associated uncertainty" [25] which emphasizes the uncertainty as an important issue in risk assessment. It is therefore interesting to consider how the proposed approach considers uncertainty.

There is limited uncertainty about the circumstances and state of parameters that may influence risk, on the day the activity is carried out. The competence of the crew is known, the weather conditions are known, technical status of protective systems (barriers) as well as any simultaneous activities are also known. There is still some uncertainty about how the activity will be carried out, whether there will be any human errors, or whether any technical systems will experience faults, but the uncertainty is nevertheless at its minimum.

When the work was planned, months or weeks in advance, a number of the above factors will be unknown. This span of uncertainty in aspects that influence the risk levels is covered by the proposed methodology through the probability distributions of the RIFs.

Another aspect of uncertainty is the extent to which there are unknown hazards or threats that have not been included in the risk assessment, so-called 'unknown unknowns' [26]. We would argue that hazards related to leaks of hydrocarbons are well known, although there may be unknown causal mechanisms that have not been revealed yet. In any case, 'unknown unknowns' must be managed in other ways than through risk analysis (e.g. by inherently safe design, robust work practices, application of precautionary principle). Searching for 'unknown unknowns' is more relevant in relation to the QRA studies, including their regular updates.

5 Conclusion

In the MIRMAP project, we have developed a conceptual method for analyzing risk in an operational setting, to provide decision support related to major accident risk for day-to-day decision making in process plants. A key to the project has been that we have taken a step back to look at the decisions that we need decision support for and what these decisions require in terms of information about major accident risk. The method that has been developed is still at the conceptual stage and has not been implemented in practice. However, based on the work done so far we believe this is a fully feasible approach that can provide useful information. There are challenges that need to be overcome, e.g. availability of data. Further, there are also uncertainties, such as aspects related to how much work is actually required to develop useful methods. Clearly, if the work is too extensive and implies too high cost, it may be questioned if this is worthwhile.

The project will continue approximately for another year, focusing on application to practical problems. This is expected to give more information about the practicality and feasibility of the method, but our preliminary conclusion is that this is a promising approach.

Acknowledgement

The authors wish to acknowledge the Norwegian Research Council for their financial support to the MIRMAP project, No.228237/E30, funded by PETROMAKS2.

6 References

[1] Rasmussen et al., Reactor Safety Study: An Assessment of Accident Risks in U.S. Commercial Nuclear Power Plants [NUREG-75/014 (WASH-1400)]. US Nuclear Regulatory Commission, 1975.
[2] NPD, 'Guidelines for safety evaluation of platform conceptual design', Oslo, 1981.
[3] NORSOK, 'Risk and emergency preparedness analysis', NOROSK Stand. Z-013 Rev, vol. 3, no. 2010, pp. 16–8, 2010.
[4] J. E. Vinnem and W. Røed, 'Root causes of hydrocarbon leaks on offshore petroleum installations', J. Loss Prev. Process Ind., vol. 36, pp. 54–62, Jul. 2015.
[5] S. Haugen, J. E. Vinnem, and J. Seljelid, 'Analysis of causes of hydrocarbon leaks from process plants', 2011.

[6] M. Knochenhauer, 'Risk expression and modelling in risk monitors - using RiskSpectrum RiskWatcher as an example', presented at the MIRMAP workshop, Trondheim, Norway, 04-Mar-2014.
[7] J. Yang, M. Yang, H. Yoshikawa, and F. Yang, 'Development of a risk monitoring system for nuclear power plants based on GO-FLOW methodology', Nucl. Eng. Des., vol. 278, pp. 255–267, Oct. 2014.
[8] T. Aven, S. Sklet, and J. E. Vinnem, 'Barrier and operational risk analysis of hydrocarbon releases (BORA-Release). Part I. Method description', J. Hazard. Mater., vol. 137, no. 2, pp. 681–691, Sep. 2006.
[9] S. Sklet, A. J. Ringstad, S. A. Steen, L. Tronstad, S. Haugen, J. Seljelid, T. Kongsvik, I. W\a erø, and others, 'Monitoring of human and organizational factors influencing the risk of major accidents', in SPE International Conference on Health Safety and Environment in Oil and Gas Exploration and Production, 2010.
[10] J. E. Vinnem, R. Bye, B. A. Gran, T. Kongsvik, O. M. Nyheim, E. H. Okstad, J. Seljelid, and J. Vatn, 'Risk modelling of maintenance work on major process equipment on offshore petroleum installations', J. Loss Prev. Process Ind., vol. 25, no. 2, pp. 274–292, Mar. 2012.
[11] S. Haugen, J. Seljelid, O. M. Nyheim, S. Sklet, and E. Jahnsen, 'A generic method for identifying major accident risk indicators', presented at the 11th International Probabilistic Safety Assessment and Management Conference and the Annual European Safety and Reliability Conference, Helsinki, Finland, 2012, pp. 5743–5752.
[12] K. Øien, 'A framework for the establishment of organizational risk indicators', Reliab. Eng. Syst. Saf., vol. 74, no. 2, pp. 147–167, Nov. 2001.
[13] S. Haugen, J. Seljelid, K. Mo, and O. M. Nyheim, 'Major Accident Indicators for Monitoring and Predicting Risk Levels', 2011.
[14] C. Wang, Hybrid causal logic methodology for risk assessment. ProQuest, 2007.
[15] W. Røed, A. Mosleh, J. E. Vinnem, and T. Aven, 'On the use of the hybrid causal logic method in offshore risk analysis', Reliab. Eng. Syst. Saf., vol. 94, no. 2, pp. 445–455, Feb. 2009.
[16] J. T. Luxhoej and D. W. Coit, 'Modeling low probability/high consequence events: an aviation safety risk model', in Reliability and Maintainability Symposium, 2006. RAMS '06. Annual, 2006, pp. 215–221.
[17] L. J. Bellamy, M. Mud, H. J. Manuel, and J. I. H. Oh, 'Analysis of underlying causes of investigated loss of containment incidents in Dutch Seveso plants using the Storybuilder method', J. Loss Prev. Process Ind., vol. 26, no. 6, pp. 1039–1059, Nov. 2013.
[18] D. Etterlid, 'iSee - visualization of HMS related factors', presented at the SINTEF IO conference, Trondheim, Norway, 2013.
[19] D. Schellings, 'Total risk approach', SPE166640, SPE Offshore Europe, September 2013, Aberdeen, UK
[20] PSAN, 'Trends in the risk level on the Norwegian Contintental Shelf (NCS) - main report (in Norwegian)', 2014.
[21] PSAN, 'Principles for barrier management in the petroleum industry'. Petroleum Safety Authority Norway, 29-Jan-2013.
[22] K. Øien, 'Risk indicators as a tool for risk control', Reliab. Eng. Syst. Saf., vol. 74, no. 2, pp. 129–145, Nov. 2001.

[23] SINTEF and NTNU, OREDA – Offshore and onshore reliability data handbook, 6th ed. Trondheim, Norway, 2015.
[24] D. Kahneman, Thinking, fast and slow. Macmillan, 2011.
[25] PSAN, Regulations relating to health, safety and the environment in the petroleum activities and at certain onshore facilities (The framework regulations). 2015.
[26] S. Haugen and J. E. Vinnem, 'Perspectives on risk and the unforeseen', Reliab. Eng. Syst. Saf., vol. 137, pp. 1–5, May 2015.

Depressurization of CO_2 Rich Mixtures: Challenges for the Safe Process Design of CCS facilities and CO_2 EOR Systems

Sathish Natarajan
Process Systems Enterprise Inc.
1200 Smith Street, Suite 1600
Houston, TX, 77002, USA
s.natarajan@psenterprise.com

Yemi Zaccheus, Apostolos Giovanoglou, James Marriott
Process System Enterprise Ltd
26-28 Hammersmith Grove
London W6 7HA, United Kingdom
y.zaccheus@psenterprise.com, a.giovanoglou@psenterprise.com,
j.marriott@psetnerprise.com

Prepared for Presentation at
American Institute of Chemical Engineers
2016 Spring Meeting
12th Global Congress on Process Safety
Houston, Texas
April 11-23, 2016

Depressurization of CO_2 Rich Mixtures: Challenges for the Safe Process Design of CCS facilities and CO_2 EOR Systems

Sathish Natarajan
Process Systems Enterprise Inc.
1200 Smith Street, Suite 1600
Houston, TX, 77002, USA
s.natarajan@psenterprise.com

Yemi Zaccheus, Apostolos Giovanoglou, James Marriott
Process Systems Enterprise Ltd

Keywords: CO_2, depressurization, blowdown, material selection, CCS, EOR

Abstract

The design of systems with high content of CO_2 in the process mixture is of increasing importance. This is particularly true for emerging technologies such as Carbon Capture and Storage (CCS); with over twenty CCS installations worldwide (built or under-construction) and many more now progressing through front-end engineering & design and then to final investment decision. The design of the safety depressurization system for both CCS facilities and CO_2 Enhanced Oil Recovery (EOR) installations is of particular importance, due to its impact on project costs. As with Oil & Gas processing facilities, the minimum metal temperatures in process equipment and piping are observed during highly transient depressurization operations ("blowdown"). The minimum metal temperature usually limits the material of construction: if metal temperatures below -46°C (-50°F) are possible then the usual requirement is to select materials that exhibit ductile behavior below this point, typically stainless steel. Such choices have a huge impact on project costs, vessel order times and ultimately project viability.

The design of the safety depressurization system for CO_2 rich mixtures is difficult; CO_2 introduces complex thermodynamic behavior, for example: physical properties that are not accurately predicted by standard equation of state methods, a narrow phase envelope and the potential formation of solid phases during depressurization. Furthermore, physical plant configurations which are sectionalized for depressurization consist of multiple interconnected vessels and significant quantities of piping low points where condensate may accumulate. These locations are shown to be significant to depressuring temperatures. The design of such systems is not handled well using conventional depressurization methodologies; which rely on the representation of an actual plant segment as a single pseudo-vessel volume.

In this paper, we present a validated methodology for analyzing accurately the depressurization of high pressure gas processing facilities with rich CO_2 mixtures. We describe the application of the methodology to the design of a CO_2 EOR process.

1 Introduction

A depressurization operation is designed to:

- Remove, in a controlled manner, all combustible hydrocarbons during an emergency situation or for a planned shutdown.
- Reduce the risk of vessel or pipe rupture by reducing pressure in the event of a fire.
- Prevent the escalation of a leak leading to a major loss of containment and potential explosions.

Typically, prior to depressurization a process plant is isolated into a number of independent systems. A full plant blowdown operation involves the simultaneous or staggered depressurization (blowdown) of all the pressurized gas (and/or in some cases liquid) in each system by routing it to one or more flare tips for controlled combustion.

The depressurization operation may itself be hazardous [1] and a number of factors must be considered in the design of these safety systems. Of particular importance here is the risk of brittle fracture in process equipment and piping due to auto-refrigeration chilling during depressurization which can lead to very low fluid, and potentially, low metal temperatures. Usually, the minimum metal temperature experienced during depressurization limits the material of construction for a facility: if metal temperatures below -46°C (-50°F) - the Lower Design Temperature, LDT [1], of low temperature carbon steel - are possible then the usual requirement is to select materials that exhibit ductile behavior below this point. Such choices have a huge impact on project costs, pressure vessel order times, project schedule and potentially project viability.

The requirements of the sixth edition of API 521 Guidelines [2] for an accurate analysis and design of blowdown systems for the general case of upstream oil and gas facilities are best satisfied by adoption of the following methodology:

1. Use of rigorous non-equilibrium modeling to accurately represent the thermodynamic behavior of fluids within the depressuring system.
2. Use of a distributed 'flowsheet' process model that represents the spatial distribution of volumes in the system, rather than the conventional practice of representing ('lumping') the entire depressurization system as a 'vessel'.
3. Coupling of process and flare network for accurate dynamic flare system analysis.

The focus of this paper is the application of the above methodology to the design of safety depressurization systems with process mixtures rich in CO_2. This is of importance to technologies such as Carbon Capture and Storage (CCS) and CO_2 Enhanced Oil Recovery (EOR) installations, primarily due to its impact on capital costs.

The main challenges introduced by the high content of CO_2 in the process mixture in design of safety depressurization systems are:

1. Prediction of physical properties: inaccurate estimation of phase equilibrium can lead to situations where formation of liquid is not predicted. This can have a significant impact on the minimum metal temperatures determined.
2. Complexity in the thermodynamic behavior that CO_2 rich mixtures usually exhibit during depressurization:
   a. Dense phase depressurization: this is the situation where single (dense) phase depressurization is followed by crossing of the two-phase envelope from the bubble point line (this is rare for common oil and gas hydrocarbon mixtures). These results in large amounts of "boiling" liquid being formed across the system, with a significant effect on metal wall temperatures. Such a behavior cannot explicitly be captured by conventional tools, which has important implications on the accuracy of the predictions.
   b. Formation of solid CO2 during depressurization which may lead to blockages.

The above challenges are addressed in the next section. The application of accurate blowdown system analysis to a CO_2 Enhanced Oil Recovery (EOR) FEED project is presented in Section 3.

2 Accurate analysis for the depressurization of CO_2 rich mixtures

The methodology proposed here is based on the new requirements introduced in the latest, sixth, edition of API 521 Guidelines [2] and focuses on [1]:

1. The accurate prediction of the thermodynamic behavior of fluids and heat transfer during depressurization.
2. The accurate representation of the blowdown system geometry taking into account all process equipment and piping, the locations where liquid may pool and boil (system low points) and liquid drainage across the system.

The application of the above principles to the analysis of depressurization systems for CO_2 rich mixtures as encountered in CCS and EOR applications is discussed next.

2.1 Physical property predictions

Blowdown of a system is predominantly a thermodynamically driven process; hence the predictions of a blowdown simulation rely heavily on accurate phase equilibrium, physical and transport property calculations. The thermodynamic models that are commonly used in oil and gas applications are cubic Equations of State (EOS) like Peng-Robinson [3] or Soave-Redlich-Kwong [4]. These models have specifically been developed and calibrated for typical hydrocarbon mixtures and conditions seen in oil and gas applications. Apart from the good predictions they offer in their range of application, part of their success is due to their simplicity which allows their widespread use in simulation software packages used by oil and gas engineers.

The presence of CO_2 in high quantities in a process mixture together with impurities like N_2, H_2, Ar and O_2, as commonly seen in CCS applications, can have an impact on the predictions of properties by cubic EOS. This impact may in certain situations be significant for the prediction of:

- Bubble point line and phase envelope: there are cases where cubic EOS tends to under-estimate the two-phase envelope.
- Liquid density: can have a significant impact in pressure drop calculations in long CO_2 pipelines during normal operation and during depressurization.

The predictions of Peng Robinson EOS are shown in Figure 1 for five CO_2-rich mixtures (Table 1) which have been studied experimentally [5]-[8]. Note that Case 4 compares density predictions, while the remaining cases compare the prediction of the phase envelope. The deviations observed in Case 1, Case 2 and Case 4 are caused by the non-ideal behavior of CO_2 mixtures with impurities (mainly H_2), which calls for more advanced and predictive thermodynamic models to be used. Two examples of such thermodynamic models are:

1. GERG EOS [9]: an industry standard high-accuracy model for mixtures of natural gas components.
2. SAFT-γ Mie EOS [10]: a variant of the general Statistical Associating Fluid Theory (SAFT) framework. The detailed molecular description employed within SAFT-γ Mie EOS results in an accurate representation of the fluid-phase behavior and derivative properties of complex fluids and mixtures, as recently demonstrated for the case of CO_2 [11].

The predictions of GERG and SAFT-γ Mie for the five experimental sets are also shown in Figure 1. These clearly exhibit very good predictions across all experimental sets. SAFT-γ Mie predictions are much closer to experimental data than GERG for Case 2, and in general offers better predictions than GERG.

Table 1: Composition of CO2-rich mixtures studied (in mole percent)

	CO2	N2	H2	Ar	O2
Case 1	89.83	3.07	0	2.05	5.05
Case 2	95	2	3	0	0
Case 3	97.45	0	0	2.55	0
Case 4	98.03	0	1.97	0	0
Case 5	97.99	2.01	0	0	0

As will be shown later in Section 2.2.2, there are situations where simple EOS can predict key results from a depressurization calculation (fluid and metal temperatures, pressure) equally well. This is partly due to the extremely fast transients encountered during blowdown operations for the case of bubble point predictions. Furthermore, pressure drops are usually negligible during depressurization of a process plant segment due to very low velocities (compared to those seen during normal operation).

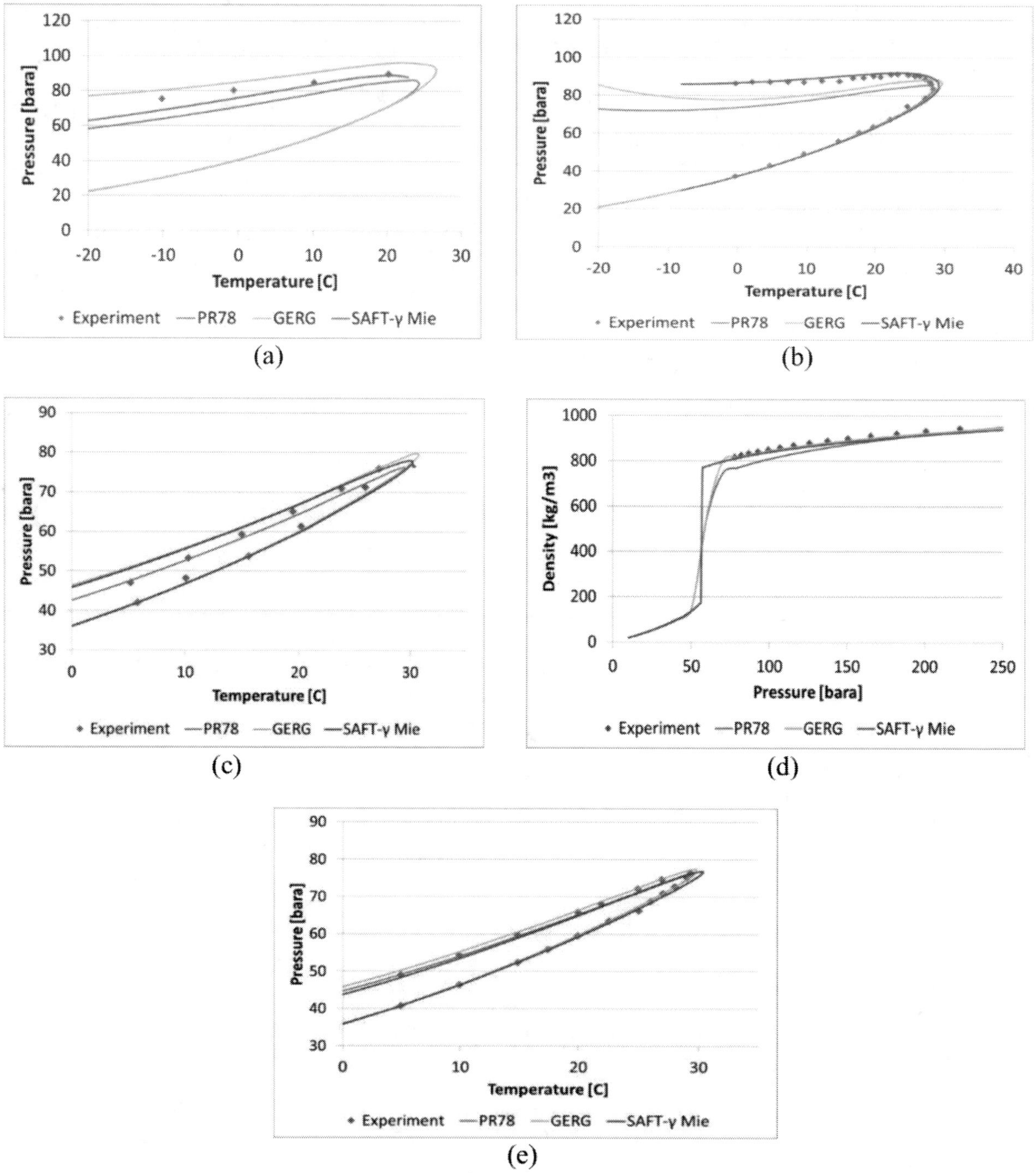

Figure 1: Comparisons between predictions of PR, GERG & SAFT-γ Mie with experimental data for (a) Case 1 (b) Case 2 (c) Case 3 (d) Case 4, and (e) Case 5

2.2 Complex thermodynamic behaviour – single vessel depressurization

The complex thermodynamic behavior that rich CO_2 mixtures exhibit during depressurization is discussed in this section through two cases of single vessel depressurization where experimental data are available:

- Case 1 is pure CO_2 depressurization and demonstrates the phenomena that occur during dense phase depressurization
- Case 2 is depressurization of a CO_2/N_2 mixture where solid CO_2 is formed during the event. A comparison of the predictions of the three thermodynamic models discussed in the previous section is also presented.

2.2.1 Case 1: Pure CO2 blowdown

This case considers depressurization of pure CO_2 from supercritical conditions (150 bara and 40°C c.f. Figure 2) in a vertical vessel with internal diameter 242 mm with tan-tan height 1.1m [12]. The outer diameter of the vessel was 315 mm. An orifice of size 17 mm^2 was used.

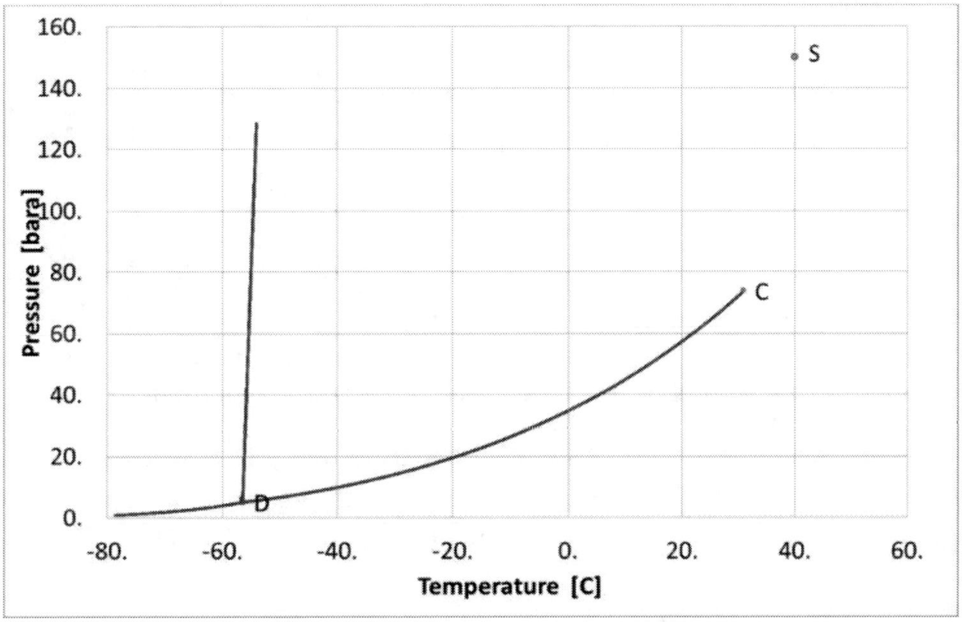

Figure 2: Phase diagram for the Pure CO2. C denotes the critical point; D is the triple point; and S is the starting point for depressurization

The initial conditions are in the supercritical region and in this experiment depressurization occurs through the dense phase region down to the bubble point line. Depressurization then follows the saturation curve. As seen by the experimental points of Figure 3 and Figure 4, there is initially a sharp decrease in pressure and temperature (due to Joule-Thomson cooling) while CO_2 is in the dense phase region. Upon reaching the bubble point line, there is significant flashing of the liquid, which results in a nearly-constant pressure profile seen at around 10 seconds (in some cases a pressure rise may also be observed at this stage). The pressure subsequently starts dropping again when the top space in the vessel is occupied by vapor and the flashing liquid is in the bottom of the vessel. The rate of depressurization is much slower now than during dense phase depressurization. The temperature of each phase drops with pressure:

- Joule-Thomson cooling is the main driving force in the vapor phase.
- Vigorous boiling is the main driving force in the liquid phase.

The metal temperatures in contact with boiling liquid are significantly colder due to the large heat transfer rates, between the liquid and metal. These cold temperatures present a real threat of brittle fracture. The large amount of liquid formed during depressurization via the bubble point line implies brittle fracture risks may be present across all equipment and piping of an actual plant blowdown system.

It should be noted that this behavior is not common during depressurization of standard hydrocarbon mixtures seen in upstream oil and gas. The critical point of such mixtures, which usually have high methane content, is at much lower temperatures such that depressurization follows the dew point curve (usually from the retrograde region). Condensation is the mechanism for liquid formation in this case and hence smaller amount of liquid is present during the event. The higher the CO_2 content in a mixture the more likely depressurization will occur via the bubble point curve.

Another important point here is the slow depressurization rates (as shown in Figure 3), after flashing, which result in much longer depressurization times; in many occasions they can exceed the allowable fire scenario rupture stress with implications on orifice sizing and violations of blowdown philosophies and rules.

Conventional tools for depressurization are not designed to represent explicitly the phenomena that occur during depressurization from the bubble point line. Instead some of them offer mechanisms to alter the thermodynamic behavior of a mixture so that a simulation can be performed. This can have a significant effect on the accuracy of minimum metal temperature and blowdown predictions based on the above discussion, and do not guarantee conservativeness.

Figure 3 and Figure 4 show the prediction of the pressure, gas and liquid temperature profiles by taking into account explicitly the phenomena that occur during CO_2 dense phase depressurization. The calculations were performed using gFLARE® Advanced Process Modelling software.

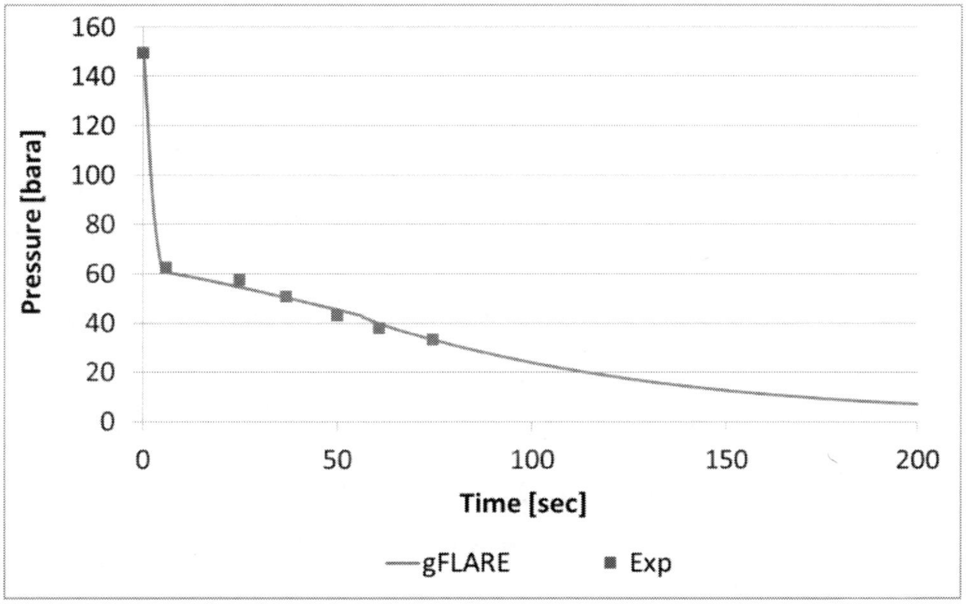

Figure 3: Pressure profile of pure CO_2 depressurization

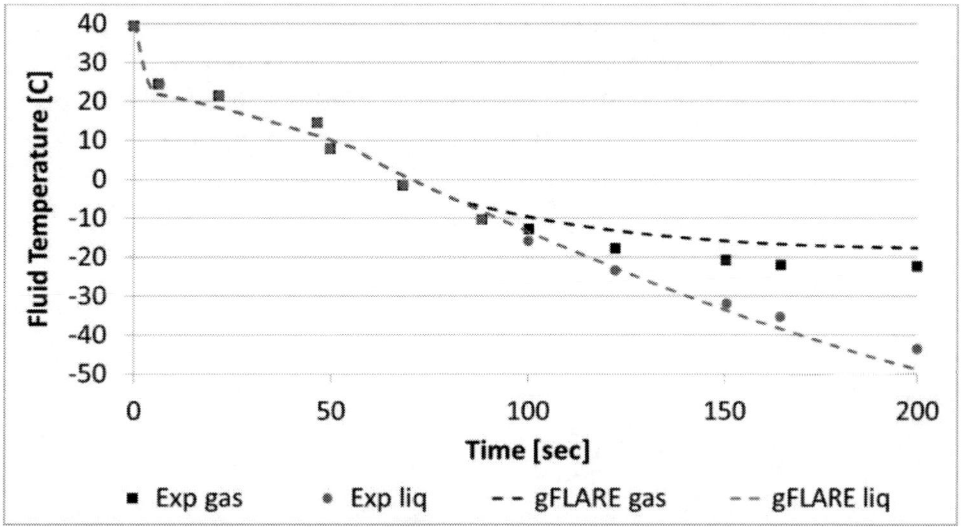

Figure 4: Pure CO_2 blowdown gas and liquid temperature profiles

2.2.2 Case 2: CO_2-N_2 mixture blowdown

Blowdown of a CO_2-N_2 mixture (30 mol% CO_2) is considered in this case from a vertical vessel with 273mm inside diameter, 1.52m tan–tan length and 25mm wall thickness. Blowdown was from the top via a 6.35mm orifice starting at 150 bara and 20°C. Figure 5 shows the phase envelope of the mixture including the CO_2 freezing line.

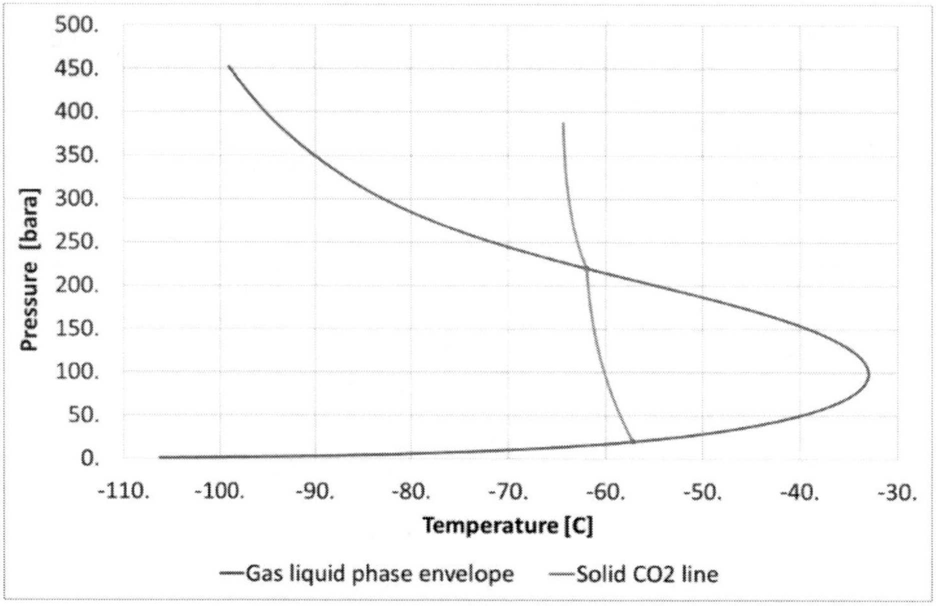

Figure 5: Phase diagram and blowdown path of gas phase for the CO_2-N_2 mixture

Due to the shape of the phase envelope of the mixture, blowdown occurs via the dew point line and a limited amount of liquid is formed via condensation. The warmer wall in this case has enough heat capacity to increase the temperature of the liquid (Figure 6). The liquid temperature subsequently drops due to boiling and as condensation continue from the colder gas. As the gas cools during blowdown the CO_2 freezing line is crossed (Figure 6). The possibility of solid CO_2 crystals forming in the vessel is observed.

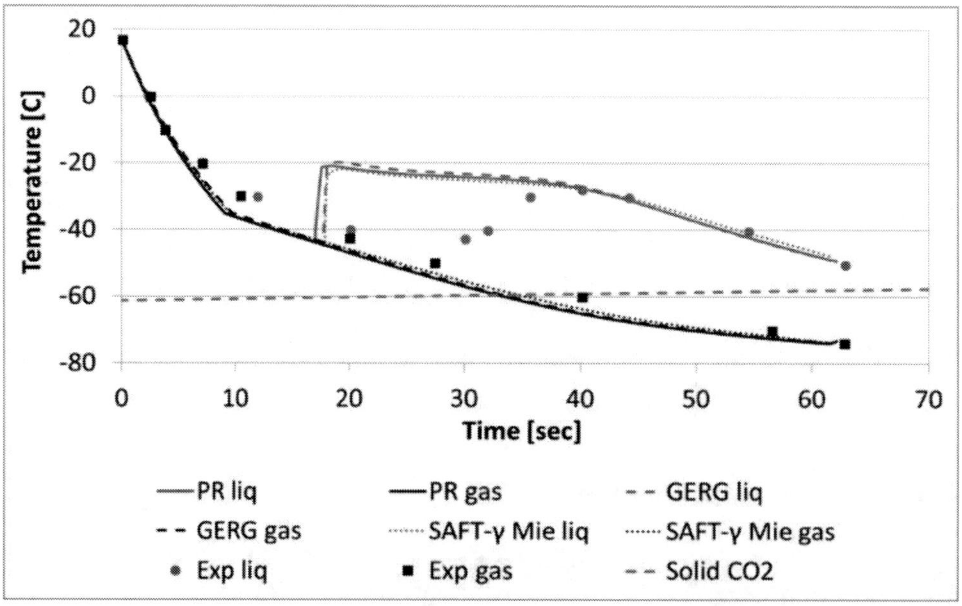

Figure 6: Gas and liquid temperatures during blowdown for the CO_2-N_2 mixture PR, GERG and SAFT-γ Mie

The blowdown event was simulated in gFLARE® Advanced Process Modelling software using the three thermodynamic models discussed in Section 2.1. A very good agreement between the experimental data and the simulation results is observed; a sharp increase in the liquid temperature when it forms and pools as it comes in contact with the warm wall is predicted. Potential formation of solid CO_2 is flagged up during the simulation. Finally, the agreement in the prediction of the three thermodynamic models should be noted as shown in Figure 6, Figure 7 and Figure 8.

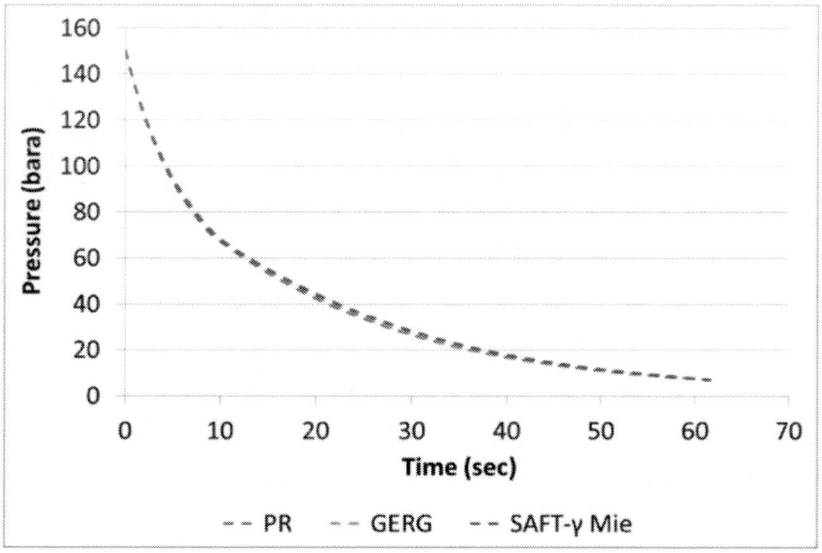

Figure 7: Pressure during blowdown for the CO_2-N_2 mixture using PR, GERG and SAFT-γ Mie

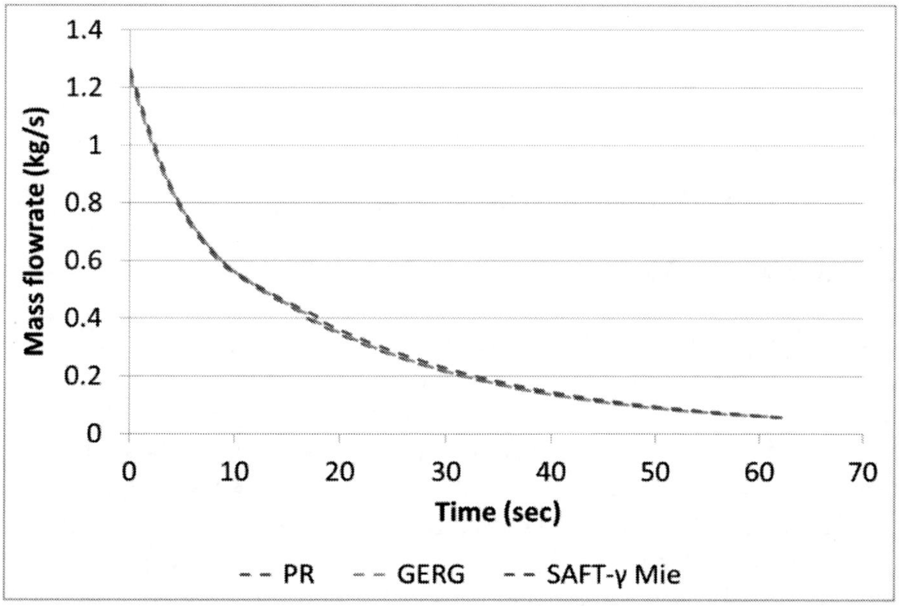

Figure 8: Flow rate during blowdown for the CO_2-N_2 mixture using PR, GERG and SAFT-γ Mie

3 System blowdown and material selection for a CO_2 EOR FEED project

Application of the approach presented in this paper to determine minimum metal temperatures in a typical CO_2 Enhanced Oil Recovery (EOR) project is described next.

This CO_2 Enhanced Oil Recovery (EOR) project is in FEED stage with the intent of having a recovery of 70% of OOIP (original oil in place) using a miscible injectant (MI) stream available in the gas gathering center. This MI stream, which is rich in CO_2 (>50% by mass), is passed through a scrubber, then compressed first in a low stage compressor and then a high stage compressor before being routed to reinjection wells (see overall PFD in Figure 9). The preferred material of construction is low temperature carbon steel (LTCS) with a lower design temperature (LDT) of -46°C. The MI compressor train is divided into three isolatable systems as shown in Figure 9. During ESD (emergency shut down), shutdown valves close isolating the train into the MI feed system, MI low stage compressor system and MI high stage compressor system, which are then simultaneously depressurized to atmospheric pressure. In this paper, depressurization of the MI feed system is evaluated in detail.

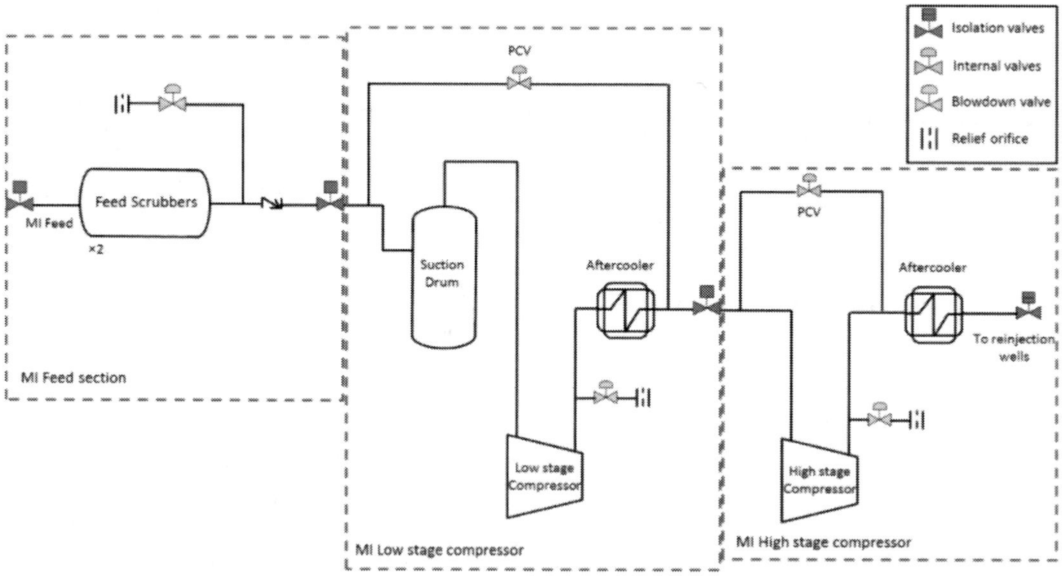

Figure 9: PFD of MI compressor train for EOR

3.1 MI injection train – Feed system

A detailed view of this system is shown in Figure 10. During FEED stage, preliminary information about pipe sizes and vessel dimensions are available. The inlet piping consisted of 24" and 28" lines which are routed to two feed scrubbers in parallel. The outlets from the two scrubbers are recombined before being passed into the compressors. As shown in Figure 10, the system consists of piping of varying sizes (28", 24", 12", 4") along with the presence of two vessels. The blowdown line is of size 4" with a restriction orifice through which depressurization occurs. The total system volume is 42m^3 with the vessel volumes accounting for 75% of the total. A conventional analysis is shown next, its drawbacks explained, followed by the detailed geometric depressurisation methodology proposed in this paper.

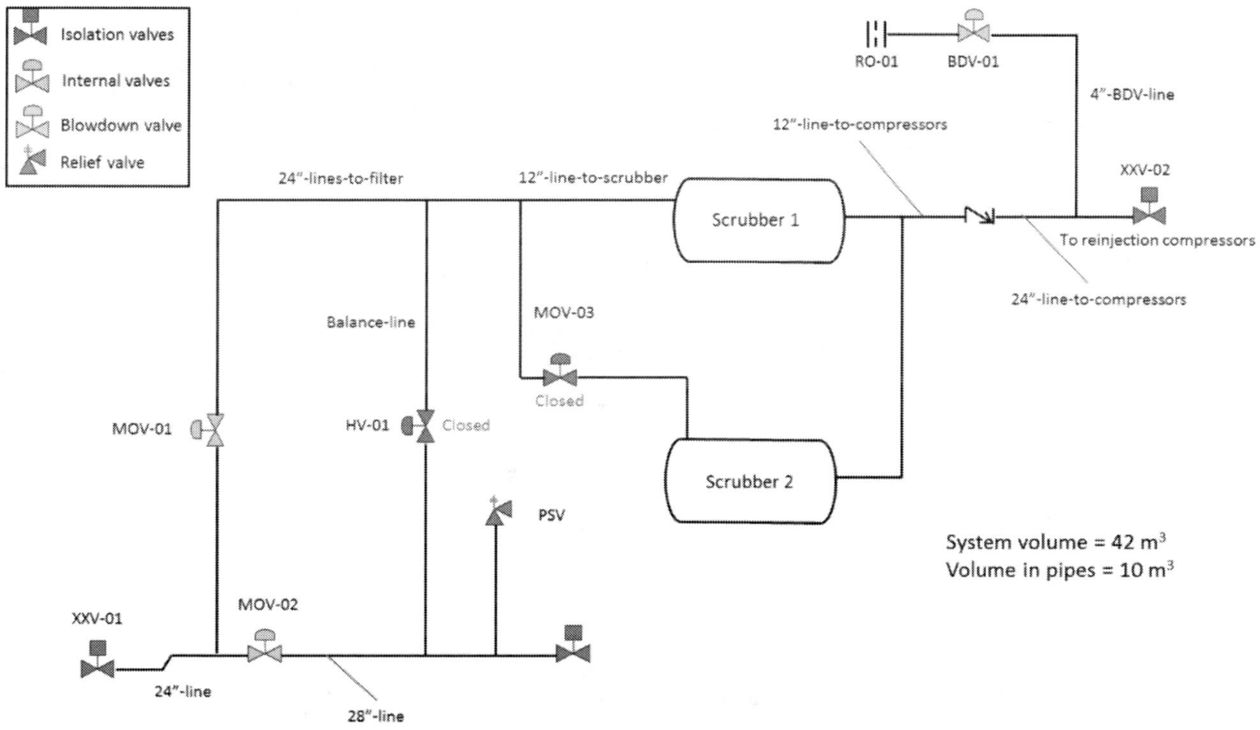

Figure 10: Schematic of the isolated MI feed system

3.2 Conventional analysis

A conventional methodology to determine minimum metal temperatures involves representing the entire blowdown system as an aggregated idealized volume and the depressurization event is simulated with an equilibrium based utility [1]. Such an analysis predicted metal wall temperatures much colder than -46°C which indicated the need for stainless steel construction of the entire system. Only a single temperature is predicted for the entire system as it is modelled in an aggregated methodology even though different aspect ratios are present in different parts of the system. In addition, conventional equilibrium based tools are not well suited to model depressurization from dense phase when depressurization is initiated from conditions to the left of the critical point and the fluid crosses the two phase envelope from the bubble point line. Results predicted by conventional tools are less reliable in such instances. Hence, a detailed system level analysis is necessary to accurately capture the behavior during depressurization and determine metal wall temperatures. Such a model configuration is described next.

3.3 Detailed depressurization model configuration

The system described above consists of multiple vessels along with piping of different sizes and wall thicknesses. The modelling approach adopted involved building a process flowsheet that represented this geometry of the system realistically. This involved analyzing P&IDs and preliminary isometrics available during this FEED stage and identifying lowpoints or liquid accumulation points. These are locations in the system where liquid formed during depressurization tends to accumulate and pool. Five lowpoints in process piping, in addition to the two vessels, were identified and configured as such in the model as shown in Figure 11. This detailed model, configured in gFLARE® Advanced Process Modelling software was used to perform the blowdown calculation in order to determine the minimum metal temperatures.

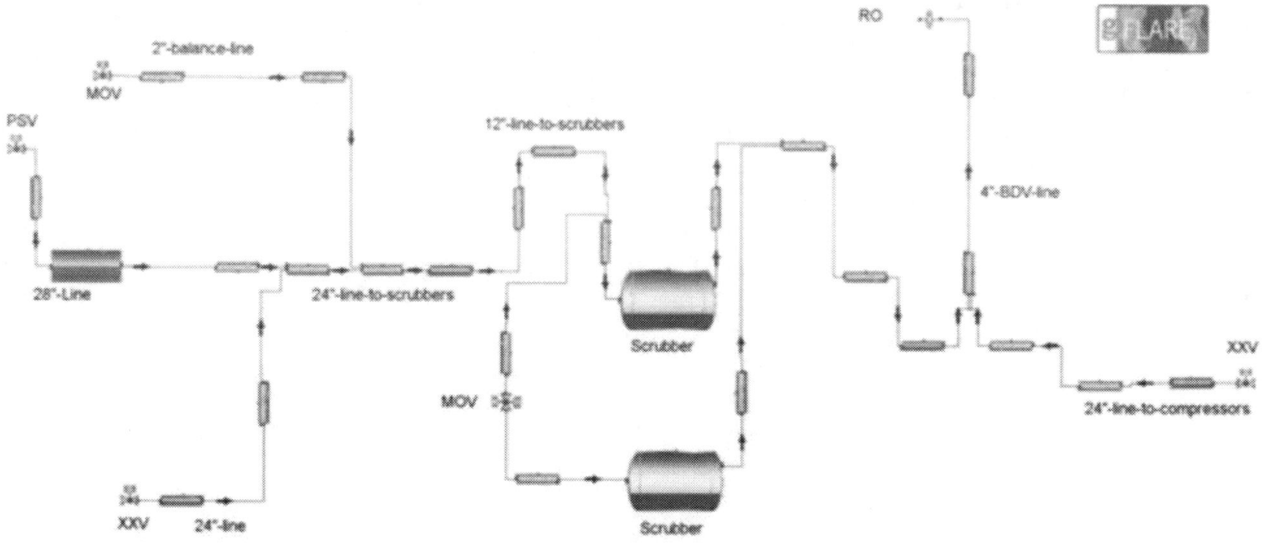

Figure 11: Geometric depressurization model for the MI feed system

3.4 Depressurization results

The geometric model configured above was used to determine the minimum metal temperatures in different parts of the system. The initial conditions for blowdown were 120 bar(g) and 40°C. The flowrate and pressure profile at the inlet of the relief orifice is shown in Figure 12. The pressure reduces to 6.9 barg in approximately 15 minutes and the depressurization is complete in 30 minutes.

Depressurization from these conditions results in the fluid crossing the two phase envelope from the left of the critical point (bubble point line) as shown in Figure 13. Thus, as described in Section 2, a large quantity of liquid is formed during depressurization. This liquid pools and evaporates which cools the metal surfaces it is in contact with. The temperature profiles in 4 key locations across the system are presented next.

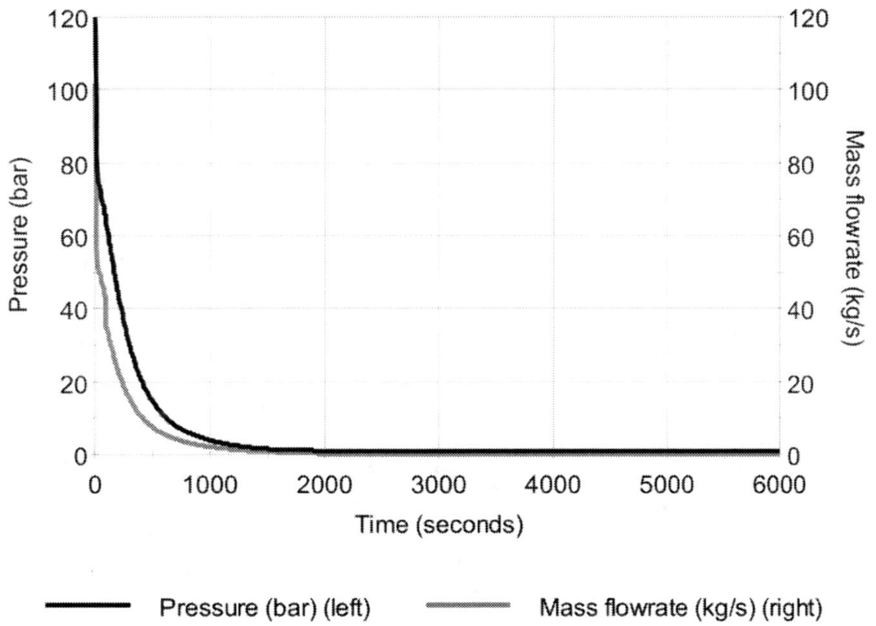

Figure 12: Pressure and flow profile at the inlet of the relief orifice

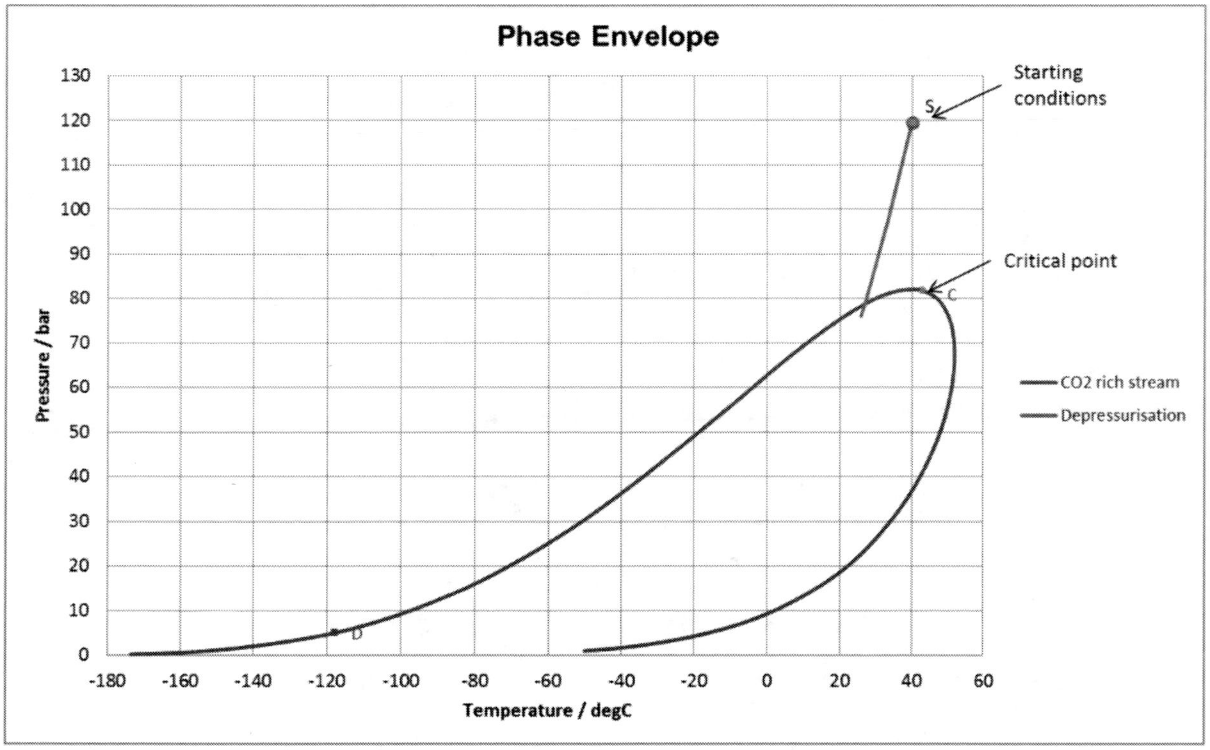

Figure 13: Phase envelope and depressurization profile

Scrubber

The temperature profile in the scrubber is shown in Figure 14. The minimum temperature observed in this vessel is -36°C which is warmer than the material LDT of -46°C. The vessel has thick walls and hence greater metal mass which keeps the temperature warm.

24"-inlet-line

The temperature profile in the 24"-inlet-line is shown in Figure 15. This line has a schedule of Sch.40. The thin wall along with the intense liquid evaporation results in minimum metal temperature of -65°C which is much colder than the material LDT of -46°C. This location presents embrittlement risks during depressurization.

12"-line

This line is at the outlet of the scrubbers leading to the compressors. The temperature profile observed in this line is shown in Figure 16. The minimum metal temperature observed is -45°C which is marginally warmer than the LDT. This line has a schedule of Sch. 80. Hence, it has a greater metal mass to internal volume ratio than the 24" line and hence experiences warmer temperatures.

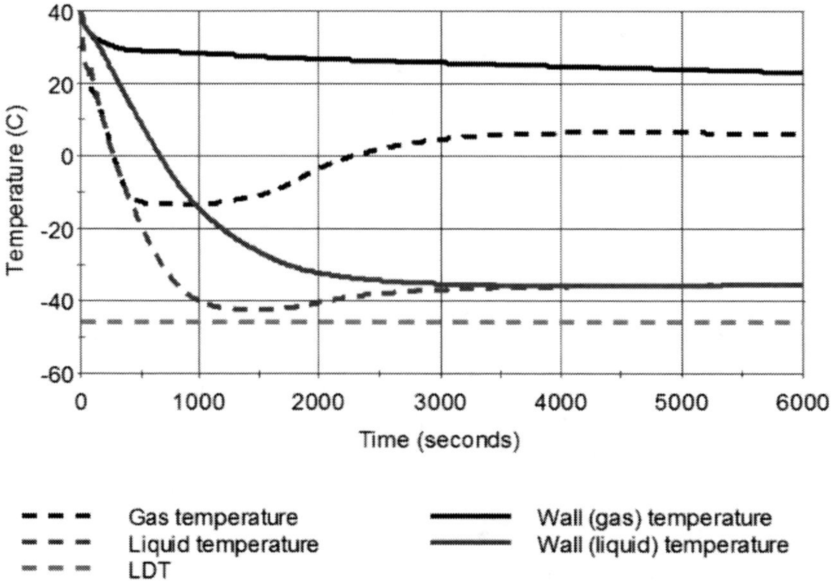

Figure 14: Temperature profile in Scrubber

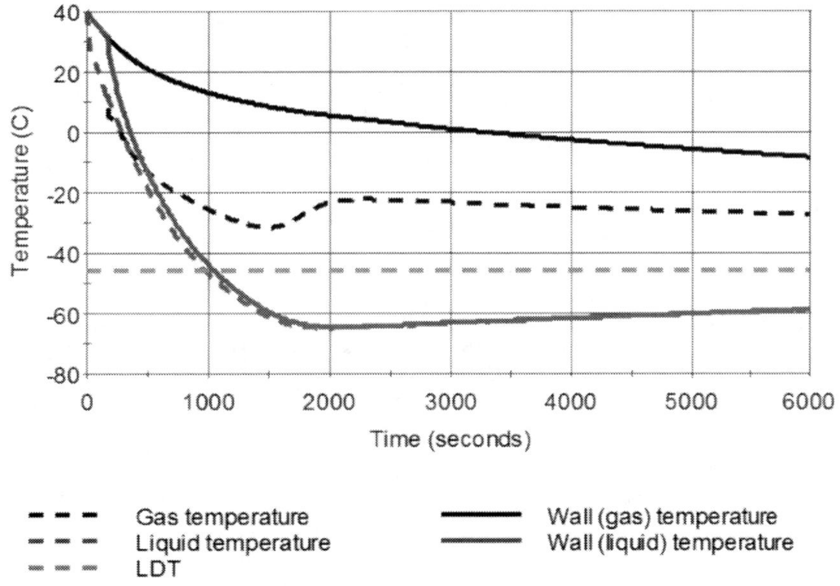

Figure 15: Temperature profile in 24”-inlet-line

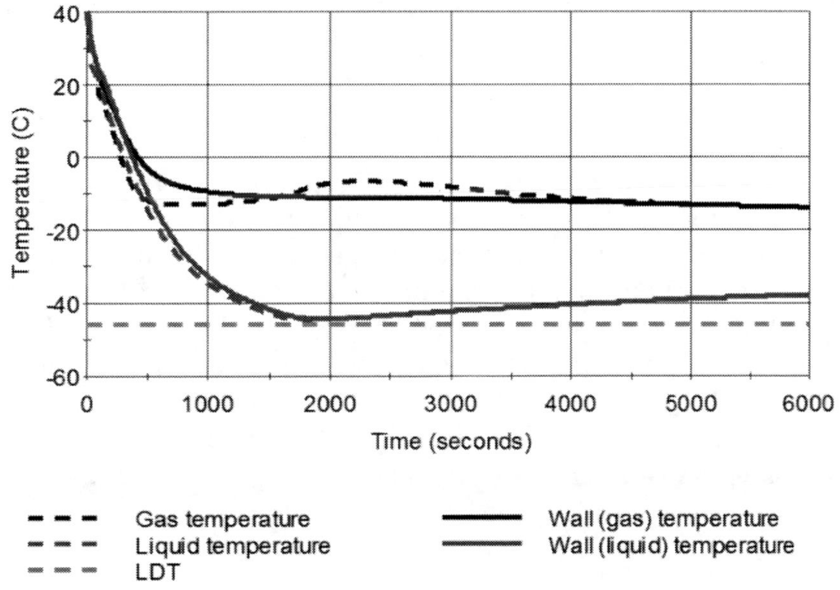

Figure 16: Temperature profiles in 12”-line

28”-line

This line is also at the inlet of the system. The temperature profile observed in this line is shown in Figure 17. The minimum metal temperature observed is -41°C. Again, this line has greater metal mass to internal volume ratio than the 24” line and hence experiences warmer temperatures.

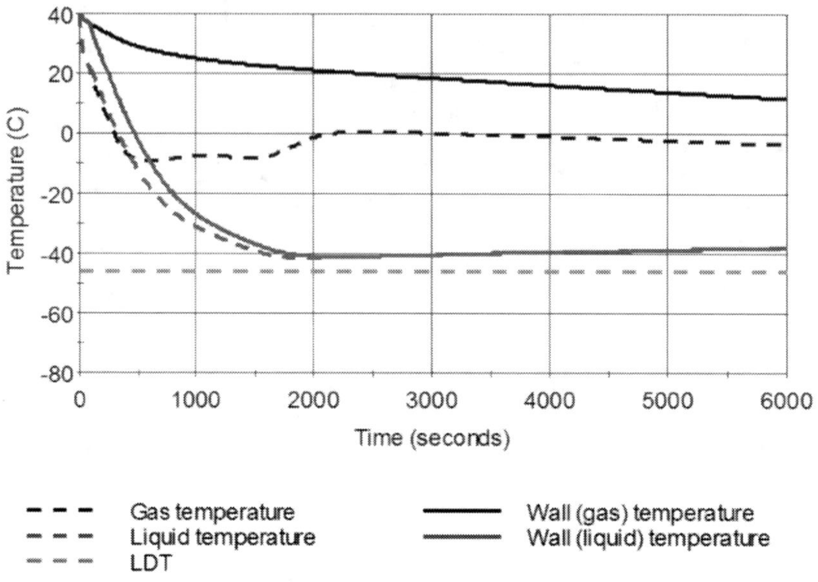

Figure 17: Temperature profiles in 28"-line

Hence in this case study, we observed that different parts of the system experienced different minimum metal temperature. A detailed geometric model allowed for accurate determination of minimum metal temperatures as well as identifying the location where these temperatures are observed.

3.5 Potential mitigation

The case study above showed that the minimum metal temperature was observed in process piping and not the vessel where most of the system volume is present. This meant that the vessels can be constructed out of low temperature carbon steel and mitigations were necessary only for the process piping. By working in collaboration with the operator engineering team during this FEED stage, the design space was explored and it was identified that there is still some flexibility with regard to choosing pipe size and wall thickness. Hence, additional calculations were performed on the 24" line with higher schedules in order to mitigate low temperature risks. Figure 18 shows the temperature profile if the schedule was increased to Sch.100. The minimum metal temperature is now marginally warmer than -46°C. Hence, the low temperature risks can be mitigated by choosing a schedule greater than Sch.100.

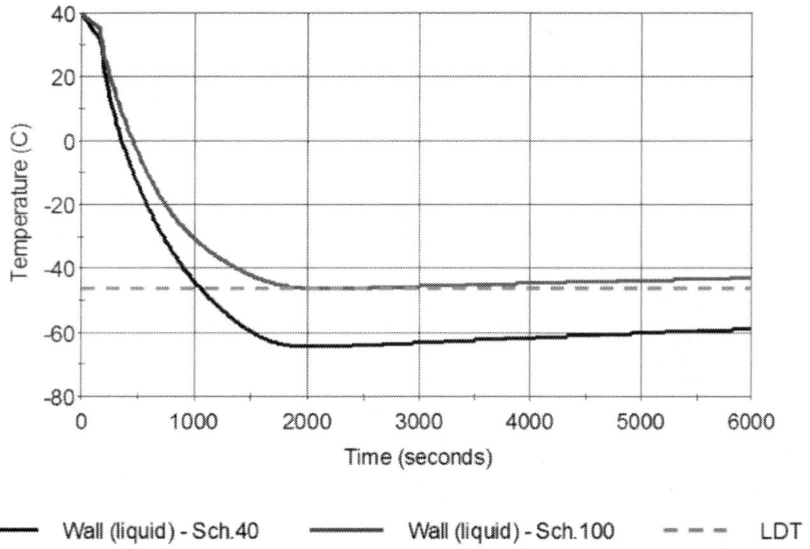

Figure 18: Temperature profile in 24”-inlet-line for two different wall thicknesses

4 Conclusion

The design of safety depressurization systems for CCS and CO_2 Enhanced Oil Recovery (EOR) installations is of particular importance due its impact on project costs. In this paper we presented a validated methodology for analyzing accurately the depressurization of high pressure gas processing facilities with rich CO_2 mixtures. The prediction of physical properties by Peng-Robinson (PR) was compared with GERG and SAFT-γ Mie EOS.

In some instances, the predictions of minimum metal temperatures between PR, GERG and SAFT-γ Mie EOS were similar, partly due to the extremely fast transients encountered during blowdown operations, from certain initial conditions. However, there can be situations where differences in predictions of phase equilibrium can result in the formation of liquid being determined incorrectly, resulting in vastly different predicted minimum temperatures. Hence, careful consideration must be given to the choice of EOS.

The complexity in the thermodynamic behavior that rich-CO_2 mixture can exhibit has been extensively discussed in this paper. It is common for such mixtures to cross the bubble point line during depressurization and it is essential to accurately capture this behavior in order to obtain accurate fluid and metal temperature predictions as well as depressurization time.

The application of the methodology was demonstrated on a typical CO_2 Enhanced Oil Recovery (EOR) project. A detailed geometric model of the system was configured and the minimum metal temperatures were accurately captured in the different parts of the system when depressurization happens from the bubble point line. It was shown that, contrary to expectation, the minimum metal temperature was observed in a process line rather than in the main vessel. Mitigations were proposed by working collaboratively with the project engineering team during this FEED stage by exploring the design space.

5 References

[1] J. Marriott, A. Giovanoglou, "Process modeling requirements for the safe design of blowdown systems – changes to industry guidelines and how this impacts current practice", *For presentation* at the Global Congress on Process Safety, 2014.

[2] API Standard 521, Pressure-relieving and Depressuring Systems, Sixth Edition, January 2014.

[3] D. Y. Peng and D. B. Robinson, "A New Two-Constant Equation of State", Ind. Eng. Chem. Fundamen., 15 (1), pp 59–64, 1976.

[4] G. Soave, "Equilibrium constants from a modified Redlich-Kwong equation of state", Chemical Engineering Science, 27(6), pp. 1197-1203, 1972.

[5] A. Chapoy, N. Mahmoud, M. Kapateh, R. Burgass, C. Coquelet and B. Tohidi. "Effect of impurities on thermophysical properties and phase behaviour of a CO2-rich system in CCS", International Journal of Greenhouse Gas Control, 19, pp. 92-100, 2014.

[6] M. J. Tenorio, A. J. Parrot, J. A. Calladine, Y. Sanchez-Vicente, A. J. Cresswell, R. S. Graham, T. C. Drage, M. Poliakoff, J. Ka and M. W. George. "Measurement of the vapour-liquid equilibrium of binary and ternary mixtures of CO2, N2 and H2, systems which are of relevance to CCS technology", International Journal of Greenhouse Gas Control, 41, pp. 68-81, 2015.

[7] M. Ahmad, J. Gernert and E. Wilbers. "Effect of impurities in captured CO2 on liquid–vapor equilibrium", Fluid Phase Equilibria, 363, pp. 149-155, 2014.

[8] Y. Sanchez-Vicente, T. C. Drage, M. Poliakoff, J. Ke and M. W. George. "Densities of the carbon dioxide + hydrogen, a system of relevance to carbon capture and storage", International Journal of Greenhouse Gas Control, 13, pp. 78-86, 2013.

[9] O. Kunz, R. Klimeck, W.Wagner, M. Jaeschke, "The GERG-2004 wide-range equation of state for natural gases and other mixtures", GERG Technical Monograph 15, 2007.

[10] V. Papaioannou, T. Lafitte, C. Avendaño, C. S. Adjiman, G. Jackson, E. A. Müller and A. Galindo. "Group contribution methodology based on the statistical associating fluid theory for heteronuclear molecules formed from Mie segments", The Journal of chemical physics, 140 (5), 054107, 2014.

[11] V. Papaioannou, F. Calado, T. Lafitte, S. Dufal, M. Sadeqzadeh, G. Jackson and A. Galindo. "Application of the SAFT-γ Mie group contribution equation of state to fluids of relevance to the oil and gas industry", Fluid Phase Equilibria, http://dx.doi.org/10.1016/j.fluid.2015.12.041, 2015.

[12] B. Gebbeken and R. Eggers. "Blowdown of carbon dioxide from initially supercritical conditions", J. Loss Prev. Process Ind., 4, pp. 285-293, 1996.

12TH GLOBAL CONGRESS ON PROCESS SAFETY

Do You Feel Lucky?

Or Do You Want to Identify and Manage Safety Critical Equipment?

Michael P. Broadribb
Baker Engineering & Risk Consultants, Inc.,
San Antonio, Texas
mbroadribb@bakerrisk.com

Eric Freiburger
Praxair Inc.,
Danbury, Connecticut
Eric_Freiburger@Praxair.com

Prepared for Presentation at
American Institute of Chemical Engineers
2016 Spring Meeting
12th Global Congress on Process Safety
Houston, Texas
April 11-23, 2016

Do You Feel Lucky?
Or Do You Want to Identify and Manage Safety Critical Equipment?

Michael P. Broadribb
Baker Engineering & Risk Consultants, Inc.,
San Antonio, Texas
mbroadribb@bakerrisk.com

Eric Freiburger
Praxair Inc.,
Danbury, Connecticut
Eric_Freiburger@Praxair.com

Keywords: safety critical equipment, safety critical element

Abstract

Today's process safety practices are invariably the result of hard lessons learned by industry from major accidents. Process safety management systems are generally credited with reducing major accident risk in process industry operations. The prevention and mitigation of major accidents relies upon appropriate layers of protection or barriers working on demand. These barriers include:

- Plant: Hardware, such as process equipment, control systems, dikes, etc.
- Process: Administrative and procedural controls
- People: Competent workforce, supervision, communication, etc.

The Plant is generally the most reliable of these barriers. To ensure that the equipment prevents and mitigates major accidents it needs to be *highly* reliable. Some companies in the process industries have started to apply criticality ranking for process safety purposes in order to identify and manage certain equipment that requires a high reliability. The identification and management of this *Safety Critical Equipment or Element (SCE)* has represented a challenge to many companies. A major issue has been an overwhelming number of SCE items, as the SCE designation has often been assigned to equipment that may not be truly safety critical. This has culminated in a failure to perform all inspection, testing and preventive maintenance (ITPM) tasks when due.

The paper will discuss a method to identify a manageable number of SCE items, and approaches for successfully performing all ITPM tasks including those tasks for SCE.

1 Introduction

Process safety management systems are commonly credited with reducing the number of major incidents in the oil & gas, and chemical industries. Those process safety practices have evolved over the years through experience and lessons learned from major incidents, which often resulted in fatalities and significant environmental and property damage. Through this experience has come a growing recognition that multiple layers of protection, as in the 'Swiss Cheese' (ref. 1) and Bow-Tie models, are necessary to prevent and/or mitigate major incidents. These layers of protection or barriers comprise, but are not limited to:

- Plant: Hardware, such as process equipment, control systems, dikes, etc.
- Process: Administrative and procedural controls
- People: Competent workforce, supervision, communication, etc.

There has also been an increasing awareness within the industry that the Process and People barriers involve human factors and interaction, which can often be the least reliable. Therefore, to prevent and mitigate a major incident, a greater onus should be placed on the equipment and systems (i.e. Plant), which must work on demand. That means that the equipment and systems need to be *highly* reliable.

Criticality analysis has been used in some industries to rank equipment and systems that require their integrity managed differently than that for other 'non-critical' equipment. Early use of Failure Modes, Effects and Criticality Analysis (FMECA) in the military, aerospace and automotive industries was focused on eliminating equipment failure to improve functionality, performance, and product quality. Increasingly the oil & gas and chemical industries have used equipment criticality ranking for improving productivity (i.e. plant availability), quality, and, more recently, some companies have started to apply criticality ranking for process safety purposes.

2 Safety Critical Equipment

When applying criticality ranking for process safety purposes, the malfunction or failure of equipment that could cause, contribute to, prevent, or mitigate the effects of a major incident is judged to be critical. This equipment has been called *safety-critical equipment* (SCE). The terminology of SCE does not fully reflect the objective of managing major hazards, and the term 'major incident critical equipment' appears more apposite.

Safety Critical Equipment
Equipment, the malfunction or failure of which could cause or contribute to a major incident, or the purpose of which is to prevent a major incident or mitigate its effects.

A similar concept and terminology, *safety-critical elements* (also abbreviated SCE), is used in UK offshore regulations (ref. 2). Following the 1988 Piper Alpha disaster in the North Sea, offshore Operators are required to identify safety-critical elements and their performance standards for managing major hazards in terms of functionality, availability, reliability, and survivability. These safety-critical elements can include equipment, systems, structures, component parts, procedures and software. Typical examples of SCE are illustrated in Table 1.

TABLE 1. Typical Examples of Safety-Critical Equipment

Pressure safety valves
Emergency depressuring and blowdown systems
Gas and fire detection systems
Emergency shutdown systems
Safety instrumented systems including combustion safeguards
Firefighting equipment; sprinkler, deluge and fire suppression systems
Emergency warning systems; other emergency communications equipment
Structures, havens and refuges for protection of personnel in an emergency
Any other equipment identified as safety-critical in HIRA studies or by other means

3 Asset Integrity

While all equipment requires appropriate inspection, testing and preventive maintenance (ITPM), a criticality rating may be used to determine the frequency of ITPM tasks and/or as a guide to which work orders require priority attention. Early attempts by some companies to meet the UK offshore regulations (ref. 2) resulted in a very large number of safety-critical elements, representing a high percentage of the master equipment list (MEL) for a specific installation. In several instances, this reached 80% of the MEL, as practically all elements of the primary containment system were captured, since any loss-of-containment of flammable hydrocarbons under pressure could result in a major incident. When such a majority of the MEL is deemed safety critical, it defeats the original philosophy of focusing on a sub-set of equipment that can be managed effectively to ensure its reliability to function on demand in the event of circumstances that could lead to a major incident.

Most companies try to level-load their maintenance programs when planning ITPM tasks in order to make the most efficient use of their resources. However, from time to time breakdowns and failures inevitably result in a peak workload that exceeds the available resources. In this unforeseen situation, it may be impossible to perform the breakdown repairs **and** all of the planned ITPM work orders on their scheduled dates. Some of the work orders will need to be deferred, and when 80% of the MEL is safety critical, this could include deferment of critical equipment maintenance, affecting its reliability.

The definition of SCE above is clearly consequence-based, which tends to be less suitable for asset integrity management (AIM) purposes, as it captures virtually all elements of the primary containment system. In reality, pressure vessels and piping systems are normally designed to recognized and generally accepted good engineering practices (RAGAGEP), and are unlikely to fail unless subject to significant corrosion or external damage (e.g. fire/explosion, dropped load).

Providing that a particular vessel or piping system has been regularly inspected and no degradation identified, it should not fail, i.e. it still has a corrosion allowance. If the design of the asset employed inherently safer design (ISD) and the provision of safety measures, the vessel or piping system should also be protected from external damage resulting in a major incident. Modern design practices include studies addressing the vulnerability (and hence survivability) of critical equipment and systems. Safety measures may include, but are not limited to, separation, blast wall, passive/active fire protection, ESD system, no heavy lift over live equipment, etc. Under these circumstances, there is **not** a *reasonable potential* of failure, and a risk-based definition for SCE is more appropriate and useful for the purposes of AIM.

Safety Critical Equipment ***(risk-based definition)***

Equipment, the malfunction or failure of which is likely to cause or contribute to a major incident, or the purpose of which is to prevent a major incident or mitigate its effects.

When situations occur that may not allow all ITPM tasks to be performed on time and according to plan, this risk-based definition allows equipment that is truly safety critical to be prioritized and their tasks accomplished to plan, while only deferring tasks on non-critical equipment. Nevertheless, any deferred task needs to be completed, but deferment by a few weeks should not significantly increase risk. Modern maintenance management systems (MMS), like Maximo and SAP, accommodate additional fields that allow SCE to be flagged in order to facilitate planning ITPM tasks.

A register of SCE should be compiled, and periodically updated. Non-critical equipment may degrade during its life and have an increased risk of failure, e.g. a corroded pressure vessel may require more frequent inspection. Under these conditions, the vessel may be elevated to the SCE category.

Ideally, the SCE in jurisdictions without specific regulations should represent up to around 20% of the equipment within the MEL. This recognizes that some critical items carry a disproportionate measure of risk, as shown in Figure 1.

The 20% number is not mandated, but significant increases in this percentage progressively limit the focus on those items of equipment that must receive preferential treatment. An AIM program built on this approach allows decisions on the deployment of available ITPM resources to be taken without significantly impacting major incident risk.

For those jurisdictions that mandate the consequence-based definition of SCE, the most expedient approach is to level-load the ITPM tasks on SCE within the MMS. Adequate resources should then be allocated to perform these tasks. Additional resources should be provided to perform breakdown repairs and ITPM tasks on non-SCE. If necessary, during periods of unforeseen high workload, tasks on the non-SCE may be rescheduled to a future date when resources become available.

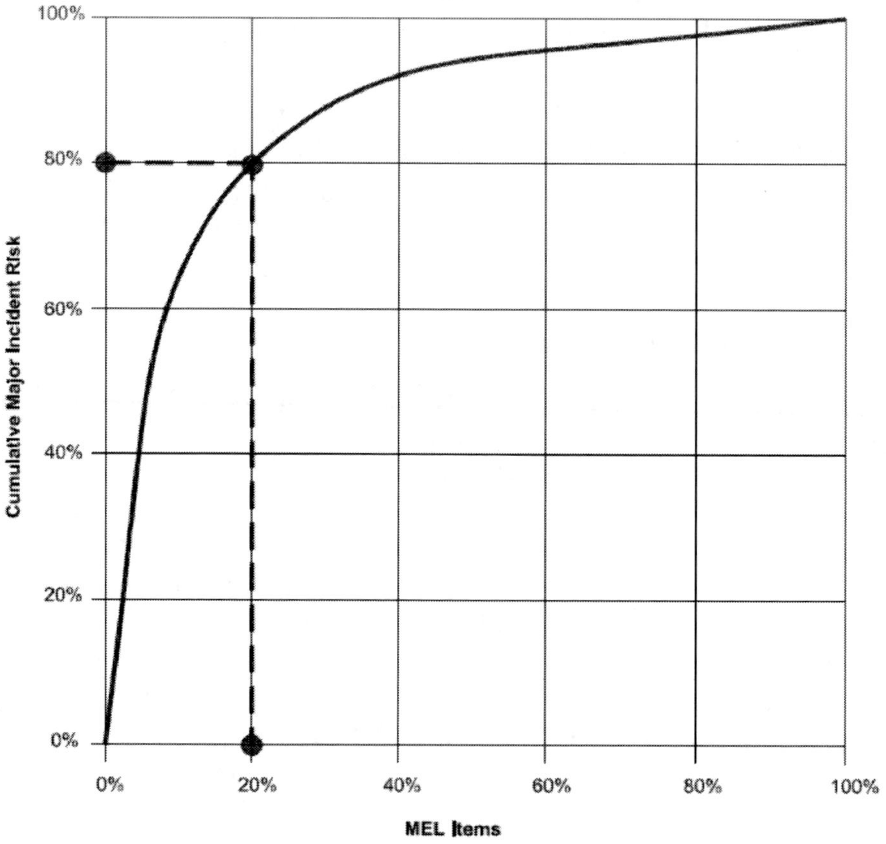

Figure 1. Major Incident Risk vs. Items on the MEL

4 Identification of SCE

Some SCE can be identified from the safeguards recorded in hazard identification and risk assessment (HIRA) studies. However, some of the safeguards in HIRA studies may be related to minor incidents or operability issues, and not major incidents. Other SCE may be identified from project ISD work and the associated design hazard management process for new facilities. Various technical studies may also identify equipment that should be categorized as SCE. For example, the following studies may also identify SCE: safety instrumented system (SIS) assessment; fire & gas detection analysis; relief, blowdown and flare study; fire water analysis; and evacuation, escape and rescue analysis.

Figure 2 is a logic tree that may be used as a comprehensive tool to identify SCE that meet the risk-based definition. Each individual item of equipment and each system should be analyzed using the logic tree. Multiple items of equipment or systems with similar functions may be analyzed as a group.

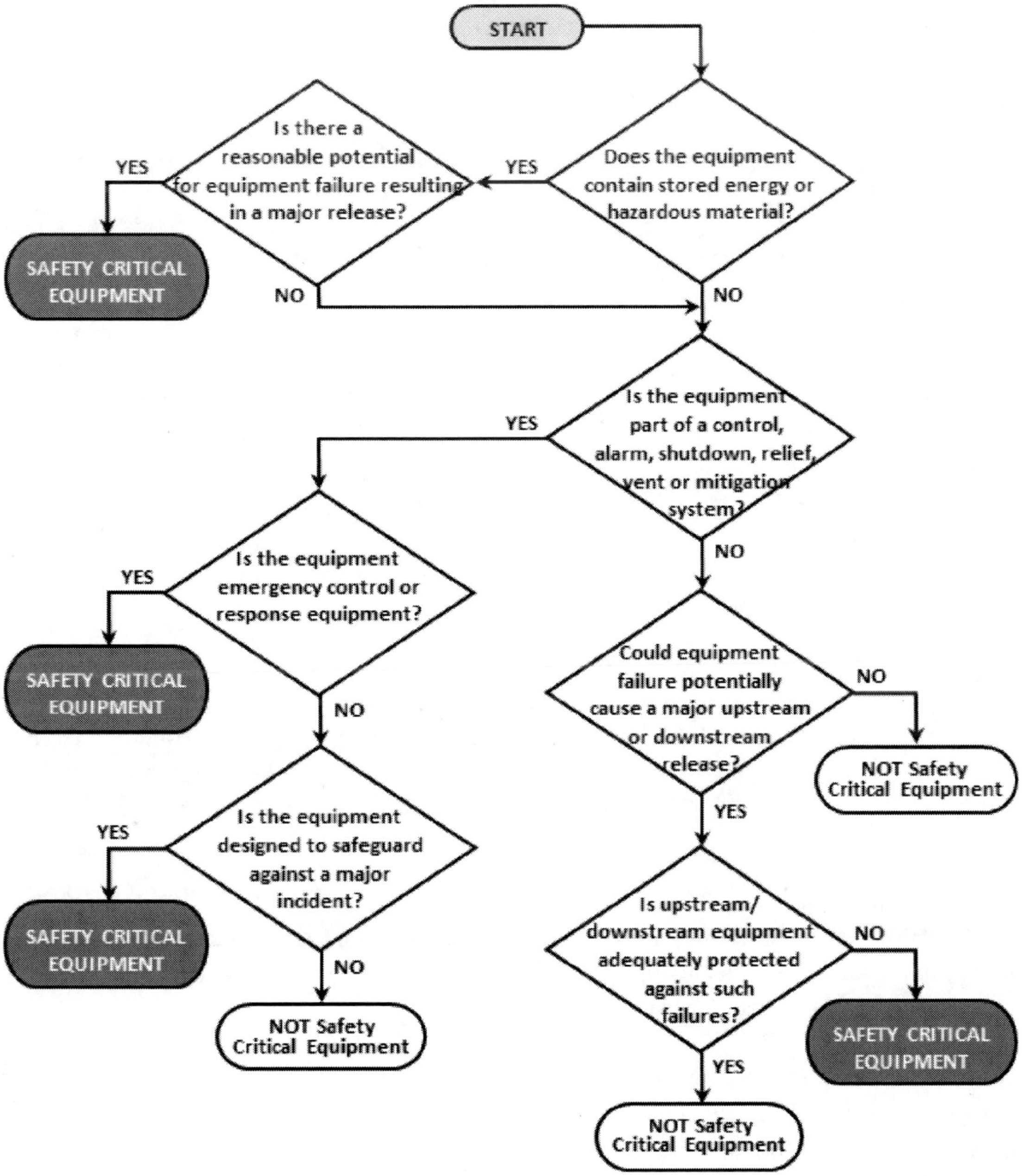

Figure 2. Logic Tree to Identify Safety Critical Equipment

4.1 Definition of Terms

Several of the terms within the logic tree (Fig. 2) should be carefully defined by the user to ensure a consistent approach. These terms include: stored energy; hazardous material; reasonable potential; major release; major incident; and adequately protected. The term 'reasonable potential' has already been discussed above.

4.1.1 Stored Energy

Stored energy may be defined as a vapor or gas above a certain high pressure with sufficient volume to cause severe injury/fatality if containment were suddenly lost. However, care should be exercised, since depending upon the process, thermal, kinetic, hydraulic, electrical and/or chemical energy may also be stored and have the potential for a major incident.

4.1.2 Hazardous Material and Major Release

The terms 'hazardous material' and 'major release' are defined by some jurisdictions, e.g. OSHA PSM (ref. 3), EPA RMP (ref. 4), and DOT (ref. 5) in the USA, but may not cover all hazards or specific chemicals (e.g. nitrogen) that may be relevant and are addressed in HIRA studies. The user will need to define all hazardous materials and their corresponding size of release that may cause a major incident.

4.1.3 Adequately Protected

Providing that equipment has layers of protection that reduce risks to a tolerable level, it is 'adequately protected'. For example, in the case of pressure, if equipment ABC fails and results in high pressure downstream, but downstream vessels and piping are protected by a properly sized pressure safety valve (PSV) or safety instrumented system (SIS), the vessels and piping are adequately protected. The PSV or SIS would be deemed SCE, but equipment ABC that failed is not classified as SCE.

4.1.4 Major Incident

The term 'major incident' has been defined by some jurisdictions. For example, the United Kingdom defines a major accident (i.e. major incident) as:

> *"...an occurrence (including in particular, a major emission, fire or explosion) resulting from uncontrolled developments in the course of the operation of any establishment and leading to serious danger to human health or the environment, immediate or delayed, inside or outside the establishment, and involving one or more dangerous substances"*
> (ref. 6).

CCPS's guidance on process safety metrics (ref. 7) provides an alternative approach. Table 2 below is an extract from Table 2 in CCPS's booklet *Process Safety Leading and Lagging Metrics* (ref. 7). Severity level 1 (i.e. multiple on-site fatalities, etc.) certainly meets the definition of major incident, but the user may elect to define it as severity level 2 (a single on-site fatality, multiple injuries, etc.).

Table 2. Process Safety Incidents & Severity Categories (ref. 7)

Severity Level	Safety/Human Health	Fire or Explosion (including overpressure)	Potential chemical impact	Community/environment impact
NA	Does not meet or exceed Level 4 threshold	Does not meet or exceed Level 4 threshold	Does not meet or exceed Level 4 threshold	Does not meet or exceed Level 4 threshold
4	OSHA recordable injury to employee or contractors (or equivalent*) associated with a process safety event	Resulting in \$25,000 to \$100,000 of direct cost	Chemical released within secondary containment, to a control device (e.g., flare or scrubber), or contained within the unit	Short -term remediation to address acute environmental impact. No long term cost or company oversight. Examples would include spill cleanup, soil and vegetation removal.
3	Lost time injury to employee or contractors associated with a process safety event	Resulting in \$100K-1M of direct cost	Chemical release outside of containment but retained on company property OR flammable release without potential for vapor cloud explosives	Minor off-site impact with precautionary shelter-in-place OR Environmental remediation required with cost less than \$1MM. No other regulatory oversight required. OR Local media coverage
2	On-site fatality - employee or contractors associated with a process safety event; multiple lost time injuries or one or more serious offsite injuries associated with a process safety event	Resulting in \$1M-10M of direct cost	Chemical release with potential for injury off site or flammable release resulting in a vapor cloud entering a building or potential explosion site (congested/confined area) with potential for damage or casualties if ignited	Substantiated shelter-in-place. OR Environmental remediation required and cost in between \$1MM - 2.5 MM. State government investigation and oversight of process. OR Regional media coverage.
1	Off-site fatality or multiple on-site fatalities associated with a process safety event	Resulting in direct cost >\$10M	Chemical release with potential for significant on-site or off-site injuries or fatalities	Significant community evacuation OR national media coverage OR Environmental remediation required and cost in excess of \$2.5 MM. Federal government investigation and oversight of process. OR other significant community impact

It should be noted that the CCPS guidance (ref. 7) and the corresponding API recommended practice (ref. 8) may be updated in 2016.

5 Case Study

Sometimes, the designation of SCE has been applied to equipment inappropriately, resulting in a lack of prioritization for competing maintenance and inspection activities. Conversely, major process safety events have occurred when equipment isn't classified as SCE as it should be. This is explored more in the following example.

For equipment to be classified as SCE, it must first be identified in the AIM program. However, one category that is often excluded is small bore piping (piping pipe components that are less than or equal to a nominal pipe size (NPS) of 2 inches). Small bore piping (SBP) typically includes bypass lines, sampling systems, PSV bleed lines, utilities (air, nitrogen), or auxiliary systems (lube or seal oil, flush or purge lines). It is understandable why much SBP is excluded as it competes with much larger and more obvious process piping, vessels, equipment, and components. However, in a recent American Fuel & Petrochemical Manufacturers (AFPM) review of process safety events, 22% of reported safety events originated from SBP (ref. 9).

In the acetylene production and packaging (cylinder) business, where the majority of the facilities do not fall under a process safety regulation, 95% of all piping is 2 inches or less, and 100% of the high pressure piping is less than 2 inches. The decision to include SBP that is classified as primary process piping is relatively obvious; however, it becomes much more difficult for very small diameter lines and tubing that are not classified as primary process piping. Examples of these include compressor bypass and unloading piping, vent lines, sample lines, reclaim lines, and condensate drain collection vessels and piping systems.

In areas where published standards are unclear or non-existent, one method to assist in the decision process is to use a simple risk review, such as the following logic tree (Fig. 3). By using Figure 3, one can assess if the failure of a SBP component would create a nuisance or reliability issue requiring maintenance according to original equipment manufacturer (OEM) recommendations, or if its failure could create a hazardous condition and should therefore be included in a more rigorous AIM program.

If included in the AIM program, the evaluation continues through the SCE logic tree (Fig. 2) to further determine if it should be classified as SCE.

An example of this process is a typical acetylene cylinder filling facility located in a 50ft by 75ft room with a 10ft ceiling. The filling process equipment is relatively simple and is comprised of a compressor, manifold, and about 40 cylinders. Considering that the process does not have emergency ventilation and shutdown/venting systems, and assuming a ¾-inch diameter manifold drain valve (i.e. a non-primary process piping item) fails/ruptures; the released acetylene would create localized pockets of flammable mixtures almost immediately and fill the entire room to the lower explosive level (LEL) in under 4 minutes.

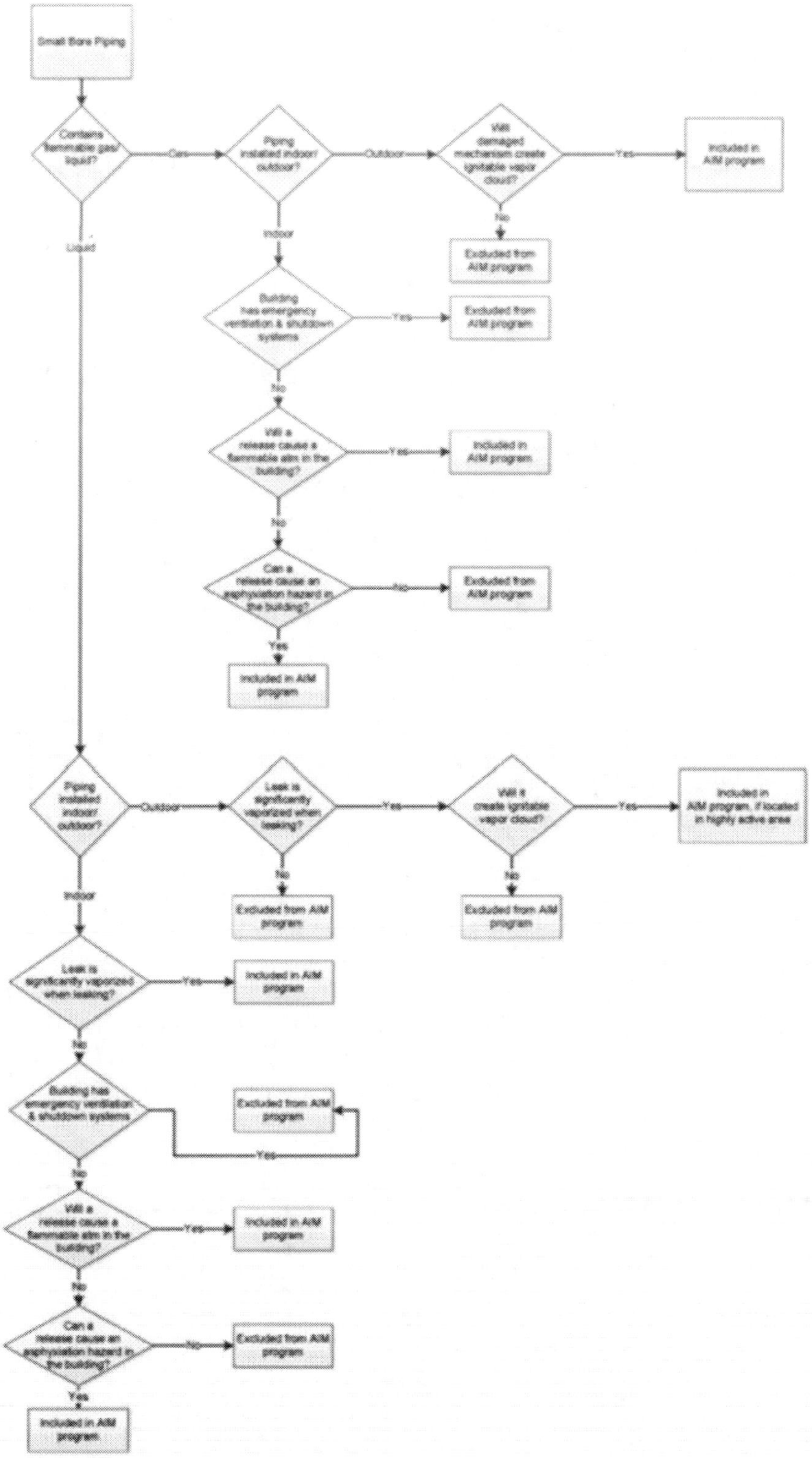

Figure 3. Logic Tree to Identify Small Bore Piping for AIM Program

So, should the drain valve be included in the AIM program and treated as SCE? Probably, but other decision factors should include whether the addition of ventilation, emergency isolation, and/or shutdown systems provide improved safety and risk reduction, and whether these additions provide wider benefits (i.e. safety, risk reduction, cost savings) to other systems as well. If the process wasn't confined in an enclosed building, a hazardous condition may not occur and the drain valve may not be considered SCE. Therefore the analysis should consider all aspects of the process and its immediate surroundings before determining which elements are SCE.

6 Conclusions

The application of criticality analysis for process safety purposes can assist with identification of equipment that requires a high reliability in order to prevent or mitigate major incidents. The management of these SCE has challenged some companies due to the large number of items that require frequent ITPM tasks to deliver that reliability. This has sometimes culminated in a failure to perform some of their ITPM tasks when due.

The use of a risk-based approach in jurisdictions without specific regulation can provide a smaller number of SCE that are truly critical to significantly reducing the risk of a major incident. A risk-based logic tree can support HIRA and ISD studies in identifying SCE.

At times of peak workload when decisions are necessary on the deployment of limited maintenance resources, the smaller number of SCE can be easily prioritized for immediate attention and their ITPM tasks performed on schedule. Some non-SCE tasks may be deferred and performed at a future date, when resources are available, without significantly increasing the risk of a major incident.

Most process plants have multiple layers of protection that represent redundant barriers (e.g. SCE) to a major incident occurring. Most of the time one or more of the barriers function as intended, and the major incident does not occur. However, a strong operating discipline is required to maintain these barriers throughout the lifecycle of the plant.

So, do you feel lucky? Or would you rather make sure by identifying and managing the SCE?

7 References

[1] J. Reason, "Human Error," Cambridge University Press, 1990.

[2] HM Government, "The Offshore Installations (Safety Case) Regulations 2005," SI 2005/ 3117.

[3] OSHA, "Process Safety Management of Highly Hazardous Chemicals," 29 CFR 1910.119, U.S. Dept. of Labor, Occupational Safety & Health Administration, Washington D.C.

[4] EPA, "Risk Management Plan (RMP) Rule," 40 CFR Part 68, U.S. Environmental Protection Agency, Washington D.C.

[5] DOT, "Hazardous Materials Regulations," 49 CFR Parts 100-185, U.S. Department of Transportation, Pipeline and Hazardous Materials Safety Administration, Washington D.C.

[6] HM Government, "Control of Major Accident Hazards Regulations 1999" (amended 2005 and 2015), UK.

[7] CCPS, "Process Safety Leading and Lagging Metrics: You Don't Improve What You Don't Measure," Center for Chemical Process Safety, American Institute of Chemical Engineers, New York, NY, 2007 (updated 2010).

[8] API, "Process Safety Performance Indicators for the Refining and Petrochemical Industries, 1st edition," ANSI/API RP 754, American Petroleum Institute, Washington D.C., 2010.

[9] AFPM, "AFPM 2013 Learnings Report, Lessons from the Event Sharing Database," American Fuel & Petrochemical Manufacturers, Washington DC, 2014.

8 Acronyms

AFPM	American Fuel and Petrochemical Manufacturers
AIM	Asset Integrity Management
API	American Petroleum Institute
CCPS	Center for Chemical Process Safety
DOT	Department of Transportation
EPA	Environmental Protection Agency
ESD	Emergency Shutdown
FMECA	Failure Modes, Effects and Criticality Analysis
HIRA	Hazard Identification and Risk Analysis
ISD	Inherently Safer Design
ITPM	Inspection, Testing and Preventive Maintenance
LEL	Lower Explosive Limit
MEL	Master Equipment List
MMS	Maintenance Management System
NPS	Nominal Pipe Size

OEM	Original Equipment Manufacturer
OSHA	Occupational Safety and Health Administration
PSM	Process Safety Management
PSV	Pressure Safety Valve
RAGAGEP	Recognized and Generally Accepted Good Engineering Practice
RMP	Risk Management Plan
SBP	Small Bore Piping
SCE	Safety Critical Equipment / Element
SIS	Safety Instrumented System

12TH GLOBAL CONGRESS ON PROCESS SAFETY

Using Past Incidents as Leading Indicators for Pipeline Integrity Management

Hao Chen
DNV GL
1400 Ravello Drive, Katy, TX 77449
hao.chen@dnvgl.com

Cynthia Spitzenberger, Tone Rice, and Deal Lindsay
DNV GL
1400 Ravello Drive, Katy, TX, 77449
Cynthia.Spitzenberger@dnvgl.com, Lindsay.Deal@dnvgl.com, Tone.Rice@dnvgl.com

Prepared for Presentation at
American Institute of Chemical Engineers
2016 Spring Meeting
12th Global Congress on Process Safety
Houston, Texas
April 11-23, 2016

Using Past Incidents as Leading Indicators for Pipeline Integrity Management

Hao Chen
DNV GL
1400 Ravello Drive, Katy, TX 77449
hao.chen@dnvgl.com

Cynthia Spitzenberger, Tone Rice, and Deal Lindsay
DNV GL

Keywords: Pipeline, Pipeline Incident, Pipeline Failure, Leading Indicator, Integrity Management

Abstract

The U.S. Department of Transportation Pipeline and Hazardous Materials Safety Administration (PHMSA) data indicate about 11,000 pipeline incidents have occurred in the U.S. in the past 20 years, causing about 400 fatalities, 1400 injuries, and over $6.3 billion in property damages. Increasing energy demand urges the expansion of pipeline networks and their flexibility and capability to deliver various products. The concerns to satisfy the increasing energy demand together with increasingly stringent regulatory requirements impose significant pressure on pipeline operators to maintain safe operation. How to allocate limited resources to the sections most susceptible to failure still remains a big challenge to most operators. One feasible preventive measure is to learn from the past pipeline failures, in which failure data analysis is essential. From effective data analysis, useful failure information such as the pipeline materials, the involved failed parts, failure types, the most frequent failure causes, and the incident location environment, can be obtained, which can help increase the effectiveness of pipeline integrity management program by optimizing the allocation of limited resources to the "weak points" most likely to fail.

Although Europe-based pipeline incident databases such as European Gas pipeline Incident data Group (EGIG) and Conservation of Clean Air and Water in Europe CONCAWE have been evaluated [1],[2], it is beneficial to understand the U.S. pipeline failure data which more accurately reflects the industry situation in the country, especially for U.S. pipeline operators. Moreover, if some similarities between the pipeline failure patterns in U.S. and in Europe can be found, the experience of effective pipeline integrity management could be shared to enhance the safety performance of the pipeline industry in both regions. An extensive public pipeline incident database is accessible from PHMSA, but it only includes raw incident data and requires a systematic analysis to generate statistically meaningful results. In this work, the raw PHMSA data (particularly liquid pipeline data) are categorized and normalized across different periods, since the reported data format varies drastically over time. Then the normalized data are systematically analyzed to reveal the failure trends and causes. In addition, the failure patterns obtained for U.S. pipelines are compared with European data to highlight the similarities and differences, aiming to provide useful information for effective decision making on pipeline integrity management.

1 Introduction

The International Energy Agency (IEA) reported the world-wide energy consumption in 2013 was about 9 million KTOE (kilo-tonne of oil equivalent), i.e., 360 quadrillion BTU [3]. According to U.S. Energy Information Administration (U.S. EIA), the total energy consumption for the U.S. in 2013 was about 97 quadrillion BTU and almost 2/3 consumed energy was from petroleum liquids and natural gas [4]. Given the fact that the supply and demand are rarely concentrated in the same region, reliable and efficient energy transportation is crucial to ensure satisfaction of the needs from transportation vehicles, building heating and industrial applications. Among various energy transportation approaches, pipeline is the primary means of moving petroleum and petroleum products to consumer markets in the U.S., with the help of over 190,000 miles of liquid pipelines and over 300,000 miles of natural gas pipelines [5].

Even though great efforts have been devoted to maintaining the integrity of the pipeline network, pipeline failures continue to occur, which have resulted in significant loss of life and property. The U.S. Department of Transportation Pipeline and Hazardous Materials Safety Administration (PHMSA) data indicate about 11,000 pipeline incidents have occurred in the past 20 years in the U.S., causing about 400 fatalities, 1400 injuries, and over $6.3 billion property damages [6]. Some recent devastating pipeline incidents in the world are as follows:

- 2013 Tioga pipeline rupture in North Dakota, U.S. caused a significant oil spill of 840,000 gallons, costing more than $20 million cost in long lasting cleanup activities [7].
- 2013 Qingdao pipeline explosion in China killed 62 people, injured 136 people, evacuated 18,000 people, and resulted in a direct economic loss of $125 million [8].
- 2014 Kaohsiung gas pipeline rupture and subsequent explosion in sewer tunnel in Taiwan killed 32 people and injured 321 people, with property damage of over $63 million [9].
- 2015 Santa Barbara pipeline in California, U.S. released 143,000 gallons of oil, introducing an environmental disaster to the local costal ecosystems and costing the operator about $257 million in response, cleanup, assessments and fines and legal settlements [10].

These incidents demonstrate that pipeline failures can have significant adverse impact on the people, environment and business operations. Meanwhile, with the growth of population and the need for higher quality of life, the energy demand is likely to keep increasing in the future, which urges the expansion of pipeline networks and increased flexibility and capability to deliver various products. Moreover, after the continuous occurrence of pipeline incidents, regulatory authorities tend to enforce increasingly stringent standards and laws on pipeline operations. All of these factors impose significant pressure on pipeline operators to maintain safe operations of their pipeline networks and establish effective pipeline integrity management programs. However, due to the limitation of time, capital investment and human resources, it is more practical and efficient to allocate the limited resources to improve the resistance of those elements which are most prone to failure.

One feasible way to identify the elements in the pipeline system susceptible to failure and improve the understanding of the failure pattern of the pipeline system is to learn from past failures/incidents by analyzing the failure data. From effective data analysis, useful failure information such as the pipeline materials, the involved failed parts, failure types, prevailing failure causes, and the incident location environment, can be obtained, which can help increase the effectiveness of a pipeline integrity management program by optimizing the allocation of limited resources to those "weak points" most likely to fail. European organizations such as the European Gas pipeline Incident data Group (EGIG) and Conservation of Clean Air and Water in Europe (CONCAWE) release periodic reports for gas pipeline incidents and oil pipeline performance respectively, in Europe [1],[2]. Although the released data are helpful, they are more reflective of the pipeline operation in Europe. Due to differences in working environment, the population along the pipeline and the enforced regulations, those data may deviate from the situation of U.S. pipeline operation. In order to better understand the failure pattern of U.S. pipelines, it is essential to analyze the failure data of those pipelines. An extensive publically available pipeline incident database is accessible from PHMSA [11], but it only includes raw incident data and requires a systematic analysis to generate statistically meaningful results.

In this work, the raw PHMSA data (particularly liquid pipeline data) are categorized and normalized across different periods, since the reported data format varies drastically over time. Then the normalized data are systematically analyzed to reveal the failure trends and causes. In addition, the failure patterns obtained for U.S. pipelines are compared with European data (CONCAWE data) to highlight the similarities and differences, aiming to provide useful information for effective decision making on pipeline integrity management.

2 Method

The annual incident raw data is obtained from the PHMSA website [11]. Over the years, the pipeline technology has changed greatly and so does the format of the accident report. Much more information is required for recent accident reports than that in early days. Take the liquid pipeline incident report as one example. The latest version of the accident report (rev 7-2014) requires 17 pages of information to be filled; while the accident report (rev 01-2001) applied in 2002-2009, only has 5 pages of information to be filled. For reports earlier than 2002, the required information was even less. Moreover, even for the common required information, different terminology was used in the report for different periods. In order to obtain the failure trend, the data was normalized according to the latest version of the accident report when the corresponding information was also available in previous versions of the report. A typical incident report and the raw database are shown in Figure 1 and Figure 2, respectively.

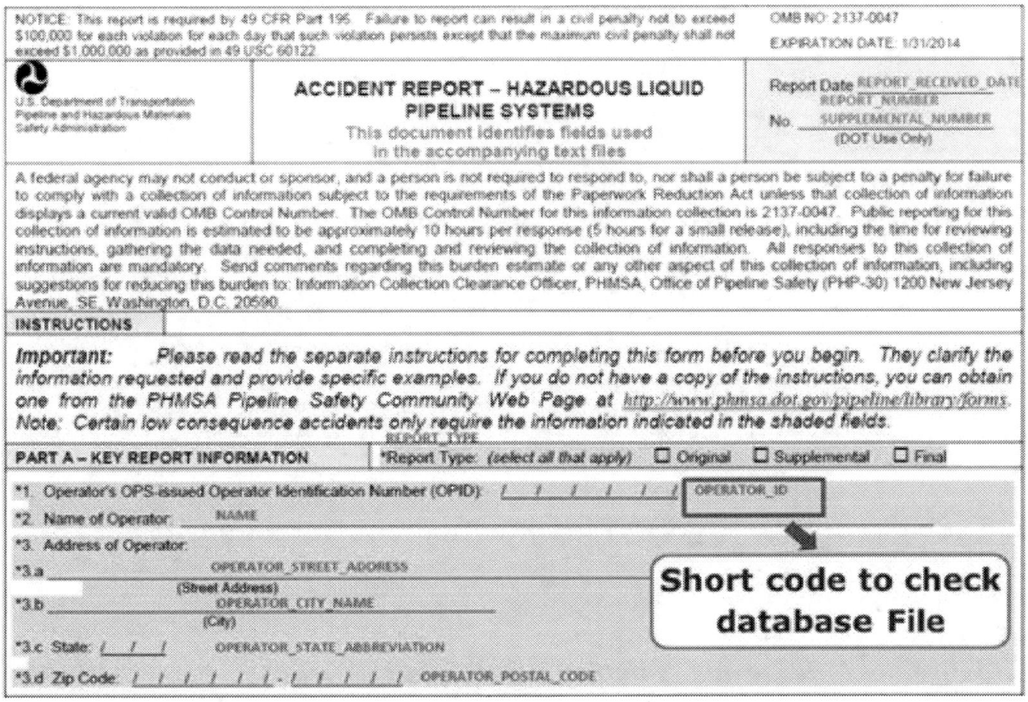
NOTICE: This report is required by 49 CFR Part 195. Failure to report can result in a civil penalty not to exceed $100,000 for each violation for each day that such violation persists except that the maximum civil penalty shall not exceed $1,000,000 as provided in 49 USC 60122.

OMB NO: 2137-0047

EXPIRATION DATE: 1/31/2014

U.S. Department of Transportation
Pipeline and Hazardous Materials
Safety Administration

ACCIDENT REPORT – HAZARDOUS LIQUID PIPELINE SYSTEMS

This document identifies fields used in the accompanying text files

Report Date REPORT_RECEIVED_DATE
REPORT_NUMBER
No. SUPPLEMENTAL_NUMBER
(DOT Use Only)

A federal agency may not conduct or sponsor, and a person is not required to respond to, nor shall a person be subject to a penalty for failure to comply with a collection of information subject to the requirements of the Paperwork Reduction Act unless that collection of information displays a current valid OMB Control Number. The OMB Control Number for this information collection is 2137-0047. Public reporting for this collection of information is estimated to be approximately 10 hours per response (5 hours for a small release), including the time for reviewing instructions, gathering the data needed, and completing and reviewing the collection of information. All responses to this collection of information are mandatory. Send comments regarding this burden estimate or any other aspect of this collection of information, including suggestions for reducing this burden to: Information Collection Clearance Officer, PHMSA, Office of Pipeline Safety (PHP-30) 1200 New Jersey Avenue, SE, Washington, D.C. 20590.

INSTRUCTIONS

Important: *Please read the separate instructions for completing this form before you begin. They clarify the information requested and provide specific examples. If you do not have a copy of the instructions, you can obtain one from the PHMSA Pipeline Safety Community Web Page at http://www.phmsa.dot.gov/pipeline/library/forms.*
Note: Certain low consequence accidents only require the information indicated in the shaded fields.

PART A – KEY REPORT INFORMATION

REPORT_TYPE
*Report Type: *(select all that apply)* ☐ Original ☐ Supplemental ☐ Final

*1. Operator's OPS-issued Operator Identification Number (OPID): /__/__/__/__/__/ OPERATOR_ID
*2. Name of Operator: NAME
*3. Address of Operator:
*3.a OPERATOR_STREET_ADDRESS (Street Address)
*3.b OPERATOR_CITY_NAME (City)
*3.c State: /__/__/ OPERATOR_STATE_ABBREVIATION
*3.d Zip Code: /__/__/__/__/__/-/__/__/__/__/ OPERATOR_POSTAL_CODE

Figure 1. Example of PHMSA Liquid Pipeline Accident Report [11].

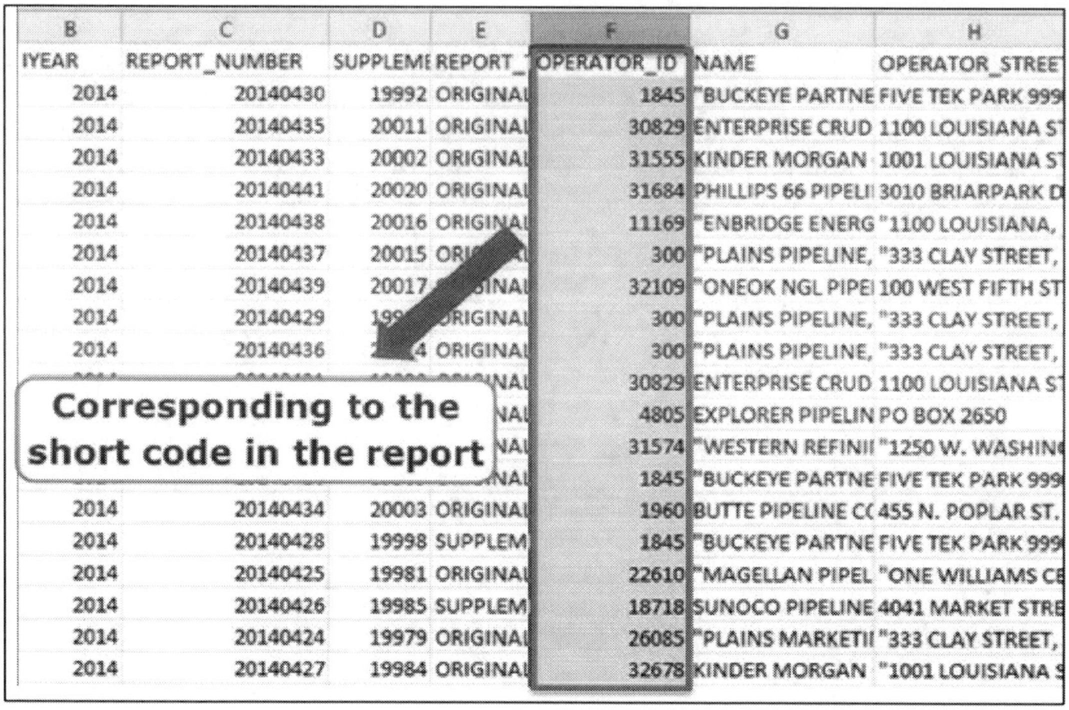

B	C	D	E	F	G	H
IYEAR	REPORT_NUMBER	SUPPLEME	REPORT_	OPERATOR_ID	NAME	OPERATOR_STREE
2014	20140430	19992	ORIGINAL	1845	"BUCKEYE PARTNE	FIVE TEK PARK 999
2014	20140435	20011	ORIGINAL	30829	ENTERPRISE CRUD	1100 LOUISIANA S
2014	20140433	20002	ORIGINAL	31555	KINDER MORGAN	1001 LOUISIANA S
2014	20140441	20020	ORIGINAL	31684	PHILLIPS 66 PIPELI	3010 BRIARPARK D
2014	20140438	20016	ORIGINAL	11169	"ENBRIDGE ENERG	"1100 LOUISIANA,
2014	20140437	20015	[illegible]	300	"PLAINS PIPELINE,	"333 CLAY STREET,
2014	20140439	20017	[illegible]	32109	"ONEOK NGL PIPE	100 WEST FIFTH ST
2014	20140429	[illegible]	[illegible]	300	"PLAINS PIPELINE,	"333 CLAY STREET,
2014	20140436	[illegible]	ORIGINAL	300	"PLAINS PIPELINE,	"333 CLAY STREET,
[illegible]	[illegible]	[illegible]	[illegible]	30829	ENTERPRISE CRUD	1100 LOUISIANA S
[illegible]	[illegible]	[illegible]	[illegible]	4805	EXPLORER PIPELIN	PO BOX 2650
[illegible]	[illegible]	[illegible]	[illegible]	31574	"WESTERN REFINII	"1250 W. WASHIN
[illegible]	[illegible]	[illegible]	[illegible]	1845	"BUCKEYE PARTNE	FIVE TEK PARK 999
2014	20140434	20003	ORIGINAL	1960	BUTTE PIPELINE C	455 N. POPLAR ST.
2014	20140428	19998	SUPPLEM	1845	"BUCKEYE PARTNE	FIVE TEK PARK 999
2014	20140425	19981	ORIGINAL	22610	"MAGELLAN PIPEL	"ONE WILLIAMS C
2014	20140426	19985	SUPPLEM	18718	SUNOCO PIPELINE	4041 MARKET STR
2014	20140424	19979	ORIGINAL	26085	"PLAINS MARKETI	"333 CLAY STREET,
2014	20140427	19984	ORIGINAL	32678	KINDER MORGAN	"1001 LOUISIANA

Figure 2. Example of Raw PHMSA Pipeline Incident Database [11].

3 Results and Discussion

3.1 Liquid Pipeline Mileage and Number of Reported Failures

3.1.1 CONCAWE Data

The CONCAWE database collects the liquid pipeline inventory in Europe since 1971. Certain criteria are applied for the pipelines recorded in the CONCAWE database, as follows **Error! Reference source not found.**:

- Pipelines transporting crude oil or petroleum products
- With at least 2 km in the public domain
- Excludes undersea pipelines and offshore pipelines
- Includes pump stations, intermediate aboveground installations and intermediate storage facilities, but excludes the origin and destination terminal facilities and tank farms

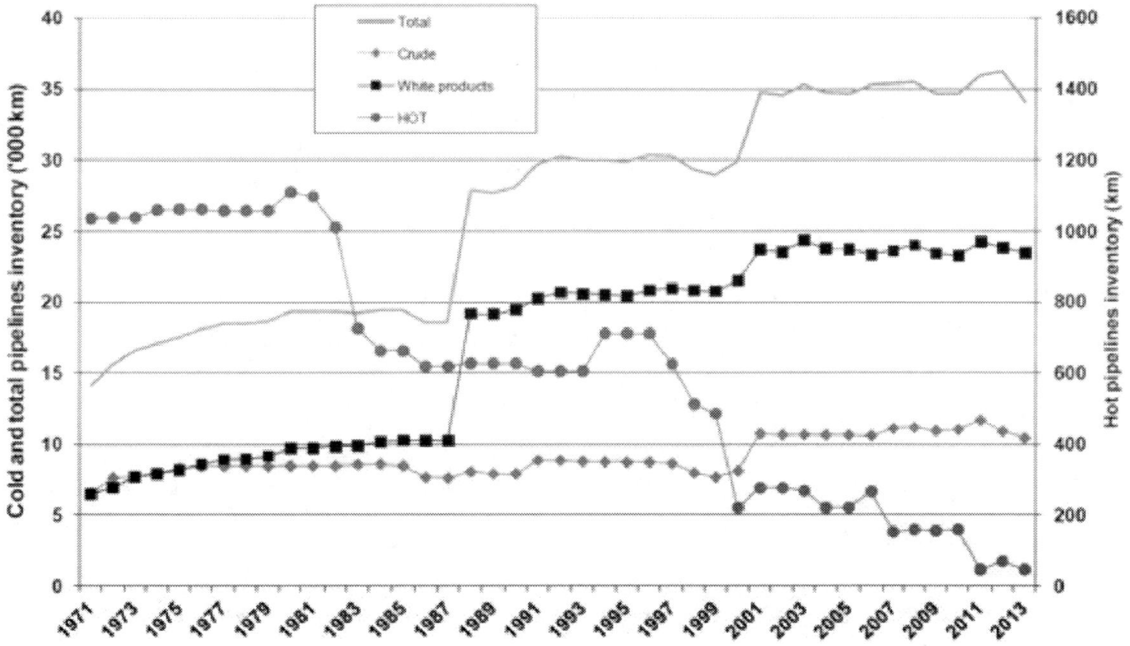

Figure 3. CONCAWE data: Liquid pipeline mileage changes over time, reproduced from [2].

In Figure 3, "Hot" indicates hot oil, i.e., fuel oil; while "White Products" indicates the products from petroleum, such as liquefied petroleum gas, gasoline, kerosene, kerosene, etc. From Figure 3, it can be seen that the liquid pipeline mileage/inventory in the CONCAWE database keeps increasing since 1971, which is mainly because of more operators becoming CONCAWE members and contributing more system data to the database, as well as the construction of new pipeline systems during the time period. Given the hot pipeline is only a very small portion of the whole pipeline system (<0.15% in 2013), the data for cold pipelines (crude and white products) will be used to compare with U.S. pipeline data.

The pipeline mileage together with the annual number of spills from liquid pipelines results in the pipeline spillage frequency. Figure 4 shows the CONCAWE liquid pipeline spillage frequency. Overall, it keeps decreasing, with an exception in 2013, due to the unusual high number of theft induced pipeline failures in that year. A better measure to use may be the 5-year moving average of spillage, which for 2013 is about 0.35E-03 per km per year.

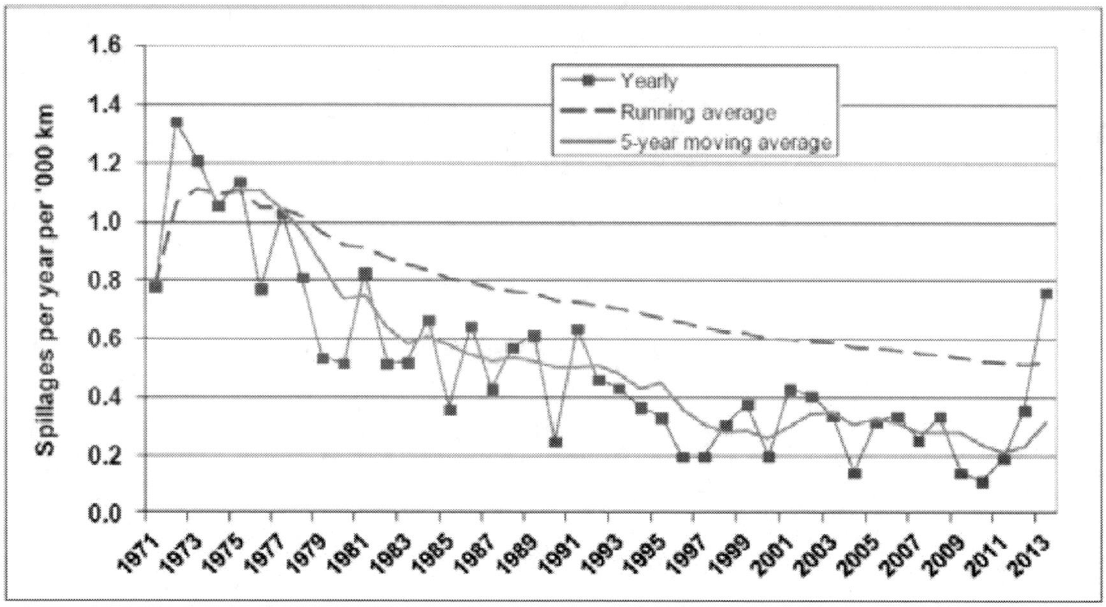

Figure 4. 43-year trend of spillage frequency for CONCAWE liquid pipeline, reproduced from [2].

3.1.2 PHMSA Data

U.S. PHMSA collects the liquid pipeline mileage data from 2004, as shown in Figure 5. The commodities transported by U.S. liquid pipeline are mainly crude and white products ("petroleum/refined products" and "high volatile liquids" were categorized into this type), with a small percentage (<3%) of CO_2, fuel grade ethanol and other types of goods. With more operators reporting pipeline data and the expansion of the existing pipeline network, the total pipeline mileage steadily increases over time. Although the collection of pipeline accident data dates back to 1986, the liquid pipeline spillage frequency can only be analyzed from 2004, due to the lack of pipeline mileage data before 2004.

Although PHMSA data cover both onshore and offshore pipelines and cover the pipelines transporting CO_2 and other types of commodities besides crude and white products, the overall U.S. pipeline failure frequency should be very close to the failure frequency for U.S. onshore crude and white products pipelines, due to the dominating potion of such pipelines in pipeline mileage (>97%) and reported incidents (>97%). The spillage frequency of U.S. liquid pipeline is shown in Figure 6.

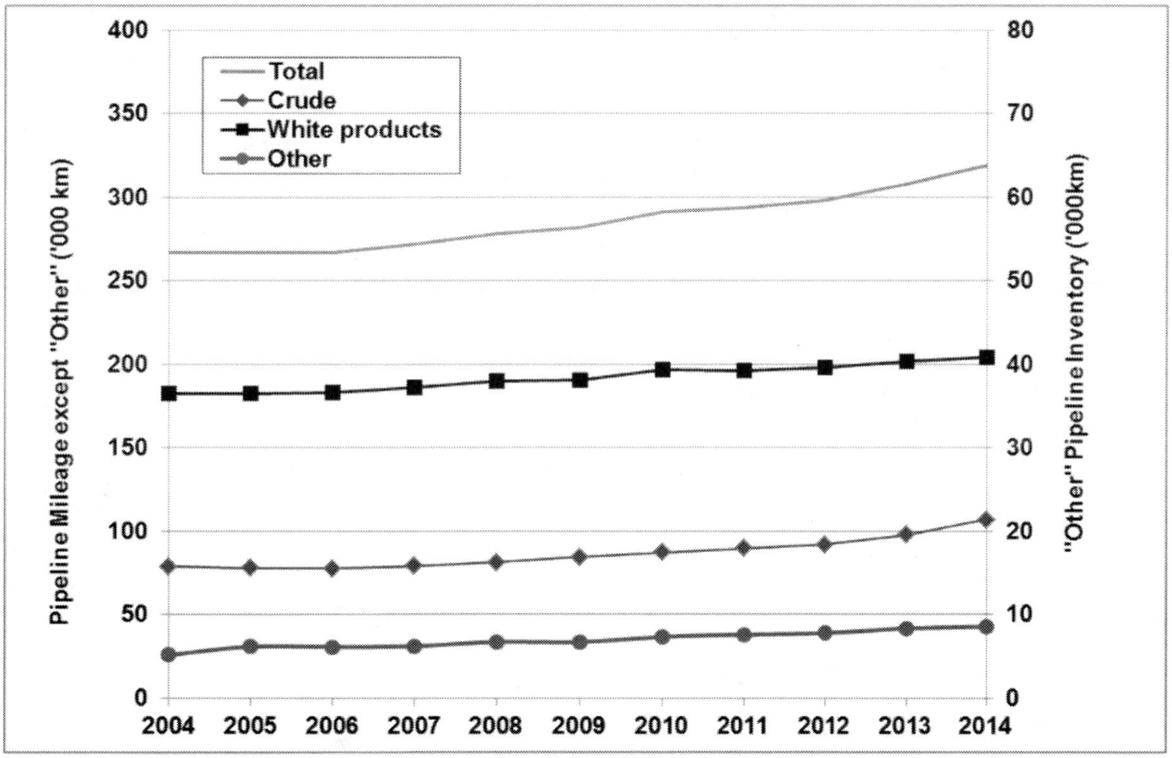

Figure 5. U.S. Liquid pipeline mileage data from 2004 to 2014, the raw data obtained from [12].

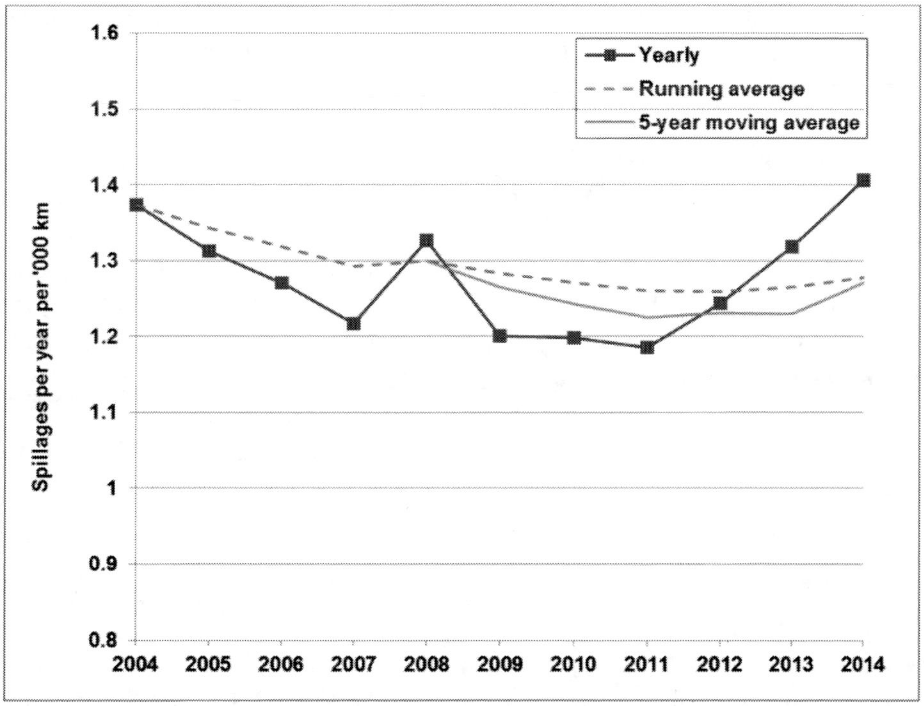

Figure 6. 10-year trend of spillage frequency for U.S. liquid pipelines, raw data obtained from [4], [12].

From Figure 6, there is no obvious trend for annual spillage frequency, but the overall running average of spillage frequency generally keeps decreasing. For 2013, the 5-year average spillage frequency is about 1.23E-03 per km per year; however, it is much higher than that of European pipelines. This may be because of the different criteria of incident reporting for PHMSA and CONCAWE. Unlike the criteria for CONCAWE mentioned above, a liquid pipeline spillage must be reported to PHMSA if the release results in any of the following [13]:

- Unintentional explosion or fire
- Release 5 gallons or more of hazardous liquid or CO_2, except the release is less than 5 barrels from a pipeline maintenance with certain restrictions
- Any fatality or injury necessitating hospitalization
- Estimated total property damage exceeding $50,000

Other factors contributing to the discrepancy between the two data sets may be different pipeline working environment, different human activities along the pipelines and different enforced regulations and standards.

3.1.3 Comparison of Pipeline Diameter and Age Distribution

Figure 7 compares the diameter (in inches) distribution of U.S. and European crude and products pipelines in 2013. The PHMSA data was grouped by applying the same category used in the CONCAWE report in order to make the comparison. There are some similarities between these two datasets. For both U.S. and European crude pipeline systems, a large portion (~30%) has a diameter of 16-24 inch; while for products pipelines, about 40% of U.S. pipelines and 50% of European pipelines are between 8 and 12 inch. However, the distribution is quite different. More than 55% of European crude pipelines are 24 inch or larger, with only about 10% pipeline smaller than 16 inch. For U.S. crude pipelines, only 20% are 24 inch or larger, while over 50% are smaller than 16 inch. From this data, it seems that Europe tends to use larger pipelines to transport crude than U.S. A similar trend is also observed for products pipelines. Much more small pipelines (<8 inch) are used in U.S. transporting products while more large pipelines (>16 inch) are used in Europe.

For age distribution, CONCAWE demonstrates the change in mileage for each age group presented by 10-year brackets (e.g., 1-10, 11-20 year old) for the whole pipeline system over the time. PHMSA also collects the mileage change for each age group by 10-year brackets, but the PHMSA data are in terms of installation year rather than the age of the pipeline as applied in CONCAWE dataset. The CONCAWE and PHMSA data are shown in Figure 8 and Figure 9.

As shown in Figure 8, European oil pipelines have been progressively ageing. In 1971, over 75% of the oil pipelines were 10 years old or less, but this ratio decreased to 5% in 2013, with about 60% over 40 years old. As shown in Figure 9, for U.S. oil pipelines, the old pipelines are being gradually replaced by new pipelines or being taken out of service, illustrated as the decreasing number of pipelines installed before 1950. In 2013, about 15% of U.S. oil pipelines were installed after 2000 (less than 13 years old), and 48%

were installed before 1970 (over 43 years old). Based on this data, U.S. oil pipelines are generally younger than European ones.

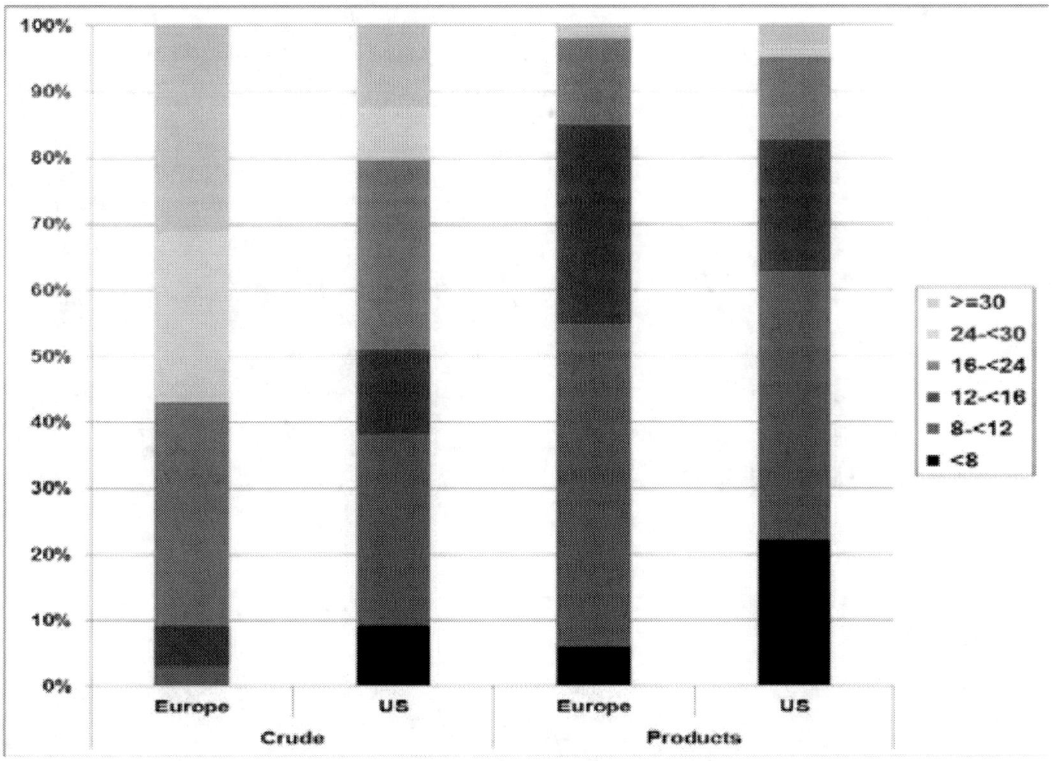

Figure 7. Comparison of U.S. and Europe crude and products pipeline diameter (in inches) distribution in 2013, raw data obtained from[2], [11].

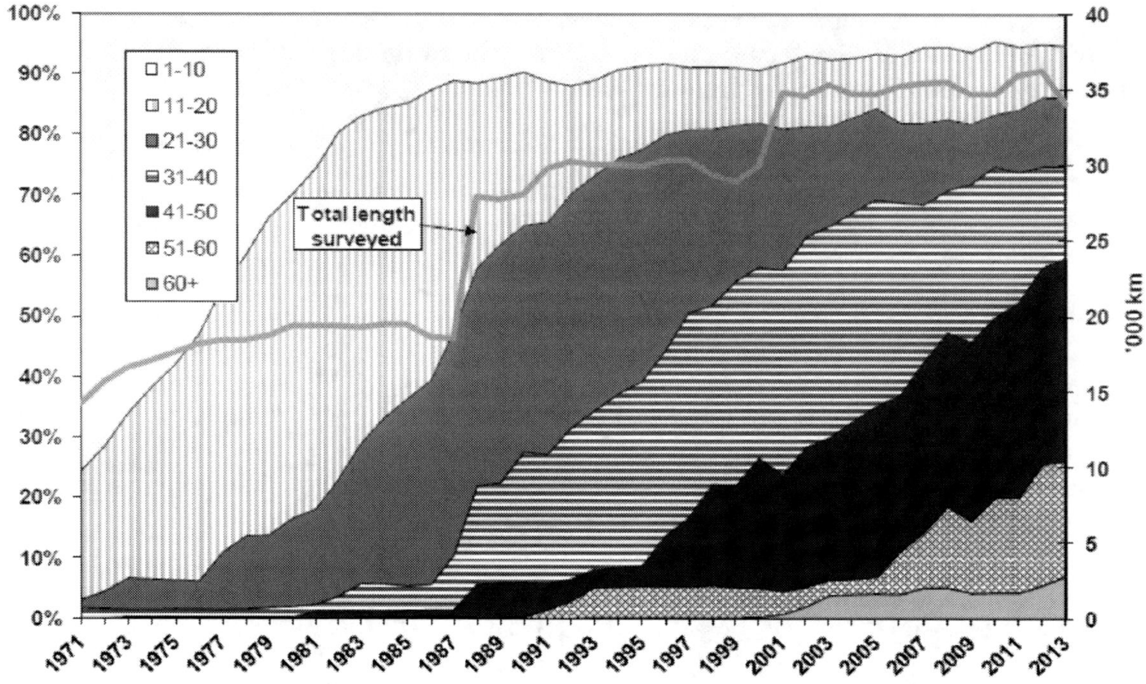

Figure 8. European oil pipeline historical age distribution (in years), reproduced from [2].

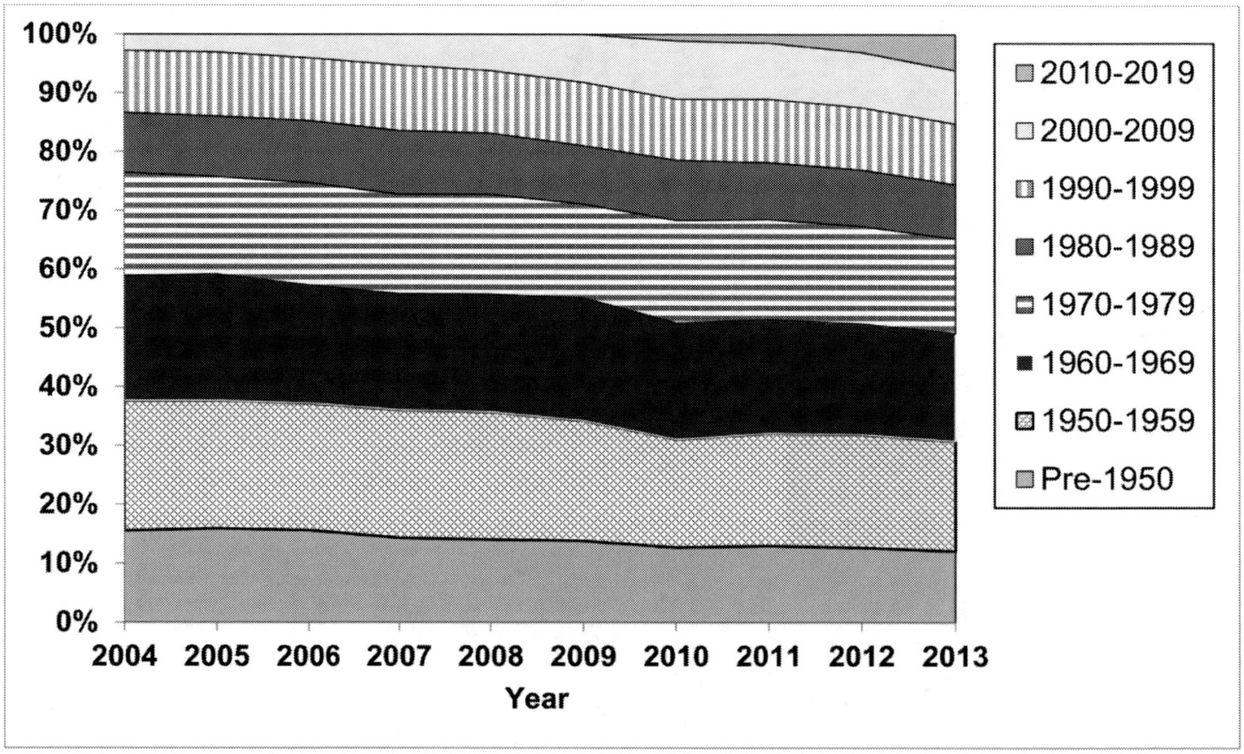

Figure 9. U.S. oil pipeline historical age distribution (in years), raw data obtained from [11],

3.2 Pipeline Failure Types and Causes

3.2.1 Failure Types

CONCAWE and PHMSA use different categorization method to present the types of pipeline accidents. For example, CONCAWE categorizes the types of pipeline accidents by hole size as "no hole, pinhole, fissure, hole, split, or rupture", while PHMSA categorizes the types as "mechanical puncture, leak, rupture, and overfill or overflow". In order to make a comparison, the types of accidents are normalized as "leak" or "rupture", by assuming "no hole, pinhole, fissure, hole and split" in the CONCAWE dataset as "leak" and assuming "mechanical puncture, leak and overfill/overflow" in the PHMSA dataset as "leak". The types of pipeline accident for PHMSA data and CONCAWE Data are shown in Figure 10 and Figure 11, respectively (Raw data from [14] and [2]). Comparing Figure 10 and Figure 11, it seems that European pipelines are more likely to experience rupture than U.S. pipelines (19% in CONCAWE data vs. 10% in PHMSA data). The pipeline configuration (e.g., diameter, thickness, materials), the age of pipeline and working conditions/environment may contribute to this discrepancy.

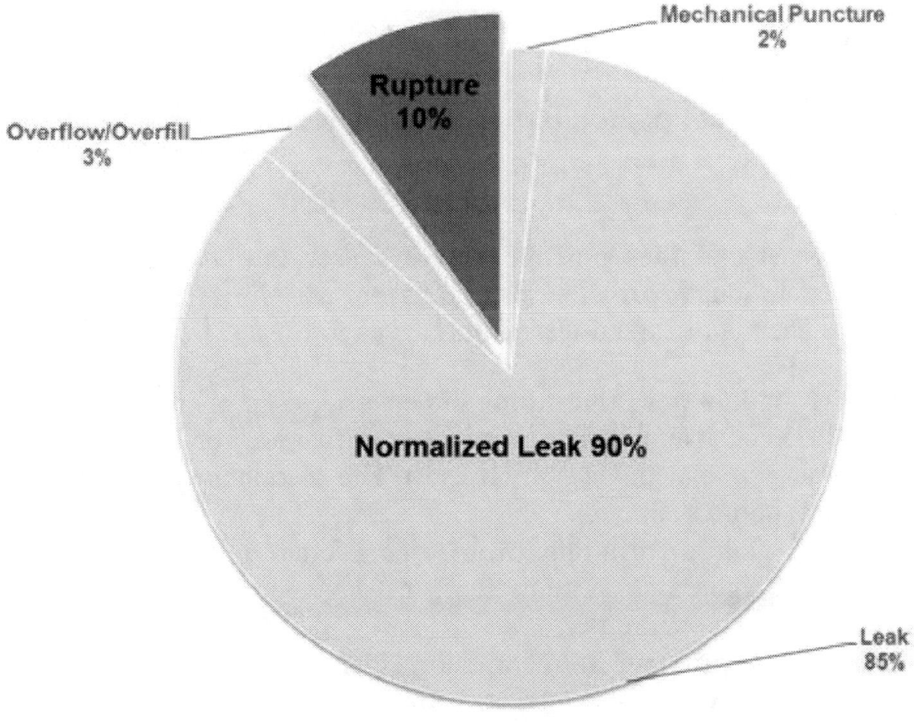

Figure 10. Type of pipeline accidents from PHMSA data 1968-2014

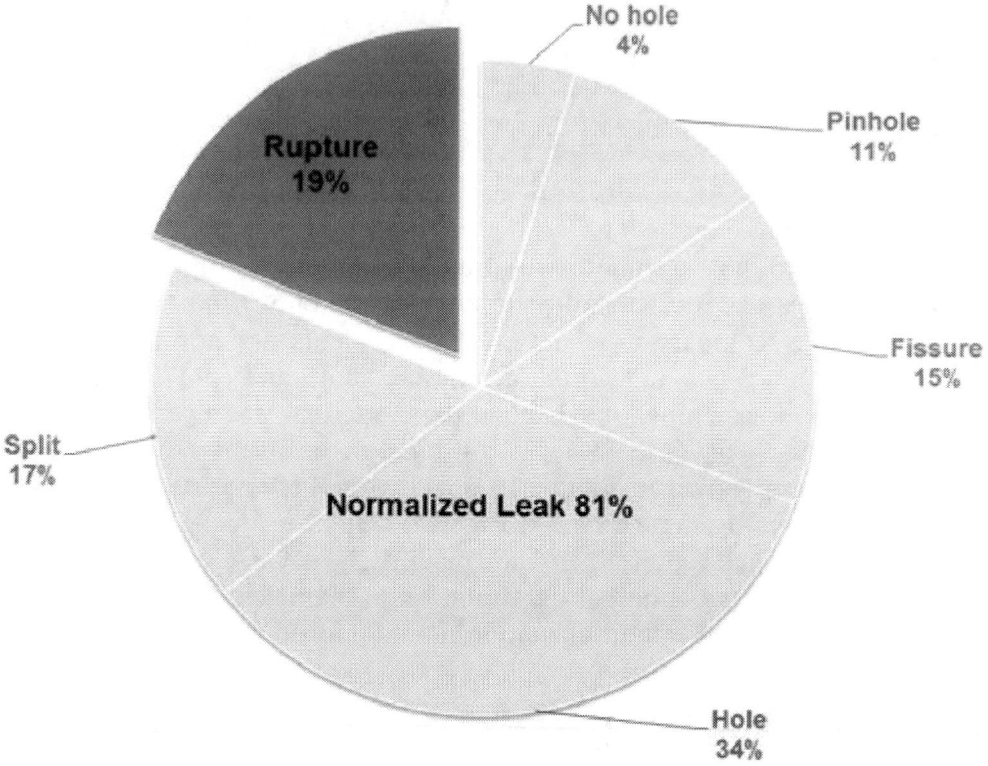

Figure 11. Type of pipeline accidents from CONCAWE data 1971-2013

3.2.2 Failure Causes

The CONCAWE dataset classifies pipeline failure causes into five major categories: mechanical failure, operational, corrosion, natural hazard and third party, each with their own sub-categories. The major categories are defined as follows [2].

- Mechanical: a failure resulting from either a design or material fault (e.g. metallurgical defect, inappropriate material specification) or a construction fault (e.g. defective weld, inadequate support, etc.). This also includes failure of sealing devices (gasket, pump seal, etc.).
- Operational: a failure resulting from operational upsets, malfunction or inadequacy of safeguarding systems (e.g. instrumentation, mechanical pressure relief system) or from operator errors.
- Corrosion: a failure resulting from corrosion either internal or external of either a pipeline or a fitting. A separate category is foreseen for stress corrosion cracking.
- Natural hazard: a failure resulting from a natural occurrence such as land movement, flooding, lightning strike, etc.
- Third party: a failure resulting from an action by a third party, either accidental or intentional (theft, terrorism, etc). This also includes "incidental" third party damage, undetected when it originally occurred but which resulted in a failure at some later point in time.

The cause distribution of CONCAWE pipeline failures during 1971-2013 are shown in Figure 12. As indicated in Figure 12, the top three prevalent causes for CONCAWE pipeline failures are 3rd party damage, mechanical failure, and corrosion.

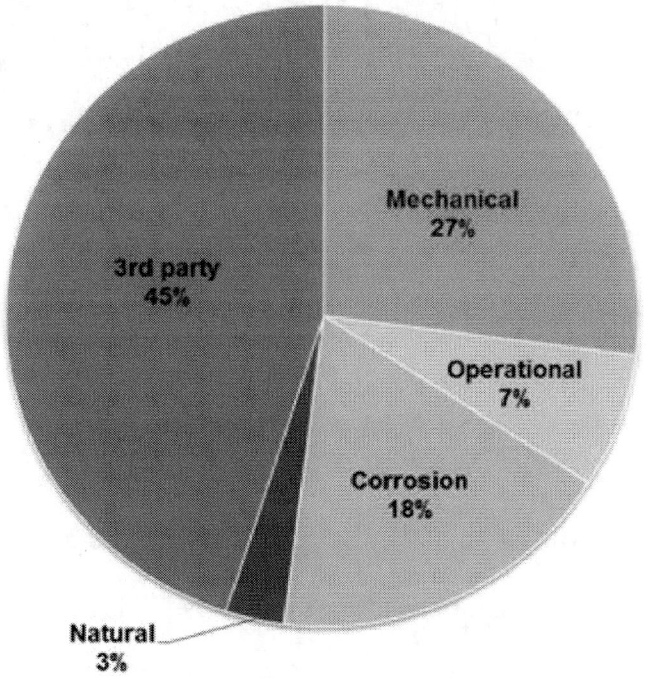

Figure 12. Major causes for CONCAWE pipeline failures, raw data from [2].

PHMSA uses a different categorization approach, which classifies the pipeline failure causes into seven major categories: corrosion failure, natural force damage, excavation damage, other outside force damage, material failure of pipe or weld, equipment failure, and incorrect operation. The cause distribution for PHMSA pipeline failures during 1968-2014 is shown in Figure 13.

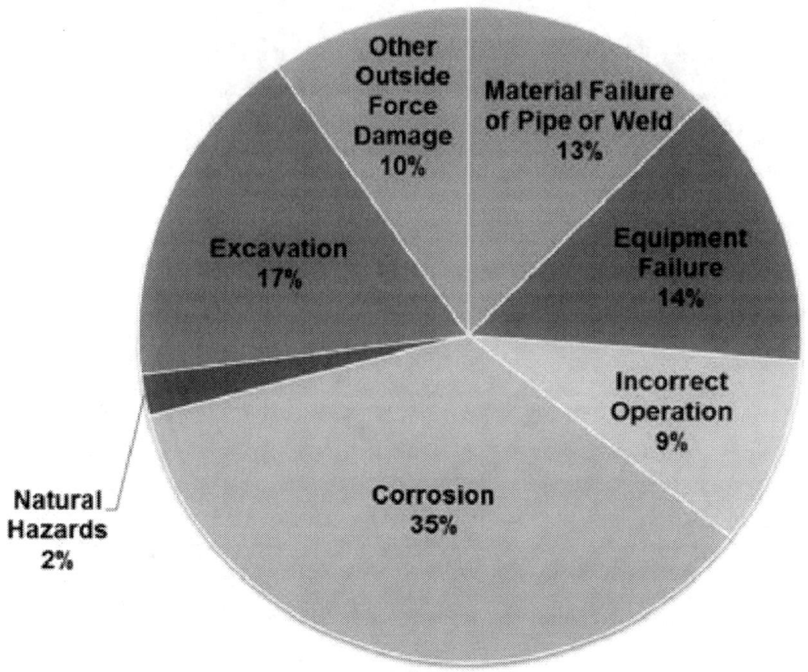

Figure 13. Major causes for PHMSA pipeline failures, raw data from [14].

The sub-categories for each major cause are as follows [11].

- Corrosion including external corrosion, and internal corrosion
- Natural force damage including earth movement (not due to heavy rains/floods), heavy rains/floods, lightning, temperature, and high winds
- Excavation including damage by operator (first party), operator's contractor (second party), third party, and previous damages due to excavation damage
- Other outside force damage including nearby fire/explosion, damage by motor vehicle/equipment (not engaged in excavation), damage by maritime equipment or asset, electrical arcing from other equipment/facility, previous mechanical damage (not related to excavation), and intentional damage (e.g., vandalism, terrorism, theft)
- Material failure of pipe or weld including failure related to construction, installation or fabrication, failure related to original manufacturing and failure related to environmental cracking
- Equipment failure including malfunction of control/relief equipment, pump related equipment failure, threaded coupling/connection failure, non-threaded connection failure, defective or loose tubing/fitting, and failure of equipment body or tank plate
- Incorrect operation including damage by operator or operator's contractor not related to excavation and motorized vehicle, tank or sump caused to overfill/overflow, valve left in wrong position but not leading to tank/vessel/sump/separator overflow or facility overpressure, pipeline/equipment over-pressurized, equipment not install properly and wrong equipment specified or installed.

Based upon the description of cause categories above, PHMSA shares some similar definitions for the causes of pipeline failures with CONCAWE, such as defining corrosion and natural damage as major causes for pipeline failures. However, there are many differences between the two datasets. Excavation damage as a cause in the PHMSA dataset not only includes the excavation damage by third party as in the CONCAWE data but also includes the excavation damage by first and second party. Furthermore, the 3rd party damage in CONCAWE dataset not only includes the excavation damage as in PHMSA dataset, but also includes some types of "other outside force damage" such as the intentional damage (theft) and previous mechanical damage as in the PHMSA dataset.

Another example showing the difference is that parts of equipment failure types such as "pump related equipment failure" together with "material failure" in the PHMSA dataset belong to the category of "mechanical failure" in the CONCAWE dataset; while other parts of equipment failure types such as "malfunction of control/relief equipment" together with "incorrect operation" failures belong to the category of "operational failure" in the CONCAWE dataset.

Therefore, it is very difficult to normalize the two datasets by using the same categories, which makes direct comparison of the pipeline failure causes between the PHMSA and CONCAWE datasets not feasible. However, certain comparison can still be made based on Figure 12 and Figure 13. For example, natural hazards only accounts for a tiny portion of pipeline failure causes in both datasets (3% in the CONCAWE data and 2% in the PHMSA data). "Mechanical failure" and "operational failure" accounts for a total of 34% in the CONCAWE dataset while a similar pattern around 36% is observed in the PHMSA dataset (as the definition of "material failure", "equipment failure" and "incorrect operation" together in the PHMSA dataset matches the definition of "mechanical failure" and "operational failure" together in CONCAWE). Corrosion plays a much more important role (35%) in U.S. pipeline failures compared with European pipeline failures (18%); while 3rd party activities caused more accidents (45%) in Europe than in U.S (<27%), which may be due to the different population distribution along the pipeline networks.

It is also interesting to see the detailed cause distribution specifically for pipeline leak and rupture, as shown in Figure 14 for the CONCAWE data and Figure 15 for the PHMSA data. For the CONCAWE data, 3rd party activities are the dominating cause of pipeline failures, contributing to over 40% pipeline leaks and almost 70% pipeline ruptures. Corrosion and mechanical failure also account for a large portion of pipeline leaks while they are not significant causes for pipeline ruptures. Unlike the CONCAWE data, corrosion is the dominating cause of leaks in the PHMSA data, while external influence ("excavation" and "other outside force damage") is the most prevalent cause of ruptures. The difference of the distribution ratios between Figure 13 and Figure 15 is because parts of the PHMSA accident reports only present the cause and do not present the type of failure (i.e., leak or rupture).

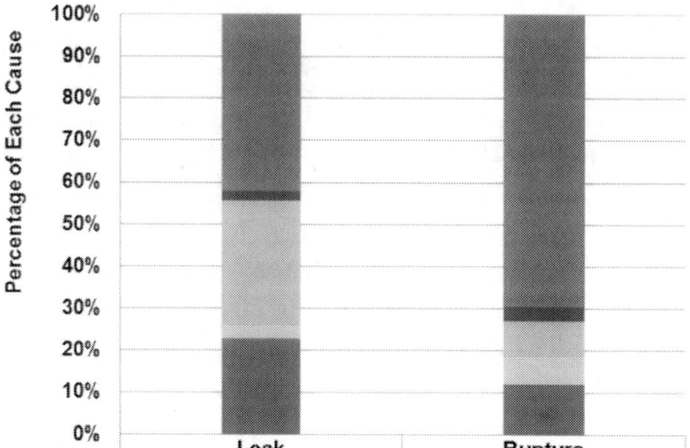

	Leak	Rupture
Third party	42%	69%
Natural hazard	2%	3%
Corrosion	30%	8%
Operational	3%	7%
Mechanical	23%	12%

Figure 14. Distribution of causes leading to leak and rupture in CONCAWE dataset, raw data from [2].

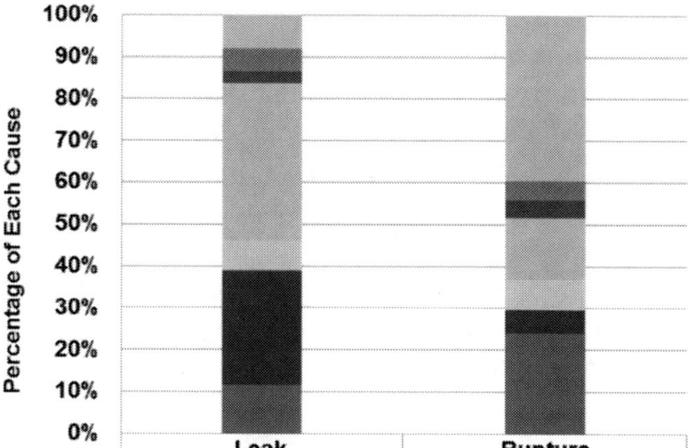

	Leak	Rupture
Other Outside Force Damage	8%	40%
Excavation	5%	5%
Natural Hazards	3%	4%
Corrosion	37%	15%
Incorrect Operation	7%	7%
Equipment Failure	27%	6%
Material Failure	12%	24%

Figure 15. Distribution of causes leading to leak and rupture in PHMSA dataset, raw data from [14].

4 Conclusions

The energy demand urges the expansion of the pipeline network, which together with the increasingly stringent regulations poses great pressure on pipeline operators to maintain pipelines under safety operation. Better understanding of the pipeline failure patterns through learning from past failures is helpful to establish an effective and efficient pipeline integrity management program to optimize the allocation of limited sources to increase the resistance of pipelines to failures.

This work systematically analyzed the PHMSA liquid pipeline accident database and compared with the European (CONCAWE) dataset, with the purpose of identifying the U.S. pipeline failure patterns and seeking common causes of pipeline failures in Europe and the U.S. Please note that all the findings in this work are only based on the liquid pipeline data, which may be different from gas pipeline data. Through the analyses, it is found that U.S. pipelines transporting crude oil and petroleum products generally are smaller than the European counterparts. The European pipeline network is gradually ageing, while more newly constructed pipelines are replacing old networks in the U.S. Regarding the type of failures, U.S. liquid pipelines tend to have a higher failure/spillage frequency (per km per year) but have fewer ruptures (percentage based) than European pipelines. The overall dominating cause of U.S. pipeline failures leading to spillage is corrosion, while 3rd party activities made the greatest contribution to European pipeline failures. More specifically, in the U.S., corrosion caused more leaks while external influence (other outside force damage plus excavation) caused more rupture than other major failure causes; in Europe, 3rd party activities are the most prevalent cause for both pipeline leak and rupture accidents. These findings indicate that the liquid pipeline failure patterns in U.S. and Europe are quite different. Therefore, when using European data for risk assessment and risk-based decision making, these findings should be considered.

The understanding of the top contributors to pipeline failures (e.g., corrosion, excavation) will help develop more targeted resolutions to increase the resistance of pipelines against those top failure contributors. More specifically, the recent incident report includes much more detailed information, which is very helpful for operators to identify the reported failures of pipelines of interest (specific size, working pressure, carried commodity, surrounding environment…). From these reported failures, operators may be able to find some common root causes for their pipelines of interest and then develop more effective integrity program to proactively safeguard the pipelines from risks. Moreover, the obtained failure frequency can be utilized in quantitative risk assessment (QRA), the results of which can help risk-based decision making on many aspects, e.g., proper material selection, proper coating type, proper exclusion zone determination. In this sense, the past failure data work as leading indicators which could prevent further occurrence of undesired events to our pipeline network.

5 References

[1] EGIG, "Gas Pipeline Incidents - 9th Report of the European Gas Pipeline Incident Data Group", Doc. Number EGIG 14.R.0403, February 2015, available at http://www.egig.eu/, Accessed on January 5, 2016.

[2] CONCAWE, ""Performance of European cross country oil pipelines - Statistical summary of reported spillages in 2013 and since 1971", Report No. 4/15, available at https://www.CONCAWE.eu, Accessed on January 5, 2016.

[3] International Energy Agency, "IEA Headline Global Energy Data (2015 Edition)", available at http://www.iea.org/statistics/ Accessed on January 5, 2016.

[4] U.S. Energy Information Administration, "Annual Energy Outlook 2015", available at https://www.eia.gov/forecasts/aeo/pdf/0383(2015).pdf Accessed on January 5, 2016.

[5] American Petroleum Institute, "Transporting Oil and Natural Gas", available at http://www.api.org/Oil-and-Natural-Gas-Overview/Transporting-Oil-and-Natural-Gas, Accessed on January 5, 2016.

[6] U.S. Department of Transportation Pipeline and Hazardous Materials Safety Administration, "Pipeline Incident 20 Year Trends", available at http://www.phmsa.dot.gov/pipeline/library/data-stats/pipelineincidenttrends, Accessed on January 5, 2016.

[7] J.S. Macpherson, "North Dakota Oil Spill Cleanup Continues, 1 Year After Tesoro Pipeline Rupture", Huff Post December 20, 2014, available at http://www.huffingtonpost.com/2014/10/20/north-dakota-oil-spill-cleanup_n_6017286.html, Accessed on January 5, 2016.

[8] NACE International, "Corrosion Failures: Sinopec Gas Pipeline Explosion", available at http://www.nace.org/CORROSION-FAILURES-Sinopec-Gas-Pipeline-Explosion.aspx, Accessed on January 5, 2016.

[9] C.F. Cheng and E. Hsu, "Compensation settled for 32 killed in Kaohsiung gas explosions", Focus Taiwan New Channel, July 7, 2015, available at http://focustaiwan.tw/news/asoc/201507170022.aspx, Accessed on January 5, 2016.

[10] J. Serna, "Refugio oil spill may have been costlier, bigger than projected", Los Angeles Times, August 5, 2015, available at http://www.latimes.com/local/lanow/la-me-ln-refugio-oil-spill-projected-company-says-20150805-story.html, Accessed on January 5, 2016.

[11] U.S. Department of Transportation Pipeline and Hazardous Materials Safety Administration, "Distribution, Transmission & Gathering, LNG, and Liquid Annual Data", available at http://www.phmsa.dot.gov/pipeline/library/data-stats/distribution-transmission-and-gathering-lng-and-liquid-annual-data, Accessed on January 5, 2016.

[12] U.S. Department of Transportation Pipeline and Hazardous Materials Safety Administration, "Annual Report Mileage for Hazardous Liquid or Carbon Dioxide Systems", available at http://www.phmsa.dot.gov/pipeline/library/data-stats/annual-report-mileage-for-hazardous-liquid-or-carbon-dioxide-systems, Accessed on January 5, 2016.

[13] U.S. Department of Transportation Pipeline and Hazardous Materials Safety Administration, "Incident Report Criteria History", May 27, 2014, available at https://hip.phmsa.dot.gov/Hip_Help/pdmpublic_incident_page_allrpt.pdf, Accessed on January 12, 2016.

[14] U.S. Department of Transportation Pipeline and Hazardous Materials Safety Administration, "Distribution, Transmission & Gathering, LNG, and Liquid Accident and Incident Data", available at http://www.phmsa.dot.gov/pipeline/library/data-stats/distribution-transmission-and-gathering-lng-and-liquid-accident-and-incident-data, Accessed on January 19, 2016.

12TH GLOBAL CONGRESS ON PROCESS SAFETY

Best Practices for Data Mining to support Asset Integrity Management

Brent G. Stanley

Principal, Asset Data Integrity Services, LLC

10631 Overbrook Lane, Houston Texas 77042

bstanley@adisllc.net

Randy Montgomery

Sr. Director Midstream and Downstream Services, ABS Consulting

2100 Space Park Drive, Suite 100, Houston Texas 77058

RMontgomery@ABSConsulting.com

Prepared for Presentation at
American Institute of Chemical Engineers
2016 Spring Meeting
12th Global Congress on Process Safety
Houston, Texas
April 11-23, 2016

Best Practices for Data Mining to support Asset Integrity Management

Brent G. Stanley

Principal, Asset Data Integrity Services, LLC

10631 Overbrook Lane, Houston Texas 77042

bstanley@adisllc.net

Randy Montgomery

Sr. Director Midstream and Downstream Services, ABS Consulting

2100 Space Park Drive, Suite 100, Houston Texas 77058

RMontgomery@ABSConsulting.com

Keywords: Data Mining; Auto-Classification; Data Extraction; Clustering

Abstract

Mechanical integrity program implementations require extensive design and operation data in order to be successful, however the standard methods for evaluating potential data sources, collecting data, reconciling to the master equipment list and locating relevant and missing data have not substantially changed in several decades, as the standard process is to send engineers and inspectors to the field with cameras and notepads (and if high-tech, with scanners in backpacks) to manually review and capture data in spreadsheets.

This presentation reviews best practices for accomplishing these tasks using modern methods which incorporate machine intelligence to auto-locate relevant data and execute data extraction, and use database rules to improve data quality prior to ingestion into predictive analytics software applications used to predict failure and better manage CAPEX and OPEX. The benefits to the operator and engineering community include reduced costs to perform the same amount of work, in shorter time frames and allowing engineering resources to focus on higher level tasks.

1.0 Data Mining Overview

In the context of integrity management, data mining is the work required to create structure from unstructured data and correlate to other existing structured data, resulting in the generation of load files or a custom database as the deliverable. Data mining requires skill sets including IT skills (including database expertise, content management, and tools knowledge), data strategy and architecture, project management and integrity management subject expertise.

Here are some simple definitions used throughout this paper to assist the reader in understanding basic concepts.

- "Structured Data" is data living in a database where the data is stored in tables which are linked together.
- "Unstructured Data" is document-based data; sometimes called "Semi-Structured" where there are forms present.
- "File Types" are the designations at the end of a file name and include a ".pdf", ".doc" and can include thousands of different types, few of which have true business value.
- "Document Type" and "Record Type" refer to the content of a file and are used as interchangeable terms here and in the industry.
- "Compound Documents" are files of file type PDF or TIFF which are aggregates of multiple records which have been combined by others for convenience in transmitting a package of information.
- "Data Extraction" refers to the process of indexing target data or data fields from records according to a prescribed target.

Data mining suggests the use of technology to power the classification and indexing of large volumes of physical and electronically stored information which are encountered in most corporations and encountered in many projects.

There is a "felt need" in the industry for data mining, as evidenced by continued focus on infrastructure safety in the midstream and downstream markets by several regulatory bodies, where records are a prime focus in order to understand as-built design of materials and testing, inspection and corrosion history, and replaced equipment design information.

Data mining is a necessity as operators have no choice but to perform the work as mandated by the regulators and is immune to fluctuations in the price of natural gas and oil. There is an opportunity for operators to improve operational efficiency and better manage CAPEX and OPEX by performing data mining, while RBI and other software applications need robust data in order to be useful and accurate. Current methods employed by operators and the engineering services community have been very manual and expensive and often use high cost engineering resources to perform basic data management tasks.

2.0 Problem Statement and Current State

2.1 Regulatory Challenges

One of the motivations which drives the need for data mining is created by the governing regulatory agencies. There is much discussion in the industry around the rule makings and best practices to support the regulatory objectives, but our observations focus on the data challenges. Given limited budgets and a desire to accomplish as much as possible with limited resources, leveraging data mining best practices allows budget dollars to go further and allows better allocation of resources, whereby skill sets are matched to the task. For example, it is wise to avoid asking engineering resources to populate spreadsheets from documents, when an algorithm and a QC resource can create the same or better output at half the cost. These resources are better utilized to analyze the data and make informed decisions.

Figure 1: Summary of Regulatory Requirements and Data Management System Impact

2.2 Typical Corporate Integrity Management Data/Records Ecosystem

Most companies have a broad mixture of records and data storage and management systems, accumulated over time and obtained through merger and acquisition. As a result, data and records live in multiple silos where are minimally connected, making the data mining process more challenging. Figure 2 illustrates the mixture of data systems existing in many corporations.

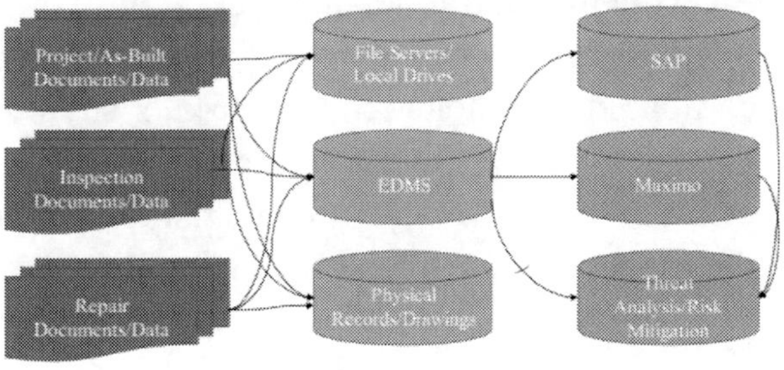

Figure 2: Example Mixture of Corporate Data Systems

2.3 Amount of Affected Infrastructure

Much of the industrial infrastructure in the United States is subject to integrity management oversight including midstream facilities and pipelines and downstream facilities. This mandated oversight necessitates an increasing need for data mining.

The typically assumed costs for the data mining effort using modern methods such as the ones suggested in this best practices summary are provided in Figure 3. However, observations on projects have shown these costs can be much higher due to inefficiencies, manual efforts, and poorly understood data and records.

Category	Miles or Facility Count	Assumed Cost/Mile or Plant for Needed Data Mining	Industry Cost
Midstream			
Gathering Lines	30000	$250	$7,500,000
Transmission	300000	$250	$75,000,000
Processing Plants	450	$20,000	$9,000,000
Liquids Gathering	21000	$250	$5,250,000
Liquids Transmission	135000	$250	$33,750,000
Downstream			
Refineries	148	$250,000	$37,000,000
Chemical Plants*	2700	$250,000	$675,000,000
			$842,500,000

*The chemical industry reports there are 13,500 facilities operating in the USA; 20% of that amount was used.

Figure 3: Amount of Infrastructure Impacted and Estimated Data Mining Costs

2.4 Typical Corporate Enterprise Management Challenges

There are common data issues affecting almost all corporations, largely due to the lack of investment over time in modernizing the IT and associated infrastructure needed to run the business and meet regulatory and risk management challenges. Energy infrastructure operators, compared to other vertical markets such as financial services, pharmaceuticals, and insurance, have yet to invest in master data management and data quality to the same level, exacerbating the effort to meet the operational, risk and regulatory challenges. The following charts summarize the most common situations generally found in industry.

Situation	Gap	Impact	Mitigation
Critical Records Recognition	Opportunity for improved cultural understanding of enterprise compliance and risk management objectives. Mixed messages from leadership. Resource and funding allocation.	Silos of activity that exhibit inconsistent business processes and have no overarching policies and controls.	Enterprise records policy and procedure with appropriate education and training program. Prioritization of effort and funding based on risk/value analysis.
New Records Generation	Departments use inconsistent business processes to generate new records.	Inability to properly categorize critical records; inconsistent and missing records; sparse and inconsistent extracted data to populate other applications.	Standardization of process, capture and avoidance of unstructured data by using database-driven forms.
Records Archiving	Records-stored locally in operations project files and regionally in corporate offices. Difficult to locate all source data, Missing critical records.	Excessive time spend finding relevant records and data. Missing records dictate expensive field mitigation for regulated facilities and pipelines.	Develop data recovery team and processes to canvass repositories for critical records and prioritize capture and indexing.
Records Access	Difficult to access source data and critical records for all end users.	Inability to provide records and extracted data to multiple applications, creating organizational inefficiencies, inability to quantify risk, and difficulty in addressing compliance.	Implement formal governance program which incorporates critical records and standardized taxonomies; leverage a common EDMS.
Data Extraction from Records	Systems lacking critical data from records; existing data has some quality issues	Reliance on default values for risk calculations. Missing repair records and their associated data causes cascading business issues.	Revise business process to input all critical data to system of record. Backfill relevant legacy data from records.
Situation	Gap	Impact	Mitigation
Electronic Records Not Available	Processing of critical records not initiated for the target projects.	Additional resources, time and costs to manually review and sort large volumes of documents comprising hundreds of thousands of pages of content.	Prioritize processing work based on compliance strategy and field location review.
Enterprise Taxonomy	Inconsistent taxonomies between GIS and other applications that are gathering critical pipeline and operational information, meaning those systems have different data hierarchies/schemas.	This condition doesn't allow for cross-system analysis of assets, nor does it allow the implementation of business process improvements.	Creation of a single enterprise taxonomy describing the assets and prioritization of the work effort needed to update current asset descriptions against the revised taxonomy.
Document Control	Documents and data associated with new construction and repairs is not managed nor integrated with other business processes requiring the data.	Incomplete data and records in applications used for enterprise risk analysis. Compromised CAPEX and OPEX planning due to poor quality and missing data.	Standardize the business process for collecting, parsing, and integrating records and data from EPC and repair activities.
Enterprise Master Data Management (MDM)	No acknowledgement or ownership of MDM as a necessary corporate function. As such, corporate records and data governance lacks appropriate stewardship.	MDM is the enterprise functional glue which ties data from various applications and processes together.	Create a formal MDM function with C-level priority.

2.5 Importance of High Quality, Adequate Data

Good data feeds an ever-increasing number of software applications which calculate specific risk (corrosion to a specific vessel, for example), cumulative risk (over time and across multiple sections or components) and/or operational risk which is measured in terms of economic impact. World class operators are able to enjoy the triumvirate benefits of risk, compliance and operational effectiveness by creating an executive-level awareness and buy-in to the investments necessary to achieve those goals.

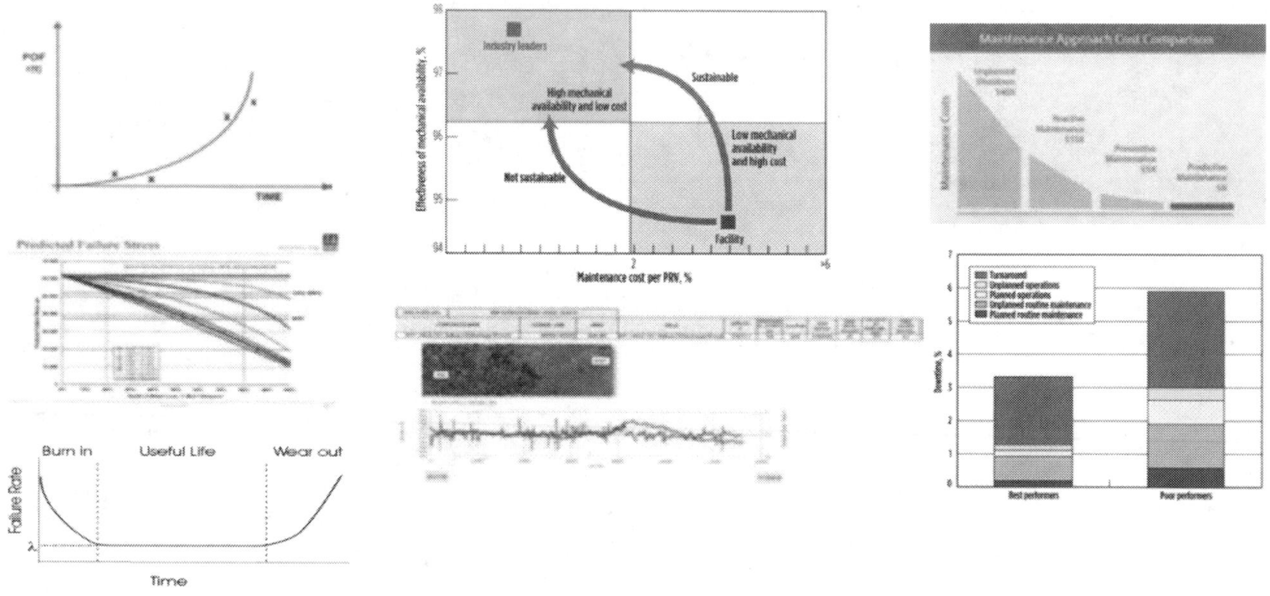

Figure 4: Collection of Diagrams Relying on Good Data

2.6 Risk of Using Default Data

In many cases, the lack of good historical data forces the operator to rely on default data for many decisions (e.g., equipment corrosion rates for RBI analysis). When this occurs, the operator can create a false sense of security since the default values might not reflect the actual materials conditions, resulting in less than accurate failure predictions. In the rush to launch RBI programs and maintenance software programs, the use of default values can create a "garbage in-garbage out" result which impacts the operator confidence in managing risk, ensuring compliance or achieving efficiency.

The following diagram is an example of default data being used for predictive failure analysis.

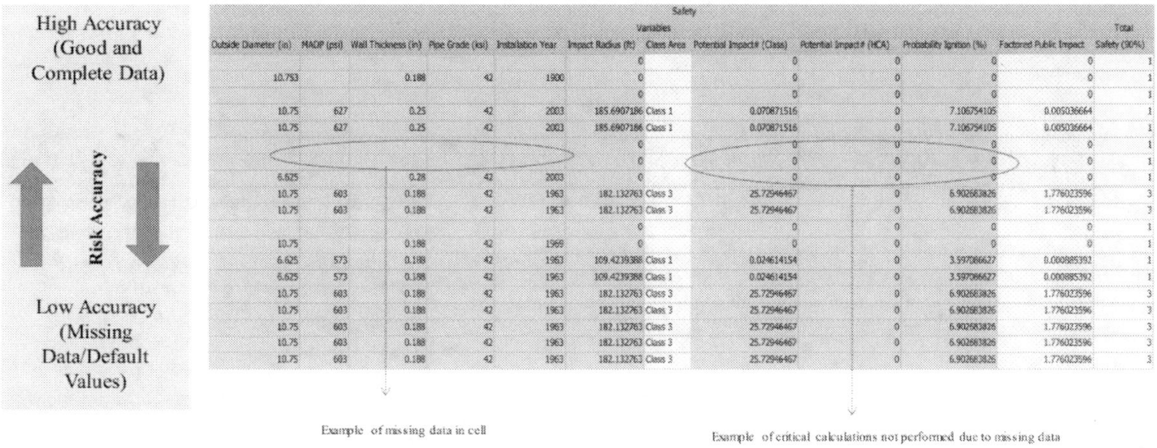

					Safety Variables						Total
Outside Diameter (in)	MAOP (psi)	Wall Thickness (in)	Pipe Grade (ksi)	Installation Year	Impact Radius (ft)	Class Area	Potential Impact# (Class)	Potential Impact# (HCA)	Probability Ignition (%)	Factored Public Impact	Safety (30%)
					0		0	0	0	0	1
10.753		0.188	42	1900	0		0	0	0	0	1
					0		0	0	0	0	1
10.75	627	0.25	42	2003	185.6907186	Class 1	0.070871516	0	7.106754105	0.005036664	1
10.75	627	0.25	42	2003	185.6907186	Class 1	0.070871516	0	7.106754105	0.005036664	1
					0		0	0	0	0	1
					0		0	0	0	0	1
6.625		0.28	42	2003	0		0	0	0	0	1
10.75	603	0.188	42	1963	182.132763	Class 3	25.72946467	0	6.902683826	1.776023596	3
10.75	603	0.188	42	1963	182.132763	Class 3	25.72946467	0	6.902683826	1.776023596	3
					0		0	0	0	0	1
10.75		0.188	42	1969	0		0	0	0	0	1
6.625	573	0.188	42	1963	109.4239388	Class 1	0.024614154	0	3.597086627	0.000885392	1
6.625	573	0.188	42	1963	109.4239388	Class 1	0.024614154	0	3.597086627	0.000885392	1
10.75	603	0.188	42	1963	182.132763	Class 3	25.72946467	0	6.902683826	1.776023596	3
10.75	603	0.188	42	1963	182.132763	Class 3	25.72946467	0	6.902683826	1.776023596	3
10.75	603	0.188	42	1963	182.132763	Class 3	25.72946467	0	6.902683826	1.776023596	3
10.75	603	0.188	42	1963	182.132763	Class 3	25.72946467	0	6.902683826	1.776023596	3
10.75	603	0.188	42	1963	182.132763	Class 3	25.72946467	0	6.902683826	1.776023596	3

2.7 ABS Group Case Study: Experience from the Field

In last 20 years of auditing/assessing mechanical integrity programs, one of challenges at most facilities is complete and accurate equipment information and data, especially design and code information. Most regulations require facilities to identify and maintain information and data related to the equipment design and its operation, and maintenance history. This is decidedly true for facilities subject to the OSHA process safety management standard (29 CFR 1910.119), and is becoming more important as OSHA continues its increased focus on demonstrated compliance with RAGAGEPs.

For a practical viewpoint, the need for this equipment data and information increased with the implementation of computerized maintenance management systems (CMMS), inspection data management systems (IDMS) and other computer-based tool used to support mechanical integrity programs. For example, the CMMS bill of material functionality use as the primary data source for spare part ordering and replacement parts information is increasing. So high quality, accurate CMMS data is needed to avoid increased downtime for repairs and possible failures if the wrong parts are installed. Another example related to fixed equipment is that incorrect or unavailable information (e.g., materials of construction, process materials, operating conditions) can result in incorrect decisions, such as the failure to identify proper damage mechanisms and then ultimately failure to determine correct inspection and non-destructive testing plans. These incorrect/insufficient plans can result in failing to identify equipment damage and subsequently equipment failures.

To date, the answer to this equipment information and data challenge has mostly been based on executing a manual "Search and Find" mission. Typically, this mission begins by identifying the types of equipment documentation believed to provide the needed information, such as pressure vessel U-1A forms, as-built vessel drawings, rotating equipment vendor drawings, mechanical seal drawings, vendor manuals, design specifications and the list goes on. Next, the "Search and Find" team identifies the likely documentation containing the information and the possible locations of the targeted documentation. This usually results in (1) pulling a number of boxes of archived records, (2) identifying personnel who might possibly have some information filed away, (3) locating server drives and folders to be searched, etc. Now the fun begins, the team must now get their hands on these information sources and manually begin going through these documents. This can be a very daunting task. The following chart provides a glimpse of the results from a recent documentation "Search and Find" mission.

Type of Information	Number of Missing Documentations	Number of Documents Found	Success Rate	Quantity of Documents Searched	Labor Expended
Pressure Vessel Code Documents (U-1As)	293	148	51%	• 12+ maintenance file cabinets • 200+ banker boxes of documents • Dozens of vendor manuals • Numerous department drive folders	More than 6 man-weeks of labor of searching and scanning, then additional 4 man-weeks to organize the scanned documents
Pressure Vessel As-built Drawings	293	245	84%		
Rotating Equipment Materials of Construction Information	158	121	76%		

While this information might not be representative of your plant (none the less), this one data point illustrates this manual process can be filled with challenges, such as

- Being highly labor intensive.
- Usually requires some level of equipment expertise (how is summer intern going to know what MAWP stands for, etc.?).
- High error/miss rate as a function of different people having different understandings of what's important or an inherent lack of consistency from person to person.
- An incomplete search of potentially valuable documents (e.g., missing information in documents not identified as an information source).

In addition, this process does not provide an effective means for capturing other important information from documents which may not have been identified as sources nor does it help retrieve valuable information which may be needed for other uses. This begs the question (or issues a challenge); is there a better and more efficient means to compile and extract information from facility documentation, which is:

- Less labor intensive
- Reduces the amount of expertise required to sort through documents
- Reduces error rate and helps eliminate missed/overlooked data
- Provides a more comprehensive search for needed information (rather a search for document which are believed to be the best source of the needed information.

2.8 Enterprise Data Requirements

Much of the critical data is needed by and may be contained in existing SAP/Maximo systems as well as documents submitted by engineering, procurement and construction (EPC) contractors and other contractors, and internal engineering and operations functions. In addition to the corporate systems, there are usually field repositories of records and in fact local databases which have been created due to an aversion to dealing with corporate IT. At the same time, the corporate initiatives are driving towards creating a "single version of the truth" with a goal of appropriately sharing the data across the various enterprise applications needing a portion of the data. In addition, knowledge workers from multiple departments need working access to the same data, further underscoring the need for high coverage and high quality data.

Some of the trends in the technology world include better tools for creating data connectors which span multiple systems to seamlessly aggregate data. These tools allow connectors to be created on the desktop without writing code. Other new technologies allow this aggregated data to be "visualized", typically in conjunction with GIS databases, and mechanical anomalies to be easily summarized in a dashboard fashion with "what if" functionality for users to drill down on conditions and even conduct analysis on potential mitigation solutions.

The following screen shot provides an example of the type of detail needed by applications such as SAP, underscoring the importance of data sharing in a typical enterprise.

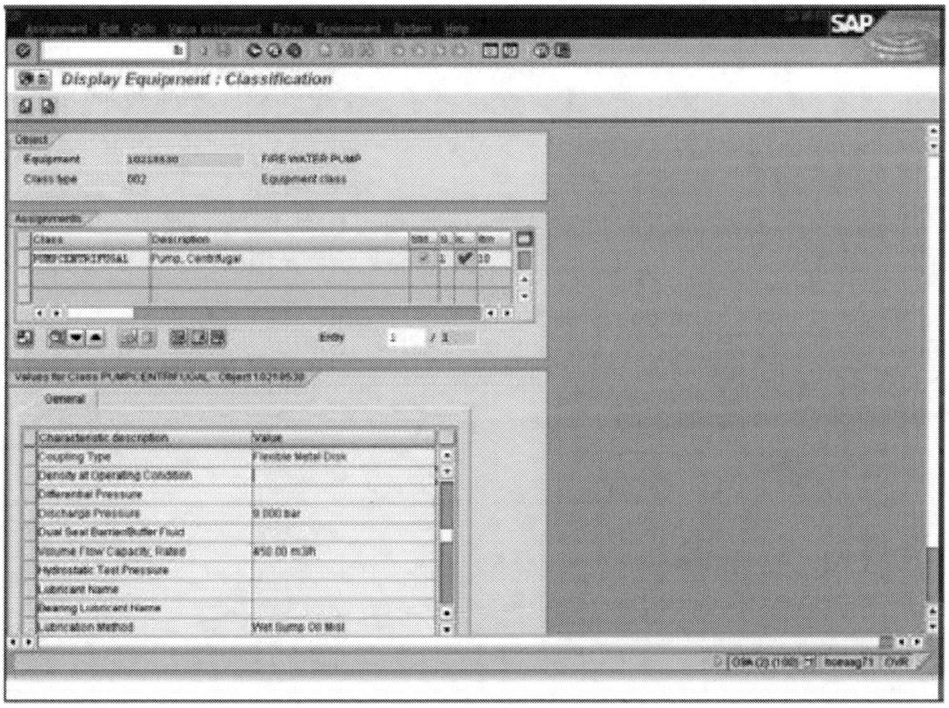

3.0 Best Practices for Performing Data Mining

3.1 Workflow Example – Facilities

A best practice workflow for facilities is shown in Figure 5 and a summary of the key steps is as follows.

- Creating a complete and accurate master equipment list (MEL): Many facilities lack a complete list of tag numbers for the current equipment. Gathering field data is often necessary to validate the MEL and then comparing to current lists provided by the operator. Digital images of name plates can be fed to a parsing algorithm to populate tag and serial numbers.
- Analysis of current databases containing equipment data: Existing operator databases usually contain some of the needed equipment data and should be the first go-to source, as it is the least expensive data to obtain. Quality checks should be performed to insure its quality and coverage against the MEL.
- Create a records strategy: Once the available database (structured) data has been analyzed, a records strategy should be created and mapped against the known repositories of records, be they physical or electronic. Often sampling of those potential repositories is needed to confirm their composition and usefulness. The records strategy should include the preparation of a data map, which is a matrix indicating what data elements are desired from the target record types.
- Execute the records strategy: Physical records will need to be scanned to a PDF file format; it is usually better to scan entire collections as opposed to cherry picking records, as the probability that additional records will be needed is high due to information gaps. Having the records electronically available makes that job easier and if automation is used to identify specific record types the location of those incremental records becomes infinitely easier.
- Apply rules to the extracted data: Various typed of rules are applied against the extracted data to normalize and transform the data. See the following section for more detail on this task.
- Prepare the load sheet: In accordance with the records strategy, a load sheet (typically a spreadsheet exported from a SQL database) is prepared.

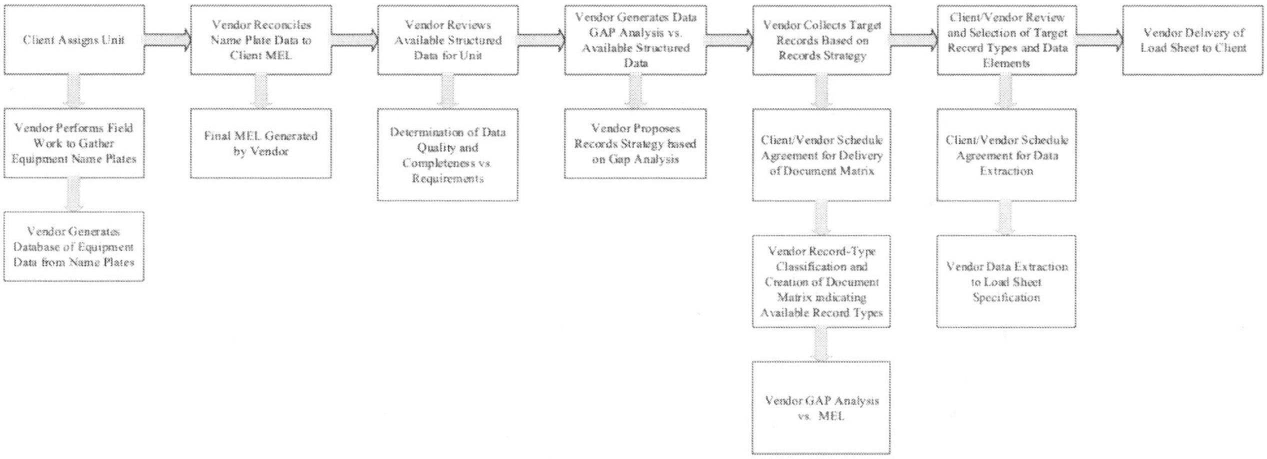

Figure 5: Example Facilities Data Mining Workflow

3.2 Workflow Example – Pipelines

Pipeline projects follow an analogous workflow, with the exception that data is typically organized by pipeline system and valve segment as opposed to an MEL. Steps are summarized as follow and illustrated in Figure 6.

- Update GIS with all repair records: Oftentimes silos of documentation of repair records have not been reconciled to the GIS database, resulting in deficiencies which cloud the "as built" nature of the system. This task prevents performing maximum operating pressure (MOP) or maximum allowable operating pressure (MAOP) and anomaly analysis on pipe which may no longer exist.
- Reconcile valve segments to the GIS database: This tasks creates a solid baseline of pipe, valves and fittings which are in scope for MOP and anomaly analysis.
- Review available structured data for useful data: Consistent with the workflow for facilities, this step digs out useful available data which is cheaper and easier to obtain before going to the records.
- Create a records strategy: Once the available database (structured) data has been analyzed, a records strategy should be created and mapped against the known repositories of records, be they physical or electronic. Often sampling of those potential repositories is needed to confirm their composition and usefulness. The records strategy should include the preparation of a data map, which is a matrix indicating what data elements are desired from the target record types.
- Execute the records strategy: Physical records will need to be scanned to a PDF file format; it is usually better to scan entire collections as opposed to cherry picking records, as the probability that additional records will be needed is high due to gaps. Having the records electronically available makes that job easier and if automation is used to identify specific record types the location of those incremental records becomes infinitely easier.

- Apply rules to the extracted data: Various typed of rules are applied against the extracted data to normalize and transform the data. See the following section for more detail on this task.
- Prepare the load sheet: In accordance with the records strategy, a load sheet (typically a spreadsheet exported from a SQL database) is prepared.

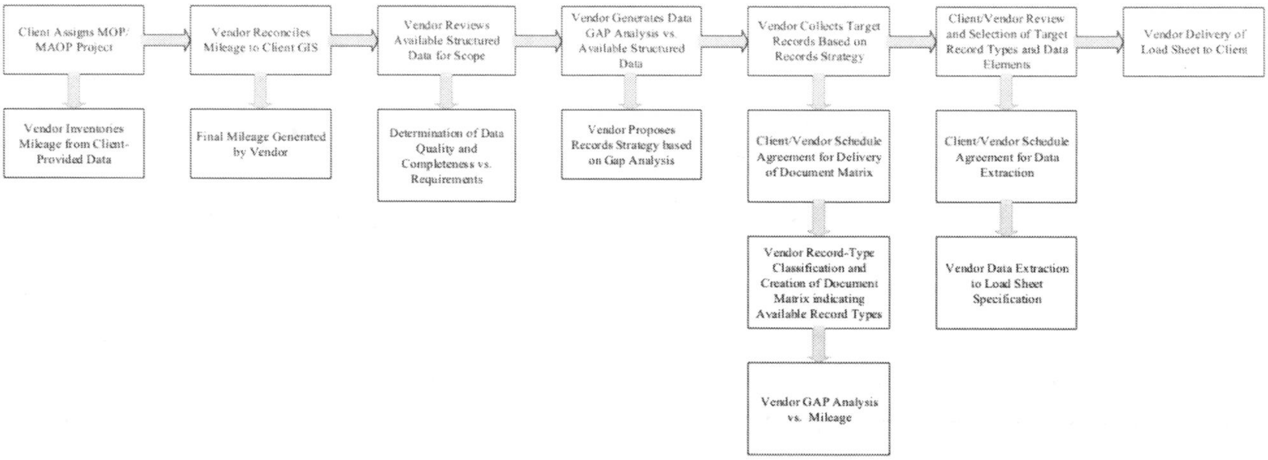

Figure 6: Example Pipeline Data Mining Workflow

3.3 Role of Data Maps

A data map is a matrix consisting of the target record types desired and the specific data elements needed from each of those record types. The data map becomes the playbook for executing the records strategy, with the extracted data residing in a SQL database. Usually there are target records not found when executing the data strategy, so having a list of alternative records is needed, often with a prioritization of value to the project. Having a data map which has been reviewed and approved avoids inefficiencies in re-work, if those alternate records have already been flagged and retrieving the records and their extracted data is machine time versus people time.

The following diagram is one example of a typical matrix listing the document type on the y-axis and the target extracted attributes on the x-axis.

Document Type	WO#	File#	Invoice#	Authorization#	Cert#	PO#	Report#	Change Order#	Test#	Test Section#	Book#	Job#	Permit#	Title	Project ID	Station/Line#	Alignment Sheet	Location	Township	Section	Range	T-S-R	Dwg#	Field Name	Well	Pipeline/System Name	Valve Section & Line
As-Built Drawings																										X	X
Authorization for Expenditures															X			X									
Budget Data	X																	X									
Completion Report	X																	X									
Completion Report for Indiana Requalification	X																	X									
Completion Report for Requalification by Replacement	X																	X									
Construction Change Order								X																			
Contruction Release																		X	X	X	X					X	
Detailed Cost Estimate	X																										
Detailed Retirement Cost Estimate	X																										
Drawing Correction																	X		X	X	X						
Field Pressure & Test Report																		X		X							

3.4 Unstructured (Document) Cleanup and Records Classification

Once all available structured data has been exhausted, a records strategy is developed. Records may be physical or electronic; if physical they should be scanned to a PDF file type and indexed to the folder, typically linked to a project or equipment ID. Other target records repositories are already in an electronic format and are usually provided as hard drives to the project team. Sometimes there is useful metadata associated with these records, especially if they are coming from a content management application such as Documentum or SharePoint. When possible, this metadata should be obtained and vetted for usefulness, which can include target data attributes themselves or pointer information. Pointer information could be a reference to work order, project number, or purchase order, as examples, which can be triangulated to other records which then lead to the specific records needed for the project.

One factor to keep in mind when processing records is the concept of lineage, which is the ability to maintain a history of where a document or page of content came from. This will be important from an audit perspective in order to validate the results down the road as well as to other parties, such as regulators.

Compound documents are documents which are aggregates of dozens or hundreds of other documents, combined into a single PDF. EPC contractors are notorious for delivering turnover data this way, often with little or no index. It is also common to find project files and completion reports packaged this way as well. Parsing these large files manually to look for specific records is time consuming and error prone, however best practices would include the following steps:

- Assigning a unique document ID to each document and to each page within using a parent-child nomenclature (for example, DOCID 123ABD_1 is the first page, DOCID 123 ABD_2 is the second page, etc.).
- Using a software utility to perform a fuzzy analysis of the document title to locate documents of interest from the records matrix. A fuzzy analysis feature will look for some percentage of the correct characters in sequence, which is useful when working with older documents which are of poor quality.
- Parsing the found records-matrix records to extract and QC the data attributes of interest.

- Some compound documents may contain a table of contents; where present, software utilities can be used to auto-create metatags which can be used as bookmarks within the PDF as well as exported to the master database.

Collections of electronic records are often provided to the project team as copies of larger document collections from server file shares. When this happens there is usually a significant amount of junk and trash, including non-relevant file types and duplicate documents. Non-relevant file types should be agreed upon as part of the project plan and can include the instruction to ignore file types such as PowerPoints, system files and other file types which are deemed not relevant by virtue of their existence. For definition, a file type is the extension associated with the file name; for example the file name ProjectAlpha.xls has a file type of ".xls", commonly known as a spreadsheet. The terms "document type" and "record type" are often used interchangeably and refer to the content of the document, and relates to the record matrix as a target document for the project, or not.

Basic processing steps include:

- File clean-up to remove out of scope file types and perform de-duplication. Duplicate files are determined using what are called hashing algorithms which compare the bit level of each document to all documents in the group. It is not uncommon for duplicates to comprise a significant percentage of the population, especially if the records are coming from a file server which has not been cleaned up in the past.
- When there are single record documents present (often determined by file name and file size), clustering software can be used to auto-group the same record types into folders; cluster folders are labeled with a record type by a content expert and that intelligence can be rolled forward to the next data set or client.

There are several software utilities available to perform the above functions; the following screen shot reflects a dashboard from one which summarizes the found file types, quantities, sizes, ages and duplicate population.

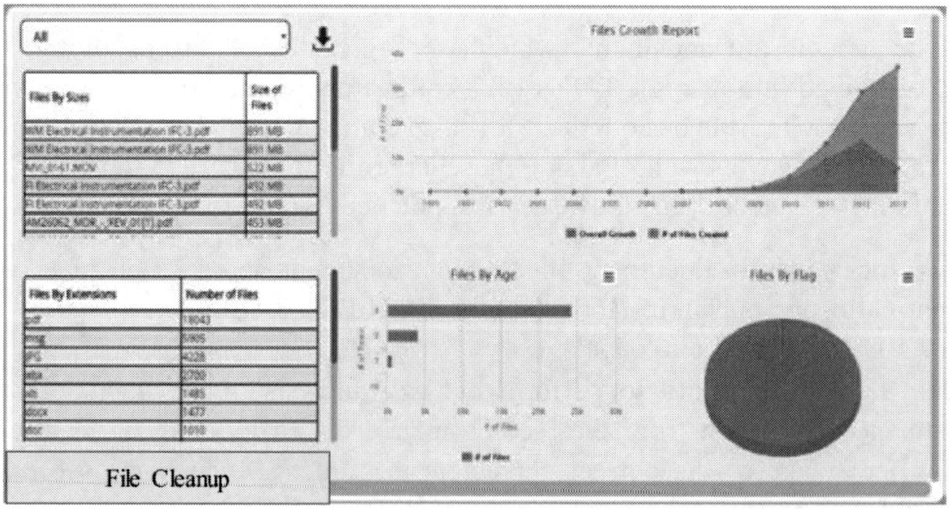

File Cleanup

One of the most useful tools is a software algorithm that automatically groups or "clusters" similar documents or pages of documents together. These tools have been used extensively in e-discovery for years, but have only recently been leveraged for records classification purposes in the engineering and operations world. There are two approaches to clustering; one uses the extracted text from optical character recognition (OCR) while the other uses what is called visual similarity to auto-group same documents or pages together. The benefit of clustering is to avoid the manual, linear review of content by people to arrive at an understanding of the available content. The results allow a faster and more cost efficient understanding of what you have, or inversely what you don't have as available documents that are needed for analysis.

Figures 7 and 8 contain the following screen shots providing a simplistic representation of the software's ability to group similar content together; this example represents the use of visual similarity to obtain high-accuracy results.

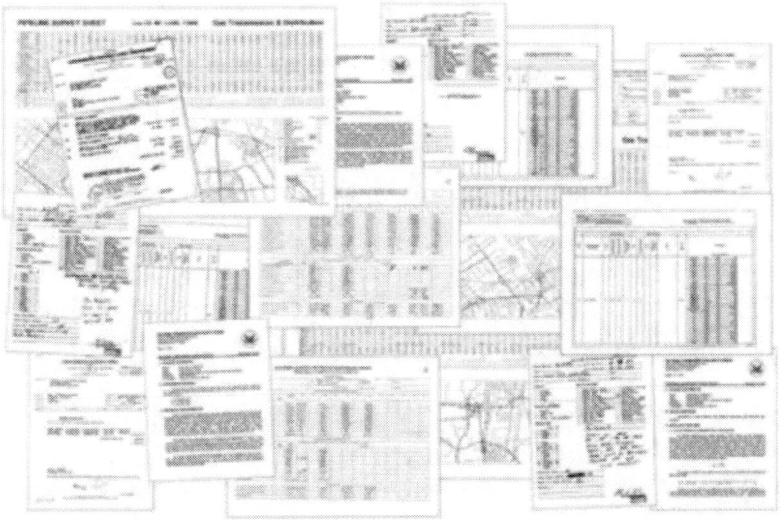

Figure 7: Unstructured Documents Before Visual Classification

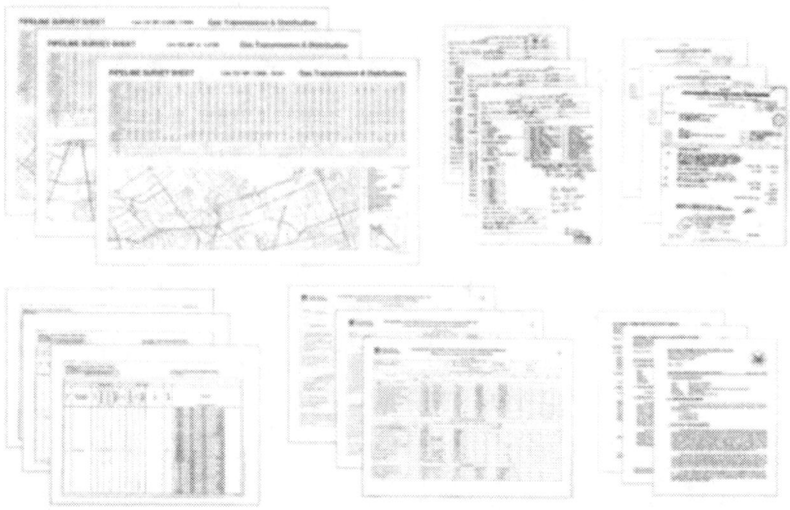

Figure 8: Structured Documents After Visual Classification

3.5 Automated Deep Data Extraction

Data extraction is performed from critical records such as U1's and drawings and placed into a SQL table; rules are then applied to transform data to output specifications. This type of software is typically pre-programmed to extract a set of target data elements from a single record type, and includes confidence-level flags as well as a quality control function to allow an operator to clean up and errors that may be present. Depending on the quality of the input document, it is possible to get extraction accuracies before quality control of 80-90%. Cursive handwriting is usually more difficult and will need to be keyed. Automation of the extraction process can save significant labor savings compared to a pure manual approach.

Figure 9 screen shot is representative of a software utility capable of being tuned to recognize and extract data from a U-1.

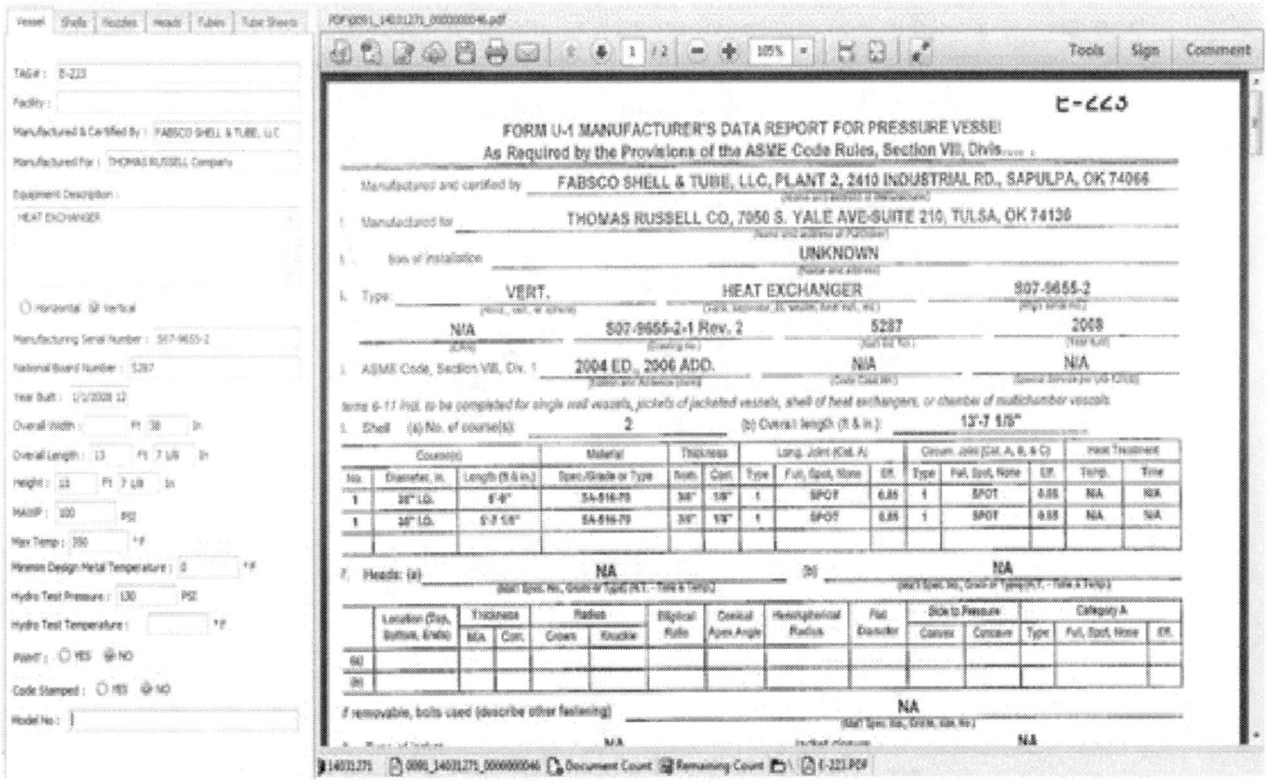

Figure 9: Example Data Extraction from U-1 Form

3.6 Applying Rules to Data using SQL

Once critical records have been found and deep data extraction performed, it is important to normalize the data to a common data dictionary, and then perform simple calculations to improve its quality prior to its use in other applications. For example, dates and tag numbers should be standardized in their form of expression, and outside diameters converted to inside diameters if that is the desired value later on.

The value of using a system-driven approach to rules application to data includes the avoidance of labor and human error to perform manually and the benefit of being able to make changes to or additions to rules quickly and with scale to large volumes of data.

```
--
Start:Update_InnerDiameter_using_OuterDiameter_IF ID
is not available for Equipments other than Drum

            Update InventoryAndCoding set
Diameter_InnerDiameter =

Case when Isnumeric(Diameter_OutsideDiameter) = 1
and Isnumeric(Furnished_Thk) = 1

then Diameter_OutsideDiameter - (2 * Case when
Furnished_Thk = 999.99 then 0 else Furnished_Thk
end)End

            where Diameter_InnerDiameter is null

                and Diameter_OutsideDiameter is
            not null

                and BatchName = @Batchname

                and ComponentCodingStatus <=2

                and ComponentType not like
                'DRUM%'

                and Component not like '%I'

--
End:Update_InnerDiameter_using_OuterDiameter_IF ID
is not available for Equipments other than Drum
```

3.7 Creating Load Sheets

One of the last tasks is to prepare the load sheet, which is structured to reflect the ingestion requirements into RBI and other software applications. The data within the load sheet represents the results from the preceding tasks and has undergone several quality checks along the way to insure data meets the high accuracy requirements of an integrity management project.

Load sheet specifications are developed in collaboration with the project team and undergo testing with sample data to insure that all data elements are present and their structure is consistent with the requirements of the downstream application(s) they are feeding. Minimizing manual re-work of the data should be avoided at all cost at this stage.

The following is an example of a portion of a load sheet for a facility.

Equipment_ID	Component_ID	RBI_Component_Family	Component_Description	Component_Type	Component_Start_Date	Operating_Pressure	Operating_Temperature
NA3-D501	NA3-D501-CS-1	Criticality RBI Component - Cylindrical Shell	NA3-D501-CS-1	Heat Exchanger-Shell	7/19/98	827	520
NA3-D501	NA3-D501-CS-2	Criticality RBI Component - Cylindrical Shell	NA3-D501-CS-2	Heat Exchanger-Channel	7/19/98	769	806
NA3-D501	NA3-D501-EB-1	Criticality RBI Component - Exchanger Bundle	NA3-D501-EB-1	Heat Exchanger-Bundle	7/19/98	769	806
NA3-D507A	NA3-D507A-CS-1	Criticality RBI Component - Cylindrical Shell	NA3-D507A-CS-1	Pressure Vessel	7/19/98	50	257
NA3-D507B	NA3-D507B-CS-1	Criticality RBI Component - Cylindrical Shell	NA3-D507B-CS-1	Pressure Vessel	7/19/98	49	257
NA3-D509	NA3-D509-CS-1	Criticality RBI Component - Cylindrical Shell	NA3-D509-CS-1	Pressure Vessel	7/19/98	76	134
NA3-E501	NA3-E501-CS-1	Criticality RBI Component - Cylindrical Shell	NA3-E501-CS-1	Heat Exchanger-Shell	7/19/98	80	298
NA3-E501	NA3-E501-CS-2	Criticality RBI Component - Cylindrical Shell	NA3-E501-CS-2	Heat Exchanger-Channel	7/19/98	266	40
NA3-E501	NA3-E501-EB-1	Criticality RBI Component - Exchanger Bundle	NA3-E501-EB-1	Heat Exchanger-Bundle	7/19/98	266	410
NA3-E502	NA3-E502-CS-1	Criticality RBI Component - Cylindrical Shell	NA3-E502-CS-1-Boot	Heat Exchanger-Shell	7/19/98	102	63
NA3-E502	NA3-E502-CS-2	Criticality RBI Component - Cylindrical Shell	NA3-E502-CS-2	Heat Exchanger-Channel	7/19/98	145	418
NA3-E502	NA3-E502-CS-3	Criticality RBI Component - Cylindrical Shell	NA3-E502-CS-3-Top	Heat Exchanger-Shell	7/19/98	102	63
NA3-E502	NA3-E502-CS-4	Criticality RBI Component - Cylindrical Shell	NA3-E502-CS-4	Heat Exchanger-Shell	7/19/98	102	63
NA3-E502	NA3-E502-EB-1	Criticality RBI Component - Exchanger Bundle	NA3-E502-EB-1	Heat Exchanger-Bundle	7/19/98	145	418
NA3-E503	NA3-E503-CS-1	Criticality RBI Component - Cylindrical Shell	NA3-E503-CS-1	Heat Exchanger-Shell	7/19/98	80	298
NA3-E503	NA3-E503-CS-2	Criticality RBI Component - Cylindrical Shell	NA3-E503-CS-2	Heat Exchanger-Channel	7/19/98	102	176
NA3-E503	NA3-E503-EB-1	Criticality RBI Component - Exchanger Bundle	NA3-E503-EB-1	Heat Exchanger-Bundle	7/19/98	102	176
NA3-E504	NA3-E504-CS-1	Criticality RBI Component - Cylindrical Shell	NA3-E504-CS-1	Heat Exchanger-Shell	7/19/98	79	806
NA3-E504	NA3-E504-CS-2	Criticality RBI Component - Cylindrical Shell	NA3-E504-CS-2	Heat Exchanger-Channel		80	1634
NA3-E504	NA3-E504-EB-1	Criticality RBI Component - Exchanger Bundle	NA3-E504-EB-1-A	Heat Exchanger-Bundle	7/19/98	147	410
NA3-E504	NA3-E504-EB-2	Criticality RBI Component - Exchanger Bundle	NA3-E504-EB-2-B	Heat Exchanger-Bundle	7/19/98	147	410
NA3-E504	NA3-E504-EB-3	Criticality RBI Component - Exchanger Bundle	NA3-E504-EB-3-C	Heat Exchanger-Bundle	7/19/98	147	410
NA3-E505	NA3-E505-CS-1	Criticality RBI Component - Cylindrical Shell	NA3-E505-CS-1	Heat Exchanger-Shell	7/19/98	159	788
NA3-E505	NA3-E505-CS-2	Criticality RBI Component - Cylindrical Shell	NA3-E505-CS-2	Heat Exchanger-Channel	7/19/98	79	857
NA3-E505	NA3-E505-EB-1	Criticality RBI Component - Exchanger Bundle	NA3-E505-EB-1	Heat Exchanger-Bundle	7/19/98	79	857
NA3-E506	NA3-E506-CS-1	Criticality RBI Component - Cylindrical Shell	NA3-E506-CS-1	Heat Exchanger-Shell	7/19/98	163	230
NA3-E506	NA3-E506-CS-2	Criticality RBI Component - Cylindrical Shell	NA3-E506-CS-2	Heat Exchanger-Channel	7/19/98	173	291
NA3-E506	NA3-E506-EB-1	Criticality RBI Component - Exchanger Bundle	NA3-E506-EB-2	Heat Exchanger-Bundle	7/19/98	173	291
NA3-E507	NA3-E507-CS-1	Criticality RBI Component - Cylindrical Shell	NA3-E507-CS-1	Heat Exchanger-Shell	7/19/98	77	367
NA3-E507	NA3-E507-CS-2	Criticality RBI Component - Cylindrical Shell	NA3-E507-CS-2	Heat Exchanger-Channel	7/19/98	77	367

Conclusion

Best practices for performing data mining for integrity management projects can be summarized in the following key points.

- Creating a data strategy at the onset of the project will insure that all important risks are understood and will reduce the probability of cost and schedule overruns.
- Identifying all potential repositories of data and records early in the process will avoid re-work once underway.
- Most companies suffer from "aging out" of critical knowledge to the project, as people retire and leave the company. Locating and interviewing these people should be included in the project plan.
- Take advantage of existing structured database data whenever possible, as it represents the cheapest and fastest means of fulfilling the data needs for the project.
- Avoid expensive manual repetitive processes by leveraging technology to classify, parse, and extract data from documents.
- Systematize the use of rules to normalize and transform data prior to creating a load sheet.
- Strive to create a "single version of the truth" which can be leveraged by multiple corporate applications.
- Instruct EPC vendors to submit more detailed and indexed turnover documents on future projects to minimize the legwork of data mining in the future.

12TH GLOBAL CONGRESS ON PROCESS SAFETY

REVITALIZING AUDITING PRACTICES

ONE COMPANY'S JOURNEY

Hope A. Luebeck &
Robert S. Romano
Veolia North America
125 Factory Lane
Middlesex, NJ 08846
robert.romano@veolia.com

Prepared for Presentation at
American Institute of Chemical Engineers
2016 Spring Meeting
12th Global Congress on Process Safety
Houston, Texas
April 11-23, 2016

Paper Revitalizing Auditing Practices - One Company's Journey

Hope A. Luebeck
Robert S. Romano
Veolia North America
125 Factory Lane
Middlesex, NJ 08846
robert.romano@veolia.com

Keywords: Audit, Auditing, Process Safety Management, PSM, OSHA

Abstract

Auditing can become a mundane and unsuccessful task when done repeatedly the same way by the same personnel. Auditors can begin to get stale in their approach and not dig deeply enough to identify the real root cause(s) of an organization's Process Safety Management (PSM) program issues. It can be beneficial to invigorate the process by utilizing innovative approaches for evaluating systems from a different perspective as well as by including new auditors. This allows for improved identification of deficiencies and opportunities to share best practices with sites. It also can help a facility to be better prepared for a potential external or regulatory group audit should one occur.

Veolia North America has re-invigorated its PSM auditing process by revamping how it plans and conducts internal audits of its PSM-covered facilities. These include aspects of pre-auditing sites, utilizing employees from sister facilities, and infusing auditing teams with third-party support, among other approaches. While the changes we have made are still works-in-progress, we believe these modifications will only strengthen our PSM programs. This paper will share examples of how Veolia has modified these vital reviews to make them more productive while training additional personnel to serve as future audit leaders and enhancing the depth of its auditing team.

1 Reasons for Auditing

Why do we audit our systems and programs? Have we ever really considered the reasons for conducting internal reviews on some periodic basis? First and foremost ... the reason for auditing is to ensure the safety of our employees. OSHA's PSM standard is rooted in employee safety. Auditing the program ensures that the facility is following OSHA's standard. Furthermore, auditing is essential for determining if the facility is in compliance with a

Federal/State regulation or a corporate program. We are essentially checking ourselves to see if we are doing what we say we are doing. We may be trying to drive program development by finding gaps and deficiencies in our systems. We may also conduct corporate or internal audits because we want to be sure that an unbiased look at a site, facility, or a group of locations takes place. We may even audit due to known or uncovered historical issues at a location. The fact of the matter is that we internally audit because we would prefer to identify problems internally rather than to have an external regulator or agency find them. The latter could result in a number of detrimental outcomes, among them the potential for: loss of company reputation, fines, negative impact on business, and other adverse possibilities.

2 Poor Audit Results/Failed Audits

Some audits are undoubtedly more productive than others. What may constitute a poor or failed audit? These may include repeat findings in subsequent audits, a continued down spiral of performance, or increasingly unsatisfactory audit results. How is this possible? Why might this be happening? Do any of the findings have a direct impact on employee safety?

Certainly management support and follow-up of audit findings and recommendations is critical for addressing audit problems. Without the support of management and the providing of resources to resolve identified issues, things are not likely to change for the better.

Other reasons for poor audit results could be the auditing practices themselves – the audit team, their focus, and their general performance. Many companies have internal audit teams, whose primary purpose is to conduct internal audits of company facilities and operations system-wide. These groups of people live on the road constantly and perform the same tasks repetitively. They can be burned out from doing the same thing and can lose their edge in seeking true root causes for audit issues.

If auditing is going to give meaningful information on effectiveness as well as compliance, the skill set and understanding of the auditor needs to match the aspirations.[1] Auditors, after all, are human beings, and can have their own weaknesses. If they are experts in certain PSM elements, they can become too comfortable in auditing what they know and possibly not dig deeply enough to identify root causes. This can come from auditing the same elements all the time, so the material gets stale with them, despite their expertise.

Audit teams may not be well-coordinated and this can cause lack of effectiveness with a team. They can have an inappropriate focus, or not drill deeply enough to find real issues with a PSM program. It can become too easy to follow a predetermined outline or form for auditing. Yes/No questions while they are necessary can stifle the true learning and uncovering of information during an audit. Asking open ended questions (i.e. How, Why, Can you explain, Can you show me) and combing through paperwork instead of only verifying the existence of specific documents are a great way to dig deeper in an audit.

Auditors may not manage their time effectively, and thus not review enough items to identify challenges within a system. While the audit team leader may be responsible for the performance of the team, the team members must also hold up their end of the bargain for effectively investigating real issues. In some cases, auditors may not even trust their instincts enough to continue looking for gaps and deficiencies in a given program. They may not look closely enough at incident history to get a real feel for what may be wrong with a local PSM program.

It can be hard to keep audits both effective and fresh. Veolia has recognized this and has taken measures to manage both aspects. In our efforts to revitalize and advance our PSM program, we also have taken a look at how we conduct our internal audits. We want to avoid some of the pitfalls which are common with auditing but yet make our audits yield better results. This does not mean that we want more findings (which sometimes can be an initial outcome of a refocusing on auditing practices), but we want better quality findings which will really contribute to driving positive change in our PSM program and to help sites be more PSM compliant. PSM Effectiveness = function [PSM performance + PSM efficiency] [2]

3 Methods to Improve Auditing Effectiveness

So what are some things which can be applied to a company's auditing practices to make them more meaningful and to help truly identify weaknesses in a company's program? Among the things which Veolia has utilized to stimulate new growth in their PSM program, while refocusing on the way it conducts its audits, are the following: Changing who audits which PSM elements, use of third party auditors, conducting "OSHA-style" interviews during the audit, overlapping audit team member approaches for multiple elements, and drilling deeper regarding proper and consistent execution/completion of forms.

Auditors may become comfortable auditing particular elements, especially topics they know best. This can certainly be a great benefit but over time they may possibly become stagnant in their approach and could miss other issues that a fresh set of eyes may see. This is why it can be beneficial to have auditors change which elements they review so they remain knowledgeable with many aspects of the PSM standard. This forces each auditor to pay particular attention to each element they review as well as stay sharp in other PSM focus areas. Auditors can also become overconfident that they know their element, and may assume some things about the site/facility/process being audited that someone newer to the game may audit more closely due to their lesser familiarity with those elements.

The use of third party auditors can be immensely beneficial. They generally audit with no biases (i.e. they don't know your facility) and they have a wealth of knowledge from industry and the variety of companies for which they have audited in the past. They are often adept in knowing common weaknesses for particular PSM elements. They also can share a variety of best practices and tools which can be most helpful for companies being audited.

Veolia has unfortunately been visited by OSHA in the recent past following an incident at its West Carrollton, OH facility in 2009. As part of that investigation, OSHA auditors spent time interviewing operators to collect facts regarding the circumstances of the blast and potential causes. From this experience, we learned how OSHA representatives may search for details by the way they pose questions and ask follow up questions. Our management thought it would be a good idea if we could mimic this inquiry style when we conduct PSM audits of our facilities. This serves a dual purpose for us. Our auditors gain the information necessary to determine if the site is PSM compliant, and the interviewees gain the valuable experience of how they might answer questions in an external audit situation should OSHA or some other regulatory group come calling. While it is hoped that this will never be necessary, this does educate personnel about how to answer questions and stay relevant to the topic.

An example of this would be an auditor interviewing an employee on a specific process. The line of questioning may include: How long have you been employed at the facility? Have you been trained on this equipment? Who trained you? Have you received any further training after your initial training? Did you read the Standard Operating Procedure (SOP) for this process? The auditor may now pull the trainer in to be interviewed. The line of questioning may include: How long have you been employed at the facility? Have you been trained on this equipment? Are you familiar with the SOP? What makes you certified to train on this equipment? Did you approve the operator on this equipment? Why did you approve this operator? The auditor may then request to see the certification for the trainer of the employee. They then may ask to see the SOP that the employee was trained on. They will also look for the employee's training documents on the process in question. Based on the information in the SOP the auditor may ask to see the referenced information such as a Personal Protective Equipment (PPE) document. As you can see, a simple line of questions and the answers that the auditor receives can open a number of different avenues for questioning and fact finding. This is a simple example with only a few questions but it does illustrate how quickly and how far an auditor can reach utilizing interview techniques.

People who have been part of their company's auditing team will certainly understand that when one specific PSM element is being audited it is not too difficult to overlap with related or complimentary elements. If this is recognized in advance as the team is planning their audit schedule, it may be possible to take a deeper dive into a site's PSM program. For example, auditors which may separately be auditing for Operating Procedures and Training might coordinate their efforts to see how the personnel being audited for each particular element might also be indirectly complying (or not) with the other element. The auditors can look for common emphases which may be able to uncover systemic issues which occur over two or more elements and indicate some sort of management or program deficiency on a larger scale.

We have also noticed that problems with completion of forms for various PSM elements (e.g. Management of Change (MOC), Hot Work, Confined Space Entry, etc.) often stem from similar issues. These may include inadequate or ineffective training, poor monitoring of contractor activities, lack of proper follow through on line items or permit/form close out and similar items. It may also indicate the degree to which a site may screen its own completed forms and permits. Are those with responsibility for those reviews completing them as intended or often enough? Is the need for retraining (or whatever the corrective action is for the deficiency) recognized and acted upon as required? Are the safety items that the form is documenting being performed properly? When there are problems with forms, it is not unusual for the problem to be systemic at a site depending on how these are managed and kept up to date at a local level.

4 Building PSM and Auditing Competency

As mentioned earlier, Veolia now adds a third party auditor to its PSM audit teams. Besides the external experience and knowledge these auditors bring to the table, they also can serve as mentors for employees learning how to audit. This provides guidance for these employees, both with auditing techniques as well as the PSM standard itself. This process also allows personnel from sister facilities to see how another Veolia site manages its PSM program. The trainee can bring both good and bad information back to their site, including learnings from the audit and whatever best practices they may have observed or been told about during the audit. In time, they will grow in PSM and auditing knowledge to eventually become a future audit leader for Veolia.

Another tool we use to build PSM and auditing competencies is to have personnel being audited help to compose their own site's findings and recommendations for PSM elements. Of course, this occurs under the guidance and supervision of an experienced audit team member. This provides insight to the auditee about what the standard requires and why an item may be deficient. It helps the auditee learn how to properly word responses. It helps a site better understand holes in their own system and hopefully provides them with a new appreciation of the PSM standard. Lastly, it can establish better ownership of individual audit recommendations by the site auditees since they were involved in determining what issues exists.

5 Resolving Findings

Naturally, once an audit is complete and the site/facility/company has accepted what the findings and recommendations may be, it is time to set a path for actions to be taken to address the noted shortcomings. The site is expected to utilize a particular software package to keep track of progress with their action items. A key to the success of any long-term initiative is the level of ongoing support it receives. This requires accountability for audit follow-up and the timely resolution of audit gaps. [3] Any action items that directly touch upon safety must be completed within an acceptable timeframe for the deficiency identified. Corporate oversight may be provided if one or more action items may require special support or intervention to resolve. Where a site has a problem which has previously been noted at a sister facility, a contact at that earlier location may be provided to the new site to provide consultation with what actions were taken. The ability to work together to prevent a facility from reinventing the wheel is paramount in resolving issues in a short period of time.

Sometimes the problems which need to be resolved reside at the corporate level and not at the site level. This is not necessarily unusual and will occur occasionally. When this is the case, a responsible person or department will be tasked with the item and it will be expected to be resolved and communicated to effected parties when completed.

What is unique to our business group is that many of its sister facilities are not grass roots plants for Veolia. Rather, they were part of numerous other owners' businesses purchased by Veolia and incorporated into its structure. As such, learnings from site to site are not always directly transferrable. Though a challenge, this at times, results in some "low hanging fruit" where issues resolved at one location are certainly applicable at a similar, but not unique other location.

We also do periodic overviews of our audit trends to identify systemic issues. Statistics of these may show that there are common problems across sister sites, a particular type of technology, or some policy that is being misapplied or misunderstood. These can be used to then identify appropriate solutions which ought to resolve common problems over multiple locations. Follow up on items of this sort may be conducted via conference calls or webinars with the effected sites as progress is made with the item and input from the site is collected along the path of progress. Outcomes of such reviews can lead to new or improved corporate policies, training or re-training emphases, or the need for additional resources to drive the needed progress.

In advance of an upcoming PSM audit, a corporate representative knowledgeable in the PSM standard will often reach out to the site and review the completion of past audit items. This often will be done several months in advance of a new audit so that if some recommendations may be weak in the degree of resolution, some guidance may be provided to help the site conduct additional activities to bring the item to a more complete and effective closure. This provides yet another opportunity for teaching and mentoring in advance of a new audit.

6 Other Factors that Affect Auditing

In addition to the auditing topics already discussed there are a number of other factors that affect the auditing of a PSM program. With so many different recognized and generally accepted good engineering practices (i.e., RAGAGEPs such as ANSI, API, NFPA) and regulatory agencies (i.e. OSHA, state DEP, EPA) touching our industries, it is sometimes difficult to sort out what standard to abide by in each situation or which standard supersedes the other. [4] It is the auditing team's job to ensure that the facility is compliant with their program, but it is the facility's job to determine what the proper regulations and RAGAGEPs are for their program. This is not typically completed in a vacuum. Information sharing between sister sites, corporate guidance, and common practices within an industry are all used to determine the proper regulations and RAGAGEPs to follow. In addition to these resources, based on the team members' education and experience, the auditing team can also play a part in aiding the facility with these decisions.

The permits that are imposed on a facility by state and federal agencies are another factor that could affect auditing. These permits (i.e. air permits, operating permit) dictate how specific operations must be carried out at the facility. Auditors must take these permits into consideration when critiquing design, procedures, and other components of PSM.

7 Beyond Compliance

In this paper we have discussed the subject of auditing for compliance with internal and external governing bodies, specifically OSHA's PSM standard. We must not forget what the root of the PSM standard is all about. On July 17, 1990, OSHA published in the Federal Register (55 FR 29150) a proposed standard,—"Process Safety Management of Highly Hazardous Chemicals"—containing requirements for the management of hazards associated with processes using highly hazardous chemicals to help assure safe and healthful workplaces. [5] All of the efforts of our auditing are to help assure a safe workplace for ourselves and our co-workers.

At times auditing can be a laborious, tedious, and painstaking job. Digging into information and looking for small details can become frustrating. Being on the receiving end of audit deficiencies can also be tough to take as well. But auditors and auditees alike must remember the true reason as to why we audit – safety. Although auditing alone does not ensure safety, it is one of the many tools we use to ensure that we are doing the right thing.

8 Conclusion

Progress in process safety programs is often asymptotic. Even the best of programs still have improvements to be made and surely the bar continues to be raised over time. Companies have to work hard to prevent their PSM practices from being non-compliant. Audits are a valuable tool to help to uncover gaps or deficiencies in a system but even these are prone to stagnation and ineffective execution. It helps to keep audits and internal audit methods fresh to get the most from them and to ensure that as much as possible, the root cause for system deficiencies are identified and addressed so we can operate our facilities safely and compliantly.

While the methods described in this paper are proving to be effective for this company, they may not be the best choice for some organizations. Each company should seek what tools and practices are optimal for their audit teams to yield the best results possible for their organization. It is generally good to change one's practices periodically to achieve the most. This keeps auditors and groups fresh and invigorated to do their best.

9 References

[1] Stephen Gill. "PSM Auditing: Thinking Beyond Compliance" Process Safety Progress, November 2015.

[2] Steve Arendt. "Continuously Improvising PSM Effectiveness – A Practical Roadmap" Process Safety Progress, June 2006.

[3] Martin C. Welch, R. Bruce Hartman. "How Sustainable are Your PSM Processes?" CEP Magazine, June 2003.

[4] www.osha.gov/SLTC/processsafetymanagement/standards

[5] https://www.osha.gov/Publications/osha3132.pdf

12TH GLOBAL CONGRESS ON PROCESS SAFETY

Beyond Compliance Auditing:
Drill 'til you find the pain points and release the pressure!

Laura Ankrom, P.E.
aeSolutions, Inc.
250 Commonwealth Drive, Ste. 200, Greenville, SC 29615
Laura.ankrom@aesolns.com

Kathy Shell, P.E.
aeSolutions, Inc.
250 Commonwealth Drive, Ste. 200, Greenville, SC 29615
Kathy.shell@aesolns.com

Prepared for Presentation at
American Institute of Chemical Engineers
2016 Spring Meeting
12th Global Congress on Process Safety
Houston, Texas
April 11-23, 2016

Beyond Compliance Auditing:
Drill 'til you find the pain points and release the pressure!

Laura Ankrom, P.E.
aeSolutions, Inc.
250 Commonwealth Drive, Ste. 200, Greenville, SC 29615
Laura.ankrom@aesolns.com

Kathy Shell, P.E.
aeSolutions, Inc.
250 Commonwealth Drive, Ste. 200, Greenville, SC 29615
Kathy.shell@aesolns.com

Keywords: Verification, Auditing, Compliance Auditing, USEPA Risk Management Program, OSHA Process Safety Management, Process Safety Lifecycle, ISA84.91, IEC 61511, Independent Protection Layer (IPL)

Abstract

In *Failure to Learn: the BP Texas City Refinery Disaster*, the author, Andrew Hopkins, explores the importance of senior managers using auditing as a tool to identify "unrecognized problems that may be lurking beneath the surface" (Hopkins, 2010, p. 114). He also goes on to state that auditors "will not set out to sample the organization. Rather, they will use their expert knowledge to zero in on areas where things might be going wrong" (Hopkins, 2010, p. 115).

Building on this concept, the authors of this paper will look beyond traditional OSHA PSM and USEPA RMP regulatory compliance auditing to explore the value of drilling down around the process safety lifecycle; locating the pain points; and releasing the pressure on the system. Compliance auditing has historically provided a "check-the-box" approach to meet regulatory requirements imposed by OSHA and USEPA. Regulatory compliance, however, is no guarantee of the prevention of major accidents. There is still a need to identify hazards, understand and manage risks. Today's auditors need to determine how to systematically identify the root cause of the "pain points" that will foster conversations around releasing the "pressure" on existing practices to achieve a vibrant integrated process safety management system.

Industry standards, such as ANSI/ISA 84.00.01 and IEC 61511, continue to drive the evolution of functional safety and meeting these requirements go far to demonstrate that a company has reduced risk to a tolerable level. The ANSI/ISA/IEC lifecycle approach actually quantifies the value of an integrated process safety program by assigning a numerical value to risk and assessing the effectiveness of the independent protection layers (IPLs) at meeting safety integrity targets. There is a growing realization in industry that having a healthy process safety management program in place is essential in meeting the full intent of IEC 61511 and ISA84.91.

The authors will present a drill down audit methodology that focuses on the organizational interfaces and the related management system procedures and practices used to meet the defined safety integrity requirements of both instrumented and non-instrumented IPLs. The approach utilizes a hazard scenario-based drill down audit methodology to expose the issues and the sources of the "pain" by digging deep into the management system processes around process hazard analysis, process safety information, mechanical integrity, operating procedures and management of change. When exposed in this manner, there is an audit trail that provides a basis for revising the work flow to "release the pressure" and accomplish the risk management objectives going forward.

1 Introduction

Organizations have been implementing the regulatory requirements of the Occupational Safety and Health Administration's (OSHA's) Process Safety Management (PSM) program and the United States Environmental Protection Agency's (USEPA's) Risk Management Program (RMP) since 1992. They have completed multiple revalidations of their Process Hazard Analyses (PHAs). Many are using the Layers of Protection Analysis (LOPA) methodology to assess their safety integrity requirements and assigning Independent Protection Layers (IPLs) to protect against their high hazard scenarios. They are also applying a risk-based focus on managing their IPLs as "critical safeguards" on their journey toward implementation of sustainable process safety management systems.

With the drive for continuous improvement and the evolution of so many aspects of PSM and RMP practices over the years, why are we still witnessing catastrophic accidents? As Andrew Hopkins has addressed in *Failure to Learn* and his subsequent book, *Disastrous Decisions*, there is a need for leadership to take a more defined role in process safety. Leaders must have an understanding of the highest hazards present in their organization and what risk barriers are in place to protect against major incidents. Assurance that these are installed, properly managed, and available to respond moves organizations one step closer to managing their hazards at the source.

In *Disastrous Decisions*, Hopkins states "If major hazards are to be managed effectively, we must first devise appropriate indicators and then make those indicators matter." (Hopkins, 2012, p.81). Figure 1.1, the Process Safety Indicator Pyramid, reprinted from the American Petroleum Institute in Recommended Practice 754, has become an industry best practice model for establishing leading and lagging indicators. Tier 1 and 2 events include any loss of containment which has significant consequences. Most organizations have very infrequent Tier 1 and 2 events, which places an emphasis on learning from Tier 3 and 4 events. A Tier 3 event "represents a challenge to the barrier system that progressed along the path to harm but is stopped short of a Tier 1 or 2 loss of primary containment" (Hopkins, 2012, p. 83). The Tier 3 events, such as an excursion from safe operating limits; inspection and test results outside acceptable limits; or a demand on a safety system, are the leading indicators that point to the problems that are "lurking beneath the surface" (Hopkins, 2012, p. 83). These indicators may point out deficiencies in an organization's process safety management systems that undermine the integrity of their IPLs. The very IPLs that are protecting against their highest hazard scenarios.

Figure 1.1: Process Safety Indicator Pyramid (API 754, p.8)

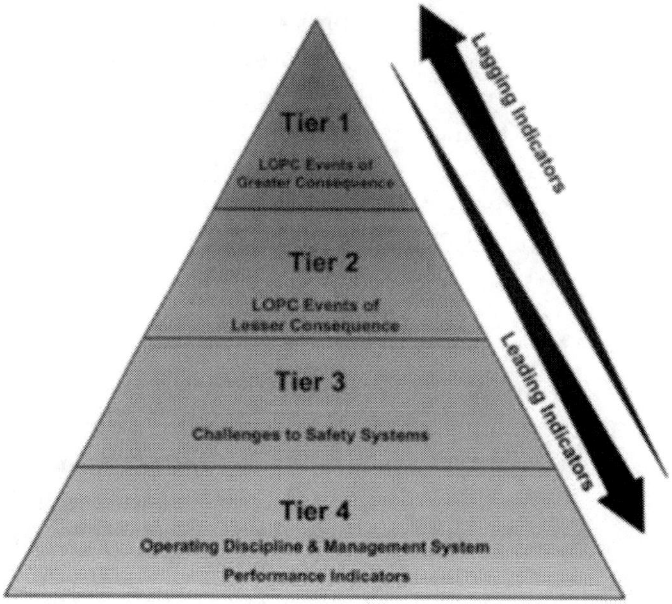

The hazard scenario-based, drill down auditing approach presented in this paper is intended to uncover weaknesses in the management systems that manifest themselves as findings when verifying the consistency of the data in the PHA and operating procedures; and the integrity of the IPLs against the requirements in International Society of Automation Standards, ANSI/ISA 84.01.01(Modified IEC 61511) and related recognized and generally accepted good engineering practices (RAGAGEP). A focused audit of this nature yields findings that are leading indicators in the same sense that Tier 3 events are; they provide actionable direction before a Tier 1 or 2 event. A hazard scenario-based, drill down audit is directed at finding issues with the handoff between management systems to ensure that the engineered and administrative protection layers in place are properly installed, managed and maintained such that they will be available to respond if called upon.

This methodology differs from a more traditional compliance audit, which is often segmented by prevention program element and reliant on the drill down on the interconnectivity of a few elements, such as the Management of Change (MOC) process, to uncover systematic problems. The author is not suggesting that the scenario-based audit supplant the element-based audit, but encourages organizations to put programs in place which draw on the most effective practices of both techniques to produce a valid indicator of the integrity of the process safety management systems in place. This form of auditing also complements the Center for Chemical Process Safety's (CCPS') recommended approach to monitor Risk Based Process Safety maturation over time through trending of relevant performance metrics to identify management system weaknesses (CCPS, p.614).

2 Hazard Scenario-Based, Drill Down Audit

The hazard scenario-based, drill down audit is focused on a trail that begins with the PHA/LOPA and the credited IPLs (or safeguards, if LOPAs have not been performed), then drills down through the management systems intended to ensure their integrity, finishing with the related human factors considerations. This approach provides a clearer view of whether the information in the PHA/LOPA has been fully integrated in the systems, procedures, and practices, as well as whether they are in place and are functioning as intended. OSHA's National Emphasis Program guidance reflects a similar drill down approach to auditing in that the emphasis of the audits (or inspections) is on implementation versus documentation. The objectives are to confirm that the facility has an integrated, functioning process safety program and to provide an indication to leadership, with a degree of confidence, that their plant is operating within the desired risk tolerance criteria.

2.1 Defined IPL Characteristics

LOPA methodology, or other quantitative risk analysis methodology, is being used to evaluate if adequate safeguards are in place for hazard events with high severity or high risk consequences to meet many organizations' risk tolerance requirements. During a LOPA, safeguards are identified that are expected to independently interrupt an initiating event from progressing to an undesired consequence. These safeguards credited for risk reduction are classified as IPLs. Publications from the CCPS and the ANSI/ISA 84.00.01 (Modified IEC 61511) and 84.91.01 specify that in order for a device, system or action to be considered an IPL it must meet the characteristics of specificity, independence, dependability, auditability and security. The IPL can be passive or active as long as the criterion in Table 2.1 are met.

Table 2.1: ISA 84 IPL Classification Characteristics

Characteristic	Description
Specificity	The IPL is designed to prevent or mitigate the consequences of the identified hazard.
Independence	An IPL shall be independent of the initiating cause and all of the other protection layers associated with the identified Hazard Event Scenario. Independence requires the performance must not be affected by the failure of another protection layer or by the conditions that caused another protection layer to fail.
Dependability	The protection provided by the IPL shall reduce the identified risk by at least ten-fold. In terms of availability, the IPL must be at least 90% available.
Auditability	The IPL must be designed to allow regular validation of the protective function.
Security	The IPL security shall be managed by design or by administrative procedure to ensure that unauthorized changes are not made that affect the integrity of the IPL, its availability, or any of its properties.

2.2 OSHA PSM Compliance and ANSI/ISA S84 Overlap

OSHA acknowledged in 2000 that ANSI/ISA S84.01, *Application of Safety Instrumented Systems for the Process Industry*, is a national consensus standard that is considered RAGAGEP for safety instrumented systems. OSHA issued a letter of interpretation stating that an organization meeting the intent of ANSI/ISA S84.01 is also meeting the PSM requirements for safety instrumented systems. In validating ANSI/ISA S84.01 requirements for an IPL, an auditor is validating many of the PSM/RMP program elements as well. In other words, execution of the safety instrumented lifecycle provides an opportunity to build and reinforce an integrated process safety management program. The overlap between ANSI/ISA S84.01 and PSM is shown below in Table 2.2.

Table 2.2: OSHA PSM Compliance and ANSI/ISA S84.01 Overlap

OSHA PSM Element	ANSI/ISA 84 Clause
Process Hazard Analysis	Clause 8 & 9 - Hazard & Risk Assessment
Process Safety Information	Clause 19 - SIS Information & Documentation
Operating Procedures	Clause 16 - Operation & Maintenance
Pre-Startup Safety Review	Clause 17 - SIS Modification
Mechanical Integrity	Clause 16 - Operation & Maintenance
Management of Change	Clause 5 - Management of Functional Safety

2.3 Drill Down Methodology

The methodology for the hazard scenario-based, drill down audit, presented herein, seeks to validate that the causes, consequences and credited IPLs are embedded in an organization's operating discipline, that the IPLs meet all of the defined characteristics in Table 2.1, and that they are fully integrated into the process safety management systems in Table 2.2.

Appendix A illustrates the suggested audit criteria for five key types of IPLs which include:

- Alarm,
- Pressure Relief Valve (PRV),
- Basic Process Control System (BPCS) Interlock,
- Safe Operating Procedure (SOP), and
- Safety Instrumented Systems (SIS).

Referencing Appendix A, the following is an example of a drill down audit trail for an Alarm IPL.

2.3.1 PHA/LOPA

A review of the PHA/LOPA should verify that the operator, alarm sensor, and final elements used by the operator are independent of the Initiating Event and other IPLs for the scenario. The auditor must also verify that the operator action will fully prevent the cause from propagating to the final consequence; either through documentation in the PHA, review of relevant process safety information (PSI), and/or interviewing an engineer knowledgeable in the process.

Example Systematic Issue(s):

If there is a lack of independence there is a potential for common mode failure which degrades the risk credit given the IPL. This may also be an indication of lack of experience of the facilitator on the LOPA methodology and the methods the organization uses to qualify their facilitators.

2.3.2 Process Safety Information (PSI)

A review of PSI would confirm that the alarm sensors are maintained on the critical IPL list and on the piping and instrumentation diagram (P&ID); the sensor data sheet and final elements are in place; and the basis for the Probability of Failure on Demand (PFD) is well documented. The auditor should verify that the alarm set point is based on the calculated time from detection to the consequence of concern (i.e. maximum calculated response time) with ample time for the operator to receive the alarm, identify the issue, and take intended action. During a field inspection, the auditor should verify that the devices are installed correctly and have open flow to the process, by physically walking down a P&ID.

Example Systematic Issue(s):

Lapses in PSI can indicate that there is a potential issue with maintaining PSI as current through the MOC process.

If the basis for the alarm set point is not readily available or found to be inadequate to support the operator action, then a lack of dependability may exist. This may reflect a lack of recognizing the need for documentation to validate the assumptions, or an issue with recognizing the significance of crediting the operator with a risk reduction factor of 10. This could also be due to not having the right team make-up in the PHA/LOPA with an engineer knowledgeable of the process (i.e. automation engineer).

2.3.3 Mechanical Integrity

Review of the mechanical integrity information should verify that calibration and proof test procedures are available; testing, calibration and inspections are scheduled at a routine frequency; and calibration and proof test records are reviewed, actioned if required, and maintained. If testing, calibration or inspection records indicate that the sensor was found to be in an undetected failed state, confirm that a near miss was initiated. Verify that records do not indicate that the field devices are prone to problems due to fouling or external environmental conditions.

Example Systematic Issue(s):

Lack of proof test procedures and routinely scheduled inspections and testing bring the reliability for the alarm sensor to function as intended in question or at risk. If there are inspection records illustrating a trend of performance issues for a device, without consideration of causes and fixes, it may represent the need for a change in the work flow where an individual is assigned responsibility for the reliability management of the instrumented functions.

2.3.4 Operating Procedures

The auditor must confirm that the alarm, along with consequences of deviation, intended operator action, and the specific parameters/authorization for bypass of the alarm are documented in the appropriate operating procedures and/or alarm summary. They must also be included in the operator training program with up-to-date documented training records.

Through operator interviews and field checks, the auditor should assess that the operator has a clear understanding of the criticality of the alarm, the consequence of concern, the urgency of responding, the intended action, and the expected automated action if it escalates. If this alarm has been activated was it captured as a demand for the revalidation process and investigated as a process safety Tier 3 event.

An auditor must validate that the operator has a means of identifying an alarm as an IPL from the console and can respond effectively in the required time. Documentation of drills, tests or simulations should be reviewed to ensure the operator has sufficient time to complete the required actions and verify that the actual response time is safely within the calculated maximum response time.

Operators should be able to tell the auditor what they would do in the event of the alarm (i.e. what their first steps would be as a response to the alarm, how they would diagnose it, and what action they would take). The auditor should also verify with the operator that their response will not place them in harm's way.

Example Systematic Issue(s):

Lapses in alarm dependability may be due to lack of proper operator training, inadequate training methods, or infrequent training on abnormal event management. The facility may not have a process in place to validate the reliability of the operator response, which results in it being considered at risk.

The facility may not have a process in place associated with demands on IPLs outside of an SIS.

If there is an inconsistency on how two operators would respond to the same alarm, the operating procedures and alarm summaries may be poorly written. It may also reflect obsolete procedures which no longer reflect practices in place, or possibly a culture of complacency where procedures are not adhered to and thus are not highly regarded as entirely necessary to learn and follow.

2.3.5 Management of Change

Much like a standard compliance audit, the auditor will also need to track an MOC down a trail to determine if changes to the alarms credited as IPLs are managed appropriately. The auditor should verify that process records exist for any changes to the DCS/PLC logic and alarm set points are securely programmed in the DCS/PLC with limited access control to the code (i.e., operators cannot adjust the set point and there is access control to the logic). The auditor should confirm with operators that they would initiate a bypass management process if they take an alarm IPL out of service. If alarms have been in a bypassed state for an extended period of time, determine if they have instituted a temporary MOC process. Any permanent changes should have resulted in an update to the PSI, Operating Procedures and Training, and Mechanical Integrity requirements, as applicable.

Example Systematic Issue(s):

If MOC issues have been identified it questions the sustainability of the operating discipline in place for continued safe operation. The auditor should continue to probe or assess where the breakdown is in the intended MOC process. This might require a need to revisit the work flow processes, roles and responsibilities, and training.

3 Conclusion

The goal of the hazard scenario-based, drill-down audit is to increase an organization's confidence that they are meeting their risk target criteria and increase their assurance that they are uncovering the systematic issues to open up opportunities for improvement that would have otherwise not been apparent. Leaders can constructively align their communications and actions around the findings of scenario-based compliance audits because they reflect degrees of achieving "operations excellence," where practices follow procedures and organizational interfaces are successfully managed day in and day out to the betterment of the whole. At the end of the day, leaders need to ensure that their hazardous processes are being operated within the risk envelope that they are intending to operate within and have a sense of assurance that they are effectively managing their risks, identifying their pain points, and relieving the pressure.

4 References

American Petroleum Institute, "Process Safety Performance Indicators for the Refining and Petrochemicals Industries," API Recommended Practice 754, API. Washington, DC. 2010.

Center for Chemical Process Safety, "Guidelines for Risk Based Process Safety," Wiley. New York. 2011.

Hopkins, Andrew, "Disastrous Decisions: The Human and Organisational Causes of the Gulf of Mexico Blowout," CCH Australia Limited. Sydney, Australia. 2012.

Hopkins, Andrew, "Failure to Learn: The BP Texas City Refinery Disaster," CCH Australia Limited. Sydney, Australia. 2010.

Appendix A: Suggested IPL Audit Criteria

IPL TYPES				
Alarm	**Pressure Relief Valve (PRV)**	**Basic Process Control System Interlock (BPCS)**	**Safe Operating Procedure (SOP)**	**Safety Instrumented System (SIS)**
Independence				
• The operator, alarm sensor, and final elements used by the operator are independent of the Initiating Event and other IPLs for the scenario. (PHA) • The operator response will not place the operator in harm's way. (OPS)	• The consequence level of the hazard scenario was determined independent of the PRV in the IPL. (PHA) • The failure of this PRV is independent of the LOPA scenario initiating cause(s). (PHA) • This PRV required is fully independent and separate from all other IPLs credited in the same LOPA scenario(s). (PHA)	• Failure modes for the BPCS interlock logic solver, sensors and control elements are independent of the LOPA scenario initiating cause(s). (PHA) • BPCS interlock's logic solver, sensors, and control elements are independent of other credited IPL(s). (PHA)	• The consequence level of the hazard scenario was determined independent of the SOP and the related operator in the IPL. (PHA) • The SOP, the related Operator Action, and the final element(s) that the operator may use are independent of the LOPA scenario initiating cause(s) and other credited IPL(s). (PHA)	• The Safety Instrumented Function's (SIF's) logic solver, sensors, and control used by the operator are independent of the Initiating Event and other IPLs for the scenario. (PHA) • The failure of this SIF's logic solver, sensors, and control elements are independent of the LOPA scenario initiating cause(s). (PHA) • The potential for Common Cause Failure between this SIF and other IPLs credited in the same LOPA scenario(s) has been ruled out, or quantified in the SIL calculations if another IPL is also a SIF. (PHA)
Specificity				
• The operator action will fully prevent the cause from propagating to the final consequence. (PHA)	• The PRV is only bypassed with heightened administrative controls and the bypass is authorized per site procedures. (OPS/MI)	• The data sheets for the BPCS Interlock's sensor(s) and control element(s) are complete and available. (PSI)	• The intended operator action(s) in the SOP are such that they fully prevent the cause from	• The logic block, sensors, and control elements are only bypassed with heightened administrative controls. (OPS)

IPL TYPES				
Alarm	**Pressure Relief Valve (PRV)**	**Basic Process Control System Interlock (BPCS)**	**Safe Operating Procedure (SOP)**	**Safety Instrumented System (SIS)**
• Sensors and final elements data sheets are complete. (PSI) • The operator has a means of identifying an alarm as an IPL from the console. (OPS) • The alarm set point provides ample time for the operator to receive the alarm, identify the issue, and take action within the maximum allowable response time. (PSI) • The alarm sensor is depicted correctly on the P&ID. (PSI)	• The data sheet is complete and available. (PSI) • The data sheet for an Associated Protective Device (APD), if applicable, is complete and available (e.g. rupture disk). (PSI) • All manual valves which can block this PRV are secured (locked or car-sealed) open. (MI) • If this PRV is in potentially fouling service, freezing service, or there are extreme environmental conditions, measures have been implemented to counter these effects (e.g., heat tracing, rupture disk) and they were taken into account in the risk reduction credit taken. (MI) • If this PRV is coupled with a Rupture Disk, a pressure gauge is installed in the interstitial space to detect leakage. (MI) • The outlet of this PRV has been designed for safe discharge (e.g., sized for adequate flow, process sewer, above congested process equipment, to a flare header). (PSI)	• When de-energized, the BPCS Interlock's control elements will transition to their "safe" states for the credited LOPA scenario(s). (PSI) • The BPCS Interlock can fully transition its control elements to their safe states upon detection and trip within the Maximum Allowable Response Time. (PSI) • The interlock's sensor(s) and control element(s) are depicted correctly on the P&ID. (PSI)	propagating to the final consequence. (PHA) • The SOP is of sufficient detail to support consistent execution by the operator. (OPS) • There is ample time for the operator to execute the SOP within the maximum allowable response time. (PSI)	• The data sheets for the logic solver, sensors and control elements are complete and available. (PSI) • This SIF Interlock is designed to fully prevent the LOPA scenario consequence(s). (PHA) • A failure of the logic block, sensors, and control elements will not initiate the LOPA scenario(s) being credited. (PHA) • When the SIF detects the hazardous process excursion, it can fully transition its control elements to their safe states within the Maximum Allowable Response Time. (PHA)

IPL TYPES				
Alarm	**Pressure Relief Valve (PRV)**	**Basic Process Control System Interlock (BPCS)**	**Safe Operating Procedure (SOP)**	**Safety Instrumented System (SIS)**
Dependability				
• The basis for the Probability of Failure on Demand (PFDavg) is well documented. (PSI/PHA) • The alarm with consequence of deviation and intended operator action is documented in an Alarm Summary or SOP. (OPS) • The operator training program includes the alarm and intended action. (TRN) • The operator response time has been established through drill or calculated basis. (PSI)	• Justification of the Probability of Failure on Demand (PFDavg) selected for this IPL is documented. (PHA/PSI) • This PRV is designed to fully prevent the LOPA scenario consequence(s). (PHA)	• The basis for the Probability of Failure on Demand (PFDavg) is well documented. (PSI/PHA) • The interlock set point with consequence of deviation and intended automated action is documented in an Alarm Summary or SOP. (OPS) • The operator training program includes the interlock and intended automated action. (TRN)	• The basis for the Probability of Failure on Demand (PFDavg) is well documented. (PSI/PHA) • The critical tasks are identified in an SOP and require positive documentation that they are executed, i.e. checklist. (OPS) • The operator training program includes the SOP. (TRN)	• The location of the SIL calculation documentation giving the Probability of Failure on Demand (PFDavg) for this IPL is recorded. (PSI)
Auditability/Validation				
• Alarm and sensors are maintained on critical IPL list. (PSI) • Calibration and proof test procedures are available. (MI) • Testing, calibration and inspection are scheduled at a routine frequency. (MI) • Calibration and Proof Test Records are reviewed, actioned if required, and maintained. (MI) • If testing, calibration or inspection records indicate that	• The PRV is inventoried on the critical equipment list for management, maintenance and auditing purposes. (PSI) • If this PRV is installed in polymerizing service, fouling service, or there are extreme external conditions, it is inspected boroscopically during a shutdown. (MI) • This PRV's proof test procedure is complete, available, and includes a	• BPCS Interlock's sensor(s) and control element(s) are maintained on the critical IPL list. (PSI) • Calibration and proof test procedures are available. (MI) • Testing, calibration and inspection are scheduled at a routine frequency. (MI) • Calibration and Proof Test Records are reviewed, actioned if required, and maintained. (MI)	• There is documentation that the procedure is executed properly, i.e. completed checklists. (OPS) • Operator training records are up to date. (TRN) • SOP is certified as current annually. (OPS) • MOC records exist for any changes to the SOP. (MOC)	• SIFs are maintained on critical IPL list. (PSI) • Calibration and proof test procedures are available. (MI) • Testing, calibration and inspection are scheduled at a routine frequency. (MI) • Calibration and Proof Test Records are reviewed, actioned if required, and maintained. (MI)

IPL TYPES				
Alarm	**Pressure Relief Valve (PRV)**	**Basic Process Control System Interlock (BPCS)**	**Safe Operating Procedure (SOP)**	**Safety Instrumented System (SIS)**
the sensor was found to be in an undetected failed state, it was recorded and investigated as a near miss. (MI/II) • Verify that records do not indicate that the field devices are prone to problems due to fouling or external environmental conditions. (MI/PSI) • Operator training records are up to date. (TRN) • SOP/Alarm Summary is certified as current annually. (OPS) • MOC records exist for any changes to the software or hardware. (MOC)	"pop test" to verify actuation pressure. (MI) • This PRV is periodically removed from service, tested, and inspected for signs of corrosion. (MI) • This PRV's inspection records are well maintained, current, available, and include both "as found" and "as left" conditions. (MI)	• If testing, calibration or inspection records indicate that the sensor(s) or control element(s) were found to be in an undetected failed state, it was recorded and investigated as a near miss. (MI/II) • Verify that records do not indicate that the field devices are prone to problems due to fouling or external environmental conditions. (MI/PSI) • Operator training records are up to date. (TRN) • SOP/Alarm Summary is certified as current annually. (OPS) • MOC records exist for any changes to the software or hardware. (MOC)		• If testing, calibration or inspection records indicate that the sensor was found to be in an undetected failed state, it was recorded and investigated as a near miss. (MI/II) • Verify records do not indicate that the field devices are prone to problems due to fouling or external environmental conditions. (MI/PSI) • MOC records exist for any changes to the software or hardware. (MOC)
Security				
• Access controlled to the hardware and software. (MOC) • Specific parameters/authorization in place for bypass of the alarm. (OPS) • Alarm set points are securely programmed in the DCS/PLC	• Block valves upstream or downstream of the relief device are adequately secured to limit access. (OPS/MI) • Relief valves are periodically bench tested by a certified individual. (MI)	• Access controlled to the hardware and software. (MOC) • Specific parameters/authorization in place for bypass of the interlock. (OPS) • Trip set points are securely programmed in the	• The operator has access to the most current version of the SOP. (OPS) • Changes to the SOP are managed through the MOC program. (MOC)	• Access controlled to the hardware and software. (MOC) • Specific parameters/ authorization in place for bypass. (OPS)

IPL TYPES				
Alarm	**Pressure Relief Valve (PRV)**	**Basic Process Control System Interlock (BPCS)**	**Safe Operating Procedure (SOP)**	**Safety Instrumented System (SIS)**
with access control to the code. (MOC)		DCS/PLC with access control to the code. (MOC)		
Field Validation				
Verify that the field devices are installed correctly, open flow to the process, by physically walking down a P&ID. (PSI/MOC)	Verify that the field devices are installed correctly, open flow to the process, by physically walking down a P&ID. (PSI/MOC)	Verify that the field devices are installed correctly, open flow to the process, by physically walking down a P&ID. (PSI/MOC)	Verify that the operator has ready access to the most current version of the SOP. (OPS/MOC)	Verify that the field devices are installed correctly, open flow to the process, by physically walking down a P&ID. (PSI/MOC)
Operator Validation				
Verify through interviewing several operators that they are knowledgeable in: • the criticality of the alarm • the consequence of concern • the urgency of responding • the intended action • the expected automated action if it escalates. (OPS/TRN)	Verify through interviewing operators that they confirm that the line-up of the block valves are maintained; and they are using the PSSR process to ensure proper alignment and all chain locks are back in place. (OPS) Verify through interviewing maintenance personnel: • the process they use to store PRVs and ensure they have a means to evaluate that they are returned to service in their proper positions. (MI) • the existence of PRV out-of-service or bypass procedures. (MI)	Verify through interviewing several operators that they are knowledgeable in: • the criticality of the interlock and its set point • the consequence of concern • the intended automated action • the expected SIS trip action or associated mitigation system (PRV, etc.), if applicable, should it continue to escalate. (OPS/TRN)	Walk through the SOP with an operator to verify that the stepwise description is current and in agreement with practice in place. (OPS)	Verify through interviewing several operators that they are knowledgeable in: • the criticality of the SIF • the consequence of concern • the expected automated action if it escalates. (OPS/TRN)

PSM Elements: II - Incident Investigation; MI – Mechanical Integrity; MOC – Management of Change; OPS – Operating Procedures; PHA-Process Hazard Analysis; PSI – Process Safety Information; TRN – Training

HOW EFFECTIVE ARE YOUR SAFETY CONTROLS, ALARMS, AND INTERLOCKS?

The Importance of Functional Safety Auditing

Monica Hochleitner, Senior SCAI Consultant
SIS-TECH Solutions, LP
12621 Featherwood Drive, Suite 120, Houston, TX 77034
mhochleitner@sis-tech.com

Eloise Roche, Senior SCAI Consultant; Angela E. Summers, President
SIS-TECH Solutions, LP
12621 Featherwood Drive, Suite 120, Houston, TX 77034
eroche@sis-tech.com, asummers@sis-tech.com

Prepared for Presentation at
American Institute of Chemical Engineers
2016 Spring Meeting
12th Global Congress on Process Safety
Houston, Texas
April 11-23, 2016

HOW EFFECTIVE ARE YOUR SAFETY CONTROLS, ALARMS, AND INTERLOCKS?

The Importance of Functional Safety Auditing

Monica Hochleitner, Eloise Roche, and Angela Summers
SIS-TECH Solutions, LP
Houston, TX

Keywords: Auditing, Management Review and Continuous Improvement, Safety Instrumentation Systems

Abstract

Some organizations invest thousands, sometimes millions, of dollars on automation systems in safety applications with the desire to minimize the risk of their enterprise. However, spending dollars does not mean that the plant will reach the desired degree of safety after implementation. Return on investment in safety controls, alarms, and interlocks (SCAI) can be negatively impacted by human error, such as inadequate design, installation, testing, maintenance, and operation of the automation systems. These human errors are systematic failures that can be reflected throughout a site. Organizational discipline and administrative controls are needed to identify and correct these failures.

This paper will discuss important aspects of process safety management and how they are connected to the effectiveness of the SCAI. Functional safety auditing specifically looks at the management systems and procedures required to keep SCAI working effectively. The case studies presented will illustrate how safety system effectiveness could have been improved if a detailed audit of the SCAI documentation and performance records had been conducted and the findings addressed in a timely fashion.

1 Introduction

Many countries have been implementing industrial incident prevention systems over the years. In 1982 the EU adopted a European Council directive, known as the Seveso Directive, containing specific requirements on Process Safety Management (PSM). In 1992 the OSHA PSM regulation was enacted in the USA. After inherently safer design strategies are applied to the process, functional safety through the application of SCAI is often used to manage some or all of the remaining risk.

SCAI projects commonly require significant investment during the analysis and implementation phases. Generally hundreds of engineering hours are used to brainstorm hazard scenarios and identify safety gaps. Even more engineering hours are dedicated to calculate the best way to resolve those gaps.

Numerous industry standards and practices have been published to address different aspects of instrumentation and controls from basic electrical safety through performance-based standards for alarm management, SCAI implementation, and safety instrumented systems (SIS). The international standard IEC 61511 [1] is widely referenced even in countries that do not have specific process safety regulations. In the USA, ANSI/ISA 84.00.01 (the USA adoption of IEC 61511) has been categorized by OSHA as a recognized and generally accepted good engineering practice (RAGAGEP) for SIS. For the purpose of clarity, all clause references made within this paper are to the normative clauses in IEC61511-1 Edition 2.0 Final Draft Industry Standard (FDIS). Similar to ANSI/ISA 84.00.01, ANSI/ISA 84.91.01 [2] and ISA 84.91.03 (draft) [3] provide requirements for safety controls, alarms, and interlocks (SCAI) in general. Functional safety audits are required for all SCAI (of which SIS is a subset), so the rest of this paper will refer to the more general term, SCAI.

Designing SCAI per these standards is not enough. Even if the risk allocation and SCAI specification are performed precisely in accordance with the industry standards and practices for safe automation, this is not a guarantee that process safety risk will remain adequately managed. If assumptions made during these early safety lifecycle phases are not consistent with reality, or if the activities that must be performed periodically are not sustained, the residual risk of catastrophic consequence will remain greater than the targets set in the facility risk criteria. Process safety management systems that assure these assumptions remain valid over time are essential to the sustainability of risk management.

Any PSM program must be subjected to a well-defined and rigorous functional audit process. Otherwise the PSM program will be at risk of losing its effectiveness. Degradation (or complete omission) of various safety system management practices is a common theme in numerous process industry incident reports, such as those published by the United States Chemical Safety Board (CSB). These reports frequently include recommendations related to implementing or improving the auditability of automation installations to ensure adherence to company standards and good practice guidelines. It is not uncommon to see defects in change management, bypass management, safety system procedure use, or in the response to findings from prior assessments or audits.

Functional safety audits are designed to ensure that the SCAI installation remains in conformance both with industry practices and with the functional specification. In this paper, we will introduce the purpose and content of these audits, briefly discuss some of the specialized SCAI management system practices of which a functional safety auditor must be aware, and present summaries of industry incidents where lack of effective functional safety audits contributed to the deterioration of the safety system effectiveness over time, causing companies to face increased health, safety, and environmental risks.

2 Functional Safety Audit vs. Functional Safety Assessment

To begin, it is important to have a clear understanding of what a functional safety audit is. For practitioners who are new to functional safety management, a common terminology error to make would be to use "functional safety audits" and "functional safety assessments" (FSA) interchangeably. However, these are not synonymous terms. As indicated in the IEC 61511 definitions below, they have distinctly different intentions.

> **3.2.24**
> **functional safety assessment**
> **FSA**
> investigation, based on evidence, to judge the functional safety achieved by one or more SIS and/or other protection layers
>
> **3.2.25**
> **functional safety audit**
> systematic and independent examination to determine whether the procedures specific to the functional safety requirements comply with the planned arrangements, are implemented effectively and are suitable to achieve the specified objectives
> Note 1 to entry: A functional safety audit may be carried out as part of a FSA.

Indeed, those familiar with financial industry terminology may realize that the use of the terms "assessment" and "audit" within that industry are essentially opposite to what is intended within the SCAI related standards. The functional safety assessor is looking at the specific details for each device and function within the SCAI. A functional safety assessment is a detailed examination of the system design and the management system procedures. In contrast, the functional safety auditor determines whether the SCAI procedures and practices are being followed consistently and whether the overall program is working effectively in a manner consistent with the required functional safety performance.

This difference in intent is further represented within the IEC 61511 safety lifecycle representation (Figure 1). Functional safety assessments (shown as Stage 1 through Stage 5) occur at discrete points during safety lifecycle, while the functional safety audit is represented as part of the functional safety management plan spanning the overall program. Those who may need to perform functional safety assessments are encouraged to read the following whitepaper on functional safety: *"Functional Safety Assessments of Safety Controls, Alarms, and Interlocks."* [4]

In short, functional safety audits look for the evidence of the effective use of the many management system procedures required to sustain performance of the SCAI across their lifecycle. Many of these practices will be familiar to any process safety management auditor, such as checking maintenance schedules and management of change policies. These process safety management practices are well documented elsewhere, and thus will not be addressed significantly in this paper [11]. However, the specialized procedures and practices required by the SCAI standards may not be as commonly understood by the process safety auditing community. For this reason, it is important that the audit team include someone who has expertise in the SCAI design and management practices.

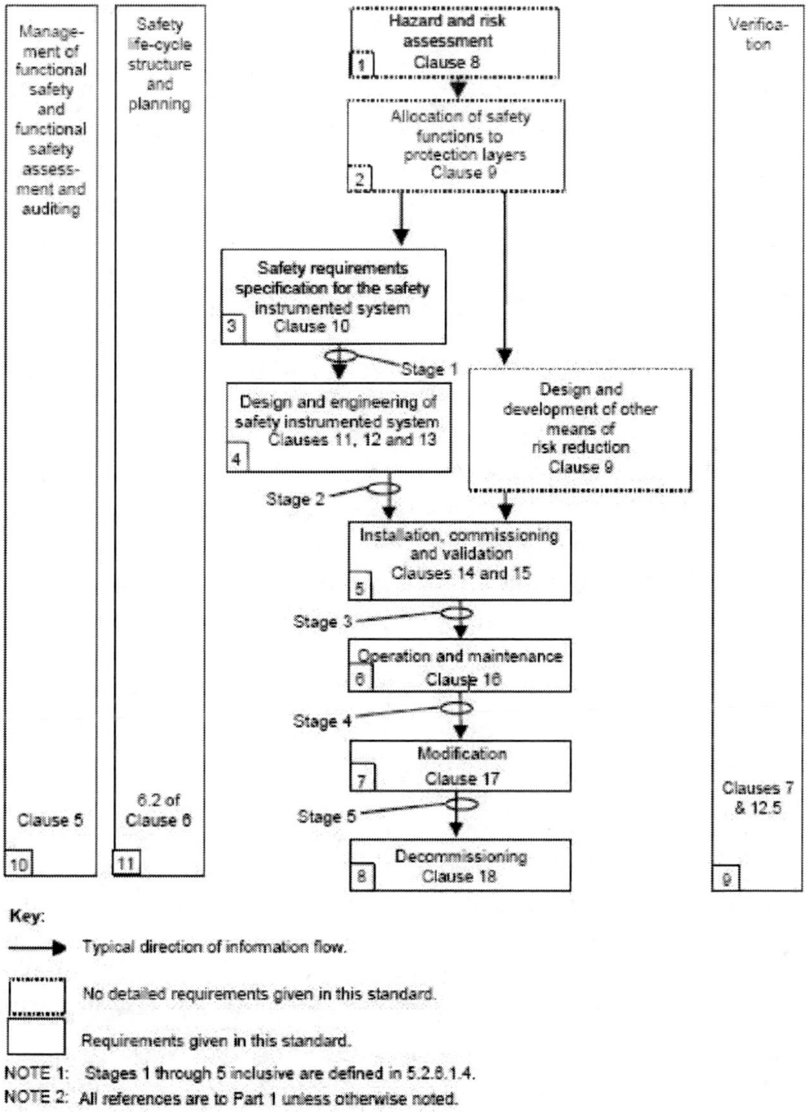

Figure 1: SIS Safety Lifecycle Phases and FSA Stages [1]

2.1 *Spot-check in auditing*

As discussed above, the purpose of functional safety auditing is to determine whether systematic failures have occurred. The primary scope of the functional safety auditor is to review documentation records for evidence that safety management systems remain active and effective. Additionally, a small sample of individual SCAI installations (field hardware and application program) is typically inspected. This "spot-check" is a technique used to detect potential errors in the design and management processes. Examples of such errors include failure of the instrument reliability program, inconsistent safety system procedure use, or a lapse in SCAI change management discipline. The recommended sample size for auditing process safety management program elements ranges from 2% for very large populations to 100% for populations of 25 or fewer [5]. In the performance of this spot-check, the functional safety auditor might use a subset of the checklists more normally used during functional safety assessment to better facilitate the detection of early signs of program degradation. Beyond this limited spot-check, however, the functional safety auditor does not generally evaluate individual SCAI installations.

2.2 Frequency of the Functional Safety Audit

> *"Safe automation depends on a quality assurance process to ensure correct operation of the process control and safety systems."* [6]

IEC 61511 establishes the recommended, and in some cases mandatory, timing of the various functional safety assessments tasks in reference to the surrounding activities within the safety lifecycle. In contrast, while the functional safety audit is a required activity, IEC 61511 does not comment further regarding the timing of the audit. The audit planner is best advised through the intended purpose of the Functional Safety Audit:

> *The purpose of the audit is to review information documents and records to determine whether the functional safety management system (FSMS) is in place, up to date, and being followed.* [Clause 5.2.6.2.1]

It follows that functional safety auditing may take place at any of the completion of any stage of the SCAI lifecycle, as well as being performed at a fixed multi-year interval in the manner of a process safety management audit. An audit at the end of a project stage would ensure the documentation and records were being generated as expected. Performing these more frequent audits may be particularly beneficial in a facility where new or unfamiliar technology is being used, where there is frequent turnover of personnel in roles essential to functional safety performance, or when the teams involved are generally less familiar with the SCAI requirements. In addition, detecting management system gaps early saves money.

2.3 Functional safety audit planning and resources

Any functional safety audit will require a degree of advanced planning to support effective execution. Functional safety audits require written procedures addressing how the audit will be carried out and how timely and satisfactory resolution of resulting recommendations will be assured. Access to all the information related to that audit will need to be arranged.

In addition, each functional safety audit must involve at least one competent auditor that is independent of the work being audited [Clause 5.2.6.2.3]. Just as for any other task in the safety lifecycle, periodic assessments are required to document the competence of the auditors with regards to this activity [Clause 5.2.2.3]. Evidence of the periodic competence assessment and effective resolution of prior assessment and audit findings are critical inputs to a functional safety audit.

2.4 Functional safety audits for other SCAI

While IEC 61511 lists specific audit requirements targeting detection and elimination of errors in the management system procedures, the guidance for other SCAI has historically been much more general. It is reasonable to expect that the auditing of other SCAI functions and devices may have been neglected. It is essential to understand, however, that even with initial robust implementation and functional safety assessment, the field hardware, application program, documentation, and human behavior associated to any SCAI are subject to the exact same degradation mechanisms.

The fundamental principle of entropy applies to all instrumented systems, regardless of the risk reduction that was assigned during allocation. Entropy applies to management systems as much as to the physical installation of hardware. It is therefore not logical to assume that SCAI installations will remain in sufficient working order without similar auditing programs. Indeed, the skills needed to perform such an audit are so similar that it is usually an efficient practice to audit both SIS and the other SCAI systems as part of the same activity.

3 Content of a Functional Safety Audit

In this section, a brief summary of what should be covered in a functional safety audit is provided. In practice, functional safety audits should be performed with more detailed checklists (or other support tools) designed for use by an audit team that in aggregate have the requisite competencies. The procedures should be designed so that all the relevant management systems and procedural practices are addressed and there is clear documented justification for the basis of each finding, whether positive or negative.

Audits that only include reviews of existing documentation and data records are likely to miss more significant systematic deficiencies, such as:

- Detailed changes being made in the instrumentation or application program without using MOC procedures

- Excessive use of bypass/manual control features
- Deterioration of field device condition through insufficient maintenance technician competency
- Misrepresentation (intentional or otherwise) within the instrument reliability program of the actual device field condition
- Approval of unsafe changes or insufficient performance caused by insufficient competence of change approvers
- Failure of the piping and instrument diagrams (P&IDs), human machine interface (HMI) drawings, or other plant safety information (PSI) documents to reflect the actual installation

A high-level review of site documentation, information and records is quite unlikely to reveal the above systematic defects. As discussed in the previous section, this is the reason that a small spot-check of the current installations is typically part of the audit scope. It is worth noting that if functional safety assessment practices have not been implemented effectively at the facility prior to the audit, significant defects may be expected and the functional safety audit team may be tempted to begin performing a functional safety assessment. In the interest of timely execution of the audit, this should be avoided. The revealed defect in the functional safety assessment practices would become a significant finding for the functional safety audit.

3.1 Key input documents

- Procedures and results for all assessment stages
    - Hazards and risk analysis (H&RA) independence and risk reduction limit review and SCAI specification consistency check
    - Detailed design review
    - Pre-startup confirmation of verification and validation results [12]
    - Operation and maintenance competency review and SCAI performance confirmation
    - Proposed SCAI change review
- Procedures and current records for resolution of findings from functional safety assessments and prior audits
- Documentation of MOCs with potential to impact H&RA, SCAI allocation, design, implementation, or procedures
- Current H&RA and Safety Requirement Specification (SRS) documentation
- Current SCAI verification and validation documentation
- Current P&IDs, HMI drawings, application program, and instrument design documentation
- All other current SCAI related procedures

3.2 Typical focus topics for functional safety audit

- Functional safety assessment procedures satisfy intended scope for that stage

- Timely execution of functional safety assessments and prior audits
- Clear documentation of timely, verified, and validated (where applicable) resolution of deficiencies from functional safety assessments and prior audits
- Completeness of documentation, including independently reviewed re-verification and re-validation (where required), of approved changes
- Clear evidence that SCAI procedures are being used in the manner consistent with the requirements of the installation and that resulting automation, instrumentation, and human system performance remains consistent with expectations
- Confirmation of effectiveness of SCAI training and competency management program
- Spot-check small % of SCAI and compensating measure device installations and program applications to audit for failure in change management, instrument reliability, or bypass control programs

4 Potential Consequences of Insufficient Functional Safety Audits

The following subsections present case studies where a functional safety audit did not appear to have been conducted or where findings were not resolved in a timely manner. In these incidents, insufficiency of the audit program likely contributed to the sequence of events that led to unwanted outcomes. As in many process industry incidents, a combination of factors led to the harmful result. This paper does not attempt to address all the contributing causes documented within the referenced reports, but instead focuses on the few deficiencies more directly related to SCAI and their management systems. In the cited events, key defects in SCAI performance or management system practices existed long enough that a robust functional safety audit should have detected them.

4.1 Managing change and procedure use - Case Study: Petrolia, 2008

This case study concerns an incident involving Oleum transfer, at Petrolia, Pennsylvania, that occurred in October 11, 2008. The internal impacts of the event included an oleum release, one person injured, and plant personnel evacuated. Externally the incident resulted in 2500 residents from three towns being evacuated and public road closures.

4.1.1 <u>Case study summary</u>

During a transfer operation, an oleum (a mixture of sulfuric acid and sulfur trioxide) process tank overflowed, filled an exhaust ventilation system, and released the oleum into a storage building. The oleum release created a cloud of sulfuric acid mist that filled the building. The sulfuric acid cloud flowed out to the facility grounds and beyond the fence line into the surrounding community. A sulfuric acid cloud is dense, visible, and slow moving. Inhaling sulfuric acid droplets can irritate the respiratory system, causing airway constriction and spasms. Severe exposure can result in fluid in the lungs, internal bleeding, and even death [7].

Twenty eight years prior to the event, a "temporary emergency" power supply was installed. The intention was to use it only during power emergencies under closely monitored conditions, so the decision was made to use an operator response to alarm safeguard instead of wiring the high level interlocks into the power circuit. The "temporary" change, with the intended limitations, was never incorporated into the facility's safety related documentation nor was any information on the new power circuit connected to the distributed control system (DCS) or other HMI.

On weekends, this particular part of the facility was lightly staffed for only a few hours. One task during this time was to transfer as much oleum into the feed tanks as possible, to support production at the start of the week.

During the weekend of the incident, operator was pumping oleum from one vessel using the primary power source and from a second vessel using the backup power source. This simultaneous transfer had become a common weekend practice since the "temporary emergency" power supply was installed. CSB report informs that "a former supervisor verbally instructed operators to use the emergency power supply to transfer oleum from the pressure vessels to the process tanks"[7]. However, the same supervisor "cautioned operators to monitor these transfers closely to prevent overfilling."

Status of the pump using backup power was not indicated on the DCS, and the automatic shutoff that was tied to a high-high level switch in the process tanks would not stop a pump powered by the emergency power supply. The pump could only be stopped locally while on backup power.

When the operator stopped pumping from the vessel using the normal power circuit via the DCS at end of the weekend shift, he inadvertently left the pump on backup power running, as shown in Figure 2. A local high level alarm beacon activated on run tank 1502, but the operator had left the building. Five minutes later, the local high high level alarm beacon activated, but no action was taken. Oleum mist (fuming sulfuric acid) was seen leaving the building an hour later by other personnel on the site. An emergency was declared and the facility was evacuated. Three nearby towns were also evacuated. About two hours later, the pump was stopped by cutting power to the oleum storage building.

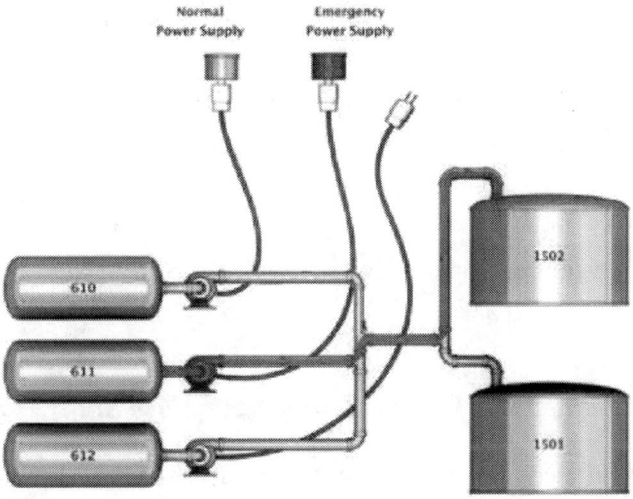

Figure 2: Weekend Power Configuration [7]

4.1.2 Instrumentation and Control Gaps

- Decision to use operator response to alarm as overfill safeguard on the emergency circuit instead of the high level interlocks that were on the primary power circuit (trade safety for project efficiency)
- Change and limitations of use were not incorporated into PHA, PSI documents, or HMI
- Over many years operators developed an undocumented practice of using the "emergency" circuit routinely on weekends, contrary to original intended use
- High level alarm used as normal fill level, and beacon not working

4.1.3 Case study conclusion

Although the CSB report does not mention the specific H&RA methodology that was used or the actual audit frequency practiced, it does mention the "temporary emergency" power supply was considered effectively permanent after a safety audit in the 1980s, replacing flexible electrical cord with wiring in conduit. However, the lack of completeness of the audit in terms of the pump use prevented observing the potential hazard being inserted to the system by this change. Assumptions made during hazard and risk analysis were actually not the same as after implementing the initial safety audit findings.

The functional safety audit recommendations should have included the upgrade going through a Management of Change (MOC) procedure. Such a process would be expected to trigger the functional safety assessments that might have revealed the inconsistency in the safeguard allocation between the two power circuits and the deficiencies in the PSI documentation associated with the "emergency" power circuit. Change documentation, verification and validation should also have been observed. Likewise, a thorough functional safety audit of SCAI procedure use should have eventually identified the operator practices during weekend that were not part of any written company procedure.

More frequent functional safety audits could have identified the lack of timely repair of the local audible alarm signal as well, although this deficiency did not directly contribute to the occurrence of this specific event.

4.2 Bypass control and audit finding resolution - Case Study: Illiopolis, 2004

This case study covers an incident that occurred in April, 2004, at a polyvinyl chloride (PVC) plant located in Illiopolis, Illinois. The internal impacts of the event included five fatalities; three hospitalized; four minor injuries; plant damage; laboratory, safety, and engineering buildings destroyed. Externally the incident resulted in 150 residents being evacuated and public road closures.

4.2.1 Case study summary

On the night of the accident, all PVC reactors were making PVC except reactor D306, which was being cleaned. After washing the reactor from the upper level, the operator went downstairs to drain out the contents of D306. Turning the wrong way coming down the stairs, the operator went to the bottom valve of reactor D310 by mistake and tried to open the bottom valve to empty the vessel (Figure 3).

The reactor pressure safety interlock prevented the reactor bottom valve from opening. The operator, presumably under the belief that he was still on D306 and that the interlock was therefore in error, connected an emergency bypass air hose to the actuator to force the interlocked valve open. He did not request permission to do the bypass or inform anyone of the bypass.

Figure 3: Cutaway of Reactor Building, from CSB Report [8]

The reactor contents rapidly emptied through the bottom valve, forming a vapor cloud. Although area release monitor alarms activated and most personnel evacuated, some operators remained at the equipment in an attempt to mitigate the situation. They were still present when the vapor cloud ignited.

4.2.2 Instrumentation and Control Gaps

- Safety Interlock bypassed with air hose, no authorization, no access controls
- Area alarms ignored by team attempting to mitigate release
- 1992 PHA identified scenario; recommendations not adopted
- 1999 PHA re-identified scenario; rationalized using administrative control
- A similar "near miss" incident had occurred at another facility the prior year and at the Illiopolis facility about 6 months prior; no corrective actions taken

4.2.3 Case study conclusion

Hazard analysis practices assume the SCAI will be operational nearly all the time. However, administrative controls (i.e. operating policies and procedures) on bypasses are subject to the same human errors as the normal operating procedures that may have initiated the event. To avoid errors like this one in Illiopolis, correctly implemented and managed access restrictions (i.e. keys and locks, passwords) may have mitigated that risk.

Access control of bypass capability depends on periodic auditing and, of course, reviews of incident and near miss investigations. Bypass control procedure use should be subjected to short term interval auditing reviewing access restrictions, authorization, communication, and how long the bypass is active, as well as the correct use of compensating measures to manage the existing risk to the target levels during the bypass. The functional safety audit should provide independent confirmation that the bypass management program is being used effectively. However, no functional safety audit or assessment program can remain effective in the absence of independent oversight that insures effective and timely resolution of the resulting recommendations. As noted by Summers and Raney, "There must also be a strong ally in upper management to support the auditing process that will be required to ensure that the EGS [engineering guidelines and standards] are used." [9]

5 Conclusion

As process industry has evolved, automation has taken a crucial role in both the normal control of the facility and in the protection of the personnel, environment, and assets of the facility in the event of loss of normal control. The creation and sustainability of safe automation is essential to the successful, productive and safe operation of these plants. As incident after incident since 1992 demonstrate on a global scale, there are a number of management systems that are critical to the continued effectiveness of safe automation design [10]. However, these management systems will likely deteriorate if not effectively audited and the results of those audits resolved in a safe and timely manner.

While general management systems, such as MOC, will be very familiar to an experienced process safety auditor, effective functional safety auditing requires additional knowledge of the policies and procedure requirements specific to instrumented safeguards. Thus, the functional safety auditor must develop a sufficient understanding of the scope, procedures, and records associated to the various functional safety assessments used throughout the SCAI safety lifecycle. Verification and validation, bypass control, device level change management, and instrument reliability programs are among the management systems whose failure appear with unfortunate frequency in process safety incident investigation reports.

Robust and timely execution of the function safety audit and of the associated functional safety assessment practices required in IEC61511 can help identify and correct such systematic failures and reduce the chance of such a failure contributing to a catastrophic event. Functional safety management gaps will potentially be detected through routine audits that may identify ways to make SCAI more effective.

6 References

[1] IEC. 2015. Functional safety: Safety instrumented systems for the process industry sector - Part 1-3, Final Draft International Standard. IEC 61511. Geneva: IEC.

[2] ANSI/ISA. 2012. Identification and Mechanical Integrity of Safety Controls, Alarms and Interlocks in the Process Industry, ANSI/ISA-84.91.01-2012. Research Triangle Park: ISA.

[3] ISA. [draft Rev 3] 2016. Functional Safety: Safety Controls, Alarms, and Interlocks for the Process Sector, ANSI/ISA-84.91.03. Research Triangle Park: ISA.

[4] Hochleitner and Roche. 2016. "*Functional Safety Assessments of Safety Controls, Alarms, and Interlocks*".

[5] CCPS. [2011]. Guideline for Auditing Process Safety Management Systems. New York, NY: Center for Chemical Process Safety.

[6] CCPS. [draft] 2016. Guidelines for Safe Automation of Chemical Processes. New York, NY: Center for Chemical Process Safety.

[7] CSB. 2009. INDSPEC Oleum Release Case Study. Case study 2009-01-I-PA. Washington, D.C.: U.S. Chemical Safety Board.

[8] CSB. 2007. Investigation report - vinyl chloride monomer explosion at Formosa Plastics Corporation. Report 2004-10-I-IL. Washington, D.C.: U.S. Chemical Safety Board.

[9] Raney, Glenn and Angela Summers. [1999]. "Common Cause and Common Sense, Designing Failure Out of Your Safety Instrumented Systems (SIS)," ISA Transactions, 38, pages 291-299

[10] Summers, A. *et all* [2015]. "Incidents that Define Safe Automation", 61st International Instrumentation Symposium, Alabama, May 11-14, 2015.

[11] Summers, A. [2008]. "Lessons Learned in Auditing Automated Systems for PSM Compliance", 1st Latin America CCPS Conference, Buenos Aires, Argentina, March 27-29, 2008.

[12] CCPS. [2007]. Guideline for Performing Effective Pre-Startup Safety Reviews. New York, NY: Center for Chemical Process Safety.

Leveraging PHA to Develop Requirements for Cyber-Security

BY EDWARD M. MARSZAL

Process plants need to be protected against physical consequences generated from cyber-attack. The starting point for developing a program for cyber-security is the assessment of the risk posed by cyber-attacks. Instead of starting with a blank sheet for paper for this analysis, use of the existing PHA documentation and process will provide optimal results with minimal additional work, and also ensuring that risk tolerance criteria and risk management methods are consistently applied. This paper will explain how existing methods for process hazards analysis (PHA) of process industry plants can be expanded with an additional "cyber review" called a Security PHA Review (SPR – pronounced 'spur'). The purpose of the SPR is to determine if there are any cyber-attack vectors that can cause significant physical damage to the facility, and if so, make recommendations for modifying one or more of the safeguards in a cyber-vulnerable vector so that they are not vulnerable to cyber-attack, or assigning an appropriate level of performance to cyber-security measures. The same way that definition of safety integrity levels (SIL) flow from HAZOP through LOPA to a safety instrumented system specifications, the definition of "security levels" as defined in IEC/ISA 62443 can flow from the PHA through a SPR to industrial control network specifications.

The approach that will be discussed includes analysis of the causes of safety incidents, the safeguards that prevent the causes from resulting in consequences, and the magnitude of the consequences that might result from the realization of these hazards. The discussion will include an example of a batch chemical reactor where a Hazards and Operability (HAZOP) study was assessed using a PHA cyber review in order to determine whether or not the facility was inherently cyber-safe, and if not, make cyber-safe recommendations and define IEC 62443 Security Level specifications.

Introduction

Managing process risk and ensuring that the risk is below a tolerable level is a difficult and resource intensive activity, made more difficult by evolving technology. New equipment may result in new hazard scenarios to be considered during design and addressed with appropriate safeguarding. The shift from controls systems that were either electronic or pneumatic to Distributed Control Systems (DCS) and Programmable Logic Controllers (PLC) is an example. These computer-based systems allowed gains in functionality over analog systems, including quick execution of complex algorithms, storage of historical process data, and quick, ubiquitous communication.

While use of computer based systems is common, most process facilities have performed little in the way of cybersecurity, leaving IT colleagues with the task of perimeter guarding around the control systems. But even with the lack of cyber-security efforts, physical damage to process plants caused by cyber-attacks is virtually quite rare. Which is surprising to those who do not understand how process plants are designed and safeguarded, but obvious to process safety professionals. Process engineers design plants to be safeguarded against scenarios that can cause significant safety consequences, and this is true whether or not the failure occurs organically or deliberately through cyber-attack. The safeguards employed are common, inexpensive, and very often inherently safe against cyber-attack because most of these devices were invented dozens or hundreds of years before the advent of the computer.

While the advantages of computerized control systems are obvious, they also introduce new hazard scenarios that did not exist in the analog days. While new scenarios generated from inherent failure modes of the new equipment items were appropriately addressed by existing process hazards analysis (PHA) methods, other new hazard scenarios were generated that are not adequately addressed by existing PHA techniques – at least not without extending those methods to address the new threats. These new threats are not adequately addressed by existing PHA methods because instead of being random failures of equipment, they are deliberate acts of humans.

Even though these new threats are not appropriately assessed with existing PHA methods, there is no reason to abandon everything that we know about process hazards analysis and start from scratch, which, unfortunately, has been proposed by many cyber-security practitioners – most of whom have no experience in process industry risk analysis. Instead of starting from scratch, methods are being adopted in industry that extend tried and true methodologies for PHA to address the problem of deliberate cyber-attack. As a result, none of the existing PHA effort is wasted, and PHA efforts are not needlessly duplicated. Instead, a small amount of additional effort is utilized to start with traditional PHA and focus in only on scenarios where cyber-attack can cause the scenario to occur and also prevent all of the safeguards from operating properly. It is these key scenarios that will result in generation of recommendations to implement safeguards that are inherently safe against cyber-attack, or to define the appropriate level of cyber-safeguarding, as defined by a Security Level (SL).

Security Level

The effectiveness of cyber-attack mitigation designed into an industrial control system is represented by its Security Level (SL), as defined in the IEC/ISA 62443 standard. Unlike Safety Integrity Levels, or SIL, of safety instrumented functions, which are quantitative measures of probability of failure on demand of safety instrumentations, SL are set of qualitative measures and techniques that are employed on industrial control systems and networks to prevent unauthorized use and access. The method for selection of SL should consider the potential consequences of scenarios that could cause loss of containment accidents in process plants by manipulating process control equipment to both cause events and prevent existing safeguards from properly operating. IEC/ISA 62443 defined four security levels, one (1) through four (4), with SL 1 being the least secure and SL 4 being the most secure. The levels are defined (in the abstract) as follows:

- SL 1 Prevent the unauthorized disclosure of information via eavesdropping or casual exposure
- SL 2 Prevent the unauthorized disclosure of information to an entity actively searching for it using simple means with low resources, generic skills, and low motivation
- SL 3 Prevent the unauthorized disclosure of information to an entity actively searching for it using sophisticated means with moderate resources, Industrial Automation and Control Systems (IACS) specific skills and high motivation

SL 4 Prevent the unauthorized disclosure of information to an entity actively searching for it using sophisticated means with extended resources, IACS specific skills and high motivation.

The above definitions of SL are quite philosophical in nature and require much more explanation. To that end, an entire document in the IEC 62443 standard set is dedicated to explaining the differences between the various SL, which is outside the scope of what can be discussed in this paper

Risk Analysis Methods for Cyber-Security

IEC/ISA 62443 cyber-security design requires risk assessment to set the required performance level. While the standard provides some guidance on risk assessment from a cyber-security perspective, practitioners of process safety can easily become confused as the standard often conflates and confuses risk analysis, vulnerability analysis, and general equipment design review. As such, to a cyber-security professional a server with out of date software patches is considered a risk (or sometimes a vulnerability). In the process industries we generally refer to complete risk scenarios that start with initiating events such as equipment failures or human error, include consideration of safeguards that might prevent that initiating event from being realized if they function correctly, and the consequences in terms of physical damage if the ultimate event occurs.

The methodologies discussed in the IEC/ISA 62443 standard do not conform to the process safety professionals' expectations of PHA. Instead, they should be considered what they are, failure modes and effects analysis analogs to be used by specialty equipment designers to identify weaknesses in their designs. Methods that have been promoted for cyber-risk analysis, such as Cyber HAZOP or CHAZOP, fail to generate complete process hazard scenarios that will define risk in a process plant. These methods all start with a "threat vector" which is a vulnerability of a computer system to cyber-attack, and then try to build out a scenario in the following framework:

1. Identify an Industrial Control System (ICS) asset
2. Identify a threat employing that asset
3. Identify a vulnerability allowing threat to occur
4. Determine likelihood of attack
5. Determine consequence
6. Calculate risk

This general process has weaknesses that make it ineffectual for the actual development of scenarios that define the risk related to cyber-attack. The primary problem is that this approach does not identify hazard scenarios, it identifies vulnerabilities in equipment design. While the new cyber-risk assessment approaches use names that imply they are similar to hazard identification techniques, they are not. All of these methods are the same in that they identify a potential failure mode of an ICS. Unfortunately for these methods, an ICS vulnerability, in and of itself, is not a scenario, and not even an initiating event for a scenario.

As an analogy, consider replacing your PHA process, such as HAZOP, with a technique that looks at the design of your safety instrumented systems as the basis for determining what hazard scenarios are present. The study might identify that a plugged exhaust port of a solenoid valve could prevent a shutoff valve from moving to a closed position. While that is an important finding, no initiating event, no safeguards, and no risk have been identified. Ultimately, the failure mode of a plugged exhaust port does not CAUSE anything! No risk scenario is generated because no initiating event is identified. Similarly, most of the cyber risk assessment methods suffer from the problem of identifying weaknesses in the design of the ICS equipment, but not being able to generate any initiating events that drive risk scenarios. Without these complete risk scenarios, which often require the simultaneous exploitation of multiple ICS vulnerabilities, the overall risk of the plant cannot be determined. Another approach must be used to identify the risk scenarios where cyber-vulnerabilities can impact the overall risk profile of the process plant.

The SPR Study

Security PHA Review or SPR (pronounced "spur") study is an evolution of PHA for assigning cyber-security performance and recommending safeguards that are inherently safe against cyber-attack. The SPR process was developed by technical safety practitioners with strong backgrounds not only in process safety, but also in the design and implementation of industrial control systems. The SPR approach was developed to tightly integrate with the process industry project lifecycle, leveraging existing engineering tasks and reports generated for general process safety. As a result, the limitations of existing cyber risk analysis approaches could be eliminated while maximizing the use of information and documentation generated in other stages of the engineering lifecycle.

The SPR study is specifically designed to generate required SL using the existing PHA as the foundation. SPR allows for the selection of SL in a manner that is analogous to the way that Layer of Protection Analysis (LOPA) allows for the selection of Safety Integrity Level (SIL) targets for Safety Instrumented Functions (SIF).

The SPR process is shown in *Figure 1*.

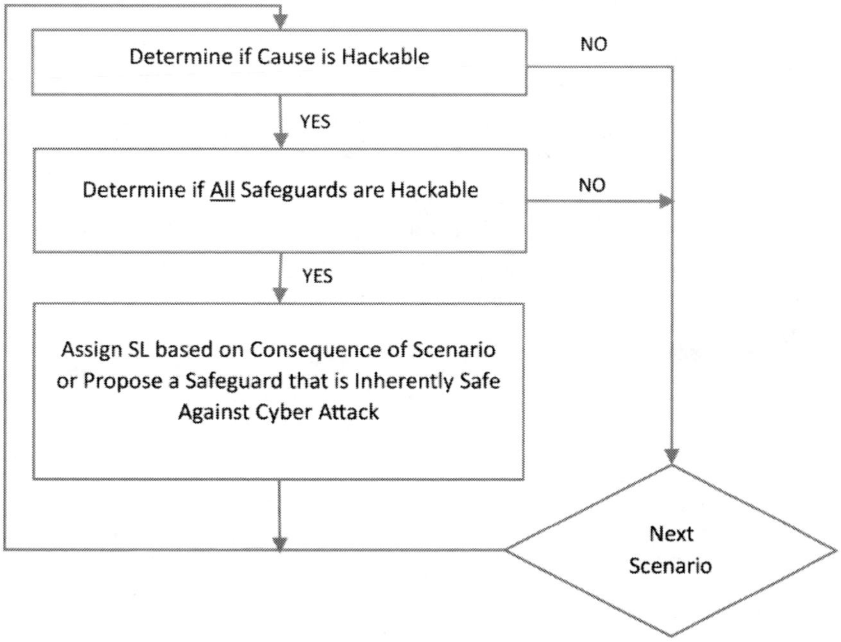

Figure 1 Simplified Security PHA Review Process

The process begins with the collection of the results of a process hazards analysis. This can either be done with the report of an existing PHA, or even as an additional step during a PHA study – while the study is in progress. Each scenario of the PHA is then reviewed to determine if it is hackable, which means that the scenario could be forced to occur by a malevolent actor who has taken control of the ICS. First, the cause or initiating event is reviewed to determine if it is hackable. Generally, this would be true for any computer control loop failure or equipment item starting or stopping. It would not be true for human interactions with mechanical process equipment not connected to a computer. If a hacker could not be responsible for the cause, the analyst moves to the next scenario.

Next, the safeguards are reviewed to determine if they are hackable. In general, all control loops, safety instrumented system functions, and operator responses to alarms are hackable, but mechanical devices such as relief valves are not. If any one of the safeguards is not hackable, the analyst moves on to the next scenario.

If the cause of a scenario can be hacked and all of the safeguards can be hacked then the overall scenarios is determined to be hackable, meaning that if a malevolent actor could take control of the ICS that person would be able to generate the scenario under consideration and realize its consequence. For each hackable scenario, the consequence category from the PHA needs to be determined. Based on the risk tolerance criteria of the process owner, an IEC/ISA SL would then be assigned to that scenario. Of course, if the consequence is severe and results in an SL that is not desirable, the analysis team has the option of recommending a safeguard that is inherently safe against cyber-attack which would then remove the scenario from consideration as a driver of the selected SL. After all of the scenarios have been reviewed in this way, the SL that is assigned to a zone is the highest of all of the SL's that were assigned to the scenarios that are associated with the ICS equipment of that zone.

Reactor Runaway SPR Example

In this example, a Continuously Stirred Tank Reactor (CSTR) is employed in the production of a highly reactive chemical compound. The reactor is charged with a fixed quantity of Reactant A. After reactant A has been completed charged, the agitator is started and cooling water is introduced into the cooling water jacket. After cooling and agitation are established, Reactant B is added to the reactor at a slow and controlled rate. After a set period of time, all of Reactant A is consumed by the addition of Reactant B, and the reactor vessel is left containing the reaction product, Product C. The product is drained from the tank reactor and sent downstream for further processing. A simplified process flow diagram of the process is shown in *Figure 2*.

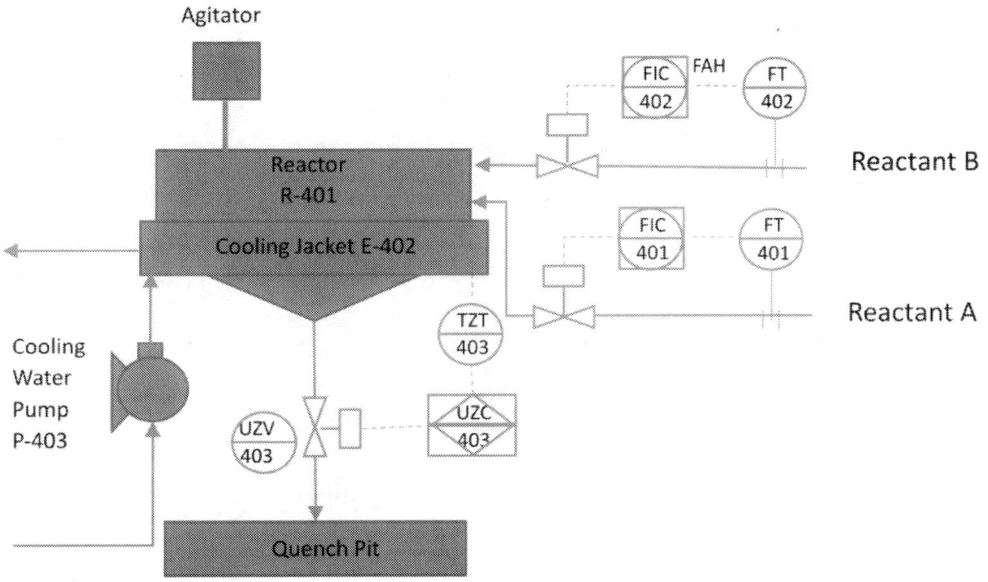

Figure 2 CSTR Simplified Process Flow Diagram

If the cooling water were to fail to be put into the cooling jacket, for instance, because the cooling water pump failed in such a way that water stopped flowing, the temperature in the reactor would begin to increase. The increased temperature would quickly result in increased reaction rates and runaway reaction. In this case, once the runaway reaction starts it can no longer be stopped by re-establishing the flow of cooling water. The runaway reaction will cause a rise in temperature, but it will also cause a significant rise in pressure as the Product C decomposes into gaseous by-products. The pressure in the reactor vessel will quickly rise in excess of the maximum allowable working pressure of the vessel causing it to rupture. The vessel rupture is also expected to result in a fire and vapor cloud explosion in addition to the physical explosion of the pressure vessel.

This scenario was considered during a HAZOP-style PHA. The worksheet for the low flow deviation for cooling water is shown in *Figure 3*

Cause	**Consequence**	**Cons Cat**	**Safeguards**	**Likelihood**	**Risk Rank**
Failure of Cooling Water Pump P-403, which results in loss of cooling water flow to Cooling Jacket E-402 of Reactor R-401	Once the flow of cooling water is stopped, the heat of reaction begins to build up in the reactor, resulting in temperatures. A runaway reaction will then start resulting in thermal decomposition of Product C into gaseous byproducts and a rapid increase in pressure in the reactor beyond the Maximum Allowable Working Pressure (MAWP). Loss of containment of process material with physical explosion. Potential fire and potential vapor cloud explosion. Potential single fatality is expected to operator or maintenance in area of reactor.	High	Safety Instrumented Function UZC-402 Dumping the reactor contents into the quench vessel which stops the reaction.	Moderate	Medium

Figure 3 Runaway Reaction PHA Worksheet

The SPR begins with an analysis of the initiating event. In this case, the initiating event is the failure of a cooling water pump such that flow of cooling water stops flowing to the reactor cooling jacket. While it is unlikely that any sort of cyber-attack could cause the pump to fail, a cyber-attack could cause the pump to stop if any start/stop functionality is included in the control system or the safety instrumented system, which is quite likely in a batch process. As such, the initiating event is determined to be hackable.

Initiating Event	Location	Hackable?
Failure of Cooling Water Pump P-403, which results in loss of cooling water flow to Cooling Jacket E-402 of Reactor R-401	DCS	Yes

Next, all of the initiating events are reviewed to determine if they are hackable. In this case there is only one safeguard which is a SIF. The SIF is determined to be hackable because it resides in a SIS that is based on a programmable logic controller. If the control system were taken over by a malevolent actor, the output of the SIF could be frozen in an energized state, preventing the dump valve from opening.

Safeguard	Location	Hackable?
Safety Instrumented Function UZC-402 Dumping the reactor contents into the quench vessel which stops the reaction.	SIS	Yes
ALL SAFEGUARDS HACKABLE?		YES

In this case, all of the safeguards were determined to be hackable. Since the initiating event and all safeguards are determined to be hackable, the next step is assign SL based on consequence category as presented in *Figure 4.*

S	Category	Safety	Environment	Commercial	TMEL	SL
0	None	No significant safety consequence	None	None	N/A	1
1	Very Low	Minor injury - first aid	Small release with minimal clean up requirements	$50,000	1E-02	1
2	Low	Lost time injury not requiring extended hospitalization	Moderate release limited to onsite damage with moderate clean up effort	$500,000	1E-03	2
3	Moderate	Severe injury (extended hospitalization, dismemberment)	Large release with limited offsite impact requires significant onsite clean up	$5 Million	1E-04	2
4	High	Single fatality	Large release offsite on extensive clean up and damage to sensitive areas	$50 Million	1E-05	2
5	Very High	Multiple fatalities	Very large release off site with extensive clean of and permanent damage to several sensitive areas	$500 Million	1E-06	3
6	Very-Very High	Multiple Offsite Fatalities	Very-Very large release offsite with extensive clean up and remediation ongoing for many years along with permanent damage to many sensitive areas	$5 Billion	1E-07	4

Figure 4 Tolerable Risk Criteria

In this case, the consequence category is high, based on the potential for a single fatality as the result of the fire and explosion that could accompany the loss of containment which results in a SL assignment of SL 2. SL 2 can be reasonably achieved by typical cyber-security measures, and accepted by the project team.

Scenario Hackable?	Consequence Category	SL Requirement
Yes	High	2

Since this scenario only has a single safeguard that is hackable it warrants some further analysis. Consider the process scenario again, but with a different set of consequences. For instance, in another similar case, the reactor could be located in an indoor building where the consequences of the explosion would be exacerbated. Instead of compromising the local area around an outdoor reactor, it could cause damage and collapse to the entire reactor building resulting in multiple fatalities of all of the people that are stationed in the normally occupied building. *Figure 5* presents a revised PHA study report excerpt for this revised situation.

Cause	Consequence	Cons Cat	Safeguards	Likelihood	Risk Rank
Failure of Cooling Water Pump P-403, which results in loss of cooling water flow to Cooling Jacket E-402 of Reactor R-401	Once the flow of cooling water is stopped, the heat of reaction begins to build up in the reactor, resulting in temperatures. A runaway reaction will then start resulting in thermal decomposition of Product C into gaseous byproducts and a rapid increase in pressure in the reactor beyond the Maximum Allowable Working Pressure (MAWP). Loss of containment of process material with physical explosion. Potential fire and potential vapor cloud explosion. Potential collapse of normally occupied reactor building. Potential for multiple fatalities of all staff occupying the reactor building.	Very High	Safety Instrumented Function UZC-402 Dumping the reactor contents into the quench vessel which stops the reaction.	Moderate	High

Figure 5 Runaway Reaction PHA Worksheet

For the new case, the SPR would proceed the same way. The initiating event analysis would show that it is hackable, and the safeguard analysis would show that all of the safeguards are hackable. In this case, instead of a consequence category of "High" that results in a SL of 2, the consequence category is "Very High" resulting in a SL of 3. An SL of 3 is a somewhat difficult target to achieve because many ICS cyber-security designs are not suitable for SL 3 without significant costly and difficult modifications. Sometimes in situations like this, the project team may recommend the use of a non-hackable safeguards so that the scenario consequence does not unduly increase the required SL.

Upon review of the common non-hackable safeguards, it is determined that no self-contained mechanical device will prevent the scenario under consideration. In addition, modifications to safe operating limits or equipment material selection in an attempt to make the design inherently safe would also not be effective. The only effective safeguard that is inherently safe against cyber-attack is an analog "mimic" of the SIF.

The analog "mimic" of the SIF UZC-403 will employ the second thermocouple of a dual element thermocouple set in the existing thermowell. The second thermocouple element will be wired to an analog temperature transmitter that will convert the temperature measurement to a 4-20 mA signal. The 4-20 Ma signal will be analyzed by an analog current monitor relay that will open a contact in the 24 VDC signal to the solenoid valve for UZV-403, de-energizing the solenoid, venting the valve's actuator, and causing the valve to go to the open position. As designed, this entire analog "mimic" is inherently safe against cyber-attack, and any cyber-attack that is waged on the digital complement (UZC-207) will be non-interfering in the safety functionality of the analog "mimic" function. The design of the "mimic is shown in more detail in *Figure 6*.

Since the scenario is no longer hackable, the SPR analysis yields a result on "no requirements" for SL as a result of this scenario.

Scenario Hackable?	Consequence Category	SL Requirement
No	N/A	None

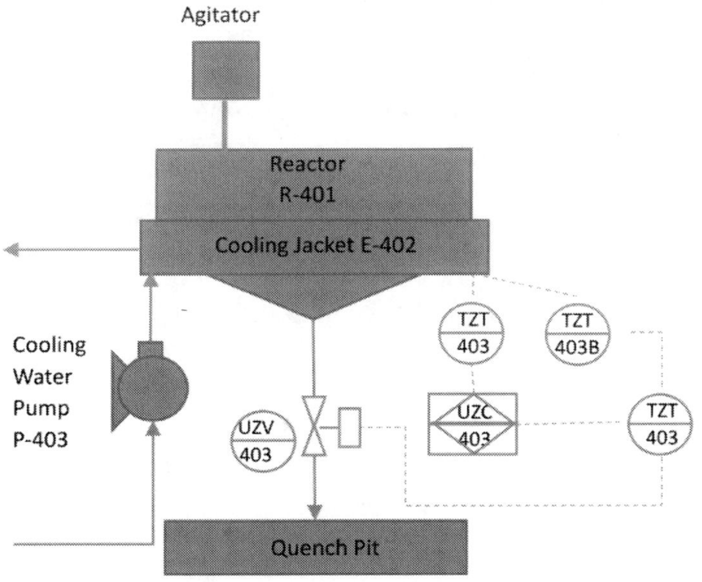

Figure 6 CSTR SIF with Analog 'Mimic'

Summary

The purpose of the SPR study is to provide an additional look at the process hazards analysis, from the specific perspective of generating the scenarios through deliberate and malicious means. The outcome of the SPR process is either recommendations on how to modify the process to prevent these scenarios from being possible – i.e., making the plant inherently safe against cyber-attack – or selecting an appropriate SL to set the design requirements of cyber-security programs, procedures, and equipment. SL are collections of design and administrative procedure attributes that effect cyber-security. The IEC 62443 standard has defined four SL, with SL 1 providing the lowest degree of security and SL 4 providing the highest degree of security. SL are assigned to security zones, which are physical or logical groupings of ICS equipment that have similar security requirements, generally because they are serving a set of process equipment with similar hazards.

While the IEC/ISA 62443 standard have established the need for risk assessment to determine SL, guidance in the standard is of a general nature. In fact, the standard's discussion of risk assessment often conflates and confuses the distinction between risk assessment of the process and failure modes and effects analysis of the ICS equipment. While analysis of the ICS equipment has its value, the overall risk of the process can only be determined by means that focus on the process.

The process industries have already established methods for risk assessment which have been successfully in use for decades. The optimal solution for assessment of cyber-security risks is not to start from scratch with new studies, but to extend the existing studies to incorporate the new objectives and requirements. In the same way that the PHA was extended through LOPA to yield SIL requirements for SIS, the PHA process is extended through SPR to yield SL requirements for cyber-security.

SPR is an extension of a PHA. The SPR process proceeds by analyzing each individual scenario of the PHA for hackability, or the ability of a malevolent actor to take control of programmable devices remotely due to cyber-vulnerabilities. First, the initiating events, or causes, are assessed. If they can be hacked, then all of the safeguards are assessed. If all of the safeguards can be hacked too, then the entire scenario can be hacked. For each hackable scenario, two options exist. Either a recommendation must be made for the employment of a non-hackable safeguard in order to make the scenario inherently safe against cyber-attack, or an enhanced level of cyber-security, as defined by the selected SL, will need to be employed, based on the consequence of the scenario.

442321 Pipeline Risk Assessment Using Artificial Intelligence: A Case from the Colombian Oil Network

Abstract:

Recently, as the world's energy consumption has increased, the number of serious accidents related with the failure of energy infrastructure has significantly grown and most of them have had a large impact on people and environment. Those serious accidents have increased public awareness about the risk of failure of certain energy infrastructure, and have also increased the concern about the risk acceptability by governments, regulatory bodies and operators.

Then, regarding the concern about the safety of energy infrastructure, operators and regulators have been performing risk assessments in attempt to identify and evaluate accurately the probability of failure of the infrastructure as well as the consequences related with that failure. However, estimating accidental risks in critical energy infrastructure involves a substantial operational expenditure due to the complexity and lack of information.

Thus, this study attempts to show an aplication case for oil pipelines of the Colombian Oil network. The case analysis is divided into three stages: 1. Fuzzification of probabilities and consequences, 2. Fuzzy inference, and 3. Defuzzyfication of risk values.

In summary, this paper presents a fuzzy risk inference system in which it considers the conversion of crisp values of probability and consequence into grades of memberships for linguistic terms of fuzzy sets, such as very low, low, medium, high and very high. After that, the inference system considers 25 conditional rules extracted from a risk matrix stated by the Department of Defense of U.S.A., MIL-STD-882D [2000]. Finally, the inference system converts the fuzzy risk values into crisp values by using a centroid method of defuzzification which allows the interaction of all conditional rules to get a numerical risk value.

445435 Effective Operational Discipline Programs

Abstract:

To achieve successful process safety performance and consistent management of process risk, effective process safety programs must include emphasis on safety culture and leadership, process safety systems, and operational discipline (OD). Effective OD programs help ensure that well-designed process safety systems are successfully implemented and help achieve consistent, desired, day-to-day performance. Failure to follow process safety system requirements can result in unsafe behaviors, unsafe acts, and ultimately, serious incidents and injuries. A focus on OD requires continued commitment by company leadership to ensure appropriate systems, procedures, and safe work practices have been established and by workers at all levels in the company to consistently follow and meet these requirements. This presentation reviews the benefits, key concepts, and practical applications of effective OD programs for evaluating and improving OD in order to help improve process safety performance.

445434 Best-in-Class MOC Implementation – Case Studies

Abstract:

So your facility has an MOC procedure, right? You hear complaints about the MOC process being burdensome, and you worry that people aren't doing MOCs when they ought to. What do companies do that are successful at MOC?

Successful MOC practitioners have several key characteristics. First, they treat Management of Change just like a project--not as a form that needs to be filled in. The project orientation engenders the formality and respect that MOCs deserve. Second, successful companies think broadly in terms of the business benefits that the MOC procedure/process can bring them. While MOC is required for PSM covered processes, the methodology can also be leveraged for replacements-in-kind, as well as smaller capital projects. And finally, successful MOC practitioners simply "get on with it". They realize that, after 24 years since the PSM regulations were promulgated, most companies are into a "sustaining" mode, complacency may have set in, and it is time to take the next steps in MOC evolution. They take action and avoid waiting for perfect solutions and universal commonality.

This presentation will present a number of case studies, from companies that have recently updated their MOC processes, and use those case studies to illustrate the key learnings.

16 AIChE
Spring Meeting, Houston, TX

Facility Integrity: Learnings from the Equipment Lifecycle

Robert F. Wasileski, III
Baker Engineering and Risk Consultants, Inc.
11011 Richmond Avenue, Suite 700, Houston, TX 77042
RWasileski@BakerRisk.com

Prepared for Presentation at
American Institute of Chemical Engineers
2016 Spring Meeting
11th Global Congress on Process Safety
Houston, Texas
April 10-14, 2016

UNPUBLISHED

Facility Integrity:
Learnings from the Equipment Lifecycle

Robert F. Wasileski, III
Baker Engineering and Risk Consultants, Inc.
11011 Richmond Avenue, Suite 700, Houston, TX 77042
RWasileski@BakerRisk.com

Keywords: Mechanical Integrity, Reliability, Lifecycle, Near Miss

Abstract

The Chemical Processing Industry (CPI) has witnessed growth in mechanical integrity (MI) programs, which have evolved from standards-based compliance, to continuous improvement programs, and on to risk-based programs. For instance, operators (i.e., manufacturers) have redesigned corporate standards, plant-level procedures and field practices to keep pace with incident learnings and published best practices, such as those produced by the American Institute of Chemical Engineers' (AIChE) Center for Chemical Process Safety (CCPS). Furthermore, the property insurance industry has duly taken note of this MI evolution, giving rise to a significantly greater focus on MI programs during insurance surveys and inspections. Meanwhile, incident investigators remain in demand as relatively large incidents with MI-related causes continue to occur in the CPI. This apparent disconnect naturally raises questions in regards to subjects such as "risk based inspection (RBI)", "reliability centered maintenance (RCM)," "inspection frequency", "industry best practice" and "inspection, testing and preventive maintenance (ITPM)".

MI programs play a considerably important role in the process safety lifecycle of equipment. The engineering design phase of a project may be on the order of just days, up to several years for a complete plant. Procurement and construction typically follow a similar timeline proportionate with the engineering design phase. However, once a facility becomes fully operational, the time devoted towards operation and maintenance will normally far out-weigh the engineering, procurement and construction (EPC) period, and may last for many decades. As such, the process safety equipment lifecycle (PSEL) is typically dominated by the MI program, relative to other lifecycle phases such as the EPC period or decommissioning and retirement.

The PSEL is explained in this article, with an in-depth examination of the central elements which should be embraced by a comprehensive MI program. While the article demonstrates the breadth and depth of MI, it aptly proposes an innovative approach towards the management system that should form the foundation of a robust MI program. The proposed framework leverages the common "Onion Skin" diagram in the context of the equipment lifecycle to create an intuitive approach to MI management. This unique management system framework is comprehensive, sound, and yet practical for implementation at facilities of most any size.

1. Glossary (Terminology, Definitions and Acronyms)

1.1 The author uses or infers to a variety of acronyms and terminology throughout this article. Where a definition does not contain a reference, the definition is that of the author. Terminology, definitions and acronyms are listed in Appendix A of this article.

2. Introduction and MI Background

> "*Operators sometimes regard tests and inspections as a nuisance, interfering with the smooth operation of the plant. Operator training should emphasize that protective equipment is there for their protection and they should 'own' it.*" – Trevor Kletz [1].

According to the Marsh Energy Practice, the world's leading property insurance broker, major accidents due to mechanical integrity related failures continue to make a significant contribution to large losses in the CPI. Marsh comments, "…loss totals in the refining industry have continued to *trend upwards over the last few years. Piping failures or leaks* (corrosion or incorrect metallurgy) and start-up and shut-down events *continue to be significant causes.*" [2] This quote may not be particularly surprising to some, given the fact that it directly references the refining industries. However, given the relative age of most refineries to other facilities such as those in petrochemicals and oil & gas, this upward trend in piping failures should be a concern to all readers from every sector of the CPI. Logically, if the age of the facility is a significant correlating factor towards piping (and other mechanical-related) failures, the future holds cause for concern.

MI programs began to receive serious attention perchance in the early 1990s, with the promulgation of the United States *Process Safety Management (PSM) Standard for Highly Hazardous Chemicals*, OSHA 29CFR1910.119. This U.S. Standard contains a relatively lengthy discourse on "Mechanical Integrity", and the performance-based requirements for those facilities covered by the Standard. Once the Standard was promulgated into law, most facilities adopted a compliance-based maintenance program to meet the requirements of the Standard. However, MI programs have evolved and matured since then, with the CPI adopting integrated approaches and philosophies to MI management such as run-to-failure maintenance, reliability-centered maintenance and, most recently, risk-based maintenance.

This article will briefly explain basic concepts such as the PSEL, the familiar "onion skin diagram", and then go on to propose a new framework for MI management programs which leverages the onion skin diagram in the context of leading-edge practices promoted by respected organizations such as the AIChE/CCPS.

3. Facility Lifecycle Explained

With respect to preventing Process Related Incidents (PRIs), perhaps one of the greatest challenges an organization may face is changing the mindset of employees from an "equipment-based" to "facility-based" mechanical integrity program. In retrospect, an arguable failure of the U.S. Process Safety Management (PSM) Standard was the over-stated emphasis on flammables and what OSHA refers to as "Highly Hazardous Chemicals". Since the early 1990s this Standard has largely shaped the CPI into a mindset that MI programs are *limited to* what is known as a "covered process" under the Standard. In practice, some companies have intentionally created artificial boundaries around "covered" equipment, in order to limit the scope of coverage of the plant's MI program. In some cases, the "covered process" literally is separated or "bounded" by a single flange on a line containing a flammable or toxic process material. The concern with this approach is that MI programs, in some cases quite robust programs, are not being applied to hazardous areas where significant process risks may be present. These artificial boundaries have potentially limited the scope of MI programs to equipment which is only required by law to have an MI program. This approach may create vulnerabilities in the plant and amplify the opportunity for PRIs beyond the user-defined "covered process" (See Table 1).

The risk of a PRI outside of the "covered process" is a very real possibility. These are process risks which involve a mindset change from "equipment-based" MI programs to a "facility-based" MI program. The scope of a facility-based MI program is clearly broader in nature. However, with practical approaches to MI such as Run-to-Failure and Risk-Based Maintenance, this increase in scope can be managed with available resources, while still controlling the risks posed by equipment outside of the "covered process" areas. Thus, the PSEL needs to be viewed as being applicable and necessary to the entire facility – not just "covered process" equipment – using a risk-based decision making process (See Figure 1).

The PSEL begins at the conceptual design phase (Phase 1). In practice, this phase may occur in a Research & Development (R&D) facility, pilot plant, or with process engineering (e.g., with an EPC). During Phases 1 through 4 it is *essential* for the Process Design Basis (PDB narrative) and Basic Design Data (BDD) to be developed and documented for process equipment, critical buildings, infrastructure and so forth. The PDB and BDD forms the foundation upon which the final plant or facility will be operated. BDD often includes chemical and physical material properties, materials of construction, reliability data, and information on operating history. However, the scope of BDD is not limited to traditional process safety information; in certain cases it may be necessary to complement the scope of data with information from other disciplines (e.g., Occupational Health, Toxicology, Product Stewardship,etc.) to ensure the data is complete and thoroughly documented [6].

Table 1: Examples of CPI Incidents *Outside* of the PSM "Covered Process" Boundaries [3], 1

EVENT NO.	EVENT DATE	EVENT TYPE	LOCATION	PRIMARY PROCESS MATERIAL INVOLVED[2]	FAILURE TYPE OR DAMAGE MECHANISM	DAMAGE (2013 U.S. Dollars)[3]
1	07/04/2012	Fire and Explosion	Bangkok, Thailand	Kerosene	Internal vessel (stripper) fire	$140,000,000
2	09/28/2011	Fire	Palau Bakom, Singapore	Blended refined products	During maintenance inside a pump house building	$150,000,000
3	04/20/2010	Fire and Explosion	Deepwater Horizon Rig (Macondo Well), Mississippi Canyon Block #252, Gulf of Mexico	Crude Oil	Well blow-out during maintenance / abandonment activities	$4,500,000,000[4]
4	11/22/2002	Fire and Explosion	Port of Mohammedia, Morocco	Waste oil	Hot surface ignition following a loss of primary containment	$200,000,000
5	09/21/2001	Explosion	Toulouse, France	Ammonium Nitrate	Internal vessel (tower) explosion	$680,000,000
6	08/14/2001	Fire	Lemont, Illinois, U.S.	Pyrophoric material	Ruptured pipework leading to air ingress	$370,000,000
7	04/09/2001	Fire	Wickland, Aruba	High flash point oil and hot surface ignition	Block valve failure (leak) during maintenance activities	$250,000,000
8	12/13/1994	Explosion	Port Neal, Iowa, U.S.	Ammonium Nitrate	Unknown	$370,000,000
9	08/07/1994	Fire	Ryazan, Russia	Crude oil	Procedural: Inadequate equipment preparation for hot work	$180,000,000
10	06/20/1991	Explosion	Dhaka, Bangladesh	Anhydrous Ammonia	Failure of a welded joint on a process vessel	$140,000,000

[1]Incidents listed in Table 1 occurred outside of the U.S. PSM "covered process" area, (a) if the incident occurred in the United States or (2) where the incident occurred outside of the United States, the incident would have been outside of the "covered process" area and/or exempt from the U.S. PSM Standard for other reasons, such as the primary process material involved.
[2] "Primary Process Material" listed is the initiating material that led to the event. Other process materials within the facility may have contributed to the total damage; however, those process materials were secondary in nature due to "domino" or knock-on effects.
[3]Damage values inflated to December 2013 U.S. Dollars using the Nelson-Farrar Petroleum Plant Cost Index. Further, estimated damage values in Table 1 are "ground up" values, limited to *property damage alone*. Other costs, such as business interruption, injuries and fatalities, and liability claims are not included in the estimated damage.

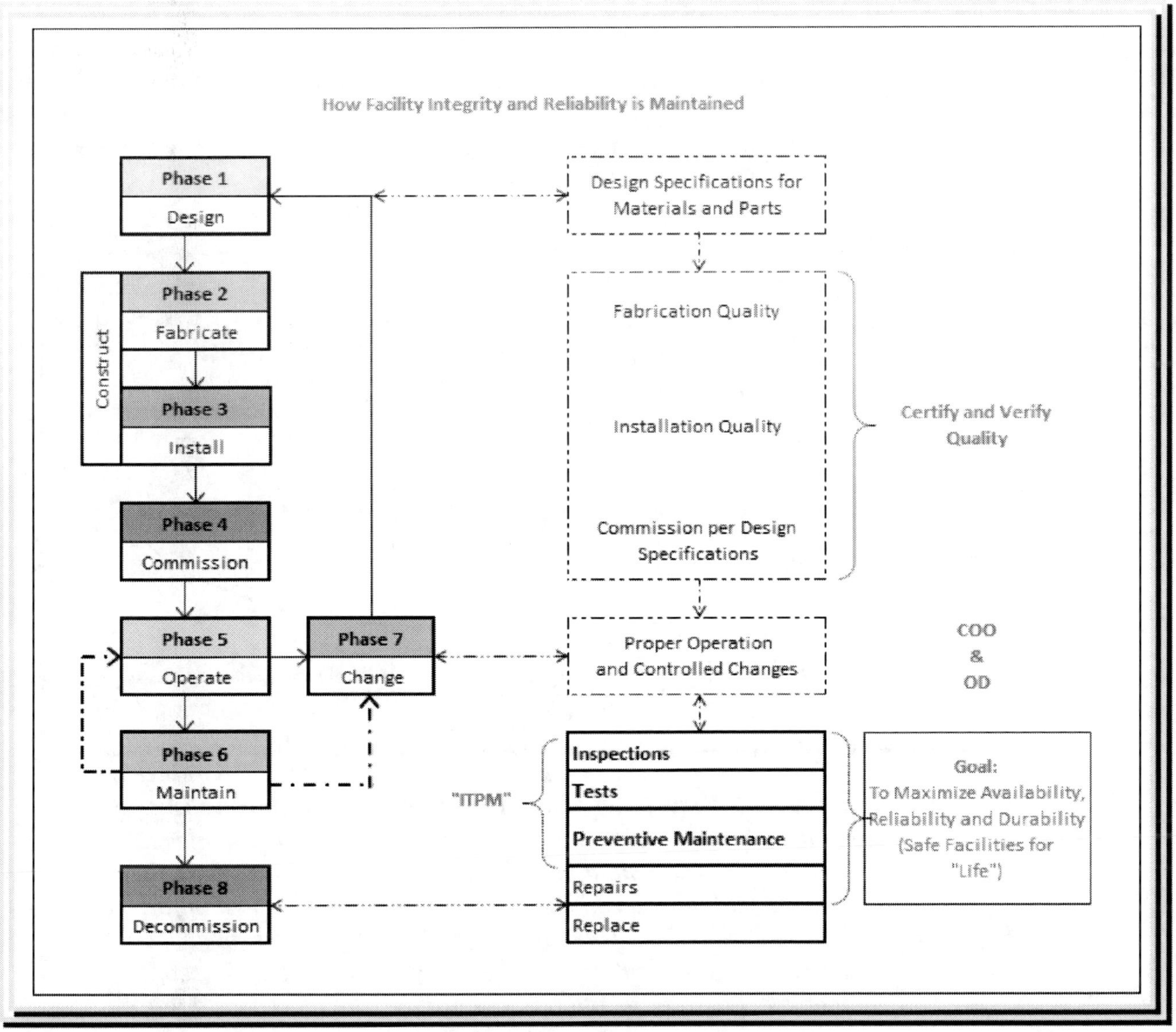

Figure 1: PSEL Diagram [5]

Phases 5 through 7 represent the *operational period* of the facility, which may last for decades and far out-weigh Phases 1 through 4 (in terms of length of time). With the operational period dominating the time element of the PSEL, it is particularly important to operate the facility in a manner which preserves the original process design basis. Preservation of the process design basis is achieved through attentive execution of the following four elements:

1. *Conduct of Operations (COO)*
2. *Operational Discipline (OD)*
3. *Management of Change (MOC)*
4. *Inspection, Testing and Preventive Maintenance (ITPM)*

Phases 5 through 8 of Figure 1 illustrate the relationship between operations, maintenance and engineering (via "Changes" to the facility) which act to preserve the original process design basis of the facility, with the support of robust management systems.

Figure 2 illustrates the PSEL with respect to time, demonstrating how the failure rate of (critical) equipment changes over time. In Figure 2, the previously explained Phases 1 through 4 are contained in the "Early Life" period. Phases 5 through 7 are within the "Intrinsic" period, and Phase 8 occurs in the "Wear Out" period of the distribution.

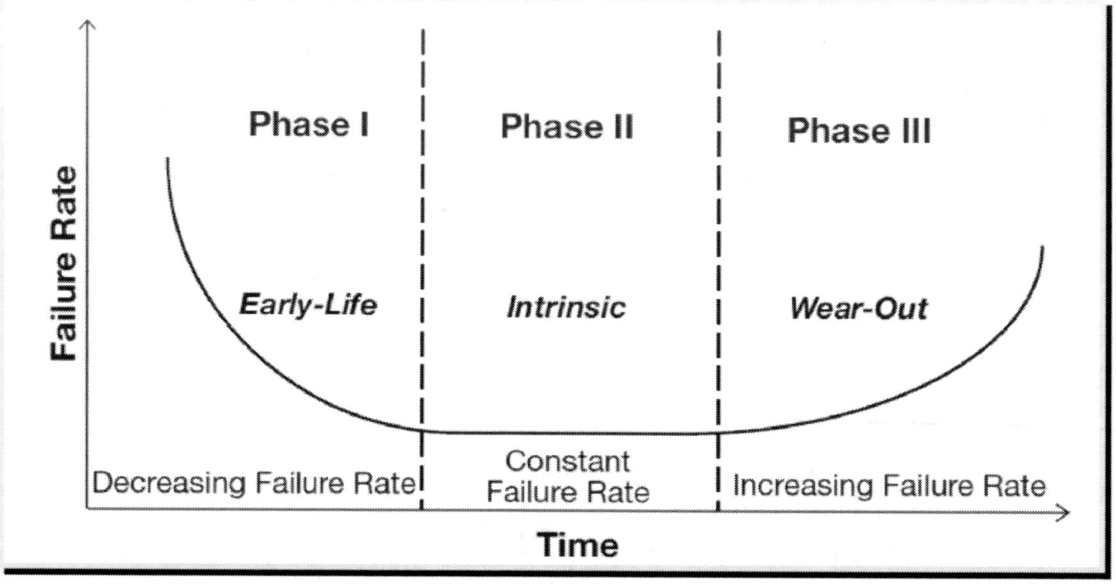

Figure 2: PSEL Expressed Using the Weibull Distribution [7]
"The Weibull distribution can be applied to mechanical component reliability, biology, weather, finance, and a broad range of other systems that obey the characteristics of the bathtub curve. Early in the lifecycle, component failures rates decrease (Phase I), eventually leveling off and remaining relatively constant (Phase II). As the system approaches the end of its life (Phase III), components begin to wear out and the failure rate increases again."

4. "Facility Integrity Program" Scope

The "onion skin Diagram" is perhaps the most useful and practical illustration of the scope of a "Facility Integrity Program." Figure 3 illustrates this scope, and further identifies the core elements and process safety barriers (i.e., layers of protection) that comprise the Facility.

Legend:

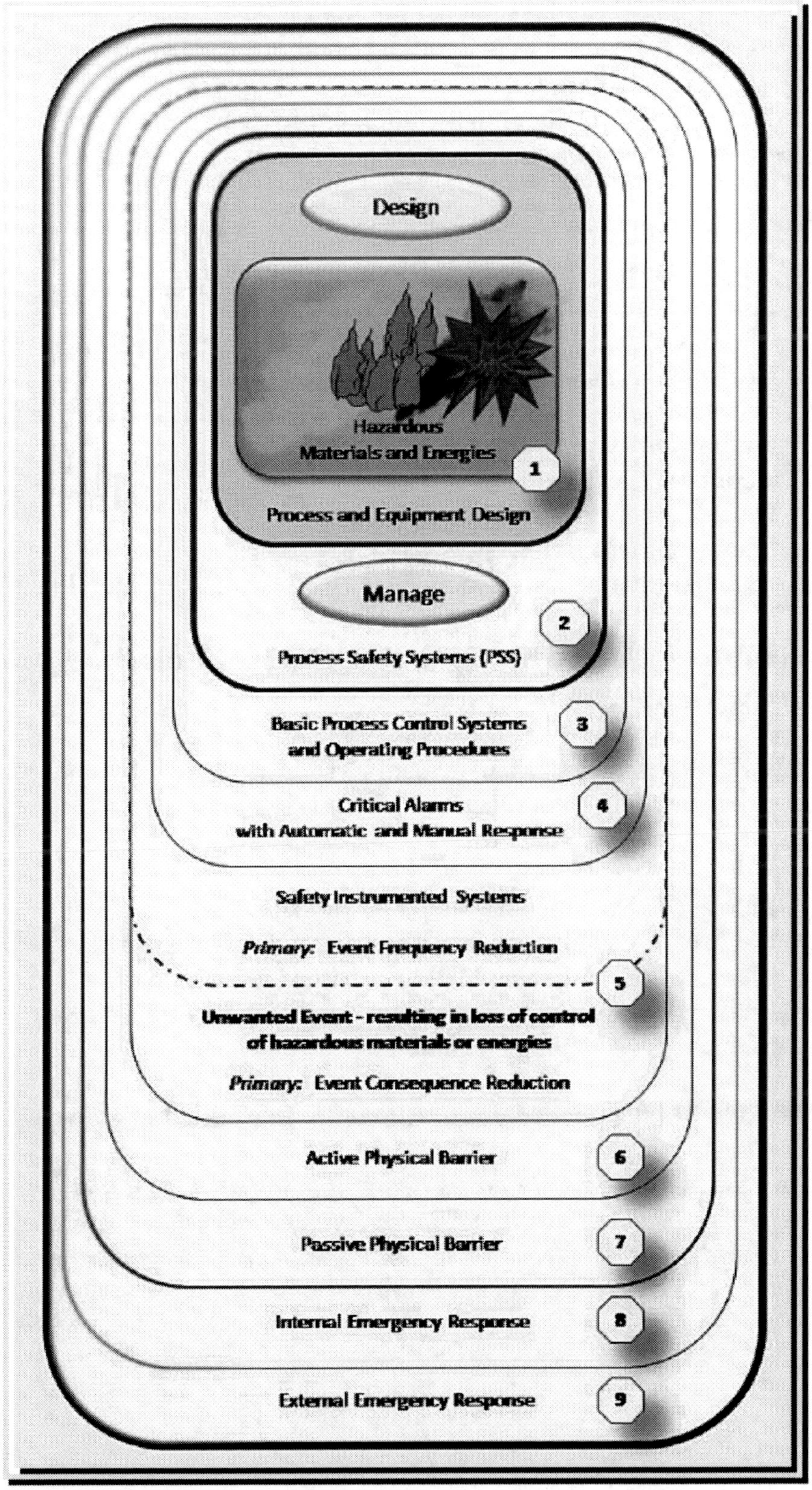

Figure 3: Onion Skin Diagram: The Scope of a "Facility Integrity Program" [8]

The barriers identified in Figure 3 are detailed below. The relevance of the barriers is that every barrier identified in the figure needs to be included in the scope of the facility integrity program. Therefore, it is *strongly recommended* for practitioners, operations and maintenance to:

a) Identify the critical equipment in their plant which is included in these barriers;
b) Fully understand the PDB, BDD and purpose of each equipment within the barrier (i.e., protection layer); and,
c) Operate and maintain that equipment in accordance with the original design basis, until which time the equipment is properly decommissioned [9].

The barriers in Figure 3 are as follows:

1. Process Design, including Inherently Safer Design (ISD) applications.
2. Process Safety Management (PSM) Systems.
3. The Basic Process Control System (BPCS) and Standard Operating Procedures (SOPs).
4. Critical Alarms **with** Operator Response.
5. Instrumented Protective Systems (IPS) and Safety Instrumented Systems (SIS) (automated systems).
6. Active Physical Protective Systems.
7. Passive Physical Protective Systems.
8. Plant (Internal) Emergency Response Teams.
9. External Emergency Responders (e.g., Local Emergency Planning Committees).

5. Coupling the PSEL and the Onion Skin Diagram: The *New* "Facility Integrity Management System"

The new Facility Integrity Management System is created by coupling the previously described PSEL with the Onion Skin Diagram. When this union is created, five critical elements of the management system are created:

1. Element 1: Documenting the PDB and BDD.
2. Element 2: Establishing and Reinforcing COO and OD.
3. Element 3: Establishing ITPM on critical equipment.
4. Element 4: Ensuring Competencies and Conducting Training.
5. Element 5: Managing Deviations.

5.1 The Five Management System Elements Explained

"A major issue for organizations is being able to learn from incidents and capture those lessons into design, procedures, training, maintenance and other programs". "But one factor that is often overlooked is the types of incidents that are tracked or investigated. [3]" -Dr. M. Sam Mannan, PE, CSP, DHC, Regents Professor and Director, Mary Kay O' Connor Process Safety Center

5.1.1 Documenting the PDB and BDD

Documentation of the PDB and BDD may include a broad range of documents which serve to support both hazard identification and risk assessment (HIRA) studies (e.g., Hazard and Operability (HazOp) studies) and Management of Change (MOC) programs. These documents must be kept *evergreen* and used as *"baseline"* for HIRA studies and MOCs, in order to thoroughly evaluate the risk of proposed changes against the original baseline design of the plant. Examples of information which is typically included in this element include:

a) Process chemistry narratives;
b) Process flow diagrams (PFDs);
c) Piping and instrumentation diagrams (P&IDs);
d) Nameplate capacities;
e) Area electrical classification diagrams;
f) Safe operating limits, and
g) Critical operating parameters.

MOCs inherently act to make changes to the facility. The PDB and BDD, when kept evergreen, allow engineering and other practitioners to evaluate the risk of changes in a meaningful manner. Conversely, a proposed facility change cannot be thoroughly evaluated for changes in process risk if the necessary baseline documents are not available, because the PDB and BDD is not an accurate reflection of the current facility.

The author has witnessed several cases where either incorrect or incomplete BDD documentation has resulted in accidents or cost increases on large capital expense projects to correct the deficiencies. For example, sizing and design data for pressure safety devices (PSDs) is normally developed during the EPC stage of a new facility. If, through MOC practices during the operational phase, the PSD data is not kept evergreen this documentation becomes increasingly unreliable over time. This slow "creeping" action of leaving PSD design basis documentation in a static, unchanged state will often become a significant issue during plant expansions, where the expansion project affects a relatively large number of PSDs in the facility. In several actual cases, the author has seen the cost of a plant expansion project dramatically increase due to outdated PSD design documentation. In order for the expansion project to proceed, all of the PSD design documentation within the project scope had to be individually evaluated, re-engineered and updated at a significant cost to the project. These evaluations, in each case, were unforeseen at the start of the project and, as such resulted in substantial scope change and cost increases to the projects.

Another real example of *incorrect* BDD resulted in an internal deflagration within a newly installed catalytic thermal oxidizer (CatOx). In an article written by the author titled "*The Role of Basic Design Data in Preventing Explosions within Fired Equipment: A Case Study* [6]", the purpose of the BDD is explained in great detail. However, in summary, a project was executed to collect various waste streams from points throughout a process in a common vent header, and effectively destroy the hydrocarbons with a CatOx in order to comply with environmental permits. During the EPC stage of the project, the method by which the BDD was developed steered the project team to incorrectly assess and document the design basis for operating the CatOx. Upon initial start-up of the CatOx unit, a mechanical failure occurred, resulting in a blow-out of a pressure relief panel – a safety critical device. A second attempt to re-start the unit allowed air ingress into the vent header, creating a flammable atmosphere that entered the combustion chamber of the CatOx. When this flammable mixture contacted the open ignition source (i.e., the flame) in the combustion chamber, a damaging internal deflagration resulted, setting the project back several months.

These examples, while not necessarily catastrophic in nature, are real events and they do in fact emphasize the importance of developing and maintaining evergreen PDB documentation and BDD.

5.1.2 Establishing and Reinforcing COO and OD

Element 2 addresses the day-to-day operations of the facility. This element goes beyond the "process equipment" itself, and includes provisions such as staffing (e.g., minimum operating complements), building use and occupancy practices (e.g., changes in building use which could impact vulnerabilities and/or design and construction requirements), spare part stocking philosophies and temporary safety system impairments. **Operations owns the risk** pertaining to day-to-day operations and these various provisions, and therefore must adhere to COO and OD practices in order to safely manage the risks in the facility on an on-going basis. Referring back to Table 1, incidents number 1, 5, 6 and 9 arguably have components of failed COO and OD which contributed to these particular losses when examined with greater detail.

MOC practices must consider not only process risks, but also a spectrum of other risks that may impact personnel, operability of the plant, environmental concerns and so forth. For example, changes in occupancy within a critical building can increase the risk profile of the facility by concentrating the risk to employees via an increase in building occupancy. While this is not an equipment change per se, it is a *change* that MOC practices should evaluate with a facility siting study in order to understand the change to personnel risk. In similar fashion, changes in staffing and workforce may adversely impact the operability of the plant. Again, this is not an equipment change but one that does warrant a MOC review. This type of change underscores the importance of establishing a minimum operating complement for the facility, particularly with the operations staff. The minimum operating complement should be considered as PDB documentation, and maintained in an evergreen status with physical changes to the plant, such as plant expansions and decommissioned units.

Wolf and Wasileski have written on the importance of establishing a robust COO and OD program in their article "*THE ILLUSION OF ATTENTION: Are There Gorillas in Your Plant?*"[7]. The failure to notice an unexpected stimulus in your field of vision while performing other attention-demanding tasks is a cognitive phenomenon known as inattentional blindness. The article presents experimental evidence which suggests that our ability to recognize hazards (in the field) is more impaired than we think. The thrust of the article goes on to present a proactive model for Organizational MOC which seeks an optimum level of awareness for operating staff. The relevance of this article with respect to

COO and OD is that MOC plays an important and integral role to successfully managing Element 2.

5.1.3 Establishing ITPM on critical equipment

Element 3 addresses maintenance and reliability, and their role in managing facility risks. Of particular importance in this element is the need for written ITPM plans for each equipment and facility included in the onion skin diagram, *especially for barriers 3 through 7* (the main protection layers used to prevent or mitigate PRIs). Written ITPM plans must include clear instructions for maintenance personnel pursuant to ITPM methods, frequencies/intervals, and **contingency measures** that must be in place to manage process risk during temporary safety system impairments which are required to carry out ITMP activities. This requires clear, close communication with operations during ITPM activities.

Perhaps no other incident exemplifies this need for close communication between operations and maintenance than the fire and explosion on July 7th, 1988 on the Piper Alpha drilling rig in the North Sea. Inflated 2003 U.S. dollars puts the property damage cost of this loss at over $1.8 Billion [3], in addition to 167 fatalities. Much has been written about this accident, in articles, books and other learning material. Possibly one of the most important lessons learned from the Piper Alpha accident has been described in an article by M. Elisabeth Pate-Cornell [10]. Pate-Cornell writes, "*...a piece of equipment (a critical pump with one redundancy) had been turned off for repair and the night crew that operated the platform had not been informed of it. This problem, in turn, was mostly a failure of the "permit-to-work system" that did not ensure proper communications. Things would have not taken catastrophic proportions...*" if not for the lack of communications between operations and maintenance. Pate-Cornell goes on in her article to describe numerous additional shortcomings in communication between operations and maintenance personnel, which are categorically described as "Human Errors". As she goes on to summarize, "*The human errors, questionable decisions, and bad judgments that have been identified above and contributed to the accident can therefore be divided into four categories: (1) design decisions; (2) production and expansion decisions; (3) personnel management; and (4) inspection, maintenance, and correction of detected problems*". This statement alone reinforces not only the importance of COO and OD, but also once again points to the importance of Element 1 which addresses process design.

5.1.4 Ensuring Competencies and Conducting Training

Engineering staff must receive training and instruction towards their responsibilities under Element 1. Element 1 is essentially the foundation of the facility integrity management system. As such, any errors or deficiencies within Element 1 can have adverse impacts towards the other four elements in the management system. For example, if BDD is not fully documented for a particular process, HIRA studies and MOCs become overly cumbersome, difficult endeavors to properly execute and evaluate risks. In similar fashion, operations and maintenance both must be properly trained such that personnel understand their responsibilities under Elements 2 and 3. Operations needs to understand they own the risk of the facility that they operate. Therefore, it is the responsibility of operations to ensure equipment has been restored to a "safe to operate" condition following ITPM activities. Furthermore, maintenance personnel must be in close communication with operations during ITPM activities such that operations is aware of maintenance activities being carried out, equipment that has been temporarily impaired for inspections, and any changes made to process equipment.

In light of these conditions, other management systems such as Permit-To-Work (PTW) and Safety System Impairments [11] become critical systems. These management systems need to clearly describe the nature of the work taking place, the location in the plant where the work will be performed, the anticipated duration of the work, and approval signatures from those with accountability for the equipment, with an escalation process to address work that lasts longer than originally anticipated.

5.1.5 Managing Deviations

The fifth element of the facility integrity management system captures deviations – the unplanned acts and undesired events that occur in a process. Unplanned acts are often undesirable acts or unsafe acts, with no readily observable negative consequences to the process (See Figure 4).

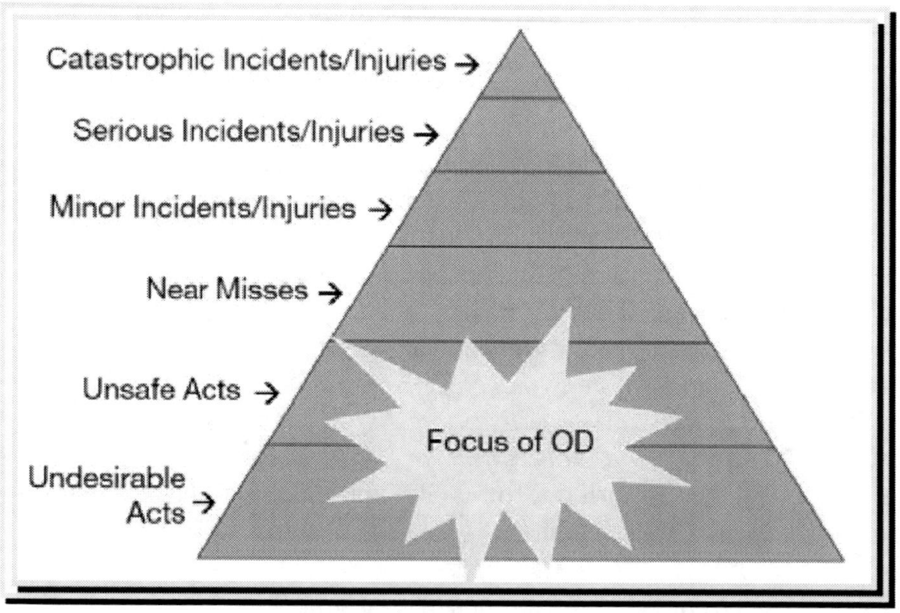

Figure 4: Process-Related Incident Pyramid [8]

"A program to reduce the OD-related problems at the bottom of the triangle helps to reduce the possibility of more serious undesired events at the top"

Conversely, undesired events are often revealed through the ITPM program, and frequently fall into the category of a "near miss". Equipment failures, both "revealed" and "unrevealed", demands placed upon safety systems (i.e., protection layers), spurious activation of safety systems, and unintentionally impaired equipment fall into this fifth element. Each of these "unplanned acts" and "undesired events" must be treated as an incident, and subjected to an incident investigation commensurate with the actual and/or potential consequences of the finding or observation. In investigating these acts and events as an incident, the facility will understand the basic causes, learn from the investigation and be prepared to put measures in place to prevent a recurrence from escalating higher into the PRI Pyramid (e.g., to a serious or catastrophic incident). This fifth element of managing deviations through incident investigation completes the PSEL by providing a circular path back to Elements 1 through 4, such that deficiencies in any element may be corrected, and thus reducing the probability of a more severe incident.

6. Conclusions

MI programs are clearly necessary for managing process risk in a manufacturing facility. While relatively short, the engineering design phase of a project is the starting point for a robust MI program. Procurement and construction typically follow a similar, relatively short timeline which is commensurate with the engineering design phase. However, once a facility becomes fully operational, the length of time devoted towards operation and maintenance normally far outweighs the EPC period, and in many cases lasts for decades. As a consequence, the PSEL is dominated by the operation of the plant and the ITPM program.

When the U.S. PSM Standard was promulgated into law, most facilities adopted a compliance-based maintenance program to meet the requirements of the Standard. However, MI programs have evolved and matured since then, with the CPI adopting blended approaches and philosophies to MI management such as run-to-failure maintenance, reliability-centered maintenance and, most recently, risk-based maintenance. While these approaches each have their own merit, the CPI needs to continue evolving and maturing by adopting a "**facility integrity**" management system approach to curtail equipment-related failures, which remain a leading cause of large-scale accidents.

7. References

[1] **Kletz, Trevor,** "*Lessons From Disaster: How Organizations Have No Memory and Accidents Recur*", Gulf Publishing Company, 1993 Institution of Chemical Engineers, ISBN: 0-88415-154-9.

[2] **Marsh Energy Practice**, "*The 100 Largest Losses: 1972-2009, 21st Edition*", Marsh Ltd, 2010.

[3] **Marsh Energy Practice**, "*The 100 Largest Losses: 1974-2013, 23rd Edition*", Marsh Ltd, 2014.

[4] **Dittrick, Paula**, "BP reaches deal with DOJ over Macondo well blowout in gulf", *Oil and Gas Journal*, November 15, 2012, Houston, TX.

[5] **J. A. Klein, and B. K. Vaughen**, *Process Safety: Key Concepts and Practical Applications*, CRCPress, Taylor & Francis, Boca Raton (to be published in 2016).

[6] **Wasileski, Robert F.**, "The Role of Basic Design Data in Preventing Explosions within Fired Equipment: A Case Study", *Process Safety Progress*, Vol. 26, No. 3, September 2007, DOI 10.1002/prs.10195.
[7] **Elliot M. Wolf, and Robert Wasileski**, "THE ILLUSION OF ATTENTION: Are there Gorillas in Your Plant?", *Chemical Engineering Progress,* July 2014, pp. 28-33.
[8] **J. A. Klein, and B. K. Vaughen, "**Implement an Operational Discipline Program to Improve Plant Process Safety", *Chemical Engineering Progress,* June 2011, pp. 48-52.
[9] **Wasileski, Robert F.**, "Retired and Dangerous: Out of Service Equipment Hazards", *Professional Safety, Journal of The American Society of Safety Engineers (ASSE)*, July 2014, pp. 33-40.
[10] **M. Elisabeth Pate,** "Learning from the Piper Alpha Accident: A Postmortem Analysis of Technical and Organizational Factors", *Risk Analysis*, Vol. 13, No. 2, 1993, pp. 215-232.
[11] **David K. Lewis and Fred Henselwood**, "Safety System Impairments and the Need to Manage Peak Risk", *Process Safety Progress*, Vol. 31, No. 4, December 2012, DOI: 10.1002/prs.11535.
[12] **AIChE/CCPS**, Process Safety Glossary, http://www.aiche.org/ccps/resources/glossary
[13] **AIChE/CCPS**, "*Guidelines for Mechanical Integrity Systems*", CCPS / John Wiley & Sons, Inc., 2006, ISBN: 13-978-0-8169-0952-0.
[14] **Benac, Daniel J.**, "Using Failure Analysis of Near-Misses to Avoid "Two-Peats" and "Three-Peats"", *Journal of Failure Analysis and Prevention*, Vol. 15, Issue 6, DOI 10.1007/s11668-015-0048-z.

Appendix A: Glossary

AIChE – American Institute of Chemical Engineers.

Availability – The fraction of time that the system is actually capable of performing its mission. [12]

CCPS – Center for Chemical Process Safety (an AIChE Technological Community).

Compliance-Based Maintenance (CBM) – A maintenance strategy whereby the principal goal of the maintenance program is to comply with the minimum regulatory requirements set forth by the governmental authority which has jurisdiction in the region of interest (e.g., *U.S. 29CFR1910.119 "Process Safety Management of Highly Hazardous Chemicals"*). [13]

Conduct of Operations (COO) – The embodiment of an organization's values and principles in management systems that are developed, implemented, and maintained to (1) structure operational tasks in a manner consistent with the organization's risk tolerance, (2) ensure that every task is performed deliberately and correctly, and (3) minimize variations in performance. [12]

Critical Equipment – Equipment, control rooms, occupied buildings and other facilities, which upon failure, may lead to a PRI with a consequence severity exceeding that of the organization's tolerable risk criteria. While this definition does not explicitly exclude knock-on and domino effects, it does assume that the damage is primarily limited to the plant and/or unit from where the PRI originates.

Deviation – An unplanned event or condition which does not comply with the design intent of the equipment and/or facility under consideration.

Equipment (and Process Equipment) – A piece of hardware which can be defined in terms of mechanical, electrical or instrumentation components contained within its boundaries. [12]

Facility – The physical location where the management system activity is performed. In early life-cycle stages, a facility may be the company's central research laboratory or the engineering offices of a technology vendor. In later stages, the facility may be a typical chemical plant, storage terminal, distribution center, or corporate office. [12] A "Facility" includes process equipment, infrastructure, buildings, process control hardware and software, and other materials used for activities such as storage, processing, manufacturing, etc. [9]

Fitness for Service (FFS) – A systematic approach for evaluating the current condition of a piece of equipment in order to determine if the equipment item is capable of operating at defined operating conditions (e.g., temperature, pressure). [12]

Hazard Identification and Risk Assessment (HIRA) - A collective term that encompasses all activities involved in identifying hazards and evaluating risk at facilities, throughout their life cycle, to make certain that risks to employees, the public, or the environment are consistently controlled within the organization's risk tolerance. [12]

Human Reliability Analysis (HRA) – A method used to evaluate whether system-required human-actions, tasks, or jobs will be completed successfully within a required time period. Also used to determine the probability that no extraneous human actions detrimental to the system will be performed. [12]

Inspection, Testing and Preventive Maintenance (ITPM) - Scheduled proactive maintenance activities intended to (1) assess the current condition and/or rate of degradation of equipment, (2) test the operation/functionality of equipment, and/or (3) prevent equipment failure by restoring equipment condition. [12]

Management of Change (MOC) - A management system to identify, review, and approve all modifications to equipment, procedures, raw materials, and processing conditions, other than replacement in kind, prior to implementation to help ensure that changes to processes are properly analyzed (for example, for potential adverse impacts), documented, and communicated to employees affected. [12]

Mechanical Integrity (MI) Program – A management system for ensuring the ongoing durability and functionality of equipment and/or facilities. [13]

Near Miss – A precursor event of negligible consequence, which may precede a deviation or process-related incident that has significant, undesirable consequences. If near-misses are identified, investigated, and the reasons for occurring are resolved, communicated, and documented, then major failures can be reduced. [14]

Onion Skin Diagram – A graphical illustration of a Facility which shows inherent process hazards at the center of the diagram, followed by successive layers of protection that are normally in place to stop the progression of a PRI. Note: The successive protection layers are intentionally ordered ("layered") in the sequence upon which a demand would be placed on that specific layer of protection.

Operational Discipline (OD) – The performance of all tasks correctly every time; OD is the execution of the COO system by individuals within the organization. [12]

OSHA – United States Occupational Safety and Health Administration

Preventive Maintenance – Maintenance that seeks to reduce the frequency and severity of unplanned shutdowns by establishing a fixed schedule of routine inspection and repairs. [12]

Process-Related Incident (PRI) – The materialization of a process hazard into an adverse event such as fire, explosion, undesirable chemical reaction, personnel exposure to hazardous materials, unnecessary complexity in the operation, and the unintended transfer of hazardous material from one operating area to another. [9]

PSM Standard – United States Standard OSHA 29CFR1910.119 "*Process Safety Management of Highly Hazardous Chemicals*"

Reliability – The probability that, when operating under stated environment conditions, process equipment will perform its intended function adequately for a specified exposure period. [12]

Reliability Centered Maintenance (RCM) – A systematic analysis approach for evaluating equipment failure impacts on system performance and determining specific strategies for managing the identified equipment failures. The failure management strategies may include preventive maintenance, predictive maintenance, inspections, testing, and/or one-time changes (e.g., design improvements, operational changes). [12]

Risk Based Inspection (RBI) – A risk assessment and management process that is focused on loss of containment of pressurized equipment in processing facilities, due to material deterioration. These risks are managed primarily through equipment inspection. [12]

Run-to-Failure Maintenance (RTFM) – A maintenance strategy whereby facility management proactively commits to equipment failures by ensuring that: (1) spare parts are obtained, (2) OEM manuals and/or repair procedures are available, and (3) applicable training is conducted to perform the repair. [13]

Safety Instrumented Function (SIF) – A system composed of sensors, logic solvers, and final control elements for the purpose of taking the process to a safe state when a process deviation occurs at a predetermined condition or state.

Safety Integrity Level (SIL) – Criterion defining the acceptable probability of failure on demand for a safety instrumented function. [13]

Safety Instrumented System (SIS) – One or more safety instrumented functions combined to protect a process or a key process component. [13]

CSB INVESTIGATION OF THE WEST (TEXAS) FERTILIZER FIRE AND EXPLOSION

Presenter:
Johnnie A. Banks, CFEI, Supervisory Investigator,
U.S. Chemical Safety and Hazard Investigation Board, Washington, DC
johnnie.banks@csb.gov

Prepared for Presentation at
American Institute of Chemical Engineers
2016 Spring Meeting
12th Global Congress on Process Safety
Houston, Texas
April 10-14, 2016

CSB INVESTIGATION OF THE WEST (TEXAS) FERTILIZER FIRE AND EXPLOSION

Presenter:
Johnnie A. Banks, CFEI, Supervisory Investigator,
U.S. Chemical Safety and Hazard Investigation Board, Washington, DC
johnnie.banks@csb.gov

Keywords: *Fertilizer Grade Ammonium Nitrate (FGAN), Hazard Awareness, Emergency Planning and Response, FGAN Storage Practices, Land Use Planning*

Abstract

On April 17, 2013, a fire and explosion occurred at the West Fertilizer Company (WFC), a fertilizer blending, retail, and distribution facility in West, Texas. The violent detonation fatally injured 12 emergency responders and three members of the public. Local hospitals treated more than 260 injured victims, many of whom required hospital admission. The blast completely destroyed the WFC facility and caused widespread damage to more than 150 offsite buildings. The WFC explosion is one of the most destructive incidents ever investigated by the U.S. Chemical Safety and Hazard Investigation Board (CSB) as measured by the loss of life among emergency responders and civilians; the many injuries sustained by people both inside and outside the facility fenceline; and the extensive damage to residences, schools, and other structures.

The CSB conducted an exhaustive investigation that lasted more than two years and which yielded 19 recommendations directed to ten recipients. The tragic loss of life to the civilians and emergency responders, in addition to issues related to community-wide property damage were central in the development of these recommendations, and cover a range of elements which include:

- Regulatory oversight
- Insurance
- Emergency response
- Emergency planning
- Land use planning
- Technical conditions

This presentation will provide an overview of the incident, consider the key findings from the investigation, and illustrate the relationship of these findings to the recommendations. Click here for the final West Fertilizer Investigation Report and related videos and information.

Process Safety Incidents: Big Picture Revelations

Paul Baybutt
Primatech Inc.
Columbus, Ohio, USA
paulb@primatech.com

Prepared for Presentation at
American Institute of Chemical Engineers
2016 Spring Meeting
12th Global Congress on Process Safety
Houston, Texas
April 11-23, 2016

Process Safety Incidents: Big Picture Revelations

Paul Baybutt
Primatech Inc.
Columbus, Ohio, USA
paulb@primatech.com

Keywords: process safety instruments; process safety management; risk management programs

Abstract

Many incidents have helped to define and develop process safety. Each has provided valuable learning opportunities. However, it is even more important to identify insights that can be obtained from an analysis of a large set of incidents that represents those that typically occur. This larger picture illuminates trends and commonalities and provides learning opportunities that are even more important than the causes of any one individual incident.

The Chemical Safety Board has published the results of numerous investigations of process safety incidents and the Center for Chemical Process Safety has published information on incidents that have helped to define process safety. These data have been analyzed to identify commonalities and trends so that measures to help protect against future incidents can be developed. Recommendations are made to address key issues identified.

1.0 Introduction

Process safety regulations have been promulgated in various parts of the world to address protecting process plants against catastrophic incidents in response to the occurrence of a number of major incidents such as Flixborough, Seveso, and Bhopal [1]. However, despite the widespread adoption of process safety practices, major incidents continue to occur. The continued occurrence of such incidents is unacceptable and the issue of what can be done to protect against them must be addressed. This paper addresses the issue of why such incidents continue to occur.

Process safety regulations contain requirements for periodic audits for programs to identify any omissions or deficiencies and some process safety programs require periodic management reviews. However, while omissions and deficiencies must be identified and corrected, knowledge of what omissions and deficiencies have actually resulted in process safety incidents is essential to help prevent further incidents.

Compilations of incident descriptions have been published by various authors, for example, [1 - 7]. Usually, incidents are organized in various ways, such as by root cause, release type, equipment involved, operation being conducted, and process safety element involved. Descriptions of individual incidents also can be found, for example, in reports from the U.S. Chemical Safety Board (CSB) [8], and in journals such as the Loss Prevention Bulletin [9]. Such publications usually focus on providing a description of the incident, identifying its causes, and providing recommendations for corrective actions. Certainly, lessons can be learned from each incident. However, it is also important to take an overall view to identify commonalities and trends across incidents so that measures can be identified that will help to protect against future incidents. In particular, the identification of issues that have contributed to multiple incidents is especially important so that future incidents may be prevented. For example, the CSB has flagged reactive hazards [10], combustible dusts [11], and hot work [12] as significant issues across multiple incidents. This paper identifies additional issues that were important contributors in numerous incidents.

CSB investigation reports have been analyzed concerning management system failures [13], process hazard analysis (PHA) [14], and the hierarchy of controls [15]. The evolution of process safety standards and legislation following nine key process safety incidents has been addressed in a recent article [16].

Section 2 identifies the incidents reviewed and the analysis approach followed. Section 3 identifies and discusses the common contributors that were found for the incidents studied and provides recommendations to address them. Section 4 provides a discussion of the results. Section 5 provides some additional noteworthy observations that were made during the review of incident descriptions. Section 6 identifies three issues that underlie all of the other issues. Each underlying issue is discussed and recommendations are made for addressing them. Conclusions are presented in section 7.

2.0 Incidents Addressed and Analysis Approach

Incidents investigated by the Chemical Safety Board (CSB) [8] and incidents described by the Center For Chemical Process Safety (CCPS) [1] were reviewed. CCPS provides descriptions of over 100 incidents. Some date back to the early 20th century but most occurred within the past 40 years. They cover a variety of industries including chemicals, oil and gas production and refining, aviation, shipping, and nuclear power generation. The CSB has issued over 60 incident investigation reports, case studies, and safety bulletins dating back to 1998 when the CSB began investigating incidents. The CSB incidents cover various sectors of the process industries. The CSB currently is investigating one incident from 2010, three from 2013, three from 2014 and one from 2015. Sixty-eight CSB incidents and 12 CCPS incidents making a total of 80 incidents were studied (see Tables 1 and 2, respectively). About 40% of the CSB incidents occurred in processes covered by the Occupational Safety and Health Administration's (OSHA's) process safety management (PSM) regulations. The descriptions of incidents and their causes were taken at face value.

Table 1. CSB Incidents Reviewed.

No.	Incident	Incident Date	Report Date	PSM[1]	PHA[2]
1	Sonat Exploration Co. Catastrophic Vessel Overpressurization	03/04/1998	09/21/2000	N	N
2	Morton International Inc. Runaway Chemical Reaction, Explosion and Fire	04/08/1998	08/16/2000	N	Y
3	Concept Sciences Hydroxylamine Explosion	02/19/1999	02/01/2002	Y	Y
4	Tosco Avon Refinery Petroleum Naphtha Fire	02/23/1999	03/21/2001	Y	N
5	Bethlehem Steel Corporation Gas Condensate Fire	02/02/2001	12/06/2001	N	N
6	BP Amoco Thermal Decomposition Incident	03/13/2001	05/20/2002	N	Y
7	Motiva Enterprises Sulfuric Acid Tank Explosion	07/17/2001	08/28/2002	N	Y
8	Georgia-Pacific Corp. Hydrogen Sulfide Leak	01/16/2002	11/20/2002	N	N
9	Third Coast Industries Petroleum Products Facility Fire	05/01/2002	03/06/2003	N	U
10	DPC Enterprises Festus Chlorine Release	08/14/2002	05/01/2003	Y	U
11	First Chemical Corp. Reactive Chemical Explosion and Fire	10/13/2002	10/15/2003	N	N
12	Environmental Enterprises Hydrogen Sulfide Release	12/11/2002	09/17/2003	N	U
13	Catalyst Systems Inc. Reactive Chemical Explosion and Fire	01/02/2003	10/29/2003	N	N
14	BLSR Operating Ltd. Vapor Cloud Deflagration and Fire	01/13/2003	09/17/2003	N	U
15	West Pharmaceutical Services Dust Explosion and Fire	01/29/2003	09/23/2004	N	N
16	Technic Inc. Ventilation System Explosion and Fire	02/07/2003	08/20/2004	N	N
17	CTA Acoustics Dust Explosion and Fire	02/20/2003	02/15/2005	N	N
18	D.D. Williamson & Co. Catastrophic Vessel Failure	04/11/2003	03/12/2004	N	N
19	Honeywell Chemical Releases (3 incidents)	a. Chlorine 07/20/2003 b. Antimony pentachloride 07/29/2003 c. Hydrogen fluoride 08/13/2003	08/08/2005	Y N Y	Y N N
20	Isotec / Sigma Aldrich Nitric Oxide Explosion	09/21/2003	08/24/2004	Y	Y
21	Hayes Lemmerz Dust Explosions and Fire	10/29/2003	09/27/2005	N	N
22	DPC Enterprises Glendale Chlorine Release	11/17/2003	02/28/2007	Y	Y
23	Giant Industries Oil Refinery Explosion and Fire	04/8/2004	10/26/2005	Y	Y
24	MFG Chemical Inc. Toxic Gas Release	04/12/2004	04/11/2006	Y	N
25	Formosa Plastics Vinyl Chloride Explosion	04/23/2004	03/06/2007	Y	Y
26	Sterigenics Ethylene Oxide Explosion	08/19/2004	03/30/2006	Y	Y
27	Marcus Oil and Chemical Tank Explosion and Fire	12/03/2004	06/06/2006	N	N

No.	Incident	Incident Date	Report Date	PSM[1]	PHA[2]
28	Acetylene Service Company Gas Explosion	01/25/2005	01/26/2006	Y	Y
29	BP Texas City Refinery Explosion and Fire	03/23/2005	03/20/2007	Y	Y
30	Formosa Plastics Propylene Fire and Explosions	10/06/2005	07/20/2006	Y	Y
31	Bethune Point Wastewater Plant Explosion and Fire	01/11/2006	03/13/2007	N	N
32	Synthron Chemical Reaction and Vapor Cloud Explosion	01/31/2006	07/31/2007	Y	N
33	Partridge Raleigh Oilfield Explosion and Fire	06/05/2006	06/12/2007	N	N
34	Universal Form Clamp Co. Explosion and Fire	06/14/2006	04/10/2007	Y	N
35	EQ Hazardous Waste Plant Explosions and Fire	10/05/2006	04/16/2008	N	U
36	CAI / Arnel Chemical Plant Explosion	11/22/2006	05/13/2008	Y	N
37	Valero Refinery Propane Fire	02/16/2007	07/09/2008	Y	Y
38	Barton Solvents Explosions and Fire	07/17/2007	06/26/2008	N	U
39	Xcel Energy Company Hydroelectric Tunnel Fire	10/02/2007	08/25/2010	N	Y
40	Barton Solvents Flammable Liquid Explosion and Fire	10/29/2007	09/18/2008	N	U
41	T2 Laboratories Inc. Reactive Chemical Explosion	12/19/2007	09/15/2009	N	N
42	Imperial Sugar Company Dust Explosion and Fire	02/07/2008	09/24/2009	N	U
43	Goodyear Heat Exchanger Rupture and Ammonia Release	06/11/2008	01/27/2011	Y	U
44	Packaging Corporation Storage Tank Explosion	07/29/2008	03/04/2010	N	N
45	Bayer CropScience Pesticide Waste Tank Explosion	08/28/2008	01/20/2011	Y	Y
46	INDSPEC Chemical Corporation Oleum Release	10/11/2008	09/30/2009	Y	Y
47	Allied Terminals Fertilizer Tank Collapse	11/11/2008	05/26/2009	N	N
48	Veolia Environmental Services Flammable Vapor Explosion and Fire	05/04/2009	07/21/2010	Y	N
49	ConAgra Natural Gas Explosion	06/09/2009	02/04/2010	N	U
50	CITGO Refinery Fire and Hydrofluoric Acid Release	07/19/2009	12/09/2009	Y	Y
51	Caribbean Petroleum Refining Tank Explosion and Fire	10/23/2009	10/21/2015	N	N
52	Silver Eagle Refinery Explosion	11/04/2009	04/14/2014	Y	U
53	NDK Crystal Inc. Explosion	12/07/2009	11/14/2013	N	U
54	DuPont Corporation Toxic Chemical Releases (3 incidents)	a. Methyl chloride 01/22/2010 b. Oleum 01/23/2010 c. Phosgene 01/23/2010	09/20/2011	U U Y	U U Y
55	Kleen Energy Natural Gas Explosion	02/07/2010	06/28/2010	N	U
56	Tesoro Refinery Explosion and Fire	04/02/2010	05/01/2014	Y	Y
57	Horsehead Holding Company Explosion and Fire	07/22/2010	03/11/2015	N	U

No.	Incident	Incident Date	Report Date	PSM[1]	PHA[2]
58	Millard Refrigerated Services Ammonia Release	08/23/2010	01/15/2015	U	U
59	E. I. DuPont De Nemours Co. Hotwork Explosion	11/09/2010	04/19/2012	Y	Y
60	AL Solutions Dust Explosion	12/09/2010	07/16/2014	N	U
61	Carbide Industries Fire and Explosion	03/21/2011	02/07/2013	N	U
62	Hoeganaes Corporation Flash Fires and Explosion	05/27/2011	01/05/2012	U	U
63	Chevron Refinery Fire	08/06/2012	01/28/2015	Y	Y
64	US Ink Fire	10/09/2012	01/15/2015	U	U

[1] Covered by OSHA's PSM standard.
[2] PHA was performed
Y – Yes
N – No
U – Unknown

Table 2. CCPS Incidents Reviewed.

No.	Incident	Incident Date
1	ELF refinery, LPG tank farm, Feyzin, France	01/04/1966
2	DSM Nypro plant, Flixborough, UK	06/01/1974
3	ICMESA, chemical plant, Seveso, Italy	07/10/1976
4	BP hydrocracker unit, Grangemouth, UK	03/21/1987
5	Shell refinery, FCC unit, Norco, LA	05/05/1988
6	Occidental Petroleum, Piper Alpha Platform, North Sea, UK	07/06/1988
7	Phillips Chemical Company, polyethylene reactor, Pasadena, TX	10/23/1989
8	Total refinery, FCC unit, La Mede, France	11/09/1992
9	Texaco Oil Refinery, FCC Unit, Milford Haven	07/24/1994
10	Napp Technologies, Lodi, NJ	04/21/1995
11	Tosco Avon Refinery, hydrocracker unit, Martinez, CA	07/21/1997
12	Sonatrach LNG plant, Skikda, Algeria	01/19/2004

All incident descriptions were first reviewed to identify commonalities. A second reading of the incident descriptions was performed to confirm the initial findings and document characteristics of the incidents. A third reading of the incident descriptions was conducted to corroborate the findings and provide further detail.

3.0 Incidents Addressed and Analysis Approach

This section identifies common contributors to the incidents. Of course, each incident has its own individual characteristics but there is a remarkable degree of similarity in some deficiencies and omissions across the incidents. The focus in this paper is on issues common to many of the incidents.

Incidents where each issue was a contributor are identified, the percentage of CSB incidents where the issue was present is given, aspects of each issue are discussed, and recommendations are made to address the issues.

3.1 Process Design

Many incidents occurred in processes that arguably had deficient designs {1, 3, 6, 15, 16, 17, 18, 21, 25, 29, 31, 34, 37, 40, 41, 42, 57, 58, 64} and {*10, 11*}[3]. This issue was present in 28% of CSB incidents and 18% of incidents covered by OSHA's PSM standard. Deficiencies arose in the initial design and as a result of modifications.

Hazards were not recognized, designs created hazards, safety systems were inadequate, process instrumentation was insufficient, human factors issues were not addressed, hazards of scale-up were not recognized, and throughput was increased without review. In some cases deficient equipment design contributed to incidents {54a, 54c}. Possibly designers were unaware of the problems and/or design reviews were inadequate but these reasons are not acceptable.

Design deficiencies can be addressed by making process design and engineering an element in process safety programs [17]. Examples of topics that should be addressed by requirements for process design and engineering include design practices, design reviews, inherent safety, hazardous area classification, pressure relief, alarm management, damage mechanisms, human factors, facility siting, competency, and training for designers. Also, process design and engineering should be subject to the same management system requirements as other elements in a process safety program such as written policies and procedures, document management, assigning roles and responsibilities, auditing, management review, and continuous improvement. Process design and engineering should be an element of process safety management systems in order to provide a specific focus on the activities involved and to ensure the application of management system practices to it. Also, the competency of process designers must be addressed.

For new designs and process modifications, proof of safety should be required rather than proof of danger. This will require a change in culture.

---

[3] Braces or curly brackets, that is, { }, are used to enclose the number designations of incidents from Table 1, using normal font, and Table 2, using italic font.

3.2 Safeguards

Many incidents involved processes that were not adequately protected by safeguards {2, 7, 8, 9, 11, 12, 13, 17,18, 20, 21, 22, 24, 25, 26, 28, 29, 30, 32, 34, 35, 36, 37, 38, 40, 41, 42, 46, 48, 49, 50, 51, 54a, 54b, 54c, 60, 62, 64} and *{1, 3, 4, 5, 6, 9, 11}*. This issue was present in 56% of CSB incidents and 57% of incidents covered by OSHA's PSM standard. Needed safeguards were not present or those that were present were insufficient, unreliable, disabled, or lacked independence. Commonly, the hierarchy of controls was not applied (see Figure 1). Often reliance was placed on procedural safeguards which are at the lower end of the hierarchy (19a, 22, 25, 29, 32, 46, 54c}. Process risks should be matched with safeguards in the hierarchy of controls. Higher consequence scenarios, such as a chlorine release in a densely populated area, require safeguards higher in the hierarchy and likely multiple, independent safeguards. Procedural safeguards typically are viewed as less reliable, and, therefore, inadequate for protection against high-consequence events.

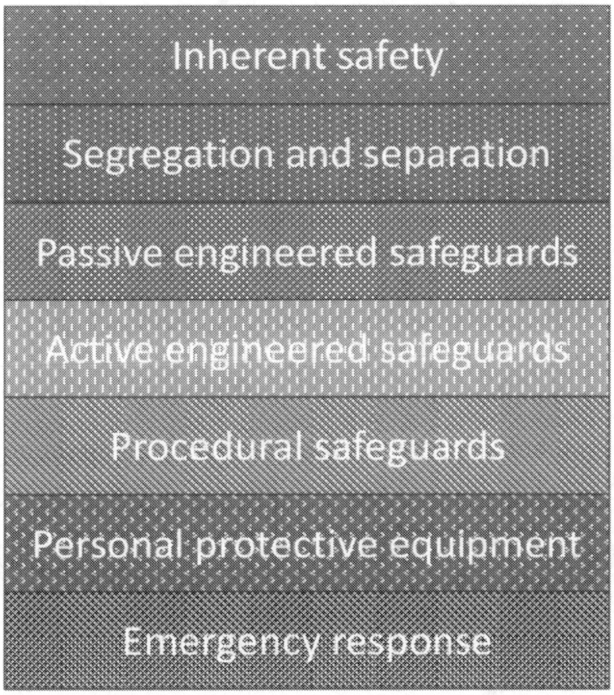

Figure 1. Hierarchy of Controls.

Inherently safer designs or practices could have prevented or mitigated some incidents {29, 39, 55, 56, 54b, 63}. Currently, there are no requirements to address inherently safer systems or the hierarchy of controls in Federal process safety regulations. Process safety regulations in New Jersey require an inherently safer technology review [18] and in Contra Costa County, California require that inherently safer systems be addressed [19]. Proposed amendments to California's process safety regulations for refineries contain a requirement to perform a hierarchy of controls analysis [20].

The operability of safeguards was an issue during several incidents. Explosions and fires disabled safeguards {15, 30, 42} and {*7*}, and toxic releases and fires prevented access to controls {19a, 30, 37, 50} and {*6*}.

Currently, there is no requirement in US process safety regulations to address the design basis for safeguards, evaluate or document the effectiveness of process safeguards, or show that safeguards adequately reduce risks. Proposed amendments to California's process safety regulations for refineries contain a requirement to perform a hierarchy of controls analysis and a safeguards protection analysis (SPA) [20]. The SPA should be performed before a PHA study and updated after the PHA study has been completed. The SPA should include the preparation of a Safety Requirements Specification (SRS) for each individual safeguard [21]. An SRS documents the functional and integrity requirements for a safeguard. Functional requirements describe what the safeguard does to prevent a hazardous state of the process and integrity requirements specify how well the safeguard should perform and what is needed to support that performance level. The SPA should validate that safeguards adequately protect against the hazards of the process, have a specified integrity level, have been installed correctly, are functional, are being maintained appropriately, etc.

Once the SPA has been performed, the PHA team can focus on the qualification of safeguards for specific hazard scenarios [21, 22]. Qualification of safeguards addresses whether safeguards are appropriate for each hazard scenario in a PHA study. Information developed in a PHA study may necessitate revisions to the SPA and SRSs. For example, it may be determined that a redundant power supply is needed to support the integrity required for a scrubber system. SPAs should be updated during management of change (MOC) reviews and when developing recommendations from incident investigations.

3.3 Operations and Maintenance

Problems with procedures were common {1, 2, 3, 5, 6, 11, 12, 13, 14, 18, 19b, 19c, 21, 22, 25, 28, 29, 32, 33, 36, 41, 42, 45, 46, 47, 51, 54c, 59, 61, 62} and {*4, 9*}. . This issue was present in 44% of CSB incidents and 43% of incidents covered by OSHA's PSM standard. In numerous incidents, procedures had not been developed, were not followed, were incomplete, or were otherwise deficient. Such situations are unacceptable. Properly written procedures must be developed following best practices for all process operations and activities before they occur and for all stages and modes of operation. Procedures are needed to deal with any situation where there is the potential for a process safety incident. Procedures must be workable in all circumstances where they apply and personnel need to be trained in the procedures and practice them regularly in drills. Facility personnel must be competent.

In six CSB investigations that identified problems with standard operating procedures (SOPs), the CSB noted that each incident involved SOP deviations that became "necessary violations" to get the job done {45}. Such situations cannot be tolerated.

Existing process safety regulations require written operating procedures. However, the regulations should be clarified to encompass all process tasks and activities and a mechanism is needed to ensure the regulatory requirements are followed.

Deficiencies in maintenance were present in a number of incidents {5, 12, 21, 29, 31, 32, 51, 54c}. Issues regarding the adequacy of inspections were common {4, 7, 47, 53, 56, 62} and {2, 8}. For some incidents, there was no or an inadequate preventive maintenance (PM) program {11, 13, 16, 41} and no or inadequate MI program for some incidents {10, 23, 29, 52, 61, 63}. Such issues were present in 34% of CSB incidents and 32% of incidents covered by OSHA's PSM standard. Existing process safety regulations address the integrity of equipment but a mechanism is needed to ensure the requirements are followed for all safety-critical equipment. Current US process safety regulations require that covered equipment comply with recognized and generally accepted good engineering practices (RAGAGEP) but the term is not defined. OSHA issued a Request for Information (RFI) in advance of rulemaking to revise the PSM standard [23]. Amongst other topics, the RFI sought comment on whether an explicit definition of RAGAGEP should be provided, and, if so, what definition should be used. The current OSHA PSM regulation provides a limited list of covered equipment [24] and the RFI sought comments on whether mechanical integrity requirements should apply to all equipment designated as safety-critical equipment.

The CCPS Risk Based Process Safety (RBPS) element, Conduct of Operations, provides requirements for operational discipline that tolerate no deviation from approved procedures, even if the outcome is inconsequential or desirable [26, 27]. Appropriate work procedures must be determined, documented, and enforced. The importance of following procedures must be emphasized to facility personnel. The adoption of an element on operational discipline would help in this regard coupled with an element that addresses process safety culture. The CCPS RBPS program contains both these elements but they are not included in existing US regulations. Additionally, deviations from approved procedures should be identified and investigated as process safety incidents.

An operational discipline element would encourage and enforce strict adherence to established procedures and imbue facility personnel with the belief that no operation or activity should be performed without a governing procedure. When people's lives are at risk it should not be acceptable to perform activities without proper preparation or by improvising.

Another deficiency present in several incidents was the acceptance of practices as normal and acceptable by facility personnel that outside observers would view as deviant and unacceptable {17, 19c, 21, 29, 42, 57, 62, 63}. This phenomenon has been called "normalization of deviance" [28]. It involves allowing operation of a process that deviates from established limits by relaxing performance standards rather than identifying and correcting the underlying causes of deviations so the deviations become the new, normal operating conditions. Hazards have been normalized in some cases {12, 27, 46, 56}. Normalization of deviance was present in 18% of CSB incidents and 18% of incidents covered by OSHA's PSM standard.

While the acceptance of a lesser performance standard in such cases may not immediately result in an incident, continued successful operation instills a sense of false security. It will be only a matter of time before uncorrected hazardous conditions eventually result in an incident.

Unfortunately, facility personnel can become so accustomed to exposure to hazardous situations that the risks are not recognized even when their lives are at stake. Just because a facility has not experienced a catastrophic incident, perhaps ever in its lifetime, does not mean one could not happen tomorrow.

Normalization of deviance can be addressed by ensuring that any deviations from established performance standards are identified and resolved as soon as they occur. Operational discipline will help in this regard.

3.4 Abnormal and Non-Routine Operations

Many incidents involved non-routine operations and situations {1, 3, 4, 5, 6, 10, 11, 16, 17, 19c, 20, 21, 23, 24, 26, 27, 29, 32, 33, 39, 41, 43, 45, 46, 47, 49, 52, 54a, 54c, 55, 56, 58, 59, 60, 61, 62, 63, 64} and {*1, 7, 9*}. In some cases, they were not recognized as non-routine operations and were not reviewed to ensure that work could proceed safely. Flaws in hot work {7, 31, 33, 44, 59} and other safe work practices {4, 5, 6, 11, 14, 23, 39, 43} and {*6*} also contributed to incidents. Lack of equipment isolation was a contributor to some incidents {11, 33, 59}. Abnormal and non-routine operations were involved in 62% of CSB incidents and 64% of incidents covered by OSHA's PSM standard.

People play an intimate role in non-routine operations. This both increases the likelihood of human error being a factor and places people closer to process hazards. Furthermore, a number of incidents occurred in processes that involved manual operations or were manually controlled {2, 12, 13, 16, 18}.

Current US process safety regulations contain requirements to develop and implement safe work practices and utilize hot work permits. However, the regulations do not specifically address non-routine operations. There are requirements for written procedures under the Operating Procedures element for specific phases of operation listed in the regulations (initial startup, normal operations, temporary operations, emergency shutdown, emergency operations, normal shutdown, and startup following a turnaround or after an emergency shutdown). However, other phases of operation are not specified including commissioning, standby, bypass, unattended operation, maintenance on live equipment, regeneration, troubleshooting, operation with disabled safety systems, cleaning, decommissioning, etc. Process safety programs must address all phases of operation that are or may be experienced by a process. Emphasis should be placed on non-steady state operations including abnormal operations because they are the process states in which most incidents have occurred.

Other elements of process safety programs besides operating procedures should include requirements to address non-routine operations, for example, mechanical integrity and PHA [29]. A separate element on non-routine operations may be merited to provide requirements in a single place. Absent specific requirements to address non-routine operations, many non-routine operations are likely not to be viewed as such and their hazards not recognized or addressed.

Procedures are needed for identifying and planning for non-routine work and abnormal situations, e.g. restart after emergency shutdown or an extended unplanned shutdown. Processes should be operated defensively and conservatively. Procedures should specify alternative courses of action if safety preconditions cannot be met. Furthermore, job hazard analysis [29] should be performed immediately prior to any non-routine activity.

3.5 Human and Organizational Factors

Human and organizational factors contributed to many incidents, for example, valve positions were not clear {23, 51}, valve operation was difficult {56}, lighting was insufficient {51}, operators were distracted {18}, operators were fatigued {29, 45}, operators had insufficient information {57}, procedures were complex and lengthy {45}, communications were problematic {25, 29, 43, 45}, layout was error prone {25}, staffing was not sufficient {25, 29, 45, 51}, operators became desensitized to an alarm due to its history of unreliability {54a}, the work environment encouraged operations personnel to deviate from procedures {29}, a computerized control system was poorly designed {29}, a new control system posed challenges {45}, and there were difficulties with organizational structure {25, 29}. Undoubtedly, problems with safety culture played an important role in many incidents, with BP Texas City being the best known {29}. Most CSB investigations have not explicitly identified contributions to incidents from human and organizational factors so it is not possible to provide meaningful statistics for their contributions to incidents. Undoubtedly, human and organizational factors played a role in most accidents.

Fundamentally, all of the incidents reviewed resulted from human failures ranging from design to operation to maintenance. Even when equipment failures such as corrosion were the basic cause of an incident, the underlying cause usually was human failure {63}.

Some incidents involved conflicts and tradeoffs between process safety and other imperatives {4, 7, 29, 39, 45, 51, 54c, 58, 63}, often financial but sometimes other issues such as noise attenuation {36}.

OSHA's PSM standard requires that "the process hazard analysis shall address human factors" [24]. However, it is not sufficient simply to require PHA studies to address human factors. Certainly, PHA teams must address the role of human factors in influencing human failures but this approach leads to insufficient consideration of human factors issues for processes. Commonly, human factors are addressed by using a checklist with questions that are asked about the entire process so that analysts can try and think of any problems. A much more detailed approach is needed. Otherwise, many aspects of human factors will be missed because of practical limits on the level of detail that can be addressed in PHA studies, limitations on the ability of PHA methods to address human factors, and time pressures on PHA teams. Also, human factors issues affect each of the elements of a process safety program including PHA itself [31]. It is not reasonable to expect a PHA team to address human factors issues that impact its own performance. The Contra Costa industrial safety ordinance requires the development and implementation of a written human factors program [19]. Proposed amendments to California's process safety regulations for refineries contain a requirement to develop, implement and maintain an effective written human factors program that goes beyond the requirements of Contra Costa's industrial safety ordinance [20].

All process safety management programs should have a distinct element that addresses human factors for processes with specific requirements to be addressed. Many aspects of human factors for process safety have been addressed by CCPS [32] and they should be covered by process safety programs.

For many incidents, it is difficult to believe that the conditions that resulted in the incident were allowed to occur. An independent observer easily can conclude that surely someone could have intervened at some stage in the sequence of events that occurred and taken action to prevent the incident. However, such a view does not take account of the psychological factors that influence decision making by people, particularly in groups and when under pressure to achieve competing goals. Factors such as group think, peer pressure, and other cognitive biases interfere with objective decision making [33]. Questionable decision making played a role in numerous incidents, for example {58, 63}. Facility personnel need to apply critical thinking in their daily work activities [34]. Critical thinking can be taught and key facility personnel should be so trained.

3.6 Process Hazard Analysis

Failures in PHA are a common finding in CSB incident investigation reports. Indeed, issues with PHA were present for all CSB incidents where a PHA study was performed. Of particular note is that CSB found that incidents were not identified by the PHA studies performed. However, many of the incidents involved deliberate operation of processes in ways that a PHA team would not consider credible. PHA teams look for ways in which a process may deviate from its design intent but usually the specified design intent identifies what should be done, not what should *not* be done. It is the latter that occurs in many incidents and, while PHA teams try to identify ways that a process may be misoperated, it is unreasonable to expect a PHA team to identify the ways in which a process may be maloperated. Moreover, there is essentially an infinite number of ways in which a process may be maloperated and it would be impractical to ask a PHA team to try and identify all of them. Maloperation must be addressed through operating discipline, proper change reviews, and designing robust processes.

In some cases, the CSB found that PHA studies did not adequately address safeguards because safeguards listed in PHA studies were non-specific or general {19a, 56, 63} and, in many cases, relied too heavily on administrative procedures. For example, safeguards were recorded as "design", "standard operating procedures", "testing and inspection", and "proper emergency response procedures". PHA studies must identify specific safety-critical equipment, for example, a specific relief valve, and specific safety-critical actions, for example, activation of an emergency quench system by an operator. PHA teams should consider the hierarchy of controls when evaluating existing safeguards and deciding on recommendations for further risk reduction. PHA team leaders and team members must be competent [35].

Some CSB incident investigation reports noted the failure to validate critical assumptions used in PHAs {45, 56, 29}. However, such validations should not be part of PHA. They should be performed before a PHA study is performed [21]. Such validations require a different skill set than that required for PHA team members. Moreover, if they were included in PHA studies they would add complexity to the studies and detract from their primary goal of scenario identification. Of course, the adequacy of safeguards for individual hazard scenarios must be addressed by the PHA team.

Some incidents have occurred when more than one activity was being performed simultaneously or sequentially and the activities conflicted {4, 7, 8, 25, 33, 43, 47, 48, 59}. Simultaneous and sequential operations (SIMOPS) reviews [36] should be performed as adjuncts to PHA studies to address such issues. However, they are not yet common and there are no regulatory requirements to perform them.

PHA studies were known to be performed for 22 of the 68 CSB incidents that were reviewed, although incident descriptions did not indicate whether or not a PHA study was performed for 21 of the incidents. However, it is likely that a PHA study was not performed for most of these incidents. Of the 22 PHA studies that were performed, only 4 studies were conducted for processes not covered by OSHA's PSM standard (see Table 1). This finding suggests that PHA studies are only performed when required by regulations.

3.7 Process Changes

Many incidents involved changes that occurred but were not reviewed for their impact on process safety {2, 4, 6, 7, 8, 16, 17, 18, 21, 25, 27, 29, 37, 45, 46, 54c, 59, 63} and {*2, 9, 11*}, and inadequate reviews were performed in some cases {7, 15, 19a, 29, 32, 45, 54a, 56, 63}. Issues with process change reviews were present in 34% of CSB incidents and 39% of incidents covered by OSHA's PSM standard. Omissions and deficiencies can arise because changes are not recognized as requiring MOC reviews, personnel lack awareness of MOC requirements, or MOC reviews are bypassed for expediency. These problems relate to the competency of facility personnel and operational discipline. If personnel are involved in unauthorized and unreviewed changes, they cannot be viewed as competent to work in the facility in the same way as if they maloperated a process or performed maintenance work incorrectly. The establishment of a competency management program should help in this regard. It should be reinforced by operational discipline.

Also, MOC reviews required by the OSHA PSM standard have a narrow definition of covered changes. They cover only changes to PSI, procedures, and facilities that affect a covered process [24]. These requirements are neither global nor sufficiently specific and they are open to interpretation. Organizational changes have played a role in several incidents {29, 45, 56, 64} but they are not currently covered by the regulations, although OSHA's RFI addresses this omission [23]. However, incidents have involved other types of changes not explicitly defined in US process safety regulations including changes in feedstock {4, 63}, addition or removal of equipment {16, 22, 37, 42, 59}, degraded equipment {59}, scale-up of recipes {32}, and prolonged operation {6}. Furthermore, MOC programs should encompass operational deviations and variances, and abnormal situations as well as preplanned changes [37]. Regulations and best practices should specify that any change that may impact on or affect a release of a highly hazardous chemical should be covered by MOC requirements.

Additionally, activities involved in making process changes can be hazardous and are not covered by current US process safety regulations. MOC reviews of changes should include the performance of job hazard analysis for the activities involved.

3.8 Previous Incidents

Information from previous incidents that could have helped to prevent incidents was not used in many cases {2, 5, 6, 11, 17, 19a, 20, 21, 25, 26, 29, 41, 42, 45, 51, 53, 54a, 54c, 60, 61, 63, 64} and {*4, 11*}. Issues with previous incidents were present for 32% of CSB incidents and 29% of incidents covered by OSHA's PSM standard. Previous incidents were not investigated properly, or at all, and near misses were ignored. The issue of retaining information from previous incidents in the corporate memory of companies is well known [38]. However, information cannot be retained if it has not first been developed.

Facility personnel may not have the time or inclination to study incident descriptions and they may not see or understand their relevance to their own facilities. Reliance cannot be placed on people to address lessons learned from previous incidents unless process safety programs contain specific requirements to do so. OSHA's PSM standard requires that previous incidents be addressed [24]. However, the regulatory requirement is part of the PHA element which simply states that the PHA shall address previous incidents. This weak requirement is likely to result in the inadequate consideration of how information from previous incidents can be used to prevent repeat or similar incidents. Process safety programs should have a separate element on previous incidents, including near misses, that requires companies to identify relevant previous incidents from within the company and from other companies in the process industries, compile pertinent information on them, analyze the information in the context of reducing risks in existing processes, and develop recommendations to do so and track them to implementation. The previous incidents review should be performed periodically, e.g. annually, and documented. It should address both the issue of ensuring that incidents are identified, reviewed, and lessons learned are implemented, and that information is retained.

3.9 Facility and Stationary Source Siting

The proximity of facilities to members of the public was an issue in multiple incidents {2, 3, 9, 10, 15, 16, 17, 18, 19a, 20, 22, 23, 24, 25, 26, 27, 29, 30, 32, 35, 36, 41, 45, 46, 48, 51, 52, 53, 63}. This issue was present for 43% of CSB incidents and 64% of incidents covered by OSHA's PSM standard. It raises concerns regarding land use planning in the US. Proximity of facility personnel to hazards in plants also was an issue in numerous incidents {11, 19a, 20, 21, 25, 26, 29, 32, 41, 48, 61} and {*2, 5, 8, 9, 12*]. This issue was present for 16% of CSB incidents and 25% of incidents covered by OSHA's PSM standard. The consequences of some incidents have been exacerbated by the presence of non-essential personnel {29, 49, 56}.

OSHA's PSM standard requires that "the process hazard analysis shall address facility siting" [24] and EPA's RMP rule requires that "the process hazard analysis shall address stationary source siting [39]. However, it is not sufficient simply to require PHA studies to address facility and stationary source siting. Certainly, PHA teams must address the impact of facility and stationary source siting on hazard scenarios but this approach likely leads to insufficient consideration of facility and stationary source siting issues for processes. Aspects of facility and stationary source siting will be missed because of practical limits on the level of detail that can be addressed in PHA studies, limitations on the ability of PHA methods to address facility and stationary source siting, and time pressures on PHA teams.

Process safety management programs should have a distinct element that addresses facility and stationary source siting for processes with specific requirements to be addressed. These requirements should include topics that often are overlooked including using designs that facilitate minimal operator presence; limiting the number of people exposed to hazards, particularly at more hazardous times, by moving unnecessary people away from non-routine operations; training and requiring people to leave when hazardous conditions or situations arise; and assessing the potential for domino events.

The problem of the proximity of people to hazards is compounded by poor emergency planning and response which was present for about 40% of these incidents (see next section).

3.10 Emergency Preparedness and Response

Deficiencies in emergency preparedness and response existed in many cases {5, 8, 10, 11, 16, 19a, 20, 22, 24, 25, 32, 34, 35, 39, 42, 43, 45, 51, 54a, 54b, 63, 64} and {*1, 3, 6, 10, 12*}. This issue was present for 32% of CSB incidents and 39% of incidents covered by OSHA's PSM standard. Deficiencies occurred in hazard communication, evacuation procedures and emergency egress, community notification, emergency preparedness, knowledge of responders, and coordination between responders. OSHA's RFI addresses requiring facilities to coordinate emergency planning with local emergency response authorities [23].

An effective way to help ensure that emergency response plans are in place and will work is to hold regular, periodic drills to test them. It is important to ensure that emergency preparedness and response correlates emergency response with actual hazard scenarios and addresses access to fire water monitors and HAZMAT equipment during incidents and provisions for containment of fire fighting water to prevent environmental pollution.

A hazard analysis should be performed for emergency response activities to identify issues before emergency response is needed. Such a hazard analysis should incorporate task analysis, job safety analysis, and a SIMOP study.

3.11 Audits and Inspections

CSB investigations found that audits by companies, regulators, insurance providers, and others often did not identify process safety omissions and deficiencies and that audits and inspections missed issues that contributed to incidents in several cases {17, 18, 22, 42, 56, 59, 64}. Inadequate audits were identified in several cases {4, 22, 29}. Issues with audits and inspections were present for 13% of CSB incidents and 18% of incidents covered by OSHA's PSM standard.

The CSB noted that in the case of a refinery incident {56} a refinery National Emphasis Program (NEP) inspection was carried out about 18 months before the incident occurred but the mechanical integrity issues that were at the heart of the incident were not identified. NEP audits are the most comprehensive ones that have been performed by regulators. This audit was conducted by a team of people that spent several weeks at the refinery and it focused on the unit that subsequently experienced the catastrophic incident investigated by the CSB.

OSHA's PSM standard requires that a company conduct a compliance audit every three years to "verify that the procedures and practices developed under the standard are adequate and are being followed" [24]. This terse requirement can make for perfunctory audits. Furthermore, process safety compliance audits do not focus on fundamental inadequacies in the design of a process safety program; rather they evaluate the design against regulatory requirements and their implementation. Audits may not properly address non-routine operations because usually they do not occur when audits are performed. Additionally, audits are conducted under time limitations and use sampling techniques.

More comprehensive audits should be performed. They should cover all aspects of a process safety program including safety culture, roles and responsibilities, Auditors must be competent and should be certified [40]. OSHA's RFI addresses requiring third-party audits, increasing the required frequency of compliance audits, and requiring specific time frames for responding to deficiencies found by audits. There is an international standard for auditing management systems [41]. Process safety regulations should require audits be conducted in compliance with the standard.

3.12 Equipment Damage Mechanisms

Equipment damage mechanisms played an important role in several incidents {4, 10, 19a, 20, 53, 56, 63}. Interestingly, all but one of these incidents occurred in a facility covered by OSHA's PSM standard. The CSB has noted the need to perform damage mechanism reviews (DMRs) {63}. Proposed amendments to California process safety regulations include requirements for conducting DMRs [20].

3.13 Process Safety Regulations

Incidents occurred in facilities both covered and not covered by US process safety regulations. Many facilities that are not covered by process safety regulations in the US pose serious process safety risks, for example, facilities handling reactive chemicals and dusts. These issues have been recognized by the CSB [10, 11].

In some cases, facilities that are covered by process safety regulations had not implemented a process safety program {24, 32, 34, 36, 48}.

3.14 Compliance with Industry and Company Standards and Practices and Regulations

Non-compliance with industry and company standards and practices were common in numerous incidents {1, 4, 8, 9, 13, 15, 16, 17, 21, 24, 27, 28, 29, 31, 33, 40, 42, 43, 47, 50, 51, 56, 54b, 54c, 62, 64} and {2, 7}. This issue was present for 38% of CSB incidents and 29% of incidents covered by OSHA's PSM standard. Also, there were failures to follow regulations for multiple incidents {8, 12, 16, 24, 25, 33, 34, 36, 39, 40, 41, 64}. This issue was present for 18% of CSB incidents and 14% of incidents covered by OSHA's PSM standard.

Non-compliance with standards, practices and regulations is a fundamental issue for process safety management. If standards, practices and regulations are ignored, the comprehensiveness of their content does not matter and incidents will continue to occur. Certainly, management review and continuous improvement may go some way to address this problem and the inclusion of this CCPS RBPS element [25] was addressed in OSHA's RFI [23]. The RFI also addressed the inclusion of the CCPS RBPS element on measurement and metrics [25] to track the effectiveness of process safety management programs. Regulators in California have proposed amendments to the state's process safety regulations for refineries that will require the use of process safety performance indicators [20]. Public disclosure of process safety performance would encourage management to pay close attention to ensuring those measures reflect favorably on the company. Thus, for example, it can be expected that there would be fewer action items that remain unclosed past their due dates. Of course, if such an approach were to be applied, it may still be ignored by some companies. Changes in regulatory enforcement may be needed.

A recent case in the UK contrasts markedly with US practices. Phosphine was released from a plant requiring 4,500 people to shelter in place for 2-3 hours and causing disruption to a motorway [42]. The U.K. Health and Safety Executive (HSE) investigated the incident and found that an equipment failure caused loss of containment. The company was fined a total of £333,000 and ordered to pay costs of £110,000 after pleading guilty to an offence under the U.K. Health and Safety at Work Act of 1974 having failed to properly assess and act upon the risk of the equipment failing. Such a penalty for an incident of relatively low consequences contrasts with smaller penalties for greater consequence incidents in the United States.

After the prosecution, an HSE inspector said, "This case should serve as a warning to other companies dealing with harmful substances that they need to get their processes absolutely right, in order to ensure the safety of the public, if they don't they will face the consequences." In virtually, every incident analyzed in this paper, the company did not get it right. There are very few, if any process safety accidents; they are preventable incidents.

3.15 Industry Standards and Guidelines

Numerous CSB incident investigation reports identified flaws and omissions in industry standards and recommended practices {9, 14, 16, 25, 26, 29, 36, 37, 48, 51, 53, 54c, 55, 56, 63} from organizations such as the American Petroleum Institute (API), the American Society of Mechanical Engineers (ASME), the Compressed Gas Association (CGA), and the National Fire Protection Association (NFPA). Issues with industry standards and guidelines were present for 22% of CSB incidents and 32% of incidents covered by OSHA's PSM standard.

Such standards and practices usually are subject to periodic review and updating as part of a process of continuous improvement so such omissions and deficiencies ultimately can be addressed. However, the CSB has criticized the permissive nature of some of these standards and recommended practices {56} which leave options open to companies not to follow some of the guidance. Also, the CSB has criticized the process by which standards are set by some organizations, for example API {56}.

This finding suggests that CSB should investigate a diversity of incidents to identify as many deficiencies and omissions as possible in industry standards and guidelines.

3.16 Product Stewardship

Some incidents involved the use of highly hazardous chemicals obtained by one company from another which did not recognize their hazards {14, 17}. In another case, a company was not informed by a supplier that process safety regulations would need to be implemented before receiving shipments {24}. While chemical supplier companies will not necessarily know the details of how purchasers intend to use the chemicals, they should at least communicate their known hazards to the purchaser. Process safety considerations should be a key part of product stewardship initiatives such as that in the ACC's Responsible Care Code of Management Practices [43]. Suppliers of highly hazardous chemicals have responsibilities in supplying such chemicals to other companies.

Some additional issues were identified for a few CSB incidents (see Table 3).

Table 3. Additional Issues Noted for CSB Incidents.

Issue	Examples
Process boundaries defined too narrowly	59
Inadequate workforce participation	56 / *24*
Inadequate process safety information	3, 6
Inaccurate or incomplete safety data sheets	2, 5, 15, 38
Loss of process design and safety information due to poor management of records	11
Lack of hazard recognition and understanding	14, 15, 21, 26, 31
Inadequate communication of plant hazards	5, 17
Need for positive materials identification during commissioning, modifications, and maintenance	10
Inadequate or no pre-startup safety review	3, 13, 29, 45
No or inadequate near miss reporting	54c
Inadequate or no incident investigation	12, 20
Inadequate contractor management	39, 59, 64
Outstanding action items	2, 4, 6, 7, 16, 25, 45, 54b, 54c, 63 / *27*
Inadequate oversight by management	4, 7, 10, 12, 29, 32

4.0 Discussion of the Analysis

A summary of the results of the analysis is provided in Table 4. Clearly the top issues are abnormal / non-routine work and safeguards which are about equally important for all incidents and for PSM incidents and stationary source siting for PSM incidents. These are followed by procedures which are about equally important for all incidents and for PSM incidents and stationary source for all incidents. Compliance with standards and practices, process changes, emergency preparedness and response, maintenance, and previous incidents come next.

Table 4. Summary of Analysis for CSB Incidents.

Issue	% Incidents	% PSM Incidents
Design	28	18
Safeguards	56	57
Procedures	44	43
Maintenance	34	32
Normalization of deviance	18	18
Abnormal and non-routine operations	62	64
Process changes	34	39
Previous incidents	32	29
Stationary source siting	43	64
Facility siting	16	25
Emergency preparedness and response	32	39
Audits and inspections	13	18
Compliance with industry and company standards and practices	38	29
Failure to follow regulations	18	14
Industry standards and guidelines	22	32

Notable differences between the issues for all incidents and PSM incidents occur in some cases. There are fewer issues for both design and compliance with standards and practices for PSM incidents than for all incidents, but there are more issues for PSM incidents than for all incidents for each of stationary source siting, facility siting, and industry standards and guidelines. The difference for stationary source siting is particularly significant.

The lowest percentage of incidents with any of the issues is 13% and the highest is 64%. This is poor process safety performance.

Time trends for the CSB incidents are provided in Table 5. The numbers of incidents displayed per year in the table largely reflects resources available and decisions made by the CSB. However, it appears that the trend for virtually all issues is for them to continue.

Table 5. Time Trends for CSB Incidents.

Issue	1998		1999		2000		2001		2002		2003		2004		2005		2006		2007		2008		2009		2010		2011		2012	
	N	P	N	P	N	P	N	P	N	P	N	P	N	P	N	P	N	P	N	P	N	P	N	P	N	P	N	P	N	P
Design	1	-	-	1	-	-	1	-	-	-	5	-	-	1	-	1	1	1	2	1	1	-	-	-	2	-	-	-	1	-
Safeguards	1	-	-	-	-	-	1	-	4	-	4	2	-	3	-	3	1	3	2	1	1	1	2	2	3	1	1	-	1	-
Procedures	2	-	-	1	-	-	1	-	3	-	5	2	-	1	-	2	1	1	1	-	2	2	1	-	-	2	2	-	-	-
Maintenance	-	-	-	1	-	-	2	-	2	1	3	-	-	1	-	1	1	1	1	-	1	-	2	1	-	2	2	-	-	1
Normalization of deviance	-	-	-	-	-	-	-	-	1	-	2	1	1	-	-	1	-	-	-	-	1	1	-	-	1	-	1	-	-	1
Abnormal and non-routine operations	1	-	-	2	-	-	3	-	1	1	4	2	1	3	-	1	2	1	2	-	2	3	1	1	4	3	2	-	1	1
Process changes	1	-	-	1	-	-	2	-	1	-	5	-	1	1	-	2	-	1	-	1	-	2	-	-	1	3	-	-	-	1
Previous incidents	1	-	-	-	-	-	2	-	1	-	2	2	-	2	-	1	-	-	1	-	1	1	2	-	2	1	1	-	1	1
Stationary source siting	1	-	-	1	-	-	-	-	1	1	4	3	1	4	-	2	1	2	1	-	-	2	2	2	-	-	-	-	-	1
Facility siting	-	-	-	-	-	-	-	-	1	-	1	2	-	2	-	1	-	1	1	-	-	-	-	1	-	-	1	-	-	-
Emergency preparedness and response	-	-	-	-	-	-	1	-	2	1	1	3	-	2	-	-	1	2	1	-	1	2	1	-	2	-	-	-	1	1
Audits and inspections	-	-	-	1	-	-	-	-	-	-	2	1	-	-	-	1	-	-	-	-	1	-	-	-	-	2	-	-	1	-
Compliance with industry and company standards and practices	1	-	-	1	-	-	-	-	2	-	5	-	1	1	-	2	2	-	1	-	2	1	1	1	1	2	1	-	1	-
Failure to follow regulations	-	-	-	-	-	-	-	-	2	-	1	-	-	2	-	-	1	2	3	-	-	-	-	-	-	-	-	-	1	-
Industry standards and guidelines	-	-	-	-	-	-	-	-	1	-	2	-	-	2	-	1	-	1	-	1	-	-	2	1	1	2	-	-	-	1

N – Non PSM Covered
P – Covered by PSM

There is an average of 5 issues per incident and as many as 13 issues. Some CSB investigations appear to be more thorough than others so the number of issues identified in the investigation reports does not necessarily reflect all of the issues that were present. Consequently, these numbers may be low. However, it is clear that most incidents involved multiple issues and this conclusion would not change. It is likely that the underlying reason for multiple issues being present for the majority of incidents is a systemic problem with the process safety management system and process safety culture.

5.0 Further Observations

This section provides some additional noteworthy observations that were made during the review of incident descriptions.

5.1 Age of Plants

Many process facilities in the US were constructed some time ago. Of course the longer a process operates the more exposure there is to risk. Plant aging factors need to be addressed by process safety programs.

Equipment deteriorates over time and this has played a role in some incidents {63}. Damage mechanism reviews can help to address this issue. Old equipment may be used, for example, a tank constructed in 1929 failed catastrophically in 2008 {47}. Use of old equipment must be subject to close review.

As time passes, industry standards and best practices evolve but the designs of processes are frozen in time. Thus, gaps between current industry standards and best practices and existing company standards and practices may exist. Use of dated standards and practices has played a role in several incidents {30}. Outdated technology also has played a role in some incidents {29}. Process safety programs should require periodic formal reviews to identify and evaluate the feasibility of adopting current standards, best practices, and technology retrospectively. Current standards, best practices, and technology should be addressed when developing recommendations to reduce risk, for example, in PHA studies, MOC reviews, and incident investigations. OSHA's PSM standard does not require employers to evaluate updates to applicable RAGAGEP or to examine new RAGAGEP after evaluating and documenting compliance with mechanical integrity requirements. OSHA's RFI sought comments on the best approach for requiring the evaluation of updates to applicable RAGAGEP. However, RAGAGEP is defined narrowly by CCPS which is the definition being advanced by OSHA. It covers generally approved ways to perform specific engineering, inspection, or mechanical integrity activities, such as fabricating a vessel, inspecting a storage tank, or servicing a relief valve. There are many more types of standards and practices that could be included in the definition of RAGAGEP.

Informal procedures may become institutionalized with the passage of time with operators passing them on verbally without others being aware {46}. Key tribal knowledge may be lost through departures of personnel including retirements {11, 21, 54}. An aging workforce may display diminishing physical and mental faculties. These are human factors issues that should be addressed by a human factors program.

Often process facilities have been built in rural areas away from the public but over time housing developments are built near the facility. This issue was present for several incidents. It must be addressed through land use planning.

5.2 Magnitude of Consequences

The consequences of some incidents could have been much worse. For example, releases occurred under favorable weather conditions {22}, with the wind blowing away from populated areas {10}, when projectiles created by explosions missed hitting adjacent processes {11}, and at a time when the number of people potentially exposed was at a minimum {3, 36}. In some cases, people exposed to serious hazards miraculously escaped harm {63} and avoided being caught in the collapse of structures {49}.

These situations were serendipitous. However, it does suggest a hazard control strategy in which riskier activities, such as non-routine work, should be timed for when the risk to receptors is at a minimum. Also, facility personnel in the vicinity of a riskier activity who are not part of the activity should be relocated for the duration of the activity. Such considerations should be an integral part of planning for non-routine work.

Some injuries occurred when people investigated process abnormalities {32, 45, 64} or tried to fix leaks {62, 63}. Such situations must be taken seriously and not considered part of normal operations. Appropriate precautions must be taken.

6.0 Underlying Issues

This section identifies three issues that underlie all of the other issues. Each underlying issue is discussed and recommendations are made for addressing it.

6.1 Competency of Facility Personnel

The CSB has not explicitly addressed the issue of competency in many incident investigations. However, it was mentioned as a contributor in at least one CSB incident {29} and two CCPS incidents {*2, 4*}. In reading incident descriptions it is hard not to reach the conclusion that the people responsible for the processes were not competent in one or more ways in many cases. This issue likely underlies most of the process safety incidents that have occurred.

The decisions made by personnel ranging from managers to operators can be questioned for many incidents. While facility personnel were not obviously incompetent, in numerous cases clearly they were not competent to perform the work required of them safely. In particular, facility personnel often lacked an awareness and/or understanding of one or more critical aspects of a process, such as its hazards {6, 8, 11, 24, 32, 36, 39, 41, 44, 54a, 54c, 62}. Inadequate operator and mechanic training was an issue in a number of incidents {1, 2, 3, 10, 11, 12, 13, 14, 18, 21, 22, 25, 28, 29, 32, 39, 42, 45, 64}.

OSHA's RFI sought comments on the addition of an element on process safety competency [23] as described by CCPS [25]. Certainly, facilities should implement a competency management program for process safety but the CCPS RBPS competency approach should be modified [44]. Critical personnel should be certified and periodically re-certified. For example, operators and mechanics should not be allowed to operate or maintain a process without demonstrating their competence, including such topics as knowledge and understanding of process hazards, the higher risks of non-steady state operations such as startup, previous incidents, the safe operating envelope for the process, handling abnormal and emergency situations, conditions under which the process must be shut down, use of permit-to-work systems, and the identification and management of changes. Responsibility for supervision of the operation of a process should be formally assigned to a single person and the responsibility formally transferred to another individual as necessary. There must be supervisory oversight of processes at all times.

In some incidents, the availability of technical expertise has been an issue {29, 32, 45}. When critical decisions are faced regarding process safety, someone must be present who understands mass and energy balances, physical and chemical phenomena that may occur, process controls and safeguards, hazards present, and the safe operating envelope, that is tolerable ranges of values of safety-critical parameters. Individuals should be designated as engineering and technical authorities who ensure the proper application of engineering standards and technical practices and adjudicate when deviations are proposed. They must be available for consultation at all times and they must have the authority to shut down processes without the need for consultation with managers and without recrimination. They must be free of production or commercial pressures and should report directly to upper management. A stop work authority that authorizes personnel to stop work in the face of imminent risk or dangerous activity and an ultimate work authority that requires designation of who has the authority for operational safety and decision making are addressed by OSHA's RFI [23].

6.2 Standards for Process Safety Elements

Process safety programs comprise multiple elements such as employee participation, PSI, PHA, etc. For example, OSHA's PSM standard has 14 elements [24] and CCPS's RBPS program has 20 elements [25]. However, it is possible for companies to address each of these elements in widely varying ways because their requirements are general in nature and do not address the effectiveness of their contribution to risk reduction. For example, PHA studies can be conducted with widely varying levels of quality. A solution to this problem is to establish minimum performance standards for each element. These standards should be established by an international organization such as ISO so they apply around the world. There is precedent for such standards in the IEC 61511 standard for safety instrumented systems [45 - 47] and the Hazard and Operability Study (HAZOP) standard [48], although the latter is outdated. These standards are still performance-based in nature, as should be new standards for other process safety elements.

One benefit of standards for process safety elements is that they can be updated more easily than regulations. Current US process safety regulations are static and unable to adapt to innovation and advances in the management of major hazard risks.

This problem could be overcome if the US process safety regulations simply established overall requirements for each process safety element and required companies to follow RAGAGEP for each element. There is already a precedent for doing so with the PSI and mechanical integrity PSM elements although RAGAGEP is interpreted more narrowly in that context. The proposal is to broaden the meaning of RAGAGEP, establish RAGAGEPs for all process safety elements, and require that RAGAGEPs be followed. The process safety regulations would effectively update automatically as RAGAGEPs evolve. Continuous improvement is an essential element of a management system for any surety such as quality, environmental protection, and occupational health and safety, and is part of CCPS's RBPS program [25].

6.3 Safety Goals and Risk Tolerance Criteria

In an least one incident, the CSB concluded that one company had knowledge of and acceptance of high risk {29}. Most companies do not really know what risk their process facilities pose because they do not evaluate risk as there are no requirements to do so under the current US process safety regulatory regime. Indeed, it is possible that similar facilities may pose quite different risks because the regulations do not contain any safety goals or risk tolerance criteria.

The goal of process safety programs is to ensure safety and control risks. Consequently, some form of safety goals or risk tolerance criteria are needed [49]. Absent some metric to judge achievement of process safety goals, resources may be allocated inappropriately. There is a current precedent for the use of numerical risk tolerance criteria provided by the New Jersey Toxic Catastrophe Prevention Act (TCPA) process safety regulations [18]. Although the particular approach used is erroneous [50], companies in the state have accepted the use of numerical criteria.

Many jurisdictions around the world employ the tolerability of risk framework developed by the UK HSE including the As Low As Reasonably Practicable (ALARP) principle [51]. Its use should be incorporated into US regulations. Although the ALARP principle usually incorporates the use of numerical criteria, it can be used without specifying such criteria but using its essential element that risk should be reduced until the incremental expenditures of resources is grossly disproportionate to the value of the incremental risk reduction achieved. This is a performance-based requirement that is consistent with the current philosophy of US process safety regulations. Use of the ALARP principle would help to accommodate improved process safety practices without the need for periodic modification of regulations.

One of the difficulties faced by PHA teams is the use of risk ranking in studies to determine if recommendations for risk reduction, that is, additional safeguards, are needed. The CSB has been critical of the subjectivity involved and the criticism is warranted [31]. The CSB has suggested that Layers of Protection Analysis (LOPA) be used for this purpose. Of course, LOPA requires the use of risk tolerance criteria that are allocated from facility criteria [52]. Proposed amendments to the California process safety regulations include requirements for such an analysis [20]. Simply put, the adequacy of safeguards cannot be determined without risk tolerance criteria.

Use of risk tolerance criteria with LOPA raises the issue of what type of risk is being addressed and should be addressed. Certainly, no loss of life is acceptable but in US society single fatalities are more tolerable than multiple fatalities as evidenced by the relative attention paid by the media to single versus multiple fatality incidents. Perhaps the focus in process safety should be on protecting groups of people, particularly members of the public.

All process safety elements would benefit from the use of appropriate performance indicators [25, 53, 54]. Taken together, their use helps to control overall facility and process risks.

7.0 Conclusions

This survey of incidents covered a range of company and facility sizes, and types of processes. Also, it included facilities covered and not covered by process safety regulations. The study results can be used by companies to focus on improving areas where other companies have fallen short.

Of the 80 incidents studied, only the explosion resulting from corrosion failure of piping at the Shell refinery in Norco, LA, that occurred in 1988 {*5*} was truly unexpected owing to a unique corrosion phenomenon that occurred under conditions that were expected to be less corrosively onerous and with pH within the tolerable range. However, the failure occurred as a result of process changes and it is possible that the incident could have been prevented. All other incidents were predictable and therefore preventable.

Process safety programs aim to identify possible incidents before they occur and implement practices to prevent them. Remarkably, many of the incidents reviewed were attributable to deficiencies or omissions in the design, operation, or maintenance of processes. Therefore, it should have come as no surprise that these processes experienced incidents.

The quote, "It is said that there is nothing new in safety - it has all happened before" was published by CCPS [1]. Virtually every incident or something like it had happened somewhere before. All the incidents involved some type of deficiency or omission in adhering to established process safety practices. In many cases there were multiple deficiencies and omissions. All of these incidents were predictable and preventable. Just because an event may be described as unthinkable does not necessarily mean it is unpredictable.

Several recommendations can be made to strengthen process safety programs and regulations.

7.1 Recommendations for Operating Companies

- Add process safety elements on:
    - Design and engineering

- Human factors review
- Facility and stationary source siting review
- Safeguards protection analysis
- Operational discipline
- Competency management
- Previous incidents review
- Best practices review

- Address non-routine operations throughout all process safety elements.
- Require the use of JSA for non-routine work.
- Expand the types of changes addressed under the MOC element to include any change that may impact on or affect a release of a highly hazardous chemical.
- Make compliance audits more comprehensive.
- Require annual emergency response drills for facilities.
- Require that process risks be managed using the ALARP principle.
- Require facilities to set safety goals and/or risk tolerance criteria and demonstrate compliance with them.

7.2 Recommendations for Standards Organizations

- Develop international standards for key process safety elements that can be referenced by process safety regulations.

7.3 Recommendations for the Chemical Safety Board

- Expand the incident attributes that are addressed in investigations to gain more insights into process safety performance, for example, explicitly address competency, human and organizational factors, safety culture, plant aging factors, and operational discipline.
- Provide an appendix to each investigation report that characterizes each attribute for the incident. Not only is information on deficiencies and omissions valuable but also knowledge of what was done correctly or that there were no problems in certain areas is valuable.
- Compile a data base of incidents and their attributes that is searchable.

- Provide periodic analyses of the full data base of incidents looking for commonalities and trends and make recommendations for improvements in process safety practices.
- Publish the criteria used for the selection of incidents that are investigated by the CSB and update as appropriate.
- Expand on the current documentation in incident investigation reports to explain how the criteria were applied in selecting the incident for investigation.

The CSB adopted a motto in its early days. It still applies:

Salus populi suprema lex esto.

(People's safety should be the highest law.)

8.0 References

[1] J. Atherton and F. Gil, Incidents That Define Process Safety, Wiley-AIChE/CCPS, New York, NY, 2008.

[2] K. Gugan, Unconfined Vapor Cloud Explosions, Gulf Publishing, Houston, TX, 1979.

[3] T. A. Kletz, Learning from Accidents, 3rd edition, Routledge, Abingdon, UK, 2001.

[4] T. A. Kletz, What Went Wrong?: Case Histories of Process Plant Disasters and How They Could Have Been Avoided, 5th edition, Elsevier, Burlington, MA, 2009.

[5] T. A. Kletz, Still Going Wrong!: Case Histories of Process Plant Disasters and How They Could Have Been Avoided, Butterworth-Heinemann, Burlington, MA, 2003.

[6] Roy E. Sanders, *Chemical Process Safety: Learning from Case Histories*, Butterworth-Heinemann; 4th edition, Butterworth-Heinemann, 5th edition, 2015.

[7] The 100 Largest Losses 1974-2013, Large Property Damage Losses in the Hydrocarbon Industry, 23rd Edition, Marsh and McLennan.

[8] http://www.csb.gov.

[9] http://www.icheme.org/lpb.aspx

[10] Hazard Investigation: Improving Reactive Hazard Management, U.S. Chemical Safety Board, Washington, DC, 2002.

[11] Combustible Dust Hazard Study, U.S. Chemical Safety Board, Washington, DC, 2006.

[12] Seven Key Lessons to Prevent Worker Deaths During Hot Work In and Around Tanks, Safety Bulletin, U.S. Chemical Safety Board, Washington, DC, 2010.

[13] A. S. Blair, Management system failures identified in incidents investigated by the U.S. Chemical Safety and Hazard Investigation Board, Process Safety Progress, Volume 23, Issue 4, pages 232–236, December 2004.

[14] M. Kaszniak, Oversights and omissions in process hazard analyses: Lessons learned from CSB investigations, Process Safety Progress, Volume 29, Issue 3, pages 264–269, September 2010.

[15] Amyotte, P. R., MacDonald, D. K. and Khan, F. I. (2011), An Analysis of CSB Investigation Reports Concerning the Hierarchy of Controls. Process Safety Progress, Volume 30, Issue 3, pages 261–265, September 2011.

[16] T. Kerin, The evolution of process safety standards and legislation following landmark events—what have we learnt?, Process Safety Progress, DOI: 10.1002/prs.11762.

[17] P. Baybutt, Process Design - A Missing Process Safety Element (letter to the editor), Process Safety Progress, Volume 35, Issue 1, March, 2016.

[18] New Jersey Department of Environmental Protection, Title 7, Chapter 31, Toxic Catastrophe Prevention Act Program, Consolidated Rule Document.

[19] Contra Costa County Ordinance, Chapter 450-8-Risk Management.

[20] Proposed General Industry Safety Order (GISO) §5189.1, Process Safety Management for Petroleum Refineries, State of California, Department of Industrial Relations, Division of Occupational Safety and Health, Version 5.0 - September 14, 2015.

[21] P. Baybutt, On the Validation of Safeguards for Process Hazard Analysis, Process Safety Progress, Volume 32, Issue 2, pps. 165–169, June 2013.

[22] P. Baybutt, "What risk reduction measures should be credited in process hazard analysis?", Process Safety Progress, Vol. 31, No. 4, pps 359–362, December, 2012.

[23] Process Safety Management and Prevention of Major Chemical Accidents, Request for Information, Federal Register Volume 78, Number 236, Monday, December 9, 2013.

[24] Process Safety Management (PSM) of Highly Hazardous Chemicals, 29 CFR 1910.119.

[25] Guidelines for Risk Based Process Safety, Wiley-AIChE/CCPS, New York, NY, 2007.

[26] Risk Management Program for Chemical Accident Prevention, 40 CFR Part 68.

[27] Conduct of Operations and Operational Discipline, Wiley-AIChE, New York, NY, 2011

[28] D. Vaughn, Risky Technology, Culture, and Deviance at NASA, 1st edition, University of Chicago Press, Chicago, IL, 1996.

[29] P. Baybutt, Process hazard analysis for phases of operation in the process life cycle, Process Safety Progress, Vol. 31, No. 3, pps 279–281, September, 2012.

[30] G. Swartz, Job Hazard Analysis: A Guide to Identifying Risks in the Workplace, Government Institutes, Rockville, MD, 2001.

[31] P. Baybutt, The Role Of People And Human Factors In Performing Process Hazard Analysis And Layers Of Protection Analysis, J. of Loss Prevention in the Process Industries, Vol. 26 pps.1352-1365, 2013.

[32] Human Factors Methods for Improving Performance in the Process Industries, Wiley-AIChE/CCPS, New York, NY, 2007.

[33] P. Baybutt, The Psychology of Decision Making in Process Hazard Analysis, 11th Global Congress on Process Safety, Austin, TX, April 26 - 30, 2015.

[34] B. N. Moore and R. Parker, Critical Thinking, Eleventh Edition, McGraw- Hill Education, New York, NY, 2015.

[35] P. Baybutt, Competency requirements for process hazard analysis (PHA) teams, Journal of Loss Prevention in the Process Industries, Volume 33, January 2015, Pages 151-158.

[36] Guidance on SIMOPS, IMCA M 203, International Marine Contractors Association, March, 2010.

[37] Management of Change, Safety Bulletin No. 2001-04-SB, U.S. Chemical Safety Board, 2001.

[38] T. A. Kletz, Lessons from Disaster: How organisations have no memory and accidents recur, Institute of Chemical Engineers, Rugby, UK, 1993.

[39] Accidental Release Prevention Requirements: Risk Management Programs Under Clean Air Act Section 112(r)(7), Federal Register / Vol. 61, No. 120 / Thursday, June 20, 1996.

[40] P. Baybutt, Competency requirements for process safety auditors, Process Safety Progress, Volume 34, Issue 2, pages 179–182, June 2015.

[41] Guidelines for auditing management systems, ISO 19011:2011, International Organization for Standardization, Geneva, Switzerland, 2011.

[42] http://press.hse.gov.uk/2016/chemical-company-fined-for-dangerous-gas-release/?ebul=hsegen&cr=8/08-feb-16, last accessed 2/8/16.

[43] Responsible Care® Process Safety Code, American Chemistry Council.

[44] P. Baybutt, The meaning and importance of process safety competency, Process Safety Progress, DOI: 10.1002/prs.11767.

[45] ANSI/ISA—84.00.01—2004 Part 1 (IEC 61511-1 Mod), Functional Safety: Safety Instrumented Systems for the Process Industry Sector — Part 1: Framework, Definitions, System, Hardware and Software Requirements.

[46] ANSI/ISA—84.00.01—2004 Part 2 (IEC 61511-2 Mod), Functional Safety: Safety Instrumented Systems for the Process Industry Sector — Part 2: Guidelines for the Application of ANSI/ISA—84.00.01—2004 Part 1 (IEC 61511-1 Mod).

[47] ANSI/ISA—84.00.01—2004 Part 3 (IEC 61511-3 Mod), Functional Safety: Safety Instrumented Systems for the Process Industry Sector — Part 3: Guidance for the Determination of the Required Safety Integrity Levels — Informative.

[48] Hazard and Operability Studies (HAZOP Studies) – Application Guide, IEC 61882:2001, International Electrotechnical Commission, Geneva, Switzerland, 2001.

[49] P. Baybutt, Letter to the Editor, Managing process risks, Process Safety Progress, Volume 34, Issue 2, page 106, June 2015.

[50] P. Baybutt, Numerical Risk Tolerance Criteria in the United States: a Critique of the Risk Criterion Used for the New Jersey Toxic Catastrophe Prevention Act, Journal of Loss Prevention in the Process Industries 32 (2014), Volume 32, November 2014, Pages 428-435.

[51] P. Baybutt, The ALARP Principle in Process Safety, Process Safety Progress, Vol. 33, Issue 1, Pages: 36–40, March 2014.

[52] P. Baybutt, Allocation of risk tolerance criteria, Process Safety Progress, Volume 33, Issue 3, pages 227–230, September 2014.

[53] Guidelines for Process Safety Metrics, Wiley-AIChE, New York, NY, 2009.

[54] Guidelines for Integrating Management Systems and Metrics to Improve Process Safety Performance, Wiley-AIChE, New York, NY, 2016.

Accidents during Turnarounds, Cleanings, and other Infrequent Operations

Michael C. Stern
Exponent, Inc.
Natick, Massachusetts
mstern@exponent.com

Vijay Somandepalli,[1] Noah Budiansky,[1] Ali Reza,[2] Timothy J. Myers[1]
Exponent, Inc.
[1]Natick, Massachusetts
[2]Los Angeles, California

Prepared for Presentation at
American Institute of Chemical Engineers
2016 Spring Meeting
12th Global Congress on Process Safety
Houston, Texas
April 11-13, 2016

Accidents during Turnarounds, Cleanings, and other Infrequent Operations

Michael C. Stern
Exponent, Inc.
Natick, Massachusetts
mstern@exponent.com

Vijay Somandepalli, Noah Budiansky, Ali Reza, Timothy J. Myers
Exponent, Inc.

Keywords: Process Safety, Turnarounds, Startup, Shutdown, Cleaning

Abstract

Infrequent operations, such as turnaround services and cleaning operations, pose unique hazards due to their lesser familiarity to operators and plant personnel. For example, cleaning chemicals and methods can have unforeseen reactions with scale material, the chemistry of which may not be well-understood. Turnaround services may create a false sense of security because the plant is not in operation, which eliminates many of the commonly encountered hazards. Turnaround services, however, can be hazardous, especially if proper precautions have not been taken when shutting down the plant. Avoidance of accidents during these infrequent operations requires similar hazard analyses to those for standard operations, but these may not be thoroughly performed if the operation is assumed safe. Multiple case studies will be presented on accidents that have occurred during infrequent operations and the lessons learned from those incidents will be discussed.

1 Introduction

Operations including startups, shutdowns, turnarounds, and cleanings can pose additional risk for facilities due to their infrequency and uniqueness. The infrequency of these types of operations provides less opportunities for operators to gain experience and understanding of the procedures. The uniqueness, such as replacing or modifying a part that has been relatively untouched for years, can lead to unintended changes in the process when operation resumes.

Variability is a common risk factor for cleaning and turnaround operations. The same cleaning procedure or chemicals that might have worked for a certain type of scale could be hazardous if used for to clean out a different contaminant. Turnarounds and cleanings may also involve the use of contractors who are less familiar with the facility and may require site specific training to safely operate.

In this paper, the heighted risks of different infrequent operations are discussed and relevant case studies are presented. The case studies, some previously published and some presented for the first time, span a wide range of industries from manufacturing facilities to chemical plants, but all experienced or observed failures during an infrequent and transient operations.

2 Cleaning Operations

Both scheduled and unscheduled cleaning operations are often required at chemical and manufacturing facilities to maintain the equipment functionality and minimize contaminants in the products. Cleaning with chemicals can pose hazards due to undesired chemical reactions between the scale material and the cleaner. In certain cases when multiple cleaners are used, the cleaners could react to form hazardous substances or gases. Table 1, shown below, provides a few examples of chemical mixtures that could form explosive, toxic, or flammable substances leading to hazardous conditions. As will be described in the Condea Vista case study below, even steam cleaning can lead to an explosion in a process vessel when contacted with an aluminum powder containing sludge.

Table 1. Potential reaction products and hazards for certain chemical mixtures

Chemical Type One	Chemical Type Two	Reaction Products	Hazards
Acetone	Peroxides	Acetone-Peroxides	Explosives
Salts with (Cl^-, F^-, CN^-) ions	Acids	Acid gases	Corrosive and toxic
Metal Powders	Water or Steam	Hydrogen	Flammable

When cleaning areas that are already hazardous locations, such as areas near equipment containing flammable liquids or combustible dusts, the cleaning process can be a potential ignition source if proper precautions are not taken. Routine maintenance or cleaning actions can also release combustible dust accumulations creating a hazardous environment in an area where ignition controls are not present, leading to a fire or explosion. The case studies of the Jahn Foundry and CTA Acoustics combustible dust explosions provide examples of how ignition of disturbed accumulations can lead to catastrophic explosions.

2.1 Condea Vista Chemical Plant Explosion[1]

On October 13, 1998, an explosion and fire occurred inside a 3000-gallon Hastelloy reactor (R-2) that was part of the linear alkyl benzene (LAB) process at the Condea Vista plant in Baltimore, Maryland. Process materials released from the various vessels and process piping by the explosion fueled a fire that took approximately 2 hours to extinguish.

The investigation revealed that prior to the explosion, the bottom of R-2 was plugged with approximately 180 gallons of a sludge-like mixture that likely consisted of a mixture of coarse aluminum powder, aluminum chloride and various hydrocarbons. The explosion occurred approximately 10 minutes after steam was introduced at the bottom of the reactor in an effort to break up the sludge accumulation.

Figure 1. Photographs showing the overview of the explosion (left) and top of the exploded R-2 reactor (right).

A series of small-scale chemical reactivity tests were performed to investigate the potential reactions between steam and manufactured sludge that was similar to the material accumulated inside the R-2 reactor. Rapid runaway chemical reactions were observed after steam was injected into this mixture, while pressures in excess of 700 psig and temperatures in excess of 400 F were recorded before the test vessels ruptured.

A review of possible chemical reactions between steam, aluminum and aluminum chloride confirms that the predominant reaction mechanism is the exothermic hydrolysis of aluminum chloride to produce HCl vapor. Significant amounts of hydrogen gas can also be generated as the aluminum reacts with the acid/steam mixture after it is stripped of its protective oxide coating. Exponent concluded that the root cause of this accident was direct and sustained contact between the steam and accumulated sludge at the bottom of the reactor.

2.2 Dust Explosion at CTA Acoustics[2,3]

At approximately 7:30 am on February 20, 2003, a fire in at the CTA Acoustics facility in Corbin, Kentucky, lead to multiple secondary explosions and fires resulting in the death of seven workers and 37 injuries.

CTA Acoustics manufactured thermal acoustic insulation. The insulation material consisted of fiberglass bound together by phenolic resin powder. The product was used in automotive applications including the insulation on the hood or floor board of a car. CTA had four fiberglass forming lines, 401, 402, 403 and 405. The CSB report on the incident stated that the explosion originated near line 405 where dust generated during cleaning was ignited by the oven on line 405.

The CSB investigation concluded that workers started a routine cleaning of the dust collection system on line 405 thirty minutes before the incident occurred. The cleaning was expected to take approximately one hour. The night shift had reported significant clogging on the line. During the cleaning, the operators observed that the transition leading to the baghouse was the source of the clog and proceeded to use compressed air to break up the clog. When the baghouse was restarted at approximately 7:30 am, an initial explosion occurred that spread to the rest of the facility through secondary explosions fueled by clouds of combustible dust that were dislodged during the initial explosion.

2.3 Dust Explosion at the Jahn Foundry[3]

Figure 2. Photographs showing dust deposits on exhaust hoods in Shell Mold department (left) and the structural damage to the foundry (right).

At approximately 3:04 PM, on February 25, 1999, a series of explosions and a subsequent fire caused extensive damage within the Shell Mold building at the Jahn Foundry Corp. in Springfield, MA. Three employees suffered fatal injuries and nine others were injured. The foundry produced gray iron cast products for various customers.

The casting process included the production of shell molds, into which the molten material was ultimately poured. The molds were produced from a 95/5 percent mixture of sand and phenolic resin, added as a thermosetting bonding agent to hold the mold together. Even though the resin was supplied in the form of a fine powder, a dust collection system was not installed in the Shell Mold building. However, an odor control system designed to remove volatile organic compounds emitted from the resin during heating of the shell molds was sufficiently powerful to entrain fine particulates. The odor control system consisted of hoods or canopies above each shell mold station that vented into duct manifolds that connected to a scrubber tower where volatile organic compounds were removed.

An engineering investigation of the incident concluded the explosions were initiated due to the ignition of resin dust dislodged from an overhead exhaust duct or inadvertently dispersed during routine maintenance. A subsequent deflagration (dust explosion) within the odor control system was fueled by excessive, resin rich deposits of dust within the exhaust ducts. A second, larger dust explosion in the upper half of the Shell Mold building caused the majority of the damage to the structure. This second explosion was fueled by resin rich dust deposited on various horizontal surfaces on the upper level of the Shell Mold building.

Personnel at Jahn Foundry were generally unaware of the specific dust explosion hazard associated with this product and the foundry did not comply with the requirements of NFPA 654, *Standard for the Prevention of Fire and Dust Explosions from the Manufacturing, Processing and Handling of Combustible Particulate Solids*

3 Startup and Shutdown Operations

Operator inexperience and unidentified operational differences can lead to additional hazards when starting up or shutting down a plant. For plants that only shut down for maintenance annually or less, the operators will have only limited experience with the tasks required for startup and shutdown due to the infrequency of the operation. As a result, lessons learned from previous experiences may be less likely to be remembered, recorded, and passed on to future operators.

Differences that occur in the process during startup can be difficult to predict and assess. In systems with dust generation, for example, the properties of the collected dust during startup can vary from the dust generated during steady-state operations. It can be difficult to quantify these differences because process sampling during startup is complicated by the inherent transience of the operation.

In processes that include recycle loops or process material recycled downstream, the feedstock during startup can vary in considerable but unidentified ways. Small process changes, equipment maintenance and replacement, and other alterations to the system can introduce small but significant changes into the performance of the system that could lead to equipment failures if the changes are not anticipated. The case of the ammoxidation reactor incident described below provides an example of a chemical reactor that was damaged during startup due to the changes that occurred while the reactor was offline.

3.1 Ammoxidation reactor failure during startup

Several startup attempts of an ammoxidation reactor led to unplanned shutdowns within hours of starting due to over-temperature alarms. The ammoxidation reactor was a packed-bed counter-current tube-in-shell reactor where the tubes were packed with catalyst and an organic heat transfer fluid was pumped through the shell to cool the tubes and remove heat from the exothermic reaction. After a two-day period involving several attempted startups that were followed with subsequent shutdowns due to over-temperature conditions, the reactor was taken offline. Inspection of the reactor revealed a significant leak of heat transfer fluid and severe localized damage to the top tubesheet. The reactor was repaired by replacement of tubesheet sections and a number of tubes.

The reactor was brought back online and was used to produce a different isomer and performed without issue. After two months, the reactor was switched back to producing the original isomer. After two unsuccessful startup attempts that led to shutdowns due to over-temperature conditions, the reactor was inspected and thermal damage was observed.

Investigation of the reactor incidents revealed that prior to the startup, new catalyst was installed to replace the old catalyst that had been in the reactor. The catalyst packing gradient had also been adjusted such that higher concentrations of catalyst were present near the beginning of the reactor tubes. The new catalyst packing gradient and the increased activity of the fresh catalyst both contributed to increased reaction rates within the reactor during operation. It was also discovered that the heat transfer fluid reacted exothermically when exposed to conditions inside the reactor tubes.

It was determined that the failure of the reactor was a direct result of thermal expansion of tubes that occurred as feed rates were increased in the reactor and over-temperature conditions occurred. The thermal expansion led to the mechanical failure of a tube-to-tubesheet connection, which allowed heat transfer fluid to enter the tube side of the reactor. Exothermic reaction of the heat transfer fluid in the tube side of the reactor during repeated startups of the reactor led to the melting and heat damage of the tubesheet.

4 Plant Turnarounds

Plant turnarounds that involve significant maintenance, cleaning, replacement of equipment, and plant modifications can pose additional hazards due to the uniqueness and unfamiliarity of the activities, the involvement of contractors who are not as intimately aware of the hazards at the plant, and the difficulty of entirely purging the process lines and equipment of hazardous chemicals. Corrosion or contaminant buildup that occurred during the operation of the process can alter the condition of the process equipment in ways unbeknownst to the operators until it is inspected during the turnaround. The case study presented below provides an example of equipment that became corroded during operation but was detected before it created hazardous conditions.

4.1 Corrosion in a Flue Gas Desulfurization Absorber

All equipment is subject to both periodic planned shutdowns to conduct routine maintenance and inspections and unplanned shutdowns to replace components or make repairs. During these shutdowns careful inspections of the equipment are important to prevent accumulating degradation of a system or fix other unforeseen issues.

During a routine shutdown to perform periodic cleanout of a flue gas desulfurization (FGD) absorber associated with an incineration plant, corrosion damage was observed in the corrosion resistant liner of the absorber. FGD systems are designed to remove sulfur dioxide and other aggressive gaseous species from power plant and incineration emissions by spraying the flue gas with a limestone slurry, producing a calcium sulfite (gypsum) or other reaction species solid that falls to the bottom of the tower. The FGD slurry environments can have pH values as low as 1 and chloride concentrations as high as 200,000 ppm. The resulting slurry can be very corrosive and highly alloyed metals (or non-metallic coatings) are required to prevent corrosion of the liner to prevent exposure of the underlying carbon steel support structure.

In this case, 2205 duplex stainless was selected as the absorber liner was selected to prevent corrosion based on the nominal operating conditions, but corrosion of the duplex stainless steel liner was observed around a cleanout door. The corrosion damage had penetrated through the stainless steel liner in the time frame since the last inspection and caused minor damage to the underlying structure. A corrosion failure analysis was conducted and it was determined that the isolated environment in the gap between the cleanout access door and the door sill was more aggressive than the nominal conditions in the absorber. Laboratory corrosion testing indicated that corrosion resistance of the stainless steel liner, that was highly resistant to corrosion under the nominal design conditions, was significantly reduced in this area due to accumulated ions resulting in a reduced pH. In this case, periodic inspection of the equipment resulting in detection of corrosion degradation that was ultimately repaired prevented more serious failures from occurring.

5 Discussion and Conclusions

A review many of the industrial failures that have been investigated by Exponent shows a significant number that occurred during infrequent and transient operations including startups, shutdowns, cleanings (scheduled and unscheduled), and chemical transfers. While the specific incidents vary substantially in their magnitude and cause, an unknown change is the system can be traced to the cause. In three cases it was accumulation (combustible dust or aluminum sludge) and in another cases it was a change in a process variable (catalyst packing) that occurred during a turnaround that caused a failure upon startup.

Preparation for these upsets require thorough process hazard analyses (PHAs) and an effective management of change (MoC) program to identify these key variables before they become an issue and develop strategies for operators to employ when these events occur. The case of the FGD unit highlights the importance of careful inspection during turnarounds to identify corrosion or other changes in equipment. In some cases, such as combustible dust incidents, significant industry and governmental focus has raised awareness of the hazards of combustible dust. An OSHA National Emphasis Program and updated NFPA standards have been developed to address these hazards. In other cases, the lessons learned from the incidents should be used in future PHAs to mitigate the likelihood of similar incidents in the future.

6 References

[1] A. Reza, A. Kemal and P. Markey. "Runaway Reactions in the Aluminum, Aluminum Chloride, HCl and Steam System. An Investigation of the 1998 CONDEA Vista Explosion in Maryland," 36th Annual Loss Prevention Symposium. AICHE Spring National Meeting, March 2002.

[2] CSB. *Combustible Dust Fire and Explosions: CTA Acoustics*. United States Chemical Safety and Hazard Investigation Board, Report No. 2003-09-I-KY, February, 2005.

[3] T. Myers and A. Ibarreta. "Investigation of the Jahn Foundry and CTA Acoustics dust explosions: similarities and differences," *Journal of Loss Prevention*, pp. 740-745, 2009.

12TH GLOBAL CONGRESS ON PROCESS SAFETY

Biomass Industry Explosion Case Study

Amy E. Theis, P.E.
Fauske & Associates, LLC
16W070 83rd Street, Burr Ridge, IL USA 60524-5802
Theis@Fauske.com

Timothy L. Cullina, M.S., P.E. and Kenneth N. Kurko
Fauske & Associates, LLC
16W070 83rd Street, Burr Ridge, IL USA 60524-5802
Cullina@Fauske.com and Kurko@Fauske.com

Prepared for Presentation at
American Institute of Chemical Engineers
2016 Spring Meeting
12th Global Congress on Process Safety
Houston, Texas
April 11-23, 2016

Biomass Industry Explosion Case Study

Amy E. Theis, P.E.
Fauske & Associates, LLC
16W070 83rd Street, Burr Ridge, IL USA 60524-5802
Theis@Fauske.com

Timothy L. Cullina, M.S., P.E. and Kenneth N. Kurko
Fauske & Associates, LLC

Keywords: Explosion, thermal stability, combustible dust, biomass, incident, dust hazard analysis (DHA), management of change (MOC)

Abstract

The focus of this case study is an explosion that occurred in a biomass processing facility. There were three pieces of equipment involved including a dryer, conveyor and a cyclone. Thankfully, there were no injuries or fatalities due to this incident. This event occurred at a biomass company, resulting in damaged equipment with one piece needing to be completely replaced. Other equipment was able to be repaired and returned to service.

Material handled in the process had been observed to smolder on prior occasions, but the cause of these near miss incidents was not investigated or identified. The material was a potential combustible dust and there was also the possibility for flammable gas to be generated due to decomposition. Through material characterization, several hazards were identified. Combustible dust and calorimetry testing were employed to evaluate the thermal stability, reactivity and combustibility of the material. In addition, off-gases were evaluated to determine if flammable gas was generated. This case study will outline the test design strategy to obtain key process safety information, identify the multiple hazards based on the test results, review existing explosion safeguards and determine a postulated series of events that could have led to the incident. Lessons learned from the incident will be reviewed including recommended risk reduction strategies that could be used to prevent future events. One example is the importance of obtaining appropriate process safety information to facilitate data driven decisions to be made regarding safe operating conditions.

1 Introduction

An explosion occurred at a biomass processing facility. Fortunately no one was injured or killed in this incident. It is believed that a fire, smoldering or other ignition source began in the conveyor. Due to a lack of explosion protection, this led to a subsequent explosion in a connected piece of equipment. Findings indicate that the source of the explosion was consistent with a combustible dust deflagration that appeared to have ignited in or near the transfer equipment and accelerated through the duct work to a cyclone.

Lessons learned presented in this paper are based on experimental test results provided by the customer as well as testing performed by Fauske & Associates, LLC after the incident. Deficiencies that yield opportunities for improvement at this facility include a lack of hazards identification for the process, lack of combustible dust management systems and lack of a risk assessment.

2 Background

2.1 Process Description

In the process, the biomass is fed to a dryer that heats the material to elevated temperatures. The biomass has some residence time in the dryer before it is transferred through the bottom to a conveyor. Forced air is fed to the dryer and then goes to a cyclone where the entrained biomass is collected downstream. See Figure 1 for a process flow diagram.

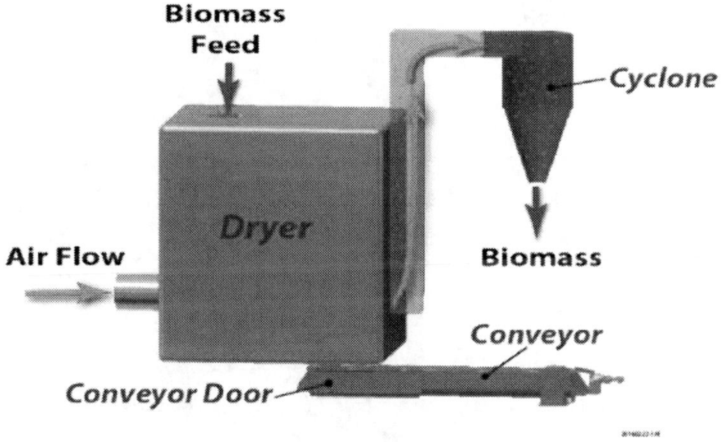

Figure 1 Biomass Process Flow Diagram

2.2 Explosion Event Description

On the day of the explosion event, a fire occurred earlier during the process. It is postulated that heated material accumulated in the conveyor, which led to self-decomposition at elevated temperatures, ultimately resulting in a fire. Employees used a water spray to suppress the flames, but perhaps this did not fully extinguish embers.

When sufficient oxygen was available, hot embers from the transfer equipment could have been drawn into the duct work and provided an ignition source for airborne biomass particulate solids. After ignition of dust, gases, or both, the rapidly expanding pressure wave and flame front moved into all of the available space in the first few milliseconds of the event.

The pressure wave blew the conveyor access door open and ejected burnt material. The flame front propagated to the dryer, fueled by airborne biomass then proceeded to the cyclone, ultimately leading to a rupture along the top of the cyclone. No personnel observed either explosive discharge. All witnesses reported hearing only one explosion. The access door opening and the cyclone failure could have occurred within less than 100 milliseconds of each other, producing a noise identifiable to witnesses as a single explosion. Pressure effects were observed downstream of the cyclone as well.

2.3 Previous Incidents

Based on history from processing similar biomass, personnel believed that smoking and fires, while not desirable, were typical and expected during the process. Fires had not only occurred during processing, but also during transportation of a by-product generated during the process.

In addition, biomass color and odor were not consistent during process runs. Operators believed there was a correlation between initial biomass color, intermediate color/odor, smoking events and fires. A lack of familiarity with combustible dust and gas hazards, especially the potential for deflagrations and explosions by both the management and operators contributed to the incident.

3 Hazard Evaluation

3.1 Material Properties

3.1.1 Biomass #1

Before the explosion, limited combustible dust and calorimetric data were available for this type of biomass. Previous combustible dust test results of a related biomass (Biomass #1) include a minimum explosive concentration (MEC) well above concentrations expected during the normal process conditions and a minimum ignition energy (MIE) value greater than 500 mJ. These results indicate that Biomass #1 was a combustible dust, but was not sensitive to ignition by electrostatic discharge.

Differential Scanning Calorimeter (DSC) testing of Biomass #1 identified one significant exotherm. This exotherm was detected above the normal operating temperature of the process and was confirmed by Carius tube testing. The Carius tube testing also indicated that non-condensable gas was formed during the event.

3.1.2 Biomass #2

After the incident, testing was performed on a sample obtained from biomass being processed at the time of the event (Biomass #2). First, DSC testing was conducted on Biomass #2 per ASTM E 537-12, *Standard Test Method for the Thermal Stability of Chemicals by Differential Scanning Calorimetry* [1] to determine the endothermic and exothermic activity of the material. A Differential Scanning Calorimeter was utilized for this test in conjunction with a 20 µl gold-plated stainless steel high pressure crucible. The DSC cell was heated at a preset rate while the differential heat flow, as a function of temperature, was measured and recorded. Five (5) DSC experiments were run on the sample at different heating rates: 1.0, 2.0, 4.0, 8.0, and 16.0°C/min.

The lowest heating rate DSC test showed that exothermic activity was detected at temperatures just less than the maximum temperature of the normal operating range for the process. Test results indicated that Biomass #2 yielded exothermic activity at lower temperatures compared to Biomass #1. The DSC results are summarized in Figure 2.

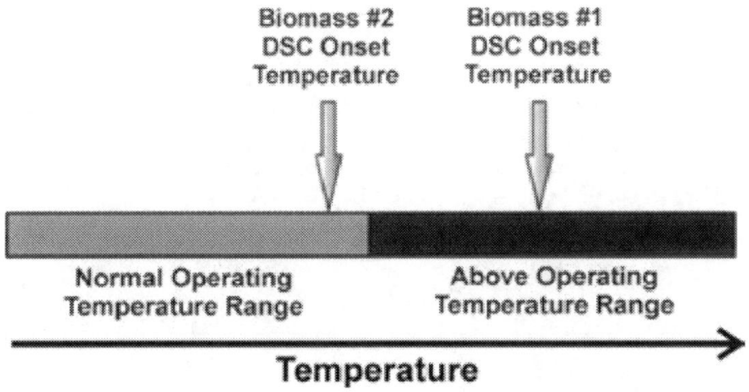

Figure 2 Thermal stability summary for biomass materials

The DSC data were evaluated to determine kinetic parameters for Biomass #2. These results showed that the material exhibited a self-accelerating decomposition temperature (SADT) value according to the United Nations Department of Transportation (UN DOT) H-1 test definition [2] that was within the normal operating temperature range. Similar evaluations were performed for smaller quantities of material (10 kg compared to 50 kg) and were also within the normal operating temperature range of the process.

Kinetic parameters were also used to determine adiabatic times to maximum rate for Biomass #2 as a function of temperature. These results indicated that the temperature that would yield a 24 hour adiabatic time to maximum rate would be in the middle of the normal operating temperature range of the process.

The next test performed on Biomass #2 was conducted using an Accelerating Rate Calorimeter (ARC), which is a high thermal inertia adiabatic calorimeter that is used to obtain data regarding the relationships between time, temperature, and pressure for exothermic reactions. The ARC utilized a 10 ml spherical test cell constructed of stainless steel with ¼" neck. In this experiment, the ARC was run in isothermal mode in order to monitor pressure generation. The instrument was programmed to track a reaction if exothermic activity was detected or maintain the desired temperature. ARC testing was conducted per ASTM E 1981-98, *"Standard Guide for Assessing Thermal Stability of Materials by Methods of Accelerating Rate Calorimetry"* [3].

The ARC isothermal test was performed on Biomass #2 at the maximum operating temperature range of the process over a time period of approximately 6 days. The temperature of the ARC test remained constant for the duration of the experiment. The evidence of non-condensable gas generated indicates that a reaction did occur. After the sample cooled, the non-condensable gas was subjected to a flame. No changes in the flame were observed, indicating that flammable gases were not present.

The last test performed on Biomass #2 was to determine the Minimum Explosible Concentration (MEC) using a 20-L Siwek chamber. The MEC is the minimum concentration of a combustible dust that is capable of propagating a deflagration through a uniform mixture of the dust in air, under the specified conditions of the test. Data obtained from this MEC test method can provide a relative measure of the concentration of a dust in the form of a uniform cloud necessary for an explosion. The test was performed per ASTM E1515, *Standard Test Method for Minimum Explosible Concentration of Combustible Dusts* [4].

Results from the MEC test on Biomass #2 showed that the MEC measured was within the range of what could be expected to occur within the process. If this concentration was present in the process, the material could be readily ignited by a burning/smoldering ember. A comparison of MEC values are shown in Figure 3.

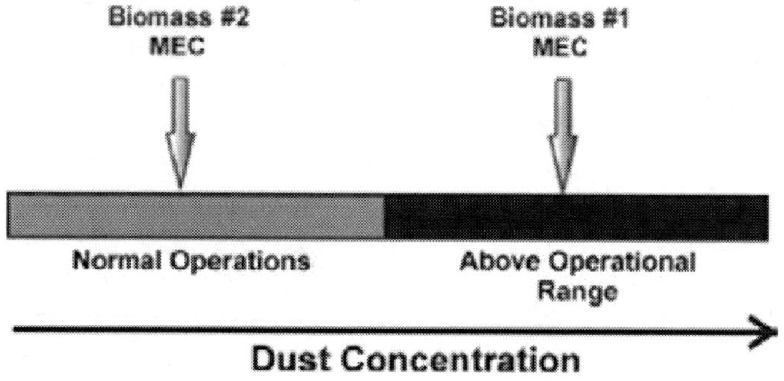

Figure 3 Minimum explosible concentrations summary for biomass materials

3.2 Incident Evaluation

It is postulated that heated material accumulated in the conveyor, which led to self-decomposition at elevated temperatures, ultimately resulting in a fire. Though fire-fighting measured were taken, it is possible that remaining material in the conveyor continued to heat or smolder. This could provide sufficient heat or an ember to serve as an ignition source for airborne particulate solids.

The combustible dust hazards of Biomass #2 handled in an existing process were not adequately identified or understood. It followed that safeguards were not in place to manage the risks associated with each piece of equipment in the process because the hazards and consequences were not characterized or evaluated. Since the hazards were not characterized at the time of the incident, employees were not made aware of the hazards associated with processing the material.

There was a concern whether flammable gas was present that could have created a hybrid mixture in addition to the combustible dust in the system. The presence of a small quantity of volatiles well below the lower flammability limit (LFL) of the gas can create a hybrid mixture with combustible dust that effectively reduces the minimum ignition energy and the minimum explosive concentration required to initiate a deflagration. Testing indicated that non-condensable gas formed during the isothermal experiment was not flammable. The gas tested may not be representative of gases formed during the deflagration event or even after the material was burned, so it is not possible to conclude whether flammable gases were present during the deflagration event. Additional testing is recommended to further evaluate this concern.

The lack of explosion prevention or mitigation devices on the equipment and absence of appropriate training are both a result of this inadequate hazard recognition. Installation of a properly designed explosion vent and installation of isolation technology could have mitigated damage to the cyclone. Figure 4 illustrates the pressure increase in a vessel both with and without the application of explosion mitigation by venting.

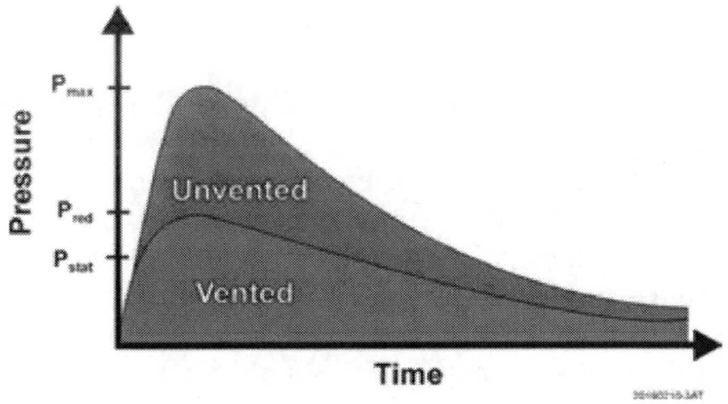

Figure 4 Progress of Explosion With and Without Explosion Protection

There was no record of a dust hazard analysis being performed for this process. Many of the required elements of a comprehensive combustible dust management program as outlined in National Fire Protection Association (NFPA) standards were not utilized to reduce the risk of dust fires and explosions.

3.3 Lessons Learned

This process included equipment covered by NFPA 654, *Standard for the Prevention of Fire and Dust Explosions from the Manufacturing, Processing, and Handling of Combustible Particulate Solids*. While processing Biomass #1, testing was performed to characterize combustible dust hazards and thermal stability for the material. Results indicated that the material was thermally stable in the normal operating temperature range and MEC values were above what would be expected during normal operations. At that time, no further protective measures were implemented. Data was gathered, but a risk basis of safety was not formally established.

When the raw material for the process changed from Biomass #1 to Biomass #2, this modification did not trigger additional testing to be performed. A management of change process, as required in NFPA 654, would have triggered the need to characterize Biomass #2 since it was different from Biomass #1. Even though the process equipment and technology remained the same, the facility changed raw materials which should have triggered Biomass #2 to be evaluated for combustible dust explosibility characteristics. This evaluation should have included thermal stability testing which were also performed previously for Biomass #1.

Fires were not uncommon in this process when processing Biomass #2 and the cause of these fires was not evaluated. It did not appear that an incident investigation process was in place at this facility, as required per NFPA 654 or the most recent NFPA 652, *Standard on the Fundamentals of Combustible Dust*. It appeared that management had allowed an acceptance of deviation to become the norm since fires were expected in the process. The response was for the operator to extinguish the fire and continue processing.

It was not evident if an inspection, cleaning and maintenance schedule was identified for pieces of equipment covered by NFPA 654 and 652. Material was allowed to accumulate in the conveyor which could have self-heated and led to fires observed in this equipment. It is not clear that inspection and access panels were in place to allow adequate cleaning of ductwork and the conveyor. It would be appropriate to establish and review cleaning and maintenance schedules per NFPA 654 and 652 requirements.

A dust hazard analysis (DHA) was not performed on this process. Given the operating temperatures for the dryer, the material should additionally be evaluated for thermal stability. This process should have triggered the need for testing to characterize the material for combustible dust hazards as well as for thermal stability. Once the hazards were identified, the equipment could then be evaluated to determine if additional protective strategies such as explosion protection and isolation were necessary.

After hazards have been identified, employees should be trained and provided guidance on required work practices in order to ensure their safety. Insufficient training contributed to the near miss of exposing employees to the deflagration. Without knowledge of the deflagration potential and the likely effects, employees were preparing to position themselves in high risk areas during the unintended explosion vent.

Data reviews after the incident revealed that employees did not understand the full intent of information recorded during the process. Design of the data forms likely contributed to the uncertainty regarding data collected. Employees did not have an awareness regarding the critical safety parameters and were not adequately trained on how to respond to deviations in the process.

Lessons learned from this case are summarized as follows:

- There was a lack of formal risk based safety documentation for the process.
- A management of change (MOC) process did not trigger data to be obtained when a change in the raw materials for the process were made.
- Sufficient data for the material/process were not obtained to establish safe operating conditions and identify potential upset scenarios.
- Performing a DHA could have identified the hazards, potential ignition sources, potential consequences and provided a framework to evaluate the adequacy of safeguards or mitigation strategies to determine if additional protection was needed.
- Proper hazard identification for the process is critical to ensure that adequate safeguards are in place and that employees receive adequate and appropriate training regarding normal operations and emergency response procedures.

- Routine fires during processing and transportation should serve as warning signs that trigger an investigation to identify preventive and corrective actions.
- The inspection, maintenance and cleaning procedures for dust handling equipment should be consistent with NFPA 654 and 652 requirements.

4 Conclusion

Testing performed after the incident indicated that the new material (Biomass #2) handled in the process was thermally unstable during normal operating conditions as well as being a combustible dust hazard.

In summarizing the lessons learned, it appears that several combustible dust management systems required by NFPA 654 and 652 did not exist, or if they existed, were not effective to prevent this incident. Key process safety information was not obtained to identify the hazards of the process. The company did not understand the warning signs for a potential combustible dust explosion in this process. Performing a DHA on this process could have identified the need to obtain relevant combustible dust and thermal stability data and understand that normal operating conditions could lead to an upset scenario. This information could then be used to propose changes to the process and determine if existing systems and equipment were adequately designed to protect against the consequences of the identified upset scenarios.

The following steps are recommended for identifying hazards associated with a process that contains equipment that could potentially handle combustible dust. These steps are consistent with the practices recommended in NFPA 652.

- Identify equipment in the process that is intended to handle particulate solids and the desired processing conditions (temperature and pressure operating range, equipment design requirements)
- Identify material to be handled in the particulate handling equipment
- Identify the hazards (safety data sheets, combustible dust hazard data, thermal stability data as warranted by process conditions, identify potential consequences due to upset conditions, perform a DHA)
- Understand the consequences in terms of severity and likelihood (determine if flash fire and/or explosion scenarios exist, determine if normal operating conditions or process deviations could lead to thermal decomposition, fire and/or explosion scenarios, identify if flammable gas could be generated)
- Determine process safety controls, safeguards and mitigation strategies (protect against overpressure situations, control and alarm set points, change in process design or procedures, explosion protection systems)
- Implement and maintain the safety measures (document safe operating ranges, perform preventive maintenance and inspections, train appropriate personnel, implement emergency response plans, employ other procedures as identified in NFPA 652)

5 References

[1] ASTM Standard E 537-12. *Standard Test Method for the Thermal Stability of Chemicals by Differential Scanning Calorimetry*, ASTM International, West Conshohocken, PA, 2012.

[2] United Nations. Recommendations on the Transport of Dangerous Goods: Manual of Tests and Criteria (6th ed.), New York and Geneva, 2015.

[3] ASTM Standard E 1981-98 e2. *Standard Guide for Assessing Thermal Stability of Materials by Methods of Accelerating Rate Calorimetry*, ASTM International, West Conshohocken, PA, 2012.

[4] ASTM E1515-14. *Standard Test Method for Minimum Explosible Concentration of Combustible Dusts*, ASTM International, West Conshohocken, PA, 2014.

[5] NFPA. NFPA 652, *Standard on the Fundamentals of Combustible Dust, 2016 Edition.* NFPA, Quincy, MA, September 2015.

[6] NFPA. NFPA 654, *Standard for the Prevention of Fire and Dust Explosions from the Manufacturing, Processing, and Handling of Combustible Particulate Solids, 2013 Edition*, Quincy, MA, June 2012.

[7] Ebadat, V. Testing to Assess Explosion Characteristics of Dust Clouds, NFPA Symposium on Dust Explosion Hazard Recognition and Control, 2009.

[8] U.S. Chemical Safety and Hazard Investigation Board. Investigation Report: Combustible Dust Hazard Study. Rep. No. 2006-H-1, 2009.

[9] Grossel, S.S. and R.G. Zalosh. Guidelines for Safe Handling of Powders and Bulk Solids. Center for Chemical Process Safety, AIChE, New York, 2005.

Atmospheric Tank Explosions: Mechanisms and an Unexpected Case Study

Michael S. Schmidt, P.E.
Principal, Bluefield Process Safety LLC
3101 Cherokee Street
Saint Louis, Missouri 63118, USA
bluefieldsafety@gmail.com

Prepared for Presentation at
American Institute of Chemical Engineers
2016 Spring Meeting
12th Global Congress on Process Safety
Houston, Texas
April 11-13, 2016

Atmospheric Tank Explosions: Mechanisms and an Unexpected Case Study

Michael S. Schmidt, P.E.
Principal, Bluefield Process Safety LLC
3101 Cherokee Street
Saint Louis, Missouri 63118, USA
bluefieldsafety@gmail.com

Keywords: storage tank, atmospheric tank, tank explosion, caustic

Abstract

Most of the literature on catastrophic failures of low pressure tanks focuses on the ignition of explosive mixtures in the vapor space of those tanks, and rightly so, because most catastrophic failures of atmospheric tanks involve explosions of such mixtures. Moreover, there is a general sense in the industry that other than explosions of flammable mixtures, the only other hazard associated with low pressure tanks is the occasional dramatic implosion of a vessel that is inadequately protected against vacuum.

In fact, there are a number of mechanisms that can lead to the catastrophic failure of a low pressure tank that have nothing to do with combustion or unprotected vacuum. Under certain circumstances, even a tank equipped with an atmospheric vent and containing nothing other than salt water can explode, with disastrous impacts.

This paper reviews the mechanisms for catastrophic failure of low pressure tanks, both implosion and explosion, and serves as a reminder for experienced process safety practitioners and as a tutorial for new process safety practitioners of what to look for during a hazard review. It also includes a case study of an atmospheric caustic tank explosion that resulted from a previously unreported mechanism that involved neither combustion nor vacuum.

1 Introduction

It is probably fair to say that atmospheric tanks are not very high on most people's list of hazards to worry about. Usually, they just sit there, storing liquids, doing what they are designed to do. Issues are typically the result of overfilling the tank and, on occasion, leaking components. Nothing sudden and catastrophic. When the words "sudden and catastrophic" are associated with atmospheric tanks, it is usually in relation to the flammable hazards of the liquids being stored. Chang and Lin [1] conducted a study of 242 storage tank incidents and found that 145 of them involved fire and another 61 involved explosion, mostly of flammable vapors from the liquids being stored. Outside of flammability issues, sudden and catastrophic incidents associated with atmospheric tanks are not what we typically worry about. Setting flammability issues aside, which have been addressed thoroughly elsewhere [2], [3], [4], [5] there are three types of atmospheric tank failures that can be sudden and catastrophic: collapses, implosions, and explosions.

Before continuing, however, it is important to first establish what is meant by the term, "atmospheric tank." The ASME boiler and pressure vessel code [6] divides vessels into two categories: pressure vessels, which operate at pressures greater than 15 psig and low pressure vessels, which operate at 15 psig or less. The BPVC applies only to pressure vessels. There are two API standards that apply to low pressure vessels. API 620 [7] applies to tanks that operate between 2½ psig and 15 psig, and at temperatures not greater than 250°F. API 650 [8] applies to tanks that operate at pressures up to 2½ psig, and at temperatures not greater than 200°F. OSHA defines an atmospheric tank as "a storage tank which has been designed to operate at pressures from atmospheric through 0.5 psig" [9].

All atmospheric tanks are low pressure tanks, but not all low pressure tanks are atmospheric tanks. There is not a consensus as to the limit—API 620 uses 2 ½ psig, OSHA uses 0.5 psig, others [10], [11] use a limit somewhere in between. For the purposes of this paper, an atmospheric tank is not intended to operate under vacuum and is not intended to operate at greater than 1 psig, or 30 inch water column. Human lungs have a greater operating pressure range.

2 Catastrophic Atmospheric Tank Failure

2.1 Collapse

One of the most famous catastrophic tank failures of the industrial era is the Great Molasses Flood of 1919 in Boston, Massachusetts [12]. The 50-foot tall, 90-foot diameter cast iron tank, built in a rush during World War I, ruptured and the wave of molasses killed 21 people. There was no fire. There was no explosion. The tank collapsed because it was under-designed for its application.

A more recent example is the failure of a liquid fertilizer storage tank at the Allied Terminals facility in Chesapeake, Virginia in 2008 [13]. The tank, originally designed for service with petroleum products, ruptured while being filled with a 32% solution of urea-ammonium nitrate (UAN). Two welders who were working on the tank as it was being filled for the first time after weld repairs were made were seriously injured when the tank collapsed.

2.1.1 Mechanisms for catastrophic failure

Atmospheric tanks fail for three reasons:

- The design or construction of the tank was inadequate.
- The tank is converted to a service that exceeds the capabilities of the initial design.
- The tank deteriorates over time.

Inadequate design or construction. While an atmospheric tank does not need to withstand high pressures, it does need to withstand the hydrostatic head of the liquid being stored. Atmospheric tanks collapse because the material of construction is not strong enough for the service, as was the case in the Great Molasses Flood, where the tank walls were 50% too thin, even for the relaxed standards of the 1910s [12]. Atmospheric tanks also collapse because the welds holding the tank together are inadequate. The Chemical Safety Board's report on their investigation into the Allied Terminals tank collapse includes an appendix of similar incidents, in which the listed cause of the incidents is "faulty welds" in more than 90% of the incidents where a cause is known [13]. Because of these issues, the CSB report notes that "tanks are more likely to fail when being filled to the maximum level for the first time."

Converted to a new service. Atmospheric storage tanks are designed for a specific service. Many tanks are designed to store petroleum products, for which a typical specific gravity is around 0.8. Simply converting the use from hydrocarbons to water increases the hydrostatic pressure on a tank by 25% when filled to the same level. Converting the use from hydrocarbons to 32% UAN, which has a specific gravity of 1.33, increases the hydrostatic pressure on a tank by two thirds. Converting the use from hydrocarbons to 50% caustic, which has a specific gravity of more than 1.5, almost doubles the hydrostatic pressure on a tank. Converting the service of atmospheric storage tanks to higher density liquids can lead to catastrophic tank failure by exceeding the design strength of the tank. Likewise, increasing the "safe fill level" has a similar impact on hydrostatic pressure, which can also lead to catastrophic tank failure.

Deterioration with time. Lastly, atmospheric storage tanks, like all equipment, deteriorate with time. Walls thin due to corrosion and erosion. Ideally, the thinning is uniform, steady, and slow. Catching this kind of deterioration during a tank inspection that includes thickness testing is straightforward. A tank with an adequate corrosion allowance will give lots of warning before the condition of the tank warrants a change in the operating conditions.

But not all deterioration is uniform. Tanks where nozzles, dip pipes, or eductors are aimed at the wall or floor of the tank are subject to localized thinning. Even a stream running down the interior wall of a tank can cause localized thinning. This weak spot then become the nucleus of a catastrophic failure. Stresses concentrate at a weak spot, the weak spot fails, and the stresses are immediately propagated through the rest of the material. The tank unzips and can collapse suddenly and catastrophically, often leading bystanders to believe there has been an explosion.

2.1.2 Protections against catastrophic failure

The causes of catastrophic tank failures also give a strong hint as to the protections that might be used against them.

Inadequate design or construction. This is best addressed with good equipment specifications according to recognized and generally accepted good engineer practice (RAGAGEP) and appropriate pre-startup safety reviews (PSSRs) to assure that the tank, as designed, constructed, and installed, meets those specifications. During commissioning, when many vulnerabilities in the tank are revealed, personnel should be kept out of harm's way, based on the assumption that the tank will fail.

Converted to a new service. Converting to a new service is a clear case for management of change, and the MOC review should consider changes in level, density, and corrosion rates, both general and local.

Deterioration with time. All equipment deteriorates, which is why a mechanical integrity program that adheres to RAGAGEP is essential. Tank inspections should test for thickness in enough spots to assure that general thinning is detected. Interior inspections should also look for specific evidence of localized thinning, especially at points where nozzles, dip pipes, or eductors are directed at the walls or floor of the tank.

2.2 Implosion

Atmospheric tanks are not designed to withstand vacuum, and they do not have to be exposed to very much vacuum to implode suddenly and catastrophically. The emptier the tank is when exposed to vacuum, the more dramatic the implosion. Tank implosions make such dramatic film that there are countless YouTube videos of tank implosions.

The lore that has been told and retold is of the maintenance team that was steaming out an atmospheric tank to prepare it for vessel entry, but did not finish before the end of the day. Rather than conduct the vessel entry at night, they closed up the tank and left. During the night, a storm front moved in, showering the tank with cold rain. The steam condensed, creating a vacuum, and when the crew returned in the morning they found the tank crumpled like an empty beer can. The crew from *Mythbusters* set out to prove or disprove this scenario using a DOT 111 tank car and decided that this story was "busted" [14]. *Mythbusters* notwithstanding, atmospheric storage tanks are not DOT 111 tank cars and the scenario should be considered credible.

Unlike explosions, implosions do not produce shock waves or flying shrapnel. Breaches may occur at the creases that form, but generally any loss of containment is not significant. The primary hazard associated with implosions are the potential injuries to personnel in the immediate area. The Center for Disease Control and Prevention reports of a worker killed while repairing a tank when it imploded [15]. Similarly, a worker in Spokane, Washington was killed in 2011 when the sewage digester he has checking imploded [16].

2.2.1 Mechanisms for catastrophic failure

Atmospheric tanks implode when they experience an internal pressure lower than they are capable of withstanding, i.e. a vacuum. Source of vacuum include the deliberate vacuum created by vacuum pumps and steam vacuum jets, as well as the incidental vacuum created by the pump suction of other pumps during pump out. Even in the absence of a pump, a gravity drain to empty a tank and other forms of emptying a tank of its liquid contents will create a vacuum. Finally, as described above, phase changes, particularly condensing vapors will create a vacuum.

2.2.2 Protections against catastrophic failure

In the absence of a source of vacuum, protections are unnecessary. However, all atmospheric tanks have a source of vacuum, if for no other reason than they are emptied. Protection include vacuum breakers, pressure/vacuum conservation vents, nitrogen pads, and most commonly, atmospheric vents. As long as these are properly sized and designed and remain fully operational, they are sufficient to protect against catastrophic implosion.

2.3 Explosion

Most atmospheric tank explosions involve the storage of flammable liquids, where it is the explosion of the flammable material that is the concern and which is outside the scope of this paper. There are also physical explosions that can occur, resulting from overpressure in the tank for reasons other than sudden combustion.

2.3.1 Mechanisms for catastrophic failure

Atmospheric tanks explode when they experience an internal pressure higher than they are capable of withstanding, which is not much. Sources of this high pressure include the pressure of inlets to the tank and heating within the tank.

Inlet line pressure can be of either gases or liquids. When the tank is open to atmosphere, the pressure drop at the inlet is complete, and the pressure in the inlet is not transferred to the interior of the tank. However, when the tank is blocked off from the atmosphere, the pressure in the tank will approach the pressure of the inlet line. Gases will flow into the tank until the pressure in the tank matches the supply pressure. Liquids will flow into the tank, compressing the headspace, until the headspace pressure matches the deadhead pressure of the liquid line. A tank does not need to be liquid full to reach the deadhead pressure of the liquid line.

Heating a vessel can result from either external heating or the heat of reaction, planned or unplanned, within the vessel. In either case, the liquid will expand and the vapor pressure will increase. Liquid expansion is only a concern when the tank is blocked off from atmosphere and the tank is completely full of liquid. In this case, since liquids are incompressible fluids, any additional expansion will sharply increase the pressure. However, atmospheric tanks are not absolutely rigid, so the effect will be to cause the tank to bulge, rather than to cause it to burst.

Increased vapor pressure is the real concern. Vapor pressure increases as temperature increases, and is independent of head space. An atmospheric tank blocked in while containing any liquid will experience an increase in pressure when heated, and atmospheric tanks are not designed to contain much increase. The vapor pressure of water, for instance, will increase by 1 psi when heated from 72°F to 113°F; it will increase by 5 psi when heated from 80°F to 165°F.

2.3.2 Protections against catastrophic failure

The key to protecting against physical explosion of atmospheric tanks is to assure that any source of pressure is released from the vessel before the pressure becomes too high. The obvious means is an atmospheric vent. Other protections include pressure/vacuum conservation vents and venting entryways or "trashcan lid."

To be useful, however, it is imperative that these protections be properly sized and designed and then remain fully operational. A vent that is too small or blocked is no protection. An atmospheric vent may be directly to atmosphere, or pass through a seal pot. Many creatures—

birds, wasps, rodents—find the outlet of an atmospheric vent a perfect place to build a nest. Bird screens, when in place, will keep out birds and rodents, but not nest-building insects. Likewise, a seal pot that has frozen solid will not vent. Vent lines that contain monomer vapor, whether directly to atmosphere or through a device can become plugged with polymer or simply glued shut.

Pressure vessels can be protected with rupture discs and pressure relief valves, but these require set pressures that are too high to be of use in atmospheric tanks. Likewise, pressure vessels are frequently protected from inlet pressure with regulators on the inlet. Unfortunately, atmospheric tanks operate at such low pressures that to regulate inlet pressures down to the operating pressure of the tank would reduce flow to unacceptably low rates.

Weak-seam roof tanks are frequently used to protect against the failure of atmospheric tank walls in the event of overpressure. They are designed so that the roof-to-shell joint will fail before any of the seams in the shell or the floor of the tank fail, thus preventing a release of the contents. However, this approach must be used with care. Accessories, such as railing or stairs, can make the roof-to-shell joint stronger than intended, as can the workmanship of an enthusiastic welder. (Welders are rarely scolded for making their welds too strong.) In general, no credit should be given for a weak-seam roof on a tank smaller than 10 meters (30 feet) in diameter because of the concern that the joint will be too strong [17]. When a weak-seam roof is used, the roof itself should be restrained with some kind of tether or hinge. Otherwise, the roof, ripped away from its weak seam, will be launched as a projectile.

3 A Case Study

While process air was blowing through the transfer line from a caustic storage tank to a caustic day tank, the caustic day tank failed catastrophically, launching the top head of the tank approximately 125 feet, where it came down through the roof of a nearby process building and triggered an impromptu evacuation in which one employee fell and had to receive medical treatment.

Figure 1. Caustic Day Tank head, after coming down through the roof of a nearby process building.

3.1 Process Background

The process consisted of a 3-inch transfer line constructed of 304 stainless steel that connected the discharge of a caustic transfer pump at the caustic storage tank to the top of a caustic day tank. A 50% solution of NaOH in water entered the caustic day tank through a dip pipe to the bottom of the tank. The transfer line was purged after each transfer of 50% caustic solution from the caustic storage tank to the caustic day tank by blowing out the line from the caustic storage tank to the caustic day tank for about 10 to 15 minutes with process air.

When the transfer line was being prepared for maintenance, there was a more extensive purge procedure. The normal procedure for purging the transfer line was to blow out the line with process air, blowing from the caustic storage tank to the caustic day tank for about 10 to 15 minutes. The next step was to flush the line with city water for 5 to 10 minutes, again flowing from the caustic storage tank to the caustic day tank. After the water flush, any residual water in the transfer line was blown out with process air, again blowing from the caustic storage tank to the caustic day tank for about 10 to 15 minutes.

The transfer line was routinely purged following transfers and was purged successfully for maintenance a month before the incident.

3.2 Incident Description

On a balmy, spring Friday morning, preparation began for maintenance on a leaking caustic transfer line from the caustic storage tank to the caustic day tank at about 9:30 am. The first step – purging the caustic from the transfer line with process air – was completed without incident.

The second step – purging residual caustic from the transfer line with city water – started at around 9:50 am. This step was also completed without incident, and at around 10 am the operator disconnected the water hose and left it to drain while he turned his attention to other matters in the unit.

At around 10:30 am, the operator returned to the caustic storage tank, disconnected the water hose, and reconnected the air line. He then opened the air supply valve in order to blow the water out of the transfer line.

At 11:10 am (the time shown on the time-stamp from the video monitoring the area), while the process air was still blowing through the transfer line from the caustic storage tank to the caustic day tank, the caustic day tank failed catastrophically. The pressure was sufficient to separate the conical head from the top of the tank and launch it approximately 125 feet, where it came down through the roof of a nearby process building to the east.

While the sides and bottom of the caustic day tank remained intact after the incident, the damage was enough to consider the caustic day tank a total loss. The piping attached to the top head of the caustic day tank was also damaged beyond repair. There were no fatalities or serious injuries associated with the incident, but one employee did require medical treatment for a sprained ankle following a fall that occurred during his evacuation from the immediate area of the incident.

3.3 Possible Issues

An explosion of a caustic tank can result from:

- ❖ Sources of pressure considerably higher than the design of the tank was capable of handling,
- ❖ Blockage in the equipment that lowered the pressure the tank was capable of handling, or
- ❖ Material failures in the tank that lowered the pressure the tank was capable of handling.

Guidance documents from suppliers of 50% caustic solution suggest the following as sources of high pressure, causes of blockage in caustic equipment, and causes of equipment material failure at pressures lower than design pressure [18], [19].

3.3.1 Possible sources of high pressure in a generic caustic tank

- ❖ Reaction of caustic solution with reactive metals (aluminum, magnesium, zinc, tin, chromium, brass, and bronzes made with zinc or tin), releasing hydrogen that forms a flammable mixture with air and subsequently ignites.
- ❖ Plant air or nitrogen used to transfer caustic solution by pressure.
- ❖ Steam used to clean out transfer lines.
- ❖ Direct steam heating of frozen caustic, exposing equipment to steam pressure.
- ❖ Pump deadhead pressure in unvented tank or line.
- ❖ Diluting strong caustic with water, rather than diluting strong caustic by adding it to water.
- ❖ Heating caustic solution while a crust of frozen caustic is formed over the top of the liquid

3.3.2 Possible causes of blockage in generic caustic equipment

- Block valves inadvertently closed.
- Lines plugged with frozen caustic solution.

3.3.3 Possible causes of generic equipment failure at lower than design pressure

- "Caustic embrittlement" – stress-corrosion failure which occurs at areas of mechanical stress or where fabricating operations produce high residual stress such as at bends in piping and at welds.
- Using iron and steel as materials of construction for equipment handling 50% caustic solutions at temperatures above 140°F. While iron and steel are the usual materials for equipment handling 50% caustic solutions below 140°F, nickel, nickel alloys, and stainless steel are required at higher temperatures and for higher concentrations.
- Using unsupported plastic pipe, which can fail because of the high density of 50% caustic solution, and using fiberglass reinforced plastic pipe except for specific conditions for which it is suited.
- Continued heating after caustic soda is melted, which may damage the special lining on the inside of railcars. The temperature of 50% caustic solution should never exceed 150°F (66°C) in lined cars.
- Wet or saturated steam above a maximum of 15 psig and 250°F. Steam at higher temperatures will damage rail car linings.

3.4 Conditions at Incident Scene

3.4.1 Noted from video

There were several dozen closed circuit video cameras positioned around the plant, three of which captured evidence of the catastrophic failure of the caustic day tank:

- The truck entrance camera, located about 600 feet southeast of the caustic day tank and aimed northwest.
- The tram gate camera, located about 275 feet southwest of the caustic day tank and aimed northeast.
- The main crosswalk camera, located about 175 feet west of the caustic day tank and aimed east.

The time stamps on the truck entrance camera and the tram gate camera coincide closely, and both showed the initial cloud release from the tank rupture at 11:10:19. The main crosswalk camera did not show the cloud release from the tank rupture, but shuddered at the time the cloud was released. Both the truck entrance camera and the main crosswalk camera captured an employee falling down and then getting up and walking away, hardhat in hand, at 14 seconds after the tank rupture.

Figure 2. Detail of screen capture from truck entrance camera at 11:10:19.647, showing launch of the caustic day tank top and associated cloud.

The main crosswalk camera, which was closest to the caustic day tank, shuddered for 0.4 sec, consistent with the passing of a shockwave. Neither of the more distant cameras showed the shudder. The more distant cameras, however, showed the initial cloud released. The cloud formed and was completely dispersed within 2 seconds. There was no evidence at all in those videos of combustion or of a secondary BLEVE. The shockwave confirms that there was an explosion. The absence of combustion or BLEVE and the quick dispersion of the cloud that was released confirm that the explosion resulted from simple overpressure, not from the ignition of a flammable mixture of hydrogen and air in the headspace or from the boil-up of the contents of the vessel.

The truck entrance camera was the only view that showed the launch of the top head. In the first 0.2 seconds, a cloud appeared above the west side of the tank. Then, 0.2 seconds after the initial release of the cloud, the conical head fully separated from the tank and began its 125-foot trajectory eastward, where it came down through the roof of a nearby process building. The video suggests that the separation began on the west side of the conical head and the break proceeded almost instantly around the tank head, propelling the conical head upward before the pressure inside the tank could be relieved.

3.4.2 Noted from field observations

The most relevant observation was that of the condition of the vent from the caustic day tank. A 3-inch gooseneck constructed from glued heavy-wall PVC pipe and elbows, the vent was attached to the top of the caustic day tank at a flanged nozzle 40 inches from the center peak of the conical head. The vent sheared off during the incident just above the flanged connection.

The internal surface of the vent (the surface that would have been visible from the outside while it was still attached to the functioning tank) was open and free of any obstruction. However, from the point the nozzle was attached to the top head, through the flange and up to the first elbow, the vent was encrusted with a white solid that tested positive for NaOH. The morphology of the solid was consistent with a slow accumulation over an extended period of time, rather than the result of an incident of overfilling the tank and then freezing in the vent. Moreover, there was no evidence of accumulation or discoloration on the head under the discharge of the vent, which would have been consistent with discharge through the vent onto the surface of the conical head.

The accumulation of solids within the vent choked it down to a diameter of about a half inch, and the rough, convoluted surface of the choked vent would have rendered it less useful than a clean half-inch pipe as a vent.

Figure 3. Vent nozzle, showing build-up of caustic.

Likewise, there were no solids built up on the interior walls of the vessel or on the interior surface of the head. This suggests that the buildup inside the vent line was the result of an entrained mist of 50% caustic solution being carried to the vent, where it collected and evaporated, leaving behind solid NaOH. The monthly vent inspections conducted on the tank since it was put into service in late 2003 would not have detected the gradual accumulation of solids.

Figure 4. View of interior of the caustic day tank after the incident. The thin white solid is caustic that dried there at the time of the incident. The interior shows some staining, but no build-up of the NaOH encrustation seen in the vent nozzle.

A thorough inspection of the vents of the two caustic tanks in the area showed no similar build-up of solids. However, one of the caustic storage tanks was put into service about 5 years before the incident and the other caustic storage tank was put into service about a year before the incident, so there would not have been the same opportunity for buildup.

The caustic day tank was specified to be constructed of 304 stainless steel. A field inspection of the vessel was consistent with that specification. The metal looked like stainless steel and was not magnetic; short of testing the metallurgy, this was considered enough to conclude that the tank was built as specified. Suppliers of 50% caustic recommend that carbon steel is a suitable material of construction for up to 140°F; over 140°F, they recommend 304 stainless steel to provide good corrosion resistance. This tank was operated at ambient temperatures. There was no reason, then, to believe that temperature or corrosion played any role in this incident.

The conical head separated completely. An inspection of the head showed that the separation was at the weld of the head to the 3"x3" angle that formed the top flange of the vertical cylinder of the tank. While the failure of an atmospheric tank is typically expected to be in the form of a split, where pressure is released without accompanying shockwave or shrapnel, an explosive separation of the head from the body of the vessel is not unheard of, e.g., Barton Solvents, Valley Center, Kansas, on July 17, 2007 [20].

Figure 5. The separation of the head from the body of the caustic day tank was at the weld of the head to the 3"x3" angle that formed the top flange of the vertical cylinder of the tank.

3.4.3 Noted from interviews

The pressure gauge on the process air was not read at the time of the incident. The production supervisor noted that the pressure should be set at around 35 to 40 psig, while the maintenance engineer stated that the process air pressure is normally around 80 psig, and that the regulator was normally wide open, with little motivation to crank it back closed. This is consistent with the pressure reading observed in the field, which indicated that the process air pressure was at 90 psig. The caustic day tank itself has a design pressure of atmospheric (0 psig) and was pneumatically tested at 5 psig at the time it was manufactured in 2003.

The supply pressure for the city water used to flush the transfer line is around 40 to 60 psig. This means that the line could have been successfully flushed into the caustic day tank, even if the pressure in the tank had risen to 35 psig during the initial purge. During the second process air purge, which lasted for around 45 minutes before the incident, there would have been more time for the pressure to build.

3.5 Contributing Factors

3.5.1 Non-contributing factors

The investigation was able to eliminate many of the generic issues identified by caustic solution suppliers.

In regard to possible sources of high pressure, the materials of construction – 304 stainless steel and PVC – make hydrogen formation as a result of reaction with reactive metals an extremely unlikely factor. This is further supported by the absence of any evidence of combustion during the incident. Likewise, neither steam nor a pump were in use, so steam pressure and pump deadhead pressure did not factor in. Temperatures were well above 65°F, the temperature at which frozen 50% caustic solution become a concern.

The purging procedure did include a step where water was added to strong caustic, a step that is normally discouraged. However, this did not create the high pressure that resulted in the incident, and overall, is a reasonable way to deal with the flush water.

In regard to possible causes of blockage in caustic equipment, there were no block valves to be inadvertently closed, and ambient temperatures were high enough to prevent lines from being plugged with frozen caustic solution.

In regard to possible causes of equipment failure at lower than design pressure, the materials of construction – 304 stainless steel and PVC – also eliminate the concerns with caustic embrittlement and with temperatures above 140°F. There was no unsupported plastic pipe to fail, and steam was not in use.

3.5.2 Direct causes

A "direct cause" is the uncontrolled release of energy or hazardous material that cannot be absorbed safely. In the case of this incident, the direct cause was high pressure in the caustic day tank, the source of which was the approximately 80 to 90 psig process air used to blow down the transfer line from the caustic storage tank to the caustic day tank.

3.5.3 Indirect causes

An "indirect cause" is the action or condition that results in direct causes. In the case of this incident, the indirect cause was the plugged tank vent on the caustic day tank. A contributing factor was probably the extended time that process air was allowed to blow into the tank. However, the vent was gradually becoming clogged and eventually, even shorter periods would have been sufficient to overpressure the vessel.

3.5.4 Basic causes

A "basic cause" is the policy, procedure, or decision that leads to indirect or direct causes. There were three basic causes. The first basic cause was the procedure that called for blowing fluids into the caustic day tank at pressures higher than the rated pressure of the tank. While pressures higher than atmospheric pressure, either as a result of pressurized fluids or pumping, are absolutely necessary to effect transfer into the tank, the higher pressure also made a functioning atmospheric vent an imperative element of normal operation.

The second basic cause was a tank design and operation that resulted in carrying entrained 50% caustic solution mist to the vent, where it could deposit and plug the vent.

The third basic cause was the vent inspection procedure. In hindsight, it is now clear that monthly visual inspection did not detect that the vent was becoming plugged. In the absence of this hindsight, there is nothing to fault the vent inspection procedure. Caustic solution is not volatile, and there is no general expectation in the industry that vents on caustic solution tanks are subject to plugging. However, now that the problem has been identified, it should be addressed.

3.6 Recommendations

3.6.1 Vent design to assure pressure relief is adequate

Dow Chemical has the following recommendation in regard to vents on caustic solution storage tanks: "In addition to a solution high-level alarm, the tank should have a liquid-level gauge and a vent. The vent should have a cross-sectional area at least four times that of the fill line (which is normally 2"), extending from the top of the tank to within about 3' of ground level. The vent, which also acts as an overflow, should be visible from the tank truck or tank car loading area so that personnel will see and/or hear caustic discharging from this overflow pipe in case the tank is overfilled. Valves should never be installed in the vent line. It must be open to the atmosphere at all times" [18].

In this particular case, the vent and fill are both 3-inch lines. The vent pipe is a gooseneck, rather than a pipe that extends to within 3 feet of ground level. Modifications to the vent piping may be helpful.

3.6.2 Vent inspection to catch any buildup

It was unexpected to discover that the vent was plugged with caustic. The likely cause of the plugging was the deposit of entrained mist in the vent. The mist, composed as it was of 50% caustic solution, would be the source of solids. The practice of blowing process air into the tank through the dip pipe assures that there for mist to be entrained after every transfer into the tank. The evidence is that process air carries the mist all the way to the vent. As a result, this is an issue that should be expected in any tank containing solutions of dissolved solids that can be entrained as mist.

In this particular case, the purpose of the vent inspection was to look for evidence of external blockage, primarily as the result of natural activity, e.g., bird nests, wasp nests, etc. In the case of tanks storing solutions of dissolved solids, a more thorough inspection of the vents is in order. This may involve the use of boroscopes and cameras, or as an alternative, the installation of inspection and clean-out ports in the vent line. The frequency of inspection should be adjusted as experience dictates, beginning with quarterly inspection, but probably settling on annual inspection.

3.6.3 Limit use of dip pipes for inlet piping

The practice of blowing process air or nitrogen into a tank through a sub-surface dip pipe assures that mist will be entrained. While the density of aqueous solutions such as 50% caustic is high enough to lead to the expectation that mist droplets will settle out before reaching a vent, especially if the level is low enough, droplets that are small enough will be carried to the vent, regardless of the level. Operating at lower liquid levels will slow down the rate of blockage, but not stop it, and the improvement would come at the cost of lost capacity.

A more effective solution is to eliminate the dip pipe so that process air is not blown through the caustic solution to create a mist in the first place. Neither Dow Chemical nor PPG Industries show dip pipes in their example configurations of storage tanks [18], [19]. Because aqueous solutions of solids are neither flammable nor non-conducting, the usual concerns regarding static discharge from free-falling liquid are simply not an issue.

In this particular case, the dip pipe also serves to avoid erosion of the head and vessel walls that result from feeding directly through a tank nozzle. Cutting the dip pipe off a few inches below the entrance to the head and fitting it with a slight elbow to direct the free-falling liquid away from the wall still avoids erosion problems but eliminates the mist that forms by sparging process air through the caustic solution.

3.6.4 Pressure regulation to reduce potential for overpressure

Atmospheric tanks are frequently connected to high pressure sources. Examples include fill lines, where the pressure can be as high as the deadhead pressure of the transfer pump; water supplies, which typically run around 40 to 60 psig; and plant air, which is often in the range of 80 to 120 psig. Any one of these can be a source of overpressure, but to regulate them down to atmospheric pressure would render them useless. While the source pressure must be above atmospheric pressure in order for these connections to work, they should not be any higher than necessary.

In this particular case, the plant air supply used to blow out the lines would have been just as effective at 20 psig as it was at 90 psig. Regulating the air supply pressure down creates no burden. The water pressure and the transfer pump pressure need to be where they are.

4 Conclusions

Atmospheric tanks are relatively fragile, especially when compared to vacuum vessels, low pressure vessels, and high pressure vessels. They are also widely used and considered relatively benign. However, because they are considered relatively benign, they are often overlooked as a hazard and neglected as a maintenance priority.

When atmospheric tanks contain flammable materials, the consequence of concern is most often a tank fire. Beyond that, the typical consequence of concern associated with atmospheric tanks is a leak. The impacts of these events are usually limited to injuries and low-consequence impacts to the environment, although there have been some notable exceptions [21], [22]. Rarely is the concern one of sudden, catastrophic failure. However, an atmospheric tank can fail suddenly and catastrophically, with life-threatening consequences. The catastrophic failure can result in a collapse, an implosion, or an explosion. Collapses result when a tank is badly installed or when a tank is misused. The best way to avoid collapses is to first make sure the tank is properly designed and installed, and then to use it within the limits of that design.

Explosions can result from the ignition of flammable vapors, but surprising to many, can also occur for purely physical reasons. Implosions result primarily when sources of vacuum are overlooked or neglected. Atmospheric tanks are called atmospheric tanks because they vent to atmosphere. When that vent becomes inoperable, it becomes all too easy for the pressure in an atmospheric tank to become lower or higher than the thin walls of the tank can withstand. So it is imperative for operators to understand what causes a vent to become inoperable, and to prevent that from happening. The response in a HazOp to "pressure-too high" or to "pressure-too low" cannot be, "That can't happen, we're open to atmosphere."

5 Recommendations

5.1 Regarding catastrophic tank collapse

To avoid catastrophic vessel collapse, apply rigorous management of change and pre-startup safety review when putting an atmospheric tank into new or different service. In particular, pay attention to the impact of using a tank for higher density liquids. When a tank is first put into service is the time when the tank is most vulnerable. Resist the temptation to fast-track atmospheric tank commissioning when it is being filled for the first time; act as though you know it will fail and deploy personnel accordingly.

Once an atmospheric tank has been successfully put into service, it should provide good service. The greatest concern will be wall thinning that compromises the strength of the vessel walls. Ideally, the thinning will be uniform and caught with a minimum of testing. However, uniform thinning is rarely the actual experience, so special attention needs to be paid to localized thinning due to erosion, which can only be detected by looking for it. Pay attention to where nozzles, dip pipes, and eductors are aimed and confirm that weak spots are not being created.

5.2 Regarding catastrophic tank implosion

Avoiding catastrophic tank implosion requires identifying and controlling sources of vacuum and assuring that measures to protect tanks from implosion remain operable. Obvious vacuum sources include vacuum pumps, pump out and other forms of emptying a tank of its liquid contents, and condensing vapors (typically, but not limited to, steam). When the presence of any of these is unavoidable, they should only be present with full awareness of their hazard and the total reliance on the protection device.

All of the devices to protect atmospheric tanks against implosion have mechanisms by which they can fail. Facilities that take credit for these devices – atmospheric vents, conservation vents, inerting pads, and vacuum breakers – need to periodically validate that they are working as designed. The means to validate the functionality of protective devices needs to be designed with the same care with which the device itself is designed.

5.3 Regarding catastrophic tank explosion

In addition to the obvious concerns of flammable vapor combustion, avoiding catastrophic tank explosion requires identifying and controlling sources of pressure. Obvious pressure sources include high inlet gas pressure, high inlet liquid pressure, and heating. Again, the presence of any of these may be unavoidable, so they should only be present with full awareness of their hazard and the importance of the protective devices.

As with any device, the devices to protect atmospheric tanks against overpressure and explosion have mechanisms by which they can fail. The term "fail-safe" does not mean they cannot fail; it means that in the absence of energy sources, they are supposed to go to the safe state. It does not mean that they cannot fail to a dangerous state. Facilities that take credit for these devices – typically atmospheric vents and conservation vents – need to periodically validate that they are working as designed. As with all protective devices, the means to validate the functionality of protective devices needs to be designed with the same care with which the device itself is designed.

For atmospheric vents in particular, the design needs to allow for easy inspection of the entire vent. While this may involve the use of boroscopes and cameras, the installation of inspection and clean-out ports in the vent line may be more cost effective and easier to implement. It is not enough to have inspection measures, however. They must be used and at a sufficient frequency to detect a developing problem in time to address it. The frequency of inspection may be adjusted as experience dictates, but two failed inspections in a row indicates that the test interval is too long.

5.4 Regarding use of dip pipes in atmospheric storage tanks

Dip pipes have two important features when used in atmospheric storage tanks. The most important is that when the liquid being stored is both non-polar and flammable, dip pipes prevent free-falling liquid, which is known to create static sparks, a source of ignition of the flammable vapors. The other feature is that dip pipes direct liquid away from areas of the tank vulnerable to erosion. When these features are not required, dip pipes should not be used.

This is of particular concern when the liquid being stored is a solution of dissolved solids. When the line and its dip pipe are blown clear, which is a fairly common practice, the gas used to blow the line clear will sparge through the liquid and carry some quantity of mist to the vent. At the vent, the droplets will collect, evaporate, and leave a deposit of solids. Unless inspections explicitly look for this deposit, it is likely to go unnoticed. When dip pipes are installed and their use with solutions of dissolved solids involves blowing them out, vent inspections for blockage should not be limited to looking for nests or other biological activity, but should also include inspection for solids build-up.

6 References

[1] Chang, J.I. and C-C Lin. "A study of storage tank accidents," *Journal of Loss Prevention*, 19 (2006) 51-59.

[2] Atherton, W. and J.W.Ash. *Review of Failures, Causes & Consequences in the Bulk Storage Industry*, National Lightning Safety Institute. 2006. http://lightningsafety.com/nlsi_lls/Causes-of-Failures-in-Bulk-Storage.pdf (accessed 02-Sep-2015).

[3] EPA. "Catastrophic Failure of Storage Tanks," *Chemical Safety Alert*, EPA 550-F-97-002b, Chemical Emergency Preparedness and Prevention Office, USEPA, Washington, DC. May 1997. www.nrt.org/production/nrt/nrtweb.nsf/AllAttachmentsByTitle/A-107tanks/$File/tanks.pdf?OpenElement (accessed 02-Sep-2015).

[4] Marsh. "Atmospheric Storage Tanks," *Risk Engineering Position Paper-01*, Rev 1.0, Marsh & McLennan Companies, February 2011. uk.marsh.com/Portals/18/Documents/Atmospheric%20Storage%20Tanks_lowres.pdf (accessed 02-Sep-2015).

[5] OGP. "Storage incident frequencies," Risk Assessment Data Directory, Report No. 434-3, International Association of Oil & Gas Producers, March 2010. http://www.ogp.org.uk/pubs/434-03.pdf (accessed 02-Sep-2015).

[6] ASME. *Boiler and Pressure Vessel Code, Section VIII-Rules for Construction of Pressure Vessels*, 2015 Ed. American Society of Mechanical Engineers, New York, NY. 2015.

[7] API. *API 620: Design and Construction of Large, Welded, Low-pressure Storage Tanks*, 12th Edition. American Petroleum Institute, Washington, DC. 2013.

[8] API. *API 650: Welded Tanks for Oil Storage*, 12th Edition. American Petroleum Institute, Washington, DC. 2013.

[9] OSHA. *29 CFR 1910.119, Process safety management of highly hazardous chemicals.*, Occupational Safety and Health Administration, U.S. Department of Labor, Washington, DC. Last revised 08-Feb-2013.

[10] API. *API 2000: Venting Atmospheric and Low-Pressure Storage Tanks*, 7th Edition. American Petroleum Institute, Washington, DC. 2014.

[11] Ennis, T. "Pressure Relief Considerations for Low-Pressure (Atmospheric) Storage Tanks," *IChemE Symposium Series No. 151*, Institution of Chemical Engineers, Rugby, UK. https://www.icheme.org/communities/subject_groups/safety%20and%20loss%20prevention/resources/hazards%20archive/~/media/Documents/Subject%20Groups/Safety_Loss_Prevention/Hazards%20Archive/XIX/XIX-Paper-63.pdf (accessed 02-Sep-2015).

[12] Schworm, P. "Nearly a century later, structural flaw in molasses tank revealed," The Boston Globe, January 14, 2015. www.bostonglobe.com/metro/2015/01/14/nearly-century-later-new-insight-into-cause-great-molasses-flood/CNqLYc0T58kNo3MxP872iM/story.html (accessed 15-Feb-2016).

[13] CSB. *Investigation Report: Allied Terminals, Inc.-Catastrophic Tank Collapse.* No. 2009-03-I-VA, U.S. Chemical Safety and Hazard Investigation Board, Washington, DC, May 2009. 49 pp.

[14] Savage, A. "On MythBusters' FIRST-EVER Implosion," *TESTED*. January 22, 2016. www.tested.com/art/makers/561346-mythbusters-first-ever-implosion/ (accessed 16-Feb-2016).

[15] FACE. "Worker Dies While Repairing a Vacuum Evaporator Tank in Virginia," FACE 8725. Fatal Accident Circumstances and Epidemiology Project, NIOSH, CDC, Atlanta, GA. October 15, 2014. www.cdc.gov/niosh/face/In-house/full8725.html (accessed 16-Feb-2016).

[16] Clouse, T. "Ruling upheld in sewage death: Supreme Court says CH2M Hill liable for $6.5 million," The Spokesman-Review, Spokane, WA, May 27, 2011. www.spokesman.com/stories/2011/may/27/ruling-upheld-in-sewage-death/ (accessed 16-Feb-2016).

[17] CCPS. *Guidelines for Engineering Design for Process Safety*, Center for Chemical Process Safety, American Institute of Chemical Engineers, New York, NY. 2010. p. 110.

[18] Dow. *Caustic Soda Solution Handbook*, Form Number 102-00011-0810XBBI, The Dow Chemical Company, Midland, Michigan. August 2010. 23 pp.

[19] PPG. *NaOH-Caustic Soda*, #0100 Rev 0406 pdf, PPG Industries, Inc., Monroeville, Pennsylvania. 2008. 63 pp.

[20] CSB. *Case Study: Barton Solvents*. No. 2007-06-I-KS, U.S. Chemical Safety and Hazard Investigation Board, Washington, D.C., June 26, 2008. 14 pp.

[21] DoJ. *Former Freedom Industries official sentenced for role in chemical spill.* U.S. Attorney's Office, Southern District of West Virginia, February 1, 2016. www.justice.gov/usao-sdwv/pr/former-freedom-industries-official-sentenced-role-chemical-spill (accessed 15-Feb-2016).

[22] Asahi. Police push for charges against TEPCO, execs over pollution problems after nuclear disaster. The Asahi Shimbun, Asia & Japan Watch. October 3, 2015. http://ajw.asahi.com/article/0311disaster/fukushima/AJ201510030052 (accessed 15-Feb-2016).

12TH
GLOBAL
CONGRESS
ON PROCESS
SAFETY

Explosions in SMPO Hydrogenation Unit

Presenter/Principal Author: Nick Gonzales
Affiliation: Shell Global Solutions
Address: 3333 Hwy 6 South, Houston TX 77082
Email address:nick.gonzales@shell.com

Prepared for Presentation at
American Institute of Chemical Engineers
2016 Spring Meeting
12th Global Congress on Process Safety
Houston, Texas
April 11-23, 2016

LEARNING FROM INCIDENTS
AWARENESS ALERT

For Industry-Wide Sharing — September 2015

Explosions in SMPO Hydrogenation Unit

Figure 1 – Hydrogenator structure after the explosion.

Target audience for this alert

- Process Safety and Process Engineers
- Production staff in Refining/Chemical and hydrogenation processes

What happened

On June 3rd 2014, a Styrene Monomer and Propylene Oxide (SMPO) unit was in the pre start-up phase after the scheduled shutdown and replacement of catalyst in the hydrogenation reactors. The reaction system contained two fixed-bed adiabatic, trickle-flow reactors. Operations were heating up the hydrogenation unit in preparation for the catalyst reduction step. The heat-up was performed by the circulation of Ethyl Benzene (EB) through the reaction system under a nitrogen blanket and continuous nitrogen purge. During this heat up, one hydrogenation reactor and an additional upstream separation vessel suddenly exploded. Two contractor workers, who were in the immediate process area, were injured with burns from hot catalyst particles. The explosion caused extensive damage requiring over a year to repair.

Why it happened

The incident investigation concluded that both the reactor and vessel had failed because of overpressure, resulting from unexpected exothermic reactions between Ethyl Benzene (EB) and copper chromite oxide catalyst which produced an excessive amount of gas.

Prior to the incident, it was generally believed that under inert nitrogen conditions EB would not react with the passivated catalyst. However, it was subsequently determined and proven in the laboratory that two exothermic reactions actually occurred during the heat up:

- EB reacted with active oxygen present in the catalyst (mainly but not limited to copper chromate $CuCr^{VI}O_4$). The catalyst contained more copper chromate than the previous catalyst used at the site. The redox reaction started when the temperature exceeded approximately 90°C. The target temperature at the end of the heat up phase was 130°C, but due to the exothermic reaction the temperature increased further in local dry zones of the trickle bed reactors.
- Copper oxide ($Cu^{II}O$) was reduced by EB when localized temperatures reached 180°C. This redox reaction generated large amounts of heat and gaseous molecules (such as CO_2 and H_2O) leading to the overpressure.

The early heat generation by the EB-chromate reaction, in combination with reduced cooling capacity in localized dry zones created localized temperatures in the reactor bed, high enough to initiate the EB-copper oxide reaction. Contributing factors include:

Low nitrogen flow: In a trickle flow reactor, sufficient nitrogen flow is required to achieve a proper liquid distribution from the distribution trays to wet the catalyst in the catalyst bed. The actual nitrogen flow was lower than needed, which led to non optimum liquid distribution in both trickle bed reactors causing dry zones in the top part of the catalyst beds.

Low EB flow: The EB flow to the hydrogenator was, at times, lower than needed for proper wetting of the catalyst and the flow variations led to preferential or rivulet flow and stagnant wet zones in the trickle flow reactor catalyst bed. These conditions led to non optimal heat removal and as a consequence high localized temperatures

Increasing Pressure: About thirty minutes before the explosion, the vent gas valve on the second separator tripped closed on high level and stayed closed. The system pressure increased up to the nitrogen supply pressure and caused the nitrogen flow to the reactor to decrease. The increasing pressure reduced the evaporative cooling of the EB from the catalyst.

Reactive hazard understanding: A Reactive Hazard Assessment (RHA) had been conducted in 2011 for this section of the SMPO unit. The RHA was based on process development work, design knowledge, literature search and application of a tool that was developed and used for identifying unintended reactions. The RHA only identified the reactions with hydrogen during the reduction phase. The reactions of EB with the catalyst in the absence of hydrogen were not identified in the RHA. This, together with the previous operational reduction experiences, reinforced the belief that EB was an inert medium in combination with the catalyst.

Lessons learned

- The passivated copper chromite hydrogenation catalyst contains bonded oxygen in the form of copper chromate and copper oxide prior to reduction. Where a hydrocarbon and oxidizing substance (i.e. both a "fuel" and reactive oxygen) are combined and in sufficient quantities, an exothermic reaction may occur and the only mitigation available may be the total amount of cooling provided.
- Heating up in EB, prior to the reduction step with hydrogen, was seen as normal and safe from a chemical reaction point of view. This influenced the choices made during the process design, process development work, the design of the trickle bed reactor, qualification of the catalyst and operational procedures.
- The purpose of a Reactive Hazard Assessment (RHA) is to identify consequences of the potential reactions, intended and unintended, between the components of the reaction streams. At the time of the Assessment, The composition of the reactants and the potential initiation temperatures were not understood adequately. Post incident investigation suggests that suitable analytical tools such as Thermogravimetric Analysis (TGA) may help to identify such exothermic reactions.
- A series of incremental and seemingly innocuous changes can allow unknown reactions to proceed and run out of control, resulting in a severe incident. Introduction of a new catalyst or catalyst supplier can introduce new components or changes even in small quantities of components that could lead to unidentified and unintended chemical reactions. The catalyst selection and subsequent change processes should address these with the focus on reactive hazards in both transient and steady state operations. Note that this may also apply to changes in process chemicals in other services (e.g. polymerization inhibitors, cleaning solutions, etc.).

AUTHOR INDEX

Acton, Sarah ... 121
Ahmed, Lubna ... 242
Akli, Andrés F. Sánchez ... 246
Allason, Daniel ... 317
Ankrom, Laura ... 777
Arnaez, Achilles ... 303
Averett, Devin ... 201
Azimov, Umid ... 244, 245, 247, 248, 249
Badam, Venkata ... 303
Banick, William R ... 46
Banks, Johnnie A. ... 838
Baum, Dick ... 341
Baybutt, Paul ... 840
Bertelsmann, Anne ... 85
Bommaganti, Praveen Kumar ... 610
Bompelly, Radhika ... 373
Brautaset, Olav ... 676
Bridges, William ... 66, 318
Brigman, James ... 423
Broadribb, Mike ... 713
Bruner, Travis ... 296
Budiansky, Noah ... 871
Bukowski, Julia V. ... 221, 233
Bukur, Dragomir B. ... 348
Bullemer, Peter ... 508
Burgan, Bassam ... 317
Busby, Kirk ... 264
Busque, Mireille ... 373
Cai, Tianxing ... 276
Carpenter, Andrew R. ... 494
Castellanos, Alejandro ... 246, 284, 285, 374
Chambliss, Sarah ... 230, 231
Chan-Urai, Polshom ... 283
Chastain, J. Wayne ... 1, 102, 493
Chastain-Knight, Denise ... 233, 460
Chaurasia, Kumud ... 123
Chen, Anqi ... 317
Chen, Hao ... 727
Cheng, Zhengdong ... 242
Cialkowski, Edward ... 103
Collins, Erin ... 349, 361
Corrales, Jesus ... 360, 425
Cox, Brenton L. ... 494
Cullina, Timothy ... 879
Dang, Wenyi ... 316
Darvishzadeh, Tohid ... 232
Deal, Lindsay ... 727
Dee, Sean J ... 641
Delanoy, Paul ... 1
Deshotels, Robert ... 171
Devlin, Chris ... 1
Doe, Natalie ... 139
Dowell, Arthur M. ... 66
Dreux, Mark ... 169
Duarte, Juan ... 360, 425
Dunjó, Jordi ... 277, 278
Edwin, Nathaniel John ... 676
Farrell, Brian E ... 629
Fecke, Mark ... 494, 641
Ferreira, Jose Filipe Michel Gagliano ... 375
Ferreira, Tatiele ... 422
Filho, Salvador Ávila ... 375, 398
Fox, Shannon B. ... 447
Freburger, Eric ... 713
Gan, Ning ... 348
Gangadharan, Preeti ... 448
Garner, Stephen W. ... 494
Ge, Chuntao ... 286
Giovanoglou, Apostolos ... 693
Giraldo, Felipe Muñoz ... 246, 284, 285, 374
Gonzales, Nick ... 910
Gonzalez, Jesus ... 360, 425
Gopalaswami, Nirupama ... 170
Grattan, Dave ... 658
Griffith, Stacey ... 449
Grosscup, Dylan ... 535
Gustafson, Richard ... 230, 231
Hamparian, Alek ... 250
Hanks, Michael ... 198
Hatanaka, Logan ... 242
Haugen, Stein ... 676
Hawkins, Steve ... 482
Hayner, Gary ... 527
Heil, Jeff ... 137
Hochleitner, Monica ... 791
Hoff, Rainer ... 819
Hong, Yizhi ... 424
Huberman, Aaron ... 570
Hull, John ... 229
Israni, Kumar ... 121, 482
Jarvis, Christian ... 535
Jin, Hui ... 35
Johnson, Michael ... 317
Johnson, Zach ... 279
Jones, David ... 279, 583
Kakosimos, Konstantinos E. ... 170
Kale, Subhash S. ... 215
Kalid, Ricardo Araujo ... 375
Kazarians, Mardy ... 250
Kill, John F. ... 373
Klein, James ... 818
Kolodji, Brian ... 122
Kumana, Zubin ... 303
Kurko, Kenneth N. ... 879
Kytomaa, Harri K. ... 120
Lea, Charles ... 535
Lee, Kwanghee ... 315
Lin, Mei-Li ... 280
Ling, Xiaodong ... 214
Liu, Yi ... 170

AUTHOR INDEX

Lopes, Jose Rafael Nascimento 398
Lou, Helen 448
Luebeck, Hope A. 768
Madden, Lauren 318
Maher, Steven T. 228
Maliakkal, George 215
Mannan, Sam M. 170, 242, 348, 424
Mariño, Oscar 285
Markowski, Adam S. 424
Marriott, James 693
Marszal, Edward 199, 806
Mayo, Michael 341
Mendez, Julian 285
Mentzer, Ray 170
Modi, Kay 583
Montgomery, Randal L. 748
Moon, Il 315
Morgan, Blair 213
Morrison, Trey 641
Mostia, William 35
Mueller, Thomas 1
Myers, Timothy J. 120, 871
Najafi, Bijan 349, 361
Natarajan, Sathish 693
Neill, Mike 150
Nguyen, Hieu 247, 248
Ocampo, Fabio 284, 285, 374
Ogle, Russell A. 641
Pape, Ronald 198
Parpacen, Gerson 360
Pasman, Hans 424
Patankar, Aniket 123
Pearson, Katherine E. 492
Prasad, Monica 610
Prophet, Neil 277, 278
Ramabhadran, Srinivasan 283
Ramlakhan, Carl D. 535
Reising, Dal Vernon 508
Reza, Ali 871
Rice, Tone 727
Richards, Amy 198
Rocha, Rafael 138
Roche, Eloise 791
Romano, Robert 768
Rowley, Jef 201
Rynkar, John 281
Sachdeva, Sonny 424
Sari, Ali 232, 244, 245, 247, 248, 249
Saud, Yaneira 121, 373
Schmidt, Michael S. 889
Sergo, Ian 137
Shaba, Kehinde 200
Shackelford, Aubry 137
Shaikh, Irfan 282
Shell, Kathy 777
Sheng, Min 229
Silva, Jean Marcel Prazeres 398
Siml, Robert 372
Simoes, Vinicius 230, 231, 243
Singh, Ravinder 448
Singh, Surendra 281
Smith, Dustin J. 139
Soczek, Charles A. 652
Sofyan, Stannia 231, 243
Somandepalli, Vijay 871
Sousa, Carlos Roberto Oliveira 375
Spitzenberger, Cynthia 727
Sposito, Nestor H. 168
Stanley, Brent 748
Stanley, Rick 426
Stauffer, Todd 460
Stern, Michael C. 871
Stewart, Loren 221
Stough, Joseph 280
Study, Karen 1
Summers, Angela 35
Summers, Angela E. 791
Tam, Vincent 317
Tambling, Todd 229
Theis, Amy 879
Turci, Laura Acevedo 168
Urbina, Alexander Guzman 817
Vianna, Sávio 422
Villalba, Nicolás 284, 285, 374
Vinnem, Jan Erik 676
Von Wald, Grant 229
Wakil, Waheed 296
Wang, Peng 316
Wang, Qingsheng 242
Wasileski, Robert 820
Wei, Cindy 46, 243
Xu, Jing 286
Xu, Qiang 276
Yu, Anfeng 316
Zaccheus, Yemi 693
Zhang, Bin 170, 242
Zhang, Jiedong 316
Zhang, Yi 214, 286
Zhao, Wen Fang 214